Biotechnology

Second Edition

Volume 11a

Environmental Processes I

Biotechnology

Second Edition

Fundamentals

Volume 1
Biological Fundamentals

Volume 2
Genetic Fundamentals and
Genetic Engineering

Volume 3
Bioprocessing

Volume 4
Measuring, Modelling and Control

Products

Volume 5a
Recombinant Proteins, Monoclonal
Antibodies and Therapeutic Genes

Volume 5b
Genomics

Volume 6
Products of Primary Metabolism

Volume 7
Products of Secondary Metabolism

Volumes 8a and b
Biotransformations I and II

Special Topics

Volume 9
Enzymes, Biomass, Food and Feed

Volume 10
Special Processes

Volumes 11a–c
Environmental Processes I–III

Volume 12
Legal, Economic and
Ethical Dimensions

All volumes are also displayed on our Biotech Website:
http://www.wiley-vch.de/home/biotech

A Multi-Volume Comprehensive Treatise

Biotechnology

Second, Completely Revised Edition

Edited by
H.-J. Rehm and G. Reed
in cooperation with
A. Pühler and P. Stadler

Volume 11a

Environmental Processes I

Wastewater Treatment

Edited by
J. Winter

Weinheim · New York · Chichester · Brisbane · Singapore · Toronto

Series Editors:
Prof. Dr. H.-J. Rehm
Institut für Mikrobiologie
Universität Münster
Corrensstraße 3
D-48149 Münster
FRG

Dr. G. Reed
1029 N. Jackson St. #501-A
Milwaukee, WI 53202-3226
USA

Prof. Dr. A. Pühler
Biologie VI (Genetik)
Universität Bielefeld
P.O. Box 100131
D-33501 Bielefeld
FRG

Prof. Dr. P. J. W. Stadler
Artemis Pharmaceuticals
Geschäftsführung
Pharmazentrum Köln
Neurather Ring
D-51063 Köln
FRG

Volume Editor:
Prof. Dr. J. Winter
Universität Karlsruhe (TH)
Institut für Ingenieurbiologie und
Biotechnologie des Abwassers
Am Fasanengarten
Postfach 6980
D-76128 Karlsruhe

Library of Congress Card No.: applied for

British Library Cataloguing-in-Publication Data:
A catalogue record for this book is available from the British Library

Die Deutsche Bibliothek – CIP-Einheitsaufnahme
Biotechnology: a multi volume comprehensive
treatise / ed. by H.-J. Rehm and G. Reed.
In cooperation with A. Pühler and P. Stadler. –
2., completely rev. ed. - VCH.
ISBN 3-527-28310-2 (Weinheim ...)

NE: Rehm, Hans-J. [Hrsg.]

Vol. 11a: Environmental Processes / ed. by J. Winter
ISBN 3-527-28321-8

Printed on acid-free and chlorine-free paper.

Composition and Printing: Zechnersche Buchdruckerei, D-67330 Speyer.
Bookbinding: J. Schäffer, D-67269 Grünstadt.
Printed in the Federal Republic of Germany

Preface

In recognition of the enormous advances in biotechnology in recent years, we are pleased to present this Second Edition of "Biotechnology" relatively soon after the introduction of the First Edition of this multi-volume comprehensive treatise. Since this series was extremely well accepted by the scientific community, we have maintained the overall goal of creating a number of volumes, each devoted to a certain topic, which provide scientists in academia, industry, and public institutions with a well-balanced and comprehensive overview of this growing field. We have fully revised the Second Edition and expanded it from ten to twelve volumes in order to take all recent developments into account.

These twelve volumes are organized into three sections. The first four volumes consider the fundamentals of biotechnology from biological, biochemical, molecular biological, and chemical engineering perspectives. The next four volumes are devoted to products of industrial relevance. Special attention is given here to products derived from genetically engineered microorganisms and mammalian cells. The last four volumes are dedicated to the description of special topics.

The new "Biotechnology" is a reference work, a comprehensive description of the state-of-the-art, and a guide to the original literature. It is specifically directed to microbiologists, biochemists, molecular biologists, bioengineers, chemical engineers, and food and pharmaceutical chemists working in industry, at universities or at public institutions.

A carefully selected and distinguished Scientific Advisory Board stands behind the series. Its members come from key institutions representing scientific input from about twenty countries.

The volume editors and the authors of the individual chapters have been chosen for their recognized expertise and their contributions to the various fields of biotechnology. Their willingness to impart this knowledge to their colleagues forms the basis of "Biotechnology" and is gratefully acknowledged. Moreover, this work could not have been brought to fruition without the foresight and the constant and diligent support of the publisher. We are grateful to VCH for publishing "Biotechnology" with their customary excellence. Special thanks are due to Dr. Hans-Joachim Kraus and Karin Dembowsky, without whose constant efforts the series could not be published. Finally, the editors wish to thank the members of the Scientific Advisory Board for their encouragement, their helpful suggestions, and their constructive criticism.

H.-J. Rehm
G. Reed
A. Pühler
P. Stadler

Scientific Advisory Board

Contributors

Dr. Rudolf Amann
MPI für Marine Mikrobiologie
Celsiusstraße 1
D-28359 Bremen
Germany
Chapter 5

Dr. Ute Austermann-Haun
Institut für Siedlungswasserwirtschaft und Abfalltechnik
Universität Hannover
Welfengarten 1
D-30167 Hannover
Germany
Chapter 10

Dr. Matthias Barjenbruch
Institut für Kulturtechnik und Siedlungswasserwirtschaft
Universität Rostock
Satower Straße 48
D-18059 Rostock
Germany
Chapter 18

Dr.-Ing. Peter Baumann
Institut für Siedlungswasserbau, Wassergüte und Abfallwirtschaft
Abt. Abwassertechnik
Bandtäle 2
D-70569 Stuttgart
Germany
Chapter 16

Prof. Dr. Eberhard Bock
Institut für Allgemeine Botanik
Abteilung Mikrobiologie
Universität Hamburg
Ohnhorststraße 18
D-22609 Hamburg
Germany
Chapter 3

Prof. Dr. Klaus Buchholz
Lehrstuhl für Technologie der Kohlenhydrate
Technische Universität Braunschweig
Langer Kamp 5
D-38106 Braunschweig
Germany
Chapter 24

Prof. Dr. Rainer Buchholz
Institut für Biotechnologie
Technische Universität Berlin
Ackerstraße 71-76
D-13355 Berlin
Germany
Chapter 21

Dr. Gerald Bunke
Institut für Biotechnologie
Technische Universität Berlin
Ackerstraße 71-76
D-13355 Berlin
Germany
Chapter 21

Dr.-Ing. Bernd Dorias
Drees & Sommer GmbH
Obere Waldplätze 13
D-70569 Stuttgart
Chapter 16

Prof. Dr. Hans-Curt Flemming
Institut für Wasserchemie und Wassertechnologie
Universität Duisburg
Moritzstraße 26
D-45476 Mülheim/Ruhr
Germany
Chapter 4

Dr. Claudia Gallert
Institut für Ingenieurbiologie und Biotechnologie des Abwassers
Universität Karlsruhe (TH)
Am Fasanengarten
Postfach 6980
D-76128 Karlsruhe
Germany
Chapter 2

Dr. Peter Götz
Institut für Biotechnologie
Technische Universität Berlin
Ackerstraße 71-76
D-13355 Berlin
Germany
Chapter 21

Prof. Dr. Ludwig Hartmann
Am neuen Berg 10
D-86673 Unterstall
Germany
Chapter 1

Prof. Dr.-Ing. Winfried Hartmeier
Lehrstuhl für Biotechnologie
RWTH Aachen
Worringerweg 1
D-52056 Aachen
Germany
Chapter 7

Prof. Dr. Mogens Henze
Department of Environmental Science and Engineering
Building 115
Technical University of Denmark
DK-2800 Lyngby
Denmark
Chapter 20

Dr. Look W. Hulshoff Pol
Department of Environmental Engineering
Agricultural University of Wageningen
P.O. Box 8129
NL-6700 EV Wageningen
The Netherlands
Chapter 25

Dr. Norbert Jardin
Ruhrverband Essen
Kronprinzenstr. 37
D-45128 Essen
Germany
Chapter 14

Dr. Hans-Joachim Jördening
Lehrstuhl für Technologie der Kohlenhydrate
Technische Universität Braunschweig
Langer Kamp 5
D-38106 Braunschweig
Germany
Chapter 24

Prof. Dr.-Ing. Rolf Kayser
Adolf-Bingel-Straße 2
D-38116 Braunschweig
Germany
Chapter 13

Prof. Dr. Paul Koppe
Obere Saarlandstraße 3
D-45470 Mülheim/Ruhr
Germany
Chapter 9

Prof. Dr. Helmut Kroiss
Institut für Wassergüte und Landschaftswasserbau
Technische Universität Wien
Karlsplatz 13/226
A-1040 Wien
Austria
Chapters 6, 23

Dr. Peter Kuschk
UFZ – Umweltforschungszentrum Leipzig-Halle GmbH
Sektion Sanierungsforschung
Permoserstraße 15
D-04318 Leipzig
Germany
Chapter 12

Prof. Dr. Gatze Lettinga
Department of Environmental Engineering
Agricultural University of Wageningen
P.O. Box 8129
NL-6700 EV Wageningen
The Netherlands
Chapter 25

Dr. Judy Libra
Institut für Verfahrenstechnik
Technische Universität Berlin
Straße des 17. Juni 135
D-10623 Berlin
Germany
Chapter 19

Prof. Dr.-Ing. Herbert Märkl
AB Bioprozeß- und Bioverfahrenstechnik
Technische Universität Hamburg-Harburg
Denickestraße 15
D-21071 Hamburg
Germany
Chapter 26

Dr. Michael J. McInerney
Department of Botany and Microbiology
University of Oklahoma
770 Van Vleet Oval
Norman, OK 73019-0245
USA
Chapter 22

Dipl.-Ing. Hartmut Meyer
Institut für Siedlungswasserwirtschaft und Abfalltechnik
Universität Hannover
Welfengarten 1
D-30167 Hannover
Germany
Chapter 10

Dr. Eberhard Morgenroth
Department of Environmental Science and Engineering
Technical University of Denmark
Building 115
DK-2800 Lyngby
Denmark
Chapter 15

Dr. Volkmar Neitzel
Ruhrverband
Kornprinzenstraße 37
D-45128 Essen
Germany
Chapter 9

Dr. Peter Nisipeanu
Ruhrverband
Kornprinzenstraße 37
D-45128 Essen
Germany
Chapter 8

Prof. Dr. Ing. Norbert Räbiger
Institut für Umweltverfahrenstechnik
Universität Bremen
Postfach 330440
D-28334 Bremen
Germany
Chapter 27

Dr. Monika Reiss
Lehrstuhl für Biotechnologie
RWTH Aachen
Worringerweg 1
D-52056 Aachen
Germany
Chapter 7

Prof. Dr.-Ing. Karl-Heinz Rosenwinkel
Institut für Siedlungswasserwirtschaft und Abfalltechnik
Universität Hannover
Welfengarten 1
D-30167 Hannover
Germany
Chapter 10

Prof. Dr. Georg Schön
Insitut für Biologie II
Universität Freiburg
Schänzlestraße
D-79104 Freiburg
Germany
Chapter 14

Dr. Andreas Schramm
MPI für Marine Mikrobiologie
Celsiusstraße 1
D-28359 Bremen
Germany
Chapter 5

Dr. Judith M. Schulz genannt Menningmann
ENVICON Kläranlagen
Postfach 100637
D-46526 Dinslaken
Germany
Chapter 17

Dr. Carin Sieker
Berliner Wasserbetriebe
Neue Jüdenstraße 1
Postfach 021098
D-10122 Berlin
Germany
Chapter 18

Prof. Dr. Ulrich Stottmeister
UFZ – Umweltforschungszentrum
Leipzig-Halle GmbH
Sektion Sanierungsforschung
Permoserstraße 15
D-04318 Leipzig
Germany
Chapter 12

Chem.-Ing. Alfred Stozek
Auf dem Loh 7
D-45289 Essen
Chapter 9

Dr. Ralf Stüven
Institut für Allgemeine Botanik
Abteilung Mikrobiologie
Universität Hamburg
Ohnhorststraße 18
D-22609 Hamburg
Germany
Chapter 3

Dr. Karl Svardal
Institut für Wassergüte und Landschaftswasserbau
Technische Universität Wien
Karlsplatz 13/226
A-1040 Wien
Austria
Chapters 6, 23

Dr. Jules B. van Lier
Department of Environmental Engineering
Agricultural University of Wageningen
P.O. Box 8129
NL-6700 EV Wageningen
The Netherlands
Chapter 25

Prof. Dr.-Ing. Peter Weiland
Bundesforschungsanstalt für Landwirtschaft
Braunschweig-Völkenrode (FAL)
Institut für Technologie
Bundesallee 50
D-38116 Braunschweig
Germany
Chapter 11

Prof. Dr.-Ing. Udo Wiesmann
Institut für Verfahrenstechnik
Technische Universität Berlin
Straße des 17. Juni 135
D-10623 Berlin
Germany
Chapter 19

Dr. Arndt Wießner
UFZ – Umweltforschungszentrum
Leipzig-Halle GmbH
Sektion Sanierungsforschung
Permoserstraße 15
D-04318 Leipzig
Germany
Chapter 12

Prof. Dr.-Ing. Peter A. Wilderer
Lehrstuhl für Wassergüte und
Abfallwirtschaft
Technische Universität München
Am Coulombwall
D-85748 Garching
Germany
Chapter 15

Dr. Jost Wingender
Institut für Wasserchemie und
Wassertechnologie
Universität Duisburg
Moritzstraße 26
D-45476 Mülheim/Ruhr
Germany
Chapter 4

Prof. Dr. Josef Winter
Institut für Ingenieurbiologie und
Biotechnologie
des Abwassers
Universität Karlsruhe (TH)
Am Fasanengarten
Postfach 69 80
D-76128 Karlsruhe
Germany
Chapter 2

Dr. Dirk Zart
Institut für Allgemeine Botanik
Abteilung Mikrobiologie
Universität Hamburg
Ohnhorststraße 18
D-22609 Hamburg
Germany
Chapter 3

Dr. Grietje Zeemann
Department of Environmental Engineering
Agricultural University of Wageningen
P.O. Box 81 29
NL-6700 EV Wageningen
The Netherlands
Chapter 25

Contents

Metal Ion Removal

Anaerobic Processes

Introduction

Josef Winter
Karlsruhe, Germany

Except for soil sanitation environmental biotechnology, including air pollution, waste and wastewater treatment processes, surface and ground water pollution and many other topics was subsumed under the title *Microbial Degradations* in Volume 8 of the First Edition of *Biotechnology*, Urbanization and industrialization, especially in developing countries, is still in progress with all negative effects on the environment.

Resulting from the accumulation of huge masses of polluted water in human settlements or in industry the limits of self-purification of surface waters are often exceeded, leading to anaerobiosis with all its deteriorating consequences for life. In industrialized countries central wastewater treatment plants have been developed to reduce the pollution freight before disposing the wastewater into the next surface water.

In the First Edition of *Biotechnology* different wastewater treatment processes contribute a major part to Volume 8. Furthermore, the volume is devoted to different processes of solid waste composting, drinking water biofiltration, exhaust gas purification, removal of pathogens and several other environmental processes.

Now, some ten years later, the biological background of aerobic or anaerobic wastewater treatment processes and of most of the other processes in environmental biotechnology (e.g., soil sanitation, waste gas purification, compost preparation, drinking water purification, etc.) has increased tremendeously and various new and differing processes are available to protect the environment. So it is the time to decribe the present state of the art of environmental biotechnological processes in a comprehensive survey.

After bringing together the most important issues that had to be covered in the Second Edition of Biotechnology, the editors immediately realized that wastewater treatment, solid waste management (also including the broad field of municipal solids composting or anaerobic fermentation), off-gas purification, biological soil remediation processes, potable water denitrification and purification and many other selected environmental processes were too broad a field to be summarized with significance in a single book.

For this reason Volume 11 *Environmental Processes* of the Second Edition of *Biotechnology* is divided into three volumes, the first of which, Volume 11a, is devoted to *Wastewater*

Treatment Processes. This first volume on environmental biotechnology summarizes the biological principles and the technical limits of all those wastewater treatment processes that are operated by municipalities and industry up to the present state to meet the legal limits for carbon, nitrogen, and phosphorus disposal into surface waters.

In the first part of the book the present status of general biological and engineering aspects of wastewater purification procedures is summarized. What does environmental legislation require for wastewater disposal of municipalities or industry into surface waters? What can biology contribute together with chemistry, physics and engineering to wastewater purification from its organic and inorganic pollutants? How can the purification efficiency be measured analytically, either off-line or – even more important for monitoring of continuous processes – on-line?

In the second part of the book the different processes for wastewater treatment are described in more detail and under the aspect of full-scale application. At first wastewater sources and variations of wastewater composition are outlined, followed by specific aerobic carbon, nitrogen and phosphate removal processes, metal ion removal and, last but not least, anaerobic wastewater treatment processes.

The volume includes well-known and practiced technologies, as well as new and only recently developed processes. Especially in the field of improved wastewater purification (N and P removal processes), which is a relatively young requirement within environmental legislation, new processes or process combinations had to be developed and applied.

It is hoped that the whole range of insights into biology and technology of wastewater treatment processes have been covered by the contributions of expert authors from Europe and America. The editors are well aware on the other hand, that not every individual system offered on the market could be described. Especially in the field of carrier-supported fixed or fluidized bed technologies not every single system could be mentioned, although carrier-supported processes may be a matter of choice for future high-rate wastewater treatment, e.g., in industry. Membrane technologies were not included, since the average lifetime of membranes is generally still too short due to membrane corrosion or biofouling.

This first volume on *Environmental Processes* should give the reader basic information on the biology of the degradation of pollutants, different processes for wastewater purification and process parameters for an optimal purification. It should be regarded as a source of overview information on frequently applied full-scale wastewater treatment processes with some more details presented for certain specific applications.

Karlsruhe, March 1999 J. Winter

I General Aspects

1 Historical Development of Wastewater Treatment Processes

LUDWIG HARTMANN

Unterstall, Germany

1 General Background

Treatment of all kinds of human wastes has a long history. Procedures depend on the lifestyle of the population and on legislation. Even today in some developing countries solid waste is dumped on little heaps in the backyard and burned once in a while. Wastewater is either generated "directly" in lakes or rivers by washing clothes or dishes or, if water is available in the houses, e.g., by single deep wells or by a public water supply system, the wastewater is disposed off untreated into the next surface water. This situation is often found in rural areas, but even today it can be observed in megacities of developing countries such as Calcutta, Bangkok, Manila, or Jakarta. Under conditions of rural life with a low population density and a rather elemental lifestyle the direct feedback of organic wastes (e.g., cattle manure) as a fertilizer on the fields and of wastewater into natural surface waters may be acceptable as long as overfertilization of soil is prevented or the capacity for self-purification of the surface water sources is not exceeded. Due to an ongoing urbanization, not only in developing countries and due to the urban lifestyle more and more waste and wastewater are generated locally in a concentrated form. Handling of huge amounts of waste or wastewater under urban conditions excludes recycling into nature for irrigation or fertilization. Instead it requires the application of highly sophisticated techniques for mechanical, chemical, and biological treatment to protect nature from permanent damage. These technically controlled instruments had to be invented as artifical ecosystems and were fitted in between the generation of huge amounts of wastes and wastewater by humans and the natural self-purification capacity. Many different technical procedures for wastewater treatment have been developed over the past 10 decades, all including biological treatment technologies at some stage, with the task of mitigating the destructive effect of man's wastes and wastewater on nature.

Concerning human and industrial wastes and wastewaters an aspect of general importance was that waste or wastewater treatment, although it was considered to be a "must" to protect the environment, costs a lot of money and reduces profit. For this reason it was always minimized to the lowest standard enforced by legislation. The philosophy was literally that of the "fire police". Only if there was a "fire" (pressure from state authorities), counteraction was necessary. Major efforts to improve the situation came mostly as a result of irreparable and no longer negligible severe environmental damage.

At this stage it should also be mentioned that the debate always started at the definition of "what is waste" and "the state of the art for its treatment". Of general importance were the accepted analytical methods for measuring pollutants or pollution, either by using so-called sum parameters, such as biological oxygen demand (BOD), chemical oxygen demand (COD), or total organic carbon (TOC), or by analyzing single substances, if they were known. Many new chemicals are still synthesized and find their way into wastes or wastewaters and finally into nature. However, their impact on the environment can often only be seen decades after their use, e.g., as wood preservatives, insecticides, pesticides, or detergents. Therefore, counteractions of course are always one step behind. The definition of chemicals in terms of BOD is uncertain and depends on the ability of microorganisms upon exposure to the xenobiotics to acquire the potential for biotransformation or – better – biodegradation. Biotransformation or biodegradation of a single chemical brings us to the general problem of waste or wastewater treatment. Disappearance of a single substance from wastewater, as measured by, e.g., gas chromatography or HPLC, does not necessarily mean complete degradation or detoxification. It may mean nothing but transformation of one substance into another, which may even be of higher environmental significance because of higher toxicity.

Finally, it has to be mentioned that progress in wastewater treatment systems was and is only slowly moving. This is mainly due to the fact that huge, central wastewater treatment plants designed for at least one generation in advance have been built and are operated by the municipalities. For this reason, it often takes almost 20–30 years until new developments will be applied.

2 The Beginnings of Waste and Wastewater Treatment

Waste and wastewater treatment reaches back to the Egyptian and Roman high cultures. In ancient Rome part of the city had a sewer channel system for collection of night soil and urine, whereas in other parts the toilets were connected to pits. The collected human excrement was sold as fertilizer for horticulture (IMHOFF, 1998).

The real history of wastewater treatment started in the second half of the previous century by the invention of the term "wastewater". The background were cholera and typhoid fever epidemics in some large cities in Central Europe. Pioneers of bacteriology and hygiene, such as PETTENKOFER (1890, 1891), worked out the scientific background for these epidemics as diseases caused by infection via contact with wastes or waste products. PETTENKOFER demanded that the wastes should be transported out of the cities. He thought that waste products were transferred into the air-filled pores of soil. From there they evaporated into the atmosphere and finally came into the houses making people ill. He calculated that men inhale daily about 9000 L of air, but only take in about 3 L of water, so the risk of an infection by air would be much higher. PETTENKOFER proposed a separation of potable water and wastewater, but his explanation of infectious diseases was wrong. Only when ROBERT KOCH in 1876, 1882, and 1983 isolated the bacteria causing anthrax, tuberculosis, and cholera infectious diseases were recognized as bacterial infections for the first time (IMHOFF, 1998).

DUNBAR (1907) found the technical answer to the problem by proposing and constructing public sewer systems for wastewater collection and transportation. By a sewer system wastes and wastewater were transported out of the cities to the next river or lake where self-purification could take place and solve the problem. For the first time a problem was solved just by exporting it from one location to another. The problem of environmental health in the settlements was transformed into an ongoing problem of river pollution, which, as will be shown, still today occupies the interest of engineers and scientists. Since most of the bigger cities were located at big rivers, it seemed to be a good solution for the time being. As a result of waste transportation out of the cities by sewer channels outbreaks of cholera and typhoid fever could be reduced and finally almost completely prevented. The first comprehensive sewer network was built in the city of Hamburg, starting in 1842. Only 25 years later other cities followed. Today the wastewater of more than 92% of population equivalents in Germany is connected to underground sewer systems for wastewater and rainwater drainage (SICKERT, 1998).

People and industry accepted the new technology and began to use the sewers to export everything that was not needed anymore. Therefore, in due time, the self-purification capacity of the wastewater-receiving natural waters was exceeded, and the water quality of rivers decreased more and more. Since river water (as a bank filtrate) was increasingly required to serve the needs for potable water supply, new actions of wastewater disposal were required.

The first pollution problem recognized in surface waters was only of optical nature. It was solved by installation of screens and sieves at the wastewater outlet into rivers (FRÜHLING, 1910; DUNBAR, 1907). Except for satisfying the psychological impression this was necessary to protect pumps for wastewater transfer into rivers or for land application. A real improvement of the situation was only achieved by the development of settling tanks to remove the settleable solids before the wastewater was released into the rivers or eventually treated further. About one third of organic pollution could be retained by this method, thus reducing the pollution freight of rivers considerably. Until the 1950s many sewage treatment plants in Germany used only mechanical treatment for removal of organic and inorganic solids (SICKERT, 1998). As a side effect huge amounts of sludge were "produced" and had to be handled.

It was found that, in analogy to aerobic self-purification, an anaerobic process existed reducing the amount of organic solids in the sludge by formation of biogas. In addition, by anaerobic treatment the water holding capacity of the sludge was reduced and, if the residence time in the digester was long enough, eggs of intestinal worms were destroyed and pathogenic bacteria were inactivated. In other words, the sludge that was removed from the settling tanks periodically could easily be dewatered, dried, and after composting be used as an agricultural fertilizer. The biogas was collected and could be integrated into the municipal gas supply systems. This was a real step forward in wastewater handling.

Since in the early days of wastewater treatment sewers and settling tanks were invented by engineers, waste handling in any form has become the domain of civil engineers. Other occupational groups were not interested. Research at that time (before and shortly after the World War I) was, therefore, concentrated on the improvement of technical installations of settling tanks with sludge fermentation. Real progress was made in the densely populated and highly industrialized regions of Germany, especially in the Ruhr and Emscher region. An outstanding pioneer of this time was KARL IMHOFF. The technologies developed at that time, the so-called Imhoff tank or the Emscher-Brunnen (Emscher well) and their variations were still in use until the 1940s and 1950s (IMHOFF, 1979; IMHOFF and IMHOFF, 1993).

Other fields of research dealt with the survival of intestinal worms or pathogenic bacteria in sludge, depending on fermentation times and conditions. Basic knowledge on the influence of fermentation time and temperature on gas production and detention time was collected. An optimal temperature for the technical process would be about 33°C to satisfy the physiological needs of mesophilic bacteria in sludge. However, this knowledge was not practically applied since heated digestion tanks were not available. Later it was found that anaerobic digestion could also be performed at thermophilic temperatures, e.g., at 55°C, the optimum of thermophilic bacteria in sewage sludge. However, except for an application in the sewage treatment plants of Moscow and Los Angeles (GARBER et al., 1975) thermophilic digesters were only operated on a laboratory or pilot plant scale (KANDLER et al., 1981; GALLERT and WINTER, 1997, PFEFFER, 1974). Today, thermophilic digestion is used in agricultural co-fermentation plants for biowaste because it kills pathogens.

3 Necessity for Further Purification of Wastewater – Development of Trickling Filters

The screens and settling tanks of the early times of wastewater treatment as the only means for wastewater purification soon turned out to be insufficient for the protection of natural water resources. This was demonstrated by a new, biological control method for pollution: the system of saprobes, introduced by KOLKWITZ and MARSSON (1902, 1908, 1909). They observed a change in the composition of the biosystem along a river upon wastewater introduction (KOLKWITZ, 1907). Many biological indicator organisms, protozoa among others – especially ciliates – and insects give information on the pollution and the progress of self-purification of rivers (FAIR et al., 1941; KOLKWITZ, 1950; ODUM, 1971). The "system of saprobes" was finally revised by LIEBMANN (1960) and made more practicable by inclusion of chemical parameters.

The saprobe index showed that pollution of rivers exceeded the capacity of self-purification, especially in densely populated areas. The distances between the different sewer inlets were too short for a full degradation of the organic pollutants. So the question was raised of how to reduce the pollution of degradable organics. The answer at that time was the installation of trickling filters, which started in England. Trickling filters have their technical origin in soil filters that served for wastewater irrigation (Royal Commission of Sewage Disposal, 1908). Instead of using soil and large areas, gravel or small rocks were piled up to a tower of 2–3 m height and the wastewater was sprinkled over the surface. After a short while a biofilm had developed on the surface of the stones. To prevent clogging, the wastewater had to pass a settling tank. A sprinkling system had already been used by CORBETT in 1893 who developed the first trickling filter (STANBRIDGE, 1976).

According to DUNBAR's theory, purification was a two-step process with (1) adsorption of organic matter to the surface of the carrier material and (2) subsequent mineralization. The adsorption theory was a result of a more physical thinking of engineers, who were not aware of the biological background. This adsorption theory for wastewater components was for quite some time the basis of treatment techniques and was even thought to explain the activated sludge process. According to the theory, the trickling filter had to be given time after a period of "adsorption" for "degradation". In other words, the technology required an intermittent operation. A period of a few minutes of wastewater application was followed by a period for biological degradation. This operation mode required rather huge filters and allowed only a low throughput of wastewater. Due to the low-rate operation, the organic pollutants were fully oxidized. Not only bacteria developed on the rocks, but also protozoa, earth worms, insects, etc. belonged to the population of a trickling filter. *Psychoda*, the so-called trickling filter fly, was a nuisance of this artifical ecosystem and attracted much scientific attention. For a permanent operation the surplus biofilm had to be washed out twice a year to avoid clogging.

According to theoretical considerations and practical observations with low-rate trickling filters no final clarifiers were required. Stabilized, clear wastewater left the treatment unit. In Germany low-rate trickling filters were still in operation after World War II, although a new understanding of the biology of purification has been gained. This resulted in the more effective technique of activated sludge systems which competed with the trickling filters. To catch up with the new development of the activated sludge technology (ARDERN and LOCKETT, 1914, 1915; LOCKETT, 1954), high-rate trickling filters with final clarifiers were constructed (HALVORSON, 1936). Much research was carried out to replace the rocks by artifical media, the height was increased, and the relationship between film formation and film removal was studied.

The time of trickling filters as the sole aerobic treatment technology, however, ended in the late 1950s. Although trickling filters could not compete with the activated sludge technology in general, they were still applied for special wastewater types or at special locations where the activated sludge technology could not be installed. Up to the present time their use as a second stage of aerobic treatment to remove the residual, more recalcitrant BOD in effluents of activated sludge treatment systems and for nitrification is of special importance. This is possible since fixed-film treatment systems can successfully host bacteria with long generation times, which would be washed out from an activated sludge system due to the limited sludge age at a short hydraulic retention time.

A special form of trickling filters are rotating disc reactors. Developed originally in the United States in the late 1920s by BUSWELL et al. (1928), many of them were built until the 1950s (FAIR et al., 1948) and were introduced in Central Europe in the late 1950s. Due to simple operation, absence of clogging, little energy consumption, and biofilm formation with a good sedimentation behavior they were the method of choice for small communities with a small amount of wastewater or for industry with special types of pollution.

4 Land Application of Wastewater and Fish Ponds

Wastewater was not only understood as a waste but also as a raw material for agricultural or aquatic production. Apart from the fertilizing effect of sewage sludge the wastewater itself could also serve as a fertilizer for agricultural soil. Especially at times of food shortages after World War I these methods were proposed and treatment plants were built. However, due to many problems, e.g., shortage of land areas in the neighborhood of big cities, integration of sewage application into agricultural practice, disinfection pretreatment to avoid epidemics, integration of wastewater application into the climatic situation, distribution of toxic substances, etc. they did not persist until the time after World War II. Only when wastewater irrigation was the major goal such methods were used for a longer time. A different situation is prevalent in the food industry, e.g., for starch production. In some parts of Germany the wastewater from potato processing is still applied by irrigation on agricultural land by use of an underground pipe distribution system.

Except for land application (e.g., USEPA, 1981), wastewater was also used as a source of nutrients in fish ponds up to World War II, e.g., in the city of Munich. However, precautions for several problems had to be taken:

(1) Pre-sedimentation of the solids to avoid sludge sedimentation in the fish pond,
(2) dilution with non-polluted water was required to avoid oxygen deficiencies, and last but not least
(3) an efficient monitoring system had to exist to prevent toxic effects. The main restraint for practical application was, however, its low efficiency during winter months. In tropical countries under constant climatic conditions fish ponds might be the method of choice to clean wastewater from small settlements.

5 Widening the Theoretical Basis

In the peak time of trickling filters in the mid-twenties a very important observation was made by STREETER and PHELPS (1925) in the United States. Studying self-purification of the Ohio river by monitoring the biological oxygen demand, they found that the degradation of organic material closely followed the characteristics of a first-order reaction. From this time on the BOD was used as the method to measure wastewater pollution as well as treatment efficiencies and the self-purification of rivers. In addition, the temperature dependence of the biological degradation was observed, leading to a standardization of the BOD analytic method. The test had to be made at 20°C. At this temperature each day roughly 20% of the remaining pollutants were oxidized. After 5 d the oxidation of organics was completed and ammonia oxidation started.

The first-order theory for biological degradation greatly enhanced basic research. Especially in the United States all types of organic materials were tested for their biodegradability in wastewater treatment plants using the BOD test. However, a real biotechnological approach was hindered at that time by the strict orientation of civil engineers to the purely physicochemical approach of the adsorption theory. To change this thinking took more than 30 years. It was only in the 1950s and 1960s that the BOD reactions were really understood. In America the research of HOOVER and BUSH (e.g., HOOVER, 1911) led to a change. In Germany improvements were made by the author and his students (HARTMANN, 1992). Bacterial proliferation was recognized as the basis for the BOD reaction. Although the bacteriological background was never really denied, the first-order reaction to describe the process was comfortable to handle for engineers. BUSH defined the plateau BOD and found that it characterized the end of those reactions that were responsible for degradation of the dissolved organic material. The plateau BOD could be reached in the BOD test in less than 24 h and followed the rules of enzyme kinetics (HARTMANN, 1992). This gave a sound basis for the analysis of degradability of wastewater components of unknown composition and of new organics that were developed by the chemical industry. The results can be expressed and handled mathematically to again permit a sound design and operation of technologies. Oxygen consumption after the plateau BOD was reached resulted from endogenous respiration of bacteria and from ciliate activity, later on from nitrification. Thus, the technical limit for removal of organics from wastewater was reached when the biological oxygen consumption reached the plateau phase.

6 The Activated Sludge Process

The activated sludge process had already been invented by ARDERN and LOCKETT (1915) at the beginning of this century and was understood as a technique for larger cities as it required a more sophisticated mode of operation. The theory for purification was taken from trickling filter systems. Purification was thought to proceed in two steps; adsorption followed by biological oxidation. The activated sludge flocs were considered as a freely floating biofilm comprised of bacteria and protozoa that did the "purification job". Contrary to the early theory it was believed later on that special physicochemical conditions were necessary to create and stabilize the flocs. Biological research concentrated on the composition of the ciliate fauna at different loading rates. The ciliates were believed to contribute greatly to the removal of colloids, thus doing the "polishing job".

A great problem of activated sludge systems was sludge bulking for which different theories were developed, but technical answers were seldom found. More important questions concerned aeration and aeration techniques, since these caused the major costs in operating activated sludge plants. Much research was done, and numerous aeration devices were invented and competed with each other. The optimal oxygen concentration to satisfy the demand of the bacteria had to be considered. It was understood that about 0.5 ppm were sufficient for degraders of carbohydrates, but about 4 ppm were required for nitrifiers.

PASVEER was one of the pioneers to define the oxygenation capacity (OC value) of aeration systems (PASVEER, 1958a), thus providing an analytical method for comparison of different aeration systems.

According to the current theory, the organics were primarily "adsorbed" at the surface of the flocs in the activated sludge basins. After separation of the sludge it "had to be cleaned from adsorbed material in re-aeration chambers before it was brought back into the plant and exposed to new pollutants for adsorption". The adsorption theory, although scientifically wrong, was still in the mind of engineers and technicians up to the 1970s: "Adsorption activated sludge plants" were designed and built, but re-aeration of return sludge was given up in the 1960s. In this context it was of importance how and where to recyle the return sludge.

Activated sludge plants were in operation under different load rates. At a high space loading rate the oxidation was incomplete, whereas at a low load rate the oxidation was complete (DOHMANN, 1998). The aeration time ranged from less than 6 h to about 12 h, which was favored in the USA.

It was already understood very early that different loading rates led to different biological systems with different ciliate communities. Low loading rates (F/M <0.2) resulted in a complete oxidation, even of ammonia, whereas high loading rates (F/M $=2$) removed only the plateau BOD.

An important side effect of the activated sludge technology was the need to also improve the technologies of sludge digestion to cope with the huge amounts of surplus sludge. A faster treatment was required, which could only be obtained in heated digesters. The retention times of heated digesters ranged from 30 d to less than 10 d. To improve dewatering of sludge which had been stabilized at a short retention time the addition of chemical flocculation agents was required. Biogas formation during anaerobic sludge stabilization served for heat and electricity generation to reduce the costs of aerobic treatment.

It was still not understood that bacterial reactions under optimal conditions are not a matter of hours, but of minutes under the conditions given in municipal wastewater, with a low concentration of pollutants and a high concentration of bacteria. Research for a better understanding was performed in Switzerland by HÖRLER (1969) and by WUHRMANN and VON BEUST (1958) as soon as in the late 1950s.

They revealed a sound, almost mathematically exact relationship between the technical conditions of operation and the treatment efficiencies. Only the biological catalyzator still had to be added. In the engineering practice, however, although understanding the activated sludge process as a biotechnological process, its theoretical background was not accepted. All this resulted from the fact that biologists, especially microbiologists, still had not found their way into this field, although some progress for a better understanding had been made by HARTMANN and his students. They bridged the gap between the BOD process and the activated sludge technology. It could be shown that the different stages of the BOD process find their technical realization at different stages of the activated sludge treatment process. The load rate is mainly responsible for different biological systems to develop.

A special form of activated sludge process, the oxidation pond, was developed in the Netherlands by PASVEER (1958b, 1964) and found wide acceptance by small communities and by industry due to its simple construction and mode of operation.

7 Detergents: An Interplay of High Significance

At the end of the 1950s a problem arose which influenced the philosophy and wastewater treatment policy more than any technical invention: surface-active substances, washing powder. This problem had to be solved rather quickly, and it found a quick answer.

The replacement of soaps by washing detergents transformed waste treatment plants, especially activated sludge plants, into lakes of foam every morning. Most of the detergents used at that time were biologically undegradable or only of low degradability and left the plants as they came in. They polluted rivers and led to foam formation everywhere. As there was no technical way to prevent foam formation, a political answer had to be found. It came from legislation demanding biodegradable detergents and outlawing others. Degradability of all newly applied chemicals in washing powder was required and degradation tests had to be performed to prove biodegradability.

This first event of successfully outlawing certain chemicals opened the way for other steps to follow. In the years to come other laws were put into operation to limit the heavy metal ion content in sludges that were used as fertilizers in order to avoid heavy metal accumulation in plants. The so-called "undegradable rest pollution" and the chlorinated hydrocarbons were also critically considered and new standards for treatment efficiencies were set (see also Chapter 8, this volume). Violating these standards led to financial fines by state authorities.

The consequences of a generally more sensitive awareness of pollution were a reduction of wastewater quantities especially in industry by changes of production techniques. Wastes and wastewater could no longer be exported into public sewers free of cost. Disposal and purification in municipal wastewater treatment was charged according to the wastewater volume or the pollution freight. It was less expensive to reduce the waste and wastewater streams in the factory and to pretreat the residual amount within the factory than to hand it over to the public treatment plant as it was. In some cases waste was no longer waste but could be recycled as a "secondary raw material" either in the factory itself or within other industrial branches. This was especially enforced in Germany by the waste recycling law (KrWAbG, 1996: Kreislaufwirtschaftsgesetz, for details see Chapter 8, this volume). New markets had to develop selling the secondary raw material, including metals, glass, and paper, for which recycling in the past had already been enforced and hence the cycles were almost closed.

8 Treatment of Secondary Pollutants

In the early 1970s the scientific background for wastewater treatment was understood and optimized technologies were developed. The ideal plant for treatment of municipal wastewater was a sequence of a high-rate activated sludge basin for removal of the carbon compounds (detention time around 1 h) followed by a trickling filter for the purpose of ammonia oxidation to nitrate, degradation of organics with low biodegradability and polishing of the bacterial turbidity by ciliates. If a higher quality of treatment was required, a final carbon absorption unit could be added. The aeration basin could be split up into several smaller units to be brought into operation or taken out as required by the wastewater flow. To improve the economy of the plant. A primary sedimentation tank was sometimes considered unnecessary as most of the primary sludge consisted of bacteria, which stabilized the operation of the aeration chamber. The surplus sludge was subjected to anaerobic digestion to produce biogas to power the aerators for oxygen introduction.

This development was stopped by new requirements, arising from new (or already known, but not taken seriously) environmental problems. In the 1950s, phosphate removal was already required for wastewater treatment plants that fed their effluents into natural still waters in order to avoid eutrophication. In most cases chemical precipitation was applied.

In the 1980s the eutrophication problem became more serious. The Baltic Sea and the North Sea, receiving most of the wastewater from England, the Netherlands, Germany, and Denmark developed dangerous algal blooms, caused by an oversupply of nitrogen and phosphate. The "new problem" had to be fought by improved technologies of wastewater treatment (e.g., ATV, 1998). The theoretical background for biological elimination of nitrogen compounds was well known. Nitrogen removal was based on oxidation followed by denitrification, using the nitrates as oxygen source for respiration.

Elimination of phosphate was originally performed by chemical precipitation, but can also be obtained by accumulation of polyphosphates in bacteria (VAN LOOSDRECHT et al., 1997). The combination of N and P removal from wastewater within the normal treatment plant led to difficulties. Both processes required carbon sources which were not available in the required amounts. Discussions on the optimal technical solution are still in progress.

9 A Second Step Forward in Anaerobic Digestion

For quite a few years no progress was achieved in anaerobic sludge treatment systems. After the installation of heated reactors, anaerobic sludge digestion seemed to be without further potential for improvement. Impulses for a change started in the 1970s and are still in progress. Detailed studies on the complexity of the sludge population followed the early work of BUSWELL and SOLLO (1948) and revealed a better understanding of the ecological requirements, the physiology and biochemistry of the three mutualistically or syntrophically interacting groups (BRYANT, 1979; WOLIN and MILLER, 1982). With an understanding of the regulatory mechanisms of interaction it was possible to develop special technologies not only for sludge treatment, but also for treatment of highly concentrated liquid wastes. Up to that time these were fed either into activated sludge plants and caused bulking sludge formation or into trickling filter systems and caused clogging.

Whereas sludge treatment was performed in **c**ompletely **s**tirred **t**ank **r**eactors (CSTR), sometimes with sludge recycling in the contact process (SCHROEPFER et al., 1955), fixed-bed and fluidized-bed anaerobic digesters were invented and used for treatment of highly polluted wastewater in the laboratory and in practice (for reviews, see SPEECE, 1983; SAHM, 1984; SWITZENBAUM, 1983; WINTER, 1984). Some types of wastewater could be stabilized by **up**flow **a**naerobic **s**ludge **b**lanket (UASB) reactors, which were developed in the Netherlands (LETTINGA et al., 1980) and which, due to the pellet or granule formation, supplied optimal conditions for syntrophic growth of all members of the anaerobic population. In principle, a similar reactor was the Clarigester of MC CARTY (1982).

10 Gaps, Lacks, and Outlook

Summing up the history of wastewater treatment to date, it is characterized by a few simple facts. For more than half of the time scientific understanding lagged behind practical knowledge. The plants worked successfully, although the operators did not know what was the basis of their success. Civil engineers had taken over the task of wastewater treatment, they developed techniques, gained experience, and the results for most of the time satisfied the needs. Wastewater treatment for all this time was more a matter of art and personal skills of the plant operator than a result of scientific understanding. Engineers also successfully developed their own ideas and basic theories on biological processes, much to the advantage of their profession.

However, increased demands on purification efficiencies required a better and more detailed understanding of the biological processes that were the basis of wastewater treatment. This was developed scientifically in the last 20–30 years by chemical engineers and biologists, and the development is still in progress (e.g., VAN LIMBERGEN et al., 1998). Today's situation is characterized by the fact that practice lags behind scientific knowledge. Not everything which is understood can be realized technologically – and even if it could be realized, there might be economic handicaps, because the costs would be too high.

The main reason for insufficient treatment lies in the object itself. In most cases wastewater and its reactions are not defined like the chemicals and their reactions in a production process. For biological treatment wastewater is a sort of nutrition broth for the organisms and they work best under constant and steady conditions. Bacteria would, therefore, need an optimal physically and chemically defined environment and nutrients of known composition. None of these requirements are fully given in wastewater. Municipal wastewater, e.g., at every moment of the day is an integral of unknowns with respect to amount, nutrients, and even toxicants or inhibitors.

The main problem of wastewater treatment arises from the quality of its nutrient composition. A good nutrition broth for bacteria should have a C/N ratio of about 12. In reality, the C/N ratio of municipal sewage is about 4, indicating a surplus amount of nitrogen. The same holds true for phosphate. Even for the simple task of biodegradation of carbohydrates conditions are not optimal. Under the new, very recent requirement of nitrogen and phosphorus removal during wastewater treatment, an appropriate carbon supply is even more deficient. No other field of biotechnology has to cope with such problems, arising from the wastewater itself which cannot easily be corrected due to the huge amounts that have to be treated.

Wastewater purification cannot be solved in treatment plants merely as an end-of-pipe technology. One has to look for solutions at an earlier stage of production. Substances that cause severe problems during wastewater treatment have to be kept out of the wastewater or must be separated and recovered at a stage where the concentration is still high enough for recovery. Only unavoidable pollutants should be released into the wastewater.

11 References

ARDERN, E., LOCKETT, W. T. (1914), Experiments on the oxidation of sewage without the aid of filters, *J. Soc. Chem. Ind.* **33**, 523–539, 1122–1124 (part I and part II).

ARDERN, E., LOCKETT, W. T. (1915), Experiments on the oxidation of sewage sludge without the aid of filters, *J. Soc. Chem. Ind.* **34**, 937–943.

ATV (1998), *Biologische und weitergehende Abwasserreinigung*, 4th Edn. Berlin: Ernst & Sohn.

BRYANT, M. P. (1979), Microbial methane production – theoretical aspects, *J. Anim. Sci.* **48**, 193–202.

BUSWELL, R. A., et al. (1928), Removal of colloids from sewage, *The Illinois Engineer* (April and May 1928).

BUSWELL, A. M., SOLLO, F. W. (1948), The mechanism of methane formation, *J. Am. Chem. Soc.* **70**, 1778–1780.

DOHMANN, M. (1998), Weitergehende Abwasserreinigung, *Korrespondenz Abwasser* **45**, 1240–1251.

DUNBAR, W. PH. (1970), *Leitfaden für die Abwasserreinigungsfrage.* München: Oldenbourg.

FAIR, G. M., MOORE, E. W., THOMAS, H. A., Jr. (1941), The natural purification of river muds and pollutional sediments, *Sewage Works J.* **13**, 270–307, 756–779, 1209–1228.

FAIR, G. M. et al. (1948), Sewage treatment at military installations, *Sewage Works J.* **20**, 52–95.

FRÜHLING, A. (1910), *Handbuch der Ingenieurwissenschaften. Die Entwässerung der Städte, Flußverunreinigung und Behandlung städtischer Abwässer.* Leipzig: Verlag W. Engelsmann.

GALLERT, C., WINTER, J. (1997), Mesophilic and thermophilic anaerobic digestion of source-sorted organic wastes: effect of ammonia on glucose degradation and methane formation, *Appl. Microbiol. Biotechnol.* **48**, 405–410.

GARBER, W. F., OHARA, G. T., COLBAUGH, J. E., RAKSIT, S. K. (1975), Thermophilic digestion at the Hyperion treatment plant, *J. Water Pollut. Contr. Fed.* **47**, 950–961.

HALVORSON, H. O. (1936), Aero-filtration of sewage and industrial wastes, *Water Works Sewage* **9**, 307–315.

HARTMANN, L. (1992), *Biologische Abwasserreinigung*. Berlin, Heidelberg, New York: Springer-Verlag.

HOOVER, C. B. (1911), A method for determining the parts per million of dissolved oxygen consumed by sewage and sewage effluents, *Columbus Sewage Works – Engineering News* **65**, 311–312.

HÖRLER, A. (1969), Entwurf Absetzbecken. *Wiener Mitteilungen Wasser, Abwasser, Gewässer* **4**, C1.

IMHOFF, K. R. (1979), Die Entwicklung der Abwasserreinigung und des Gewässerschutzes seit 1868, *GWF Wasser Abwasser* **120**, 563–576.

IMHOFF, K. R. (1998), Geschichte der Abwasserentsorgung, *Korrespondenz Abwasser* **45**, 32–38.

IMHOFF, K., IMHOFF, K. R. (1993), *Taschenbuch der Stadtentwässerung*. 28th Edn. München, Wien: Oldenbourg.

KANDLER, O., WINTER, J., TEMPER, U. (1981), Methane fermentation in the thermophilic range, in: *Energy from Biomass*, 1st E.C. Conf. (PALZ, W., CHARTIER, P., HALL, D. O., Eds.), pp. 472–477. London: Appl. Sci. Publ.

KOLKWITZ, R. (1907), Über die biologische Selbstreinigung und Beurteilung der Gewässer, *Hyg. Rundsch.* **17**, 143–150.

KOLKWITZ, R. (1950), Die Ökologie der Saprobien, *Schriftenr. Ver. Wasser-, Boden- und Lufthyg.* **4**, 1–64.

KOLKWITZ, R., MARSSON, M. (1902), Grundsätze für die biologische Beurteilung des Wassers nach seiner Flora und Fauna, *Mitteilungen der königlichen Prüfungsanstalt für Wasserversorgung und Abwässerbeseitigung zu Berlin* **1**, 33–72.

KOLKWITZ, R., MARSSON, M. (1908), Ökologie der pflanzlichen Saprobien, *Ber. Dtsch. Bot. Ges.* **26a**, 505–519.

KOLKWITZ, R., MARSSON, M. (1909), Ökologie der tierischen Saprobien, *Int. Rev. Ges. Hydrobiol.* **2**, 126–152.

KrWAbG (1996), *Kreislaufwirtschaftsgesetz:* Gesetz zur Förderung der Kreislaufwirtschaft und Sicherung der umweltverträglichen Beseitigung von Abfällen. Bundestagsdrucksache 12/8084, Bundesratsdrucksache 654/94, zuletzt geändert am 25. 08. 1998, Bundesgesetzblatt I, p. 2455.

LETTINGA, G., VAN FELSON, A. F. M., HOBMA, S. W., DE ZEEUM, W., KLAPWIJK, A. (1980), Use of the upflow sludge blanket (UASB) reactor concept for the biological wastewater treatment, especially for anaerobic treatment, *Biotech. Bioeng.* **22**, 299–334.

LIEBMANN, H. (1960), *Handbuch der Frischwasser- und Abwasserbiologie*, Vol. 1. München: Oldenbourg.

LOCKETT, W. T. (1954), The evolution of the activated sludge process, *J. Proc. Ind. Sewage Purif.* **19**, 19–23.

MC CARTY, P. L. (1982), One hundred years of anaerobic treatment, in: *Proc. 2nd Int. Symp. Anaerobic Digestion* (HUGHES et al., Eds.), pp. 3–22. Amsterdam: Elsevier Biomedical.

ODUM, E. P. (1971), *Fundamentals of Ecology.* Philadelphia: Sauders.

PASVEER, A. (1958a), Über den Begriff OC/Load, *Münchener Beiträge zur Abwasser-, Fisch- und Flußbiologie* **5**, 240ff.

PASVEER, A. (1958b), Abwasserreinigung im Oxydationsgraben, *Bauamt und Gemeindebau* **31**, 78ff.

PASVEER, A. (1964), Über den Oxydationsgraben, *Schweiz. Z. Hydrol.* **26**, 466ff.

PETTENKOFER, M. (1890), Über Verunreinigung und Selbstreinigung der Flüsse, *Schilling's Journal für Gasbeleuchtung und Wasserversorgung* **33**, 415–421.

PETTENKOFER, M. (1891), Die Untersuchung der Isar auf Flußverunreinigung von München bis Ismaning und über die Selbstreinigung der Flüsse, *Deutsche Bauzeitung* **25**, 109–112.

PFEFFER, J. T. (1974), Temperature effects on anaerobic digestion of domestic refuse, *Biotechnol. Bioeng.* **16**, 771–787.

Royal Commission of Sewage Disposal (1908), *First Report, Part III:* Purification of sewage by treatment on land, pp. 137–158.

SAHM, H. (1984), Anaerobic wastewater treatment, in: *Advances in Biochemical Engineering/Biotechnology* **29** (FIECHTER, A., Ed.), pp. 83–115. Berlin, Heidelberg, New York: Springer-Verlag.

SCHROEPFER, G. J., FULLEN, W. J., JOHNSON, A. S., ZIEMKE, N. R., ANDERSON, J. J. (1955), The anaerobic contact process as applied to packing house wastes, *Sewage Ind. Wastes* **27**, 460–496.

SICKERT, E. (1998), Kanalisationen im Wandel der Zeit, *Korrespondenz Abwasser* **45**, 220–246.

SPEECE, R. E. (1983), Anaerobic biotechnology for industrial wastewater treatment, *Environ. Sci. Technol.* **17**, 416A–427A.

STANBRIDGE, H. H. (1976), *History of Sewage Treatment in Britain. 5. Land Treatment.* The Institute of Water Pollution Control, Maidstone.

STREETER, H. W., PHELPS, E. B. (1925), A study of the pollution and natural purification of the Ohio river. III. Factors concerned in the phenomena of oxidation and reaeration. *Public Health Service Bull.* **146**, Washington, DC.

SWITZENBAUM, M. S. (1983), Anaerobic fixed-film wastewater treatment, *Enzyme Microb. Technol.* **5**, 242–250.

USEPA (1981), *Process Design Manual for Land Treatment of Municipal Wastewater.* Environmental Protection Agency, Washington, DC.

van Limbergen, H., Top, E. M., Verstraete, W. (1998), Bioaugmentation in activated sludge: Current features and future aspects, *Appl. Microbiol. Biotechnol.* **50**, 16–23.

van Loosdrecht, M. C. M., Hooijmans, C. M., Brdjanovic, D., Heijnen, J. J. (1997), Biological phosphate removal processes, *Appl. Microbiol. Biotechnol.* **48**, 289–296.

Winter, J. (1984), Anaerobic waste stabilization, *Biotechnol. Adv.* **2**, 75–99.

Wolin, M. J., Miller, T. L. (1982), Interspecies hydrogen transfer – 15 years later, *ASM-News* **48**, 561–565.

Wuhrmann, K., von Beust, F. (1958), Zur Theorie des Belebtschlammverfahrens. II. Über den Mechanismus der Elimination gelöster organischer Stoffe aus dem Abwasser bei der biologischen Reinigung, *Schweiz. Z. für Hydrol.* **20**, 311–330.

2 Bacterial Metabolism in Wastewater Treatment Systems

CLAUDIA GALLERT
JOSEF WINTER
Karlsruhe, Germany

List of Abbreviations

MW	molecular weight
BOD	biological oxygen demand
COD	chemical oxygen demand
TOC	total organic carbon
TCA	tricarboxylic acid
HRT	hydraulic retention (residence) time
$NADH_2$	reduced nicotinamide adenosine dinucleotide
FdH_2	reduced ferredoxin
atm	atmospheres
K_s	half saturation constant

1 Introduction

Water that has been used by man and is disposed into a receiving water body with altered physical and/or chemical parameters is per definition referred to as wastewater. If only the physical parameters of the water were changed, e.g., resulting in an elevated temperature after use as a coolant, treatment before final disposal into a surface water may require only cooling close to its initial temperature. If the water, however, has been contaminated with soluble or insoluble organic or inorganic material, a combination of mechanical, chemical, and/or biological purification procedures may be required to protect the environment from periodic or permanent pollution or damage. For this reason legislation in industrialized and in many developing countries has reinforced environmental laws that regulate the maximum of allowed residual concentrations of carbon, nitrogen, and phosphorous compounds in the purified wastewater, before it is disposed into a river or into any other receiving water body (for details, see Chapter 3, 10, 13, 14, this volume). However, the reinforcement of these laws is not always very strict. It seems to be related to the economy of the respective country and thus differs significantly between wealthy industrialized and poor developing countries. In this chapter basic processes for biological treatment of waste or wastewater to eliminate organic and inorganic pollutants are summarized.

2 Decomposition of Organic Carbon Compounds in Natural and Man-Made Ecosystems

Catabolic processes of microorganisms, algae, yeasts, and lower fungi are the main pathways for a total or at least a partial mineralization/decomposition of bioorganic and organic compounds in natural or man-made environments. Most of this material is derived directly or indirectly from recent plant or animal biomass. It originates from carbon dioxide fixation via photosynthesis (→plant biomass), from plants that served as animal feed (→detritus, feces, urine, etc.) or from fossil, biologically or geochemically transformed biomass (→peat, coal, oil, natural gas). Even the carbon portion of some xenobiotics may be tracked back to a biological origin, namely if these substances were produced from oil, natural gas, or coal. Only due to the fact that the mineralization process for carbonaceous material of decaying plant and animal biomass in nature under anaerobic conditions with a shortage of water was incomplete, the formation of fossil oil, natural gas, and coal deposits from biomass through biological and/or geochemical transformation occurred. The fossil carbon of natural gas, coal, and oil enters the atmospheric CO_2 cycle again, as soon as these compounds are incinerated as fuels or for energy generation in industry and private households.

The biological degradation of recent biomass and of organic chemicals during solid waste or wastewater treatment proceeds either in the presence of molecular oxygen by respiration, under anoxic conditions by denitrification, or under anaerobic conditions by methanogenesis or sulfidogenesis. Respiration of soluble organic compounds or of extracellularly solubilized biopolymers such as carbohydrates, proteins, fats or lipids in activated sludge systems leads to the formation of carbon dioxide, water, and a significant amount of surplus sludge. Some ammonia and H_2S may be formed during degradation of sulfur-containing amino acids or heterocyclic compounds. Oxygen must either be supplied by aeration or by injection of pure oxygen. The two process variants differ mainly in their capacity for oxygen transfer and the "stripping efficiency" for carbon dioxide from respiration. Stripping of carbon dioxide is necessary to prevent a drop of pH and to carry out heat energy.

Respiration with "chemically bound oxygen" supplied in the form of nitrate or nitrite in the denitrification process abundantly yields dinitrogen. However, some nitrate escapes the reduction to dinitrogen in wastewater treatment plants and contributes about 2% of the total N_2O emission in Germany (SCHÖN et al., 1994; Chapter 14, this volume). Denitrifiers are aerobic organisms that switch their respiratory metabolism to the utilization of nitrate or nitrite as terminal electron acceptors, if grown under anoxic conditions. Only if the nitrate in the bulk mass has been used completely the redox potential will be low enough for growth of strictly anaerobic organsims, such as methanogens or sulfate reducers. If in sludge flocs of an activated sludge system anaerobic zones were allowed to form, e.g., by limitation of the oxygen supply, methanogens and sulfate reducers may develop in the center of sludge flocs and form the traces of methane and hydrogen sulfide found in the off-gas.

Under strictly anaerobic conditions soluble carbon compounds of wastes and wastewater are degraded step by step to methane, CO_2, NH_3, and H_2S via a syntrophic interaction of fermentative and acetogenic bacteria with methanogens or sulfate reducers. The complete methanogenic degradation of biopolymers or of monomers via hydrolysis/fermentation, acetogenesis, and methanogenesis can proceed only at a low H_2 partial pressure, which is maintained mainly by interspecies hydrogen transfer. Interspecies hydrogen transfer is facilitated, if acetogens and hydrogenolytic methanogenic bacteria are arranged in a close spatial neigborhood in flocs or in a biofilm at short diffusion distances. The reducing equivalents for carbon dioxide reduction to methane or sulfate reduction to sulfide are derived from the fermentative metabolism, e.g., of clostridia or *Eubacterium* sp., from β-oxidation of fatty acids, or the oxidation of alcohols. Methane and CO_2 are the main products in anaerobic environments where sulfate is absent, whereas sulfide and CO_2 are the main products if sulfate is present.

2.1 Basic Biology, Mass and Energy Balance of Aerobic Biopolymer Degradation

In order to make soluble and insoluble biopolymers – mainly carbohydrates, proteins, or lipids – accessible for respiration by bacteria, the macromolecules must be hydrolyzed by exoenzymes, which often are only produced and excreted after contact with respective inductors. The exoenzymes adsorb to the biopolymers and hydrolyze them to monomers or at least to oligomers. Only soluble, low molecular-weight compounds (e.g., sugars, disaccharides, amino acids, oligopeptides, glycerol, fatty acids) can be taken up by microorganisms and are metabolized to serve for energy production and cell multiplication.

Once taken up, degradation via glycolysis (sugars, disaccharides, glycerol), hydrolysis and deamination (amino acids, oligopeptides), or hydrolysis and β-oxidation (phospholipids, long-chain fatty acids) proceeds in the cells. Metabolism of almost all organic compounds leads to the formation of acetyl-CoA as the central intermediate, which is either used for biosyntheses, excreted as acetate, or oxidized to CO_2 and reducing equivalents in the tricarboxylic acid (TCA) cycle. The reducing equivalents are respired with molecular oxygen in the respiration chain. Only the energy of a maximum of 2 mol of anhydridic phosphate bonds of ATP is conserved during glycolysis of 1 mol of glucose through substrate chain phosphorylation. Further 2 mol of ATP are formed during oxidation of 2 mol of acetate in the TCA cycle, whereas 34 mol ATP are formed by electron transport chain phosphorylation in the respiration chain with oxygen as the terminal electron acceptor. During oxygen respiration reducing equivalents react with molecular oxygen in a controlled "Knallgas reaction".

When carbohydrates are respired by aerobic bacteria overall about one third of the initial energy content is lost as heat, and two thirds are conserved biochemically in 38 phosphoanhydride bonds of ATP. In activated sludge reactors or in wastewater treatment ponds, which are not loaded with highly concentrated wastewater, wall irradiation and heat losses with the off-gas stream of aeration into the atmosphere prevent self-heating. In activated sludge reactors for treatment of highly concentrated wastewater, however, self-heating up to the thermophilic temperature range may occur if the wastewater is warm in the beginning, the hydraulic retention time for biological treatment is short (short aeration time), and the air or oxygen stream for aeration is restricted to just supply sufficient oxygen for a complete oxidation of the pollutants (small aeration volume).

The conserved energy in the terminal phospho-anhydride bond of ATP, formed during substrate chain and oxidative phosphorylation of proliferating bacteria is partially used for maintenance metabolism of the existing cells and partially serves for cell multiplication. Partitioning between both is not constant, but depends on the nutritional state. In highly loaded activated sludge reactors with a surplus or at least a non-growth-limiting substrate supply approximately 50% of the substrate are respired in the energy metabolism of the cells and 50% serve as a carbon source for cell growth (Tab. 1). The biochemically conserved energy must be dissipated to serve for the maintenance metabolism of existing cells and for cell growth.

If the substrate supply is growth limiting, e.g., in a low-loaded aerobic treatment system a higher proportion of ATP is consumed for maintenance, representing the energy proportion that bacteria must spend for non-growth-associated cell survival metabolism, and less energy is available for growth. Overall, more of the substrate carbon is respired and the proportion of respiration products to surplus sludge formation is higher, e.g., around 70:30% (Tab. 1). In a trickling filter system apparently an even higher proportion of the substrate seems to be respired. This might be due to protozoa grazing off part of the biofilm.

For comparison, Tab. 1 also summarizes the carbon dissipation for anaerobic methanogenic degradation. Only about 5% of the fermentable substrate are used for cell growth (surplus sludge formation) in anaerobic reactors, whereas 95% are converted to methane and CO_2, and most of the energy of the substrates is conserved in the fermentation products.

2.1.1 Mass and Energy Balance for Aerobic Glucose Respiration and Sewage Sludge Stabilization

In most textbooks of microbiology respiration of organic matter is described exemplarily by Eq. 1, with glucose used as a model substance. Except for an exact reaction stoichiometry of the oxidative metabolism, mass and energy dissipation, if mentioned at all, is not quantified. Both parameters are, however, very important for activated sludge treatment plants. The surplus sludge formed during wastewater stabilization requires further treatment, causes disposal costs, and – in the long run – may be an environmental risk, whereas heat evolution during unevenly high-loaded aerobic treatment may shift the population towards more thermotolerant or thermophilic species and thus, at least for some time, may decrease the process efficiency.

$$\begin{array}{l}1\,\text{Mol}\ C_6H_{12}O_6 + 6\,\text{Mol}\ O_2 \rightarrow \\ 6\ \text{Mol}\ CO_2 + 6\,\text{Mol}\ H_2O + \text{Heat Energy}\end{array} \qquad (1)$$

If 1 mol of glucose (MW=180 g) is degraded in an activated sludge system at a high BOD loading rate (e.g., >0.6 kg m^{-3} d^{-1} BOD), approximately 0.5 mol (90 g) are respired to CO_2 and water by consumption of 3 mol of O_2 (96 g), releasing 19 mol of ATP (Fig. 1). The other 0.5 mol of glucose (90 g) are converted to pyruvate via one of three glycolytic pathways, accompanied by the formation of 0.5–1 mol ATP. Pyruvate or its subsequent metabolic products, e.g., acetate or dicarboxylic acids, are directly taken as carbon substrates for cell multiplication and surplus biomass formation.

Tab. 1. Carbon Flow during Aerobic Degradation in an Activated Sludge System under a) Saturating or b) Limiting Substrate Supply[a] and during Anaerobic Degradation

(A) **Aerobic degradation:**
- (a) Saturating substrate supply = high-load condition
  1 Unit Substrate Carbon → 0.5 Units CO_2 Carbon + 0.5 Units Cell Carbon
- (b) Limiting substrate supply = low-load condition
  1 Unit Substrate Carbon → 0.7 Units CO_2 Carbon + 0.3 Units Cell Carbon

(B) **Anaerobic degradation:**
  1 Unit Substrate Carbon → 0.95 Units (CO_2+CH_4) Carbon + 0.05 Units Cell Carbon

[a] Estimated from surplus sludge formation in different wastewater treatment plants

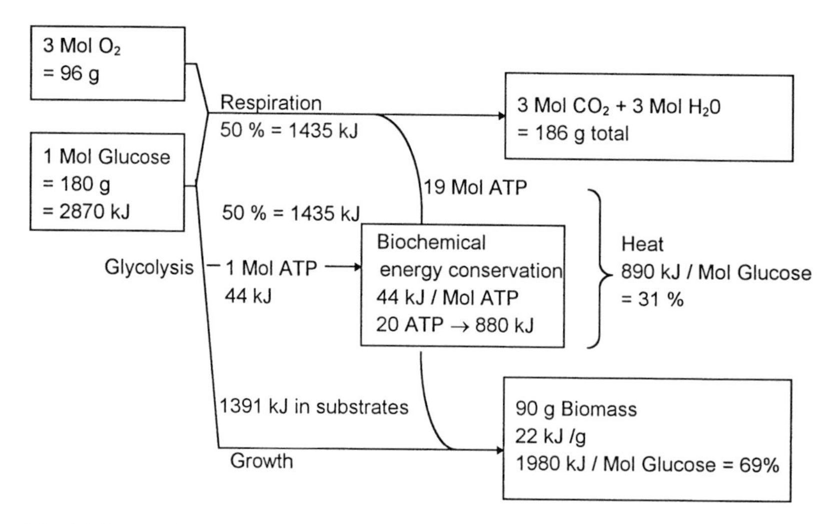

Fig. 1. Mass and energy dissipation during glucose respiration at pH 7.

A maximum amount of 20 mol ATP is thus available for growth and maintenance (Fig. 1). At a pH of 7 about 44 kJ of energy are available for growth per mol of ATP hydrolyzed to ADP and inorganic phosphate (THAUER et al., 1977). For an average molar growth yield of aerobes of 4.5 g per mol ATP (LUI, 1998) 90 g biomass can be generated from 180 g glucose. If the incineration energy per g of cell dry mass was 22 kJ, about 890 kJ (2,870 – 980 kJ) are lost as heat during respiration (Fig. 1). The energy loss is the sum of heat losses during respiration and cell growth.

At a low BOD loading rate the proportion of glucose respired in relation to the proportion of glucose fixed as surplus biomass may be shifted. Up to 0.7 mol (126 g) of glucose may be oxidized to CO_2, requiring 4.2 mol of oxygen (134.4 g O_2). Thus, for respiration of 1 mol of glucose different amounts of oxygen may be consumed, depending on the loading rate of the wastewater treatment system and different amounts of carbon dioxide and of surplus sludge formed (Fig. 1, Tab. 1).

The energy and carbon balance deduced above can be analogously transferred to aerobic stabilization of raw sewage sludge. If the initial dry matter content is around 36 g L^{-1} (average organic dry matter content of sewage sludge) and if a biodegradability of 50% within the residence time in the sludge reactor is obtained, about 9 g L^{-1} of new biomass are formed and thus 27 g L^{-1} (36 – 18 + 9) remain in the effluent.

The released heat energy is approximately 89 kJ per L reactor content. For an estimation of the theoretical temperature rise this amount of heat energy must be divided by 4,185 kJ (specific energy requirement for heating 1 L of H_2O from 14.5–15.5 °C). Thus, by respiration of 18 g L^{-1} organic dry matter the reactor temperature would increase by 21.3 °C within the residence time required for degradation ($\leq$16 h), provided that no heat energy is lost. A great proportion of the heat energy is, however, transferred via the liquid phase to the aeration gas and stripped out, whereas a smaller proportion is lost via irradiation from the reactor walls. Since air with almost 80% "ballast nitrogen" is normally used as a source of oxygen in aeration ponds or activated sludge reactors, the heat transfer capacity of the off-gas is high enough to prevent a significant increase of the wastewater temperature. Thus, ambient or at least mesophilic temperatures can be maintained. An increasing temperature of several °C would lead to a shift of the population in the reactor and – at least temporarily – would result in reduced process stability, whereas an only slightly increased temperature of a few °C might simply stimulate the metabolic activity of the prevalent mesophilic population. In practice in activated sewage sludge systems no self-heating is observed due to only about 50% degradability and a complete heat transfer with the off-gas into the atmosphere at a retention time of more than 0.5 d. If, however, wastewater from a dairy plant or from a brewery with a similar COD concentration, but with almost 100% biodegradable constituents would be stabilized with pure oxygen twice as much heat would evolve, leading to a theoretical temperature rise of 57 °C. Self-heating is observed, since the heat loss via much less off-gas is significantly lower and, due to higher reaction rates than with sewage sludge, the heat is generated during a shorter time span (shorter retention time).

2.1.2 Mass and Energy Balance for Anaerobic Glucose Degradation and Sewage Sludge Stabilization

For anaerobic wastewater or sludge treatment oxygen must be excluded to maintain the low redox potential that is required for survival and metabolic activity of the acetogenic, sulfidogenic, and methanogenic population. Hydrolysis of polymers, uptake of soluble or solubilized carbon sources and the primary metabolic reactions of glycolysis up to pyruvate and acetate formation seem to proceed identically or are at least analogous in aerobic and anaerobic bacteria. Whereas aerobes oxidize acetate in the TCA cycle and respire the reducing equivalents with oxygen, anaerobes, such as *Ruminococcus* sp., *Clostridium* sp., or *Eubacterium* sp., either release molecular hydrogen or transform pyruvate or acetate to highly reduced metabolites, such as lactate, succinate, ethanol, propionate, *n*-butyrate, etc. For further degradation within the anaerobic food chain these reduced metabolites must be oxidized anaerobically by acetogenic bacteria. Since the anaerobic oxidation of propionate or *n*-butyrate by acetogenic bacteria is obligately accompanied by hydrogen production, but is only slightly exergonic under conditions of a low H_2 partial pressure (BRYANT, 1979), acetogens can only grow when the hydrogen is consumed by hydrogen-scavenging organisms, such as methanogens or sulfate reducers.

During anaerobic degradation of 1 mol glucose approximately 95% of the glucose carbon is used for biogas formation (171 g $\triangleq$ 127.7 L CH_4+CO_2) and only about 5% of the substrate carbon (9 g) converted to biomass (Tab. 1). Much less heat energy is released during the anaerobic metabolism as compared to aerobic respiration (131 kJ mol^{-1} versus 890 kJ mol^{-1} during respiration) and the biogas contains almost 90% of the energy of the

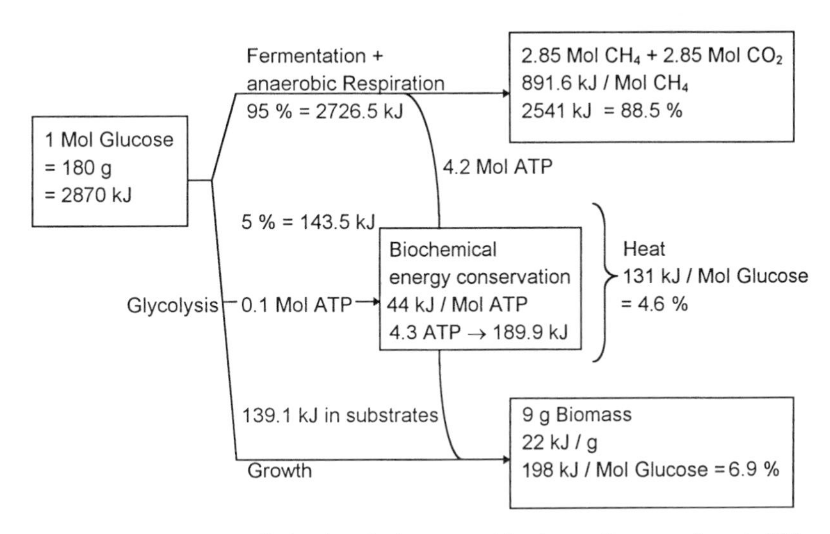

Fig. 2. Mass and energy dissipation during anaerobic glucose fermentation at pH 7.

fermented substrate (Fig. 2). Due to the heat energy requirement to warm up the wastewater and due to heat losses via irradiation from pipes and reactor walls, heat generation is by far not sufficient to maintain a constant mesophilic fermentation temperature. For this reason, anaerobic digesters must be heated.

In sewage sludge with 36 g L^{-1} organic dry matter content and 50% degradability, 0.9 g surplus sludge and 17.1 g biogas (equivalent to 12.75 L) would be formed during anaerobic stabilization. Only 13.1 kJ heat energy per mol of glucose are released, leading to a self-heating potential of 3.1 °C. Since the heat energy is only released within the hydraulic residence time of the wastewater in the reactor, which is usually more than 10 d (except for high-rate industrial wastewater treatment, where in special cases the HRT may be shorter than one day), much more heat energy is lost by irradiation via the reactor walls than would be required to maintain the actual temperature. If highly concentrated wastewater streams of the food and beverage industry are stabilized anaerobically at hydraulic retention times of <1 d (high space loading), more heat energy is generated within much shorter time. But even then process energy from external sources would have to be supplied to maintain a mesophilic environment at temperatures of 30–37 °C. Proper insulation of anaerobic reactors could minimize but not replace external heating of the reactor. The methane in the biogas, generated during anaerobic sludge or wastewater stabilization processes, contains about 90% of the energy of the fermented substrate. Since methane is a climate-relevant gas it may not be emitted into the atmosphere, but must be combusted to CO_2. Methane of anaerobic reactors may be used as fuel for gas engines to generate electricity and/or heat energy (Eq. 2).

$$CH_4 + 2\,O_2 \rightarrow CO_2 + 2\,H_2O + \text{Heat Energy} \quad (\Delta G^{\circ\prime} = -891.6\ \text{kJ mol}^{-1}\ CH_4) \qquad (2)$$

2.2 General Considerations for the Choice of Aerobic or Anaerobic Wastewater Treatment Systems

If a producer of wastewater has to make decision whether to install an aerobic or an anaerobic waste or wastewater treatment system, several points should be taken into consideration:

(1) Anaerobic treatment in general does not lead to the low pollution standards of COD, BOD_5, or TOC which can be met with aerobic systems and which are required by environmental laws (see Chapter 8, this volume). Anaerobic treatment of wastes and wastewater is often considered a pre-treatment process to minimize the oxygen demand and surplus sludge formation in a subsequent aerobic post-treatment stage. Only after a final aerobic treatment the COD, BOD_5, or TOC concentration limits stated in the environmental laws can be met. If boundary concentrations for nitrogen and phosphate have to be obeyed in addition, further treatment steps such as nitrification, denitrification, and biological or chemical phosphate removal must be considered.

(2) Highly concentrated wastewater should in general be treated anaerobically due to the possibility of energy recovery in biogas and due to much lower amounts of surplus sludge to be disposed. For aerobic treatment a high aeration rate would be necessary and much surplus sludge would be generated. Aeration would cause aerosol formation and eventually require off-gas purification.

(3) The efficiency of COD degradation for the bulk mass in concentrated wastewater or sludges (degradability of organic pollutants) by aerobic or anaerobic bacteria generally seems to be about similar. However, the degradation rates may be faster for aerobic treatment procedures than for anaerobic treatment procedures.

(4) Wastewater with a low concentration of organic pollutants should be treated

aerobically due to higher process stability at low pollutant concentrations, although aerobic treatment is more expensive and more sludge remains for disposal. If mineralized sludge is required, aerobic treatment at a low loading or at prolonged hydraulic retention times is necessary to reinforce respiration of all endogenous reserve material.

(5) Anaerobic treatment systems are more expensive to construct, but less expensive to operate than aerobic treatment systems.

2.3 Aerobic or Anaerobic Hydrolysis of Biopolymers – Kinetic Aspects

Hydrolysis of biopolymers and fermentation or respiration of the monomers may be catalyzed by strictly anaerobic, facultatively anaerobic, or aerobic microorganisms. With some exceptions (e.g., small protein molecules, dextran) biopolymers are insoluble and form fibers (cellulose), grains (starch), or globules (casein after enzymatic precipitation) or can be melted or emulsified like fat. HENZE et al. (1997) reported hydrolysis constants k_h for dissolved organic polymers of 3–20 d^{-1} under aerobic conditions and of 2–20 d^{-1} under anaerobic conditions, whereas for suspended solids the hydrolysis constants k_h were 0.6 to 1.4 d^{-1} under aerobic conditions and 0.3 to 0.7 d^{-1} under anaerobic conditions. For a kinetic description of hydrolysis and fermentation a substrate-limited first-order reaction was assumed by BUCHAUER (1997). It was deduced that the temperature-dependent reaction rate for hydrolysis is little lower than the reaction rate for fermentation of the hydrolysis products. The rate-limiting step is, therefore, hydrolysis of particles and not fermentation of solubilized material (BUCHAUER, 1997). Since hydrolysis is not only catalyzed by freely soluble exoenzymes, diluted in the bulk mass of liquid, but to a much higher extent by enzymes which are excreted in the neighborhood of bacterial colonies growing attached to the surface of the particles, the above mentioned description of complex fermentation processes is not always valid. Cellulases, may be arranged in cellulosomes, that attach to the particles, which serve as carrier material until they themselves are solubilized. For this reason, VAVILIN et al. (1997) included biomass in their description of hydrolysis of cellulose, cattle manure, and sludge. SHIN and SONG (1995) determined the maximum rates of acidification and methanation for several substrates. For hydrolysis of particulate organic matter the ratio of surface area to particle size is of importance. They found that for glucose, starch, carboxymethyl cellulose and casein or food waste from a restaurant hydrolysis proceeded faster than methanogenesis, whereas for newspaper and leaves hydrolysis was the rate-limiting step.

2.4 Hydrolysis of Cellulose by Aerobic and Anaerobic Microorganisms – Biological Aspects

Cellulose and lignin are the main shape- and structure-giving compounds of plants. Both substances are the most abundant biopolymers on earth. Cellulose fibers are formed of linear chains of 100–1,400 glucose units, linked together by β-1,4-glycosidic bonds. Inter- and intramolecular hydrogen bonds and van-der-Waal forces arrange the highly organized fibrillous regions (crystalline region), which alternate with less organized amorphous regions in the cellulose fibers. The fibers are embedded in a matrix of hemicelluloses, pectin, or lignin. The hemicelluloses consist mainly of xylans or glucomannans, which have side chains of acetyl, gluconuryl, or arabinofuranosyl units. To make cellulose fibers accessible for microorganisms, the hemicellulose, pectin, or lignin matrix must be degraded microbiologically or solubilized chemically. Cellulose degradation in the presence of oxygen in soil or in the absence of oxygen in the rumen of ruminants, in swamps, or in anaerobic digesters is the most important step in mineralization of decaying plant material. Cellulolytic organisms can be found among fungi, e.g., within the genera *Trichoderma* or *Phanaerochaete*, or bacteria, e.g., within the genera *Cellulomonas, Pseudomonas, Thermomonospora* (aerobic cellulose degraders) and *Clostridium*, *Fibrobacter*, *Bacteroides*, or *Ruminococcus* (anaerobic cellulose degraders). For more details on cellulolytic bacteria and the mechanism of cellulose cleavage the reader is referred to COUGHLAN and MAYER (1991).

In order to hydrolyze the insoluble cellulose polymers, cellulases – in some organisms organized in cellulosomes – must be excreted into the medium. Cellulases are complex biocatalysts and contain a catalytic site and a substrate-binding site. The presence of a non-catalytic substrate binding site permits tight attachment onto the different forms of the cellulose substrate without being washed off and keeps the enzyme close to its cleaving sites. Substrate binding is reversible to allow the enzyme to "hike" along the fibers in order to obtain total solubilization. Many aerobic fungi and some bacteria excrete endoglucanases that hydrolyze the amorphous region of cellulose (degradation within the chain), whereas exoglucanases would hydrolyze cellulose from the ends of the glucose chains. Cellobiose is cleaved off by cellobiohydrolases from the non-reducing ends in the amorphous region, but finally the cystalline region is also hydrolyzed.

Cleavage of cellobiose by β-glucosidases to glucose units is a necessity to avoid cellobiose accumulation, which inhibits the cellobiohydrolases (BE'GUIN and AUBERT, 1994).

Anaerobic bacteria such as *Clostridium thermocellum* form a stable enzyme complex, the cellulosome at the cell surface (LAMED and BAYER, 1988) Cellulosomes are very active in degrading crystalline cellulose. The catalytic subunits of a cellulosome, endoglucanases and xylanases, cleave the cellulose fiber into fragments, which are simultaneously degraded further by β-glucosidases. Cellulosome-like proteins are found also in *Ruminococcus* sp. and *Fibrobacter* sp. cultures. The cell-bound enzymes are associated with the capsule or the outer membrane. Other bacteria have not developed a mechanism to adhere to cellulose fibers, but excrete cellulases into the medium.

Adsorption of bacteria on cellulose fibers via cellulosomes offers the advantage of close spatial contact with the substrate, which is hydrolyzed mainly to glucose, that is taken up and metabolized. Small amounts of cellobiose are formed that induce cellulase expression. The contact of bacteria with the solid substrate surface keeps them in the neighborhood of cellobiose and thus keeps cellulase activities high. On the other hand, accumulation of hydrolysis products like glucose would repress cellulase activity.

In nature most cellulose is degraded aerobically. Only about 5–10% are supposed to be degraded anaerobically, which may be underestimated. Since most ecosystems are rich in carbonaceous substances, but deficient in nitrogen compounds, many cellulolytic bacteria can also fix nitrogen. This is of advantage for them and for other syntrophic or symbiontic partner organisms (LESCHINE, 1995). Other examples of mutualistic interactions between organisms of an ecosystem are: interspecies hydrogen transfer between anaerobes, transfer of growth factors, or production of fermentable substrates for the partner organisms.

Hydrolysis of biological structural components such as cellulose, lignin, or other shape-determining or reserve polymers (Tab. 2) is difficult. The limiting step of hydrolysis seems to be the liberation of the cleavage products. Contrary to the slow hydrolysis rates of celluloses, mainly due to lignin incrustation of naturally occurring celluloses, starch can be more easily hydrolyzed. Branching and the helical structure of starch facilitate hydrolysis (WARREN, 1996). Whereas cellulose forms fibers with a large surface, covered with lignin, starch forms grains with an unfavorable surface-to-volume ratio for enzymatic cleavage. Thus, although amylases may be present in high concentrations, the hydrolysis rate is limited due to the limited access of the enzymes to the substrate.

Whereas cellulose and starch are biodegradable, other carbohydrate-derived cellular compounds are not biodegradable and – after reaction with proteins – form humic acid-like residues in the Maillard reaction.

Tab. 2. Polysaccharides and Derivatives Occurring in Nature

Compound	Bond	Unit	Occurrance
Cellulose	β-1,4-	glucose	plant cell wall
Chitin	β-1,4-	N-acetyl glucosamine	fungal and insect cell wall
Murein	β-1,4-	N-acetyl-glucosamine N-acetyl-muramic acid	bacterial cell wall
Chitosan	β-1,4-	N-acetyl-glucosamine, substituted with acetyl residues	fungal and insect material
Mannans	β-1,4-	mannose	plant material
Xylans	β-1,4-	xylopyranose, substituted with acetyl-arabinofuranoside residues	plant material
Starch	α-1,4-	glucose amylose (contains few α-1,6-branches) amylopectin (contains many α-1,6-branches)	reserve material in plant
Glycogen	α-1,4-	glucose highly branched with α-1,6-bonds	reserve material in bacteria
Dextran	α-1,2- α-1,3- α-1,4- α-1,6-	glucose, glucose, glucose, glucose	reserve material of yeasts, exopolymer of bacteria
Laminarin	β-1,3-	glucose	reserve material in algae

2.5 Biomass Degradation in the Presence of Inorganic Electron Acceptors and by an Anaerobic Food Chain

In ecosystems in which molecular oxygen is available, plant and animal biomass is degraded to CO_2 and H_2O, catalyzed either by single species of aerobic microorganisms or the whole population of a respective ecosystem, competing for the substrates. A single organism may be able to hydrolyze the polymers and oxidize the monomers to CO_2 and H_2O with oxygen. In ecosystems where molecular oxygen is deficient – such as swamps, wet soil, the rumen of animals, the digestive tract of man, or in river and lake sediments – oxidation of the dead biomass proceeds anoxically by reduction of electron acceptors such as nitrate and nitrite or anaerobically by reduction of sulfate, Fe^{3+}, Mn^{4+}, or CO_2. The oxidation of the carbon source is either complete or incomplete with acetate excretion. In the absence of inorganic electron acceptors oxidized metabolites such as pyruvate or acetate are reduced to lactate or ethanol or biotransformed to, e.g., *n*-butyrate or *n*-butanol. In permanently anaerobic ecosystems with seasonal overfeeding, periodic accumulation of such metabolites might occur, e.g., in fall after the non-perennial plants drop their leaves or decay completely. The biopolymers of the leaves or the plants themselves decompose by extracellular enzymatic hydrolysis. The monomers are fermented and the fermentation products may be degraded further to biogas by acetogenic and methanogenic bacteria. Whereas single cultures of aerobes may catalyze the whole mineralization process to finally CO_2 and H_2O, single cultures of strictly anaerobic bacteria are not capable of degrading biopolymers to CH_4 and CO_2. Under anaerobic conditions biopolymers must be degraded by a food chain via depolymerization (hydrolysis), fermentation (acidogenesis), β-oxidation of fatty acids (acetogenesis), and biogas formation (methanogenesis) as the last step (McINERNEY, 1988). In an intitial exoenzyme-catalyzed reaction the biopolymers are hydrolyzed to soluble mono-, di-, or oligomers. These are taken up by the bacteria and fermented to CO_2, H_2, formate, acetate, propionate, butyrate, lactate, etc. If fatty acid isomers are produced, they mainly were derived from degradation of amino acids after proteolysis. The fatty acids are further oxidized by acetogenic bacteria, before the cleavage products, CO_2, H_2, and acetate, can be taken up by methanogens and are converted to methane and CO_2. Lactate is oxidized to pyruvate and the latter is decarboxylated to yield acetate, CO_2, and H_2. If ethanol is present, it is oxidized to acetate and hydrogen and the hydrogen is used for CO_2 reduction.

Tab. 3 summarizes the reactions that might be catalyzed by methanogens and that might contribute to methane emission in different ecosystems. In sewage digesters about two thirds of the methane were derived from acetate cleavage and one third from CO_2 reduction with H_2. If hexoses were the substrates and glycolysis was the main degradation pathway, then the 2 mol pyruvate might be decarboxylated by a pyruvate:ferredoxin oxidoreductase to yield 2 mol acetate and 2 mol CO_2. The hydrogen of the 2 mol $NADH_2$ from glycolysis and the 2 mol FdH_2 from pyruvate decarboxylation are then released as molecular hydrogen at low H_2 partial pressure (Eq. 3). 2 mol of CH_4 are then formed from acetate and 1 mol of CH_4 by CO_2 reduction (reactions 1 and 2 of Tab. 3).

$$1\,\text{Mol Glucose} \rightarrow 2\,\text{Mol Acetate} + 2\,\text{Mol } CO_2 + 4\,\text{Mol } H_2 \quad (\text{at low } pH_2) \tag{3}$$

In complex ecosystems formate is formed if high concentrations of hydrogen accumulate. Evidence for an interspecies formate transfer was reported (THIELE et al., 1988, Chapter 22, this volume). Due to a 30-times higher diffusion coefficient for formate than for hydrogen, interspecies formate transfer is supposed to play a major role for syntrophic butyrate (BOONE et al., 1989) and propionate degradation (STAMS, 1994; SCHINK, 1997). However, at an increased H_2 partial pressure formate is also produced by methanogens themselves, either in pure cultures or in a sewage sludge population (BLEICHER and WINTER, 1994) and this might also contribute to increasing formate concentrations. Other substrates for methanogenic bacteria (Tab. 3), such as methanol (derived e.g., from methoxy groups of lignin monomers) or methylamines and dimethylsulfide (e.g., from methylsulfonopropionate of algae; FRITSCHE, 1998) are only relevant in ecosystems where these substances are produced during microbial decay. A few methanogens might also use reduced products such as pri-

Tab. 3. Reactions Catalyzed by Methanogens and Standard Changes in Free Energy

Reaction Substrates [Mol]		Products [Mol]	$\Delta G^{0'}$ [kJ per mol of methane]
Acetate	⇒	$CH_4 + CO_2$	−31.0
$4H_2 + CO_2$	⇒	$CH_4 + 2H_2O$	−135.6
4HCOOH	⇒	$CH_4 + 3CO_2 + 2H_2O$	−130.1
$4CO + 2H_2O$	⇒	$CH_4 + 3CO_2$	−211.0
4 Methanol	⇒	$3CH_4 + CO_2 + 2H_2O$	−104.9
Methanol + H_2	⇒	$CH_4 + H_2O$	−112.5
2 Ethanol[a] + CO_2	⇒	CH_4 + 2 Acetate	−116.3
4 2-Propanol[b] + CO_2	⇒	CH_4 + 4 Acetone + $2H_2O$	−36.5
4 Methylamine + $2H_2O$	⇒	$3CH_4 + CO_2 + 4NH_3$	−75.0
2 Dimethylamine + $2H_2O$	⇒	$3CH_4 + CO_2 + 2NH_3$	−73.2
4 Trimethylamine + $6H_2O$	⇒	$9CH_4 + 3CO_2 + 4NH_3$	−74.3
2 Dimethylsulfide + $2H_2O$	⇒	$3CH_4 + CO_2 + 2H_2S$	−73.8

[a] Other primary alcohols serving as hydrogen donors for CO_2 reduction are 1-propanol and 1-butanol.

[b] Other secondary alcohols serving as hydrogen donors for CO_2 reduction are 2-butanol, 1,3-butanediol, cyclopentanol, and cyclohexanol ([a,b] restricted to a few species).

Compiled from WHITMAN et al. (1992) and WINTER (1984).

mary, secondary, and cyclic alcohols as a source of electrons for CO_2 reduction (WIDDEL, 1986; ZELLNER and WINTER, 1987a; BLEICHER et al., 1989; ZELLNER et al., 1989).

2.6 The Role of Molecular Hydrogen and of Acetate During Anaerobic Biopolymer Degradation

Molecular hydrogen is produced during different stages of anaerobic degradation. In the fermentative stage organisms like *Clostridium* sp. or *Eubacterium* sp. produce fatty acids, CO_2, and hydrogen from carbohydrates. In the acetogenic stage acetogens such as *Syntrophobacter wolinii* or *Syntrophomonas wolfei* produce acetate, CO_2, and hydrogen or acetate and hydrogen by anaerobic oxidation of propionate and *n*-butyrate (MCINERNEY, 1988). Whereas the fermentative bacteria may release molecular hydrogen even at a high H_2 partial pressure and simultaneously excrete reduced products (e.g., clostridia, *Ruminococcus, Eubacterium* sp.), the release of molecular hydrogen during acetogenesis of fatty acids or of other reduced metabolites for thermodynamic reasons may only function, if hydrogen does not accumulate, but is consumed by methanogens (Tab. 4, reaction 1) or, alternatively, by sulfate reducers (Tab. 4, reaction 2) via interspecies hydrogen transfer. In the rumen and in sewage sludge digesters the hydrogen concentration may be decreased by acetate formation from CO_2 and H_2 (Tab. 4, reaction 3). Some additional reactions consuming hydrogen to decrease its concentration are also listed in Tab. 4 (reactions 4–6).

To maintain a low H_2 partial pressure a syntrophism of acetogenic, hydrogen-producing and methanogenic, hydrogen-utilizing bacteria is essential (IANOTTI et al., 1973). Complete anaerobic degradation of fatty acids with hydrogen formation by obligate proton-reducing acetogenic bacteria is only possible at H_2 partial pressures of $<10^{-4}$ (*n*-butyrate) or 10^{-5} atm (propionate), which cannot be maintained by methanogens or sulfate reducers. However, by reversed electron transport electrons may be shifted to a lower redox potential suitable for proton reduction (SCHINK, 1997). If hydrogen accumulates beyond this threshold concentration, the anaerobic oxidation of fatty acids will get endergonic and would not proceed (for details, see Chapter 22, this volume). Whereas hydrogen prevents β-oxidation of fatty acids by acetogens already at very low H_2 partial pressure, much higher concentrations of acetate in the millimolar range are required for the same effect.

The fermentative metabolism of acidogenic bacteria is exergonic still at H_2 partial pressures $>10^{-4}$ atm. Whereas acetogenic bacteria apparently depend mainly on ATP generation by chemiosmotic phosphorylation, fermentative bacteria gain a great portion of ATP from substrate chain phosphorylation. This may be the reason why fermentative bacteria do not depend on a syntrophic interaction with electron-thriving bacteria, such as methanogens or sulfate reducers.

Tab. 4. Hydrogen-Consuming Reactions in Anaerobic Ecosystems (SCHINK, 1997)

Reaction			$\Delta G^{0'}$ [kJ per mol]
(1)	$4H_2+CO_2$	⇒ CH_4+2H_2O	−131.0
(2)	$4H_2+SO_4^{2-}$	⇒ $S^{2-}+4H_2O$	−151.0
(3)	$4H_2+2CO_2$	⇒ $CH_3COO^-+H^++2H_2O$	−94.9
(4)	H_2+S^0	⇒ H_2S	−33.9
(5)	$H_2C(NH_3^+)COO^-+H_2$	⇒ $CH_3COO^-+NH_4^+$	−78.0
(6)	$COOH-CH=CH-COOH+H_2$	⇒ $COOH-CH_2-CH_2-COOH$	−86.0

2.7 Anaerobic Conversion of Biopolymers to Methane and CO_2

The principle of the anaerobic metabolism of biopolymers was outlined by BRYANT (1979). There was an essential requirement for a syntrophic interaction between different metabolic groups for a complete anaerobic degradation (WOLIN, 1976, 1982). The most sensitive "switch" of the carbon flow of substrates to biogas was the H_2 partial pressure. The substrate supply for biomethanation processes should be limited in such a way that the most slowly growing group of the food chain, the obligately proton-reducing acetogens, were still able to excrete hydrogen at a maximum rate, but at the same time hydrogen accumulation $>10^{-5}$ atm was prevented by an active methanogenesis or sulfate reduction (BRYANT, 1979). Whereas hydrogen seemed to be the most sensitive regulator of anaerobic degradation formate (BLEICHER and WINTER, 1994), acetate, or other fatty acids could accumulate to much higher concentrations (MCINERNEY, 1988) and still did not repress anaerobic degradation reactions. If hydrogen accumulated due to an oversupply or an inhibition of methanogens, anaerobic biodegradation was disturbed successively at different stages: Initially, at the acetogenic stage β-oxidation of fatty acids and alcohols failed, leading to an accumulation of these acid metabolites. Later on the spectrum of metabolites of the fermentative flora changed towards "more reduced products" like ethanol, lactate, propionate, and *n*-butyrate, leading to an even higher concentration of volatile fatty acids and a further decreasing pH. At a pH below 6.5 methanogenic reactions were almost completely prevented. At this stage acidification with a rich spectrum of reduced products may still proceed.

2.7.1 The Anaerobic Degradation of Carbohydrates in Wastewater

Carbohydrates are homo- or heteropolymers of hexoses, pentoses, or sugar derivatives, which occur in soluble form or as particles, forming grains of different size or fibers. In some plants starch forms grains of up to 1 mm size, which is 1,000-fold the diameter of bacteria. Starch metabolism by bacteria requires hydrolytic cleavage by amylases to soluble mono- or dimers, since only soluble substrates can be taken up and are metabolized.

The anaerobic degradation of biopolymers in general or of cellulose in particular can be divided into a hydrolytic, fermentative, acetogenic, and methanogenic phase (Fig. 3). Hydrolysis and fermentation of the hydrolysis products may be catalyzed by the same trophic group of microorganisms. The distinction of both phases is more of theoretical than of practical relevance. Concerning reaction rates in a methane fermenter which is fed with a particulate substrate, the rate-limiting step is hydrolysis rather than the subsequent fermentation of the monomers, if acetogenesis and methanogenesis still proceed faster. Hydrolysis rates of polymers may be very different. Hemicellulose or pectin are hydrolyzed ten times faster than lignin-encrusted cellulose (BUCHHOLZ et al., 1986, 1988). In the acidification reactor of a two-stage anaerobic process hydrolysis of polymers to monomers normally is slower than fermentation of monomers to fatty acids and other fermentation products. For this reason no sugar monomers can be detected during steady state operation. In the methane reactor β-oxidation of fatty acids, especially of propionate or *n*-butyrate, is the rate-limiting step (BUCHHOLZ et al., 1986). Fatty acid degradation is the slowest reaction overall in a two-stage methane reactor fed carbohydrate-containing wastewater from sugar production. Thus the methane reactor has to be sized bigger than the acidification reactor to permit longer hydraulic retention times.

The rate of cellulose degradation is highly dependent on the state in which cellulose is present in the wastewater. If cellulose is lignin-encrusted, lignin prevents access of cellulases to hydrolyze the cellulose fibers. If cellulose is prevalent in a crystalline form, cellulases can easily attach to it, and then hydrolysis may be a relatively fast process. At increasing loading in an anaerobic reactor fed crystalline cellulose acetogenesis first became the rate-limiting process, leading to propionate and butyrate formation (WINTER and COONEY, 1980). In decaying plant material cellulose very often is

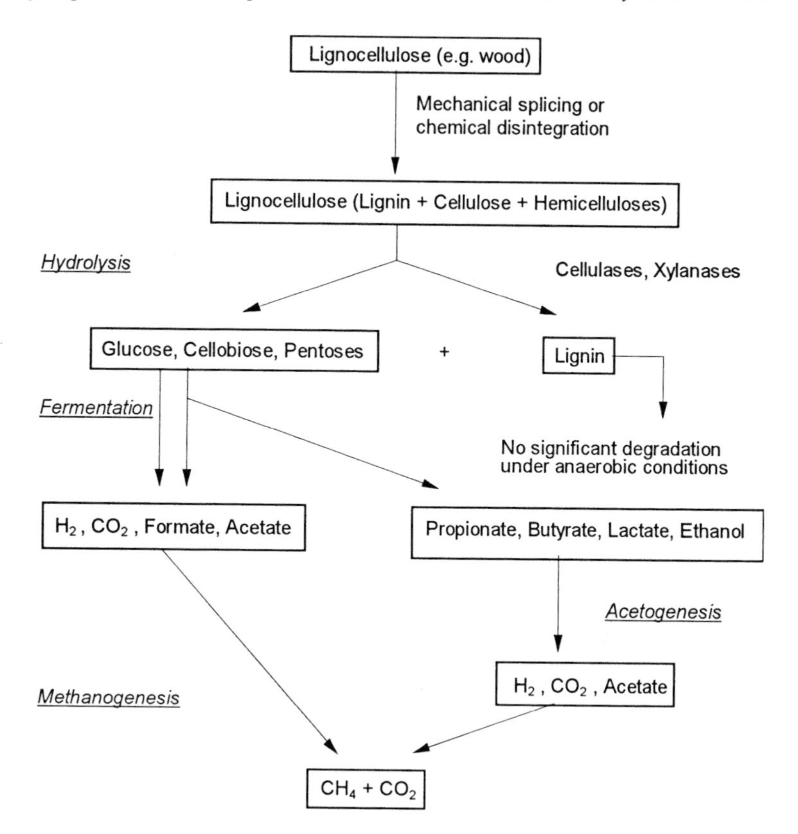

Fig. 3. Anaerobic degradation of lignocellulose or cellulose to methane and CO_2 (according to ATV, 1994).

lignin-encrusted. Due to the highly restricted access of these complexes for cellulases, hydrolysis of cellulose becomes the rate-limiting step of degradation to methane and CO_2.

Whether microorganisms are capable of degrading lignin under anaerobic conditions is still under discussion. In a natural environment without time limitation lignin was reported to be degraded anaerobically (COLBERG and YOUNG, 1985; COLBERG, 1988). However, since these results were based on long-term experiments performed *in situ* and anaerobiosis was not controlled, it remains doubtful, whether the little amount of lignin which disappeared was really degraded under strictly anaerobic conditions. The occurrence of coal and fossil oil more or less indicates that lignin compounds are highly resistant to microbial attack.

During anaerobic degradation of starch, hydrolysis by amylases proceeds with high velocity if a good contact between starch grains and amylases is maintained. Whereas in an anaerobic reactor at low loading starch degradation may proceed without the requirement of acetogenic bacteria, as pointed out in Fig. 4 route a, at high loading volatile fatty acids are formed and acetogens are essential for a total degradation (Fig. 4, route b). Fig. 4 illustrates that the requirement for the rate-limiting acetogenic reactions may be avoided by a limitation of the substrates in such a way that hydrolysis and fermentation are maximally as fast as methanogenesis. The rationale behind this is, that many fermentative bacteria produce only acetate, formate, CO_2, and hydrogen, if H_2-scavenging methanogens or sulfate reducers are able to maintain a sufficiently low H_2 partial pressure, whereas a wide spectrum of fermentation products, typical for the metabolism of the respective bacterium in pure culture, is produced at a higher H_2 partial pressure (WINTER, 1983, 1984). A disturbance of methanogenesis in continuous syntrophic methanogenic cultures may be caused by high spike concentrations of sugars or – at low concentrations of sugars – by inhibiting substances like NH_3, H_2S, antibiotics (HAMMES et al., 1979; HILPERT et al., 1981), or xenobiotics. As a consequence, the H_2 partial pressure increases and volatile fatty acids are generated (WINTER, 1984; WINTER et al., 1989; WILDENAUER and WINTER, 1985; ZELLNER and WINTER, 1987b). Once propionate or *n*-butyrate are produced, anaerobic degradation requires acetogens for β-oxidation (Fig. 4, route b).

2.7.2 The Anaerobic Degradation of Protein

Proteins are biological macromolecules, either soluble or solid (e.g., feathers, hair, nails). Outside the cell at an acid pH or in the presence of enzymes soluble proteins form precipitates, e.g., precipitation of casein by addition of rennet enzyme. The reaction sequences necessary for protein degradation in a methanogenic ecosystem are outlined in Fig. 5. Hydrolysis of precipitated or soluble protein is catalyzed by several types of proteases that cleave off membrane-permeable amino acids, dipeptides or oligopeptides. In contrast to the hydrolysis of carbohydrates, that proceeds favorably at a slightly acid pH, optimal hydrolysis of protein requires a neutral or weak alkaline pH (MCINERNEY, 1988). In contrast to the fermentation of carbohydrates, which lowers the pH due to volatile fatty acid formation, fermentation of amino acids in wastewater reactors does not lead to a significant pH change due to acid and ammonia formation. Acidification of protein-containing wastewater proceeds optimally at pH values of 7 or higher (WINTERBERG and SAHM, 1992) and ammonium ions together with the CO_2/bicarbonate/carbonate buffer system stabilize the pH. Alcetogenesis of fatty acids from deamination of amino acids requires a low H_2 partial pressure for the same

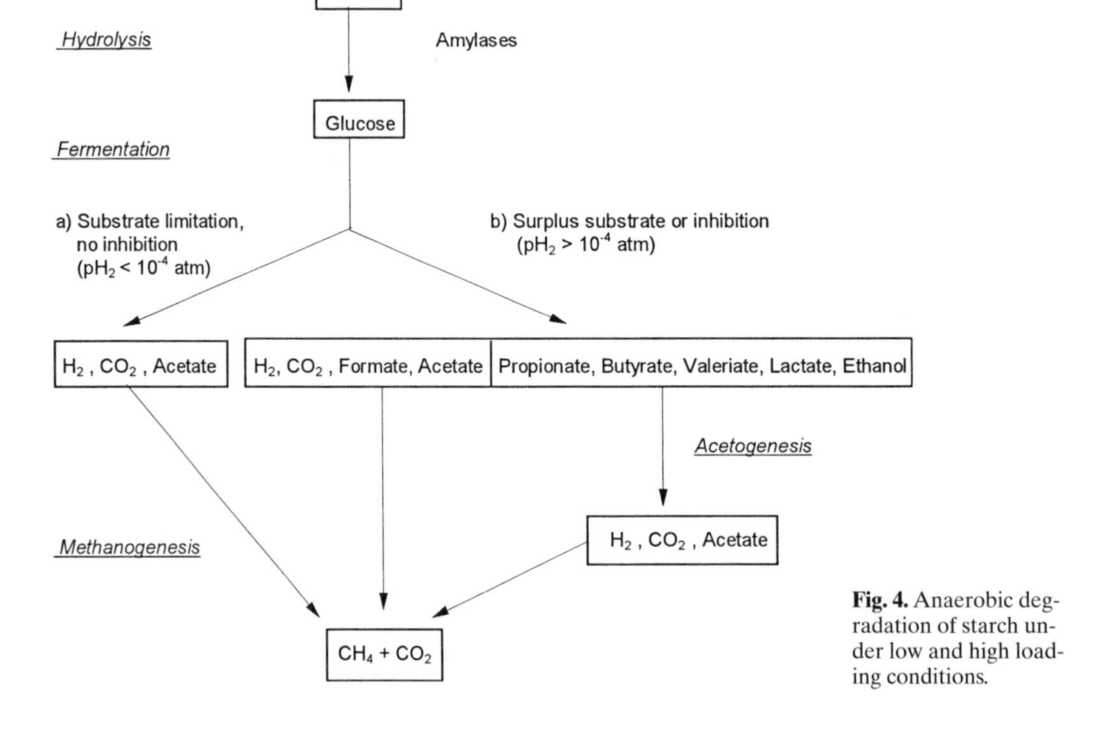

Fig. 4. Anaerobic degradation of starch under low and high loading conditions.

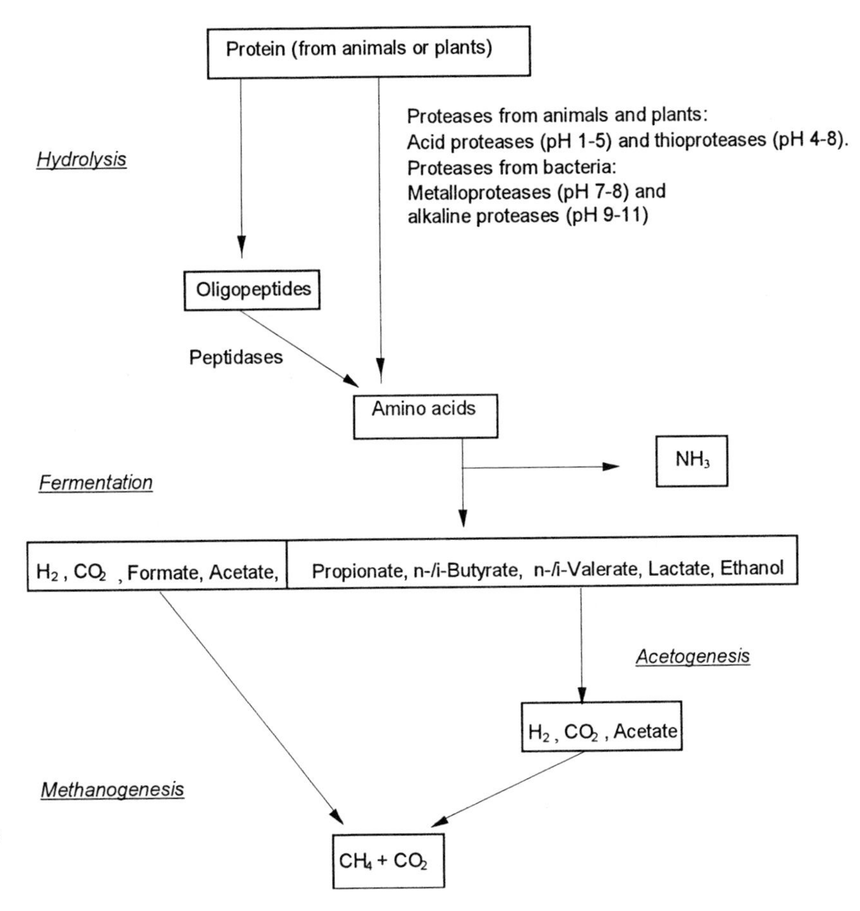

Fig. 5. Anaerobic degradation of protein.

reasons as mentioned for carbohydrate degradation. This can be maintained by a syntrophic interaction of fermentative, protein-degrading bacteria, acetogenic and methanogenic or sulfate-reducing bacteria. Except for syntrophic interaction of amino acid-degrading bacteria with methanogens for maintenance of a low H_2 partial pressure, clostridia and presumably also some other sludge bacteria may couple an oxidative and a reductive amino acid conversion via the Stickland reaction. One amino acid, e.g., alanine (Eq. 4) is oxidatively decarboxylated and the hydrogen or reducing equivalents produced during this reaction serve to reductively convert another amino acid, e.g., glycine to acetate and ammonia (Eq. 5).

$$CH_3-CHNH_2-COOH+2H_2O \rightarrow CH_3-COOH+CO_2+NH_3+2H_2 \quad \Delta G^{o\prime} = +7.5 \text{ kJ mol}^{-1} \tag{4}$$

$$2CH_2NH_2-COOH+2H_2 \rightarrow 2CH_3COOH + 2NH_3 \quad \Delta G^{o\prime} = -38.9 \text{ kJ mol}^{-1} \tag{5}$$

For complete degradation of amino acids in an anaerobic system, therefore, a syntrophism of amino acid-fermenting anaerobic bacteria with methanogens or sulfate reducers is required (WILDENAUER and WINTER, 1986; WINTER et al., 1987; ÖRLYGSSON et al., 1995). If long-chain amino acids are deaminated (Eqs. 6–9) fatty acids such as propionate, *i*-butyrate, or *i*-valerate are formed directly by deamination. The fatty acids require acetogenic bacteria for their degradation.

$$\text{Valine} + 2\,H_2O \rightarrow i\text{-Butyrate} + CO_2 + NH_3 + 2\,H_2 \quad (6)$$

$$\text{Leucine} + 2\,H_2O \rightarrow i\text{-Valerate} + CO_2 + NH_3 + 2\,H_2 \quad (7)$$

$$i\text{-Leucine} + 2\,H_2O \rightarrow \text{2-Methylbutyrate} + CO_2 + NH_3 + 2\,H_2 \quad (8)$$

$$\text{Glutamate} + 2\,H_2O \rightarrow \text{Propionate} + 2\,CO_2 + NH_3 + 2\,H_2 \quad (9)$$

In contrast to carbohydrate degradation, where the necessity for propionate- and butyrate-degrading acetogenic bacteria can be circumvented by substrate limitation as outlined in Fig. 4a, during protein degradation these fatty acids are a product of deamination, and their formation cannot be avoided by maintaining a low H_2 partial pressure. In the methanogenic phase there is no difference whether carbohydrates or proteins are fermented, except that the methanogens in a reactor fed protein should be more tolerant to ammonia and a higher pH.

2.7.3 The Anaerobic Degradation of Neutral Fat and Lipids

Fat and lipids are another group of biopolymers that contribute significantly to the COD in sewage sludge, cattle and swine manures, or wastewater from the food industry, e.g., slaughterhouses, potato chip factories (WINTER et al., 1992; BROUGHTON et al., 1998). To provide a maximum surface for hydrolytic cleavage by lipases or phospholipases, solid fat, lipids, or oil must be emulsified. Glycerol, saturated and unsaturated fatty acids (palmitic acid, linolic acid, linolenic acid, stearic acid, etc.) are formed from neutral fat. Lipolysis of phospholipids generates fatty acids, glycerol, higher alcohols (serine, ethanolamine, choline, inositol), and phosphate. Lipolysis of sphingolipids generates fatty acids and amino alcohols (e.g., sphingosine), and lipolysis of glycolipids fatty acids, amino alcohols, and hexoses (glucose, galactose). A scheme for anaerobic degradation of fat is shown as an example in Fig. 6. Sugar moieties and glycerol may be degraded to methane and CO_2 by an interaction of fermentative and methanogenic bacteria in low-loaded systems or by fermentative, acetogenic, and methanogenic bacteria in high-loaded systems. The long-chain fatty acids are degraded by acetogenic bacteria via β-oxidation to acetate and molecular hydrogen. If they accumulate, the anaerobic digestion process will be inhibited (HANAKI et al., 1981). Odd-numbered fatty acids are degraded to acetate, propionate, and hydrogen, even-numbered fatty acids to acetate and hydrogen (BRYANT, 1979). Only at a very low H_2 partial pressure, that can be maintained by hydrogen-utilizing methanogens or sulfate reducers, β-oxidation of at least *n*-butyrate or propionate is exergonic. Methanol, ethanol, and ammonia are formed from choline (Fig. 6). After hydrolysis, fermentation, and acetogenesis of the fat components in the methanogenic phase acetate, CO_2, and hydrogen are converted to biogas. All subsequent intracellular reactions may be influenced by syntrophic interaction via interspecies hydrogen transfer except for the extracellular initial lipase reaction.

Carbohydrates, protein, and fat or biogenic oil may also be degraded anaerobically under thermophilic incubation conditions. The overall degradation scheme is the same, but populations are different (e.g., WINTER and ZELLNER, 1990). Thermophilic fat degradation practically is getting more important, since waste fat from fat separators and fat flotates of the food industry are often co-fermented in agricultural biogas plants. Since for hygienic reasons the input material must be autoclaved in any case, a thermophilic process should be applied, keeping the fat in a melted, soluble form for a more effective metabolism. Biogas plants with co-fermentation of waste fat residues are considered waste treatment systems (Chapter 11, this volume) and must be designed to meet the hygienic demands relevant for treatment of the respective waste.

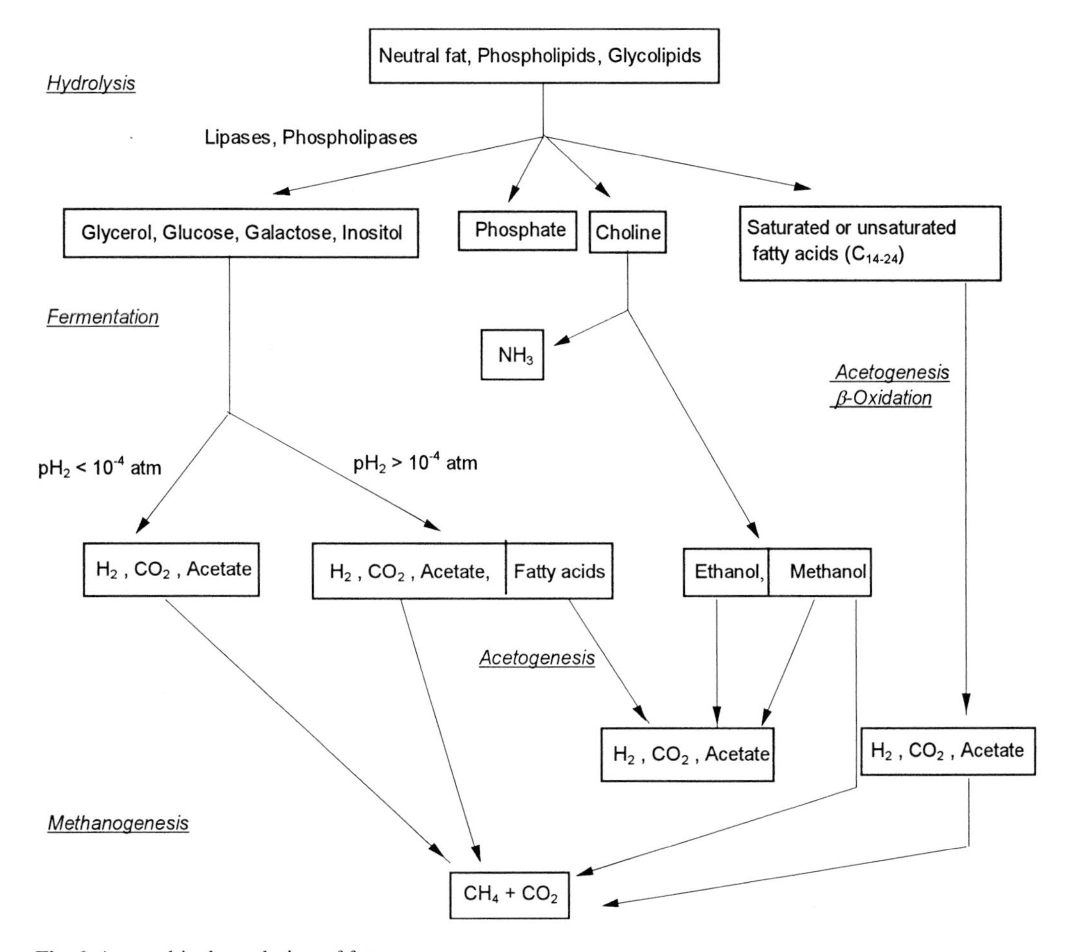

Fig. 6. Anaerobic degradation of fat.

2.8 Competition of Sulfate Reducers with Methanogens in Methane Reactors

Municipal wastewater or wastewater from sugar production, slaughterhouses, breweries, etc., normally contain less than 200 mg sulfate per liter. If sulfuric acid is used, e.g., for cleaning stainless steel containers and pipes in the dairy industry, to maintain an acid pH in bioreactors for bakers' yeast or citric acid production or if ammonia sulfate is used to inhibit metabolic routes in bakers' yeast for the production of biochemicals, the wastewater contains large amounts of sulfate. Furthermore, sulfite-containing wastewater is generated by the starch or cellulose industry during bleaching of the raw products.

If these wastewaters are subjected to an anaerobic treatment the methanogenic bacteria must compete with sulfate reducers for the hydrogen equivalents from COD degradation (Omil et al., 1998; Chapter 22, this volume). In anaerobic digesters sulfate reduction is the favored reaction due to a higher affinity of sulfate reducers for reducing equivalents or hydrogen leading to sulfide production at the expense of reduced biogas formation. Sulfide subsequently forms heavy metal precipitates and, if still present in larger amounts, remains in solution and may be toxic for acetogens and methanogens. Some hydrogen sulfide leaves the reactor with the biogas, which may contain up to a few percent of H_2S. Then gas purification is necessary before the gas can be used as fuel for gas engines. The higher affinity of sulfate reducers for reducing equivalents as compared to the methanogens can technically be used for sulfate and heavy metal ion removal from wastewater in the first stage of a two-stage anaerobic process (ELFERINK et al., 1994).

In comparison to methanogens, which have a rather restricted substrate spectrum (Tab. 3), sulfate reducers are metabolically more versatile. Sulfate reducers may utilize polymers such as starch, monomers such as sugars, fatty acids, formate, aliphatic and aromatic compounds as well as molecular hydrogen (WIDDEL, 1988) to generate reducing equivalents for sulfate reduction (Eq. 10).

$$8(H)+2H^{+}+SO_4^{2-}\rightarrow H_2S+4H_2O \quad (10)$$

Undissociated hydrogen sulfide is toxic for methanogens and sulfate reducers as well. A 50% inhibition of methanogenesis was observed at a total sulfide concentration of 270 mg L^{-1} (OLESZKIEWICZ et al., 1989), whereas 85 mg L^{-1} sulfide inhibited sulfate reducers (MCCARTNEY and OLESZKIEWICZ, 1991). For stable methanogenesis not more than 150 mg L^{-1} sulfide should be accumulated (SPEECE et al., 1986). H_2S toxicity can be avoided by intensive flushing with biogas to strip the H_2S for adsorption onto $Fe_2(OH)_3$. If large amounts of H_2S are formed during anaerobic treatment, essential heavy metal ions for methanogens (Fe, Ni, Mo, Co, etc.) may be precipitated as sulfides, and this may lead to deficiencies in the heavy metal bioavailability for the wastewater population.

The metabolites of the fermentative and acetogenic phase of anaerobic wastewater treatment systems, mainly acetate and CO_2+H_2, are substrates for methanogens and for sulfate reducers. If wastewater with a high sulfate concentration is treated in a methane reactor, the population may gradually shift from hydrogenotrophic methanogens towards hydrogenotrophic sulfate reducers, due to a more favorable K_s value for hydrogen of the sulfate reducers. For pure cultures of *Desulfovibrio* sp. the K_s value for hydrogen was 1 µM whereas for hydrogenotrophic methanogens it was only 6 µM (KRISTJANSSON et al., 1982). For acetate such a dramatic disadvantage of methanogens was not observed. The affinities of sulfate reducers and methanogenic bacteria (valid at least for *Methanosaeta* sp.) were in the same range, between 0.2–0.4 µM (ELFERINK et al., 1994).

2.9 Biogas Amounts and Composition of Biogas During Fermentation of Carbohydrates, Protein, and Fat

From an economic point of view, the specific biogas amounts and the biogas quality from anaerobic treatment of wastewater and sludge are very important process parameters. For this reason, a basis for prediction of amount and composition of biogas was elaborated already in the 1930s (BUSWELL and NEAVE, 1930; BUSWELL and SOLLO 1948; BUSWELL and MUELLER, 1952). If the elemental composition of a wastewater is known, the theoretical amount and the composition of the biogas can be predicted by using the Buswell equation (Eq. 11). The actual biogas amount is lower and can be calculated by including a correction factor for the degree of degradability, the pH, which influences CO_2 absorption and a 5–10% discount for biomass formation.

$$\begin{aligned}&C_cH_hO_oN_nS_s+1/4(4c-h-2o+3n+2s)H_2O\rightarrow\\&1/8(4c-h+2o+3n+2s)CO_2+\\&1/8(4c+h-2o-3n-2s)CH_4+\\&nNH_3+sH_2S\end{aligned} \quad (11)$$

Applying the Buswell equation (Eq. 11) for the anaerobic treatment of a wastewater with carbohydrates as pollutants the gas composition should theoretically be 50% methane and 50% CO_2 (Eq. 12).

$$1\,Mol\;C_6H_{12}O_6\rightarrow 3\,Mol\;CH_4+3\,Mol\;CO_2 \quad (12)$$

Since CO_2 is increasingly soluble in water with decreasing temperature and with an increasing pH, CO_2 reacts to bicarbonate/carbonate, the biogas may contain more than 80% methane. The total amount of gas is then diminished by the amount of CO_2 which has been absorbed and solubilized in the liquid. From a fat- and protein-containing wastewater theoretically more than 50% methane will be generated (Tab. 5).

3 Nitrogen Removal During Wastewater Treatment

Nitrogen-containing substances in wastewater are of inorganic or of organic nature. Together with phosphates they represent the main source for eutrophication of surface water. For this reason they must be eliminated together with the organic carbon during wastewater treatment. Whereas phosphates form insoluble precipitates with many heavy metal ions and the precipitates can be separated by sedimentation or flotation, all nitrogen compounds, except for magnesium ammonium phosphate, are easily soluble in water and thus cannot be removed chemically by precipitation. For biological removal of amino nitrogen and of heterocyclic nitrogen compounds a conversion to ammonia in an aerobic or anaerobic treatment process is the first step. Then ammonia must be nitrified, and the nitrate denitrified to yield nitrogen. Thus, depending on the kind of nitrogen compounds present in wastewater, nitrogen removal requires up to three processes in sequence: ammonification, nitrification, and denitrification.

The major portion of nitrogen compounds in municipal wastewater are reduced nitrogen compounds such as ammonia, urea, amines, amino acids, or protein. Oxidized nitrogen compounds such as nitrate or nitrite normally are not present at all or not in relevant amounts. Nitrate and nitrite may, however, represent the main nitrogen load in wastewater of certain food or metal industry branches (GENSICKE et al., 1998; ZAYED and WINTER, 1998).

The ammonia of raw municipal wastewater is mainly derived from urine and is formed in the sewer system by enzymatic cleavage of urea with ureases (Eq. 13).

$$NH_2CONH_2 + H_2O \rightarrow CO_2 + 2NH_3 \quad (13)$$

The residence time of the wastewater in the sewer system normally is not long enough for a significant contribution of ammonia from other sources, e.g., proteolysis and deamination of the amino acids.

3.1 Ammonification

The main organic nitrogen compounds of municipal wastewater are heterocyclic compounds (e.g., nucleic acids) and proteins (see also Chapter 10, this volume). Proteolysis and degradation of amino acids leads to the liberation of ammonia by the different mechanisms of ammonification (RHEINHEIMER et al., 1988), including hydrolytic, oxidative, reductive, and desaturative deamination (Eqs. 14–17, respectively)

$$R-NH_2 + H_2O \rightarrow R-OH + NH_3 \quad (14)$$

$$R-CHNH_2COOH + H_2O \rightleftharpoons R-CO-COOH + 2(H) + NH_3 \quad (15)$$

$$R-CHNH_2-COOH + 2(H) \rightarrow R-CH_2-COOH + NH_3 \quad (16)$$

$$R-CH_2-CHNH_2-COOH \rightleftharpoons R-CH{=}CH-COOH + NH_3 \quad (17)$$

Tab. 5. Amount of Biogas, Biogas Composition, and Energy Content

Substrate	Amount [$cm^3\ g^{-1}$]	Composition % CH_4	% CO_2	Energy content [$Wh\ g^{-1}$]
Carbohydrate[a]	746.7	50	50	3.78
Fat[b]	1,434	71	29	8.58
Proteins[c]	636	60	40	4.96

[a] Calculated for hexoses
[b] Calculated for triglycerides containing glycerol plus 3 mol palmitic acid
[c] Calculated for polyalanine and reaction of ammonia to $(NH_4)_2CO_3$

A significant amount of ammonia from urea cleavage or from ammonification of amino acids is assimilated in aerobic treatment processes for growth of bacteria (surplus sludge formation). It can be estimated that bacteria roughly consist of 50% protein and that the nitrogen content of protein is about 16%. Thus, for synthesis of 1 g bacterial biomass about 0.08 g ammonia-N are required. For elimination of ammonia that is not used for cell growth during wastewater treatment, it must first be nitrified and then denitrified to molecular nitrogen.

3.2 Nitrification of Ammonia

3.2.1 Autotrophic Nitrification

Autotrophic nitrifiers are aerobic microorganisms oxidizing ammonia via nitrite (Eq. 18) to nitrate (Eq. 19). Organisms catalyzing nitritation (Eq. 18) belong to the genera *Nitrosomonas, Nitrosococcus, Nitrosolobus, Nitrosospira*, or *Nitrosovibrio*, organisms catalyzing nitration (Eq. 19) are, e.g., members of the genera *Nitrobacter, Nitrococcus*, and *Nitrospira*.

$$NH_4^+ + 1.5\,O_2 \rightarrow NO_2^- + 2H^+ + H_2O \quad (18)$$

$$NO_2^- + 0.5\,O_2 \rightarrow NO_3^- \quad (19)$$

Ammonia oxidation to nitrite or nitrite oxidation to nitrate are the energy-yielding processes for autotrophic growth of the nitrifying bacteria. CO_2 is assimilated via the Calvin cycle. Since the positive redox potential of the oxidizable nitrogen compounds is not low enough to form $NADH_2$ for CO_2 reduction, it must be formed via an energy-consuming reversed electron transport. For this reason, the growth yield of nitrifying bacteria is very low. *Nitrosomonas* sp., e.g., must oxidize 30 g NH_3 to form 1 g of cell dry mass (SCHLEGEL, 1992).

The oxidation of ammonia by nitrifiers is initiated by an energy-neutral monoxygenase reaction that yields hydroxylamine (Eq. 20).

$$NH_3 + XH_2 + O_2 \rightarrow NH_2OH + X + H_2O \quad (20)$$

Then hydroxylamine is further oxidized, presumably via nitroxyl (Eq. 21) to nitrite (Eq. 22), which is the energy-yielding reaction during nitritation. For microbial oxidation of 1 mg NH_4^+-nitrogen to nitrite 3.42 mg O_2, and for the oxidation of 1 mg NO_2^--nitrogen to nitrate (Eq. 19) 1.14 mg O_2 are required.

$$NH_2OH + X \rightarrow XH_2 + (NOH) \quad (21)$$

$$(NOH) + 0.5\,O_2 \rightarrow HNO_2 + \text{Energy} \quad (22)$$

Due to their slow growth autotrophic nitrifiers cannot successfully compete with heterotrophic bacteria for oxygen. In a highly loaded activated sludge system the autotrophic nitrifiers will be overgrown by the heterotrophic sludge flora, which consumes the oxygen. Ammonia oxidation starts only if the BOD_5 concentration of the wastewater is below 110 mg L^{-1} (WILD et al., 1971) During nitrification of ammonia the alkalinity of the wastewater increases slightly, due to CO_2 consumption for autotrophic growth (pH increase), but in a counterreaction it drastically decreases due to nitric acid formation from ammonia (pH decrease from above neutral to acid values). If the buffer capacity of a wastewater is weak, the pH will drop far below 7 and thus prevent further nitrification by autotrophic nitrifiers (RHEINHEIMER et al., 1988).

3.2.2 Heterotrophic Nitrification

Some bacteria of the genera *Arthrobacter, Flavobacterium*, or *Thiosphaera* are able to catalyze heterotrophic nitrification of nitrogen-containing organic substances (Eq. 23).

$$R-NH_2 \rightarrow R-NHOH \rightarrow R-NO \rightarrow NO_3^- \quad (23)$$

Heterotrophic nitrifiers oxidize reduced nitrogen compounds such as hydroxylamine, aliphatic and aromatic nitrogen-containing compounds, but in contrast to autotrophic nitrification no energy is gained by nitrate formation. For this reason an organic substrate must be respired to satisfy the energy metabolism (SCHLEGEL, 1992) .

Some of the heterotrophic nitrifiers are able to denitrify nitrate or nitrite under aerobic growth conditions. The nitrogen metabolism of *Thiosphaera pantotropha* or of *Paracoccus de-*

nitrificans are well-documented examples (STOUTHAMER et al., 1997). These organisms express a membrane-bound nitrate reductase under anoxic growth conditions, which only works in the absence of molecular oxygen. Under aerobic growth conditions a periplasmic nitrate reductase is expressed which catalyzes nitrate reduction at least to the state of nitrous oxide.

3.3 Denitrification – Nitrate Removal from Wastewater

Many aerobic bacteria seem to be able to switch their oxidative metabolism to nitrate respiration. Similar to oxygen respiration the heterotrophic bacteria require a complex carbon source as an electron source for denitrification (e.g., Eqs. 24 and 25). Denitrification starts with the reduction of nitrate to nitrite by membrane-bound nitrate reductase A (a). Then a membrane-bound nitrite reductase (b) catalyzes NO formation. Finally the NO reductase (c) and the N_2O reductase (d) form N_2 (Eq. 26). The theoretical stoichiometry of denitrification with methanol or acetate as a carbon source is shown in Eqs. 27 and 28. For practical application a surplus of carbon source must be supplied, since the wastewater is not free of oxygen, and part of the carbon source is respired until anoxic conditions are obtained.

To evaluate the stoichiometry of nitrate to organic compounds for denitrification with a complex carbon source the oxidation/reduction state of the carbon substrates and the oxygen concentration in the wastewater should be known. In wastewater treatment plants more than 2.85 g COD are required for the reduction of 1 g NO_3^-–N (BERNET et al., 1996).

$$CH_3OH + H_2O \rightarrow CO_2 + 6H^+ + 6e^- \tag{24}$$

$$CH_3COOH + 2H_2O \rightarrow 2CO_2 + 8H^+ + 8e^- \tag{25}$$

$$NO_3^- \xrightarrow[a]{2e^-} NO_2^- \xrightarrow[b]{e^-} NO \xrightarrow[c]{e^-} 0.5N_2O \xrightarrow[d]{e^-} 0.5N_2 \tag{26}$$

$$5CH_3OH + 6HNO_3 \rightarrow 5CO_2 + 3N_2 + 13H_2O \tag{27}$$

$$5CH_3{-}COOH + 8HNO_3 \rightarrow 10CO_2 + 4N_2 + 14H_2O \tag{28}$$

Instead of nitrate many denitrifying bacteria are able to use NO_2^-, NO, or N_2O as terminal electron acceptors. *Vice versa* they may release these intermediates during denitrification of nitrate under unfavorable conditions as it was observed in soil (CONRAD, 1996). If, e.g., surplus nitrate is supplied and hydrogen donors are not sufficiently available NO and N_2O may be formed (SCHÖN et al., 1994). Another condition for N_2O formation would be a pH below 7.3, at which the nitrogen oxidoreductase is inhibited (KNOWLES, 1982).

Except for dissimilatory nitrate reduction many aerobic and anaerobic bacteria are capable of assimilatory nitrate reduction to supply the cells with ammonia for growth (Eqs. 29 and 30). However, the enzymes for nitrate assimilation are only expressed at concentrations of ammonia of <1 mM.

$$NO_3^- \xrightarrow[a]{2e^-} NO_2^- \xrightarrow[b]{2e^-} HNO \xrightarrow[b]{2e^-} NH_2OH \xrightarrow[b]{2e^-} NH_3 \tag{29}$$

$$HNO_3 + 8(H) \rightarrow NH_4^+OH^- + 2H_2O \tag{30}$$

In this case nitrate reductase B (enzyme a in Eq. 29) reduces nitrate to nitrite, which is then reduced to ammonia by a complex nitrite reductase (b). Whereas nitrate reduction by the oxygen-sensitive, membrane-bound enzyme nitrate reductase A conserves energy, no energy conservation is possible in the reaction of the soluble enzyme nitrate reductase B, which is not repressed by oxygen. For details on the cell biology and the molecular basis of denitrification the reader is referred to ZUMFT (1997).

3.4 Combined Nitrification and Denitrification

A strict separation of the reactions participating in aerobic, autotrophic nitrification and in anoxic, heterotrophic denitrification is not required, as judged from N^{15}-tracer experiments (KUENEN and ROBERTSON, 1994). Autotrophic ammonia oxidizers seem to be able to produce NO, N_2O, or N_2 if oxygen is limited and ammonia as well as nitrite oxidizers can be isolated from anaerobic reactors (KUENEN and ROBERTSON, 1994). *Nitrosomonas europaea* is able to use nitrite as an electron acceptor and pyruvate as an energy source under anaerobic growth conditions (ABELIOVICH and VONSHAK, 1992). Some of the nitrite oxidizers may be mixotrophic and may denitrify during heterotrophic growth (BOCK et al., 1986).

Aerobic denitrification by *Thiosphaera pantotropha* was first described by ROBERTSON and KUENEN (1984). *T. pantotropha* respires molecular oxygen and denitrifies nitrate simultaneously, provided that suitable electron acceptors are available. The conversion rate of acetate as electron donor with nitrate as electron acceptor was twice as high, if the concentration of molecular oxygen was $<30\%$ of air saturation compared to 30–80% air saturation (ROBERTSON et al., 1988). Respiration, simultaneous nitrification, and denitrification was observed in the presence of oxygen and ammonia. During heterotrophic denitrification by *T. pantotropha* ammonia was in the initial step oxidized to hydroxylamine by an ammonia monoxygenase with ubiquinone serving as electron donor. Hydroxylamine was subsequently oxidized to nitrite by a hydroxylamine oxidoreductase. During coupled nitrification/denitrification 3 of the 4 electrons from the oxidation of hydroxylamine were used for reduction of nitrite to nitrogen and were not available for an electron transport chain reaction catalyzed by cytochrome oxidase. The regeneration of ubiquinone was mediated by electrons that were generated during oxidation of an organic substrate (heterotrophic nitrification).

Conversion rates of ammonia by heterotrophic nitrifiers such as *T. pantotropha* (KUENEN and ROBERTSON, 1994) were smaller than those of autotrophic nitrifiers (35.4 for *T. pantotropha* versus 130–1,550 nmol NH_3 min^{-1} mg^{-1} dry weight for *Nitrosomonas* sp.). If, however, the higher population density of heterotrophic nitrifiers, resulting from higher growth rates was taken into consideration, the specific conversion rates were in a similar range. Heterotrophic nitrifiers can, in addition to their nitrifying capability, denitrify nitrite or nitrate to molecular nitrogen. In many wastewater treatment plants autotrophic nitrifiers may exist producing nitrite from ammonia under moderate aerobic conditions, which then is converted to nitrate and/or reduced to nitrogen by heterotrophic nitrifiers in the presence of a suitable carbon source. In practice, the only disadvantage of heterotrophic nitrification would be more surplus sludge generation for final disposal.

3.5 Anaerobic Ammonia Oxidation (Anammox®)

In the Anammox process ammonia is oxidized to nitrogen under anaerobic conditions with nitrate or nitrite serving as electron acceptors (VAN DE GRAAF et al., 1995). By using N^{15}-isotopes it could be shown that nitrite was reduced to hydroxylamine (Eq. 31), which then reacted with ammonia to yield hydrazine (N_2H_4, Eq. 32). By oxidation of hydrazine to molecular nitrogen (Eq. 33) 4 reducing equivalents were generated which were required for nitrite reduction to hydroxylamine (VAN DE GRAAF et al., 1997).

$$2\,HNO_2 + 4\,XH_2 \rightarrow 2\,NH_2OH + 2\,H_2O + 4\,X \quad (31)$$

$$2\,NH_2OH + 2\,NH_3 \rightarrow 2\,N_2H_4 + 2\,H_2O \quad (32)$$

$$2\,N_2H_4 + 4\,X \rightarrow 2\,N_2 + 4\,XH_2 \quad (33)$$

Since the redox state is balanced by the above mentioned reactions, reducing equivalents for CO_2 reduction by the autotrophic organisms must be generated by oxidation of nitrite to nitrate (Eq. 34).

$$HNO_2 + H_2O + NAD \rightarrow HNO_3 + NADH_2 \quad (34)$$

Per mol of ammonia 0.2 mol nitrate were generated and 20 mg biomass were produced (VAN DE GRAAF et al., 1996). The Anammox process seems to be suitable for nitrogen removal in ammonia-rich effluents of anaerobic reactors, fed with wastewater rich in TKN compounds. Nitrogenous compounds may be eliminated from this wastewater in a combination of nitrification for nitrite supply and anaerobic ammonia oxidation (STROUS et al., 1997).

4 Biological Phosphate Removal

Based on the early observation that microorganisms take up more phosphate than required for cell growth, it was found that single cells accumulated polyphosphate in granules, containing a few up to several thousand phosphate units (EGLI and ZEHNDER, 1994). Phosphate constituted up to 12% of the cell weight of polyphosphate-accumulating bacteria, whereas bacteria that did not accumulate polyphosphate contained only 1–3% phosphate (VAN LOOSDRECHT et al., 1997b). For biological phosphate removal by "luxury phosphate uptake" the polyphosphate-accumulating bacteria in wastewater have to be kept subsequently in an anaerobic and an aerobic environment. All former attempts to isolate polyphosphate-accumulating bacteria led to *Acinetobacter* sp., which served for investigations of the polyphosphate accumulation mechanism found in textbooks. However, pure cultures of the polyphosphate-accumulating *Acinetobacter* sp. and wastewater communities seem to behave differently (VAN LOOSDRECHT et al., 1997a). A major discrepancy is that *Acinetobacter* sp. do not play a dominant role in activated sludge systems, as deduced from a survey using specific gene probes (WAGNER et al., 1994).

In *Acinetobacter* sp. polyphosphate accumulation is closely interconnected with poly-β-hydroxybutyrate (PHB) and glycogen metabolism (VAN LOOSDRECHT et al., 1997b). Under anaerobic conditions in the absence of nitrate acetate was taken up by *Acinetobacter* cells, metabolized to β-hydroxybutyrate, polymerized to PHB, and stored intracellularly in inclusion bodies. Reducing equivalents for PHB formation from acetate were made available by glycolysis of glucose units from stored glycogen, whereas the energy for polymerization came from hydrolysis of anhydride bonds of ATP and of polyphosphate. To supply both, the maintenance metabolism and the storage metabolism of *Acinetobacter* with sufficient energy under anaerobic conditions, anhydride bonds in polyphosphate are hydrolyzed, and inorganic phosphate is excreted ("phosphate resolubilization") (SMOLDERS et al., 1994). For storage of 1 mg fatty acids about 0.6 mg orthophosphate were released (DANESH and OLESZKIEWICZ, 1997). Under aerobic conditions two situations may be prevalent. In wastewater that does not contain a suitable carbon source for respiration *Acinetobacter* sp. hydrolyze PHB which is stored in the granules and respire β-hydroxybutyrate to gain energy for growth, maintenance, formation of glycogen, and polymerization of phosphate, which is taken up from the wastewater. In the presence of oxygen much more P_i is taken up by *Acinetobacter* sp. of the activated sludge flora than has been released under anaerobic conditions ("luxury uptake"). DANESH and OLESZKIEWICZ (1997) reported that approximately 6 to 9 mg volatile fatty acids are required for biological removal of 1 mg phosphorus.

In wastewater that contains suitable carbon sources for *Acinetobacter* sp. these are partially oxidized via the TCA cycle and partially serve for PHB formation (KUBA et al., 1994). Since reducing equivalents were necessary for PHB formation less ATP could be formed by oxidative phosphorylation. The lower ATP yield was compensated by energy from hydrolysis of polyphosphate to P_i, which served the cells to form ADP from AMP. 2 mol ADP were disproportionated to AMP and ATP by the adenylate kinase reaction, and only little P_i was released into the medium.

Except for obligately aerobic bacteria of the genus *Acinetobacter* nitrate-reducing bacteria were also capable of eliminating phosphate from the aqueous environment (KERRN-JESPERSEN and HENZE, 1993). The simultaneous presence of obligately aerobic bacteria and of nitrate-reducing bacteria in activated sludge might explain why in the presence of nitrate in the anaerobic phosphate re-solubilization phase (oxygen absent) there was no accumulation of P_i in the medium. Nitrate-reducing bacteria apparently did not inhibit P_i excretion by *Acinetobacter*, but they took up phosphate and accumulated it as polyphosphate. This might explain why re-solubilization of phosphate by degradation of stored polyphosphate is not always observed in wastewater treatment plants for biological phosphate removal.

For biological phosphate accumulation the ability of cells to store reserve material plays an essential role. Polymerization of substrates and storage in intracellular granules might offer the advantage that at times of substrate shortage or in an environment with low concentrations of substrates (at high k_m values) such organisms can survive much better than other heterotrophic bacteria that have not developed effective strategies for substrate utilization.

Biological phosphate elimination is a highly effective process leading to final concentrations in the wastewater of <0.1 mg L^{-1} phosphate. For an optimal biological phosphate removal the COD:P ratio in the wastewater should be about 20 g COD g^{-1} phosphate to permit good growth of the polyphosphate-accumulating bacteria. If biological phosphate elimination is applied in combination with chemical precipitation, the minimum COD:P ratio should be 2 g COD g^{-1} phosphate (SMOLDERS et al., 1996) in processes with or without separation of primary sludge. If a great portion of the particulate COD is separated in a primary sedimentation pond, ferrous ions should be supplemented to support the biological phosphate removal by chemical precipitation. If no pre-sedimentation of sludge is available or the sedimentation efficiency is low, no ferrous ions have to be added (VAN LOOSDRECHT et al., 1997b). A detailed description of the state of the art concerning biological phosphate elimination is given by SCHÖN (Chapter 14, this volume).

5 Biological Removal, Biotransformation, and Biosorption of Metal Ions from Contaminated Wastewater

Whereas solid organic or inorganic material in wastewater or sludge can be removed by sedimentation, soluble organic pollutants or xenobiotics should be eliminated from the aqueous environment by microbial mineralization or anaerobic degradation to gaseous products, with a varying portion (5–50%) serving as substrates for bacterial growth. Most of the inorganic components present in wastewater are soluble and are ionized as cations and anions. Traces of many cations (e.g., Na^+, K^+, Ca^{++}, Fe^{++}, Ni^{++}, Co^{++}, Mn^{++}, Zn^{++}, etc.) or anions (e.g., PO_4^{3-}, Cl^-, S^{2-}) are essential micronutrients for bacterial growth. Other cations such as ammonium may also be required for bacterial growth, but the surplus amount must be oxidized to nitrate, and the nitrate denitrified to gaseous nitrogen for N-elimination of wastewater. Sulfate is reduced to sulfide under anaerobic conditions, low amounts of which are required for growth of bacteria. The sulfide not required for growth is toxic for bacteria if present as H_2S in high concentrations. In anaerobic reactors at a slightly alkaline pH most of the sulfide is precipitated as heavy metal sulfide; then it is harmless for microorganisms and their environment. To support detoxification heavy metal ions may also be precipitated chemically, in addition to for precipitation of heavy metal ions from wastewater by sulfide formed by sulfate reduction.

Metal ion contaminations in wastewater can be removed by microorganisms either by a direct or indirect influence on the redox state of the metal ions or through biosorption of metal ions on the cell surface (LOVLEY and COATES, 1997). Certain bacteria, yeasts, fungi, or algae are able to actively accumulate metal ions in the cells against a gradient. The process of bioaccumulation of metal ions depends on living, metabolically active cells, whereas biosorption is a passive, energy-independent process that can be mediated also by inactive cell material. Biosorption of metal ions includes mechanisms such as ion exchange, chelation, matrix entrapment, and surface sorption (UNZ and SHUTTLEWORTH, 1996). After biosorption or active removal of metal ions from wastewater or contaminated soil the heavy metal containing biomass must be separated and incinerated or regenerated by desorption or remobilization of the metals. As an example of successful biosorption wastewater from the galvanizing industry that contained 29 mg L^{-1} Zn and 10.5 mg L^{-1} Fe was purified in a three-stage stirred reactor cascade. With 15 g biomass 7.5 L of wastewater were decontaminated

(BRAUCKMANN, 1997). The removal of metal ions by biologically catalyzed changes of the redox state is an alternative. The altered speciation of metals can lead to precipitation, solubilization, or volatilization of the metal ions.

Many bacteria are known to use metal ions as electron acceptors for anaerobic respiration. Examples are the reduction of Fe^{3+}, Cr^{6+}, Mn^{4+}, Se^{6+}, As^{5+}, Hg^{2+}, Pb^{2+}, or U^{6+} (Tab. 6). The dissimilatory metal-reducing bacteria can use H_2 or organic pollutants as electron donors and are capable of simultaneously removing organic and inorganic contaminations. Metal ions are reduced and precipitated with sulfide which is generated by sulfate-reducing bacteria.

Solubilization of most heavy metal precipitates is favored at an acid pH, which is the favorable pH for soil or sludge decontamination, whereas an alkaline pH is favorable for precipitation of heavy metal ions to decontaminate wastewater.

An example for the formation of precipitates or soluble compounds at different redox states are ferrous ions. Under aerobic conditions they exist in the Fe^{3+}-stage and form insoluble Fe^{3+}-oxide or -hydroxide, whereas under anaerobic conditions they form soluble Fe^{2+}-ions. Since Fe is an essential element for microoganisms, aerobic bacteria must excrete siderophores, which bind Fe^{3+} to their phenolate or hydroxamate moiety and supply the cells with soluble Fe^{2+}. To accelerate Fe^{3+} reduction in biotechnological processes the chelate former nitrilotriacetic acid may be added. Except for synthesis of cell components (e.g., cytochromes, ferredoxin, etc.) Fe^{2+} salts may be electron donors for nitrate reduction (STRAUB et al., 1996).

The reduction of Hg^{2+} to metallic Hg by *Escherichia coli* or *Thiobacillus ferrooxidans* facilitates Hg separation and prevents methylation reactions under aerobic or sulfate-reducing conditions. The mechanism includes a nonenzymatic transfer of methyl groups from methylcobalamine to Hg^{2+} to form methylmercury or dimethylmercury, which both are neurotoxins and are enriched in the food chain. Except for mercury, selenium and arsenic could be transformed microbiologically by methylation to dimethyl selenide or di- or trimethyl arsine into a volatile form. The methylated arsenic compounds are less toxic than the non-methylated arsenic compounds (WHITE et al., 1997).

In contrast to bacteria, fungi are capable of leaching soluble as well as insoluble metal salts. For this purpose fungi excrete organic acids such as citric acid, fumaric acid, lactic acid, gluconic acid, oxalic acid, or malic acid, which solubilize metal salts and form complexes with the metal ions. The leaching efficiency is dependent on the soil microflora. Some soil microorganisms seem to be able to degrade the carbon skeleton of the metal-organic complex and thus immobilize the metal ions again

Tab. 6. Metals as Electron Acceptors for Anaerobic Respiration

Reaction	Microorganisms	Reference
$2Fe^{3+} + H_2 \leftrightarrow 2Fe^{2+} + 2H^+$	*Geobacter metallireducens*	LOVLEY and LONERGAN (1990)
	Pelobacter carbinolicus	LOVLEY et al. (1995)
$Mn^{4+} + H_2 \leftrightarrow Mn^{2+} + 2H^+$	*Geobacter metallireducens*	LOVLEY (1991)
	mixed culture	LANGENHOFF et al. (1997)
$2Cr^{6+} + 3H_2 \leftrightarrow 2Cr^{3+} + 6H^+$	*Desulfovibrio vulgaris*	LOVLEY and PHILLIPS (1994)
	Bacillus strain QC1-2	CAMPOS et al. (1995)
$Se^{6+} + H_2 \leftrightarrow Se^{4+} + 2H^+$	*Thauera selenatis*	MACY et al. (1993)
$Se^{6+} + 3H_2 \leftrightarrow Se^0 + 6H^+$	strain SES-1; SES-3	OREMLAND et al. (1989)
$Te^{4+} + 2H_2 \leftrightarrow Te^0 + 4H^+$	*Schizosaccharomyces pombe*	SMITH (1974)
$Pb^{2+} + H_2 \leftrightarrow Pb^0 + 2H^+$	*Pseudomonas maltophila*	LOVLEY (1995)
$As^{5+} + H_2 \leftrightarrow As^{3+} + 2H^+$	*Geospirillum arsenophilus*	AHMANN et al. (1994)
$Hg^{2+} + H_2 \leftrightarrow Hg^0 + 2H^+$	*Escherichia coli*	ROBINSON and
	Thiobacillus ferrooxidans	TUOVINEN (1984)
$U^{6+} + H_2 \leftrightarrow U^{4+} + 2H^+$	*Shewanella putrefaciens*	LOVLEY et al. (1991)

(BRYNHILDSEN and ROSSWALL, 1997). Mixed bacterial cultures or *Wolinella succinogenes* use perchlorate or chlorate as electron acceptors for respiration (WALLACE et al., 1996; VAN GINKEL et al., 1995) and thus detoxify these chemicals.

5.1 Sulfate Reduction under the Aspect of Metal Ion Precipitation

Sulfate-reducing bacteria are biotechnologically relevant for sulfate removal or heavy metal precipitation in wastewater or waste or for the elimination of SO_2 during off-gas purification. Sulfate is the terminal electron acceptor and is reduced to sulfide with reducing equivalents derived from the degradation of lactic acid or many other organic acids or compounds (WIDDEL, 1988). Alternatively, some sulfate reducers can also use molecular hydrogen. Sulfate reducers gain energy in an anaerobic electron transport chain (HANSEN, 1994), leading to sulfide, a weak two-basic acid, which dissociates according to Eq. 35.

$$H_2S \overset{K_1}{\longleftrightarrow} H^+ + HS^- \overset{K_2}{\longleftrightarrow} 2\,H^+ + S^{2-}$$
$$K_1 = 1.02 \cdot 10^{-7}, K_2 = 1.3 \cdot 10^{-13}\ (25\,°C) \quad (35)$$

The total dissociation is described by Eq. 36:

$$K = \frac{[H^+]^2 \cdot [S^{2-}]}{[H_2S]} = K_1 \cdot K_2 = 1.3 \cdot 10^{-20} \quad (36)$$

For precipitation of heavy metal ions sulfide ions are necessary (Eq. 37).

$$Me^{2+} + S^{2-} \leftrightarrow MeS$$
$$\text{or: } 2\,Me^+ + S^{2-} \leftrightarrow Me_2S \quad (37)$$

The concentration of sulfide is pH-dependent. At acid pH only those metal sulfides with a very low solubility can be precipitated. Thus, at acid pH HgS, As_2S_3, CdS, CuS, or PbS form precipitates, whereas at a more alkaline pH ZnS, FeS, NiS, or MnS form precipitates. Al_2S_3 or Cr_2S_3 are water soluble and cannot be removed by precipitation and sedimentation. Zinc removal from zinc-contaminated groundwater by microbial sulfate reduction and zinc sulfide precipitation in a 9 m^3 sludge blanket reactor was demonstrated and has been transferred into full scale of 1,800 m^3 (WHITE and GADD, 1996).

6 Aerobic and Anaerobic Degradation of Xenobiotic Substances

Except for pesticides and insecticides that are widely used in horticulture and farming, of which small amounts are washed into the groundwater, many xenobiotics are spilled in the environment, either by spot contamination or by contamination of large areas. Biological remediation of soil and groundwater from gasoline (e.g., YERUSHALMI and GUIOT, 1998) and from the volatile fractions of diesel oil (e.g., GREIFF et al., 1998), even at low temperatures (MARGESIN and SCHINNER, 1998), seems to be no problem. Results of many laboratory and field studies with a variety of substances are available (e.g., ARENDT et al., 1995; HINCHEE et al., 1995a, b; KREYSA and WIESNER, 1996). Except for studies in complex field or wastewater environments, many xenobiotics have been degraded by aerobic or anaerobic pure or defined mixed cultures (Tab. 7). These include monoaromatic or polyaromatic substances, either with or without chloro-substituents. Whereas the two-ring compound naphthalene was relatively easily degradable by *Pseudomonas* sp. or *Rhodococcus* sp., the four-ring compounds fluoranthene and pyrene were much less degradable by *Rhodococcus* sp. (BOUCHEZ et al., 1996). Aromatic compounds such as phenol or benzoic acid were degradable by mixed consortia with a rate of up to 1 g L^{-1} d^{-1}, either in the presence or absence of oxygen (MÖRSEN and REHM, 1987; KNOLL and WINTER, 1987; KOBAYASHI et al., 1989). A potent population was required, which could be obtained by pre-incubation. Even PCP was shown to be biodegradable by a methanogenic mixed culture from UASB granules (JUTEAU et al., 1995; WU et al., 1993; KENNES et al., 1996). Some pure cultures or

Tab. 7. Aerobic and Anerobic Degradation or Dechlorination of Xenobiotics

Substance	Rate [mg L^{-1} d^{-1}]	Microorganisms Reactor	Reference
Aerobic degradation			
2-Hydroxy-benzothiazole	138	*Rhodococcus rhodochrous*	DE WEVER et al. (1997)
Naphtalene	57	*Pseudomonas* sp. *Rhodococcus* sp.	BOUCHEZ et al. (1996)
Fluoranthene	6.6	*Rhodococcus* sp.	BOUCHEZ et al. (1996)
Pyrene	6.6	*Rhodococcus* sp.	BOUCHEZ et al. (1996)
Pyrene	0.56	*Mycobacterium flavescens*	DEAN-ROSS and CERNIGLIA (1996)
Toluene	57	*Pseudomonas putida*	HEALD and JENKINS (1996)
Phenol	188	*Bacillus* sp. A2	MUTZEL et al. (1996)
Phenol	1,000	mixed immobilized culture	MÖRSEN and REHM (1987)
Cresol (o,m,p)	259	mixed immobilized culture	MÖRSEN and REHM (1987)
2,4-Diphenoxy-acetic acid	33	*Alcaligenes eutrophus*[a] JMP134(pJP4)[b]	VALENZUELA et al. (1997)
2,4,6-Trichlorophenol	15	JMP134(pJP4)[b]	VALENZUELA et al. (1997)
Anaerobic degradation			
Pentachlorophenol (PCP)	107	methanogenic mixed culture fixed film reactor	JUTEAU et al. (1995)
PCP	90	methanogenic mixed culture UASB	WU et al. (1993)
PCP	4.4	methanogenic granules	KENNES et al. (1996)
PCP	22.7	methanogenic mixed culture	JUTEAU et al. (1995)
Benzene	0.029	sulfate-reducing mixed culture	EDWARDS and GRIBIĆ-GALIC (1992)
Phenol	1,000	methanogenic mixed culture	KNOLL and WINTER (1987)
Phenol	31	syntrophic mixed culture	KNOLL and WINTER (1989)
Phenol	200	syntrophic mixed culture	KOBAYASHI et al. (1989)
Benzoic acid	600	syntrophic culture	KOBAYASHI et al. (1989)
Toluene	0.1–1.5	sulfate-reducing mixed culture	EDWARDS et al. (1992)
Toluene	4.6	methanogenic mixed culture	EDWARDS and GRIBIĆ-GALIC (1994)
Xylene	0.1–1.5	sulfate-reducing mixed culture	EDWARDS et al. (1992)
Xylene	5.3	methanogenic mixed culture	EDWARDS and GRIBIĆ-GALIC (1992)
Aerobic Dechlorination			
2-Chlorophenol	102	*Pseudomonas pickettii*	KAFKEWITZ et al. (1996)
4-Chlorophenol	41	*Pseudomonas pickettii*	KAFKEWITZ et al. (1996)
1,3-Dichloro-2-propanol	671	*Pseudomonas pickettii*	KAFKEWITZ et al. (1996)
Tetrachloroethene	35.8 total[c]	anaerobic mixed culture	WU et al. (1995)
Anaerobic Dechlorination			
2,6-Dichlorophenol	38.4	methanogenic mixed culture	DIETRICH and WINTER (1990)
4-Chlorophenol	0.43	sulfate-reducing consortium	HÄGGBLOM (1998)
2-Chlorophenol	128	methanogenic mixed culture	DIETRICH and WINTER (1990)
2-Chlorophenol	1.66	anaerobic mixed culture	KUO and SHARAK GENTHNER (1996)

Tab. 7. Continued

Substance	Rate [mg L^{-1} d^{-1}]	Microorganisms Reactor	Reference
3-Chlorobenzoate	6.08	anaerobic mixed culture	KUO and SHARAK GENTHNER (1996)
3-Chloro-4-OH-benzoate	29.9	*Desulfitobacterium chlororespirans*	SANFORD et al. (1996)
Polychlorinated Biphenyls (PCB) 2,3,4,5,6-CB[d]	0.24 total	methanogenic granules	NATARAJAN et al. (1996)
2,3,4,5-CB	0.39 3,5-CB[c]	anaerobic sediments	BERKAW et al. (1996)
2,3,4,6-CB	13.3 tri-CB[c]	anaerobic sediment	WU et al. (1996)
Tetrachloroethene	6.13	methanogenic granules	CHRISTIANSEN et al. (1997)
Tetrachloroethene	1.64 dichloro-ethene[c]	strain TT 4B	KRUMHOLZ et al. (1996)
Tetrachloroethene	2.05 dichloro-ethene[c]	strain MS-1	SHARMA and MCCARTY (1996)
DCB, TCB, TeCB[e]	1.24	methanogenic mixed culture	MIDDELDORP et al. (1997)

[a] *Alcaligenes eutrophus* = *Ralstonia eutropha*
[b] pJP4 = 2,4-dichlorophenoxy acetic acid-degrading plasmid
[c] Dehalogenation product: total dehalogenation or partial dehalogenation to the corresponding dehalogenation product
[d] 2,3,4,5,6-CB = 2,3,4,5,6-chlorinated biphenyls
[e] DCB = di-chlorobenzene, TCB = tri-chlorobenzene, TeCB = tetra-chlorobenzene

mixed cultures were capable of dechlorinating aromatic or aliphatic compounds (Tab. 7, second part). As for the aerobic and anaerobic degradation of phenol the dechlorination rate for 2-chlorophenol under aerobic or anaerobic conditions was of the same order of magnitude, 102 versus 128 mg L^{-1} d^{-1} (KAFKEWITZ et al., 1996; DIETRICH and WINTER, 1990). A pure culture of *Pseudomonas pickettii* was used for aerobic dechlorination, whereas for anaerobic degradation a mixed culture was used. The main problem for degradation of xenobiotic compounds in wastewater was a too short residence time in the reactors, which did not allow selection or adaptation of bacteria for dechlorination or degradation. Only if a permanent pollution is prevalent, the degradation potential may develop with time. Alternatively, biofilm reactors should be used to enrich dechlorinating and xenobiotic degrading bacteria.

7 Bioaugmentation in Wastewater Treatment Plants for the Degradation of Xenobiotics

In biotechnology or medicine mutants of bacteria or fungi or genetically engineered organisms are widely applied for the production of citric acid, gluconic acid, ascorbate, or pharmaceuticals such as penicillin, insulin, or blood coagulation factors. For soil remediation bacteria and fungi have also been adapted or genetically transformed (ATLAS, 1981; MARGESIN and SCHINNER, 1997, KORDA et al., 1997; MEGHARAI et al., 1997). For wastewater or sludge stabilization the successful application of genetically modified bacteria or of bacteria that could serve as donors for plasmids encoding for degradative enzymes were rather rare (VAN LIMBERGEN et al., 1998). In most cases a natural selection of the most suitable microorganisms from a complex flora, simply by adapting process parameters, was enforced. The limited reports on bioaugmentation for wastewater treatment may be due to one of several reasons:

(1) The plasmids were unstable or the genes are not expressed in the new environment.
(2) Inoculated strains do not survive or, if they survive, metabolic activity is too low for successful competition with autochthonous strains.
(3) Inoculated strains, serving as a "gene pool", survive, but others are not competent for gene transfer.
(4) Wastewaters normally contain a complex spectrum of "better" carbon sources than xenobiotics and so organisms do not express genes for xenobiotics degradation.

To increase the survival potential in wastewater a selected flora with special metabolic features should be pre-adapted to the conditions that are prevalent in the new environment. Another possibility would be the addition of organisms containing broad host range plasmids, permitting conjugation and DNA exchange between different species or genera of bacteria.

To enhance degradation of organic compounds in activated sludge or by a biofilm, either selected specialized bacteria, genetically modified bacteria, or bacteria as plasmid donors for degradative pathways can be added (MCCLURE et al., 1991; FRANK et al., 1996). The probability of plasmid exchange from donor to recipient cells would be of advantage for the recipient bacteria, because they harbor pathways for xenobiotic degradation and thus extend their substrate spectrum. The transfer of naturally occurring mercury resistance plasmids between *Pseudomonas* strains in biofilms was shown to occur rapidly (BALE et al., 1988).

SELVARATNAM et al. (1997) added a phenol-degrading strain of *Pseudomonas putida* to an activated sludge SBR reactor, which removed 170 mg L^{-1} phenol before the *Pseudomonas putida* strain was augmented. Whereas the original phenol-degrading activity was partially lost in the non-augmented reactor upon further operation, the augmented reactor was capable of degrading phenol almost completely. A more convincing approach would have been to use non-phenol-degrading activated sludge for this experiment although the survival of the catabolic plasmid *dmpN* of *Pseudomonas putida* and its expression by the reactor biomass was demonstrated for 44 d by techniques of molecular biology at steady state conditions in the laboratory.

Successful bioaugmentation experiments in an upflow anaerobic sludge blanket reactor were reported by AHRING et al. (1992), who introduced a suspension of a pure culture of *Desulfomonile tiedjei* or a three-member consortium to an UASB reactor. They observed a rapid increase of the dehalogenating activity for 3-chlorobenzoate, whereas non-amended parallel incubations had no dehalogenating activity. Even after several months at 0.5 d hydraulic residence time, which was shorter than the generation time of *Desulfomonile tiedjei*, the dehalogenating activity was still observed, and *Desulfomonile tiedjei* could be traced as a member of the biofilm by the use of antibody probes. More recently an UASB reactor was supplemented with *Dehalospirillum multivorans* to improve the dehalogenating activity (HÖRBER et al., 1998). In contrast, MARGESIN and SCHINNER (1998) found for biological decontamination of fuel-contaminated wastewater that biostimulation of the autochthonous flora by a mineral mix enhanced biodegradation to a larger extent than bioaugmentation with a cold-adapted mixed inoculum containing *Pseudomonas* sp. and *Arthrobacter* sp.

Degradation of many xenobiotics in anaerobic or aerobic pure cultures or complex ecosystems has been shown. Whereas for "intrinsic sanitation" of polluted soil the time was not limited, as long as the pollutants were adsorbed tightly to the soil matrix, degradation of xenobiotics in wastewater must be completed during the residence time in the treatment system. Thus, merely the presence of a degradation potential within a wastewater ecosystem is not sufficient, but degradation rates must be high and degradation must be faster than the residence time of bacteria in suspended systems. Supplementation of potential microorganisms as a tool to increase the degradation speed or to increase the degradation potential in wastewater in general cannot be considered as a state-of-the-art procedure yet.

8 References

ABELIOVICH, A., VONSHAK, A. (1992), Anaerobic metabolism of *Nitrosomonas europaea, Arch. Microbiol.* **158**, 267–270.

AHMANN, D., ROBERTS, A. L., KRUMHOLZ, L. R., MOREL, F. M. M. (1994), Microbe grows by reducing arsenic, *Nature* **371**, 750.

AHRING, B. K., CHRISTIANSEN, N., MATHARI, I., HENDRIKSEN, H. V., MACARIO, A. J. L., DEMACARIO, E. C. (1992), Introduction of *de novo* bioremediation ability, aryl reductive dechlorination, into anaerobic granular sludge by inoculation of sludge with *Desulfomonile tiedjei, Appl. Environ. Microbiol.* **58**, 3677–3682.

ARENDT, F., BOSMANN, R., VAN DEN BRINK, W. J. (Eds.) (1995), *Contaminated Soil '95*, Vol. 1. Dordrecht: Kluwer Academic Publishers.

ATLAS, R. M. (1981), Microbial degradation of petroleum hydrocarbons: an environmental perspective, *Microbiol. Rev.* **45**, 180–209.

ATV (1994), Arbeitsbericht Fachausschuss 7.5: Geschwindigkeitsbestimmende Schritte beim anaeroben Abbau von organischen Verbindungen in Abwässern, *Korrespondenz Abwasser* **1**, 101–107.

BALE, M. J., DAY, M. J., FRY, J. C. (1988), Novel method for studying plasmid transfer in undisturbed river epilithon, *Appl. Environ. Microbiol.* **54**, 2756–2758.

BE'GUIN, P., AUBERT, J.-P. (1994), The biological degradation of cellulose, *FEMS Microbiol. Rev.* **13**, 25–58.

BERKAW, M., SOWERS, K. R., MAY, H. D. (1996), Anaerobic *ortho* dechlorination of polychlorinated biphenyls by estuarine sediment from Baltimore harbor, *Appl. Environ. Microbiol.* **62**, 2534–2539.

BERNET, N., HABOUZIT, F., MOLETTA, R. (1996), Use of an industrial effluent as a carbon source for denitrification of a high-strength wastewater, *Appl. Microbiol. Biotechnol.* **46**, 92–97.

BLEICHER, K., WINTER, J. (1994), Formate production and utilization by methanogens and by sludge consortia – interference with the concept of interspecies formate transfer, *Appl. Microbiol. Biotechnol.* **40**, 910–915.

BLEICHER, K., ZELLNER, G., WINTER, J. (1989), Growth of methanogens on cyclopentanol/CO_2 and specificity of alcohol dehydrogenase, *FEMS Microbiol. Lett.* **59**, 307–312.

BOCK, E., KOOPS, H.-P., HARMS, H. (1986), Cell biology of nitrifying bacteria, in: *Nitrification* (PROSSER, J. I., Ed.), pp. 17–38. Oxford: IRL Press.

BOONE, D. R., JOHNSON, R. L., LIU, Y. (1989), Diffusion of the interspecies electron carriers H_2 and formate in methanogenic ecosystems, and applications in the measurement of K_M for H_2 and formate uptake, *Appl. Environ. Microbiol.* **55**, 1735–1741.

BOUCHEZ, M., BLANCHET, D., VANDECASTEELE, J.-P. (1996), The microbial fate of polycyclic aromatic hydrocarbons: carbon and oxygen balances for bacterial degradation of model compounds, *Appl. Microbiol. Biotechnol.* **45**, 556–561.

BRAUCKMANN, B. (1997), Mikrobielle Extraktion von Schwermetallen aus Industrieabwässern, *Wasser Boden* **49**, 55–58.

BROUGHTON, M. J., THIELE, J., BIRCH, E. J., COHEN, A. (1998), Anaerobic batch digestion of sheep tallow, *Water Res.* **32**, 1323–1428.

BRYANT, M. P. (1979), Microbial methane production – theoretical aspects, *J. Anim. Sci.* **48**, 193–201.

BRYNHILDSEN, L., ROSSWALL, T. (1997), Effects of metals on the microbial mineralization of organic acids, *Water Air Soil Poll.* **94**, 45–57.

BUCHAUER, K. (1997), Zur Kinetik der anaeroben Hydrolyse und Fermentation von Abwasser, *Österr. Wasser- und Abfallwirtsch.* **49**, 69–75.

BUCHHOLZ, K., STOPPOCK, E., EMMERICH, R. (1986), Untersuchungen zur Biogasgewinnung aus Rübenpreßschnitzeln, *Zuckerindustrie* **111**, 873–845.

BUCHHOLZ, K., STOPPOCK, E., EMMERICH, R. (1988), Kinetics of anaerobic hydrolysis of solid material, *5th Int. Symp. Anaerobic Digestion*, Poster papers 15–18, (TILCHE, A., ROZZI, A., Eds.). Bologna: Monduzzi Editore.

BUSWELL, A. M., NEAVE, S. L. (1930), Laboratory studies on sludge digestion III, *State Water Surv. Bull.* **30**, 1–84.

BUSWELL, A. M., MUELLER, H. F. (1952), Mechanism of methane fermentation, *Ind. Eng. Chem.* **44**, 550–552.

BUSWELL, A. M., SOLLO, F. W. (1948), The mechanism of methane fermentation, *J. Am. Chem. Soc.* **7**, 1778–1780.

CAMPOS, J., MARTINEZ-PACHECO, M., CERVANTES, C. (1995), Hexavalent-chromium reduction by a chromate-resistant *Bacillus* sp. strain, *Antonie von Leeuwenhoek* **68**, 203–208.

CHRISTIANSEN, N., CHRISTENSEN, S. R., ARVIN, E., AHRING, B. K. (1997), Transformation of tetrachloroethene in an upflow anaerobic sludge blanket reactor, *Appl. Microbiol. Biotechnol.* **47**, 91–94.

COLBERG, P. J. (1988), Anaerobic microbial degradation of cellulose, lignin, oligolignols and monoaromatic lignin derivatives, in: *Biology of Anaerobic Microorganisms*. Wiley Series in Ecological and Applied Microbiology (ZEHNDER, A. J. B., Ed.). New York, Chichester: John Wiley & Sons.

COLBERG, P. J., YOUNG, L. J. (1985), Anaerobic degradation of soluble fractions of (^{14}C-Lignin)-Lignocellulose, *Appl. Microbiol. Biotechnol.* **49**, 345–349.

CONRAD, R. (1996), Soil microorganisms as controllers of the atmospheric trace gases (H_2, CO, CH_4, OCS, N_2O, and NO), *Microbiol. Rev.* **60**, 609–640.

COUGHLAN, M. P., MAYER, F. (1991), The cellulose-decomposing bacteria and their enzymes, in: *The Prokaryotes – A Handbook on the Biology of Bacteria: Ecophysiology, Isolation, Identification, Applications*, (Vol. 1) (BALLOWS, A., TRÜPER, H. G., DWORKIN, M., HARDER, W., SCHLEIFER, K.-H., Eds.), pp. 461–516. New York, Berlin, Heidelberg: Spinger-Verlag.

DANESH, S., OLESZKIEWICZ, J. A. (1997), Volatile fatty acid production and uptake in biological nutrient removal systems with process separation, *Water Environ. Res.* **69**, 1106–1111.

DEAN-ROSS, D., CERNIGLIA, C. E. (1996), Degradation of pyrene by *Mycobacterium flavescens*, *Appl. Microbiol. Biotechnol.* **46**, 307–312.

DE WEVER, H., DE CORT, S., NOOTS, I., VERACHTERT, H. (1997), Isolation and characterization of *Rhodococcus rhodochrous* for the degradation of the wastewater component 2-hydroxybenzothiazole, *Appl. Microbiol. Biotechnol.* **47**, 458–461.

DIETRICH, G., WINTER, J. (1990), Anaerobic degradation of chlorophenol by an enrichment culture, *Appl. Microbiol. Biotechnol.* **34**, 253–258.

EDWARDS, E. A., GRIBIĆ-GALIC, D. (1992), Complete mineralization of benzene by aquifer microorganisms under strictly anaerobic conditions, *Appl. Environ. Microbiol.* **58**, 2663–2666.

EDWARDS, E. A., GRIBIĆ-GALIC, D. (1994), Anaerobic degradation of toluene and *o*-xylene by a methanogenic consortium, *Appl. Environ. Microbiol.* **60**, 313–322.

EDWARDS, E. A., WILLS, L. E., REINHARD, M., GRIBIĆ-GALIC, D. (1992), Anaerobic degradation of toluene and xylene by aquifer microorganisms and sulfate-reducing conditions, *Appl. Environ. Microbiol.* **58**, 794–800.

EGLI, T., ZEHNDER, A. J. B. (1994), Phosphate and nitrate removal, *Curr. Opin. Biotechnol.* **5**, 275–284.

ELFERINK, S. J. W. H. O., VISSER, A., POL, L. W. H., STAMS, A. J. M. (1994), Sulfate reduction in methanogenic bioreactors, *FEMS Microbiol. Rev.* **15**, 119–136.

FRANK, N., SIMAO-BEAUNOIR, A. M., DOLLARD, M. A., BAUDA, P. (1996), Recombinant plasmid DNA mobilization by activated sludge strains grown in fixed-bed or sequenced-batch reactors, *FEMS Microbiol. Ecol.* **21**, 139–148.

FRITSCHE, W. (1998), *Umwelt-Mikrobiologie*. Jena, Stuttgart: Gustav Fischer Verlag.

GENSICKE, R., MERKEL, K., SCHUCH, R., WINTER, J. (1998), Biologische Behandlung von Permeaten aus der Ultrafiltration zusammen mit nitrathaltigen Abwässern aus der elektrochemischen Entgratung in submersen Festbettreaktoren, *Korrespondenz Abwasser* **1/98**, 86–91.

GREIFF, K., LEIDIG, E., WINTER, J. (1998), Biologische Aufbereitung eines BTEX-belasteten Grundwassers in einem Festbettreaktor, *Acta Hydrochim. Hydrobiol.* **26**, 95–103.

HÄGGBLOM, M. M. (1998), Reductive dechlorination of halogenated phenols by a sulfate-reducing consortium, *FEMS Microbiol. Ecol.* **26**, 35–41.

HAMMES, W., WINTER, J., KANDLER, O. (1979), The sensitivity of pseudomurein-containing genus *Methanobacterium* to inhibitors of murein synthesis, *Arch. Microbiol.* **123**, 275–279.

HANAKI, K., MATSUO, T., NAGASE, M. (1981), Mechanism of inhibition caused by long-chain fatty acids in anaerobic digestion process, *Biotechnol. Bioeng.* **23**, 1591–1610.

HANSEN, T. A. (1994), Metabolism of sulfate-reducing prokaryotes, *Antonie van Leeuwenhoek* **66**, 165–185.

HEALD, S. C., JENKINS, R. O. (1996), Expression and substrate specifity of the toluene dioxygenase of *Pseudomonas putida* NCIMB 11767, *Appl. Microbiol. Biotechnol.* **45**, 56–62.

HENZE, M., HARREMOES, P., JANSEN, J., ARVIN, E. (1997), *Wastewater treatment: Biological and Chemical Processes*, 2nd Edn. Heidelberg, New York: Springer-Verlag.

HILPERT, R., WINTER, J., HAMMES, W., KANDLER, O. (1981), The sensitivity of archaebacteria to antibiotics, *Zentralbl. Bakteriol. Mikrobiol. Hyg.* (Abt. 1 Orig. C) **2**, 11–20.

HINCHEE, R. E., MÜLLER, R. N., JOHNSON, P. C. (Eds.) (1995a), *Bioremediation 3, Vol. 2: In-Situ Aeration, Air Sparging, Bioventing and Related Processes.* Columbus, OH: Batelle Press.

HINCHEE, R. E., VOGEL, C. M., BROCKMAN, F. J. (Eds.) (1995b), *Bioremediation 3, Vol. 8: Microbial Processes for Bioremediation.* Columbus, OH: Batelle Press.

HÖRBER, C., CHRISTIANSEN, N., ARVIN, E., AHRING, B. K. (1998), Improved dechlorination performance of upflow anaerobic sludge blanket reactors by incorporation of *Dehalospirillum multivorans* into granular sludge, *Appl. Environ. Microbiol.* **64**, 1860–1863.

IANOTTI, E. L., KAFKEWITZ, P., WOLIN, M. J., BRYANT, M. P. (1973), Glucose fermentation products of *Ruminococcus albus* grown in continuous culture with *Vibrio succinogenes:* changes caused by interspecies transfer of hydrogen, *J. Bacteriol.* **114**, 1231–1240.

JUTEAU, P., BEAUDET, R., MCSWEEN, G., LÉPINE, F., MILOT, S., BISAILLON, J.-G. (1995), Anaerobic biodegradation of pentachlorophenol by a methanogenic consortium, *Appl. Microbiol. Biotechnol.* **44**, 218–224.

KAFKEWITZ, D., FAVA, F., ARMENANTE, P. M. (1996), Effect of vitamins on the aerobic degradation of 2-chlorophenol, 4-chlorophenol, and 4-chlorobiphenyl, *Appl. Microbiol. Biotechnol.* **46**, 414–421.

KENNES, C., WU, W.-M., BHATHNAGAR, L., ZEIKUS, J. G. (1996), Anaerobic dechlorination and mineralization of pentachlorophenol and 2,4,6-trichlorophenol by methanogenic pentachlorophenol degrading granules, *Appl. Microbiol. Biotechnol.* **44**, 801–806.

KERRN-JESPERSEN, J. P., HENZE, M. (1993), Biological phosphorus uptake under anoxic and aerobic conditions, *Water Res.* **27**, 617–624.

KNOLL, G., WINTER, J. (1987), Anaerobic degradation of phenol in sewage sludge. Benzoate formation from phenol and CO_2 in the presence of hydrogen, *Appl. Microbiol. Biotechnol.* **25**, 384–391.

KNOLL, G., WINTER, J. (1989), Degradation of phenol via carboxylation to benzoate by a defined, obligate syntrophic consortrium of anaerobic bacteria, *Appl. Microbiol. Biotechnol.* **30**, 318–324.

KNOWLES, R. (1982), Denitrification, *Microbiol. Rev.* **46**, 43–70.

KOBAYASHI, T., HASHINAGE, T., MIKAMI, E., SUZUKI, T. (1989), Methanogenic degradation of phenol and benzoate in acclimated sludges, *Water Sci. Technol.* **21**, 55–65.

KORDA, A., SANTAS, P., TENENTE, A., SANTAS, R. (1997), Petroleum hydrocarbon bioremediation: sampling and analytical techniques, *in situ* treatments and commercial microorganisms currently used, *Appl. Microbiol. Biotechnol.* **48**, 677–686.

KREYSA, G., WIESNER, J. (Eds.), (1996), *In-situ-Sanierung von Böden*, Resümee und Beiträge des 11. DECHEMA-Fachgespräches Umweltschutz. Frankfurt am Main: DECHEMA, Deutsche Gesellschaft für Chemisches Apparatewesen.

KRISTJANSSON, J. K., SCHÖNHEIT, P., THAUER, R. K. (1982), Different K_s-values for hydrogen of methanogenic bacteria and sulfate reducing bacteria: an explanation for the apparent inhibition of methanogenesis by sulfate, *Arch. Microbiol.* **131**, 278–282.

KRUMHOLZ, L. R., SHARP, R., FISHBAIN, S. S. (1996), A freshwater anaerobic coupling acetate oxidation to tetrachloroethene dehalogenation, *Appl. Environ. Microbiol.* **62**, 4108–4113.

KUBA, T., WACHTMEISTER, A., VAN LOOSDRECHT, M. C. M., HEIJNEN, J. J. (1994), Effect of nitrate on phosphorus release in biological phosphorus removal systems, *Water Sci. Technol.* **30**, 263–269.

KUENEN, J. G., ROBERTSON, L. A. (1994), Combined nitrification–denitrification processes, *FEMS Microbiol. Rev.* **15**, 109–117.

KUO, C.-W., SHARAK GENTHNER, B. R. (1996), Effect of added heavy metal ions on biotransformation and biodegradation of 2-chlorophenol and 3-chlorobenzoate in anaerobic bacterial consortia, *Appl. Environ. Biotechnol.* **62**, 2317-2323.

LAMED, R., BAYER, E. A. (1988), The cellulosome of *Clostridium thermocellum, Adv. Appl. Microbiol.* **33**, 1–46.

LANGENHOFF, A. A. M., BROUWERS-CEILER, D. L., ENGELBERTING, J. H. L., QUIST, J. J., WOLKENFELT, J. G. P. N. et al. (1997), Microbial reduction of manganese coupled to toluene oxidation, *FEMS Microbiol. Ecol.* **22**, 119–127.

LESCHINE, S. B. (1995), Cellulose degradation in anaerobic environments, *Annu. Rev. Microbiol.* **49**, 399–426.

LOVLEY, D. R. (1991), Dissimilatory Fe(III) and Mn(IV) reduction, *Microbiol. Rev.* **55**, 259–387.

LOVLEY, D. R. (1995), Bioremediation of organic and metal contaminants with dissimilatory metal reduction, *J. Ind. Microbiol.* **14**, 85–93.

LOVLEY, D. R., COATES, J. D. (1997), Bioremediation of metal contamination, *Curr. Opin. Biotechnol.* **8**, 285–289.

LOVLEY, D. R., LONERGAN, D. J. (1990), Anaerobic oxidation of toluene, phenol and *p*-cresol by the dissimilatory iron-reducing organism GS-15, *Appl. Environ. Microbiol.* **56**, 1858–1864.

LOVLEY, D. R., PHILLIPS, E. J. P. (1994), Reduction of chromate by *Desulfovibrio vulgaris* (Hildenborough) and its C_3 cytochrome, *Appl. Environ. Microbiol.* **60**, 726–728.

LOVLEY, D. R., PHILLIPS, E. J. P., GORBY, Y. A., LANDA, E. R. (1991), Microbial reduction of uranium, *Nature* **350**, 413–416.

LOVLEY, D. R., PHILLIPS, E. J. P., LONERGAN, D. J., WIDMAN, P. K. (1995), Fe(III) and S^0 reduction by *Pelobacter carbinolicus, Appl. Environ. Microbiol.* **61**, 2132–2138.

LUI, Y. (1998), Energy uncoupling in microbial growth under substrate-sufficient conditions, *Appl. Microbiol. Biotechnol.* **49**, 500–505.

MACY, J. M., LAWSON, S., DEMOLL-DECKER, H. (1993), Bioremediation of selenium oxyanions in San Joaquin drainage water using *Thauera selenatis* in a biological reactor system, *Appl. Microbiol. Biotechnol.* **40**, 588–594.

MARGESIN, R., SCHINNER, F. (1997), Bioremediation of diesel-oil-contaminated alpine soils at low temperatures, *Appl. Microbiol. Biotechnol.* **47**, 462–468.

MARGESIN, R., SCHINNER, F. (1998), Low-temperature bioremediation of a waste water contaminated with anionic surfactants and fuel oil, *Appl. Microbiol. Biotechnol.* **49**, 482–486.

McCartney, D. M., Oleszkiewicz, J. A. (1991), Sulfide inhibition of anaerobic degradation of lactate and acetate, *Water Res.* **25**, 203–209.

McClure, N. C., Weightman, A. J., Fry, J. C. (1991), Survival and catabolic activity of natural and genetically engineered bacteria in a laboratory-scale activated sludge unit, *Appl. Environ. Microbiol.* **57**, 366–373.

McInerney, M. J. (1988), Anaerobic hydrolysis and fermentation of fats and proteins, in: *Biology of Anaerobic Microorganisms* (Zehnder, A. J. B., Ed.), pp. 373–415. New York, Chichester: John Wiley & Sons.

Megharaj, M., Wittich, R.-M., Blasco, R., Pieper, D. H., Timmis, K. N. (1997), Superior survival and degradation of dibenzo-*p*-dioxin and dibenzofuran in soil by soil-adapted *Sphingomonas* sp. strain RW1, *Appl. Microbiol. Biotechnol.* **48**, 109–114.

Middeldorp, P. J. M., de Wolf, J., Zehnder, A. J. B., Schraa, G. (1997), Enrichment and properties of a 1,2,4-trichlorobenzene-dechlorinating methanogenic microbial consortium, *Appl. Environ. Microbiol.* **63**, 1225–1229.

Mörsen, A., Rehm, H. J. (1987), Degradation of phenol by a mixed culture of *Pseudomonas putida* and *Cryptococcus elinovii* adsorbed on activated carbon, *Appl. Microbiol. Biotechnol.* **26**, 283–288.

Mutzel, A., Reinscheid, U. M., Antranikian, G., Müller, R. (1996), Isolation and characterization of a thermophilic *Bacillus* strain, that degrades phenol and cresols as sole carbon source at 70 °C, *Appl. Microbiol. Biotechnol.* **46**, 593–596.

Natarajan, M. R., Wu, W.-M., Nye, J., Wang, H., Bhatnagar, L., Jain, M. K. (1996), Dechlorination of polychlorinated biphenyl congeners by an anaerobic microbial consortium, *Appl. Microbiol. Biotechnol.* **46**, 673–677.

Oleszkiewicz, J. A., Mastaller, T., McCartney, D. M. (1989), Effects of pH on sulfide toxicity to anaerobic processes, *Environ. Technol. Lett.* **10**, 815–822.

Omil, F., Lens, P., Visser, A., Hulshoff, L. W., Lettinga, G. (1998), Long-term competition between sulfate reducing and methanogenic bacteria in UASB reactors treating volatile fatty acids, *Biotechnol. Bioeng.* **57**, 676–685.

Oremland, R. W., Hollibaugh, J. T., Maest, A. S., Presser, T. S., Miller, L. G., Culbertson, C. W. (1989), Selenate reduction to elemental selenium by anaerobic bacteria in sediments and culture: biogeochemical significance of a novel, sulfate-independent respiration, *Appl. Environ. Microbiol.* **55**, 2333–2343.

Örlygsson, J., Houwen, F. P., Svensson, B. H. (1995), Thermophilic anaerobic amino acid degradation: deamination rates and end-product formation, *Appl. Microbiol. Biotechnol.* **43**, 235–241.

Rheinheimer, G., Hegemann, W., Raff, J., Sekoulov, I. (1988), *Stickstoffkreislauf im Wasser*. München: R. Oldenbourg Verlag.

Robertson, L. A., Kuenen, J. G. (1984), Aerobic denitrification: a controversy revived, *Arch. Microbiol.* **139**, 351–354.

Robertson, L. A., van Niel, W. W. J., Torremans, R. A. M., Kuenen, J. G. (1988), Simultaneous nitrification and denitrification in aerobic chemostat cultures of *Thiosphera pantotropha, Appl. Environ. Microbiol.* **54**, 2812–2818.

Robinson, J. B., Tuovinen, O. H. (1984), Mechanisms of microbial resistance and detoxification of mercury and organomercury compounds: physiological, biochemical, and genetic analyses, *Microbiol. Rev.* **48**, 95–124.

Sanford, R. A., Cole, J. R., Löffler, F. E., Tiedje, J. M. (1996), Characterization of *Desulfitobacterium chlororespirans* sp. nov., which grows by coupling the oxidation of lactate to the reductive dechlorination of 3-chloro-4-hydroxybenzoate, *Appl. Environ. Biotechnol.* **62**, 3800–3808.

Schink, B. (1997), Energetics of syntrophic cooperation in methanogenic degradation, *Microbiol. Mol. Biol. Rev.* **61**, 262–280.

Schlegel, H. G. (1992), *Allgemeine Mikrobiologie*, 7th Edn. Stuttgart: Georg Thieme Verlag.

Schön, G., Bußmann, M., Geywitz-Hetz, S. (1994), Bildung on Lachgas (N_2O) im belebten Schlamm aus Kläranlagen, *GWF Wasser Abwasser* **135**, 293–301.

Selvaratnam, C., Schoedel, B. A., McFarland, B. L., Kulpa, C. F. (1997), Application of the polymerase chain reaction (PCR) and the reverse transcriptase/PCR for determining the fate of phenol-degrading *Pseudomonas putida* ATCC 11172 in a bioaugmented sequencing batch reactor, *Appl. Microbiol. Biotechnol.* **47**, 236–240.

Sharma, P. K., McCarty, P. L. (1996), Isolation and characterization of a facultatively aerobic bacterium that reductively dehalogenates tetrachloroethene to *cis*-1,2-dichloroethene, *Appl. Environ. Microbiol.* **62**, 761–765.

Shin, H.-S., Song, Y.-C. (1995), A model for evaluation of anaerobic degradation characteristics of organic waste: Focusing on kinetics, rate-limiting step, *Environ. Technol.* **16**, 775–784.

Smith, D. G. (1974), Tellurite reduction in *Schizosaccharomyces pombe, J. Gen. Microbiol.* **83**, 389–392.

Smolders, G. J. F., van der Meij, J., van Loosdrecht, M. C. M., Heijnen, J. J. (1994), Model of the anaerobic metabolism of the biological phosphorus removal process; stoichiometry and pH influence, *Biotechnol. Bioeng.* **43**, 461–470.

Smolders, G. J. F., van Loosdrecht, M. C. M., Heijnen, J. J. (1996), Steady state analysis to evaluate the phosphate removal capacity and acetate requirement of biological phosphorus removing mainstream and sidestream process configurations, *Water Res.* **30**, 2748–2760.

SPEECE, R. E., PARKIN, G. F., BHATTACHARYA, S., TAKASHIMA, S. (1986), Trace nutrient requirements of anaerobic digestion, in: *Proc. EWPCA Conf. Anaerobic Treatment, a Grown Up Technology*, Amsterdam, Industrial Presentations (Europe) B.V. 's-Gravelandseweg 284–296, Schiedam, The Netherlands, pp. 175–188.

STAMS, A. J. M. (1994), Metabolic interactions between anaerobic bacteria in methanogenic environments, *Antonie von Leeuwenhoek* **66**, 271–294.

STOUTHAMER, A. H., DE BOER, A. P. N., VAN DER OOST, J., VAN SPANNING, R. J. M. (1997), Emerging principles of inorganic nitrogen metabolism in *Paracoccus denitrificans* and related bacteria, *Antonie von Leeuwenhoek* **71**, 33–41.

STRAUB, K. L., BENZ, M., SCHINK, B., WIDDEL, F. (1996), Anaerobic, nitrate-dependent microbial oxidation of ferrous iron, *Appl. Environ. Microbiol.* **62**, 1458–1460.

STROUS, M., VAN GERVEN, E., ZHENG, P., KUENEN, J. G., JETTEN, M. S. M. (1997), Ammonium removal from concentrated waste streams with the anaerobic ammonium oxidation (Anamox) process in different reactor configurations, *Water Res.* **8**, 1955–1962.

THAUER, R. K., JUNGERMANN, K., DECKER, K. (1977), Energy conservation in chemotrophic anaerobic bacteria, *Bacteriol. Rev.* **41**, 100–180.

THIELE, J. H., CHARTRAIN, M., ZEIKUS, J. G. (1988), Control of interspecies electron flow during anaerobic digestion: role of the floc formation, *Appl. Environ. Microbiol.* **54**, 10–19.

UNZ, R. F., SHUTTLEWORTH, K. L. (1996), Microbial mobilization and immobilization of heavy metals, *Curr. Opin. Biotechnol.* **7**, 307–310.

VALENZUELA, J., BUMANN, U., CESPEDES, R., PADILLA, L., GONZALEZ, B. (1997), Degradation of chlorophenols by *Alcaligenes eutrophus* JMP134 (pJP4) in bleached kraft mill effluent, *Appl. Environ. Microbiol.* **63**, 227–232.

VAN DE GRAAF, A. A., MULDER, A., DE BRUIJN, P., JETTEN, M. S. M., ROBERTSON, L. A., KUENEN, J. G. (1995), Anaerobic oxidation of ammonium is a biologically mediated process, *Appl. Environ. Microbiol.* **61**, 1246–1251.

VAN DE GRAAF, A. A., DE BRUIJN, P., ROBERTSON, L. A., JETTEN, M. S. M., KUENEN, J. G. (1996), Autotrophic growth of anaerobic, ammonium-oxidizing microorganisms in a fluidized bed reactor, *Microbiology* **142**, 2187–2196.

VAN DE GRAAF, A. A., DE BRUIJN, P., ROBERTSON, L. A., JETTEN, M. S. M., KUENEN, J. G. (1997), Metabolic pathway of anaerobic ammonium oxidation on the basis of ^{15}N studies in a fluidized bed reactor, *Microbiology* **143**, 2415–2421.

VAN GINKEL, C. G., PLUGGE, C. M., STROO, C. A. (1995), Reduction of chlorate with various energy substrates and inocula under anaerobic conditions, *Chemosphere* **31**, 4057–4066.

VAN LIMBERGEN, H., TOP, E. M., VERSTRAETE, W. (1998), Bioaugmentation inactivated sludge: current features and future perspectives, *Appl. Microbiol. Biotechnol.* **50**, 16–23.

VAN LOOSDRECHT, M. C. M., SMOLDERS, G. J., KUBA, T., HEIJNEN, J. J. (1997a), Metabolism of microorganisms responsible for enhanced biological phosphorus removal from wastewater, *Antonie van Leeuwenhoek* **71**, 109–116.

VAN LOOSDRECHT, M. C. M., HOOIJMANS, C. M., BRDJANOVIC, D., HEIJNEN, J. J. (1997b), Biological phosphate removal processes, *Appl. Microbiol. Biotechnol.* **48**, 289–296.

VAVILIN, V. A., RYTOV, S. V., LOKSHINA, L. Y. (1997), Two-phase model of hydrolysis kinetics and its application to anaerobic degradation of particulate organic matter, *Appl. Biochem. Biotechnol.* **63–65**, 45–58.

WAGNER, M., ERHART, R., MANZ, W., AMMAN, R., LEMMER, H. et al. (1994), Development of an rRNA-targeted oligonucleotide probe specific for the genus *Acinetobacter* and its application for *in situ* monitoring in activated sludge, *Appl. Environ. Microbiol.* **60**, 792–800.

WALLACE, W., WARD, T., BREEN, A., ATTAWAY, H. (1996), Identification of an anaerobic bacterium which reduces perchlorate and chlorate as *Wolinella succinogenes, J. Ind. Microbiol.* **16**, 68–72.

WARREN, R. A. J. (1996), Microbial hydrolysis of polysaccharides, *Annu. Rev. Microbiol.* **50**, 183–212.

WHITE, C., GADD, G. M. (1996), Mixed sulphate-reducing bacterial cultures for bioprecipitation of toxic metals: factorial and response-surface analysis of the effects of dilution rate, sulphate and substrate concentration, *Microbiology* **142**, 2197–2205.

WHITE, C., SAYER, J. A., GADD, G. M. (1997), Microbial solubilization and immobilization of toxic metals: key biogeochemical processes for treatment of contamination, *FEMS Microbiol. Rev.* **20**, 503–516.

WHITMAN, W. B., BOWEN, T. L., BOONE, D. R. (1992), The methanogenic bacteria, in: *The Prokaryotes* (BALOWS, A., TRÜPER, H. G., DWORKIN, M., HARDER, W., SCHLEIFER, K.-H. (Eds.), pp. 719–767. New York, Berlin: Springer-Verlag.

WIDDEL, F. (1986), Growth of methanogenic bacteria in pure culture with 2-propanol and other alcohols as hydrogen donors, *Appl. Environ. Microbiol.* **51**, 1056–1062.

WIDDEL, F. (1988), Microbiology and ecology of sulfate- and sulfur-reducing bacteria, in: *Biology of Anaerobic Microorganisms* (ZEHNDER, A. J. B., Ed.), pp. 469–585. New York, Chichester: John Wiley & Sons.

WILD, H. E., SANYER, C. N., MCMAHON, T. C. (1971), Factors affecting nitrification kinetics, *J. Water Pollut. Control Fed.* **43**, 1845–1854.

WILDENAUER, F. X., WINTER, J. (1985), Anaerobic digestion of high-strength acidic whey in a pH-controlled up-flow fixed film loop reactor, *Appl. Microbiol. Biotechnol.* **22**, 367–372.

WILDENAUER, F. X., WINTER, J. (1986), Fermentation of isoleucine and arginine by pure and syntrophic cultures of *Clostridium sporogenes, FEMS Microbiol. Ecol.* **38**, 373–379.

WINTER, J. (1983), Energie aus Biomasse, *Umschau* **25, 26**, 774–779.

WINTER, J. (1984), Anaerobic waste stabilization, *Biotechnol. Adv.* **2**, 75–99.

WINTER, J. U., COONEY, C. L. (1980), Fermentation of cellulose and fatty acids with enrichments from sewage sludge, *Eur. J. Appl. Microbiol. Biotechnol.* **11**, 60–66.

WINTER, J., ZELLNER, G. (1990), Thermophilic anaerobic degradation of carbohydrates – metabolic properties of microorganisms from the different phases, *FEMS Microbiol. Rev.* **75**, 139–154.

WINTER, J., SCHINDLER, F., WILDENAUER, F. (1987), Fermentation of alanine and glycine by pure and syntrophic cultures of *Clostridium sporogenes, FEMS Microbiol. Ecol.* **45**, 153–161.

WINTER, J., KNOLL, G., SEMBIRING, T., VOGEL, P., DIETRICH, G. (1989), Mikrobiologie des anaeroben Abbaus von Biopolymeren und von aromatischen und halogenaromatischen Verbindungen, in: *Biogas – Anaerobtechnik in der Abfallwirtschaft* (THOMÉ-KOZMIENSKY, K. J., Ed.). Berlin: EF-Verlag für Energie und Umweltschutz.

WINTER, J., HILPERT, H., SCHMITZ, H. (1992), Treatment of animal manures and wastes for ultimate disposal – Review, *Asian Aust. J. Anim. Sci.* **5**, 199–215.

WINTERBERG, R., SAHM, H. (1992), Untersuchungen zum anaeroben Proteinabbau bei der zweistufigen anaeroben Abwasserreinigung, *Lecture:* DECHEMA Arbeitsausschuß Umweltbiotechnologie, January 1992.

WOLIN, M. J. (1976), Interactions between H_2-producing and methane-producing species, in: *Microbial Formation and Utilization of Gases (H_2, CH_4, CO)* (SCHLEGEL, H. G., GOTTSCHALK, G., PFENNIG, N., Eds.), pp. 14–15. Göttingen: Göltze.

WOLIN, M. J. (1982), Hydrogen transfer in microbial communities, in: *Microbial Interactions and Communities*, Vol. 1 (BULL, A. T., SLATER, J. H., Eds.), pp. 323–356. London: Academic Press.

WU, W. M., BHATNAGAR, L., ZEIKUS, J. G. (1993), Performance of anaerobic granules for degradation of pentachlorophenol, *Appl. Environ. Microbiol.* **59**, 389–397.

WU, M.-W., NYE, J., HICKEY, R. F., JAIN, M. K., ZEIKUS, J. G. (1995), Dechlorination of PCE and TCE to ethene using anaerobic microbial consortium, in: *Bioremediation of Chlorinated Solvents* (HINCHEE, R. E., LEESON, A., SEMPRINI, L., Eds.), pp. 45–52. Columbus Richland: Batelle Press.

WU, Q., BEDARD, D. L., WIEGEL, J. (1996), Influence of incubation temperature on the microbial reductive dechlorination of 2, 3, 4, 6, tetrachlorobiphenyl in two freshwater sediments, *Appl. Environ. Microbiol.* **62**, 4174–4179.

YERUSHALMI, L., GUIOT, S. R. (1998), Kinetics of biodegradation of gasoline and its hydrocarbon constituents, *Appl. Microbiol. Biotechnol.* **49**, 475–481.

ZAYED, G., WINTER, J. (1998), Removal of organic pollutants and of nitrate from wastewater from the dairy industry by denitrification, *Appl. Microbiol. Biotechnol.* **49**, 469–474.

ZELLNER, G., WINTER, J. (1987a), Secondary alcohols as hydrogen donors for CO_2-reduction by methanogens, *FEMS Microbiol. Lett.* **44**, 323–328.

ZELLNER, G., WINTER, J. (1987b), Analysis of a highly efficient methanogenic consortium producing biogas from whey, *Syst. Appl. Microbiol.* **9**, 284–292.

ZELLNER, G., BLEICHER, K., BRAUN, E., KNEIFEL, H., TINDALL, B. J. et al. (1989), Characterization of a new mesophilic, secondary alcohol-utilizing methanogen, *Methanobacterium palustre* spec. nov. from a peat bog, *Arch. Microbiol.* **151**, 1–9.

ZUMFT, W. G. (1997), Cell biology and molecular basis of denitrifications, *Microbiol. Molec. Biol. Rev.* **61**, 533–616.

3 Nitrification and Denitrification – Microbial Fundamentals and Consequences for Application

DIRK ZART
RALF STÜVEN
EBERHARD BOCK
Hamburg, Germany

1 Introduction

The elimination of inorganic nitrogen compounds plays an important role in the course of wastewater treatment since these compounds contribute significantly to the eutrophication of aquatic environments and act as a strong fish poison (LA COUR JANSEN et al., 1997a). Ammonium occurs as the main nitrogen component of untreated wastewater. Organically bound nitrogen is also released as ammonium (US Environmental Protection Agency, 1994). Generally, ammonium-N is removed by means of biological treatment such as activated sludge systems or trickling filters (FOCHT and CHANG, 1975). However, there are also several chemical or physical methods like precipitation, stripping, or ion exchange in use (MAEKAWA et al., 1995; GREEN et al., 1996; SIEGRIST, 1996).

The first step of biological ammonium elimination is the nitrification, in detail the aerobic oxidation of ammonium to nitrite by ammonia oxidizing bacteria and the subsequent oxidation of the produced nitrite to nitrate by nitrite oxidizing bacteria. In the course of denitrification, the second step of biological nitrogen removal, nitrite and nitrate are reduced to gaseous N-compounds such as nitric oxide, nitrous oxide, or dinitrogen.

2 Ammonia Oxidation

2.1 Physiology of Ammonia Oxidation

Ammonia oxidizers, e.g., members of the genera *Nitrosomonas*, *Nitrosolobus*, *Nitrosococcus* (WATSON et al., 1989) are lithoautotrophic organisms using carbon dioxide as the main carbon source (BOCK et al., 1991). Their only way to gain energy is the two-step oxidation of ammonia to nitrite (HOOPER, 1969). First, ammonia is oxidized to hydroxylamine by the ammonia monooxygenase (AMO) (HOLLOCHER et al., 1981). This enzyme is unspecific and oxidizes also several apolar compounds such as methane, carbon monoxide, or some aliphatic and aromatic hydrocarbons (HOOPER et al., 1997). These compounds can act as competitive inhibitors of ammonia oxidation (HYMAN et al., 1988; KEENER and ARP, 1993). The second step in this metabolism is performed by the hydroxylamine oxidoreductase (HAO). This enzyme oxidizes hydroxylamine to nitrite (WOOD, 1986). Two of the four electrons released from this reaction (ANDERSSON and HOOPER, 1983) are required for the AMO reaction (TSANG and SUZUKI, 1982) while the remaining ones are used for the generation of a proton motive force (HOLLOCHER et al., 1982) in order to regenerate ATP and NADH (WHEELIS, 1984; WOOD, 1986).

2.2 Anoxic Ammonia Oxidation

In the absence of dissolved oxygen, ammonia oxidation has recently been observed in cultures of *Nitrosomonas eutropha* (SCHMIDT and BOCK, 1997). These cells were able to replace molecular oxygen by nitrogen dioxide or dinitrogen tetroxide, respectively, in the course of ammonia monooxygenase reaction. Hydroxylamine and nitric oxide were formed in this reaction. While nitric oxide was not further metabolized, hydroxylamine was oxidized to nitrite as shown under oxic conditions. The nitrite produced was partly used as an electron acceptor leading to the formation of dinitrogen. However, anoxic ammonia oxidation with nitrogen dioxide was nearly 10-fold slower than oxic ammonia oxidation and is, therefore, hardly suited for municipal wastewater treatment.

Another kind of anaerobic ammonia oxidation has been patented by MULDER (1989). The so-called "Anammox" process is based on the oxidation of ammonium with nitrate or nitrite serving as electron acceptors (MULDER et al., 1995; VAN DE GRAAF et al., 1995). Dinitrogen was shown to be the main end product of this anaerobic ammonia oxidation. Additionally, small amounts of nitrate were formed, when nitrite was used as an electron acceptor. No intermediates such as hydroxylamine, nitric oxide, or nitrous oxide were detectable. The organisms probably involved in "Anammox" were shown to be slowly growing, lithoautotrophic, irregularly shaped cells. Their metabolic activity was strongly inhibited by oxygen, phosphate, acetylene, or shock loading. Even organic electron donors or high concentrations of nitrite were inhibitory (VAN DE GRAAF et al., 1996). However, up to now the organisms have not been taxonomically classified. The metabolic pathway proposed for anaerobic ammonia oxidation differs significantly from the known pathway of aerobic, lithotrophic ammonia oxidizers since ammonium is supposed to be oxidized with hydroxylamine to form hydrazine, which is subsequently oxidized to dinitrogen. Hydroxylamine is assumed to be formed by reduction of nitrite (VAN DE GRAAF et al., 1997). "Anammox" was shown to exhibit rather efficient ammonium elimination rates of up to 3 kg NH_4^+ m^{-3} d^{-1} (VAN DE GRAAF et al., 1996) but unfortunately maximum anoxic ammonia oxidation rates were only obtained after very long lag phases (MULDER et al., 1995; JETTEN et al., 1997), which might be a great disadvantage for applications.

2.3 Denitrification by Ammonia Oxidizers

During oxic hydroxylamine oxidation by ammonia oxidizers small amounts of nitric and nitrous oxide are released (ANDERSON, 1965; HOOPER and TERRY, 1979). Both gases are also produced in the course of aerobic denitrification by ammonia oxidizing bacteria (HOOPER, 1968; GOREAU et al., 1980; REMDE and CONRAD, 1990; STÜVEN et al., 1992). Additionally, the formation of dinitrogen was observed (POTH, 1986; BOCK et al., 1995). In the absence of dissolved oxygen *Nitrosomonas* is capable of anoxic denitrification with molecular hydrogen (BOCK et al., 1995) or simple organic compounds (ABELIOVICH and VONSHAK, 1992) serving as electron donors while nitrite is used as electron acceptor. High aerobic denitrification activities are only obtained when the cells are cultivated under extremely oxygen-limited conditions. But under these conditions ammonia oxidation rates are low (ZART et al., 1996). For this reason, the denitrifying potentials of ammonia oxidizers have not been intentionally used in the treatment of wastewater. However, evidence was given that ammonia oxidizers of sewage sludge generally possess aerobic denitrifying activity (MULLER et al., 1995). The removal of nitrite already in the nitrification step would decrease costs of the denitrification process because the denitrification tank could be smaller in dimension and the demand for an organic substrate would be lower. Consequently, the use of nitrifier denitrification can distinctly contribute to cost reduction in the wastewater treatment.

2.4 Effect of Nitrogenous Oxides on Ammonia Oxidation

Ammonia oxidizing bacteria of the genus *Nitrosomonas* were distinctly inhibited when gaseous nitric oxide was removed from laboratory-scale cultures by means of intensive aeration. Nitrification started again when nitric oxide was added to the gas inlet of the culture vessels (ZART, unpublished data). Using a laboratory-scale fermentor with complete biomass retention it could be shown that nitrogenous oxides such as nitric oxide and especially nitrogen dioxide have a significant promoting effect on pure cultures of *Nitrosomonas eutropha* (ZART and BOCK, 1998). Compared to cultures grown without these externally added nitrogenous oxides their addition resulted in a pronounced increase in nitrification rate, specific activity of ammonia oxidation, growth rate, maximum cell density, and aerobic denitrification capacity. Maximum cell numbers amounted to $2 \cdot 10^{10}$ *Nitrosomonas* cells per mL and nitrification rates up to 3,100 mg $NH_4-N \cdot L^{-1}\,d^{-1}$ were accessible. Furthermore, approximately 50% of the nitrite produced was aerobically denitrified to dinitrogen when nitrogen dioxide was present. Hence, the addition of nitrogen dioxide could significantly improve the purification especially of high strength wastewaters, e.g., from sludge treatment.

3 Nitrite Oxidation

3.1 Physiology of Nitrite Oxidation

The second step of nitrification, the oxidation of nitrite to nitrate, is performed by nitrite oxidizing bacteria, e.g., members of the genera *Nitrobacter*, *Nitrospira*, or *Nitrococcus* (PROSSER, 1989). Several strains of *Nitrobacter* and one strain of *Nitrospira* are the only nitrite oxidizers which are not restricted to marine environments (BOCK et al., 1991; EHRICH et al., 1995).

The key enzyme of nitrite oxidizing bacteria is the membrane bound nitrite oxidoreductase (TANAKA et al., 1983) which oxidizes nitrite with water as the source of oxygen to form nitrate (ALEEM et al., 1965). The electrons released from this reaction are transferred via a- and c-type cytochromes to a cytochrome oxidase of the aa_3-type. However, the mechanism of energy conservation in nitrite oxidizers is still unclear. Neither HOLLOCHER et al. (1982) nor SONE et al. (1983) were able to find any electron transport chain-linked proton translocation which is necessary to maintain a proton motive force for ATP regeneration. Thus, NADH is thought to be produced as the first step of energy conservation (SUNDERMEYER and BOCK, 1981). The overall process deserves further research for complete elucidation.

3.2 Growth Characteristics of Nitrite Oxidizers

Nitrite oxidizers are generally lithoautotrophic organisms. Carbon dioxide is fixed as the main carbon source using RuBisCO, which is in parts carboxysome-bound (BOCK et al., 1974). Higher growth rates are obtained, when the cells are maintained mixotrophically (STEINMÜLLER and BOCK, 1976; WATSON et al., 1986). In contrast to ammonia oxidizing bacteria, several strains of *Nitrobacter* are capable of heterotrophic growth under oxic as well as anoxic conditions (BOCK, 1976; STEINMÜLLER and BOCK, 1977). Heterotrophic growth is significantly slower than lithoautotrophic (BOCK, 1976), although, 10- to 50-fold higher cell densities are obtained (BOCK et al., 1991).

Nitrite oxidation occurs obligately under oxic conditions. The organisms involved are much more sensitive to oxygen limitation than ammonia oxidizers are. Already at dissolved oxygen concentrations of about 0.5 mg L^{-1} nitrite oxidation is completely inhibited (HANAKI et al., 1990). Additionally, *Nitrobacter* is inhibited at high oxygen concentrations (EIGNER and BOCK, 1972). Thus, the oxygen content of a nitrite oxidizing nitrification tank has to be maintained carefully to avoid accumulation of nitrite. With sufficient oxygen supply nitrite oxidation proceeds at a faster rate than conversion of ammonia to nitrite. Therefore, high nitrite concentrations are rarely found in natural environments or in wastewater treatment plants (BOCK et al., 1991).

3.3 Denitrification by Nitrite Oxidizers

Some strains of *Nitrobacter* were shown to be denitrifying organisms as well. Nitrate can be used as an acceptor for electrons derived from organic compounds to promote anoxic growth (BOCK et al., 1988). Since the oxidation of nitrite is a reversible process, the nitrite oxidoreductase can reduce nitrate to nitrite in the absence of oxygen (SUNDERMEYER-KLINGER et al., 1984). Furthermore, the nitrite oxidoreductase copurifies with a nitrite reductase which reduces nitrite to nitric oxide (AHLERS et al., 1990). Evidence is given that subsequent oxidation of nitric oxide to nitrite is involved in NAD reduction in *Nitrobacter hamburgensis* (FREITAG and BOCK, 1990; BOCK et al., 1991). However, since denitrifying cells of *Nitrobacter* grow very slowly, it seems so far improbable that denitrification of nitrite oxidizers might have any significance in the treatment of wastewater.

4 Heterotrophic Nitrification

The oxidation of ammonia (VAN NIEL et al., 1993), hydroxylamine (RALT et al., 1981), or organic nitrogen compounds, e.g., oximes (Castignetti and Hollocher, 1984) to nitrite and nitrate by various chemoorganotrophic microorganisms is called heterotrophic nitrification. The latter is a co-metabolism which is not coupled to energy conservation (Wood, 1988). Heterotrophic nitrifiers are found among algae (Spiller et al., 1976), fungi (Stams et al., 1990), and bacteria (Robertson and Kuenen, 1990). Compared to the nitrification rates of autotrophic nitrifiers those of heterotrophic nitrifiers are small (Robertson and Kuenen, 1988). Therefore, heterotrophic nitrification was thought to occur preferentially under conditions which are not favorable for autotrophic nitrification, e.g., acidic environments (Killham, 1986). But recent research revealed that heterotrophic nitrification contributes to overall nitrate production only to a minor extent, even in acid soils (Stams et al., 1990; Barraclough and Puri, 1995). Over the last years, attention has been focused on heterotrophic nitrification because it occurs frequently accompanied by aerobic denitrification, leading to the complete elimination of dissolved nitrogen compounds (Castignetti and Hollocher, 1984; Robertson et al., 1989).

5 Denitrification

5.1 General Aspects of Denitrification

Denitrification is the reduction of oxidized nitrogen compounds like nitrite or nitrate to gaseous nitrogen compounds. This process is performed by various chemoorganotrophic, lithoautotrophic, and phototrophic bacteria and some fungi (Shoun and Tanimoto, 1991; Zumft, 1997), especially under oxygen-reduced or anoxic conditions (Focht and Chang, 1975). Denitrification can be described as a kind of anoxic respiration. Electrons originated from organic matter, reduced sulfur compounds, molecular hydrogen, etc., are transferred to oxidized nitrogen compounds instead of oxygen in order to build up a proton motive force usable for ATP regeneration. Involved enzymes are the nitrate reductase, the nitrite reductase, the nitric oxide reductase, and finally the nitrous oxide reductase (Zumft et al., 1988; Zumft and Körner, 1997). Dinitrogen is the main end product of denitrification while the nitrogenous gases nitric oxide and nitrous oxide occur as intermediates in low concentration (Zumft, 1992). However, these gases are also released as end products when denitrification enzymes are not completely expressed, e.g., when the concentration of dissolved oxygen is too high (Körner and Zumft, 1989).

5.2 Substrate Requirements of Denitrifiers

The common denitrifiers in municipal wastewater treatment plants are chemoorganotrophic bacteria, e.g., *Paracoccus denitrificans* or *Alcaligenes eutrophus*, which need an electron donor of an organic nature. Hence, complete denitrification requires a sufficient supply of organic matter. In detail, a specific C:N ratio is required to provide an adequate amount of electron donors for the reduction of a distinct amount of nitrogen oxides (la Cour Jansen et al., 1997b). Wastewater with an inadequate COD loading should therefore, be supplemented with an external organic electron donor like acetate or methanol to avoid exceeding outlet concentrations of nitrate or nitrite (Her and Huang, 1995).

5.3 Aerobic Denitrification

Denitrification also occurs in the presence of oxygen. The range of oxygen concentration permitting aerobic denitrification is broad and differs from one organism to another (Robertson and Kuenen, 1990). The onset of aerobic denitrification is not depending on oxygen sensitivity of the corresponding enzymes but rather on regulation of oxygen- or redox-sensing factors involved in the regulation on a transcriptional level. This is further indicated by denitrifying enzymes expressed under anoxic conditions which remain active in the presence of oxygen (Zumft, 1997). Generally, the ability to denitrify under oxic conditions seems to be the rule rather than the exception among denitrifiers (LLOYD et al., 1987). However, in the treatment of wastewater denitrification commonly proceeds under anoxic and not under oxic conditions because in the presence of oxygen large amounts of organic matter would be "wasted" on respiration. Hence, denitrification would cease or run incompletely as a result of lack of electron donors (LA COUR JANSEN et al., 1997b).

6 Combined Nitrification/ Denitrification Process – an Example of Application

The nitrogen content of wastewater from the sludge treatment might range from 200 to 2,000 mg N L^{-1}. The small volume of this so-called sludge liquor (about 2% of the total wastewater stream) contributes roughly 10–30% of the overall nitrogen load of municipal wastewater treatment plants (KOLLBACH and GRÖMPING, 1996). Biological treatment of these highly loaded wastewaters is feasible with good efficiency because of the high ammonium concentration. Furthermore, the small bypass streams can be purified under optimal conditions for the microorganisms involved. This refers especially to tempering of the process, which is important for maximal microbial activity. As an example, a biological treatment process (NITRA process) will be described, which uses nitrification and denitrification reaction sequences to eliminate nitrogen. A flow chart of this process is presented in Fig. 1. The sludge liquor is fed into the aerated nitrification reactor, which is connected to a stirred but not aerated denitrification tank. Due to a high internal circulation between the two biological reactors, the nitrified wastewater is nearly simultaneously denitrified. The activated sludge is recirculated from a sedimentation tank (clarifier). Temperature, pH value, and oxygen supply are kept in an optimal range. As depicted in Fig. 2A more than 90% of the ammonium load was eliminated when the process had been optimized.

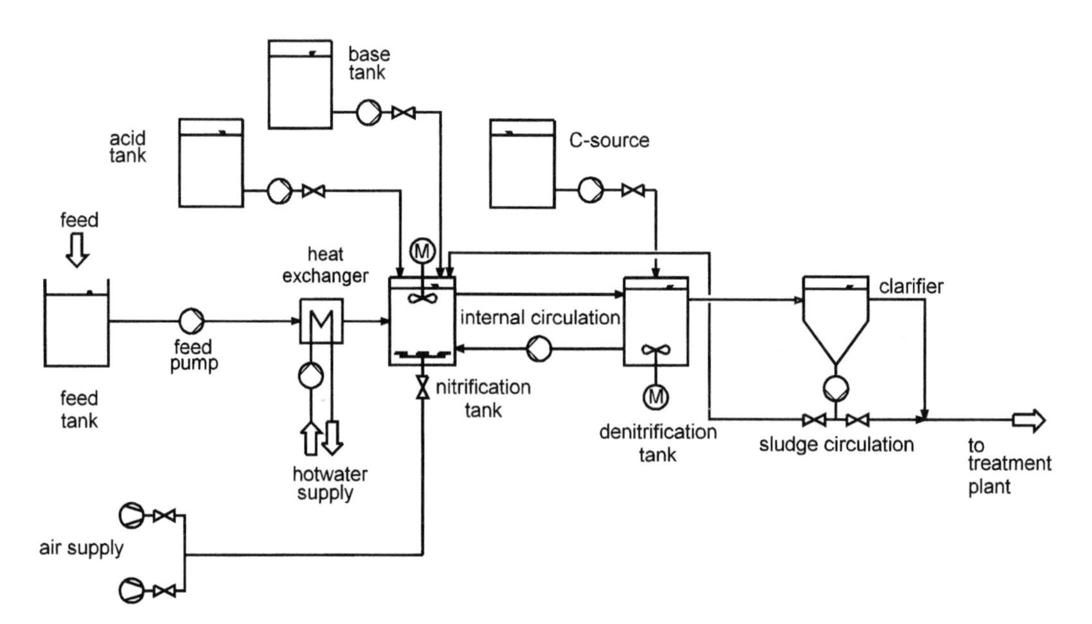

Fig. 1. Simplified flow chart of a biological treatment process for N-elimination in sludge liquor (NITRA process).

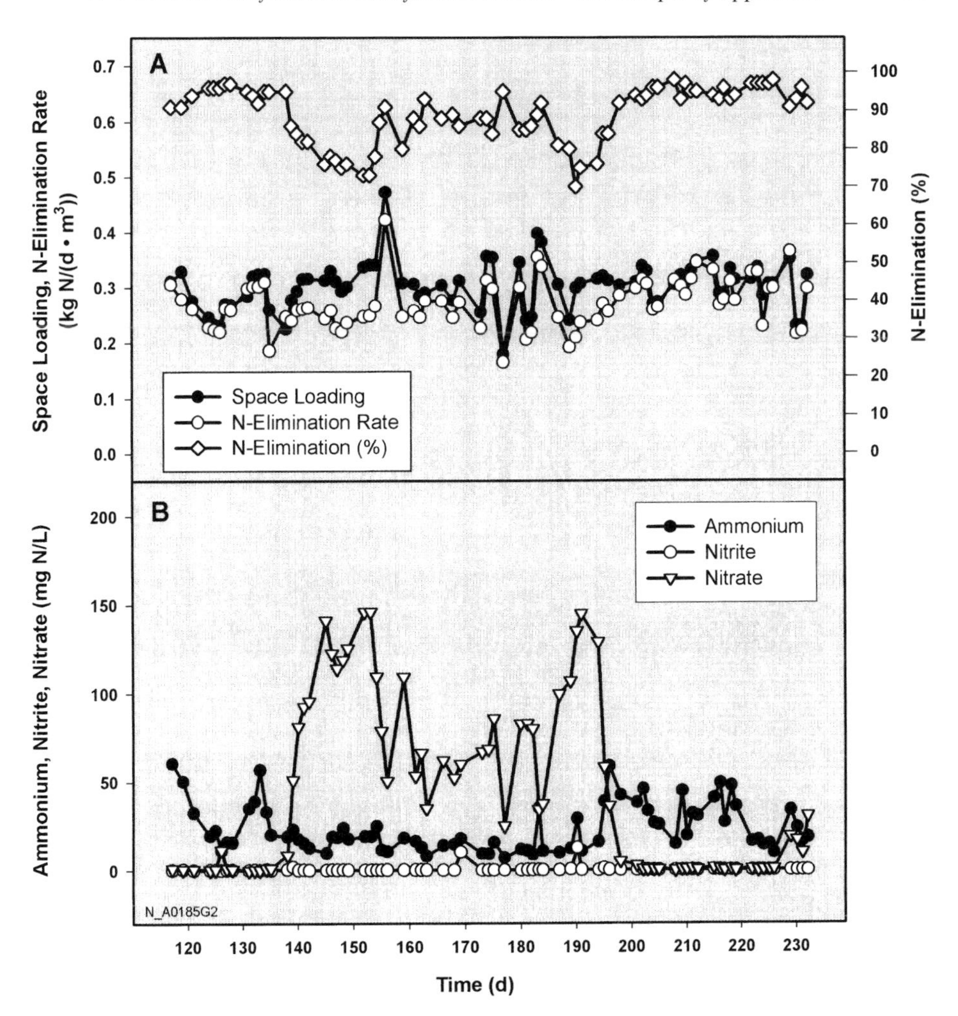

Fig. 2. N-elimination during biological treatment of sludge liquor with high ammonium content (average: 650 mg N L^{-1}). **A** Nitrogen space loading, nitrogen elimination rate, and nitrogen elimination given as a percentage of the nitrogen load. **B** Concentrations of nitrogenous compounds in the sludge liquor after treatment with the NITRA process.

Nitrate only accumulated when the dosage of the external carbon source for denitrification was too low (Fig. 2B). Furthermore, it was possible to avoid the formation of nitrate completely, when the ammonium level was kept high enough for active ammonia oxidation (not shown). Under these conditions, nitrite was the only end product of nitrification and was denitrified directly. This might be due to the formation of hydroxylamine during ammonia oxidation, which was shown to inhibit nitrite oxidation at very low concentrations (STÜVEN et al., 1992). Consequently, the carbon source supply was reduced and aeration was lowered.

7 References

ABELIOVICH, A., VONSHAK, A. (1992), Anaerobic metabolism of *Nitrosomonas europaea*, *Arch. Microbiol.* **158**, 267–270.

AHLERS, B., KÖNIG, W., BOCK, E. (1990), Nitrite reductase activity in *Nitrobacter vulgaris*, *FEMS Microbiol. Lett.* **67**, 121–126.

ALEEM, M. I. H., HOCH, G. E., VARNER, J. E. (1965), Water as the source of oxidant and reductant in bacterial chemosynthesis, *Proc. Nat. Acad. Sci. USA* **54**, 869–873.

ANDERSON, J. H. (1965), Studies on the formation of nitrogenous gas from hydroxylamine by *Nitrosomonas*, *Biochim. Biophys. Acta* **97**, 337–339.

ANDERSSON, K. K., HOOPER, A. B. (1983), O_2 and H_2O are each the source of one O in HNO_2 produced from NH_3 by *Nitrosomonas*: ^{15}N-NMR evidence, *FEBS Lett.* **164**, 236–240.

BARRACLOUGH, D., PURI, G. (1995), The use of ^{15}N pool dilution and enrichment to separate the heterotrophic and autotrophic pathways of nitrification, *Soil Biol. Biochem.* **27**, 17–22.

BOCK, E. (1976), Growth of *Nitrobacter* in the presence of organic matter. II. Chemoorganic growth of *Nitrobacter agilis*, *Arch. Microbiol.* **108**, 305–312.

BOCK, E., DÜVEL, D., PETERS, K.-R. (1974), Characterization of a phage-like particle from cells of *Nitrobacter*. I. Host-particle correlation and particle isolation, *Arch. Microbiol.* **97**, 115–127.

BOCK, E., WILDERER, P. A., FREITAG, A. (1988), Growth of *Nitrobacter* in the absence of dissolved oxygen, *Water Res.* **22**, 245–250.

BOCK, E., KOOPS, H.-P., HARMS, H., AHLERS, B. (1991), The biochemistry of nitrifying organisms, in: *Variations of Autotrophic Life* (SHIVELY, J. M., Ed.), pp. 171–200. London: Academic Press.

BOCK, E., SCHMIDT, I., STÜVEN, R., ZART, D. (1995), Nitrogen loss caused by denitrifying *Nitrosomonas* cells using ammonium or hydrogen as electron donors and nitrite as electron acceptor, *Arch. Microbiol.* **163**, 16–20.

CASTIGNETTI, D., HOLLOCHER, T. C. (1984), Heterotrophic nitrification among denitrifiers, *Appl. Environ. Microbiol.* **47**, 620–623.

EHRICH, S., BEHRENS, D., LEBEDEVA, E., LUDWIG, W., BOCK, E. (1995), A new obligately chemolithoautotrophic, nitrite-oxidizing bacterium, *Nitrospira moscoviensis* sp. nov. and its phylogenetic relationship, *Arch. Microbiol.* **164**, 16–23.

EIGNER, U., BOCK, E. (1972), Synthesis and breakdown of polyphosphate fraction in cells of *Nitrobacter winogradskyi* Buch, *Arch. Microbiol.* **81**, 367–378.

FOCHT, D. D., CHANG, A. C. (1975), Nitrification and denitrification process related to wastewater treatment, *Adv. Appl. Microbiol.* **19**, 153–186.

FREITAG, A., BOCK, E. (1990), Energy conservation in *Nitrobacter*, *FEMS Microbiol. Lett.* **66**, 157–162.

GOREAU, T. J., KAPLAN, W. A., WOFSY, S. C., MCELROY, M. B., VALOIS, F. W., WATSON, S. W. (1980), Production of NO_2^- and N_2O by nitrifying bacteria at reduced concentration of oxygen, *Appl. Environ. Microbiol.* **40**, 526–532.

GREEN, M., MELS, A., LAHAV, O., TARRE, S. (1996), Biological ion exchange process for ammonium removal from secondary effluent, *Water Sci. Technol.* **34**, 449–458.

HANAKI, K., WANTAWIN, C., OGAKI, S. (1990), Nitrification at low levels of dissolved oxygen with and without organic loading in a suspended-growth reactor, *Water Res.* **24**, 297–302.

HER, J.-J., HUANG, J.-S. (1995), Influence of carbon source and C/N ratio on nitrate/nitrite denitrification and carbon breakthrough, *Biores. Technol.* **54**, 45–51.

HOLLOCHER, T. C., TATE, M. E., NICHOLAS, D. J. D. (1981), Oxidation of ammonia by *Nitrosomonas europaea*: definitive ^{18}O-tracer evidence that hydroxylamine formation involves a monooxygenase, *J. Biol. Chem.* **256**, 10834–10836.

HOLLOCHER, T. C., KUMAR, S., NICHOLAS, D. J. D. (1982), Respiration-dependent proton translocation in *Nitrosomonas europaea* and its apparent absence in *Nitrobacter agilis* during inorganic oxidations, *J. Bacteriol.* **149**, 1013–1020.

HOOPER, A. B. (1968), A nitrite-reducing enzyme from *Nitrosomonas europaea* – preliminary characterization with hydroxylamine as electron donor, *Biochim. Biophys. Acta* **162**, 49–65.

HOOPER, A. B. (1969), Biochemical basis of obligate autotrophy in *Nitrosomonas europaea*, *J. Bacteriol.* **97**, 776–779.

HOOPER, A. B., TERRY, K. R. (1979), Hydroxylamine oxidoreductase of *Nitrosomonas* – production of nitric oxide from hydroxylamine, *Biochim. Biophys. Acta* **571**, 12–20.

HOOPER, A. B., VANNELLI, T., BERGMANN, D. J., ARCIERO, D. M. (1997), Enzymology of the oxidation of ammonia to nitrite by bacteria, *Antonie van Leeuwenhoek* **71**, 59–67.

HYMAN, M. R., MURTON, I. B., ARP, D. J. (1988), Interaction of ammonia monooxygenase from *Nitrosomonas europaea* with alkanes, alkenes, and alkynes, *Appl. Environ. Microbiol.* **54**, 3187–3190.

JETTEN, M. S. M., LOGEMANN, S., MUYZER, G., ROBERTSON, L. A., DE VRIES, S., VAN LOOSDRECHT, M. C. M., KUENEN, J. G. (1997), Novel principles in the microbial conversion of nitrogen compounds, *Antonie van Leeuwenhoek* **71**, 75–93.

KEENER, W. K., ARP, D. J. (1993), Kinetic studies of ammonia monooxygenase inhibition in *Nitrosomonas europaea* by hydrocarbons and halogenated hydrocarbons in an optimized whole-cell assay, *Appl. Environ. Microbiol.* **59**, 2501–2510.

KILLHAM, K. (1986), Heterotrophic nitrification, in: *Nitrification* (PROSSER, J. I., Ed.), pp. 117–126. Oxford: IRL Press.

KOLLBACH, J. S., GRÖMPING, M. (Eds.) (1996), *Stickstoffrückbelastung: Stand der Technik 1996/97 – zukünftige Entwicklungen.* Neuruppin: TK Verlag Karl Thomé-Kozmiensky.

KÖRNER, H., ZUMFT, W. G. (1989), Expression of denitrification enzymes in response to the dissolved oxygen level and respiratory substrate in continuous culture of *Pseudomonas stuzeri*, *Appl. Environ. Microbiol.* **55**, 1670–1676.

LA COUR JANSEN, J., HARREMOËS, P., HENZE, M. (1997a), Treatment plants for nitrification, in: *Wastewater Treatment: Biological and Chemical Processes*, 2nd Edn. (HENZE, M., HARREMOËS, P., LA COUR JANSEN, J., ARVIN, E., Eds.), pp. 195–228. Berlin: Springer-Verlag.

LA COUR JANSEN, J., HARREMOËS, P., HENZE, M. (1997b), Treatment plants for denitrification, in: *Wastewater Treatment: Biological and Chemical Processes*, 2nd Edn. (HENZE, M., HARREMOËS, P., LA COUR JANSEN, J., ARVIN, E., Eds.), pp. 229–272. Berlin: Springer-Verlag.

LLOYD, D., BODDY, L., DAVIES, K. J. P. (1987), Persistence of bacterial denitrification under aerobic conditions: the rule rather than the exception, *FEMS Microbiol. Ecol.* **45**, 185–190.

MAEKAWA, T., LIAO, C.-L., FENG, X.-D. (1995), Nitrogen and phosphorus removal from swine wastewater using intermittent aeration batch reactor followed by ammonium crystallization process, *Water Res.* **29**, 2643–2650.

MULDER, A. (1989), *European Patent* 0 327 184 A1.

MULDER, A., VAN DE GRAAF, A. A., ROBERTSON, L. A., KUENEN, J. G. (1995), Anaerobic ammonia oxidation discovered in a denitrifying fluidized bed reactor, *FEMS Microbiol. Ecol.* **16**, 177–184.

MULLER, E. B., STOUTHAMER, A. H., VAN VERSEFELD, H. W. (1995), Simultaneous NH_3-oxidation and N_2-production at reduced O_2-tensions by sewage sludge subcultured with chemolithotrophic medium, *Biodegradation* **6**, 339–349.

POTH, M. (1986), Dinitrogen production from nitrite by a *Nitrosomonas* isolate, *Appl. Environ. Microbiol.* **52**, 957–959.

PROSSER, J. I. (1989), Autotrophic nitrification in bacteria, in: *Advances in Microbial Physiology*, Vol. 30 (ROSE, A. H., TEMPEST, D. W., Eds.), pp. 125–181. London: Academic Press.

RALT, D., GOMEZ, R. F., TANNERBAUM, S. R. (1981), Conversion of acetohydroxamate and hydroxylamine to nitrite by intestinal microorganisms, *Eur. J. Appl. Microbiol. Biotechnol.* **12**, 226–230.

REMDE, A., CONRAD, R. (1990), Production of nitric oxide in *Nitrosomonas europaea* by reduction of nitrite, *Arch. Microbiol.* **154**, 187–191.

ROBERTSON, L. A., KUENEN, J. G. (1988), Heterotrophic nitrification in *Thiosphaera pantotropha*: oxygen uptake and enzyme studies, *J. Gen. Microbiol.* **134**, 857–863.

ROBERTSON, L. A., KUENEN, J. G. (1990), Combined heterotrophic nitrification and aerobic denitrification in *Thiosphaera pantotropha* and other bacteria, *Antonie van Leeuwenhoek* **57**,139–152.

ROBERTSON, L.A., CORNELISSE, R., DE VOS, P., HADIOETOMO, R., KUENEN, J. G. (1989), Aerobic denitrification in various heterotrophic nitrifiers, *Antonie van Leeuwenhoek* **56**, 289–300.

SCHMIDT, I., BOCK, E. (1997), Anaerobic ammonia oxidation with nitrogen dioxide by *Nitrosomonas eutropha*, *Arch. Microbiol.* **167**, 106–111.

SHOUN, H., TANIMOTO, T. (1991), Denitrification by the fungus *Fusarium oxysporum* and involvement of cytochrome P-450 in the respiratory nitrite reduction, *J. Biol. Chem.* **25**, 1527–1536.

SIEGRIST, H. (1996), Nitrogen removal from digester supernatant – comparison of chemical and biological methods, *Water Sci. Technol.* **34**, 399–406.

SONE, N., YANAGITA, Y., HON-NAMI, K., FUKUMORI, Y., YAMANAKA, T. (1983), Proton-pump activity of *Nitrobacter agilis* and *Thermus thermophilus* cytochrome c oxidases, *FEBS Lett.* **155**, 150–153.

SPILLER, H., DIETSCH, E., KESSLER, E. (1976), Intracellular appearance of nitrite and nitrate in nitrogen-starved cells of *Ankistrodesmus braunii*, *Planta* **129**, 175–181.

STAMS, A. J. M., FLAMELING, E. M., MARNETTE, E. C. L. (1990), The importance of autotrophic *versus* heterotrophic oxidation of atmospheric ammonium in forest ecosystems with acid soil, *FEMS Microbiol. Ecol.* **74**, 337–344.

STEINMÜLLER, W., BOCK, E. (1976), Growth of *Nitrobacter* in the presence of organic matter. I. Mixotrophic growth, *Arch. Microbiol.* **108**, 299–304.

STEINMÜLLER, W., BOCK, E. (1977), Enzymatic studies on autotrophically, mixotrophically and heterotrophically grown *Nitrobacter agilis* with special references to nitrite oxidase, *Arch. Microbiol.* **115**, 51–54.

STÜVEN, R., VOLLMER, M., BOCK, E. (1992), The impact of organic matter on nitric oxide formation by *Nitrosomonas europaea*, *Arch. Microbiol.* **158**, 439–443.

SUNDERMEYER, H., BOCK, E. (1981), Energy metabolism of autotrophically and heterotrophically grown cells of *Nitrobacter winogradskyi*, *Arch. Microbiol.* **130**, 250–254.

SUNDERMEYER-KLINGER, H., MEYER, W., WARNINGHOFF, B., BOCK, E. (1984), Membrane-bound nitrite oxidoreductase of *Nitrobacter*: evidence for a nitrate reductase system, *Arch. Microbiol.* **140**, 153–158.

TANAKA, Y., FUKUMORI, Y., YAMANAKA, T. (1983), Purification of cytochrome a_1c_1 from *Nitrobacter agilis* and characterization of nitrite oxidation system of the bacterium, *Arch. Microbiol.* **135**, 265–271.

TSANG, D. C. Y., SUZUKI, I. (1982), Cytochrome c554 as a possible electron donor in the hydroxylation of ammonia and carbon monoxide in *Nitrosomonas europaea*, *Can. J. Biochem.* **60**, 1018–1024.

US Environmental Protection Agency (1994), *Nitrogen Control.* Lancaster: Technomic Publishing Company.

VAN DE GRAAF, A. A., MULDER, A., DE BRUIJN, P., JETTEN, M. S. M., ROBERTSON, L. A., KUENEN, J. G. (1995), Anaerobic oxidation of ammonium is a biologically mediated process, *Appl. Environ. Microbiol.* **61**, 1246–1251.

VAN DE GRAAF, A. A., DE BRUIJN, P., ROBERTSON, L. A., KUENEN, J. G. (1996), Autotrophic growth of anaerobic ammonium-oxidizing microorganisms in a fluidized bed reactor, *Microbiology* **142**, 2187–2196.

VAN DE GRAAF, A. A., DE BRUIJN, P., ROBERTSON, L. A., JETTEN, M. S. M., KUENEN, J. G. (1997), Metabolic pathway of anaerobic ammonium oxidation on the basis of ^{15}N studies in a fluidized bed reactor, *Microbiology* **143**, 2415–2421.

VAN NIEL, E. W. J., ARTS, P. A. M., WESSELINK, B. J., ROBERTSON, L. A., KUENEN, J. G. (1993), Competition between heterotrophic and autotrophic nitrifiers for ammonia in chemostat cultures, *FEMS Microbiol. Ecol.* **102**, 109–118.

WATSON, S. W., BOCK, E., VALOIS, F. W., WATERBURY, J. B., SCHLOSSER, U. (1986), *Nitrospira marina* gen. nov. sp. nov.: a chemolithotrophic nitrite oxidizing bacterium, *Arch. Microbiol.* **144**, 1–7.

WATSON, S. W., BOCK, E., HARMS, H., KOOPS, H.-P., HOOPER, A. B. (1989), Genera of ammonia-oxidizing bacteria, in: *Bergey's Manual of Systematic Bacteriology* (STALEY, J. T., BRYANT, M. P., PFENNIG, N., HOLT, J. G., Eds.), pp. 1822–1834. Baltimore, MD: Williams & Wilkins.

WHEELIS, M. (1984), Energy conservation and pyridine nucleotide reduction in chemoautotrophic bacteria: a thermodynamic analysis, *Arch. Microbiol.* **138**, 166–169.

WOOD, P. M. (1986), Nitrification as a bacterial energy source, in: *Nitrification* (PROSSER, J. L., Ed.), pp. 63–78. Oxford: IRL Press.

WOOD, P. M. (1988), Monooxygenase and free radical mechanism for biological ammonia oxidation, in: *The Nitrogen and Sulfur Cycles* (COLE, J. A., Ed.), pp. 217–243. Cambridge: Cambridge University Press.

ZART, D., BOCK, E. (1998), High rate of aerobic nitrification and denitrification by *Nitrosomonas eutropha* grown in a fermentor with complete biomass retention in the presence of gaseous NO_2 or NO, *Arch. Microbiol.* **169**, 282–286.

ZART, D., SCHMIDT, I., BOCK, E. (1996), Neue Wege vom Ammonium zum Stickstoff, in: *Ökologie der Abwasserorganismen* (LEMMER, H., GRIEBE, T., FLEMMING, H.-C., Eds.), pp. 183–192. Berlin, Heidelberg: Springer-Verlag.

ZUMFT, W. G. (1992), The denitrifying prokaryotes, in: *The Prokaryotes*, 2nd Edn. (BALOWS, A., TRÜPER, H. G., DWORKIN, M., HARDER, W., SCHLEIFER, K.-H., Eds.), pp. 554–582. New York: Springer-Verlag.

ZUMFT, W. G. (1997), Cell biology and molecular basis of denitrification, *Microbiol. Mol. Biol. Rev.* **61**, 533–616.

ZUMFT, W. G., KÖRNER, H. (1997), Enzyme diversity and mosaic gene organization in denitrification, *Antonie van Leeuwenhoek* **71**, 43–58.

ZUMFT, W. G., VIEBROCK, A., KÖRNER, H. (1988), Biochemical and physiological aspects of denitrification, in: *The Nitrogen and Sulfur Cycles* (COLE, J. A., FERGUSON, S. J., Eds.), pp. 245–279. Cambridge: Cambridge University Press.

4 Autoaggregation of Microorganisms: Flocs and Biofilms

JOST WINGENDER
HANS-CURT FLEMMING
Mülheim/Ruhr, Germany

1 Introduction

In aquatic environments microorganisms preferably grow and live in aggregated forms, in which mixed species exist in close proximity. It is estimated that more than 99% of all microorganisms on earth live in aggregates such as in biofilms and flocs (COSTERTON et al., 1987). Biofilms develop at interfaces, well known adherent to a solid surface (substratum) at the solid–water interface, but also at water–air interfaces, where hydrophobic substances and microorganisms accumulate (KJELLEBERG, 1985), and at solid–air interfaces such as rocks and building materials where they give rise to biological weathering as well as to biodeterioration (SAND et al., 1989).

For historical reasons, the term "biofilm" has been introduced for most of these microbial aggregates, and it is assumed that structure and function of flocs and biofilms are similar so that the characteristics of biofilms also grossly apply to flocs. It is acknowledged that there may be differences between various forms of biofilms as well as between flocs and biofilms. However, as a working definition, "biofilm" may be a useful generic term, because it applies to the common feature of all microbial aggregates, namely the mode of life in a slime matrix. Thus, the expression embraces flocs in the sense of "planktonic biofilms", sessile microcolonies, thin and thick layers of microorganisms, and sludges of microbial origin. All these aggregates are accumulations of microorganisms (prokaryotic and eukaryotic unicellular organisms), extracellular polymeric substances (EPS), multivalent cations, biogenic and inorganic particles as well as colloidal and dissolved compounds (Tab. 1). The common aspect of the various appearances of microbial aggregates is that the retention time of the constituent cells in a fixed position to each other is much longer than for single cells in suspension, allowing to establish synergistic microconsortia of different species, which may be able to degrade complex substrates otherwise inaccessible to single strains. Complex trophic levels can evolve, including protozoa and metazoa, whose existence is based on the consumption of biomass accumulated in the form of microorganisms. Recent research has revealed that even in primary biofilms cell-to-cell signals are exchanged as a form of bacterial communication (DAVIES et al., 1998), and it is speculated that bacteria in biofilms may be modular organisms (ANDREWS, 1995).

Flocs and biofilms have in common that microbial cells are kept together by high molecular-weight EPS, often enforced by divalent cations, among which Ca^{2+} plays an important role. EPS are key components of microbial aggregates, because they provide three-dimensional gel-like networks, which keep the cells together, allow the evolution of stable microconsortia, and create an environment, where the microbial cells are protected from adverse environmental conditions. Flocs and biofilms display physicochemical properties which are of fundamental importance to their ecological function in nature, their effects on technical systems, their use in biotechnological applications, and the efficacy of countermeasures to prevent or control undesirable biofilm formation (biofouling) and microbially influenced corrosion (MIC) of materials (biocorrosion), which affects metals, minerals, and synthetic polymers (FLEMMING, 1996).

Biofilms represent the oldest form of life recorded and date back 3.5 billion years (SCHOPF et al., 1983); since then, biofilms are ubiquitous in nature. Their occurrence marks the extremes of environmental conditions, under

Tab. 1. Composition of Flocs and Biofilms in Aquatic Environments

Component	Examples
Microorganisms	heterotrophic and autotrophic bacteria, algae, protozoa, fungi
EPS	polysaccharides, proteins, glycoproteins, nucleic acids, lipids, humic substances
Biogenic particles	cellular debris, detritus
Inorganic particles	clay, sand, silt, corrosion products
Inorganic ions	multivalent ions (e.g., Ca^{2+}, Mg^{2+}, SO_4^{2-}, Fe^{3+})
Dissolved organic molecules	amino acids, sugars, benzene, toluene, xylene
Water	free, bound

which life on earth can exist, including the entire redox stability of water, temperatures between −12°C and +118°C, pH values between 0.5 and 14, and even growth on radiation sources as well as on the wall of biocide distribution pipes (compiled by FLEMMING, 1991). Suspended aggregates ("planktonic biofilms") have been described in marine and freshwater environments, e.g., riverine, lacustrine, and marine flocs referred to as "marine snow" (ALLREDGE and SILVER, 1988).

Biological wastewater treatment makes use of mixed microbial populations in the form of both engineered flocs and biofilms. Floc formation is the basis of the activated sludge process; fixed-biofilm processes are employed in biofilm reactors with attached microbial biomass such as in trickling filters, rotating biological contactors, fluidized-bed reactors, or submerged fixed-bed reactors (BRYERS and CHARACKLIS, 1990; BITTON, 1994). In wastewater treatment these reactor systems are used for the biological oxidation of organic matter, nitrification and denitrification processes, or anaerobic digestion of wastewater. From a process engineering standpoint, the major advantage is that dissolved organic substances are sequestered from the bulk water phase and partially converted into locally immobilized biomass, which can be handled separately. This is the process which underlies "biological filtration". Exactly the same process can occur in the wrong place, where it cannot be easily controlled any more, e.g., in heat exchangers or process water treatment systems. Then, it is called biofouling (FLEMMING and SCHAULE, 1996).

2 Composition and Development of Flocs and Biofilms

2.1 Spatial Arrangements

Microbial aggregates vary in size from flocs in microscopic dimensions in the micrometer range to macroaggregates of many centimeters in diameter and extended microbial mats. Under conditions of extensive mucilage production and minimal turbulence, marine snow may agglomerate to large macroscopic flocs yielding threads up to 30 cm and clouds up to 300 m across (LEPPARD, 1995). Biofilms on submerged surfaces occur as thin layers of bacteria or can develop into massive accumulation of biomass as macroscopic deposits or slimes, e.g., in organically polluted surface waters or in biofilm reactors for wastewater treatment (GRAY, 1985; BITTON, 1994). Conventional microbiological methods are not well suited to the investigation of microbial aggregates.

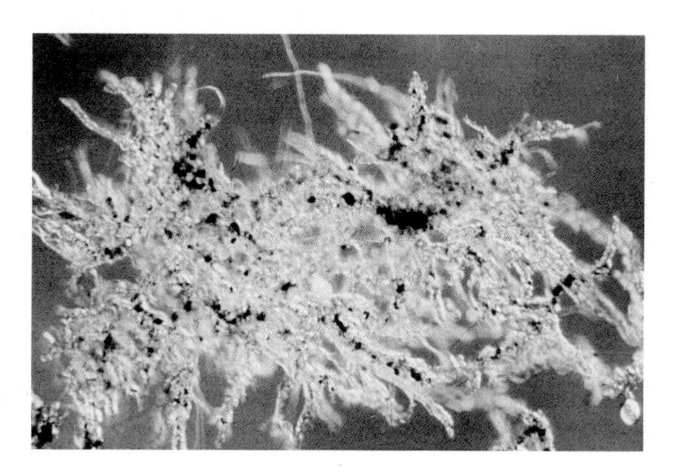

Fig. 1. Light microscopical picture of a floc in an activated sludge system with enclosed anthracite powder particles (courtesy of F. OPALLA).

Thus, during the last two decades various new techniques have been developed to reveal the structure, dynamics, and processes of biofilms. Combinations of different microscopic methods, including light microscopy, epifluorescence microscopy, confocal laser scanning microscopy (CLSM) (LAWRENCE et al., 1991), transmission electron microscopy (TEM), scanning electron microscopy, and atomic force microscopy (BEECH et al., 1994) have been applied in parallel for the study of floc and biofilm structure in marine and freshwater environments as well as in drinking and wastewater systems (DROPPO et al., 1996; LISS et al., 1996; STEWART et al., 1995; WALKER et al., 1995). This correlative microscopy in combination with nondestructive stabilization techniques of microbial flocs (e.g., embedding in agarose) allowed the analysis of the same sample over a range of different magnifications bridging the resolution gaps that exist with single microscopic techniques (DROPPO et al., 1996). The application of CSLM and the environmental scanning electron microscope (ESEM) (LITTLE et al., 1998), the use of microelectrodes for measurement of chemical species (e.g., dissolved oxygen, ammonium) (REVSBECH, 1989), and the use of fluorescent probes for the study of the penetration behavior of defined molecules and for determining physiological activities allowed examination of fully hydrated flocs and biofilms with minimal influence on the internal structure (COSTERTON et al., 1994; DE BEER et al., 1993). Digitized microtome sections of activated sludge flocs were processed with a discrete smooth interpolation (DSI) technique, which was adapted from medical applications such as reconstruction of embryo organ images from histological sections (ZARTARIAN et al., 1997). This method allowed not only the three-dimensional visualization of the floc, but also the manipulation of the image as a mathematical object. Application of all the methods mentioned above showed that the overall structure of flocs and biofilms was not uniform, but generally characterized by a heterogeneous distribution of the components was typical of these microbial aggregates. Fig. 1 shows a light microsopic picture of a floc in a wastewater treatment system supported by powdered anthracite. Fig. 2 depicts a scanning electron micrograph of a thick biofilm (ca. 100 μm), which colonized a reverse osmosis membrane and survived many cleaning cycles, eventually leading to biofouling of that system (FLEMMING and SCHAULE, 1989). Fig. 3 is a vertical cut through a biofilm composed by image processing of single horizontal optical sections that were obtained by CSLM. The biofilm was grown in a nutrient solution on a glass surface. The space between the cells is filled with EPS gel.

Under the light microscope activated sludge flocs appear as aggregates with irregular boun-

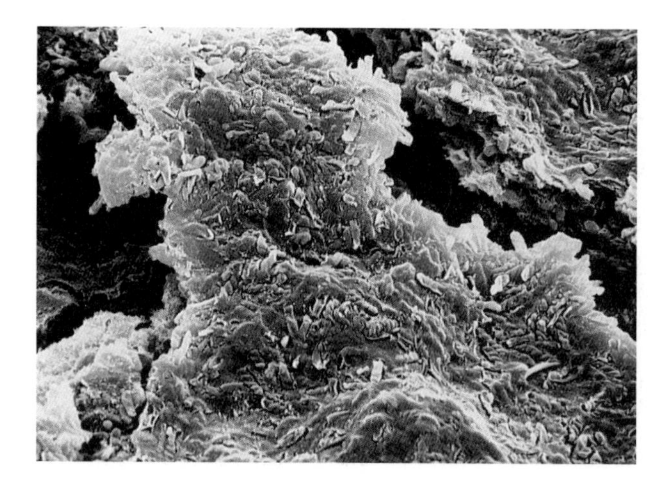

Fig. 2. Biofouling layer on a reverse osmosis membrane (FLEMMING and SCHAULE, 1989).

Fig. 3. Vertical cut through a biofilm originating from image processing of horizontal sections obtained by CSLM technique (SCHINDLER, KÜHN, and FLEMMING, unpublished).

daries containing microorganisms (bacteria of different shapes, protozoa) and undefined organic and inorganic particles (EIKELBOOM and VAN BUIJSEN, 1981; JORAND et al., 1995). Within the flocs, smaller aggregates of microbial cells were observed as distinct microcolonies of bacterial cells with similar morphology, when activated sludge flocs were studied at higher magnifications using TEM (ZARTARIAN et al., 1994; JORAND et al., 1995). Generally, microcolonies and single cells within flocs and biofilms are separated by transparent voids. However, specific staining (e.g., Thiery's stain, ruthenium red) allowed microscopic detection of polysaccharidic material between microcolonies and single cells indicating that the intercellular space was filled with EPS connecting the cells (JORAND et al., 1995). This was confirmed when flocs from natural riverine systems and from wastewater were viewed by high-resolution TEM (LISS et al., 1996). Pores devoid of physical structures under optical microscopes were found to be filled with complex matrices of extracellular polymeric fibrils (4–6 nm diameter). These fibrils provided the dominant bridging mechanism between organic and inorganic components of the flocs and contributed to the extensive surface area per volume unit of the flocs. It is thought that this mechanism is also valid for biofilms.

The heterogeneous arrangement of floc components shown by microscopic analysis was confirmed by sonication experiments with activated sludge (JORAND et al., 1995). Size distribution studies demonstrated that activated sludge flocs varied in size between about 1 μm to 1,000 μm or more (BITTON, 1994; LI and GANCZARCZYK, 1991; JORAND et al., 1995). Flocs smaller than 5 μm in diameter (dispersed microorganisms, aggregates of a few microorganisms) predominated by number, whereas most of the surface area, the volume, and the biomass was made up of larger flocs (64–256 μm) (LI and GANCZARCZYK, 1991; JORAND et al., 1995). Activated sludge flocs proved to be relatively resistant to mechanical agitation (stirring), but could easily be dispersed into smaller flocs by short periods of ultrasound (30–60 s at 37 W) with minimal bacterial cell lysis (JORAND et al., 1995). Ultrasonic disruption resulted in the appearance of subunits with sizes of 2.5 μm and 13 μm from larger flocs of 125 μm with concomitant release of EPS (carbohydrates, proteins, DNA). The 13 μm aggregates proved particularly resistant to sonication. Similarly, bacterial microcolonies with a diameter of 10–20 μm embedded in EPS were shown to remain unaffected, when activated sludge flocs were disintegrated by sulfide treatment (NIELSEN and KEIDING, 1998). On the basis of the microscopic observations and size distribution studies a model of the aggregation mechanisms and structure of activated sludge flocs was suggested (JORAND et al., 1995). Primary particles, 2.5 μm in size, form secondary particles (microcolonies), 13 μm in size. The secondary particles make up tertiary particles with a mean diameter of 125 μm. Some of the 2.5 μm particles may lie between the 13 μm aggregates. EPS form links between the 13 μm aggregates and between microorganisms within the 13 μm and 2.5 μm aggregates. Since only the EPS between the 13 μm flocs were extracted by sonication, it was assumed that these polymers ("type I polymers") must be different from those within the 13 μm and 2.5 μm flocs ("type II polymers"), which could not be removed by sonication.

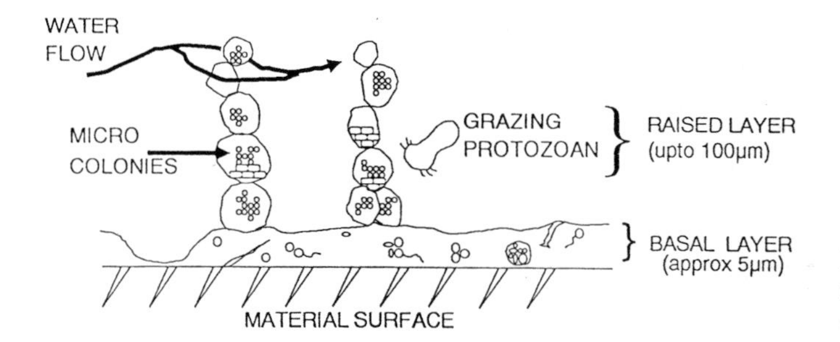

Fig. 4. Heterogenous mosaic biofilm model (according to WALKER et al., 1995).

There is no such thing as a "general biofilm model", as biofilms are formed by various organisms under a wide range of conditions. A number of conceptual models exist for the structure of microbial biofilms, reflecting the various biofilm phenomena observed. Among these models are: (1) the heterogeneous mosaic biofilm model (Fig. 4) (WALKER et al., 1995), (2) the penetrated water-channel mushroom-like biofilm model (Fig. 5) (COSTERTON et al., 1994), and (3) the dense confluent biofilm model (Fig. 6) (MARSHALL and BLAINEY, 1991; MARSH, 1995). Examples for all three models can be found in natural and technical environments. Observations of natural and laboratory biofilms as well as computer modeling of biofilm growth suggest that biofilm structure is largely determined by the prevailing substrate concentration (WIMPENNY and COLASANTI, 1997) and shear forces (VAN LOODSRECHT et al., 1995). The heterogeneous mosaic biofilm model was described on the basis of the examination of the structure of biofilms in oligotrophic drinking water environments (WALKER et al., 1995). In this model the biofilm consists of a thin (≈ 5 µm) basal layer of microor-

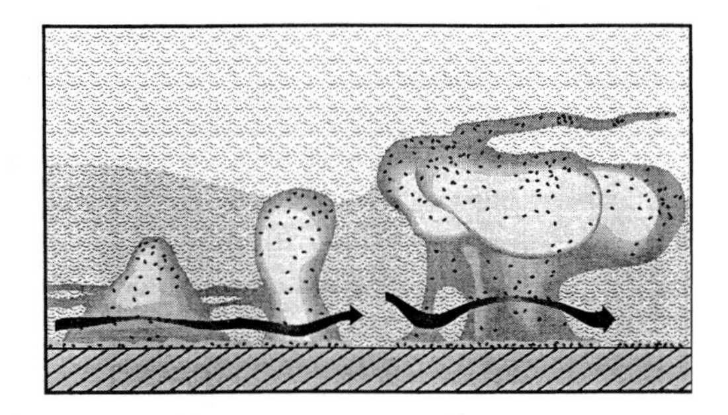

Fig. 5. Penetrated water-channel mushroom-like biofilm model (according to COSTERTON et al., 1994).

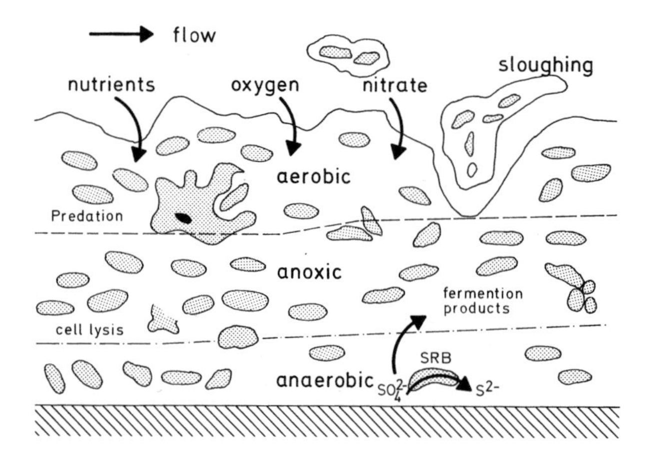

Fig. 6. Dense confluent biofilm model (according to MARSHALL and BLAINEY, 1991).

ganisms with tall stacks (up to 100 μm) of microcolonies rising from the surfaces and well separated from each other. In the water-channel model, microcolonies form mushroom-like structures attached by stalks of EPS and microorganisms which may merge into each other and are penetrated by water channels (COSTERTON et al., 1994). Smaller branching channels can also be observed within the microcolonies. This model was proposed on the basis of observations of natural living biofilms observed under a range of environmental conditions. The dense confluent biofilm model has been described for some medically important biofilms such as oral biofilms (dental plaque) (MARSH, 1995) as well as for aquatic environments (MARSHALL and BLAINEY, 1991). These confluent biofilms showed no evidence of water channels or porous structures, but consisted of numerous microcolonies and other specific associations between differently shaped bacteria. In addition to nutrient concentration and shear forces, other factors also control development, composition, and structural heterogeneity of microbial aggregates as summarized for biofilms in Tab. 2. Complex interactions between a number of physical, chemical, and biological processes determine the temporal and spatial changes during the development and aging of flocs and biofilms and are the underlying cause for the dynamic properties of microbial aggregates (BRADING et al., 1995).

Tab. 2. Major Factors that Control Biofilm Development, Composition, and Structure

Surface properties of substratum (e.g., roughness, hydrophobicity)
Surface properties of microorganisms
Physicochemical conditions of the bulk liquid phase (temperature, pH, salinity, ions, organic matter)
Concentration of available organic substrates, measured as assimilable organic carbon (AOC), biodegradable dissolved organic carbon (BDOC), or biological oxygen demand (BOD)
Morphology of microorganisms (e.g., filaments)
Physiological activity of microorganisms
Lysis of biofilm organisms
Grazing by protozoa
Activity of invertebrates
Formation of gas bubbles in anoxic and anaerobic zones (e.g., N_2, CH_4)
Continuous detachment of small particles (erosion) or sporadic detachment of larger fragments of biofilms (sloughing)
Age of biofilm
Adsorption of exogenous material from the bulk water phase
Hydraulic conditions (flow rate, shear stress)
Presence of antimicrobial agents

2.2 Development of Biofilms

The development of biofilms can be divided into three phases of an overall sigmoidal growth curve (Fig. 7) (CHARACKLIS, 1990):

(1) During the induction phase the "race for the surface" takes place, usually leading first to a conditioning film of traces of adsorbed organic macromolecules such as humic acids, proteins, and polysaccharides (BAIER, 1980). Microorganisms are transported more slowly than macromolecules and reach the surface either by their own mobility or by Brownian motion. In flowing systems a laminar sublayer exists even under highly turbulent conditions, which prevents cell transport directly

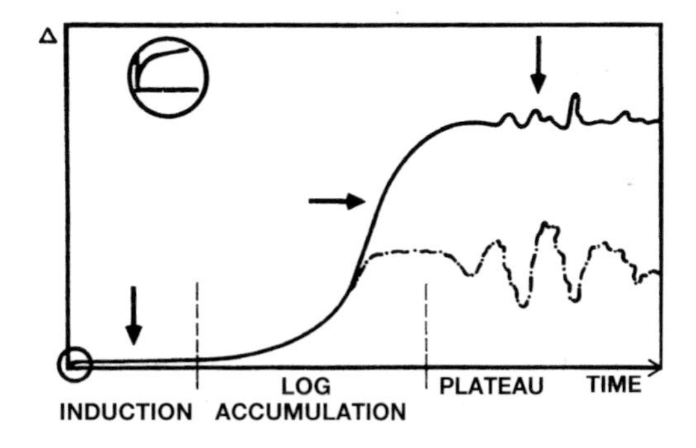

Fig. 7. Development of biofilms (according to CHARACKLIS, 1990).

to the surface by advection (CHARACKLIS, 1990). The induction phase is initiated by the reversible attachment of microbial cells to surfaces (primary adhesion). Some cells are able to scavenge biodegradable material from surfaces without staying irreversibly adhered (MARSHALL et al., 1994; MARSHALL, 1996). Sometimes, irreversible adhesion occurs with concomitant phenotypic changes such as increased EPS production (VANDEVIVERE and KIRCHMAN, 1993; DAVIES et al., 1993). How microorganisms sense contact to a surface is still unclear. In the induction phase cells are described as living particles (MARSHALL et al., 1971) and primary adhesion is attributed to weak physicochemical interactions such as electrostatic and dispersion interactions and hydrogen bonding (VAN LOODSRECHT et al., 1990). During this phase, accumulation of cells on the surface is a function of adhesion of cells from the water phase and, thus, dependent on suspended cell concentration.

(2) The logarithmic phase with subsequent multiplication of the sessile bacteria results in the formation of single-species microcolonies with the daughter cells bound within the EPS matrix. In this phase biofilm accumulation is due to the growth of adhering cells and becomes independent from cell concentration in the water phase.

(3) The plateau phase is reached after further cell division and new recruitment of microorganisms from the bulk fluid phase, leading to a more or less continuous biofilm that finally consists of coalesced microcolonies and single cells embedded in a highly hydrated matrix of EPS and trapped macromolecules as well as particulate matter from the water phase (COSTERTON et al., 1987; KORBER et al., 1995; PALMER and WHITE, 1997). The level of the plateau is controlled by the mechanical stability of the biofilm towards shear forces and by the availability of nutrients. The external structure of biofilms has been described to be the result of a balance between growth processes and detachment mainly resulting from shear forces (VAN LOODSRECHT et al., 1995); under high shear stresses biofilms become more smooth and less susceptible to shear forces. The organisms can detach from the biofilm by means of different mechanisms such as erosion of cells and sloughing off of larger biofilm fragments (RITTMANN, 1989) as well as actively leaving the biofilm by swarming (SZEWZYK and SCHINK, 1988) or by change of cell surface hydrophobicity (ASCON-CABERA et al., 1995).

2.3 Role of Organisms in Flocs and Biofilms

A diversity of life forms can be found in flocs and biofilms. They usually contain mixed populations of bacteria, but also represent a habitat for unicellular eukaryotic organisms (fungi, algae, protozoa such as flagellates, ciliates, and amoebae). Biofilms in nutrient-rich and aerobic environments (e.g., trickling filters) have often been found to be colonized by metazoa like rotifers, nematodes, annelids, insect larvae, or snails that feed on biofilm microorganisms and help control biofilm thickness (BITTON, 1994). Heterotrophic polymer degrading bacteria are often numerically dominant, but autotrophic organisms such as nitrifying bacteria or photosynthetic bacteria and algae may be present in considerable amounts depending on the physicochemical conditions of the environment (GEESEY et al., 1978; LEMMER et al., 1994). The concentrations of microorganisms within flocs and biofilms exceed those in the water by several orders of magnitude (CARON et al., 1982; GEESEY et al., 1978; LEWIS and GATTIE, 1990). Maximal concentrations of bacteria in laboratory-grown aggregates and in activated sludge ranged from about 10^{10} to 10^{12} cells mL^{-1} (LEWIS and GATTIE, 1990; URBAIN et al., 1993); highest population densities in sewer biofilms and activated sludges were in the range of 10^9 to 10^{11} culturable organisms per gram of dry weight determined for heterotrophic bacteria, but also for ammonifying and denitrifying bacteria (LEM-

MER et al., 1994). JAHN and NIELSEN (1998) reported values of approximately 10^{12} cells per g of volatile solids in sewer biofilms.

There may also be differences in the distribution of morphological types of bacteria between aggregates and the bulk aqueous phase and between flocs and biofilms from different origins. In samples of oceanic aggregates (marine snow) non-coccoid forms (straight and curved rods, spirals, filaments) constituted 54% of the bacteria, but only 21% of the bacterioplankton samples (CARON et al., 1982). Microbial morphologies influence physical properties of flocs and biofilms such as structural heterogeneity, density, and porosity of the aggregates that may interfere with technical processes. Filamentous bacteria are ubiquitous in activated sludge flocs. They have been assumed to provide a backbone for activated sludge flocs, to which flocculent microorganisms attach providing the macrostrucure of strong flocs (SEZGIN et al., 1978; LAU et al., 1984; URBAIN et al., 1993). If there are not enough filaments, the floc will be mechanically unstable and disaggregate into smaller particles referred to as pinpoint flocs in turbulent environments. However, excessive growth of filamentous microorganisms in activated sludge flocs results in filamentous bulking phenomena that consist of slow settling and poor compaction of solids in the clarifier of activated sludge systems (BITTON, 1994; SEZGIN et al., 1978). Thus, balanced growth of floc forming and filamentous organisms may be desirable for the activated sludge process to promote the formation of flocs with good settling properties.

Flocs and biofilms as characterized by high population densities result in direct and indirect interactions between the organisms. Although the direct inter- and intraspecific interaction between microbial cells (coaggregation) has been studied extensively with respect to dental plaque formation (MARSH, 1995), little information exists on the coaggregation of microorganisms in aquatic environments. Coaggregation of oral bacteria contribute to the overall structure of dental biofilms. The cellular interaction can be inhibited or reversed by addition of certain sugars or protease treatment suggesting that proteins with properties of lectins (non-enzymatic proteins with multiple binding sites for sugar residues) on the cell surface of oral bacteria were involved in coaggregation. Some observations indicate that coaggregation phenomena are also effective in the development of flocs and biofilms in environments other than the oral cavity. Non-floc forming *Cytophaga* bacteria normally growing as freely dispersed cells grew in close and stable association with floc forming *Zoogloea*-like bacteria, when grazing protozoa were present (GÜDE, 1982). These observations suggest that microbial aggregates are composed not solely of floc forming organisms, but may also harbor non-floc organisms, offering a protective habitat for those bacteria under selective pressures. Coaggregation has also been described for bacteria isolated from activated sludge (BOSSIER and VERSTRAETE, 1996) and from a water biofilm (BUSWELL et al., 1997). Pairs of different isolates demonstrated coaggregation giving rise to visible flocs (BUSWELL et al., 1997). The cell surface structures involved in the coaggregation process appeared to be similar to the lectin-like molecules implicated in the interactions between bacteria in oral biofilms. Some bacteria may act as bridging organisms by mediating coaggregations between two non-aggregating cell types (KOLENBRANDER et al., 1985; BUSWELL et al., 1997). The different rate and extent of coaggregation reactions among aquatic biofilm bacteria suggested a role of this phenomenon in determining the sequence of bacterial succession in the biofilm and its architecture (BUSWELL et al., 1997). These cell-to-cell interactions seem to make an active contribution to the development of microbial aggregates.

A major problem is the characterization of biofilm organisms as the majority of them cannot be cultured by conventional microbiological methods (WAGNER et al., 1993). The authors could demonstrate that many species abundant in activated sludge flocs could not be grown on agar media. Using fluorescently labeled gene probes in combination with CSLM enabled these authors to reveal the three-dimensional structure of functional groups of microorganisms within activated sludge flocs (WAGNER et al., 1994).

2.4 Role of Extracellular Polymeric Substances in Floc and Biofilm Structures

EPS consist mainly of polysaccharides and proteins, but other macromolecules such as DNA, lipids, and humic substances can also be found in EPS from wastewater biofilms and activated sludges (NEU, 1992; URBAIN et al., 1993; JAHN and NIELSEN, 1996; NIELSEN et al., 1997; see Tab. 1). Most bacteria are able to produce EPS, whether they grow planktonically or in biofilms. Cell surface polymers and EPS are of major importance for the development and structural integrity of flocs and biofilms. They mediate interactions between microorganisms and form the matrix in which the microorganisms are immobilized and kept in a three-dimensional arrangement. In general, the proportion of EPS in biofilms can vary between roughly 50% and 90% of the total organic matter (CHRISTENSEN and CHARACKLIS, 1990; NEU, 1992; NIELSEN et al., 1997). In activated sludges and sewer biofilms 85–90% and 70–98%, respectively, of total organic carbon were found to be extracellular, indicating that cell biomass can constitute only a minor fraction of the organic matter in microbial aggregates (JAHN and NIELSEN, 1998; FRØLUND et al., 1996). The quantification of EPS is generally performed after extraction of the polymers from microbial aggregates by physical and/or chemical extraction procedures (e.g., heat, blending, centrifugation, alkali addition, extraction of divalent cations with EDTA or with cation exchange resin). Quantification of EPS usually involves gravimetric or standard biochemical methods to analyze specific constituents of the EPS (LAZAROWA and MANEM,, 1995). EPS have different origins. They are actively secreted by floc and biofilm organisms, are released after cell death and cell lysis, or are adsorbed from the bulk water phase (e.g., cellulose or humic acids in wastewater).

Three types of noncovalent interactions have to be considered as cohesive forces between the components within the EPS matrix of microbial aggregates: London (dispersion) forces, electrostatic interactions, and hydrogen bonds (FLEMMING et al., in press).

Dispersion forces mainly act between hydrophobic regions of molecules (e.g., within and between proteins) and are not dependent on functional groups. The binding energy is about 2.5 kJ mol^{-1}. *Electrostatic interactions* occur between charged molecules (ions) and between permanent or induced dipoles. Repulsion is expected between functional groups such as carboxyl groups of proteins and polysaccharides; however, divalent cations such as Ca^{2+} can act as bridges contributing significantly to the overall binding force. Positively charged groups of amino sugars in polysaccharides or of amino acids in proteins can also interact with negatively charged groups providing cohesion forces. Electrostatic interactions seem to be of major importance for the stability of the EPS matrix. The binding energy of nonionic electrostatical interactions ranges between 12 and 29 kJ mol^{-1}. The binding force is strongly dependent on the distance between the partners of the bond and the water concentration. *Hydrogen bonds* form mainly between hydrogen atoms of hydroxyl groups and the more electronegative oxygen or nitrogen atoms that are abundant in polysaccharides and proteins. Hydrogen bonds are active within (e.g., maintenance of the secondary and tertiary structure of proteins) and between macromolecules, but are also involved in binding of water to EPS. The binding energy ranges between 10 and 30 kJ mol^{-1} and reaches only a short distance.

The individual binding force of any type of these interactions is relatively small compared to a covalent C—C bond (about 250 kJ mol^{-1}). However, if an EPS molecule possesses 10^6 functional groups and only 10% of them are involved in bonding, the total binding energy adds up to values in the range of several covalent C—C bonds. All three types of binding forces contribute to the overall stability of floc and biofilm matrices, probably to various extents.

EPS are the key components in a number of models explaining the aggregation of microorganisms as well as the physicochemical properties of the extracellular matrix in flocs and biofilms (e.g., PAVONI et al., 1972; HARRIS and MITCHELL, 1973; NIELSEN et al., 1997). In activated sludge flocs EPS have been implicated in determining floc structure, floc charge, the flocculation process, floc settleability, and dewatering properties. In the polymer-bridging model, floc formation is considered as the result of the interaction of high-molecular-weight, long-chain EPS with microbial cells and other particles as well as with other EPS molecules, so that EPS bridge the cells in a three-dimensional matrix. Flocculation is associated with the formation of EPS. Cellular aggregation was found to depend on the physiological state of the microorganisms; flocculation of cultures of mixed populations from domestic wastewater did not occur until they entered into a restricted state of growth (PAVONI et al., 1972). There was a direct correlation between microbial aggregation and EPS accumulation; the ratio of EPS to microorganism mass rapidly increased during culture aggregation. The major EPS components were polysaccharides, proteins, and nucleic acids (RNA, DNA). Surface charge was not considered a necessary prerequisite for flocculation, since it remained constant throughout all growth phases regardless of the flocculability of the culture. Bacteria washed free of EPS formed stable dispersions, but re-addition of extracted EPS again resulted in flocculation. In batch cultures with *Zoogloea* it was also shown that production of an extracellular polysaccharide was accompanied by flocculation of the bacteria (UNZ and FARRAH, 1976). The polymer formation was initiated in the mid-logarithmic growth phase and the quantity produced appeared to be influenced by the level of carbon and nitrogen in the medium. The detection of extracellular polymeric fibrils in natural and wastewater flocs by high-resolution TEM confirmed the role of EPS as structural support to the microbial aggregates (LISS et al., 1996).

In addition to EPS, divalent cations are regarded as important constituents of microbial aggregates, since they bind to negatively charged groups present on bacterial surfaces, in EPS molecules, and on inorganic particles entrapped in flocs and biofilms. It has been reported that extraction of Ca^{2+} from flocs and biofilms by displacement with monovalent cations or by chelation with the more general complexing agent EDTA or the more Ca^{2+}-specific chelant EGTA resulted in the destabilization of flocs (BRUUS et al., 1992; HIGGINS and NOVAK, 1997) and biofilms (TURAKHIA et al. 1983). Practical implications are that weakening of activated sludge structure by removal of Ca^{2+} leads to an increase in the number of small particles with subsequent decrease of filterability and dewaterability. These observations suggest that divalent cations may be particularly important for the maintenance of floc and biofilm structure by acting as bridging agents within the three-dimensional EPS matrix. BRUUS et al. (1992) also integrated the role of divalent cations into their sludge floc model. The floc structure was proposed to be considered as a three-dimensional EPS matrix kept together by divalent cations with varying selectivity to the matrix ($Cu^{2+}>Ca^{2+}>Mg^{2+}$). It was argued that approximately half of the Ca^{2+} pool was associated with EPS forming a matrix that resembled gels of carboxylate containing alginates. Fe^{3+} ions may also be of importance in floc stabilization. Specific removal of Fe^{3+} from activated sludge flocs caused a weakening of floc strength resulting in release of particles to bulk water, dissolution of EPS, and partial floc disintegration (NIELSEN and KEIDING, 1998). On the basis of investigations on laboratory-scale activated sludge reactors, HIGGINS and NOVAK (1997) emphasized the role of structural proteins in conjunction with divalent cations in flocculation. Increasing the concentrations of Ca^{2+} or Mg^{2+} resulted in an increase of bound protein, whereas there was little effect on bound polysaccharides. Addition of high concentrations of Na^{+} led to a decrease of the proportion of bound protein. It was supposed that the monovalent sodium ions displaced divalent cations within the flocs. This displacement would reduce binding of protein within the floc and result in solubilization of protein. Further support for the involvement of extracellular protein in the aggregation of bacteria into flocs came from the observation that treatment of activated sludge flocs with a proteolytic enzyme (Pronase) resulted in deflocculation, in a shift to smaller particles in the 5–40 μm range, and in a release of polysaccharide. Gel electrophoretic analysis of extracted EPS from municipal, industrial, and laboratory activated sludge revealed the presence of a protein with a molecular mass of approximately 15,000 Da. Analysis of amino acid composition and sequence indicated that this protein displayed similarities to lectins; binding site inhibition studies demonstrated lectinlike activity of the 15,000 Da protein (HIGGINS and NOVAK, 1997). On the basis of these results a model of flocculation was proposed: lectin-like proteins bind polysaccharides that are cross-linked to adjacent proteins. DIGNAC et al. (1998) also postulated that extracellular proteins play a role as structural elements. Divalent cations bridge negatively charged functional groups on the EPS molecules. The cross-linking of EPS and cation bridges leads to the stabilization of the biopolymer network mediating the immobilization of microbial cells. URBAIN et al. (1993) concluded from their studies on 16 activated sludge samples from different origins that internal hydrophobic bondings were involved in flocculation mechanisms and their balance with hydrophilic interactions determined the sludge settling properties. Hydrophobic areas in between the cells were considered as essential adhesives within the floc structure. Cell surface hydrophobicity was shown to be important for adhesion of bacteria to activated sludge flocs (OLOFSSON et al., 1998). Cells with high cell surface hydrophobicity attached in higher numbers to the flocs than bacteria with a more hydrophilic surface. The hydrophobic cells attached not only on the surface of the flocs, but also penetrated the flocs through channels and pores, whereas hydrophilic cells did not. It was assumed that adhesion of hydrophobic bacteria within flocs would increase the potential of the flocs to clear free-living cells from the water phase (OLOFSSON et al., 1998).

As described for flocs, EPS are also considered as essential matrix polymers responsible for the integrity of the three-dimensional structure of biofilms. In addition, EPS may be involved in the interaction between microbial cells and the substratum leading to irreversible adhesion and surface colonization. The chemical composition of EPS largely determines the physical properties of biofilms (CHRISTENSEN and CHARACKLIS, 1990). Much information has been gathered on the chemical and physical properties of extracellular polysaccharides, since they are abundant in bacterial EPS. Specific polysaccharides (e.g., xanthan) are only produced by individual bacterial strains, whereas nonspecific polysaccharides (e.g., levan, dextran, or alginate) are found in a variety of bacterial strains or species (CHRISTENSEN and CHARACKLIS, 1990). Noncarbohydrate substituents like acetyl, pyruvyl, and succinyl groups can greatly alter the physical properties of extracellular polysaccharides and the way in which the polymers interact with one another, with other polysaccharides and proteins, and with inorganic cations (SUTHERLAND, 1994). The network of microbial polysaccharides displays a relatively high water binding capacity and is mainly responsible for aquisition and retention of water with the generation of a highly hydrated environment within flocs and biofilms (CHAMBERLAIN, 1997).

As mentioned above, secreted polysaccharides, but also proteins, are supposed to have mainly structural functions in forming and stabilizing the floc and biofilm matrix. In addition, part of the extracellular proteins have been identified as enzymes. Enzyme activities in flocs and biofilms include the activities of aminopeptidases, glycosidases, esterases, lipases, phosphatases, and oxidoreductases (FRØLUND et al., 1995; LEMMER et al., 1994; GRIEBE et al., 1997). Most of these enzymes are an integrated part of the EPS matrix (FRØLUND et al., 1995). They are supposed to function in the extracellular degradation of macromolecules to low-molecular-weight products which can be transported into the cells and are available for microbial metabolism.

3 Life in Microbial Aggregates

Flocs and biofilms are microenvironments that continuously interact with their surroundings. In general, there are complex interdependencies between the different physical, chemical, and biological factors of the aquatic environment and microbial aggregates, which control the temporal and spatial changes during development and maturation of flocs and bio-

Tab. 3. Characteristics of Life in Microbial Aggregates

Immobilization of cells in the EPS matrix
High cell density
Formation of stable microconsortia
Long retention time
Facilitated gene exchange; preservation of genes
Cell-to-cell communication
Recycling of nutrients, e.g., phosphorus
Restriction of mass transport in comparison to the bulk water phase
Formation of gradients in pH value, concentration of oxygen, salts, nutrients
Aerobic, microaerophilic, and anaerobic habitats in close proximity due to gradients
Tolerance to many toxic substances such as chlorine, heavy metals, phenol, etc.
Interaction between extracellular enzymes and other EPS components
Utilization of particulate material

films (Tab. 3). Aggregates provide survival benefits to otherwise uncompetitive microorganisms and act as important reservoirs of microbial diversity (LEWIS and GATTIE, 1990). The EPS matrix of flocs and biofilms provides the possibility that microorganisms form stable aggregates of diverse microbial species leading to synergistic microconsortia. Examples for floc and biofilm communities acting as functional consortia are the ability to perform the sequential degradation of recalcitrant organic substances (e.g., certain xenobiotic compounds) that are not readily biodegradable by the single constituent species or the two-step oxidation of ammonia via nitrite to nitrate by nitrifying bacteria such as members of the genera *Nitrosomonas* (ammonia oxidizers) and *Nitrobacter* (nitrite oxidizers).

Sessile bacteria are inherently different from their planktonic counterparts. In the sessile state bacteria express different genes, grow at different rates, and produce EPS in large amounts. Genetic responses of bacteria at solid surfaces have been demonstrated using reporter gene techniques (GOODMAN and MARSHALL, 1995). In *Pseudomonas aeruginosa*, attachment to surfaces resulted in increased expression of genes involved in the biosynthesis of the major extracellular polysaccharide alginate (DAVIES et al., 1993; HOYLE et al., 1993). However, the factors that regulate the expression of biofilm phenotypes are still poorly understood. Decreased growth rates and/or EPS production have been associated with increased resistance to antimicrobial agents (GILBERT and BROWN, 1995). Aggregation may also afford some protection of floc and biofilm microorganims from predation by protozoa (KORBER et al., 1995). Another protective effect is due to enhanced water retention by the EPS matrix, preventing or delaying desiccation and increasing survival of EPS producing micoorganisms (ROBERSON and FIRESTONE, 1992; OPHIR and GUTNICK, 1994). The EPS matrix is supposed to play a significant role in nutrition of immobilized bacteria by adsorbing organic nutrients and metal ions from the surrounding medium. This may represent an important mechanism for aquisition of carbon and energy sources in oligotrophic environments. Accumulated organic matter can be utilized by extracellular enzymes, which are secreted or entrapped from the environment and concentrated within the EPS matrix. The formation of complexes of extracellular enzymes such as lipases with alginates and concomitant stabilization of the enzymes has been demonstrated (WINGENDER, 1990). Interaction of secreted enzymes with polysaccharides of the gel matrix may prevent loss of digestive enzymes into the bulk fluid leading to a localized reservoir of high catabolic enzyme activity in close proximity to the immobilized cells within flocs and biofilms and allowing the cells to maintain a certain level of control over enzymes in their microenvironment.

Microelectrode studies and the use of fluorescently labeled rRNA-targeted oligonucleotide probes have shown that an important characteristic of microbial aggregation is the heterogeneous spatial arrangement of microorganisms giving rise to gradients of solutes such as protons, oxygen, ammonia, nitrite, nitrate, sulfide, organic substrates, and microbial products within the dimensions of flocs and biofilms (KÜHL and JØRGENSEN, 1992; DE BEER et al., 1993; COSTERTON et al., 1994; WIMPENNY and KINNIMENT, 1995; SCHRAMM et al., 1997). In aerobic environments, oxygen is consumed in outer layers of flocs and biofilms through respiration so that in deeper regions fermentative species or anaerobic organisms such as sulfate reducers or methanogens can develop and be active. A vertical zonation of oxygen respiration, sulfide oxidation, and sulfate reduction has been demonstrated in aerobic trickling filter biofilms (KÜHL and JØRGENSEN, 1992). Thus, aerobic and anaerobic habitats allow the coexistence of different physiological groups of microorganisms in close proximity, which may lead to metabolic versatility of biofilm communities. Transport processes may be irregular due to the structural heterogeneity of biofilms. COSTERTON et al. (1994) have shown that water channels appeared to facilitate transport of oxygen into the depth of biofilms, but diffusion limitation and oxygen utilization still results in very low oxygen levels at the centers of cellular microcolonies of the biofilm. Thus, in heterogeneous biofilm systems, convective flow through channels and pores combined with molecular diffusion into and out of constituent microcolonies determine solute transport, whereas in more confluent microbial biofilms the penetration of solutes is dominated by molecular diffusion. Although it was commonly assumed that mass transport in biofilms was only due to diffusion, convectional transport within biofilms has been demonstrated by LEWANDOWSKI et al. (1994). Although flow velocity was about three decimals lower than in the adjacent water phase, mass transport was much higher than by diffusion only. Observations by transmission electron microscopy of steady state biofilms of *P. aeruginosa* or dental plaque organisms suggested that nutrient diffusion limitations in lower parts of biofilms may lead to starvation, cell death, and lysis of microorganisms (WIMPENNY and KINNIMENT, 1995). In general, lysis of bacteria and generation of gas by anaerobes in deeper zones of biofilms can lead to destabilization and sloughing of biofilm fragments. DE BEER et al. (1993) determined vertical microprofiles of ammonium, oxygen, nitrate, and pH within nitrifying aggregates (approximately 100 to 120 μm in diameter) grown in a fluidized-bed reactor. Both ammonium consumption and nitrate production occurred in the outer parts of the aggregates, even under conditions in which ammonium and oxygen penetrated the whole aggregate suggesting that nitrifying bacteria were localized in distinct areas within the aggregates. In trickling filter biofilms oxygen consumption and nitrite/nitrate production were restricted to the upper 50–100 μm of the biofilm (SCHRAMM et al., 1997). Ammonia and nitrite oxidizers were found in a dense layer of cells and cell clusters in the upper part of the biofilm in close vicinity to each other favoring the sequential two-step metabolism from ammonia to nitrate. Limited penetration into microbial aggregates may be among one of several resistance mechanism of floc and biofilm organisms against the action of biocides and antibiotics (GILBERT and BROWN, 1995). It depends greatly on the nature of the antimicrobial agent, whether the EPS matrix constitutes a penetration barrier and protective structure. In mucoid strains of *P. aeruginosa*, slime formation afforded protection against chlorine, but not against hydrogen peroxide (WINGENDER et al., in press). Chlorine penetration into biofilms was found to be a function of simultaneous diffusion into the matrix and reaction with EPS resulting in the neutralization of the biocide (DE BEER et al., 1994). Generally, it is assumed that mass transfer of antimicrobial agents to microorganisms within aggregates may be restricted by boundary layers surrounding floc and biofilm surfaces, by the EPS matrix acting as diffusion barrier, and by the interaction with matrix components. However, this does not always apply to small molecules and ions, which can rapidly penetrate the matrix (DE BEER et al., 1994).

High cell densities characteristic of flocs and biofilms promote communication processes between the constituent cells. Horizontal gene transfer may be facilitated due to the close proximity of aggregated cells and the accumulation of DNA in the EPS matrix. Several mechanisms of gene transfer are likely to occur in the natural environment: conjugation, transduction, and natural transformation (LORENZ and WACKERNAGEL, 1994). Little is known about the exchange of genetic information within biofilms, although a few studies have demonstrated gene exchange in pure culture biofilms (ANGLES et al., 1993; LISLE and ROSE, 1995) and plasmid transfer in river epilithon (BALE et al., 1988). These observations suggest that gene transfer can occur in microbial aggregates, possibly contributing to phenotypic changes of constituent microorganisms.

Another form of information transfer resides in the phenomenon of quorum sensing which is a signaling mechanism in response to population density (SWIFT et al., 1996). This kind of cell-to-cell communication is mediated by low-molecular-weight diffusible signaling molecules (autoinducers), which are chemically N-acyl-L-homoserine lactones (AHLs). Extracellular accumulation of AHLs above a critical threshold level results in transcriptional activation of a range of different genes with concomitant expression of new phenotypes. This may be one reason why populations with high cell densities characteristic of microbial aggregates display biological properties which are different from populations with lower cell densities typical of microbial suspensions in aqueous environments. AHLs allow bacteria to sense when cell densities in their surroundings reach the minimal level for a coordinate population response to be initiated. AHLs control gene expression of diverse processes. They are involved in the regulation of bioluminescence, swarming, conjugation, and the production of extracellular products such as antibiotics, surfactants, and enzymes (SWIFT et al., 1996). AHLs have been implicated in the shortened recovery process of *Nitrosomonas europaea* biofilm cells from nitrogen starvation due to accumulation of the signaling molecules within the biofilm to levels unobtainable in planktonic populations which showed a prolonged phase of recovery (BATCHELOR et al., 1997). The ecological significance lies in the ability of high cell densities to respond immediately to increased nutrient concentrations after starvation periods. With respect to biofilm formation AHLs were considered as mediators of adhesion by switching cells to an attachment phenotype through expression of adhesive polymers, and they were supposed to facilitate induction of other genes essential for the maintenance of the biofilm mode of growth (HEYS et al., 1997). Cell-to-cell communication via AHLs seems to be of fundamental importance for aggregated bacteria to adapt dynamically in response to prevailing environmental conditions. DAVIES et al. (1998) suggested that the involvement of an intercellular AHL in the development of *P. aeruginosa* biofilms might provide possible targets to control biofilm growth, e.g., on catheters, in the lungs of cystic fibrosis patients and in other environments, where *P. aeruginosa* represents a persistent problem. STEINBERG et al. (1997) reported that halogenated furanones excreted by seaweeds prevented their colonization by bacteria, probably by interfering with AHL functions that were necessary for bacterial adhesion and/or biofilm formation. From a practical point of view, this finding offers the perspective to use these inhibitors of bacterial AHL activity in anti-fouling strategies.

Microbial aggregates are not random accumulations of microbial cells, but seem to be organized and dynamic systems with respect to structure, composition, and activity as a result of complex developmental events. Due to interactive processes coordination exists between members of the microbial community based on regulatory mechanisms, which are only possible under high cell density conditions characteristic of microbial aggregates. These observations indicate that microbial processes are strongly influenced by continued movement of substances (nutrients, metabolites, ions, genetic material, and signaling molecules) through the floc and biofilm matrices controlling the spatial and temporal dynamics of these structures. Thus, in life flocs and biofilms offers survival advantages, because they allow effective adaptation of microorganisms to diverse environmental conditions. It is justified to consider flocs and biofilms as precursors of multicellular organisms, as they communicate and interact in a way which makes them superior to isolated single cells. That means that living organisms possibly "invented" cooperation as early as 3.5 billion years ago.

4 References

ALLREDGE, A. L., SILVER, M. W. (1988), Characteristics, dynamics and significance of marine snow, *Prog. Oceanogr.* **20**, 41–82.

ANDREWS, J. H. (1995), What if bacteria are modular organisms? *ASM News* **61**, 627–632.

ANGLES, M. L., MARSHALL, K. C., GOODMAN, A. E. (1993), Plasmid transfer between marine bacteria in the aqueous phase and biofilms in reactor microcosms, *Appl. Environ. Microbiol.* **59**, 843–850.

ASCON-CABRERA, M. A., THOMAS, D., LEBEAULT, J.-M. (1995), Activity of synchronized cells of a steady-state biofilm recirculated reactor during xenobiotic biodegradation, *Appl. Environ. Microbiol.* **61**, 920–925.

BAIER, R. E. (1980), Substrata influences on adhesion of microorganisms and their resultant new surface properties, in: *Adsorption of Microorganisms to Surfaces* (BITTON, G., MARSHALL, K. C., Eds.), pp. 59–104. New York: John Wiley & Sons.

Bale, M. J., Fry, J. C., Day, M. J. (1988), Transfer and occurrence of large mercury resistance plasmids in river epilithon, *Appl. Environ. Microbiol.* **54**, 972–978.

BATCHELOR, S. E., COOPER, M., CHHABRA, S. R., GLOVER, L. A., STEWART, G. S. A. B. et al. (1997), Cell density-regulated recovery of starved biofilm populations of ammonia-oxidizing bacteria, *Appl. Environ. Microbiol.* **63**, 2281–2286.

BEECH, I. B. (1994), The use of atomic force microscopy for studying biodeterioration of metals due to the formation of biofilms, *Abstract 2nd COST 511 Workshop*, IFREMER, France, 30 June–1 July 1994, pp. 1–14.

BITTON, G. (1994), *Wastewater Microbiology.* New York: Wiley-Liss.

BOSSIER, P., VERSTRAETE, W. (1996), *Comamonas testosteroni* colony phenotype influences exopolysaccharide production and coaggregation with yeast cells, *Appl. Environ. Microbiol.* **62**, 2687–2691.

BRADING, M. G., JASS, J., LAPPIN-SCOTT, H. M. (1995), Dynamics of bacterial biofilm formation, in: *Microbial Biofilms* (LAPPIN-SCOTT, H. M., COSTERTON, J. W., Eds.), pp. 46–63. Cambridge: Cambridge University Press.

BRUUS, J. H., NIELSEN, P. H., KEIDING, K. (1992), On the stability of activated sludge flocs with implications to dewatering, *Water Res.* **26**, 1597–1604.

BRYERS, J. D., CHARACKLIS, W. G. (1990), Biofilms in water and wastewater treatment, in: *Biofilms* (CHARACKLIS, W. G., MARSHALL, K. C., Eds.), pp. 671–696. New York: John Wiley & Sons.

BUSWELL, C. M., HERLIHY, Y. M., MARSH, P. D., KEEVIL, C. W., LEACH, S. A. (1997), Coaggregation amongst aquatic biofilm bacteria, *J. Appl. Microbiol.* **83**, 477–484.

CARON, D. A., DAVIS, P. G., MADIN, L. P., SIEBURTH, J. MCN. (1982), Heterotrophic bacteria and bacteriovorous protozoa in oceanic macroaggregates, *Science* **218**, 795–797.

CHAMBERLAIN, A. H. L. (1997), Matrix polymers: the key to biofilm processes, in: *Biofilms: Community Interactions and Control* (WIMPENNY, J., HANDLEY, P., GILBERT, P., LAPPIN-SCOTT, H., JONES, M., Eds.), pp. 41–46. Cardiff: BioLine.

CHARACKLIS, W. G. (1990), Biofilm processes, in: *Biofilms* (CHARACKLIS, W. G., MARSHALL, K. C., Eds.), pp. 195–232. New York: John Wiley & Sons.

CHRISTENSEN, B. E., CHARACKLIS, W. G. (1990), Physical and chemical properties of biofilms, in: *Biofilms* (CHARACKLIS, W. G., MARSHALL, K. C., Eds.), pp. 93–130. New York: John Wiley & Sons.

COSTERTON, J. W., CHENG, K.-J., GEESEY, G. G., LADD, T. I., NICKEL, J. C. et al. (1987), Bacterial biofilms in nature and disease, *Annu. Rev. Microbiol.* **41**, 435–464.

COSTERTON, J. W., LEWANDOWSKI, Z., DEBEER, D., CALDWELL, D., KORBER, D., JAMES, G. (1994), Biofilms, the customized microniche, *J. Bacteriol.* **176**, 2137–2142.

DAVIES, D. G., CHAKRABARTY, A. M., GEESEY, G. G. (1993), Exopolysaccharide production in biofilms: substratum activation of alginate gene expression by *Pseudomonas aeruginosa*, *Appl. Environ. Microbiol.* **59**, 1181–1186.

DAVIES, D. G., PARSEK, M. R., PEARSON, J. P., IGLEWSKI, B. H., COSTERTON, J. W., GREENBERG, E. P. (1998), The involvement of cell-to-cell signals in the development of a bacterial biofilm, *Science* **280**, 295–298.

DE BEER, D., VAN DEN HEUVEL, J. C., OTTENGRAF, S. P. P. (1993), Microelectrode measurements of the activity distribution in nitrifying bacterial aggregates, *Appl. Environ. Microbiol.* **59**, 573–579.

DE BEER, D., SRINIVASAN, R., STEWART, P. S. (1994), Direct measurement of chlorine penetration into biofilms during disinfection, *Appl. Environ. Microbiol.* **60**, 4339–4344.

DIGNAC, M.-F., URBAIN, V., RYBACKI, D., BRUCHET, A., SNIDARO, D. et al. (1998), Chemical description of extracellular polymers: Implication on activated sludge floc structure, *Water Sci. Technol.* **38**, 45–33.

DROPPO, I. G., FLANNIGAN, D. T., LEPPARD, G. G., JASKOT, C., LISS, S. N. (1996), Floc stabilization for multiple microscopic techniques, *Appl. Environ. Microbiol.* **62**, 3508–3515.

EIKELBOOM, D. H., VAN BUIJSEN, H. J. J. (1981), Microscopic sludge investigation manual, *Report No. A94a*, TNO Research Institute, Delft, The Netherlands.

FLEMMING, H.-C. (1991), Biofilms as a particular form of microbial life, in: *Biofouling and Biocorrosion in Industrial Water Systems* (FLEMMING, H.-C., GEESEY, G. G., Eds.), pp. 3–9. Heidelberg: Springer-Verlag.

FLEMMING, H.-C. (1996), Biofouling and microbially influenced corrosion (MIC) – an economical and technical overview, in: *Microbial Deterioration of Materials* (HEITZ, E., SAND, W., FLEMMING, H.-C., Eds.), pp. 5–14. Heidelberg: Springer-Verlag.

FLEMMING, H.-C., SCHAULE, G. (1989), Biofouling auf Umkehrosmose- und Ultrafiltrationsmembranen. Teil II: Analyse und Entfernung des Belages, *Vom Wasser* **73**, 287–301.

FLEMMING, H.-C., SCHAULE, G. (1996), Measures against biofouling, in: *Microbial Deterioration of Materials* (HEITZ, E., SAND, W., FLEMMING, H.-C., Eds.), pp. 121–139. Heidelberg: Springer-Verlag.

FLEMMING, H.-C., WINGENDER, J., MORITZ, R., BORCHARD, W., MAYER, C. (in press), Physicochemical properties of biofilms – a short review, in: *Biofilms in Aquatic Systems* (KEEVIL, C., HOLT, D., DOW, C., GODFREE, A., Eds.), Cambridge: Royal Society of Chemistry.

FRØLUND, B., GRIEBE, T., NIELSEN, P. H. (1995), Enzymatic activity in the activated-sludge floc matrix, *Appl. Microbiol. Biotechnol.* **43**, 755–761.

FRØLUND, B., PALMGREN, R., KEIDING, K., NIELSEN, P. H. (1996), Extraction of extracellular polymers from activated sludge using a cation exchange resin, *Water Res.* **30**, 1749–1758.

GEESEY, G. G., MUTCH, R., COSTERTON, J. W., GREEN, R. B. (1978), Sessile bacteria: an important component of the microbial population in small mountain streams, *Limnol. Oceanogr.* **23**, 1214–1223.

GILBERT, P., BROWN, M. R. W. (1995), Mechanisms of the protection of bacterial biofilms from antimicrobial agents, in: *Microbial Biofilms* (LAPPIN-SCOTT, H. M., COSTERTON, J. W., Eds.), pp. 118–130. Cambridge: Cambridge University Press.

GOODMAN, A. E., MARSHALL, K. C. (1995), Genetic responses of bacteria at surfaces, in: *Microbial Biofilms* (LAPPIN-SCOTT, H. M., COSTERTON, J. W., Eds.), pp. 80–98. Cambridge: Cambridge University Press.

GRAY, N. F. (1985), Heterotrophic slimes in flowing waters, *Biol. Rev.* **60**, 499–548.

GRIEBE, T., SCHAULE, G., WUERTZ, S. (1997), Determination of microbial respiratory and redox activity in activated sludge, *J. Ind. Microbiol.* **19**, 118–122.

GÜDE, H. (1982), Interactions between floc-forming and nonfloc-forming bacterial populations from activated sludge, *Curr. Microbiol.* **7**, 347–350.

HARRIS, R. H., MITCHELL, R. (1973), The role of polymers in microbial aggregation, *Annu. Rev. Microbiol.* **27**, 27–50.

HEYS, S. J. D., GILBERT, P., EBERHARD, A., ALLISON, D. G. (1997), Homoserine lactones and bacterial biofilms, in: *Biofilms: Community Interactions and Control* (WIMPENNY, J., HANDLEY P., GILBERT, P., LAPPIN-SCOTT, H., JONES, M., Eds.), pp. 103–112. Cardiff: BioLine.

HIGGINS, M. J., NOVAK, J. T. (1997), Characterization of exocellular protein and its role in bioflocculation, *J. Environ. Eng.* **123**, 479–485.

HOYLE, B. D., WILLIAMS, L. J., COSTERTON, J. W. (1993), Production of mucoid exopolysaccharide during development of *Pseudomonas aeruginosa* biofilms, *Appl. Environ. Microbiol.* **61**, 777–780.

JAHN, A., NIELSEN, P. H. (1996), Extraction of extracellular polymeric substances (EPS) from biofilms using a cation exchange resin, *Water Sci. Technol.* **32**, 157–164.

JAHN, A., NIELSEN, P. H. (1998), Cell biomass and exopolymer composition in sewer biofilms, *Water Sci. Technol.* **37**, 17–24.

JORAND, F., ZARTARIAN, F., THOMAS, F., BLOCK, J. C., BOTTERO, J. Y. et al. (1995), Chemical and structural (2D) linkage between bacteria within activated sludge flocs, *Water Res.* **29**, 1639–1647.

KJELLEBERG, S. (1985), Mechanisms of bacterial adhesion at gas–liquid interfaces, in: *Bacterial Adhesion* (SAVAGE, D. C., FLETCHER, M. M., Eds.), pp. 163–194. New York: Plenum Press.

KOLENBRANDER, P. E., ANDERSEN, R. N., HOLDEMAN, L. V. (1985), Coaggregation of oral *Bacteroides* species with other bacteria: central role in coaggregation bridges and competitions, *Infect. Immun.* **48**, 741–746.

KORBER, D. R., LAWRENCE, J. R., LAPPIN-SCOTT, H. M., COSTERTON, J. W. (1995), Growth of microorganisms on surfaces, in: *Microbial Biofilms* (LAPPIN-SCOTT, H. M., COSTERTON, J. W., Eds.), pp. 15–45. Cambridge: Cambridge University Press.

KÜHL, M., JØRGENSEN, B. B. (1992), Microsensor measurements of sulfate reduction and sulfide oxidation in compact microbial communities of aerobic biofilms, *Appl. Environ. Microbiol.* **58**, 1164–1174

LAU, A. O., STROM, P. F., JENKINS, D. (1984), Growth kinetics of *Sphaerotilus natans* and a floc former in pure and dual continuous culture, *J. Water Pollut. Control Fed.* **56**, 41–51.

LAWRENCE, J. R., KORBER, D. R., HOYLE, B. D., COSTERTON, J. W., CALDWELL, D. E. (1991), Optical sectioning of microbial biofilms, *J. Bacteriol.* **173**, 6558–6567.

LAZAROWA, V., MANEM, J. (1995), Biofilm characterization and activity analysis in water and wastewater treatment, *Water Res.* **29**, 2227–2245.

LEMMER, H., ROTH, D., SCHADE, M. (1994), Population density and enzyme activities of heterotrophic bacteria in sewer biofilms and activated sludge, *Water Res.* **28**, 1341–1346.

LEPPARD, G. G. (1995), The characterization of algal and microbial mucilages and their aggregates in aquatic ecosystems, *Sci. Total Environ.* **165**, 103–131.

LEWANDOWSKI, Z., STOODLEY, P., ALTOBELLI, S., FUKUSHIMA, E. (1994), Hydrodynamics and kinetics in biofilm systems – recent advances and new problems, *Water Sci. Technol.* **29**, 223–229.

LEWIS, D. L., GATTIE, D. K. (1990), Effects of cellular aggregation on the ecology of microorganisms, *ASM News* **56**, 263–268.

LI, D., GANCZARCZYK, J. (1991), Size distribution of activated sludge flocs, *J. Water Pollut. Control Fed.* **63**, 806–814.

LISLE, J. T., ROSE, J. B. (1995), Gene exchange in drinking water and biofilms by natural transformation, *Water Sci. Technol.* **31**, 41–46.

LISS, S. N., DROPPO, I. G., FLANNIGAN, D. T., LEPPARD, G. G. (1996), Floc architecture in wastewater and natural riverine systems, *Environ. Sci. Technol.* **30**, 680–686.

LITTLE, B., WAGNER, P. A., LEWANDOWSKI, Z. (1998), Spatial relationships between bacteria and mineral surfaces, *Rev. Mineralog.* **35**, 123–159.

LORENZ, M. G., WACKERNAGEL, W. (1994), Bacterial gene transfer by natural genetic transformation in the environment, *Microbiol. Rev.* **58**, 563–602.

MARSH, P. D. (1995), Dental plaque, in: *Microbial Biofilms* (LAPPIN-SCOTT, H. M., COSTERTON, J. W., Eds.), pp. 282–300. Cambridge: Cambridge University Press.

MARSHALL, K. C. (1996), Adhesion as a strategy for access to nutrients, in: *Bacterial Adhesion: Molecular and Ecological Diversity* (FLETCHER, M., Ed.), pp. 59–87. New York: John Wiley & Sons.

MARSHALL, K. C., BLAINEY, B. (1991), Role of bacterial adhesion in biofilm formation and biocorrosion, in: *Biofouling and Biocorrosion in Industrial Water Systems* (FLEMMING, H.-C., GEESEY, G. G., Eds.), pp. 8–45. Heidelberg: Springer-Verlag.

MARSHALL, K. C., STOUT, R., MITCHELL, R. (1971), Mechanism of the initial events in the sorption of marine bacteria to surfaces, *J. Gen. Microbiol.* **68**, 337–348.

MARSHALL, K. C., POWER, K. N., ANGLES, M. L., SCHNEIDER, R. P., GOODMAN, A. E. (1994), Analysis of bacterial behaviour during biofouling of surfaces, in: *Biofouling and Biocorrosion* (GEESEY, G. G., LEWANDOWSKI, Z., FLEMMING, H.-C., Eds.), pp. 15–26. New York: Lewis Publishers.

NEU, T. R. (1992), Polysaccharides in biofilms, in: *Jahrbuch Biotechnologie 4* (PRÄVE, P., SCHLINGMANN, M., ESSER, K., THAUER, R., WAGNER, F., Eds.), pp. 73–101. Munich: Carl Hanser Verlag.

NIELSEN, P. H., KEIDING, K. (1998), Disintegration of activated sludge flocs in presence of sulfide, *Water Res.* **32**, 313–320.

NIELSEN, P. H., JAHN, A., PALMGREN, R. (1997), Conceptual model for production and composition of exopolymers in biofilms, *Water Sci. Technol.* **36**, 11–19.

OLOFSSON, A.-C., ZITA, A., HERMANSSON, M. (1998), Floc stability and adhesion of green-fluorescent-protein-marked bacteria to flocs in activated sludge, *Microbiology* **144**, 519–528.

OPHIR, T., GUTNICK, D. L. (1994), A role for exopolysaccharides in the protection of microorganisms from desiccation, *Appl. Environ. Microbiol.* **60**, 740–745.

PALMER, R. J., Jr., WHITE, D. C. (1997), Developmental biology of biofilms: implications for treatment and control, *Trends Microbiol.* **5**, 435–440.

PAVONI, J. L., TENNEY, M. W., ECHELBERGER, W. F. (1972), Bacterial exocellular polymers and biological flocculation, *J. Water Pollut. Control Fed.* **44**, 414–431.

REVSBECH, N. P. (1989), Microsensors: spatial gradients in biofilms, in: *Structure and Function of Biofilms* (CHARACKLIS, W. G., WILDERER, P. A., Eds.), pp. 129–144. Chichester: John Wiley & Sons.

RITTMANN, B. E. (1989), Detachment from biofilms, in: *Structure and Function of Biofilms* (CHARACKLIS, W. G., WILDERER, P. A., Eds.), pp. 49–58. Chichester: John Wiley & Sons.

ROBERSON, E. B., FIRESTONE, M. K. (1992), Relationship between desiccation and exopolysaccharide production in a soil *Pseudomonas* sp., *Appl. Environ. Microbiol.* **58**, 1284–1291.

SAND, W., AHLERS, B., KRAUSE-KUPSCH, T., MEINCKE, M., KRIEG, E. et al. (1989), Mikroorganismen und ihre Bedeutung für die Zerstörung von mineralischen Baustoffen, *Z. Umweltchem. Ökotox.* **3**, 36–40.

SCHOPF, J. W., HAYES, J. M., WALTER, M. R. (1983), Evolution on earth's earliest ecosystems: recent progress and unsolved problems, in: *Earth's Earliest Biosphere* (SCHOPF, J. W., Ed.), pp. 361–384. New Jersey: Princeton University Press.

SCHRAMM, A., LARSEN, L. H., REVSBACH, N. P., AMANN, R. I. (1997), Structure and function of a nitrifying biofilm as determined by microelectrodes and fluorescent oligonucleotide probes, *Water Sci. Technol.* **36**, 263–270.

SEZGIN, M., JENKINS, D., PARKER, D. S. (1978), A unified theory of filamentous activated sludge bulking, *J. Water Pollut. Control Fed.* **50**, 362–381.

STEINBERG, P. D., DE NYS, R., KJELLEBERG, S. (1997), Chemical inhibition of epibiota by Australian seaweeds, *Biofouling* **12**, 227–244.

STEWART, P. S., MURGA, R., SRINIVASAN, R., DEBEER, D. (1995), Biofilm structural heterogeneity visualized by three microsopic methods, *Water Res.* **29**, 2006–2009.

SUTHERLAND, I. W. (1994), Structure–function relationships in microbial exopolysaccharides, *Biotech. Adv.* **12**, 393–448.

SWIFT, S., THROUP, J. P., WILLIAMS, P., SALMOND, G. P. C., STEWART, G. S. A. B. (1996), Quorum sensing: a population-density component in the determination of bacterial phenotype, *Trends Biochem. Sci.* **21**, 214–219.

SZEWZYK, U., SCHINK, B. (1988), Surface colonization by and life cyclus of *Pelobacter acidigalli* studied in a continuous-flow microchamber, *J. Gen. Microbiol.* **134**, 183–190.

TURAKHIA, M. H., COOKSEY, K. E., CHARACKLIS, W. G. (1983), Influence of a calcium-specific chelant on biofilm removal, *Appl. Environ. Microbiol.* **46**, 1236–1238.

UNZ, R. F., FARRAH, S. R. (1976), Exopolymer production and flocculation by *Zoogloea* MP6, *Appl. Environ. Microbiol.* **31**, 623–626.

URBAIN, V., BLOCK, J. C., MANEM, J. (1993), Bioflocculation in activated sludge: an analytical approach, *Water Res.* **27**, 829–838.

VANDEVIVERE, P., KIRCHMAN, D. L. (1993), Attachment stimulates exopolysaccharide synthesis by a bacterium, *Appl. Environ. Microbiol.* **59**, 3280–3286.

VAN LOODSRECHT, M. C. M., LYKLEMA, J., NORDE, W., ZEHNDER, A. (1990), Influence of interfaces on microbial activity, *Microbiol. Rev.* **54**, 75–87.

VAN LOODSRECHT, M. C. M., EIKELBOOM, D., GJALTEMA, A., MULDER, A., TIJHUIS, L., HEIJNEN, J. J. (1995), Biofilm structures, *Water Sci. Technol.* **32**, 35–43.

WAGNER, M., AMANN, R. J., LEMMER, H., SCHLEIFER, K.-H. (1993), Probing activated sludge with proteobacteria-specific oligonucleotides: inadequacy of culture-dependent methods for describing microbial community structure, *Appl. Environ. Microbiol.* **59**, 1520–1525.

WAGNER, M., ASSMUS, B., AMANN, R. J., HUTZLER, P., HARTMAN, A. (1994), *In-situ* analysis of microbial consortia in activated sludge using fluorescently labeled rRNA-targeted oligonucleotide probes and scanning confocal laser microscopy *J. Microsc.* **176**, 181–187.

WALKER, J. T., MACKERNESS, C. W., ROGERS, J., KEEVIL, C. W. (1995), Heterogeneous mosaic biofilm – a haven for waterborne pathogens, in: *Microbial Biofilms* (LAPPIN-SCOTT, H. M., COSTERTON, J. W., Eds.), pp. 196–204. Cambridge: Cambridge University Press.

WIMPENNY, J. W. T., COLASANTI, R. (1997), A unifying hypothesis for the structure of microbial biofilms based on cellular automaton models, *FEMS Microbiol. Ecol.* **22**, 1–16.

WIMPENNY, J. W. T., KINNIMENT, S. L. (1995), Biochemical reactions and the establishment of gradients within biofilms, in: *Microbial Biofilms* (LAPPIN-SCOTT, H. M., COSTERTON, J. W., Eds.), pp. 99–117. Cambridge: Cambridge University Press.

WINGENDER, J. (1990), Interactions of alginate with exoenzymes, in: *Pseudomonas Infection and Alginates* (GACESA, P., RUSSELL, N. J., Eds.), pp. 160–180. London: Chapman & Hall.

WINGENDER, J., GROBE, S., FIEDLER, S., FLEMMING, H.-C. (in press), The effect of extracellular polysaccharides on the resistance of *Pseudomonas aeruginosa* to chlorine and hydrogen peroxide, in: *Biofilms in Aquatic Systems* (KEEVIL, C., HOLT, D., DOW, C., GOODFREE, A., Eds.). Cambridge: Royal Society of Chemistry.

ZARTARIAN, F., MUSTIN, C., BOTTERO, J. Y., VILLEMIN, G., THOMAS, F. et al. (1994), Spatial arrangement of the components of activated sludge flocs, *Water Sci. Technol.* **30**, 243–250.

ZARTARIAN, F., MUSTIN, C., VILLEMIN, G., AIT-ETTAGER, T., THILL, A. et al. (1997), Three-dimensional modeling of an activated sludge floc, *Langmuir* **13**, 35–40.

5 Nucleic Acid-Based Techniques for Analyzing the Diversity, Structure, and Dynamics of Microbial Communities in Wastewater Treatment

ANDREAS SCHRAMM
RUDOLF AMANN
Bremen, Germany

1 Introduction

Biological wastewater treatment systems, like activated sludge basins, trickling filters, or anaerobic digesters, essentially can be interpreted as specialized aquatic ecosystems, where microorganisms are the main players. In order to fully characterize, understand, and, in the long run, control those microbial communities, knowledge of both their structure and function is necessary. Attention should, therefore, be given to identification, enumeration, and spatial distribution (structural parameters) as well as to *in situ* activities (functional parameter) of the community members. Furthermore, from an ecological and from an engineering point of view stability or dynamics of these microbial communities are important both for theory and practice. Until recently, identification of microorganisms required the isolation of pure cultures and the investigation of physiological and biochemical traits. Enumeration had to be done by plate counts or most probable number (MPN) techniques. However, since all cultivation-dependent techniques are not only time-consuming and labor-intensive but select for certain organisms, they are inadequate for determining reliable cell numbers (STALEY and KONOPKA, 1985), and in many cases even for the identification of the main catalysts of a system (WAGNER et al., 1993, 1994c). Questions like micro-scale distribution and *in situ* activity of microorganisms are almost impossible to address by classical methods.

To circumvent these limitations identification techniques based on nucleic acids have been developed and successfully applied to wastewater treatment systems during the last five years. This chapter intends to summarize the potential, current applications, and limitations of nucleic acid-based techniques for analyzing the microbial communities present in wastewater treatment systems.

2 Ribosomal Ribonucleic Acid (rRNA)-Based Methods

Macromolecules like rRNA or proteins can be used as "molecular clocks" for evolutionary history (ZUCKERKANDL and PAULING, 1965). By comparative sequence analysis, reconstruction of evolution and classification of organisms based on their phylogeny is possible. For several reasons 16S rRNA sequence comparison is currently considered the most powerful tool for the classification of microorganisms (WOESE, 1987):

- Ribosomes and consequently rRNA molecules are present in all organisms. As an essential component of the protein synthesis apparatus they have a homologous origin and show functional constancy. No lateral gene transfer has been shown for rRNA genes so far. Therefore it is a valid assumption to reconstruct the phylogeny of the organisms based on these molecules.
- Some positions of the rRNA molecules are evolutionary more conserved than others. Consequently, sequence regions can be found that allow differentiation at any taxonomic level from species and genera up to kingdoms or domains.
- 16S rRNA sequences have been determined for many of the described bacterial species and deposited in public databases (e.g., MAIDAK et al., 1997).
- The natural amplification of rRNA within microbial cells (usually more than 1,000, frequently several 10,000 copies) makes it easier and more sensitive to assay this molecule and gives, e.g., the opportunity for identification of single bacteria by fluorescent oligonucleotide hybridization (see Sect. 2.3.3).
- The presence and abundance of ribosomes and, consequently, rRNA in individual cells is connected to their viability and general metabolic activity, or at least their metabolic potential. Cells in a rapidly growing *E. coli* culture need and

have more ribosomes than those in a slowly growing culture (SCHAECHTER et al., 1958).

The applications of rRNA-based nucleic acid techniques to the analysis of wastewater treatment systems today range from a simple identification of isolates over the detection of bacterial diversity and population dynamics to attempts at fully and quantitatively describing the complex microbial communities.

2.1 Molecular Characterization of Isolates

Despite the success of molecular methods in analyzing microbial communities classical isolation of bacteria is still absolutely necessary for, e.g., investigations on the metabolic potential of the community members. However, nucleic acid-based techniques can support this classical approach by providing tools for screening, characterization, and identification of isolates. Compared with microbiological or biochemical methods molecular methods are more rapid and more reliable since they are not affected by growth conditions and culture media.

2.1.1 Restriction Fragment Length Polymorphism (RFLP) and Amplified Ribosomal DNA Restriction Analysis (ARDRA)

A common methodology for a rapid molecular characterization of isolates is based on the generation of so-called "genetic fingerprints". DNA of an organism is digested by rare-cutting restriction enzymes, and the resulting fragments are separated by length using pulsed-field gel electrophoresis (SWAMINATHAN and MATAR, 1993). Different species will have differing fragment lengths due to mutations of the restriction sites, insertions, or deletions, and these restriction fragment length polymorphisms (RFLPs) are used for comparison of the isolates.

Since cutting the whole genome requires high quality DNA extraction prior to RFLP analysis and time-consuming, complicated separation of the rather large fragments by pulsed-field gel electrophoresis, most recent applications of RFLP are based only on part of the genome. For example, amplified ribosomal DNA restriction analysis (ARDRA) starts with the amplification of the genes encoding 16S rRNA by polymerase chain reaction (PCR), followed by restriction digestion and analysis of the fragment lengths by standard gel electrophoresis (HEYNDRICKX et al., 1996). The resulting patterns or fingerprints are compared to reference strains and can be used for cluster analysis of different isolates. In principle, a database of restriction patterns of reference organisms can be established to facilitate rapid identification of isolates by ARDRA.

ARDRA allows the processing of numerous isolates in a very short time, and can yield valuable information on the similarity of isolates. Consequently, appropriate isolates can be selected for further physiologic and phylogenetic investigations. On the other hand, reliable identification using ARDRA is hampered by the enormous amount of reference patterns required. Furthermore, the information content in an ARDRA pattern is limited (frequently only few bands) and this could cause false-positive identification. ARDRA was, e.g., applied by MASSOL-DEYA et al. (1997) to study aerobic, hydrocarbon degrading biofilms.

2.1.2 Other Methods

Besides ARDRA, the methods described in the following sections can also be applied. Denaturing gradient gel electrophoresis (DGGE, explained in detail in Sect. 2.2.1) is ideal to screen a large set of isolates for redundancy. Hybridization (Sect. 2.3) with sets of rRNA-targeted probes also allows rapid screening of isolates for their phylogenetic affiliation. By this technique, e.g., the majority of isolates from a municipal activated sludge was characterized as members of the gamma subclass of Proteobacteria (WAGNER et al., 1993; KÄMPFER et al., 1996). In a second example, genus- or species-specific probes were used for confirming the identification of *Paracoccus* sp. initially based on physiological tests (NEEF et al., 1996).

Finally, the full description of a new species identified to be important in a given system should always include the determination of its 16S rRNA sequence for valid phylogenetic classification and, if applicable, the DNA–DNA hybridization with related species or strains.

2.2 Diversity and Dynamics of Microbial Communities

Microbial diversity and population dynamics can be rapidly monitored using rRNA-based pattern or fingerprinting methods. These encompass extraction of total nucleic acids (DNA and/or RNA) from an environmental sample (like activated sludge or biofilms), amplification of part of the genes encoding 16S rRNA by PCR, and subsequent separation of the resulting gene fragments on a gel to form a pattern or fingerprint of the community.

2.2.1 Denaturing Gradient Gel Electrophoresis (DGGE)

2.2.1.1 Method

By DGGE DNA fragments of the same length but with different sequences can be separated (Muyzer et al., 1995a, recently reviewed by Muyzer and Smalla, 1998). 16S rRNA gene fragments of a length of typically 200–500 bp are amplified by PCR with an additional 40 bp GC-rich sequence at the 5′ end of one of the primers. When analyzed on a polyacrylamide gel containing an increasing gradient of DNA denaturants (a mixture of urea and formamide), for each of these DNA molecules (ds DNA) a transition from a double-stranded, helical to a partially single-stranded secondary structure will occur at a certain position in the gel. This will stop, or at least strongly slow down, the migration of the respective gene fragment. The GC-rich region acts as a "clamp" to prevent formation of single-stranded DNA (ssDNA). As sequence variations cause a difference in the melting behavior of ds DNA, sequence variants of particular fragments will stop at different positions in the denaturing gradient gel and hence can be separated effectively (Fig. 1). The result of DGGE analysis of PCR products obtained from a microbial community is a band pattern, and the number of bands is a rough estimate for the microbial diversity of a given system.

2.2.1.2 Applications

The patterns are frequently used for comparisons of different systems, e.g., aerobic and anaerobic biofilms (Muyzer et al., 1993) or different activated sludge plants (Curtis and Craine, 1997). DGGE is particularly useful to detect population changes addressing the question of stability and dynamics of microbial communities (Fig. 2). The method can be modified by using rRNA instead of DNA as a template for 16S rDNA amplification. For that purpose, extracted rRNA is transcribed into ribosomal copy DNA (rcDNA) by the enzyme reverse transcriptase prior to the PCR amplification. While the rDNA-based DGGE pattern is solely determined by the presence of DNA and, therefore, in first approximation by the abundance of populations, the rRNA-based DGGE pattern should more strongly represent the metabolically active and, therefore, rRNA-rich parts of the community (Teske et al., 1996).

A band at a given site of a DGGE gel has *per se* no biological meaning. Therefore, to learn more about the population represented by a certain band, this band needs to be further characterized, which can be achieved, in principle, by two techniques. The DGGE gel can be blotted to a nylon membrane and the pattern can be examined by hybridization analysis with taxon-specific probes. Alternatively, bands can be retrieved from the gel and subsequently sequenced; comparative sequence analysis then allows identification or at least affiliation with the closest relative (Muyzer et al., 1993, 1995b).

2.2.1.3 Limitations

Whereas DGGE analysis of rDNA PCR products is a powerful tool to analyze diversity

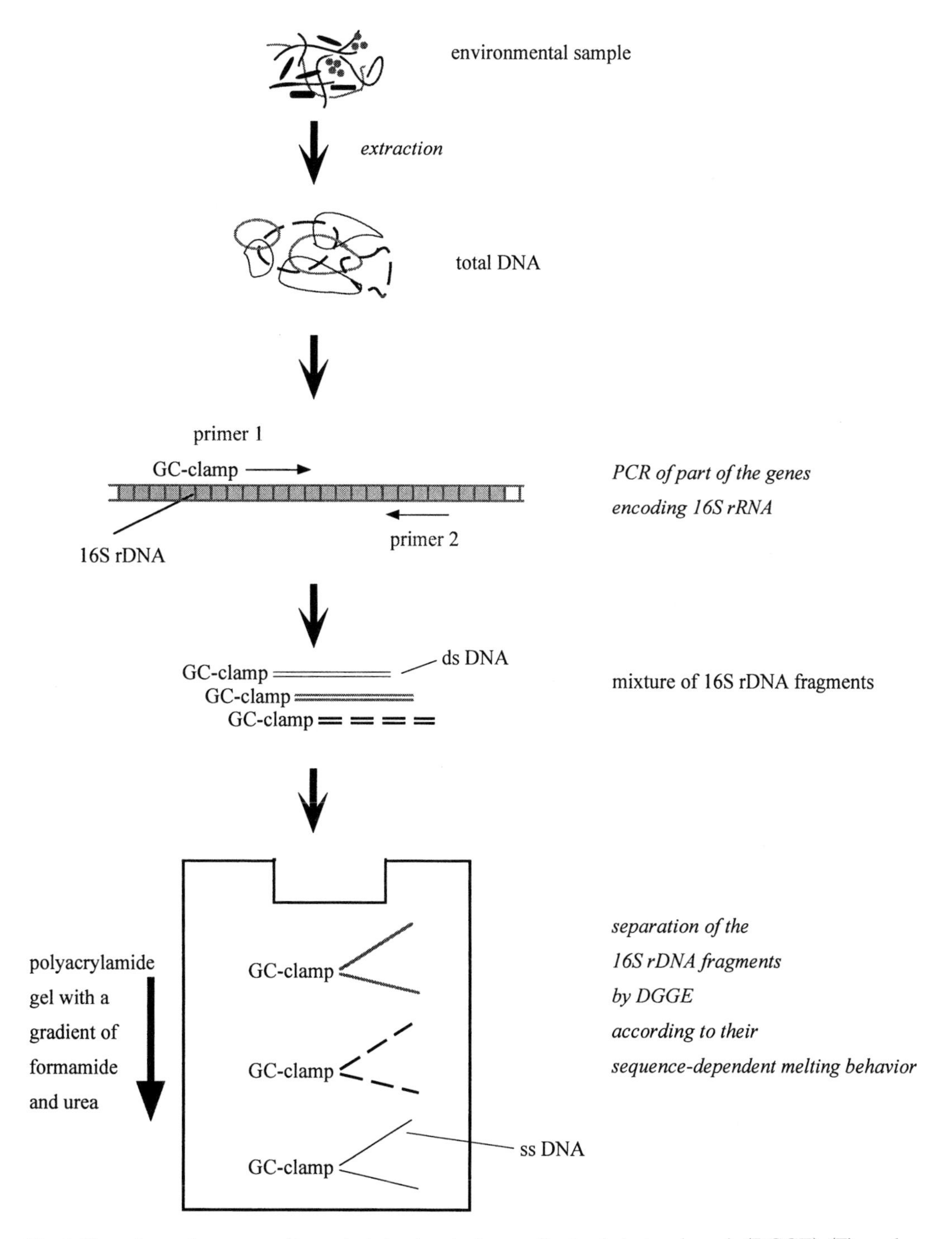

Fig. 1. Flow chart of a community analysis by denaturing gradient gel electrophoresis (DGGE). (Figure by A. SCHRAMM and G. MUYZER).

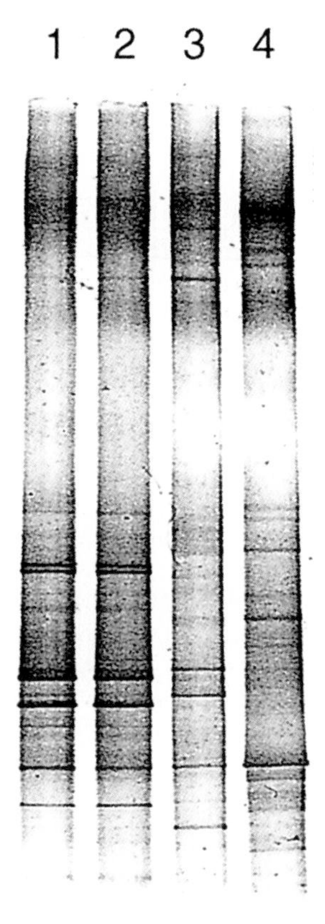

Fig. 2. DGGE patterns of 16S rDNA fragments from activated sludge. No population change is apparent during an anoxic–oxic cycle of a sequencing batch reactor (SBR) (lane 1: anoxic, lane 2: oxic phase), but a change is detectable between the original inoculum (aeration basin wastewater treatment plant Prague, lane 3) and the established biocoenosis of the SBR. Lane 4 displays the differing community pattern of another aeration basin, wastewater treatment plant Bremen (SANTEGOEDS, SCHRAMM and MUYZER, unpublished data).

and dynamics of microbial communities, it has severe limitations in the analysis of community structures, and is like any other method prone to specific biases:

(1) The method involves extraction of nucleic acids and subsequent PCR, which may both cause some bias: Not all cells lyse under the same conditions (WARD, 1989), and preferential amplification of certain templates can occur (SUZUKI and GIOVANNONI, 1996). Therefore, different intensities of DGGE bands must not be interpreted as quantitative measures of the abundance of species relative to each other. For qualitative statements on the development of a certain band (population) over time, i.e., to monitor its appearance or disappearance, identical treatment of all samples has to be ensured.

(2) Separation of DNA fragments with high resolution is restricted to a maximum size of about 500 bp. Consequently, the phylogenetic information that can be retrieved by sequencing is relatively little. In case of full identity with an rRNA sequence in a database it might be sufficient for an identification, but in cases in which only distantly related sequences are available classification becomes difficult if not impossible. Also characterization of bands by hybridization analysis with rRNA-targeted oligonucleotide probes is only possible if the probe target region is within the amplified fragment, which is only the case for a small fraction of the full set of available probes.

(3) The main difficulty, however, is the "one band – one species" hypothesis. Especially in complex communities bands might originate from two or more fragments that co-migrate on the denaturing gradient gel. Furthermore, single species might result in two or more DGGE bands due to inter-operon microheterogeneity (NÜBEL et al., 1996).

2.2.2 Terminal Restriction Fragment Length Polymorphisms (T-RFLP)

Recently, RFLP analysis (see Sect. 2.1.1) has been modified for application to microbial communities (LIU et al., 1997): After extraction of total DNA from an environmental sample 16S rDNA is amplified by PCR with one of the two primers being fluorescently labeled.

The fluorescent PCR products are then digested with frequently cutting restriction enzymes, and analyzed by a standard high resolution gel electrophoresis in which the restriction fragments are separated solely by size. The abundance and length of only the fluorescent terminal fragments is determined in automated sequencing devices or by fluorimetry. This again yields a pattern or "community fingerprint" like in the case of DGGE. The method may be suitable for analyzing diversity and dynamics of microbial communities as demonstrated for activated sludge (LIU et al., 1997). Besides similar advantages it suffers even more from similar limitations as DGGE: The length of the terminal rDNA fragments ranges from 60–500 bp, too little information for valid phylogenetic affiliation or application of specific oligonucleotide probes. Different species may have in many cases the same terminal restriction fragment length, hence leading to an underestimation of the actual bacterial diversity. T-RFLP seems to be an easier, but also a less sensitive alternative to DGGE.

2.3 Community Structure

The structure of a microbial community is mainly defined by two parameters: Identity and abundance of its members. A broader definition of structure would also include the spatial arrangement of species relative to each other, an information which might be especially important in stratified habitats like biofilms. Since both cultivation and fingerprint methods are not sufficient to address these questions, hybridization techniques applying rRNA-targeted oligonucleotide probes have been developed (AMANN et al., 1995).

Oligodeoxynucleotides are single-stranded pieces of DNA with a length of 15–25 nucleotides. When such oligonucleotides are labeled, e.g., with radioisotopes or fluorescent dye molecules they become so-called probes that allow detection of complementary target sites by specific base pairing. For the reasons outlined before, target molecules are rRNAs, extracted and immobilized on a membrane (dot blot hybridization), or maintained in fixed cells (whole cell hybridization), respectively. In both cases, quantification is possible, either relative to total extracted rRNA, or relative to total cell counts.

2.3.1 Probe Selection and Design

An important question for users of the hybridization approach is of course how to find the right probe sequence. Currently, there are already numerous published probes available specific for microorganisms involved in wastewater treatment processes (Tab. 1). For the detection or quantification of well known wastewater bacteria one should first refer to this established probe pool before considering the design of new probes. Public probe databases have been established and can be accessed by internet [*http://www.cme.msu.edu/ OPD* (ALM et al., 1996) or *http://www.mikro. biologie.tu-muenchen.de*].

When a new probe has to be designed based on sequences accessible in the updated 16S rRNA databases, software packages like ARB, developed at the Department of Microbiology of the Technical University München (STRUNK et al., ARB: a software environment for sequence data, *http://www.mikro.biologie.tu-muenchen.de*), facilitate searching for diagnostic sequences and comparing probes with the complete database. With these tools probes can be tailor-made for the needs of the respective study. A detailed review on the theoretical aspects and practical experiences of the development of oligonucleotide probes is given by STAHL and AMANN (1991).

Considering the high microbial diversity in many wastewater treatment plants the risk for false-positive or false-negative assignments based on single probes is rather high. Consequently, "one probe is no probe" and probes should always be applied as sets of probes. One possibility is the so called "top-to-bottom" approach (AMANN et al., 1995). As it is rather laborious and time-consuming to evaluate the abundance of all the bacteria that may be important in a certain sample by specific probes on the species or genus level, a preselection of the necessary probes has to be performed. In this approach, analysis starts with probes specific for the highest taxonomic levels, the three domains or kingdoms of life, Archaea, Bacteria, and Eukarya, followed by

Tab. 1. Oligonucleotide Probes Specific for Microorganisms or Groups of Microorganisms Involved in Important Processes of Wastewater Treatment

Probe	Sequence 5′-3′	Target Position	Specificity	Phylog. Affiliat.[a]	Dot Blot	*in situ*	Reference
Nitrifying bacteria							
NSO190	CGATCCCCTGCTTTTCTCC	16S, 190–208	ammonia oxidizing β-Proteobacteria	β	+[b]	+[b]	Mobarry et al. (1996)
NSO1225	CGCCATTGTATTACGTGTGA	16S, 1225–1244	ammonia oxidizing β-Proteobacteria	β	+	+	Mobarry et al. (1996)
NEU	CCCCTCTGCTGCACTCTA	16S, 653–670	halophilic and halotolerant members of the genus *Nitrosomonas*	β		+	Wagner et al. (1995)
Nsm156	TATTAGCACATCTTTCGAT	16S, 156–174	genus *Nitrosomonas*, *Nitrosococcus mobilis*	β	+	+	Mobarry et al. (1996)
Nsv443	CCGTGACCGTTTCGTTCCG	16S, 444–462	*Nitrosolobus multiformis*, *Nitrosospira briensis*, *Nitrosovibrio tenuis*	β	+	+	Mobarry et al. (1996)
NIT3	CCTGTGCTCCATGCTCCG	16S, 1035–1048	genus *Nitrobacter*	α		+	Wagner et al. (1996)
Nb1000	TGCGACCGGTCATGG	16S, 1000–1014	genus *Nitrobacter*	α	+		Mobarry et al. (1996)
Denitrifying bacteria							
PAR651	ACCTCTCTCGAACTCCAG	16S, 651–668	genus *Paracoccus*	α		+	Neef et al. (1996)
PAR1457	CTACCGTGGTCCGCTGCC	16S, 1457–1474	genus *Paracoccus*	α		+	Neef et al. (1996)
Pdv198	CTAATCCTTTGGCGATAAATC	16S, 198–232	*Paracoccus denitrificans*, *P. versutus*	α		+	Neef et al. (1996)
Pdv1031	CCTGTCTCCAGGTCACCG	16S, 1031–1051	*Paracoccus denitrificans*, *P. versutus*	α		+	Neef et al. (1996)
Hvu1034	GCACCTGTCCCACTGCCT	16S, 1034–1051	*Hyphomicrobium vulgare*	α		+	Neef et al. (1996)
Ps	GCTGGCCTAGCCTTC	23S, 1432–1446	most rRNA group I pseudomonads	γ		+	Schleifer et al. (1992)
Pst67	AAGCTCTCTTCATCCG	16S, 67–82	*Pseudomonas stutzeri*, *P. mendocina*	γ		+	Amann et al. (1996a)
Pae997	TCTGGAAAGTTCTCAGCA	16S, 997–1014	*Pseudomonas aeruginosa*, *P. alcaligenes*, *P. aureofaciens*, *P. fluorescens*, *P. mendocina*, *P. oleovorans*, *P. pseudoalcaligenes*	γ		+	Amann et al. (1996a)
AEU	CTGACACACTCTAGCCTT	16S, 646–663	*Alcaligenes eutrophus*	β		+	Rothemund et al. (1996)

Tab. 1. (Continued)

Probe	Sequence 5′-3′	Target Position	Specificity	Phylog. Affiliat.[a]	Dot Blot	*in situ*	Reference
Sulfate reducing bacteria							
SRB385	CGGCGTCGCTGCGTCAGG	16S, 385–402	most sulfate reducing δ-Proteobacteria, Myxobacteria, few gram-positive bacteria	δ		+	AMANN et al. (1992a)
804	CAACGTTTACTGCGTGGA	16S, 804–821	*Desulfococcus multivorans*, *Desulfosarcina variabilis*, *Desulfobotulus sapovorans*, genera *Desulfobacterium*, *Desulfobacter*	δ	+	–	DEVEREUX et al. (1992)
814	ACCTAGTGATCAACGTTT	16S, 814–883	*Desulfococcus multivorans*, *Desulfosarcina variabilis*, *Desulfobotulus sapovorans*	δ	+	–	DEVEREUX et al. (1992), RAMSING et al. (1996)
129	CAGGCTTGAAGGCAGATT	16S, 129–146	genus *Desulfobacter*	δ	+	+	DEVEREUX et al. (1992), RAMSING et al. (1996)
221	TGCGCGGACTCATCTTCAAA	16S, 221–240	genus *Desulfobacterium*	δ	+	+	DEVEREUX et al. (1992), RAMSING et al. (1996)
660	GAATTCCACTTTCCCCTCTG	16S, 660–679	genus *Desulfobulbus*	δ	+	+	DEVEREUX et al. (1992), RAMSING et al. (1996)
687	TACGGATTTCACTCCT	16S, 687–702	Desulfovibrionaceae	δ	+	+	DEVEREUX et al. (1992), RAMSING et al. (1996)
Methanotrophs							
9-α	CCCTGAGTTATTCCGAAC	16S, 142–159	genera *Methylocystis*, *Methylosinus*, *Methylobacterium*	α	+	+	TSIEN et al. (1990)
1034-Ser	CCATACCGGACATGTCAAAAGC	16S, 987–1008	genera *Methylocystis*, *Methylosinus*	α	+		BRUSSEAU et al. (1994)
1038-Ser-M	GGTAACATGCCATGTCCAG	16S, 990–1008	genus *Methylobacterium*	α	+		BRUSSEAU et al. (1994)
10-γ	GGTCCGAAGATCCCCCGCTT	16S, 197–216	genera *Methylobacter*, *Methylococcus*, *Methylomonas*, *Methylobacillus*, *Methylophilus*	γ	+	+	TSIEN et al. (1990)

Tab. 1. (Continued)

Probe	Sequence 5′-3′	Target Position	Specificity	Phylog. Affiliat.[a]	Dot Blot	*in situ*	Reference
Methanogens							
MB310	CTTGTCTCAGGTTCCATCTCCG	16S, 310–331	Methanobacteriales	Eury	+	+	Raskin et al. (1994b), Harmsen et al. (1996a)
MB1174	TACCGTCGTCCACTCCTTCCTC	16S, 1174–1195	Methanobacteriales	Eury	+		Raskin et al. (1994b)
MC1109	GCAACATAGGGCACGGGTCT	16S, 1109–1128	Methanococcales	Eury	+		Raskin et al. (1994b)
MG1200	CGGATAATTCGGGGCATGCTG	16S, 1200–1220	Methanomicrobiaceae, Methano-corpusculaceae, Methanoplanaceae	Eury	+	+	Raskin et al. (1994b), Harmsen et al. (1996a)
MSMX860	GGCTCGCTTCACGGCTTCCCT	16S, 860–880	Methanosarcinaceae	Eury	+		Raskin et al. (1994b)
MS1414	CTCACCCATACCTCACTCGGG	16S, 1414–1434	genera *Methanosarcina*, *Methanococcoides*, *Methanolobus*, *Methanohalophilus*	Eury	+		Raskin et al. (1994b)
MS821	CGCCATGCCTGACACCTAGCGAGC	16S, 821–844	genus *Methanosarcina*	Eury	+		Raskin et al. (1994b)
MX825	TCGCACCGTGGCCGACACCTAGC	16S, 825–847	genus *Methanosaeta*	Eury	+	+	Raskin et al. (1994b), Harmsen et al. (1996a)
Bacteria involved in floc formation, bulking, foaming							
ZRA	CTGCCGTACTCTAGTTAT	16S, 647–664	*Zoogloea ramigera*	β	+	+	Rossello-Mora et al. (1995)
SNA	CATCCCCCTCTACCGTAC	16S, 656–673	*Sphaerotilus natans*, few *Leptothrix* spp., Eikelboom type 1701	β	+	+	Wagner et al. (1994a)
LDI	CTCTGCCGCACTCCAGCT	16S, 649–666	"*Leptothrix discophora*", *Aquaspirillum metamorphum*	β	+	+	Wagner et al. (1994a)
LMU	CCCCTCTCCCAAACTCTA	16S, 652–669	*Leucothrix mucor*	β	+	+	Wagner et al. (1994a)

Tab. 1. (Continued)

Probe	Sequence 5′-3′	Target Position	Specificity	Phylog. Affiliat.[a]	Dot Blot	*in situ*	Reference
HHY	GCCTACCTCAACCTGATT	16S, 655–672	genus *Haliscomenobacter*	β	+	+	WAGNER et al. (1994a)
TNI	CTCCTCTCCCACATTCTA	16S, 652–669	*Thiothrix nivea*	β	+	+	WAGNER et al. (1994a)
021N	TCCCTCTCCCAAATTCTA	16S, 652–669	Eikelboom type 021N	β	+	+	WAGNER et al. (1994a)
MPA60	GGATGGCCGCGTTCGACT	16S, 60–77	"*Microthrix parvicella*"	LGC		+	ERHART et al. (1997)
MPA223	GCCGCGAGACCCTCCTAG	16S, 223–240	"*Microthrix parvicella*"	LGC		+	ERHART et al. (1997)
MPA645	CCGGACTCTAGTCAGAGC	16S, 645–661	"*Microthrix parvicella*"	LGC		+	ERHART et al. (1997)
S-*-Myb-0736-a-A-22	CAGCGTCAGTTACTACCCAGAG	16S, 736–757	*Mycobacterium* complex	HGC	+	+	DE LOS REYES et al. (1997b)
S-G-Gor-0596-a-A-22	TGCAGAATTTCACAGACGACGC	16S, 596–617	genus *Gordona*	HGC	+	+	DE LOS REYES et al. (1997b)
S-S-G.am-0192-a-A-18	CACCCACCCCCATGCAGG	16S, 192–209	*Gordona amarae*	HGC	+	+	DE LOS REYES et al. (1997b)

[a] Phylogenetic affiliation of target bacteria: α, β, γ, δ: alpha, beta, gamma, and delta subclass of Proteobacteria, respectively; LGC: gram-positive bacteria with low G + C DNA content; HGC: gram-positive bacteria with high G + C DNA content; Eury: phylum Euryarchaeota.

[b] + indicates successful application, – the failure of dot blot or *in situ* hybridization, respectively, with the listed probe.

intermediate levels such as the alpha, beta, and gamma subclasses of Proteobacteria (MANZ et al., 1992), the *Cytophaga–Flavobacterium* and *Bacteroides* cluster (MANZ et al., 1996), the gram-positive bacteria with high DNA G+C content (ROLLER et al., 1994), etc. (Tab. 2). Genus-, species-, and subspecies-specific probes are then selected based on the information obtained with the more general probes. Readers unfamiliar with modern bacterial taxonomy are referred to WOESE (1987) or BALOWS et al. (1992). Using this approach increasingly refined information on the community structure can be obtained rapidly by both dot blot and whole-cell hybridization.

In order to achieve real complete community structure analysis, also including hitherto uncultured or unsequenced organisms, the full cycle rRNA approach (Fig. 3) has to be performed (OLSEN et al., 1986; AMANN, 1995b). This approach starts with the extraction of DNA from an environmental sample, followed by the PCR-amplification of the almost full length 16S rDNA, and the construction of a clone library of these rDNA fragments. Clones can be screened for redundancy by DGGE (see Sect. 2.2.1), colony or dot blot hybridization, and the 16S rRNA of interest can be sequenced and compared with databases to yield information about the identity or relatedness of new sequences. This gives insight into the diversity and, to a much lesser degree, some rough estimate of community structure, as observed by BOND et al. (1995) for a comparison of phosphate-removing and non-phosphate-removing activated sludge. Since the retrieved 16S rDNA sequences might have been derived from naked DNA, dead cells, or contaminants, it is important to quantify the populations behind the sequences by oligonucleotide probing of either extracted total RNA or of intact fixed cells. The full cycle rRNA approach has recently been applied by SNAIDR et al. (1997) to activated sludge. They found the majority of

Tab. 2. Oligonucleotides Specific for Higher Taxa Like Domains, Kingdoms, Classes, or Subclasses

Probe	Sequence 5′-3′[a]	Target Position	Specificity	Reference
ARC915	GTGCTCCCCCGCCAATTCCT	16S, 915–934	domain Archaea	STAHL and AMANN (1991)
EUB338	GCTGCCTCCCGTAGGAGT	16S, 338–355	domain Bacteria	AMANN et al. (1990b)
ALF1b	CGTTCGYTCTGAGCCAG	16S, 19–35	alpha subclass of Proteobacteria	MANZ et al. (1992)
BET42a	GCCTTCCCACTTCGTTT	23S, 1027–1043	beta subclass of Proteobacteria	MANZ et al. (1992)
GAM42a	GCCTTCCCACATCGTTT	23S, 1027–1043	gamma subclass of Proteobacteria	MANZ et al. (1992)
SRB385	CGGCGTCGCTGCGTCAGG	16S, 385–402	most members of delta subclass of Proteobacteria	AMANN et al. (1992a)
CF319a	TGGTCCGTGTCTCAGTAC	16S, 303–319	*Cytophaga–Flavobacterium* cluster of CFB phylum	MANZ et al. (1996)
HGC69a	TATAGTTACCACCGCCGT	23S, 1901–1918	gram-positive bacteria with high DNA G+C content	ROLLER et al. (1994)
EURY499	CTTGCCCRGCCCTT	16S, 498–510	most Euryarchaeota	BURGGRAF et al. (1994)
CREN498	CCAGRCTTGCCCCCCGCT	16S, 499–515	most Crenarchaeota	BURGGRAF et al. (1994)

[a] Y=C or T, R=A or G

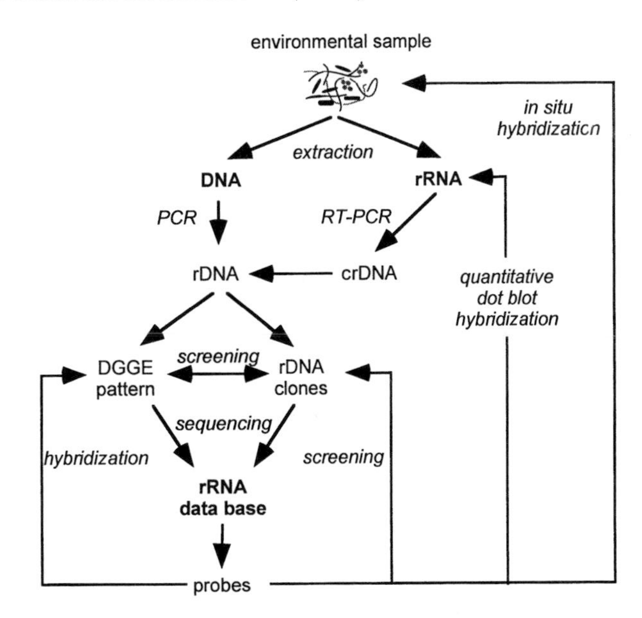

Fig. 3. The rRNA approach. See text for details.

clones, and the dominant population detected by *in situ* hybridization, to be affiliated with the beta 1 group of Proteobacteria, which encompasses many former pseudomonads such as members of the genera *Comamonas*, *Hydrogenophaga*, and *Acidovorax*. On the other hand, members of the *Cytophaga–Flavobacterium* cluster and of the gram-positive bacteria with high G + C DNA content, were abundant *in situ* at about 12%, but absent in the clone library. This again shows that community structure should not be described solely based on a clone library.

2.3.2 Dot Blot Hybridization

2.3.2.1 Method

For dot blot or slot blot hybridization total RNA is extracted from an environmental sample and immobilized ("blotted") on a nylon membrane (STAHL et al., 1988). DNA oligonucleotide probes are labeled with ^{32}P (STAHL et al., 1988), or – with a significant loss of sensitivity – non-radioactively with digoxygenin (DIG) (MANZ et al., 1992). Hybridization conditions are optimized by adjusting the final wash temperature to provide adequate sensitivity and specificity relative to RNA extracted from reference organisms. Processing of the membrane including prehybridization, hybridization, and washing takes several hours. Subsequently, bound probe is quantified by phosphorimaging or densitometry after autoradiography in case of ^{32}P or enzymatic antibody detection of probe-conferred DIG, respectively. The relative abundance of a certain 16S rRNA is expressed as a fraction of total 16S rRNA in the sample, which is determined by hybridization of a universal oligonucleotide probe (STAHL et al., 1988).

2.3.2.2 Applications

Quantitative dot blot hybridization was successfully applied to several population analyses of wastewater treatment systems: WAGNER et al. (1993) observed the dominating role of proteobacterial rRNA from the alpha, beta, or gamma subclass in activated sludge samples from aeration tanks, and a severe population shift towards the gamma subclass during cultivation-dependent analysis. Similar results were obtained by MANZ et al. (1994), additionally emphasizing the significance of the *Cytophaga–Flavobacterium* cluster (CF) and of the gram-positive bacteria with high G+C content (HGC) in activated sludge. Anaerobic systems like sewage sludge digesters (RASKIN et al., 1994a, 1995b) and anaerobic fixed-bed biofilm reactors (RASKIN et al., 1995a, 1996) have been analyzed focusing on the methanogenic populations and their competition and coexistence with sulfate reducing bacteria. The studies showed among other things the importance of Methanosarcinaceae, Methanobacteriales, and *Desulfovibrio* sp. in anaerobic processes. In nitrifying biofilms MOBARRY et al. (1996) identified rRNA of the genus *Nitrosomonas* to be the main fraction (70%) of ammonia oxidizer rRNA. In contrast, only a weak hybridization signal was observed with probes for the genera *Nitrosovibrio-Nitrosospira-Nitrosolobus* and *Nitrobacter*, indicating their minor contribution to nitrification in this system and leaving space for yet uncharacterized nitrifying bacteria. Finally, levels of mycobacterial rRNA, i.e., of the genus *Gordona* and some strains of the species *Gordona* (formerly *Nocardia*) *amarae* were compared in foam and mixed liquor from municipal and industrial activated sludge and anaerobic digester systems (DE LOS REYES et al., 1997b, c). A relatively high percentage (10–25%) of the rRNA in the foams could be attributed to the *Mycobacterium* complex, with the majority consisting of *Gordona* rRNA, thus confirming the participation of this group in the foaming process.

2.3.2.3 Advantages and Limitations

The detection limit of a microbial population is with approximately 0.1% relative abundance rather low. The method is like most other methods in molecular biology based on nucleic acid extraction. It is of advantage that for RNA isolation procedures can be used that are harsher and consequently less selective than those required for isolation of good quality DNA. However, extraction always bears the potential for some bias, at least requiring a careful evaluation whether the extraction protocol is suitable for the respective environmental sample. The spatial resolution of the method is restricted to a few hundred micrometer by the need to apply enough material for the extraction step. Furthermore, relative cell numbers cannot directly be extrapolated from the data of relative rRNA abundance (WAGNER et al., 1993). The ribosome content of one single cell ranges roughly between 1,000 and 100,000, changing with growth rate, cell size, and largely varying between different species. However, the relative rRNA abundance should represent a valid measurement for the contribution of a certain population to the total metabolic activity (STAHL et al., 1988), since it is the product of their number of detected cells and the average rRNA content.

2.3.3 *In situ* Hybridization

In contrast to quantitative dot blot hybridization, the detection of a certain rRNA sequence within morphologically intact cells in their natural environment ("*in situ* hybridization") can provide numbers and spatial distribution of microorganisms on the cell level. Quantification of the probe-conferred signal should even allow the estimation of *in situ* growth rates of individual cells (POULSEN et al., 1993). Due to advantages like superior spatial resolution, immediate detectability by epifluorescence microscopy, and easier handling, fluorescent dyes like fluorescein, carboxytetramethylrhodamine, or recently CY3 have commonly replaced the initially used radioisotopes as labels for *in situ* hybridization probes. The application of two probes with different fluorescent labels to the same sample allows the simultaneous detection of two different populations. Using three different labeled probes for different taxonomic levels even up to seven distinct populations can be visualized due to the occurrence of additive colors resulting from the simultaneous binding of two or three probes to one cell (AMANN et al., 1996b).

2.3.3.1 Method

The *in situ* hybridization procedure (Fig. 4) has recently been described in detail (AMANN, 1995b). Shortly, the permeabilization of microorganisms is a prerequisite for successful *in situ* hybridization with oligonucleotide probes. This is, in most cases, achieved by fixation of

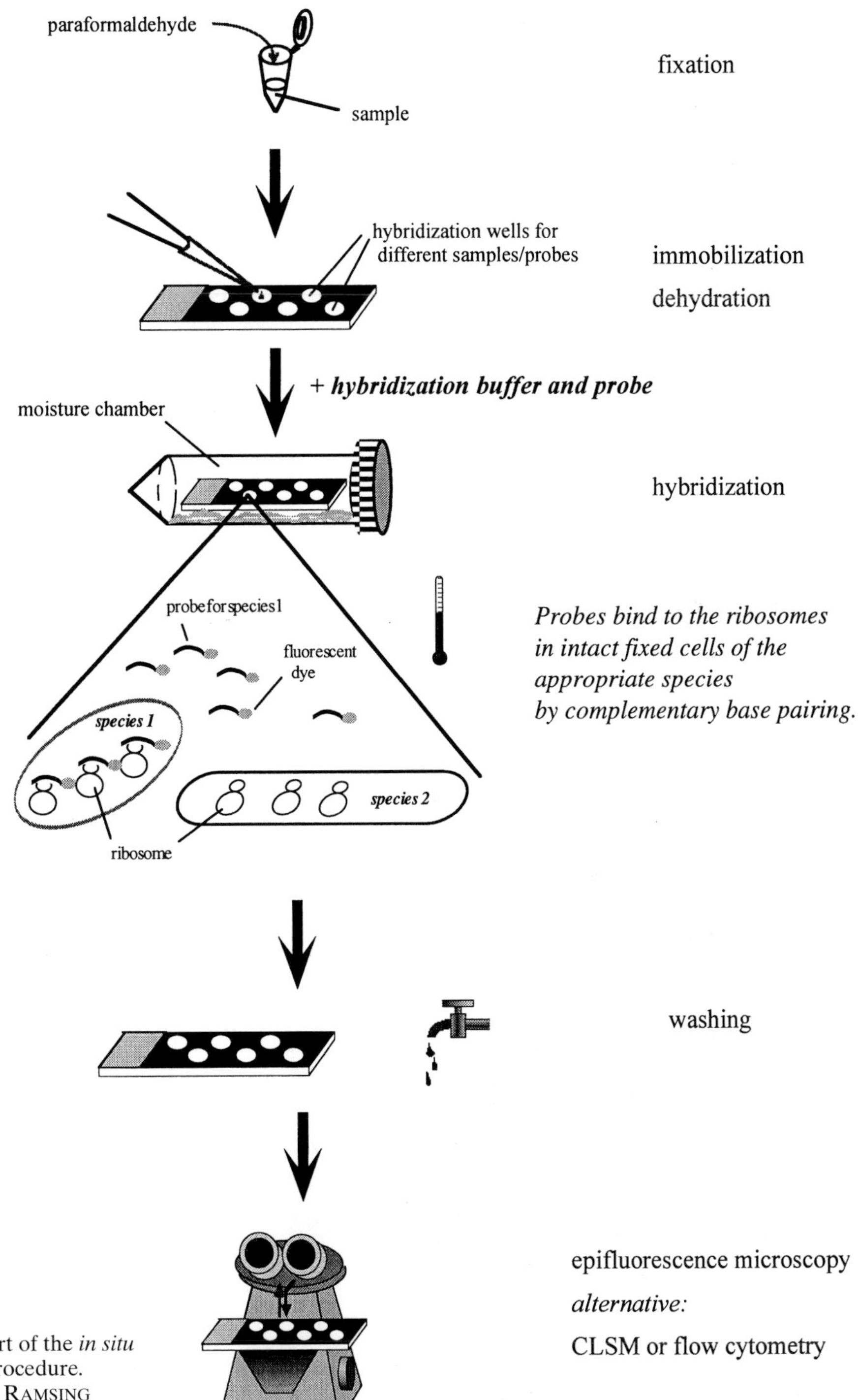

Fig. 4. Flow chart of the *in situ* hybridization procedure. (Figure by N. B. RAMSING and A. SCHRAMM).

the sample with paraformaldehyde (for gram-negative bacteria) or ethanol (for gram-positive bacteria). Occasionally, additional pretreatment with lysozyme (HAHN et al., 1993) or mutalysine (ERHART et al., 1997) may be required. Samples are then immobilized on microscopic slides coated with gelatin (or alternatively with poly-L-lysine) to improve adhesion to the glass surface, and dehydrated in an ethanol dilution series. Hybridization is done in a moisture chamber at an appropriate hybridization temperature (STAHL and AMANN, 1991) by covering the slides with hybridization solution containing the labeled probes. After a stringent washing step to remove unbound or unspecifically bound probes the slides are air dried, mounted in antibleaching buffer, and examined with an epifluorescence microscope. Using suitable filter sets individual bacterial cells can be identified and their morphology, spatial distribution, and number can be determined (for an example see Figs. 5 and 6).

2.3.3.2 Analytical Tools

Conventional epifluorescence microscopy is just one of several possibilities to examine a hybridized sample. In combination with highly sensitive charge-coupled device (CCD) cameras weak signals can be visualized, and image analysis can be used for signal quantification (POULSEN et al., 1993) or background correction (RAMSING et al., 1996) of digitized images. Another option is confocal laser scanning microscopy (CLSM). This technique of optical sectioning removes out-of-focus fluorescence and leads to a sharp and clear image (WAGNER

a

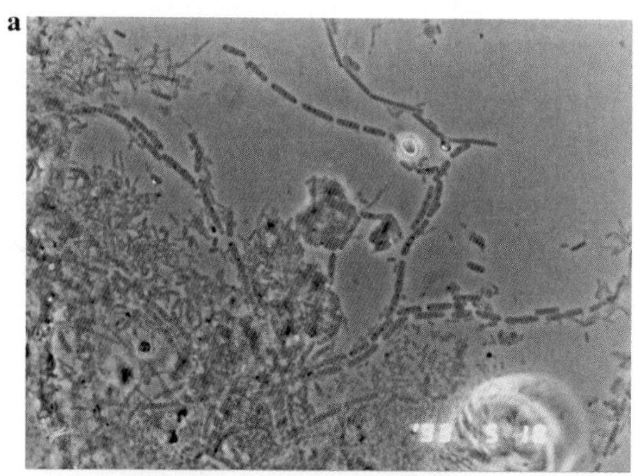

b

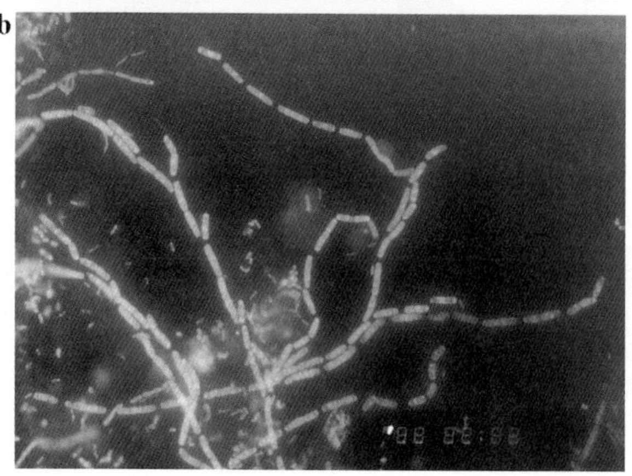

Fig. 5. *In situ* hybridization of activated sludge with a fluorescently labeled oligonucleotide probe specific for the beta subclass of Proteobacteria. **a** Phase contrast and **b** epifluorescence micrograph are shown of the same microscopic field.

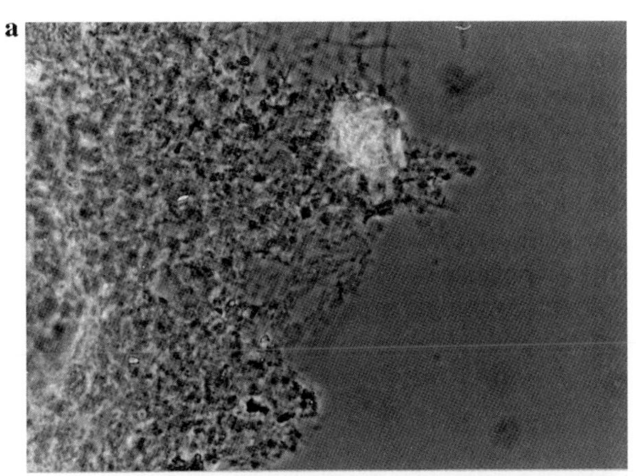

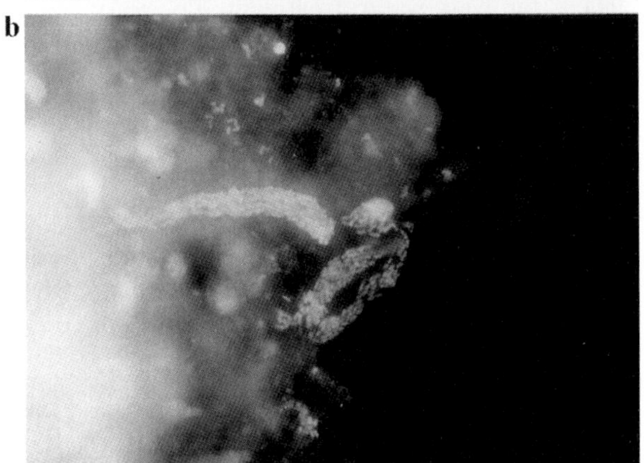

Fig. 6. Detection of *Zoogloea ramigera* "fingers" by fluorescence *in situ* hybridization with probe ZRA. **a** Phase contrast and **b** epifluorescence micrograph are shown of the same microscopic field.

et al., 1994b). Especially in samples with high background fluorescence or extremely dense cell clusters like biofilms it is helpful to restrict the collected signal to thin sections of about 1 μm. From these optical sections 3-D images can be reconstructed and displayed as 3-D red-green anaglyph, depth profiles, or 3-D space-filling reconstruction to visualize directly the spatial distribution of defined bacterial populations, e.g., in an activated sludge floc (WAGNER et al., 1994b) or in a biofilm sample (MØLLER et al., 1996).

Rapid and quantitative analysis of suspended samples can be achieved using flow cytometry (AMANN et al., 1990a; WALLNER et al., 1996). This allows not only the automation of the *in situ* hybridization approach but also an easier and more rapid quantification of relative cell numbers, signal intensity (rRNA content), and cell size. The applicability of this promising combination to activated sludge was demonstrated by WALLNER et al. (1995).

2.3.3.3 Applications

Numerous studies were performed using *in situ* hybridization for the analysis of activated sludge, anaerobic digesters, and biofilms. The spectrum ranges from population analyses at higher taxonomic levels (e.g., subclasses of Proteobacteria, the *Cytophaga–Flavobacterium* cluster, or the gram-positive bacteria with high DNA G+C content (WAGNER et al., 1993; MANZ et al., 1994; WALLNER et al., 1995; KÄMPFER et al., 1996) to the identification and enumeration of functional important monophyletic groups like ammonia oxidizing bacteria (WAGNER et al., 1995, 1996; MOBARRY et al., 1996; SCHRAMM et al., 1996), or sulfate reducing bacteria (RAMSING et al., 1993; RASKIN et al., 1995a). It encompasses the quantification of single key genera or species like *Acinetobacter* sp. (WAGNER et al., 1994c), *Zoogloea ramigera* (ROSSELLO-MORA et al., 1995; Fig. 6), *Nitrobacter* sp. (WAGNER et al., 1996), different filamentous bacteria (WAGNER et al., 1994a; ERHART et al., 1997; NIELSEN et al., 1997), or *Gordona amarae* (DE LOS REYES et al., 1997a). Significant findings were, e.g., that the importance of gamma subclass Proteobacteria, in particular the genera *Aeromonas* (KÄMPFER et al., 1996) and *Acinetobacter* (WAGNER et al., 1994c), for wastewater treatment has been overestimated by cultivation-dependent methods. The same is true for the commonly isolated nitrite oxidizers of the genus *Nitrobacter* that can hardly be detected *in situ* in many activated sludge plants showing complete nitrification and in certain biofilm communities (WAGNER et al., 1996). Recent molecular data (BURRELL et al., 1998; SCHRAMM et al., 1998; JURETSCHKO et al., 1998) indicate that members of the genus *Nitrospira* seem to perform the bulk of the nitrite oxidization in these plants. *In situ* hybridization in contrast to cultivation studies also demonstrated that beta Proteobacteria (Fig. 5) are often the dominant group of bacteria in low-load municipal activated sludge plants (WAGNER et al., 1993).

It is also possible to combine classical cultivation techniques and *in situ* hybridization in studies on the structure and function of wastewater treatment processes. Along these lines NEEF et al. (1996) showed that *Paracoccus* sp. and *Hyphomicrobium* sp. are important denitrifying species in methanol-fed biofilms established in a sand filter. Classical screening for denitrifying bacteria had yielded numerous isolates of the genus *Paracoccus.* Probes designed for this genus based on published 16S rRNA sequences did indeed hybridize to approximately 3.5% of all cells in the biofilm. The detected cells were arranged in dense clusters and had a relatively high ribosome content indicating high metabolic activity. Hyphomicrobia could also be detected in significant numbers. The smaller cells had a lower ribosome content and were distributed more evenly over the biofilm. In studies like this, the physiology of isolates can be determined classically and the molecular techniques serve as a way to get accurate enumeration and information on the spatial distribution. Insights are obtained into both the structure and the function of the communities. It is also appealing to consider a backward approach: If a new and so far unknown population is localized *in situ* by molecular means, probes could help in the screening for the right isolate which might either be of slow growth or difficult to culture by standard techniques.

Fluorescent oligonucleotide probes have been developed as a tool for basic research, but offer high promise as a relatively simple and inexpensive monitoring technique. In the future, important sludge properties like floc formation, bulking, or foaming could be related, e.g., to the abundance of *Zoogloea ramigera*, certain filaments, or *Gordona amarae*, respectively, and thereby to the causative agents. Sensitive and specific molecular techniques could allow to detect upcoming malfunction of wastewater treatment systems in an early stage.

Whereas monitoring techniques might ultimately not rely on microscopic localization of single bacterial cells *in situ* hybridization will always be the technique necessary to understand the 3-D structure of immobilized microbial communities, like biofilms. Also the correct functioning of other commercially important wastewater treatment processes such as the anaerobic granular sludge process (HARMSEN et al., 1996a, b) is dependent on a long-term stratification of bacterial populations and on correct positioning of the various community members relative to each other. Microorganisms are currently thought to inhabit in these systems zones in relation to physicochemical gradients that are optimal for their metabolism. The spatial distributions can only be resolved by *in situ* techniques such as immunostaining (for a review, see BOHLOOL and SCHMIDT, 1980) or *in situ* hybridization. Several studies have therefore recently focused on the *in situ* hybridization of immobilized systems (AMANN et al., 1992a; RAMSING et al., 1993; MANZ et al., 1993; SZEWZYK et al., 1994; ROTHEMUND et al., 1996; HARMSEN et al., 1996a, b; MØLLER et al., 1996). For nitrifying biofilms it was, e.g., shown that ammonia oxidizers (*Nitrosomonas* sp.) and nitrite oxidizers (*Nitrobacter* sp.) are arranged in clusters of close vicinity facilitating an effective inter-species transport of nitrite (MOBARRY et al., 1996; SCHRAMM et al., 1996).

2.3.3.4 Combination with Other Methods

In situ hybridization can also be combined with measurements of the *in situ* activity of microorganisms to address both, structure and function of a microbial population. For that purpose, microsensors for oxygen, sulfide, and nitrate have been used to quantify sulfate reduction (RAMSING et al., 1993) and nitrification (SCHRAMM et al., 1996) with a spatial resolution of 20–50 μm in trickling filter biofilms. The microsensor data were then compared with *in situ* detection of sulfate reducing and nitrifying bacteria, respectively, in vertical biofilm sections, displaying a highly stratified community. In case of a high load nitrifying biofilm treating the wastewater of an aquaculture (SCHRAMM et al., 1996) *in situ* hybridization visualized densely clustered aggregates of *Nitrosomonas* and *Nitrobacter* in high abundance in the upper aerobic nitrifying zone. Individual cells of the two nitrifying genera could also be detected in the lower anoxic zones. Here, the findings almost perfectly matched the theory, however, they yielded new insights regarding the relative localization of ammonia and nitrite oxidizers. These were present in the same zone, however, in dense monospecies clusters rather than randomly distributed.

The development of new and the improvement of known microsensors (e.g., for nitrite or ammonia) (DE BEER et al., 1997) will in the future promote combined studies in immobilized systems, and may be especially valuable in testing the increasing number of biofilm models.

Recently, microautoradiography has been combined with *in situ* hybridization (NIELSEN et al., 1997). With this approach, the uptake of radioactively labeled substrate (e.g., ^{14}C acetate) by individual cells can be detected, and the cells are subsequently identified by *in situ* hybridization.

2.3.3.5 Limitations

Although *in situ* hybridization is well established and proved to work for many wastewater treatment systems, several difficulties can occur:

(1) rRNA accessibility may be limited by two factors: (a) diffusion of a labeled probe into a morphologically intact cell is dependent on the permeability of the cell periphery (e.g., cell walls, membranes, capsules). Whereas fixation with paraformaldehyde usually is sufficient for permeabilization of gram-negative bacteria, different treatments had to be developed for gram-positive bacteria (e.g., ROLLER et al., 1994). Currently, ethanol is used as a standard fixative for gram-positive bacteria. In general, it should be kept in mind that fixation is an integral part of *in situ* hybridization (AMANN et al., 1995). (b) Even if the probe diffuses freely into the cell, the target sites on the 16S or 23S rRNA are not necessarily available for hybridization. Probably due to higher order structures of rRNA or the ribosome, i.e., RNA–RNA or RNA–protein interactions, some positions are inaccessible *in situ*. A good hint for suitable probing sites can be retrieved from an *in situ* accessibility map recently completed for *E. coli* 16S rRNA (FUCHS et al., 1998), or from probing sites can be retrieved from the target positions of successful whole-cell hybridization probes (e.g., Tab. 1; AMANN et al., 1995).

(2) As mentioned above, there is a positive correlation between rRNA content and growth rate of bacterial cells. Since the signal intensity after fluorescence *in situ* hybridization is directly correlated to the cellular rRNA content it can be used as a rough estimate of the *in situ* growth rates of individual cells. On the other hand, slowly growing or starving cells will yield poor signals and, in some cases, will not even be detectable by *in situ* hybridization. In eutrophic environments like activated sludge, however, this should be the exception.

(3) Another frequent problem is the autofluorescence of cells and surrounding material, that may lead to false-positive cells or prevent reliable detection of hybridized cells. Several attempts were made to overcome this limitation, including new brighter fluorescent dyes like CY3 or CY5 in combination with more sensitive band pass high quality filters (GLÖCKNER et al., 1996), image analysis (RAMSING et al., 1996), or the application of enzyme-labeled oligonucleotides (AMANN et al., 1992b, SCHÖNHUBER et al., 1997).

Finally, it should be considered for all hybridization methods involving specific oligonucleotides that any probe is, at best, as good as the according database. The number of sequences deposited in public databases is steadily increasing, and before applying an existing probe its specificity should be reevaluated with an updated rRNA database. Keeping in mind that only a minor fraction of the microbial diversity is currently known, and only for part of it rRNA sequences have been determined, one should be cautious in relying too much on the specificity of one single probe. Again, the application of probes in sets is strongly recommended (see Sect. 2.3.1), either for identification at different taxonomic levels or as multiple probes of the same specificity (AMANN, 1995a).

3 Detection of Functional Genes or Gene Products

Nucleic acid techniques based on rRNA offer a wide range of tools for analyzing the structural parameters of microbial communities active in wastewater treatment. We are getting close to an almost complete and high resolution structural description of complex systems. What is now required is an increasing correlation of these structural data to functional parameters on a macro- and microscale. Studies like those performed by NEEF et al. (1996), RAMSING et al. (1993), and SCHRAMM et al. (1996) are important steps in this direction.

Another approach to analyze *in situ* functions of bacteria would be the detection of specific gene products involved in key steps of the metabolism. The *in situ* detection of messenger RNA (mRNA) in single bacterial cells is currently anything but routine (HONERLAGE et al., 1995; WAGNER et al., 1998) but it bears great potential for the analysis of processes in wastewater treatment. For example, sequences of ammonia monooxygenase, methane monooxygenase, or nitrite reductase have recently been determined and used for the development of specific PCR approaches or polynucleotide probes (SINIGALLIANO et al., 1995; KLOOS et al., 1995; NIELSEN et al., 1996). However, their application is at the moment restricted to the analysis of DNA or mRNA extracted from isolates (BAUMANN et al., 1996) or bulk samples (WAWER et al., 1997) due to the low sensitivity of *in situ* hybridization. In the future, detection of lower-copy-number nucleic acids like mRNA will hopefully become possible by the development of increasingly sensitive hybridization techniques. The combination of *in situ* hybridization for rRNA and mRNA would then allow to simultaneously address the identity and function of individual wastewater bacteria.

Acknowledgements
The original work of the authors was supported by grants of the Körber Stiftung, Deutsche Forschungsgemeinschaft, and the Max-Planck-Gesellschaft. We thank Dr. GERARD MUYZER for his comments on the manuscript.

4 References

ALM, E. W., OERTHER, D. B., LARSEN, N., STAHL, D. A., RASKIN, L. (1996), The oligonucleotide probe database, *Appl. Environ. Microbiol.* **62**, 3557–3559.

AMANN, R. I. (1995a), Fluorescently labelled, rRNA-targeted oligonucleotide probes in the study of microbial ecology, *Mol. Ecol.* **4**, 543–554.

AMANN, R. I. (1995b), *In situ* identification of microorganisms by whole cell hybridization with rRNA-targeted nucleic acid probes, *Mol. Microb. Ecol. Man.* **3.3.6**, 1–15.

AMANN, R. I., BINDER, B. J., OLSON, R. J., CHISHOLM, S. W., DEVEREUX, R., STAHL, D. A. (1990a), Combination of 16S rRNA-targeted oligonucleotide probes with flow cytometry for analyzing mixed microbial populations, *Appl. Environ. Microbiol.* **56**, 1919–1925.

AMANN, R. I., KRUMHOLZ, L., STAHL, D. A. (1990b), Fluorescent-oligonucleotide probing of whole cells for determinative, phylogenetic, and environmental studies in microbiology, *J. Bacteriol.* **172**, 762–770.

AMANN, R. I., STROMLEY, J., DEVEREUX, R., KEY, R., STAHL, D. A. (1992a), Molecular and microscopic identification of sulfate-reducing bacteria in multispecies biofilms, *Appl. Environ. Microbiol.* **58**, 614–623.

AMANN, R. I., ZARDA, B., STAHL, D. A., SCHLEIFER, K.-H. (1992b), Identification of individual prokaryotic cells by using enzyme-labeled, rRNA-targeted oligonucleotide probes, *Appl. Environ. Microbiol.* **58**, 3007–3011.

AMANN, R. I., LUDWIG, W., SCHLEIFER, K.-H. (1995), Phylogenetic identification and *in situ* detection of individual microbial cells without cultivation, *Microb. Rev.* **59**, 143–169.

AMANN, R., LUDWIG, W., SCHULZE, R., SPRING, S., MOORE, E., SCHLEIFER, K.-H. (1996a), rRNA-targeted oligonucleotide probes for the identification of genuine and former pseudomonads, *Syst. Appl. Microbiol.* **19**, 501–509.

AMANN, R., SNAIDR, J., WAGNER, M., LUDWIG, W., SCHLEIFER, K.-H. (1996b), *In situ* visualization of high genetic diversity in a natural microbial community, *J. Bacteriol.* **178**, 3496–3500.

BALOWS, A., TRÜPER, H. G., DWORKIN, M., HARDER, W., SCHLEIFER, K.-H. (Eds.) (1992), *The Prokaryotes*, 2nd Edn. New York: Springer-Verlag.

BAUMANN, B., SNOZZI, M., ZEHNDER, A. J. B., VAN DER MEER, J. R. (1996), Dynamics of denitrification activity of *Paracoccus denitrificans* in continuous culture during aerobic–anaerobic changes, *J. Bacteriol.* **178**, 4367–4374.

BOHLOOL, B. B., SCHMIDT, E. L. (1980), The immunofluorescence approach in microbial ecology, *Adv. Microb. Ecol.* **4**, 203–241.

BOND, P. L., HUGENHOLTZ, P., KELLER, J., BLACKALL, L. L. (1995), Bacterial community structure of phosphate-removing and non-phosphate-removing activated sludges from sequencing batch reactors, *Appl. Environ. Microbiol.* **61**, 1910–1916.

BRUSSEAU, G. A., BULYGINA, E. S., HANSON, R. S. (1994), Phylogenetic analysis and development of probes for differentiating methylotrophic bacteria, *Appl. Environ. Microbiol.* **60**, 626–636.

BURGGRAF, S., MAYER, T., AMANN, R., SCHADHAUSER, S., WOESE, C. R., STETTER, K. O. (1994), Identifying Archaea with rRNA-targeted oligonucleotide probes, *Appl. Environ. Microbiol.* **60**, 3112–3119.

BURREL, P. C., KELLER, J., BLACKALL, L. L. (1998), Microbiology of a nitrite-oxidizing bioreactor, *Appl. Environ. Microbiol.* **64**, 1878–1883.

CURTIS, T. P., CRAINE, N. G. (1997), The Comparison of the Diversity of Activated Sludge Plants, in: *Proc. 2nd Int. Conf. Microorganisms in Activated Sludge and Biofilm Processes* (JENKINS, D., HERMANOWICZ, S. W., Eds.), pp. 65–72. Berkeley, CA: IAWQ.

DE BEER, D., SCHRAMM, A., SANTEGOEDS, C. M., KÜHL, M. (1997), A nitrite microsensor for profiling environmental biofilms, *Appl. Environ. Microbiol.* **63**, 973–977.

DE LOS REYES, F. L., OERTHER, D. B., DE LOS REYES, M. F., HERNANDEZ, M., RASKIN, L. (1997a), Characterization of filamentous foaming in activated sludge systems using oligonucleotide hybridization probes and antibody probes, in: *Proc. 2nd Int. Conf. Microorganisms in Activated Sludge and Biofilm Processes* (JENKINS, D., HERMANOWICZ, S. W., Eds.), pp. 283–284. Berkeley, CA: IAWQ.

DE LOS REYES, F. L., RITTER, W., RASKIN, L. (1997b), Group-specific small-subunit rRNA hybridization probes to characterize filamentous foaming in activated sludge systems, *Appl. Environ. Microbiol.* **63**, 1107–1117.

DE LOS REYES, M. F., DE LOS REYES, F. L., HERNANDEZ, M., RASKIN, L. (1997c), Identification and quantification of *Gordona amarae* strains in activated sludge systems using comparative rRNA sequence analysis and phylogenetic hybridization probes, in: *Proc. 2nd Int. Conf. Microorganisms in Activated Sludge and Biofilm Processes* (JENKINS, D., HERMANOWICZ, S. W., Eds.), pp. 677–680. Berkeley, CA: IAWQ.

DEVEREUX, R., KANE, M. D., WINFREY, J., STAHL, D. A. (1992), Genus- and group-specific hybridization probes for determinative and environmental studies of sulfate-reducing bacteria, *Syst. Appl. Microbiol.* **15**, 601–609.

ERHART, R., BRADFORD, D., SEVIOUR, R. J., AMANN, R., BLACKALL, L. L. (1997), Development and use of fluorescent *in situ* hybridization probes for the detection and identification of "*Microthrix parvicella*" in activated sludge, *Syst. Appl. Microbiol.* **20**, 310–318.

FUCHS, B. M., WALLNER, G., BEISKER, W., SCHWIPPL, I., LUDWIG, W., AMANN, R. (1998), Flow cytometric analysis of the *in situ* accessibility of *Escherichia coli* 16S rRNA for fluorescently labeled oligonucleotide probes, *Appl. Environ. Microbiol.* **64**.

GLÖCKNER, F. O., AMANN, R., ALFREIDER, A., PERNTHALER, J., PSENNER, R. et al. (1996), An *in situ* hybridization protocol for detection and identification of planktonic bacteria, *Syst. Appl. Microbiol.* **19**, 403–406.

HAHN, D., AMANN, R. I., ZEYER, J. (1993), Whole-cell hybridization of *Frankia* strains with fluorescence- or digoxigenin-labeled 16S rRNA-targeted oligonucleotide probes, *Appl. Environ. Microbiol.* **59**, 1709–1716.

HARMSEN, H. J. M., AKKERMANS, A. D. L., STAMS, A. J. M., DE VOS, W. M. (1996a), Population dynamics of propionate-oxidizing bacteria under methanogenic and sulfidogenic conditions in anaerobic granular sludge, *Appl. Environ. Microbiol.* **62**, 2163–2168.

HARMSEN, H. J. M., KENGEN, H. M., AKKERMANS, A. D. L., STAMS, A. J. M., DE VOS, W. M. (1996b), Detection and localization of syntrophic propionate-oxidizing bacteria in granular sludge by *in situ* hybridization using 16S rRNA-based oligonucleotide probes, *Appl. Environ. Microbiol.* **62**, 1656–1663.

HEYNDRICKX, M., VAUTERIN, L., VANDAMME, P., KERSTERS, K., DE VOS, P. (1996), Applicability of combined amplified ribosomal DNA restriction analysis (ARDRA) patterns in bacterial phylogeny and taxonomy, *J. Microbiol. Methods* **26**, 247–259.

HONERLAGE, W., HAHN, D., ZEYER, J. (1995), Detection of mRNA of nprM in *Bacillus megaterium* ATCC 14581 grown in soil by whole-cell hybridization, *Arch. Microbiol.* **163**, 235–241.

JURETSCHKO, S., TIMMERMANN, G., SCHMIDT, M., SCHLEIFER, K.-H., POMMERENING-RÖSER, A. et al. (1998), Combined molecular and conventional analysis of nitrifying bacterial diversity in activated sludge: *Nitrosococcus mobilis* and *Nitrospira*-like bacteria as dominant populations, *Appl. Environ. Microbiol.* **64**, 3042–3051.

KÄMPFER, P., ERHART, R., BEIMFOHR, C., BÖHRINGER, J., WAGNER, M., AMANN, R. (1996), Characterization of bacterial communities from activated sludge: culture-dependent numerical identification versus *in situ* identification using group- and genus-specific rRNA-targeted oligonucleotide probes, *Microb. Ecol.* **32**, 101–121.

KLOOS, K., FESEFELDT, A., GLIESCHE, C. G., BOTHE, H. (1995), DNA-probing indicates the occurrence of denitrification and nitrogen fixation genes in *Hyphomicrobium*. Distribution of denitrifying and nitrogen fixing isolates of *Hyphomicrobium* in a sewage treatment plant, *FEMS Microbiol. Ecol.* **18**, 205–213.

LIU, W.-T., MARSH, T. L., FORNEY, L. J. (1997), Determination of the microbial diversity of anaerobic–aerobic activated sludge by a novel molecular biological technique, in: *Proc. 2nd Int. Conf. Microorganisms in Activated Sludge and Biofilm Processes* (JENKINS, D., HERMANOWICZ, S. W., Eds.), pp. 247–254. Berkeley, CA: IAWQ.

MAIDAK, B. L., OLSEN, G. J., LARSEN, N., OVERBEEK, R., MCCAUGHEY, M. J., WOESE, C. R. (1997), The RDP (ribosomal database project), *Nucleic Acids Res.* **25**, 109–111.

MANZ, W., AMANN, R., LUDWIG, W., WAGNER, M., SCHLEIFER, K.-H. (1992), Phylogenetic oligodeoxynucleotide probes for the major subclasses of Proteobacteria: problems and solutions, *Syst. Appl. Microbiol.* **15**, 593–600.

MANZ, W., SZEWZYK, U., ERICSSON, P., AMANN, R., SCHLEIFER, K.-H., STENSTRÖM, T.-A. (1993), *In situ* identification of bacteria in drinking water and adjoining biofilms by hybridization with 16S and 23S rRNA-directed fluorescent oligonucleotide probes, *Appl. Environ. Microbiol.* **59**, 2293–2298.

MANZ, W., WAGNER, M., AMANN, R., SCHLEIFER, K.-H. (1994), *In situ* characterization of the microbial consortia active in two wastewater treatment plants, *Water Res.* **28**, 1715–1723.

MANZ, W., AMANN, R., LUDWIG, W., VANCANNEYT, M., SCHLEIFER, K.-H. (1996), Application of a suite of 16S rRNA-specific oligonucleotide probes designed to investigate bacteria of the phylum cytophaga–flavobacter–bacteroides in the natural environment, *Microbiology* **142**, 1097–1106.

MASSOL-DEYA, A., WELLER, R., RIOS-HERNANDEZ, L., ZHOU, J.-Z., HICKEY, R. F., TIEDJE, J. M. (1997), Succession and convergence of biofilm communities in fixed-film reactors treating aromatic hydrocarbons in groundwater, *Appl. Environ. Microbiol.* **63**, 270–276.

MOBARRY, B. K., WAGNER, M., URBAIN, V., RITTMANN, B. E., STAHL, D. A. (1996), Phylogenetic probes for analyzing abundance and spatial organization of nitrifying bacteria, *Appl. Environ. Microbiol.* **62**, 2156–2162.

MØLLER, S., PEDERSEN, A. R., POULSEN, L. K., ARVIN, E., MOLIN, S. (1996), Activity and three-dimensional distribution of toluene-degrading *Pseudomonas putida* in a multispecies biofilm assessed by quantitative *in situ* hybridization and scanning confocal laser microscopy, *Appl. Environ. Microbiol.* **62**, 4632–4640.

MUYZER, G., SMALLA, K. (1998), Application of denaturing gradient gel electrophoresis (DGGE) and temperature gradient gel electrophoresis (TGGE) in microbial ecology, *Antonie van Leeuwenhoek* **73**, 127–141.

MUYZER, G., DE WAAL, E. C., UITTERLINDEN, A. G. (1993), Profiling of complex microbial populations by denaturing gradient gel electrophoresis analysis of polymerase chain reaction-amplified genes coding for 16S rRNA, *Appl. Environ. Microbiol.* **59**, 695–700.

MUYZER, G., HOTTENTRÄGER, S., TESKE, A., WAWER, C. (1995a), Denaturing gradient gel electrophoresis of PCR-amplified 16S rDNA – a new molecular approach to analyse the genetic diversity of mixed microbial communities, *Mol. Microb. Ecol. Man.* **3.4.4**, 1–22.

MUYZER, G., TESKE, A., WIRSEN, C. O., JANNASCH, H. W. (1995b), Phylogenetic relationships of *Thiomicrospira* species and their identification in deep-sea hydrothermal vent samples by denaturing gradient gel electrophoresis of 16S rDNA fragments, *Arch. Microbiol.* **164**, 165–172.

NEEF, A., ZAGLAUER, A., MEIER, H., AMANN, R., LEMMER, H., SCHLEIFER, K.-H. (1996), Population analysis in a denitrifying sand filter: conventional and *in situ* identification of *Paracoccus* spp. in methanol-fed biofilms, *Appl. Environ. Microbiol.* **62**, 4329–4339.

NIELSEN, A. K., GERDES, K., DEGN, H., MURELL, J. C. (1996), Regulation of bacterial methane oxidation: transcription of the soluble methane monooxygenase operon of *Methylococcus capsulatus* (Bath) is repressed by copper ions, *Microbiology* **142**, 1289–1296.

NIELSEN, P. H., ANDREASEN, K., WAGNER, M., BLACKALL, L. L., LEMMER, H., SEVIOUR, R. J. (1997), Variability of type 021N in activated sludge as determined by *in situ* substrate uptake pattern and *in situ* hybridization with fluorescent rRNA targeted probes, in: *Proc. 2nd Int. Conf. Microorganisms in Activated Sludge and Biofilm Processes* (JENKINS, D., HERMANOWICZ, S. W., Eds.), pp. 255–262. Berkeley, CA: IAWQ.

NÜBEL, U., ENGELEN, B., FELSKE, A., SNAIDR, J., WIESHUBER, A. et al. (1996), Sequence heterogeneities of genes encoding 16S rRNAs in *Paenibacillus polymyxa* detected by temperature gradient gel electrophoresis, *J. Bacteriol.* **178**, 5636–5643.

OLSEN, G. J., LANE, D. J., GIOVANNONI, S. J., PACE, N. R., STAHL, D. A. (1986), Microbial ecology and evolution: a ribosomal RNA approach, *Annu. Rev. Microbiol.* **40**, 337–365.

POULSEN, L. K., BALLARD, G., STAHL, D. A. (1993), Use of rRNA fluorescence *in situ* hybridization for measuring the activity of single cells in young and established biofilms, *Appl. Environ. Microbiol.* **59**, 1354–1360.

RAMSING, N. B., KÜHL, M., JORGENSEN, B. B. (1993), Distribution of sulfate-reducing baeteria, O_2 and H_2S in photosynthetic biofilms determined by oligonucleotide probes and microelectrodes, *Appl. Environ. Microbiol.* **59**, 3840–3849.

RAMSING, N. B., FOSSING, H., FERDELMAN, T. G., ANDERSEN, F., THAMDRUP, B. (1996), Distribution of bacterial populations in a stratified fjord (Mariager Fjord, Denmark) quantified by *in situ* hybridization and related to chemical gradients in the water column, *Appl. Environ. Microbiol.* **62**, 1391–1404.

RASKIN, L., POULSEN, L., NOGUERA, D. R., RITTMAN, B. E., STAHL, D. A. (1994a), Quantification of methanogenic groups in anaerobic biological reactors by oligonucleotide probe hybridization, *Appl. Environ. Microbiol.* **60**, 1241–1248.

RASKIN, L., STROMLEY, J. M., RITTMAN, B. E., STAHL, D. A. (1994b), Group-specific 16S rRNA hybridization probes to describe natural communities of methanogens, *Appl. Environ. Microbiol.* **60**, 1232–1240.

RASKIN, L., AMANN, R. I., POULSEN, L. K., RITTMANN, B. E., STAHL, D. A. (1995a), Use of ribosomal RNA-based molecular probes for characterization of complex microbial communities in anaerobic biofilms, *Water Sci. Technol.* **31**, 261–272.

RASKIN, L., ZHENG, D., GRIFFIN, M. E., STROOT, P. G., MISRA, P. (1995b), Characterization of microbial communities in anaerobic bioreactors using molecular probes, *Antonie van Leeuwenhoek* **68**, 297–308.

RASKIN, L., RITTMAN, B. E., STAHL, D. A. (1996), Competition and coexistence of sulfate-reducing and methanogenic populations in anaerobic biofilms, *Appl. Environ. Microbiol.* **62**, 3847–3857.

ROLLER, C., WAGNER, M., AMANN, R., LUDWIG, W., SCHLEIFER, K.-H. (1994), *In situ* probing of gram-positive bacteria with high DNA G+C content using 23S rRNA-targeted oligonucleotides, *Microbiology* **140**, 2849–2858.

ROSSELLO-MORA, R. A., WAGNER, M., AMANN, R., SCHLEIFER, K.-H. (1995), The abundance of *Zoogloea ramigera* in sewage treatment plants, *Appl. Environ. Microbiol.* **61**, 702–707.

ROTHEMUND, C., AMANN, R., KLUGBAUER, S., MANZ, W., BIEBER, C. et al. (1996), Microflora of 2,4-dichlorophenoxyacetic acid degrading biofilms on gas permeable membranes, *Syst. Appl. Microbiol.* **19**, 608–615.

SCHAECHTER, M., MAALOE, O., KJELDGAARD, N. O. (1958), Dependency on medium and temperature of cell size and chemical composition during balanced growth of *Salmonella typhimurium*, *J. Gen. Microbiol.* **19**, 592–606.

SCHLEIFER, K.-H., AMANN, R., LUDWIG, W., ROTHEMUND, C., SPRINGER, N., DORN, S. (1992), Nucleic acids for the identification and *in situ* detection of pseudomonads, in: *Pseudomonas: Molecular Biology and Biotechnology* (GALLI, E., SILVER, S., WITHOLT, B., Eds.), pp. 127–134. Washington, DC: American Society for Microbiology.

SCHÖNHUBER, W., FUCHS, B., JURETSCHKO, S., AMANN, R. (1997), Improved sensitivity of whole-cell hybridization by the combination of horseradish peroxidase-labeled oligonucleotides and tyramide signal amplification, *Appl. Environ. Microbiol.* **63**, 3268–3273.

SCHRAMM, A., LARSEN, L. H., REVSBECH, N. P., RAMSING, N. B., AMANN, R., SCHLEIFER, K.-H. (1996), Structure and function of a nitrifying biofilm as determined by *in situ* hybridization and the use of microelectrodes, *Appl. Environ. Microbiol.* **62**, 4641–4647.

SCHRAMM, A., DE BEER, D., WAGNER, M., AMANN, R. (1998), *Nitrosospira* sp. and *Nitrospira* sp. as dominant populations in a nitrifying fluidized bed reactor: identification and activity *in situ*, *Appl. Environ. Microbiol.* **64**, 3480–3485.

SINIGALLIANO, C. D., KUHN, D. N., JONES, R. D. (1995), Amplification of the *amoA* gene from diverse species of ammonium-oxidizing bacteria and from an indigenous bacterial population from seawater, *Appl. Environ. Microbiol.* **61**, 2702–2706.

SNAIDR, J., AMANN, R., HUBER, I., LUDWIG, W., SCHLEIFER, K.-H. (1997), Phylogenetic analysis and *in situ* identification of bacteria in activated sludge, *Appl. Environ. Microbiol.* **63**, 2884–2896.

STAHL, D. A., AMANN, R. (1991), Development and application of nucleic acid probes, in: *Nucleic Acid Techniques in Bacterial Systematics* (STACKEBRANDT, E., GOODFELLOW, M., Eds.), pp. 205–248. Chichester: John Wiley & Sons.

STAHL, D. A., FLESHER, B., MANSFIELD, H. R., MONTGOMERY, L. (1988), Use of phylogenetically based hybridization probes for studies of ruminal microbial ecology, *Appl. Environ. Microbiol.* **54**, 1079–1084.

STALEY, J. T., KONOPKA, A. (1985), Measurement of *in situ* activities of nonphotosynthetic microorganisms in aquatic and terrestrial habitats, *Annu. Rev. Microbiol.* **39**, 321–346

SUZUKI, M. T., GIOVANNONI, S. J. (1996), Bias caused by template annealing in the amplification of mixtures of 16S rRNA genes by PCR, *Appl. Environ. Microbiol.* **62**, 625–630.

SWAMINATHAN, B., MATAR, G. M. (1993), Molecular typing methods, in: *Diagnostic Molecular Microbiology* (PERSING, D. H., SMITH, T. F., TENOVER, F. C., WHITE, T. J., Eds.), pp. 26–50. Washington, DC: American Society for Microbiology.

SZEWZYK, U., MANZ, W., AMANN, R., SCHLEIFER, K.-H., STENSTRÖM, T.-A. (1994), Growth and *in situ* detection of a pathogenic *Escherichia coli* in biofilms of a heterotrophic water-bacterium by use of 16S- and 23S-rRNA-directed fluorescent oligonucleotide probes, *FEMS Microbiol. Ecol.* **13**, 169–176.

TESKE, A., WAWER, C., MUYZER, G., RAMSING, N. B. (1996), Distribution of sulfate-reducing bacteria in a stratified fjord (Mariager Fjord, Denmark) as evaluated by most-probable-number counts and denaturing gradient gel electrophoresis of PCR-amplified ribosomal DNA fragments, *Appl. Environ. Microbiol.* **62**, 1405–1415.

TSIEN, H. C., BRATINA, B. J., TSUJI, K., HANSON, R. S. (1990), Use of oligodeoxynucleotide signature probes for identification of physiological groups of methylotrophic bacteria, *Appl. Environ. Microbiol.* **56**, 2858–2865.

WAGNER, M., AMANN, R., LEMMER, H., SCHLEIFER, K.-H. (1993), Probing activated sludge with oligonucleotides specific for Proteobacteria: inadequacy of culture-dependent methods for describing microbial community structure, *Appl. Environ. Microbiol.* **59**, 1520–1525.

WAGNER, M., AMANN, R., KÄMPFER, P., ASSMUS, B., HARTMANN, A. et al. (1994a), Identification and *in situ* detection of gram-negative filamentous bacteria in activated sludge, *Syst. Appl. Microbiol.* **17**, 405–417.

WAGNER, M., ASSMUS, B., HARTMANN, A., HUTZLER, P., AMANN, R. (1994b), *In situ* analysis of microbial consortia in activated sludge using fluorescently labelled, rRNA-targeted oligonucleotide probes and confocal scanning laser microscopy, *J. Microscopy* **176**, 181–187.

WAGNER, M., ERHART, R., MANZ, W., AMANN, R., LEMMER, H. et al. (1994c), Development of an rRNA-targeted oligonucleotide probe specific for the genus *Acinetobacter* and its application for *in situ* monitoring in activated sludge, *Appl. Environ. Microbiol.* **60**, 792–800.

WAGNER, M., RATH, G., AMANN, R., KOOPS, H.-P., SCHLEIFER, K.-H. (1995), *In situ* identification of ammonia-oxidizing bacteria, *Syst. Appl. Microbiol.* **18**, 251–264.

WAGNER, M., RATH, G., KOOPS, H.-P., FLOOD, J., AMANN, R. (1996), *In situ* analysis of nitrifying bacteria in sewage treatment plants, *Water Sci. Technol.* **34**, 237–244.

WAGNER, M., SCHMID, M., JURETSCHKO, S., TREBESIUS, K.-H., BUBERT, A. et al. (1998), *In situ* detection of a virulence factor mRNA and 16S rRNA in *Listeria monocytogenes*, *FEMS Microbiol. Lett.* **160**, 159–168.

WALLNER, G., ERHART, R., AMANN, R. (1995), Flow cytometric analysis of activated sludge with rRNA-targeted probes, *Appl. Environ. Microbiol.* **61**, 1859–1866.

WALLNER, G., STEINMETZ, I., BITTER-SUERMANN, D., AMANN, R. (1996), Combination of rRNA-targeted hybridization probes and immuno-probes for the identification of bacteria by flow cytometry, *Syst. Appl. Microbiol.* **19**, 569–576.

WARD, D. M. (1989), Molecular probes for analysis of microbial communities, in: *Structure and Function of Biofilms* (CHARACKLIS, W. G., WILDERER, P. A., Eds.), pp. 145–163. Chichester: John Wiley & Sons.

WAWER, C., JETTEN, M. S. M., MUYZER, G. (1997), Genetic diversity and expression of the [NiFe] hydrogenase large-subunit gene of *Desulfovibrio* spp. in environmental samples, *Appl. Environ. Microbiol.* **63**, 4360–4369.

WOESE, C. R. (1987), Bacterial evolution, *Microb. Rev.* **51**, 221–271.

ZUCKERKANDL, E., PAULING, L. (1965), Molecules as documents of evolutionary history, *J. Theor. Biol.* **8**, 357–366.

6 Analytical Parameters for Monitoring of Wastewater Treatment Plants

HELMUT KROISS
KARL SVARDAL
Wien, Austria

List of Abbreviations

AOX adsorbable organic halogenated compounds
BOD biochemical oxygen demand
COD chemical oxygen demand
DEV Deutsche Einheitsverfahren
DN denitrification
DO dissolved oxygen
GAC granulated activated carbon
MAP magnesium ammonium phosphate
MCRT mean cell residence time (sludge age)
OUC oxygen uptake for carbon removal
OUN uptake for nitrification
PAC powdered activated carbon
PAH polycyclic aromatic hydrocarbons
POX purgeable organic halogenated compounds
rH redox potential
S_0 dissolved substrate
SS_e suspended solids in the effluent
SS_o suspended solids in the influent
SST secondary sedimentation tank
SVI sludge volume index
T temperature
TC total carbon
TN total nitrogen
TOC total organic carbon
TP total phosphorus
X_0 particulate degradable pollution
Y yield coefficient

1 Introduction

Wastewater is a complex mixture of thousands of different compounds most of them cannot be determined in detail, and wastewater treatment is a complex combination of mechanical, chemical, and biological processes which causes problems for monitoring. Flow and composition of wastewater vary continuously over time even when coming from the same source (municipal, industry, and trade effluents). This is the main difference as compared to industrial production process control.

Monitoring comprises the information needed for control and supervision of the different processes as well as of the result of the treatment: i.e., the required quality of the discharged or reused effluent. There are many physical, chemical and biochemical analytical procedures which can be applied to characterize the wastewater composition and the process performance. In practice the economical constraints have to be considered. The consequence is that it is necessary to find an optimum between a minimum of analytical effort and a maximum of reliability of process performance. The key role in this respect plays our knowledge on the relationship between wastewater flow and composition and the process parameters. Another important influence on the monitoring system derives from the accuracy and reliability of analytical procedures starting from sampling, sample treatment and the analytical problems associated with the complex matrix of wastewater compounds ranging from dissolved to coarse particulate matter.

Mechanical and chemical processes in wastewater treatment in most of the cases can be described by relationships which are not dependent on the specific local situation and its history (i.e., loading situation during the last days or weeks). This is basically not the case for biological processes. Microorganisms live in communities with their own genetically fixed monitoring system and within certain limits adapt to the situation which also depends on the historic development. This can be interpreted as a great advantage of biological processes as long as we can rely on their abilities. It creates a lot of problems if we exceed their adaptability. Associated with the previous statements it turns out that the time scale for monitoring differs between seconds and weeks which shall be demonstrated by the following example: Nitrifying bacteria adapt their respiration rate (oxygen consumption) according to the ammonia concentration in the liquid phase within seconds, the number of active nitrifiers in a biological reactor needed to get constant low effluent concentrations depends on the ammonia load of the last days or weeks which has been nitrified and a series of other conditions (temperature, sludge production, etc.) during this period. The consequence is that decision making in wastewater treatment plant operation has many different aspects and needs many different parameters for monitoring in order to get the important information for the solution of the problems.

Legal requirements for the documentation of treatment performance often have another background than the requirements for process monitoring. They reflect a compromise between the multitude of different wastewaters and treatment processes, the specific local situations and conditions related to the receiving waters and an effective administration of the compliance with the legal requirements. This compromise can be seen as a continuous challenge for all parties involved in order to optimize cost efficiency.

2 Basic Tools for Monitoring and Data Assessment

As already mentioned there are two major challenges with process monitoring, i.e.,

(1) reliability of data,
(2) understanding of the process behavior in order to be able to make use of the data for process control.

The two challenges are linked to a large extent and it seems to be important to present some of the most important tools to improve the situation.

The first and one of the most efficient tools for data assessment is the application of the conservation law of mass and energy (1st law of thermodynamics). Even in the case the behavior of the processes is unknown (black box) input and output load of mass must be equal as long as the parameters are conservative. Such mass balances must fit for, e.g., total carbon (TC), total phosphorus (TP), total nitrogen (TN), but not for total organic carbon (TOC). For COD (chemical oxygen demand) the mass balance concept can also be applied as COD can be interpreted as an electrical charge balance for organic carbon, while BOD (biochemical oxygen demand) balances cannot be used. Of course such mass balances can be made for any chemical element, which can be very interesting in industrial wastewater treatment.

Data assessment is only possible if all input and all output mass flows can be measured with comparable accuracy, which in practice often poses problems. In most of the cases TP and COD balances can be used while for TN the analytical determination of denitrified N loads which leave the process with the exhaust gas is very difficult. TC balances have to incorporate CO_2 production and the equilibria between carbonate and bicarbonate.

With "closed" mass balances systematic errors in sampling, sample treatment and analytical procedures can be detected. This tool becomes more effective using mass balances over longer periods of time (days, weeks, months) as the changes in the mass content of the reactors, also a source of erroneous data, can be neglected in many cases or have little effect on the mass balance. The period of time for mass balances has to be adapted to the process for which they are applied. For aerobic and anaerobic biological processes the important mass balances are well described in the literature (NOWAK et al., 1998; KROISS, 1985; SVARDAL, 1991; SVARDAL et al., 1998).

A similar tool which also refers to the conservation laws is electro-neutrality, which is of great importance in the understanding of chemical and biochemical processes, where, e.g., pH or alkalinity play an important role. COD balances are also balances of electrical charges. Electro-neutrality means that the sum of all cation concentrations multiplied with their charge number must be equal to the sum of all anion concentrations multiplied with their respective charge number. As normally not all ions present in a wastewater can be measured in practice the application of this tool needs experience and results in an approximation which in many cases is sufficient for data assessment and process understanding of, e.g., anaerobic biological wastewater treatment (SVARDAL, 1991).

Chemical equilibria in liquid solutions (wastewaters) are rather complex and are often directly linked to the pH and solubility of different compounds. As biological processes are very sensible to changes in the equilibrium between the undissociated and the dissociated form of a compound, pH often plays an important role. As wastewater treatment normally influences on these equilibria pH and solubility will change during the different treatment steps. pH can be controlled by the addition of chemicals or other process parameters (e.g., oxygen supply with aerobic biological treatment). Both influences can result in the precipitation of dissolved wastewater compounds (Ca, P, Mg, H_2S) causing improved quality in the effluent, but – on the other hand – severe scaling or sedimentation problems in pipes and reactors.

For biological processes there exist basic relationships between the reaction rate of bacteria expressed as growth rate μ or compound removal rate v and substrate concentration S of a biodegradable compound. These relationships have been developed by MONOD and MICHAELIS-MENTEN, respectively (Fig. 1).

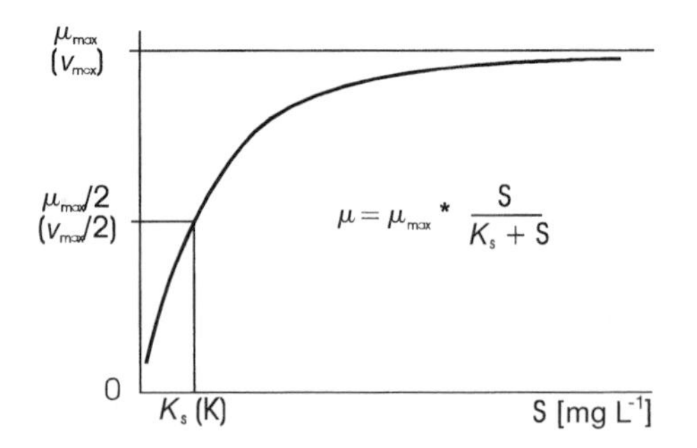

Fig. 1. MONOD and MICHAELIS-MENTEN relationships between growth rate μ, removal rate v, and substrate concentration S.

This means that pollution load variations of a wastewater constituent in the influent affect the bacterial activity via a variation of concentration. The maximum removal rate v_{max} (maximum growth rate μ_{max}) of bacteria will only be reached at high substrate (effluent) concentrations S ($S \gg K_s$). Low effluent concentrations can only be achieved if the bacteria have low growth rates which means that they operate far from their maximum capacity ($S < K_s$). The half saturation constants K_s and K for many bacteria are in the range of 1 mg L^{-1} and less. The determination of the actual maximum growth rate of the bacteria in a reactor (determining, e.g., the maximum oxygen uptake rate, the maximum methane production rate, etc.) gives the best information on the maximum removal capacity of a biological treatment system which can be compared to the actual loading situation.

As long as the yield coefficient Y (g biomass produced per g substrate removed) is a constant, which can be assumed in most of the cases, the two equations (MONOD and MICHAELIS-MENTEN) contain similar information, e.g., oxygen uptake for nitrification is proportional to the growth rate of nitrifiers. It is important to note that these relationships only apply if the substrate S is rate limiting while all the other necessary substrates for the bacterial life are not rate limiting (e.g., N, P, O_2, trace, elements, etc.). It has to be checked which compound in a biological reactor is rate limiting for the bacterial activity to avoid wrong operational decisions. This is especially important for industrial wastewater treatment.

3 Wastewater Characterization

3.1 General Aspects

Wastewater characterization can be achieved by:

(1) physical,
(2) chemical, and
(3) biochemical parameters.

Physical parameters are treated in Sect. 3.2, chemical and biochemical parameters will be discussed in the following sections.

The need for wastewater characterization using different parameters is mainly depending on:

(1) the source of the wastewater and its variability over the year (season),
(2) the treatment process to be applied,
(3) the standards of treatment efficiency to be met (effluent concentration).

As treatment systems have a long lifetime also the future development of wastewater flow and composition has to be projected for a proper design.

It is a general characteristic of wastewater that flow, composition and none of the parameters for its characterization are constant and that in practice it is not possible to determine all polluting compounds present in the influent, during the treatment and in the effluent. The consequence is that parameters as TN or TP are used or sum parameters representative for the organic pollution as COD, TOC and the biochemical parameter BOD_5. For some of the wastewater compounds specific analytical procedures are applied, e.g., for inorganic nitrogen compounds (nitrate nitrogen, nitrite nitrogen, ammonia nitrogen). It is common practice to refer the concentrations to the elements (N, P, O_2). In this way it is much easier to calculate mass balances as compared to the concentrations based on molecular weight of the different compounds (ions).

3.2 Physical Parameters

3.2.1 Flow

The mass flow of different qualities of water (raw water influent, recirculated water from sludge handling, etc.) represents a basis for the calculation of pollution loads (kg d^{-1}) and for the hydraulic parameters as flow velocity, hydraulic detention time, mixing conditions, etc. There are many different ways to reliably measure flow, nevertheless flow data have to be checked before use in order to avoid erroneous mass balance results.

3.2.2 Temperature

Temperature of the wastewater and its variations over time represent a very important information for design and operation of treatment processes. The temperature influences on important physical parameters such as density, viscosity, etc. Most of the chemical and biochemical reactions are strongly dependent and often very sensitive to temperature changes and ranges. As chemical or biochemical reactions can be endothermic or exothermic they can contribute to changes in temperature in the wastewater reactors. Climatic conditions as air temperature, wind velocity, air humidity and solar irradiation also can cause temperature variations in outdoor reactors.

Balances for all energy inputs and losses are necessary to estimate the influence of the wastewater temperature on the treatment processes. The high values for the specific heat and the evaporation energy of water should always be kept in mind, especially if a water temperature adjustment is aimed at by heating or cooling.

Temperatures will have to be monitored at nearly every type of wastewater treatment process.

3.2.3 Suspended and Settleable Solids

The suspended solids concentration or load in the wastewater (determined with a membrane filtration, pore size of 45 μm) is an important parameter, but does not contain enough information for design and operational purposes, except for the biological treatment of normal domestic sewage, where enough experience is available.

The settleable solids are of interest if primary settling is applied or sediment accumulation in mixed reactors has to be avoided.

For a better characterization of the solids the following fractions can be distinguished:

(1) inorganic solids which remain inert during the treatment process,
(2) inorganic solids which are dissolved during the treatment process,
(3) organic solids which are not degraded during the treatment process,
(4) organic solids which are degraded (or dissolved) during the treatment process,
(5) organic solids which are produced during the treatment process (biomass),
(6) inorganic solids which are produced by precipitation of dissolved matter during the treatment (e.g., carbonates, sulfides).

There is a close relationship between the suspended solids load entering a treatment system and the produced sludge loads which have to be removed from the process (and disposed off). As many biological treatment processes are designed using the sludge age (mean cell residence time) as the decisive parameter, sludge production is directly linked to the treatment efficiency. For industrial and mixed wastewaters it is difficult to predict the behavior of sludge production from the solids without detailed investigations on lab or pilot scale. While suspended solids in the wastewater in many cases behave similar in anaerobic and aerobic treatment systems, solids production from dissolved matter is quite different. Inorganic solids (e.g., quartz sand) can result in erosion problems for pipes and pumps. Precipitation of dissolved inorganic salts can cause severe operational problems such as scaling, clogging and sediment accumulation in reactors and pipes. The most common precipitates are calcium carbonate and magnesium ammonium phosphate (MAP).

The density and size of solids in reactors influence the mixing energy ($W m^{-3}$) required for complete mixing in order to avoid accumulation of sediments in reactors and to have good contact between solids and bacteria. The size of particles in the influent of (biological) treatment processes can be limited by screens and sieves (<1 mm in diameter). The density and size of the particles determine the efficiency of sand traps, grit chambers, and primary settling tanks.

4 Monitoring of Treatment Plants during Operation

4.1 General Remarks

Monitoring of treatment plants has at least the following goals:

(1) to have enough information for correct decisions during operation in order to avoid malfunction,
(2) to meet the standards laid down in the discharge permits or the requirements for reuse (depending on the purpose of reuse) including documentation,
(3) to optimize cost efficiency.

In many respects the data acquisition can be the same for all goals but normally the data for operational decisions exceed the number of permit values and often less accurate information can be sufficient. For the second goal normally standard analytical methods (DEV, American Standard Methods) have to be used as exceeding of limit values can have legal (criminal law) and financial consequences.

The following sections concentrate on the operation of treatment plants with different processes. As there are numerous processes and process combinations applied for different wastewaters and treatment efficiency requirements it will be necessary to concentrate on the most common unit processes and treatment standards. Descriptions are limited to those directly related to treatment (removal) processes. Equipment reliability problems are not discussed. Emphasis is put on the understanding of the most important processes in biological wastewater treatment.

4.2 Mechanical Treatment Processes

The most common mechanical treatment steps are:

(1) screening,
(2) grit removal, and
(3) primary sedimentation.

Grit removal and primary sedimentation efficiency can easily be checked by the determination of settleable solids using the Imhoff cone and 2 h settling time. Sand in the effluent of the grit chamber is directly visible ($>0{,}2$ mm in diameter). The effluent of primary settling tanks should contain less than 1 $mL\,L^{-1}$ of settleable solids. Normally a daily to weekly check of the mechanical step is sufficient to avoid malfunction.

4.3 Aerobic Biological Treatment Processes

4.3.1 Process Application

Most of the wastewaters containing organic carbon, nitrogen and phosphorus compounds can be treated using aerobic biological processes. Bacteria degrade the organic carbonaceous pollution, oxidize ammonia to nitrate, and phosphorus can be removed by luxury uptake. The nutrients nitrogen and phosphorus can be reduced in biological treatment plants in order to avoid eutrophication in the receiving waters as lakes, inland seas or bays.

Bacteria are living organisms with sophisticated control systems which are genetically fixed. The bacterial population in the biological reactors develops adaptation to the wastewater composition, the process configuration (layout), loading pattern, process control and environmental conditions (mainly temperature, pH).

4.3.2 Biological Processes

4.3.2.1 Degradation of Organic Carbonaceous Pollution

Fig. 2 shows the basic elements of aerobic carbon degradation.

It is important to clearly differentiate between reduction efficiency (reduction) and degradation. The pollution components in the influent and in the effluent are different to a large extent, which cannot be detected using the common parameters BOD, COD and TOC

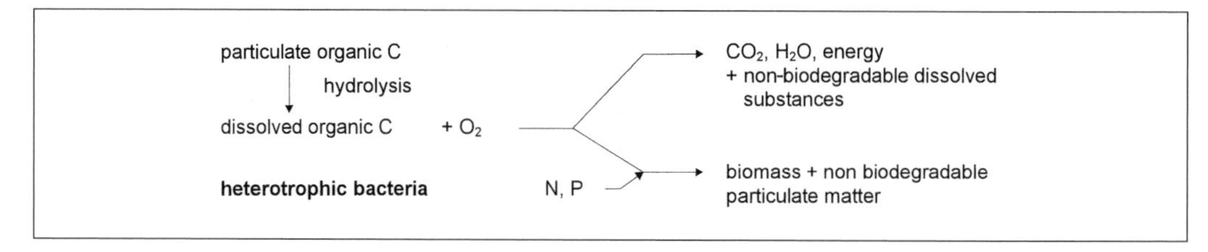

Fig. 2. Carbon removal process.

for the characterization of the organic pollution. This is especially true for BOD_5, where the concentrations and loads alone cannot characterize the effect to the receiving water adequately. Normally the degradation of raw water compounds is much higher than the measured reduction expressed in terms of the common parameters. BOD_5 being available only 5 d after sampling can only be used for documentation but not for control purposes. COD is the only parameter which can be used for balances which is important for the check of the correctness of the data. Continuous monitoring of TOC is possible but it needs special attention to correctly include the solids into the analytical procedure.

A relatively new development is the use of UV probes directly submerged into the flow of the wastewater. Despite the fact that the method is rather unspecific with regard to TOC or COD it is possible to get signals which correspond to COD (influent and effluent). As reported (MATSCHÉ and RUIDER, 1982) for domestic and other wastewaters containing aromatic compounds UV probes can be used with great success for continuous monitoring needing only little maintenance. This method cannot be used for wastewaters mainly polluted with, e.g., carbohydrates (MATSCHÉ and STUMWÖHRER, 1996),

Oxygen is the important substrate in order to achieve organic carbon removal from wastewater. Oxygen supply by aeration systems, therefore, is a key element of process control which needs the basic understanding of the different biological processes involved.

Heterotrophic aerobic bacteria reach nearly their maximum growth rate (removal rate) at dissolved oxygen (DO) concentrations of ca. 2 mg L^{-1}. At 0 mg L^{-1} the aerobic degradation process will stop (see Fig. 1). If carbon (BOD, COD) removal is the only requirement for the biological treatment a DO concentration of 2 mg L^{-1} should be maintained in order to prevent oxygen limitation. The oxygen supply will have to be adapted to the oxygen uptake for carbon removal (OUC) in order to minimize energy consumption for aeration. The OUC is directly related to the availability of dissolved biodegradable substrates. Variations of dissolved substrate (S_0) in the influent will immediately result in variations of OUC. But as heterotrophic bacteria are able to store dissolved substrate there is not a linear correlation (PRENDL and KROISS, 1998) between the variations. At treatment plants for BOD removal only (MCRT <4 d) this storage capacity can be neglected; this is not the case at MCRT >5 d.

Particulate degradable pollution (X_0) has to undergo hydrolysis before it can be removed by heterotrophic bacteria. As hydrolysis is the rate limiting factor, the OUC for particulate matter remains relatively constant over time even if there are strong variations in the influent. The consequence is that, e.g., for the control of oxygen supply continuous TOC influent monitoring cannot be used. There are indications that the DO concentration also influences hydrolysis which is mainly of interest in the case of aerobic sludge stabilization. At low DO concentrations (<0.5 mg L^{-1}) diffusion limitations will play an important role even in dispersed growth reactors (activated sludge process).

Heterotrophic bacteria need nitrogen (N) and phosphorus (P) for their growth. In domestic wastewater N and P are abundant but industrial wastewater often has a lack of nutrients. In such a case nutrients have to be added. For rough estimates a BOD:N:P ratio of 100:5:1 can be used as the basis for nutrient addition. In the case low nutrient effluent standards have to be achieved or in order to minimize the costs for nutrient addition continuous monitoring of low concentrations (<0.5 mg L^{-1}) of NH_4-N and PO_4-P in the biological reactors can be used. A regular check of nutrient deficiency will be required. It can be detected by adding short shock loads of nutrients to the biological reactor. An immediate response in increased OUC (drop in DO or increase of aeration capacity) is a clear indication that nutrients have been a limiting factor. If there is no response enough nutrients are added.

4.3.2.2 Nitrification

Fig. 3 shows the basic scheme of the biological nitrification process.

Nitrifying bacteria are strictly aerobic. For the oxidation of 1 g NH_4-N an oxygen load of 4.3 g O_2 will be consumed. Nitrification completely stops at a DO concentration of 0 mg L^{-1}. It reaches nearly maximum capacity at DO concentrations of 2 mg L^{-1}.

Nitrifiers have a low maximum growth rate $\mu_{N,max}$ as compared to most of the heterotrophic bacteria in treatment plants. Therefore, they can only accumulate in a treatment system if the MCRT in the aerobic reactor volume is greater than $\mu_{N,max}$ [$MCRT_a \gg (\mu_{N,max})^{-1}$.

The maximum growth rate $\mu_{N,max}$ is strongly dependent on temperature (doubling for an increase in temperature of 7 °C). The required aerobic volume, therefore, will strongly depend on temperature.

Full nitrification in a treatment plant is an indicator of excellent receiving water protection as three goals are achieved:

(1) oxygen consumption in the receiving water due to the wastewater discharge is nearly completely avoided,
(2) ammonia toxicity is avoided,
(3) excellent removal of carbonaceous pollution is a (compulsory) side effect.

Nitrification, therefore, is an excellent tool for monitoring precautionary water protection. As long as ammonia concentrations in the effluent are low, which can easily be monitored even continuously, carbon removal needs no additional monitoring.

In order to maintain nitrification the MCRT in the aerobic volume has to be monitored and adjusted to the temperature. The aeration has to be adapted to the oxygen uptake for nitrification (OUN). As bacteria cannot store ammonia the variations of ammonia load over time result in a linear response of OUN. Nitrifying bacteria are present in domestic sewage and in the natural environment. They convert ammonia to nitrite (*Nitrosomonas*) and nitrate (*Nitrobacter*) using CO_2 as a carbon source. For 1 mol of ammonia 2 mol of H^+ are produced, which results in a decrease of alkalinity and a drop of pH which depends on the wastewater composition (buffer capacity).

Oxygen supply can be controlled best by continuous monitoring of NH_4-N concentration in the aerobic reactor volume (outlet). Even on a full scale it could be proved that control of oxygen supply can effectively be achieved with DO monitoring at one point in circular flow aeration tanks (FRANZ, 1996). As long as only nitrification has to be achieved a constant DO of 2 mg L^{-1} in the aerated volume would be sufficient. If only nitrogen removal by denitrification has to be achieved or is aimed at, a minimization of oxygen supply is of great importance.

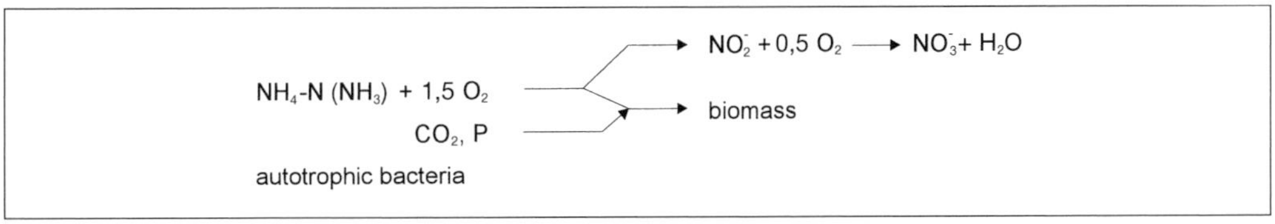

NH_4-N (NH_3) + 1,5 O_2 → NO_2^- + 0,5 O_2 → NO_3^- + H_2O
CO_2, P → biomass
autotrophic bacteria

Fig. 3. Nitrification process.

In the case the wastewater to be treated has a low buffer capacity (alkalinity, hardness of raw water) the pH also has to be monitored, since at a pH $<6{,}8$ nitrification will be inhibited. The drop in pH due to nitrification can be reduced by denitrification or by chemical (alkalinity) addition.

4.3.2.3 Denitrification

Fig. 4 shows the basic scheme of nitrogen removal from wastewater by denitrification, which is a respiratory carbon removal process in the presence of NO_x-N and the absence of DO (anoxic conditions).

Most of the heterotrophic bacteria in biological treatment plants are able to use nitrogen as electron acceptor instead of oxygen. Nitrogen is released from the wastewater as N_2 gas.

A prerequisite of denitrification is that nitrogenous compounds are already converted to nitrate (nitrite) by nitrification. The denitrification process can be limited by:

(1) availability of biodegradable organic carbon,
(2) availability of nitrate (NO_x),
(3) denitrification rate (temperature dependent).

As the hydrolysis rate is reduced in anoxic zones, it is efficient to make use of the dissolved substrate in the influent for denitrification (predenitrification). For increased denitrification efficiency particulate substrate can also be used as a carbon source. The rate of hydrolysis will then determine the denitrification. Simultaneous nitrification–denitrification combined with pre-denitrification offers the best opportunities for maximum N removal. If denitrification is efficient, oxygen uptake in the aerated volume will become rather constant over time. The denitrification process can be monitored by the nitrate concentration in the anoxic zone. This analytical information can be used for the recirculation of nitrate from the nitrification zone to the anoxic volume or for the addition of an easily biodegradable carbon source (methanol, acetate, ethanol). For the optimization of nitrogen removal plants it has to be clearly stated whether nitrification (i.e., low ammonia effluent concentrations) or nitrogen removal (i.e., low total nitrogen removal) has priority. In every case a minimum MCRT in the aerobic zone has to be maintained in order to keep enough nitrifiers in the system.

Whenever complete removal of NO_x can be achieved in the denitrification (non-aerated) volume the redox potential (rH) can be used for monitoring (PLISSON-SAUNÉ, 1996). If NO_x and DO are absent anaerobic conditions start which results in biological H_2S production causing a sharp drop of rH. This can be detected and used for aeration control in intermittent denitrification systems. The ratio of $BOD(N)^{-1}$ has to be high enough not to limit denitrification by a lack of biodegradable substrate.

The overall nitrogen removal efficiency can be monitored by the analysis of nitrate (NO_3-N) in the effluent. At high temperatures in the reactors and with industrial wastewaters nitrite, a necessary intermediate product of nitrification and denitrification, in the effluent can cause problems as it is very toxic to the receiving water biocenosis. This fact results in very low effluent or in stream standards for nitrite.

The optimization of nitrification–denitrification plant aeration is a challenging operational problem as oxygen is preferred by the bacteria as an electron acceptor as compared to nitrogen. The goal, therefore, must be to

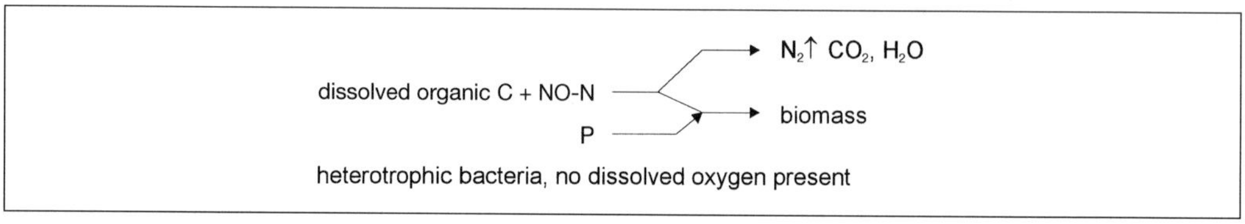

dissolved organic C + NO-N
P
$N_2\uparrow$ CO_2, H_2O
biomass
heterotrophic bacteria, no dissolved oxygen present

Fig. 4. Denitrification process.

minimize oxygen transfer to the anoxic zones and to avoid anaerobic conditions as they can affect nitrifying bacterial activity.

Sludge treatment, especially biological sludge stabilization, normally results in important flows of nitrogen within the plant. In order to achieve low ammonia and total nitrogen effluent concentrations these internal nitrogen flows will have to be monitored (NH_4-N) and controlled. It is even possible to have separate supernatant treatment for enhanced N removal (KOLLBACH and GRÖMPING, 1996; WETT et al., 1988).

4.3.2.4 Phosphorus Removal

The removal of phosphorus can be achieved by:

(1) chemical precipitation,
(2) enhanced biological phosphorus removal.

The two processes can easily be combined. In domestic wastewater about 1/3 of the phosphorus (if only P free detergents are used) will be removed with the excess sludge produced by the carbon removal process.

P precipitation by the addition of metal salts (mainly Fe, Al) is a very simple and cheap process. The chemical addition required to obtain about 1 mg L^{-1} total phosphorus in the effluent corresponds to a molar ratio of about Me:P = 1:1. The correct dosing of precipitants can be monitored by PO_4-P effluent analysis. Pre-precipitation (combined with primary sedimentation) also has to be controlled in order to avoid P deficiency in the following biological steps.

The efficiency of enhanced biological P removal (BARNARD, 1974) is depends on many wastewater and process parameters and is not as reliable as chemical precipitation. In any case the bacteria will have to be subject to strict anaerobic conditions for a certain period of time. The transfer of oxygen and NO_x to the anaerobic zone (tank) has to be minimized. If enhanced biological P removal is aimed at the addition of precipitating chemicals should be located at the end of the nitrification zone in order to avoid competition between the chemical and the biochemical processes.

Enhanced biological P removal capacity can be monitored by analyzing the increase of PO_4-P load in the effluent of the anaerobic zone as compared to the P load in the influent and in the return sludge.

Due to the low cost of P precipitation it is not economic to use continuous monitoring systems for optimization of enhanced biological P removal, except for very large plants. The key elements for biological P removal are good plant configurations (design) and optimized nitrogen removal by nitrification at N–DN processes. In any case it has to be avoided that P stored in the biomass is released during sludge treatment or in the secondary clarifier due to anaerobic conditions.

4.3.2.5 Role of Biological Treatment Process Selection

Even though all aerobic processes make use of bacteria present in the wastewater and the environment process control and monitoring are greatly influenced by the treatment system. We distinguish at least two different concepts:

(1) reactors with dispersed growth (activated sludge process),
(2) reactors with fixed films (trickling filters, biofilters, moving bed biofilters).

A detailed comparison of the process behavior of the two different concepts can be found in the literature (KROISS, 1994).

For the reactors with dispersed growth it can be assumed, that diffusion limitation for substrate transport to the bacteria plays a minor role. This means that the concentration in the liquid phase of completely mixed systems is related to the growth rate of the bacteria. Only if the oxygen concentration is close to 0 mg L^{-1} simultaneous nitrification and denitrification will occur as oxygen diffusion into the bacterial flocs is rate limiting for nitrification in the flocs and for denitrification on the surface of the floc. The large completely mixed reactors also act as equalization tanks.

The changes in the concentration of all parameters are slow, the control systems can also slowly respond to the changes.

In fixed film systems diffusion limits the process rates as the diffusion of substrate into the biofilm determines the amount of biomass involved in degradation. If the carrier material is fixed (trickling filters, RBC, biofilters) these reactors act similar to plug flow systems with very low detention times (minutes). This means that control systems will have to react very fast and the control needs instant determination of the relevant parameters. Using, e.g., continuous measurement of NH_4-N, NO_3-N, PO_4-P the time lag between sampling and data output has to be minimized, otherwise the control system will react too late.

Due to diffusion limitation the concentrations measured in the free liquid phase of a fixed film reactor cannot be used for the relevant relationship between the substrate concentration and the growth rate of most of the bacteria as described in Fig. 1. HARREMOES (1982) showed that this relationship can be described by a half order kinetics for fixed film reactors. This means that, e.g., the nitrification capacity of a reactor can be doubled by increasing the DO concentration by a factor of 4. For aeration control this means that the removal capacity of a fixed film reactor can be increased by increased DO concentrations (even $>10 \text{ mg L}^{-1}$) in order to avoid oxygen diffusion limitations. DO control, therefore, will have to be quite different between an activated sludge and an aerated biofilm reactor. Aeration in the aerobic reactor volume of an activated sludge system can be controlled by continuous DO measurements (constant DO $\sim 2 \text{ mg L}^{-1}$). For fixed film reactors substrate concentrations (NH_4-N) in the effluent are suitable control parameters for aeration.

4.3.2.6 Liquid–Solid Separation

Fixed film reactor performance is not directly linked to liquid–solid separation, it is only necessary to retain the solids for receiving water protection. The excess sludge of fixed film reactors normally has good settling properties and is separated from the treated effluent by sedimentation or by simultaneous filtration in biofilm systems. The excess sludge is withdrawn from the treatment system by the backwash process. Monitoring the suspended solids in the effluent can be performed using turbidity meters.

For dispersed growth systems like the activated sludge process liquid–solid separation and thickening in the secondary settling tanks as well as sludge recirculation are part of the treatment process.

Monitoring of the liquid–solid separation process is essential in order to achieve the required treatment efficiency. The following parameters can be used for this purpose:

(1) sludge volume (diluted or stirred), SSV,
(2) sludge volume index (together with microscopic investigation) SVI,
(3) settling velocity,
(4) sludge level in the secondary clarifier,
(5) turbidity in the effluent (Secchi depth, photometric determination).

The performance of secondary clarifiers depends on the sludge volume loading rate q_{sv} [$m^3 \, m^{-2} \, h^{-1}$] (ATV, 1991). Daily determination of VSV gives enough information to detect relevant changes in secondary clarifier loading. The thickening properties of activated sludge are responsible to maintain a stable mass balance for the secondary clarifiers. The thickening properties can be monitored by the sludge volume index (SVI). The dominating influence on SVI comes from filamentous bacteria. Excessive growth of filamentous bacteria causes bulking (SVI $>150 \text{ m g}^{-1}$) and can easily lead to process failure due to sludge losses via the effluent.

Continuous monitoring of the sludge level in the secondary clarifiers can be used for effective control of the mass balance. Under normal flow conditions (dry weather) the sludge level should be at least 2 m below the surface of a clarifier. Regular microscopic investigation of the sludge allows to differentiate between different causes of SVI changes and to decide on the correct prevention strategy.

In rare cases the settling velocity of the sludge and not the thickening properties are responsible for high suspended solid concentrations in the effluent. It can be determined in stirred cylinders and compared with the hydraulic loading of the secondary clarifier in order to check the cause of the problems. Turbidity can be used as a parameter to control suspended solids in the effluent which is of importance to meet effluent standards mainly for BOD and total P. For the early detection of bulking it is not reliable.

4.3.2.7 Special Parameters for Industrial Wastewater Treatment

Depending on the type of industrial production and the wastewater composition additional parameters can be necessary for monitoring as:

(1) adsorbable organic halogenated compounds (AOX),
(2) purgeable organic halogenated compounds (POX),
(3) color (extinction at certain wavelengths),
(4) heavy metals,
(5) polycyclic aromatic hydrocarbons (PAH).

These and many other parameters can be included in the standards for effluent quality, but most of them can only marginally be controlled by biological treatment processes. They have to be controlled at their source or by physical chemical processes like:

(1) granulated activated carbon (GAC) filtration,
(2) addition of powdered activated carbon (PAC),
(3) chemical oxidation (O_3, UV, Fenton),
(4) membrane separation processes,
(5) evaporation, incineration.

Monitoring of these processes has to be adapted to the specific situation and is beyond the scope of this chapter.

4.4 Parameters for Operational Decisions

The preceeding sections have tried to explain the different processes involved in wastewater treatment based on the relationship of the parameters of importance and the process behavior. Tabs. 1 and 2 show a condensed scheme of the processes and the most important parameters for process monitoring. They do not reflect the data requirements for documentation regarding the limit values contained

Tab. 1. Monitoring Parameters for the Operation of Anaerobic Biological Treatment Plants

Process	Parameters	
	Activated Sludge	Fixed Film
C removal (only)		
Treatment efficiency	BOD_5, COD, TOC	BOD_5, COD, TOC
Oxygen supply	DO ~2 mg L^{-1}	TOC, DOC (DO)
Nitrification	T	T
Treatment efficiency	NH_4-N	NH_4-N
Oxygen supply	NH_4-N, (DO)	NH_4-N
Denitrification	T	T
Treatment efficiency	NO_3-N	NO_3-N
Recirculation rate	(NO_3-N), (DO)	(NO_3-N)
(oxygen supply)	rH	DO in recirculation
P removal		
Precipitation	PO_4-P	PO_4-P
Enhanced bio P	PO_4-P	
Liquid–solid separation		
Separation efficiency	SS_e, turbidity	SS_e, turbidity
Mass balance	sludge level (SST), SSV, SVI	

Tab. 2. Monitoring Parameters for the Operation of Anaerobic Biological Treatment of (Industrial) Wastewaters

Process	Parameters
C removal efficiency	COD, TOC, (BOD_5)
Stability of C-removal (inhibition by organic acids)	organic acid concentrations and spectrum, pH, gas composition (CO_2/CH_4 ratio)
Ammonia inhibition	NH_4–N, pH
Sulfide inhibition	Partial pressure of H_2S in the biogas (sulfide concentration in the liquid phase, pH)

in discharge permits. These tables are restricted to the main elements involved in the treatment processes and do not reflect special problems with industrial wastewaters containing hazardous or inhibiting compounds.

Anaerobic biological processes (see Chapter 23, this volume) are mainly used for the removal of organic carbonaceous pollution and only the basic operational problem is process stability. This is reflected in Tab. 2. The background for this table is published in ATV (1994) and KROISS (1985).

5 Quality Control

Monitoring will only be successful if the collected data are correct. Wrong data can easily lead to wrong decisions. Continuous quality control is necessary and comprises a regular check of:

(1) sampling point and procedure (is the sample representative for the process and the purpose?),
(2) sample treatment (does the sample treatment influence on its composition and analysis?),
(3) analytical procedure (is the procedure sensitive enough and are the results correct?),
(4) maintenance and calibration (especially of continuous measuring instruments).

For each treatment plant it is necessary to develop a quality control procedure which is adapted to the specific needs for operation and for documentation. For some of the data the mass balance concept (conservation law for elements) can be used to check the correctness of data. This is very important for the documentation of data needed for at least the following purposes:

(1) proving legal compliance to avoid prosecution,
(2) optimization of cost efficiency; at industrial plants the optimization process will have to include the production processes and pollution prevention at the source,
(3) long-term decisions on future investments (e.g., extension design).

For those parameters which require on-line data for control purposes the aim is to use either two sensors at the same location or the cross check of one information source with other available on-line data using the relationship between them. By this it should at least be possible to distinguish between changes in the process behavior and changes in the data production chain from sampling until the electronic transfer to the operator.

6 References

ATV (1991), *Arbeitslbatt A 131*, Bemessung von einstufigen Belebungsanlagen ab 5,000 Einwohnerwerten. GFA, Hennef.

ATV (1994), Geschwindigkeitsbestimmende Schritte beim anaeroben Abbau von organischen Verbindungen in Abwässern, *Korrespondenz Abwasser* **41**, 101–107.

BARNARD, J. L. (1974), Cut N and P without chemicals, *Water Waste Eng.* **11**, 33–36.

FRANZ, A. (1997), Ein Beitrag zur Beurteilung der Wechselwirkungen zwischen Kläranlagentechnik, -betrieb und Gewässergüte des Vorfluters, der an der Kläranlage entspringt, *Wiener Mitt.* **140** (*Thesis*, Institute for Water Quality and Waste Management, Vienna University of Technology, Austria).

HARREMOËS, P. (1982), Criteria for nitrification in fixed film reactors, *Water Sci. Technol.* **14**, 167–187

KOLLBACH, J., GRÖMPING, M. (1996), *Stickstoffrückbelastung*. Neuruppin: TK-Verlag Karl Thomé-Kozmiensky.

KROISS, H. (1985), Anaerobe Abwasserreinigung, *Wiener Mitt.* **62** (publication of the Institute for Water Quality and Waste Management, Vienna University of Technology, Austria).

KROISS, H. (1994), Vergleichende Betrachtung von Belebungs- und Festbettverfahren für die Biologische Abwasserreinigung, *awt* **4**, 51–57.

MATSCHÉ, N., RUIDER, E. (1982), UV-Absorption, ein aussagekräftiger Parameter zur Erfassung der Restverschmutzung von biologisch gereinigtem Abwasser, *Wiener Mitt.* **49**, 241–260.

MATSCHÉ, N., STUMWÖHRER, K. (1996), UV absorption as control parameter for biological treatment plants, *Water Sci. Technol.* **33**, 211–218.

NOWAK, O., SCHWEIGHOFER, P., NICOLAVCIC, B. (1998), Aspekte zweistufiger Verfahren und von Biofilmverfahren, *Wiener Mitt.* **145**, pp. 411–466 (publication of the Institute for Water Quality and Waste Management, Vienna University of Technology, Austria).

PLISSON-SAUNÉ, S. (1996), Application de la mesure du potential d'électrode de platine en contrôle dynamique et à l'optimisation des stations d'épuration à boues activées de faible charge éliminant l'azote, *These* INSA Toulouse No. 405.

PRENDL, D., KROISS, H. (1998), Bulking sludge prevention by an aerobic selector, *Water Sci. Technol.* **38**, 19–27.

SVARDAL, K. (1991), Anaerobe Abwasserreinigung – Ein Modell zur Berechnung und Darstellung der maßgebenden chemischen Parameter, *Wiener Mitt.* **95** (publication of the Institute for Water Quality and Waste Management, Vienna University of Technology, Austria).

SVARDAL, K., NOWAK, O., SCHWEIGHOFER, P. (1998), Datendokumentation und Auswertung; Plausibilitätsanalyse von Meßwerten, *Wiener Mitt.* **147**, pp. 439–475 (publication of the Institute for Water Quality and Waste Management, Vienna University of Technology, Austria).

WETT, R., ROSTEK, W., RAUCH, W., INGERLE, K. (1998), pH-controlled reject-water treatment, *Water Sci. Technol.* **37**, 165–172.

7 Monitoring of Environmental Processes with Biosensors

MONIKA REISS
WINFRIED HARTMEIER
Aachen, Germany

1 Introduction

Environmental biotechnology is a rapidly growing sector with efforts into many different directions not only comprising removal of unwanted wastes and pollutants but also the production of wanted material using biological principles. In order to be successful and safe, all these fields need careful analytical control and monitoring, and many analytical systems may be used to meet these needs. Development and application of biosensors offer one of the most promising possibilities.

Biosensors are defined as devices having a biological recognition element combined with a suitable transducer, e.g., an electrode, a piezoelectric crystal, a photosensitive element, or a thermistor (Fig. 1). The transducer element can also be used in miniaturized form as a microelectrode or an ion-sensitive field effect transistor (ISFET). The biological component reacts with the analyte and leads to a specific change, e.g., in pH-value, mass, heat, or ion concentration which is registered by the transducer element, converted mostly into an electrical signal, and often transferred to a computer (SCHELLER and SCHUBERT, 1989; HALL, 1995).

The biologically active components such as enzymes, antibodies, nucleic acids, organelles, tissue slices, or whole microbial cells are responsible for the selectivity and sensitivity and also for the range of substances which may be analyzed by biosensors. Due to the large variety of possible analytes, biosensors have found applications in medicine, food and fermentation technology, environmental monitoring, and the defence and security industry. Biosensors for environmental monitoring have been described by NEUJAHR (1984), MATTIASSON (1991), and DENNISON and TURNER (1995).

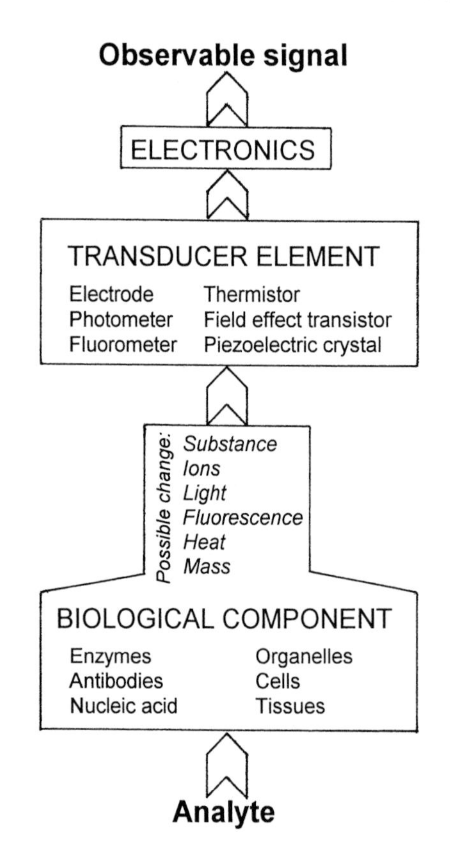

Fig. 1. Scheme of a biosensor.

2 Principles of Monitoring with Biosensors

As can be seen from the examples in Tab. 1, various biological reactions can be used to detect pollutants. There are two different approaches in environmental monitoring: one registers specific substances or, at least, groups of pollutants while the other monitors sum parameters (NEUJAHR, 1984; MATTIASSON, 1991). Immobilized enzymes or other biomolecules with high specificity, e.g., antibodies, nucleic acids, or lectins, are used as biological recognition elements if a high selectivity is required (WENDZINSKI et al., 1997). Thus, even enantioselective analysis is possible (SCHÜGERL et al., 1996).

Increasing complexity of the biological compound often results in a lower selectivity allowing the measurement of sum parameters, as performed by sensors containing mutants of *Salmonella typhimurium* to determine mutagens (KARUBE et al., 1989) or biosensors for BOD measurement and toxicity control (KONG et al., 1996). Microorganisms used in devices for the determination of bioavailable

Tab. 1. Some Approaches to Detect Environmental Pollutants

Mode of Detection	Pollutants Detected	Reference
Enzymatic reaction or microbial metabolism naphthalene	pesticides, nitrate/nitrite, phenols, halogenated hydrocarbons,	RIEDEL et al. (1992); MATTIASSON (1991); SASAKI et al. (1998); NEUJAHR and KJELLEN (1979); EREMENKO et al. (1995); PETER et al. (1996); KÖNIG et al. (1996)
Enzyme inhibition	pesticides, metals	MATTIASSON et al. (1978); MARTY et al. (1993)
Respiration inhibition	toxic materials	IKEBUKURO et al. (1996); SCHOWANEK et al. (1987)
Bacterial luminescence	metals, organic compounds	HYUN et al. (1993); TESCIONE and BELFORT (1993)
Photosynthetic activity	herbicides	VAN HOOF et al. (1992); JOCKERS and SCHMID (1993)
Molecularly imprinted membranes	pesticides	PILETSKY et al. (1995); SCHELLER et al. (1997)
Immunochemistry	herbicides, polychlorinated biphenyls	KARUBE (1988); CHEMNITIUS et al. (1996)

organic carbon or biological oxygen demand (BOD) should have a broad spectrum of substrates. In those cases the bacterium *Bacillus cereus* (SUN and LIU, 1992) or the yeasts *Trichosporon cutaneum* (HIKUMA et al., 1979; YANG et al., 1996) and *Rhodotorula mucilaginosa* (NEUDÖRFER and MEYER-REIL, 1997) are often used.

Specific monitoring implies two kinds of analytes. One group consists of synthetic organic compounds derived from industrial wastes, biocides, and detergents. Phenols belong to this group, but many phenols also occur naturally, arising mostly from the lignins of decaying wood. The second group consists of compounds which normally occur in nature and may be – at certain concentrations – essential for biochemical processes. This group includes nitrite, nitrate, sulfate, phosphate, and even some heavy metals. However, when such compounds exceed a certain concentration they become an environmental risk. In the following survey examples of biosensor detection are given for both groups.

3 Monitoring of Inorganic Substances

3.1 Heavy Metals

Heavy metals are ubiquitous throughout the environment and are well known for their toxicity, polluting air, water, and food. Thermometric measurements of heavy metals were performed with immobilized urease in an enzyme thermistor for the determination of mercury and copper concentrations in water solutions (MATTIASSON et al., 1978). The inhibition of urease was expressed as the ratio of the temperature peaks as the result of urea addition obtained after and before the presence of the inhibiting metal ions. In another device, ÖGREN and JOHANSSON (1978) also used immobilized urease but in a reactor connected with an ammonia-sensitive electrode. Thus, mercury could be determined in the range of 0 to 0.7 nmol. Only silver and copper led to interferences.

Heavy metal biosensors using immobilized oxidases and dehydrogenases have been developed based on the degree of enzyme inactivation for the detection of mercury, zinc, silver, cadmium, and copper salts (GAYET et al., 1993). These make use of immobilized enzymes coupled with an oxygen electrode to register the decline in oxygen levels which results from the incubation with the substrate. Addition of heavy metals inhibits these enzymes and results in lower consumption rates of oxygen. Another enzyme used to detect heavy metals is β-glucosidase. This has been covalently coupled to Eupergit C (Röhm, Weiterstadt), filled in a glass column, and used in a flow injection system for mercury determination (REISS et al., 1993).

Detection via bacterial luminescence has been realized in a bioluminescent sensor for the detection of mercury constructed with *Escherichia coli*, genetically engineered with a mercury (II)-sensitive promoter and the *lux* genes from *Vibrio fischeri* (TESCIONE and BELFORT, 1993). The gene product of the *mer* operon, inactive because the *lux* genes have been fused into the middle of the operon, detoxifies the mercuric ion by reducing it to Hg^0. The mercuric ions activate gene transcription producing luciferase and cell luminescence is observed, the level of which can be related to the concentration of mercury. The sensitivity of the minisensor to the mercuric ions ranged from 10 nM to 4 µM, when the system was aerated.

3.2 Nitrate and Nitrite

The toxicity of nitrite is based on the formation of methemoglobin and of carcinogenic N-nitroso compounds by the reaction between nitrite and secondary or tertiary amines and amides. Nitrate as such is not toxic but it is reduced to the toxic nitrite by several microorganisms. The detection of nitrite and nitrate can be performed using an electrochemical assay with nitrate and nitrite reductase combined with the reduced form of methylviologen as an electron donor. A column with immobilized nitrite reductase serves for nitrite assay, a second column with immobilized nitrate and nitrite reductase indicates the amount of nitrate plus nitrite (Fig. 2). The nitrate concentration can be calculated by subtracting the value of the first column from the value of the second column (KIANG et al., 1978). By the use of an amperometric measurement in combination with a mediator nitrite could be determined with a detection limit of 0.01 mg L^{-1} (SASAKI et al., 1998). A microsensor for nitrate based on denitrifying bacteria has recently been described by LARSEN et al. (1996).

3.3 Cyanide

Immobilized rhodanase and injectase can be applied in an enzyme thermistor for cyanide detection. The heat signal, generated by the conversion of cyanide catalyzed by the immobilized enzymes, leads to linear relationships of 20–600 µM cyanide with injectase and 20–1,000 µM with rhodanase (MATTIASSON et al., 1977). Besides thermistors, electrochemical sensors are also in laboratory use for cyanide detection together with rhodanase as biological component (GROOM and LUONG, 1991).

3.4 Sulfite and Sulfide

A sulfite ion sensor has been built using immobilized organelles and an oxygen electrode (KARUBE et al., 1983). In this device a linear relationship between the steady state current and the sulfite ion concentration was obtained up to 340 µM. The time necessary for determination was about 10 min.

A photomicrobial electrode for selective determination of sulfide has been tested by MATSUNAGA et al. (1984). They used the photosynthetic bacterium *Chromatium* sp. on a hydrogen-sensitive electrode. This led to a sensor with a response time between 5 and 10 min and a linear range up to concentrations of 3.5 mM. The minimum detectable sulfide concentration was 0.5 mM.

Sulfide could be monitored with an oxygen electrode after fixing of *Thiobacillus thiooxidans* cells by a gas permeable membrane onto the probe (KUROSAWA et al., 1994). The response had a standard deviation of 2.5% when

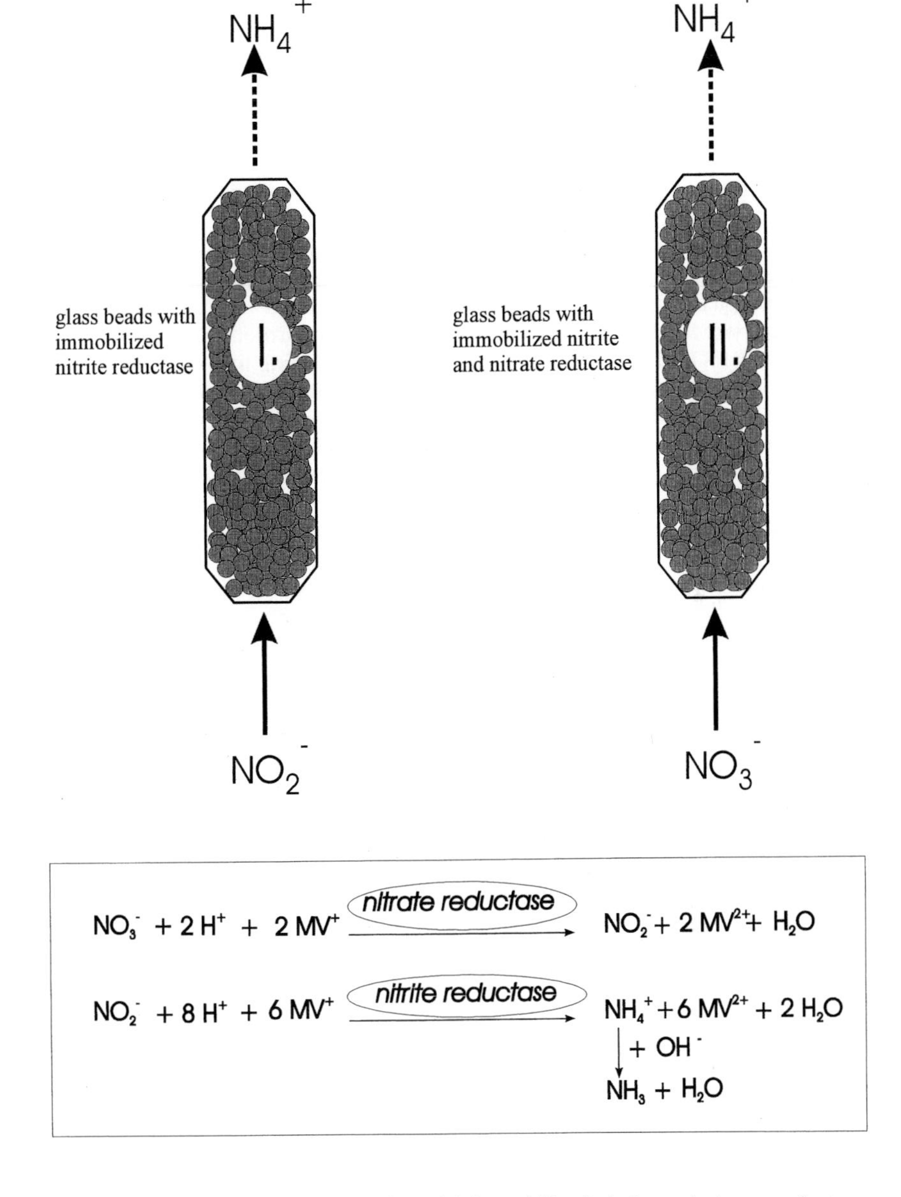

Fig. 2. Nitrite and nitrate monitoring with immobilized nitrite and nitrate reductase and ammonium detection (MV: methyl viologen).

the sample contained 200 μM sulfide. A linear relationship between sensor signal and concentration of sodium sulfide resulted below 400 μM. The stability of the sensor was satisfactory and it could be applied for the determination of sulfide in springwater.

4 Monitoring of Organic Substances

4.1 Phenolic Compounds

Environmental pollution by phenolic substances is widespread, because of their intensive use as industrial chemicals, e.g., for the production of polymeric resins or pesticides. Many biological compounds have been tested for the construction of phenol biosensors. Microbial sensors have been developed with *Trichosporon cutaneum* by NEUJAHR and KJELLEN (1979) with a linear range of 0–15 mg L^{-1} phenol, and with two different strains of *Pseudomonas putida* by IGNATOV and KOZEL (1995), where linear calibration curves of 2–11.7 µM phenol and 17–34 µM *p*-nitro-phenol resulted. The combination of an *Azotobacter* species with an oxygen electrode was used for the determination of phenol and hydroquinones, whereby linearity was shown between 2 and 16 mg L^{-1} (REISS et al., 1995). ORUPOLD et al. (1996) investigated two genetically engineered *Pseudomonas putida* strains. A linear relationship between sensor response and phenol concentrations up to 0.1 or 0.2 mM, respectively, was observed; a sensor response was also obtained by catechol, resorcinol, and 2,4-dinitrophenol. A biosensor for 2-ethoxyphenol constructed with *Rhodococcus rhodochrous* and an oxygen electrode also gave a linear range of detection between 0.05 and 0.4 mM (BEYERSDORF-RADECK et al., 1994).

Many enzymes which catalyze the oxidation of phenolic substances exist in nature, e.g., in organisms that produce polyphenolic compounds such as lignins or other secondary metabolites such as alkaloids or tannins, and these, therefore, can be utilized in the construction of enzymatic phenol sensors. Tyrosinase, e.g., catalyzes the *ortho*-hydroxylation of phenol to produce catechol, and the dehydrogenation of catechol to *o*-quinone. This enzyme has a broad substrate specificity (PETER and WOLLENBERGER, 1997). In addition to phenol and catechol, it can also catalyze the oxidation of *para*-substituted phenols. Tyrosinase has been used in combination with an oxygen electrode. However, lower detection limits can be achieved by amperometric detection of the biocatalytically generated quinone products, e.g., 0.02 µM by DENG et al. (1996) or 30 ppb by a gas phase sensing tyrosinase sensor of DENNISON et al. (1995).

Amperometric detection can also be performed via redox mediators, such as ferrocyanide (SCHILLER et al., 1978), tetracyanoquinodimethane (KULYS and SCHMID, 1990), or 5-methyl-phenazonium (NMP^+). In the presence of the reduced mediator, the enzymatically formed quinone is converted into diphenol leading to a signal amplification (Fig. 3). Using tyrosinase with NMP^+ a detection limit for phenol of 0.25 nM can be obtained (KOTTE et al., 1995). Tyrosinase was also applied in organic solvents. Operating a biosensor in an organic solvent simplifies the immobilization procedure since the enzyme, insoluble in the organic phase, remains in the aqueous phase and is retained by a hydrophilic support. A tyrosinase electrode for the determination of phenols in chloroform was developed by HALL et al. (1988).

Laccase oxidizes a wide variety of phenols and aromatic amines. In combination with a fiber-optic luminescent oxygen sensor, laccase can be used for sensing of polyphenols, producing a good linearity within the range of

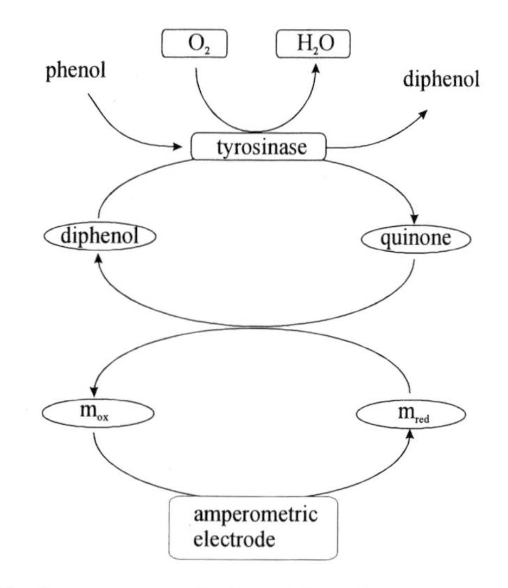

Fig. 3. Amperometric phenol detection using mediators.

0–5 mM (PAPKOVSKY et al., 1993). Laccase has also been used in a bi-enzyme electrode, where signal amplification is performed by shuttling the analyte between two enzymes acting in a reaction cycle (EREMENKO et al., 1995). Diphenols are oxidized by laccase to generate quinones, the second enzyme, glucose dehydrogenase (GDH), reduces these quinones in the presence of glucose, thereby regenerating the original diphenols. These can be oxidized by laccase again leading to a signal amplification. The consumption of oxygen or the pH-drift produced by the GDH reaction can be measured (Fig. 4). With this amplification EREMENKO et al. (1995) measured detection limits of 1 mM for norepinephrine and *p*-aminophenol.

Another enzyme used for phenol detection, phenol hydroxylase, catalyzes the oxidation of phenol to the corresponding diphenol. NADPH is used as cofactor in this case. The maximum rate of oxygen consumption shows a linear dependency on phenol concentration over a range of 0.5–50 μM (KJELLEN and NEUJAHR, 1980). Without added phenol this enzyme acts as an NADPH oxidase with H_2O_2 as the product of O_2 reduction. Since both oxidation of NADPH and the hydroxylation reactions occur concurrently, oxygen consumption is a result of both phenol and NADPH oxidation. Because NADPH diffuses slower than oxygen and phenol into the immobilization matrix of the enzyme electrode, the biosensor has to be incubated in a buffer solution with NADPH before phenol is added. Phenol hydroxylase also reacts with several mono-substituted phenols carrying halogen, hydroxyl, or methyl groups, but not carboxyl groups (NEUJAHR and GAAL, 1973). In general, the activity towards these is considerably lower compared with that towards unsubstituted phenol.

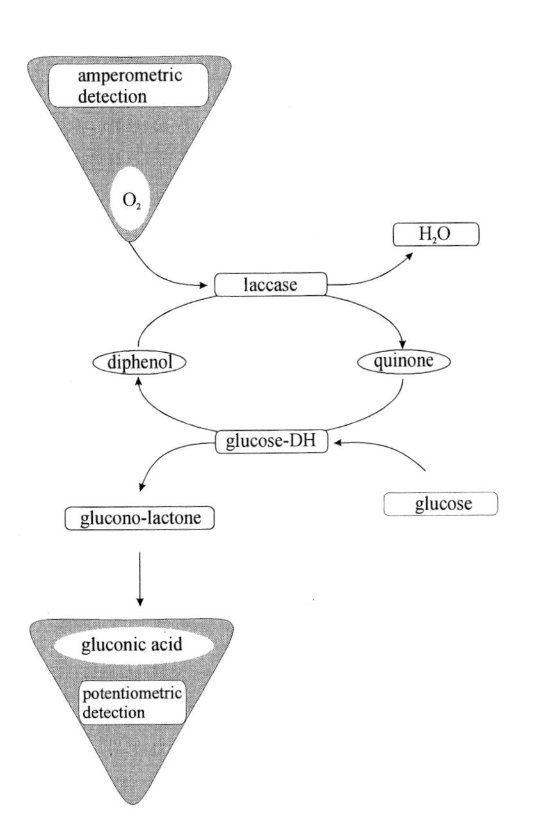

Fig. 4. Principle of enzymatic signal amplification for the detection of di- or polyphenols.

Tissue material for use in a biosensor has the advantage of good stability and activity associated with the presence of large amounts of the enzymes in their natural environment. The first tissue biosensors were developed with plant or animal tissues placed on the top of the electrochemical transducer, which show longer response times than biosensors in which the tissue is directly incorporated into a carbon paste matrix (WIJESURIYA and RECHNITZ, 1993). These mixed tissue-carbon paste electrodes have been shown to be useful for the detection of analytes both in static and flow systems. They combine a fast response with a good reusability, are easy to miniaturize, and can be used in organic phases. Mushroom or banana tissues were spread in a thin layer onto a rough graphite disk for the determination of phenolic compounds in chloroform. Reaction times of only a few seconds and a detection limit of $3 \cdot 10^{-5}$ M were reached for *p*-chlorophenol detection (WANG et al., 1992).

4.2 Pesticides

Due to industrial manufacturing processes and intensive agriculture there is an increasing number of pesticides in the environment. Various families of pesticides, e.g., fungicides, herbicides, and insecticides, can be determined by enzymes. For the identification and quantification of pesticides enzyme inhibition can be used for analytical methods with a high selectivity and sensitivity. Cholinesterases (acetyl or butyryl cholinesterase) play an important role in the detection of pesticides (DUMSCHAT et al., 1991; SCHWEDT and HAUCK, 1988; MARTY et al., 1993). Organophosphorous and carbamate insecticides act as inhibitors of cholinesterases thus leading to a decreased pH change after addition of the corresponding choline ester. The inhibition by various pesticides differs strongly and is, in the presence of a mixture of different pesticides, not proportional to the sum of pesticides (HERZSPRUNG et al., 1989). Cholinesterase biosensors can be used for a screening test to establish the presence of cholinesterase inhibitors and indicate a more detailed investigation. Acetylcholinesterase has also been used in organic solvents for the detection of pesticides (MIONETTO et al., 1994).

Several biosensors for the detection of organophosphorous and carbamate insecticides have been developed with potentiometric (TRAN-MINH, 1993), piezoelectric (GUILBAULT and NGEH-NGWAINBI, 1988), semiconductor (DUMSCHAT et al., 1991), or amperometric transducers (SKLADAL and MASCINI, 1992; MARTY et al., 1993; WOLLENBERGER et al., 1994). Two amperometric approaches were studied depending on the substrate being utilized. When acetylcholine was used as a substrate, choline esterase was co-immobilized with choline oxidase on a hydrogen peroxide measuring platinum anode (MARTY et al., 1992). Choline, as product of acetylcholinesterase, is oxidized by choline oxidase consuming oxygen and producing hydrogen peroxide. Acetylthiocholine as substrate is hydrolyzed by acetylcholine esterase to acetate and thiocholine, which can be oxidized electrochemically. Since a large overvoltage is necessary for thiocholine oxidation with a platinum anode, carbon paste electrodes with mediators such as tetracyano-*p*-quinodimethane (TCNQ), tetrathiafulvalene, 1,1′-dimethylferrocene, or cobalt phthalocyanine were developed (Fig. 5). In this way, sensitive and rapidly responding sensors were constructed, being more stable and cheaper than systems with choline oxidase (SKLADAL and MASCINI, 1992).

Many herbicides used in agriculture affect photosynthetic electron transport chains, which can be monitored with cyanobacteria and a suitable chemical mediator, e.g., potassium ferricyanide. The mediator is reduced by the photosynthetic activity of the cyanobacteria and subsequently reoxidized at a working electrode. By disturbing the photosynthetic electron transport within the cells, a decreased mediator reduction can be observed and rapid detection of concentrations of 50 µg L^{-1} is possible for linuron and atrazine (VAN HOOF

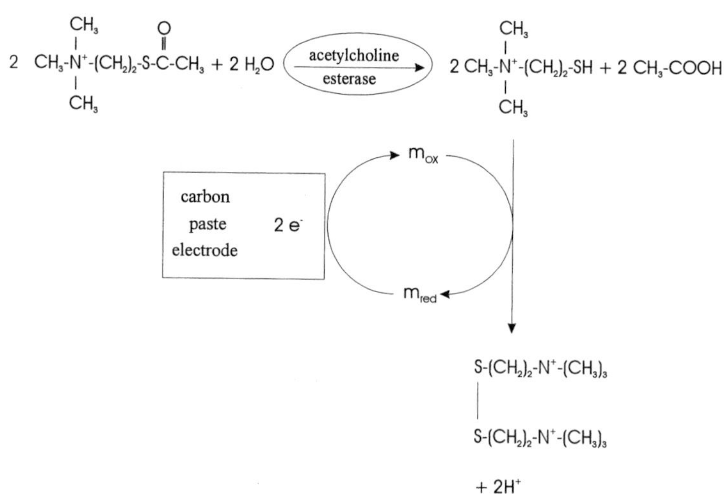

Fig. 5. Amperometric approach to determine choline esterase inhibitors.

et al., 1992). Bacterial luciferase combined, e.g., with the photoreaction center from phototrophic bacteria can be used for sensitive detection of herbicides (JOCKERS and SCHMID, 1993).

In addition to enzyme inhibition, it is also possible to use the inhibition of the respiration of microbial cells for environmental analysis. IKEBUKURO et al. (1996) developed a reactor employing immobilized yeast to measure cyanide. In the presence of cyanide, the yeast's respiration is inhibited and a decrease in oxygen consumption is consequently observed. Microbial sensors of this type are simple, rapid, and inexpensive biosensors for the measurement of cyanide in river water. They could be affected by other toxic compounds in river water, e.g., pesticides, but they were not affected by metal ions. The influence of other compounds led to another application of devices with immobilized microbial cells, the detection of toxicity, where the differences in oxygen consumption were used to quantify inhibiting substances in wastewater. These applications will be discussed below.

Environmental substances can inhibit enzyme activity or microbial respiration, thus leading to sensitive detection methods, but enzymatic or microbial metabolism of these substances also offers a great potential for the detection of environmental compounds (RIEDEL et al., 1992; MATTIASSON, 1991). KAUFFMANN and MANDELBAUM (1996) investigated the entrapment of cell-free crude extracts of a *Pseudomonas* strain in sol–gel glass which could rapidly degrade the herbicide atrazine. Industrial effluents from atrazine manufacturing plants could be treated by *Pseudomonas* and could be analyzed by a biosensor containing this strain.

Biosensors can be constructed by mimicking biological receptors through an imprinting procedure. Polymeric membranes containing molecular recognition sites for atrazine have been prepared by molecular imprinting technology (PILETSKY et al., 1995). The membrane is formed by radical polymerization of diethyl aminoethyl methacrylate and ethylene glycol dimethacrylate in the presence of atrazine as template. When the template molecules are removed, the recognition of atrazine in drinking water, food, or environment can be performed by a combination of reversible binding and shape complementarity. This technique has also been used for the detection of D- and L-amino acids, cholesterol, ephedrine, diazepam, and morphine (SCHELLER et al., 1997).

When dealing with very small amounts of analyte, as is often the case with rests of environmental pollutants, immunochemistry can be applied to increase the sensitivity of detection (KARUBE, 1988). Immuno-FIA(flow-injection-analysis)systems have the advantage of achieving low detection limits, use small sample volumes, and there is no necessity to pretreat samples (CHEMNITIUS et al., 1996). Using a column with immobilized monoclonal antibodies against atrazine and atrazine labeled with alkaline phosphatase, atrazine can be detected under continuous flow conditions in the range of 9–180 $\mu g\ L^{-1}$ (VIANELLO et al., 1998).

4.3 Halogenated Organics

Cells of *Rhodococcus* sp. contain the enzyme alkyl-halidohydrolase, which transforms appropriate halogenated hydrocarbons into the corresponding alcohols and halogen ions. Alginate-entrapped cells, placed on ion-sensitive potentiometric electrodes for both chloride and bromide ions, lead to a compact biosensor for halogenated hydrocarbons in water samples with detection limits for ethylene bromide and 1-chloro butane of 0.04 and 0.22 $mg\ L^{-1}$, respectively (PETER et al., 1996).

An amperometric biosensor for the determination of benzoate and 3-chlorobenzoate using *Pseudomonas putida* was developed by RIEDEL et al. (1991). A further development of the same group (RIEDEL et al., 1995) led to an amperometric biosensor for phenol and chlorophenols with *Trichosporon beigelii* (identical with *T. cutaneum*) as biological component. This sensor has much more sensitivity for chlorophenols than for phenols and is insensitive to benzoate. The detection limit is 2 μM, and the linear range for 4-chlorophenol is observed up to 40 μM.

In a system developed by ZHAO et al. (1995) partially purified polyclonal anti-PCB antibodies (ABs), in combination with a quartz fiber, were used to detect polychlorinated biphenyls (PCBs). PCBs are very stable compounds with paired phenyl rings and various degrees of chlorination. The fiber, coated with AB, binds the fluorescein conjugate of 2,4,5-trichlorophenoxybutyrate and the level of fluorescence can be monitored. PCBs compete with these conjugates for binding and a decreased fluorescence is observed.

4.4 Naphthalene

For quantification of naphthalene in aqueous solutions immobilized *Sphingomonas* sp. or *Pseudomonas fluorescens* cells have been used by KÖNIG et al. (1996). An oxygen electrode was first covered with a gas-permeable membrane on which the microorganisms were immobilized and then with a capillary membrane. Both *Sphingomonas* sp. and *P. fluorescens* express the same behavior in the detection of naphthalene in aqueous solutions. In a flow-through system a linear range between 0.03 and 2.0 mg L^{-1} and a response time of 2 min can be reached. *Pseudomonas fluorescens* genetically supplemented with the luciferase gene has also been applied for naphthalene detection by KING et al. (1991) using a photomultiplier as transducer.

5 Monitoring of Sum Parameters

5.1 Biological Oxygen Demand

Due to the time-consuming classical determination method for biological oxygen demand (BOD) it is almost impossible to control wastewater plants. A dramatic reduction in time, from five days to several minutes, is possible using microbial BOD-sensors. A linear relationship between the sensor response and BOD_5-value is given up to maximum BOD-values of 10 to 300 mg L^{-1}, depending on the microorganisms used. Since BOD-biosensors should detect a broad spectrum of substrates, activated sludges obtained from wastewater treatment plants (KARUBE et al., 1977) or mixed cultures of *Bacillus* strains were used (GALINDO et al., 1992; LI et al., 1994), however, the reproducibility was quite poor. This aspect was improved by BOD-biosensors with pure cultures of yeasts or bacteria. Sensors were constructed with *Bacillus cereus* (SUN and LIU, 1992), *Clostridium butyricum* (KARUBE et al., 1977), *Torulopsis candida* (SANGEETHA et al., 1996), *Hansenula anomala* (KULYS and KADZIAUSKIENE, 1980), and *Trichosporon cutaneum* (HIKUMA et al., 1979; YANG et al., 1996). The BOD-module produced by the Prüfgerätewerk Medingen (Germany) using the yeast *Trichosporon cutaneum*, has a measuring range of 2 to 22 mg L^{-1} BOD. Samples with a higher concentration are analyzed after dilution, which is suggested by the module (SZWEDA and RENNEBERG, 1994).

Most BOD-sensors employing stationary state measurements exhibit response times of 15 to 20 min. By the use of the kinetic method, a rapid determination, taking between 15 to 30 s, can be achieved (RIEDEL et al., 1989). Even microorganisms with a broad spectrum of substrates cannot metabolize all compounds in wastewater. One possibility to improve BOD-measurements is the incubation of the sensors in the respective wastewater (RIEDEL et al., 1990). Another possibility is the co-immobilization of the corresponding enzymes. The yeast *Lipomyces kononenkoae* can be used for BOD-determination of lactose containing wastewater after co-immobilization of β-galactosidase (HARTMEIER et al., 1993). The immobilization of the microorganisms often introduces diffusion barriers so that biosensors cannot sufficiently register macromolecular substances. The rapidly increasing response time of the sensor would make continuous measurements impossible. Therefore, for continuous BOD-measurements of polymer containing liquids, it is advantageous to break down the polymers (e.g., starch, cellulose, pectin) to mono- and oligomers, which are more easily metabolized by microorganisms and thus detectable by a biosensor. One method using this approach is shown in Fig. 6, where a BOD-device is combined with enzyme columns containing immobilized α-amylase and glucoamylase for improved BOD-determinations of starch containing wastewaters.

In most cases BOD-sensors measure the decrease of dissolved oxygen caused by the

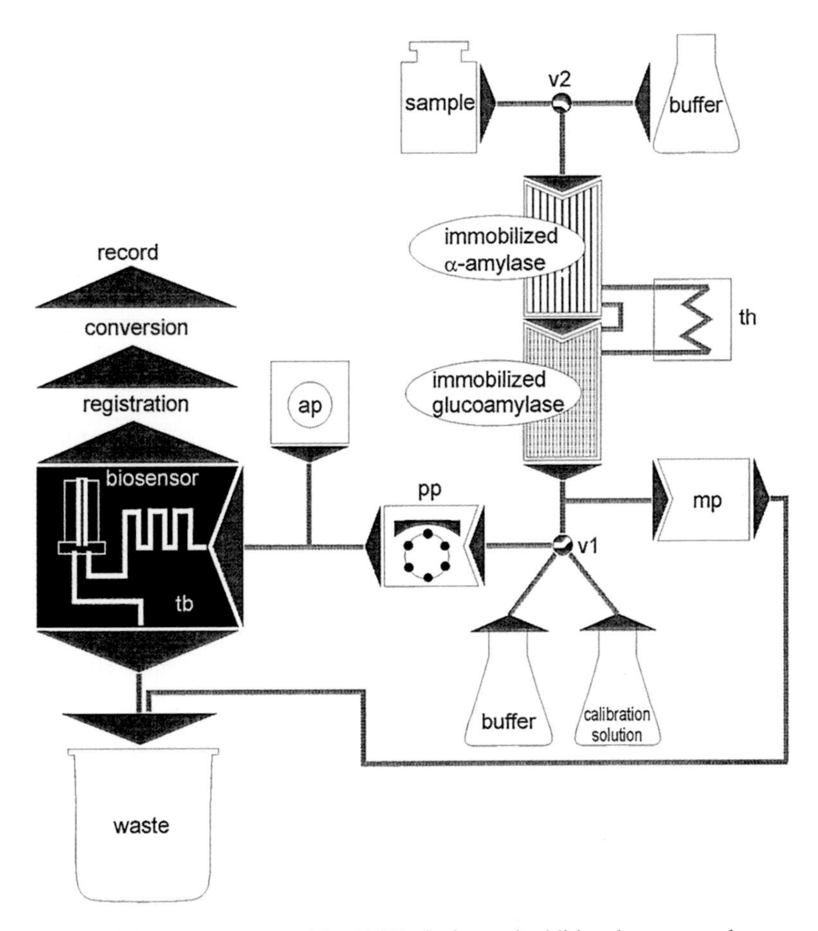

Fig. 6. BOD-measurement with a BOD-device and additional enzyme columns; ap: air pump, mp: membrane pump, pp: peristaltic pump, tb: thermoblock, th: thermostate, v1, v2: valves.

oxygen consumption of the microorganisms, but BOD-sensors based on bacterial luminescence can also be obtained. Luminous bacteria emit light as a result of normal metabolism. By the metabolism of organic substances in wastewater the level of light emission by luminous bacteria increases. With *Photobacterium phosphoreum*, a photo diode, and an amplifier good agreement can be achieved between BOD_5-values and the biosensor measurements (HYUN et al., 1993).

5.2 Toxicity

By measuring the consumption of oxygen it is also possible to register inhibiting substances in wastewater. If inhibiting substances are present in the wastewater, a reduction of the metabolic activity of the microorganisms used in the biosensor is observed leading to a reduction of oxygen consumption. These sensors can be used for toxicity control in plants treating industrial effluents. Toxicity detection based on acetic acid catabolism in a stirred and aerated reactor filled with activated sludge was developed by SCHOWANEK et al. (1987). Automated toxicity control is possible giving an alarm signal if the oxygen consumption rate decreases below a defined level. Another respirometer based on luminescence quenching by molecular oxygen was developed by WONG et al. (1997). The toxicity of several heavy metals on activated sludge was tested using the respirometer.

Metabolic ultraviolet spectroscopy response of bacterial cells immobilized in an agarose membrane was used for the detection of toxic compounds in water by BAINS (1994). The UV absorbance of the membrane was measured at 200 nm and compared to a membrane without cells. The UV spectra of the cells changed in presence of toxins as a result of a reduction of energy metabolism within 15 s exposure to the toxin.

6 Perspectives

Continuous and *in situ* monitoring technologies have a high priority for industries and for government agencies responsible for environmental monitoring (ROGERS, 1995). Driven by the need for fast and cost-effective methods for environmental monitoring, many field analytical methods have been developed. Among these, biosensors are a highly interesting sector. They principally have the potential to meet many of the analytical and monitoring requirements. Research and development in the last decades led to the construction of a broad spectrum of biosensors. However, most of these developments have only been tested under well-controlled and mild laboratory conditions. Consequently, until now, only very few biosensors are really commercially produced and applied on a large scale.

Major problems arise from the biological material in biosensors. This organic material is by nature not stable so that the desire for sensors with long-time stability cannot be sufficiently met in most cases. Possibilities to improve the stability are still limited but under progress with techniques of protein engineering and of enzyme crystallization and intramolecular cross-linking (FERNANDEZ-LAFUENTE et al., 1995) Self-regenerating whole cell systems can overcome stability problems to a certain extent, but those systems are only thinkable for very few applications and even then a change of the biochemical properties within weeks and months cannot be avoided.

The diversity of wastes and environmental pollutants and the often small market associated with the detection of one single analyte implies that the most competitive biosensors can be constructed for the detection of a variety of analytes. If such a multifunctionality of sensors is desired, miniaturization of biosensors is an attractive goal. Thin-film electrodes, ion-sensitive field effect transistors (ISFETs), and gas-sensitive metal oxide semiconductors (MOS) play an important role within these developments, since these elements can be made by microelectronic production technologies. Small devices make it possible to include sensors for different analytes in the same sensor so that one exposure of the sensing device gives informations on different analytes. Current developments in this field are very promising.

7 References

BAINS, W. (1994), A spectroscopically interrogated flow-through type toxicity biosensor, *Biosens. Bioelectron.* **9**, 111–117.

BEYERSDORF-RADECK, B., KARLSON, U., SCHMID, R. D. (1994), A microbial sensor for 2-ethoxyphenol, *Anal. Lett.* **27**, 285–298.

CHEMNITIUS, G., MEUSEL, M., ZABOROSCH, C., KNOLL, M., SPENER, F., CAMMAN, K. (1996), Highly sensitive electrochemical biosensors for water monitoring, *Food Technol. Biotechnol.* **34**, 23–29.

DENG, Q., GUO, Y., DONG, S. (1996), Cryo-hydrogel for the construction of a tyrosinase-based biosensor, *Anal. Chim. Acta* **319**, 71–77.

DENNISON, M. J., TURNER, A. P. F. (1995), Biosensors for environmental monitoring, *Biotechnol. Adv.* **13**, 1–12.

DENNISON, M. J., HALL, J. M., TURNER, A. P. F. (1995), Gas-phase microbiosensor for monitoring phenol vapor at ppb levels, *Anal. Chem.* **67**, 3922–3927.

DUMSCHAT, C., MÜLLER, H., STEIN, K., SCHWEDT, G. (1991), Pesticide-sensitive ISFET based on enzyme inhibition, *Anal. Chim. Acta* **252**, 7–9.

EREMENKO, A. V., MAKOWER, A., SCHELLER, F. W. (1995), Measurement of nanomolar diphenols by substrate recycling coupled to a pH-sensitive electrode, *Fresenius J. Anal. Chem.* **351**, 729–731.

FERNANDEZ-LAFUENTE, R., ROSELL, C. M., RODRIGUEZ, V., GUISAN, J. M. (1995), Strategies for enzyme stabilization by intramolecular crosslinking with bifunctional reagents, *Enzyme Microb. Technol.* **17**, 517–523.

GALINDO, E., GARCIA, J. L., TORRSE, L. G., QUINTERO, R. (1992), Characterization of microbial membranes used for the estimation of biochemical oxygen demand with a biosensor, *Biotechnol. Techniques* **6**, 399–404.

GAYET, J.-C., HAOUZ, A., GELOSO-MEYER, A., BURSTEIN, C. (1993), Detection of heavy metal salts with biosensors built with an oxygen electrode coupled to various immobilized oxidases and dehydrogenases, *Biosens. Bioelectron.* **8**, 177–183.

GROOM, C. A., LUONG, J. H. T. (1991), A flow through analysis biosensor system for cyanide, *J. Biotechnol.* **21**, 161–172.

GUILBAULT, G. G., NGEH-NGWAINBI, J. (1988), Use of protein coatings on piezoelectric crystals for assay of gaseous pollutants, *Biotec* **2**, 17–22.

HALL, E. A. H. (1995), *Biosensoren*. Berlin: Springer-Verlag.

HALL, G. F., BEST, D. J., TURNER, A. P. F. (1988), Amperometric enzyme electrode for the determination of phenols in chloroform, *Enzyme Microb. Technol.* **10**, 543–546.

HARTMEIER, W., TARI, A., REISS, M. (1993), BOD-biosensor basing on *Lipomyces* cells and β-galactosidase from mold origin, *Med. Fac. Landbouww. Univ. Gent* **58**, 1799–1801.

HERZSPRUNG, P., WEIL, L., QUENTIN, K.-E. (1989), Bestimmung von Phosphorpestiziden und insektiziden Carbamaten mittels Cholinesterasehemmung, *Z. Wasser-Abwasser-Forsch.* **22**, 67–72.

HIKUMA, M., SUZUKI, H., YASUDA, T., KARUBE, I., SUZUKI, S. (1979), Amperometric estimation of BOD by using living immobilized yeast, *Eur. J. Appl. Microbiol. Biotechnol.* **8**, 289–297.

HYUN, C.-K., TAMIYA, E., TAKEUCHI, T., KARUBE, I. (1993), A novel BOD sensor based on bacterial luminescence, *Biotechnol. Bioeng.* **41**, 1107–1111.

IGNATOV, O. V., KOZEL, A. B. (1995), The determination of aromatic compounds by microbial biosensor. in: *Environmental Biotechnology: Principles and Applications*, (MOO-YOUNG, M., Ed.), pp. 656–674. Dordrecht: Kluwer Academic Publishers.

IKEBUKURO, K., MIYATA, A., CHO, S. J., NOMURA, Y., CHANG, S. M. et al. (1996), Microbial cyanide sensor for monitoring river water, *J. Biotechnol.* **48**, 73–80.

JOCKERS, R., SCHMID, R. D. (1993), Detection of herbicides via a bacterial photoreaction centre and bacterial luciferase, *Biosens. Bioelectron.* **8**, 281–289.

KARUBE, I. (1988), Novel immunosensor systems, *Biotec* **2**, 37–47.

KARUBE, I., MATSUNAGA, T., MITSUDA, S., SUZUKI, S. (1977), Microbial electrode BOD sensors, *Biotechnol. Bioeng.* **19**, 1535–1547.

KARUBE, I., SOGABE, S., MATSUNAGA, T., SUZUKI, S. (1983), Sulfite ion sensor with use of immobilized organelle, *Eur. J. Appl. Microbiol. Biotechnol.* **17**, 216–220.

KARUBE, I., SODE, K., SUZUKI, M., NAKAHARA, T. (1989), Microbial sensor for preliminary screening of mutagens utilizing a phage induction test, *Anal. Chem.* **61**, 2388–2391.

KAUFFMANN, C. G., MANDELBAUM, R. T. (1996), Entrapment of atrazine-degrading enzymes in sol–gel glass, *J. Biotechnol.* **51**, 219–225.

KIANG, C.-H., KUAN, S. S., GUILBAULT, G. G. (1978), Enzymatic determination of nitrate: electrochemical detection after reduction with nitrate reductase and nitrite reductase, *Anal. Chem.* **50**, 1319–1322.

KING, P. M., DIGRAZIA, P. M., MITTELMAN, M. W., WHITE, D. C., SAYLER, G. S. (1991), Bioluminescent reporter technology for *in situ* monitoring of bacterial populations, *Proc. Int. Symp. Environmental Biotechnology*, pp. 61–63. Ostend: Koninglijke Vlaamse Ingenieursvereniging.

KJELLEN, K. G., NEUJAHR, H. Y. (1980), Enzyme electrode for phenol, *Biotechnol. Bioeng.* **22**, 299–310.

KONG, Z., VAEREWIJCK, M., VERSTRAETE, W. (1996), On-line BOD measurement and toxicity control of waste waters with a respirographic biosensor, *Environ. Technol.* **17**, 399–406.

KÖNIG, A., ZABOROSCH, C., MUSCAT, A., VORLOP, K.-D., SPENER, F. (1996), Microbial sensors for naphthalene using *Sphingomonas* sp. B1 or *Pseudomonas fluorescens* WW4, *Appl. Microbiol. Biotechnol.* **45**, 844–850.

KOTTE, H., GRÜNDIG, B., VORLOP, K.-D., STREHLITZ, B., STOTTMEISTER, U. (1995), Methylphenazonium-modified enzyme sensor based on polymer thick films for subnanomolar detection of phenols, *Anal. Chem.* **67**, 65–70.

KULYS, J., KADZIAUSKIENE, K. (1980), Yeast BOD sensor, *Biotechnol. Bioeng.* **22**, 221–226.

KULYS, J., SCHMID, R. D. (1990), A sensitive enzyme electrode for phenol monitoring, *Anal. Lett.* **23**, 589–597.

KUROSAWA, H., HIRANO, T., NAKAMURA, K., AMANO, Y. (1994), Microbial sensor for selective determination of sulphide, *Appl. Microbiol. Biotechnol.* **41**, 556–559.

LARSEN, L. H., BEVSBECH, N. P., BINNERUP, S. J. (1996), A microsensor for nitrate based on immobilized denitrifying bacteria, *Appl. Environ. Microbiol.* **62**, 1248–1252.

LI, F., TAN, T. C., LEE, Y. K. (1994), Effects of pre-conditioning and microbial composition on the sensing efficacy of a BOD biosensor, *Biosens. Bioelectron.* **9**, 197–205.

MARTY, J.-L., MIONETTO, N., ROUILLON, R. (1992), Entrapped enzymes in photocrosslinkable gel for enzyme electrodes, *Anal. Lett.* **25**, 1389–1398.

MARTY, J.-L., MIONETTO, N., NOGUER, T., ORTEGA, F., ROUX, C. (1993), Enzyme sensor for the detection of pesticides, *Biosens. Bioelectron.* **8**, 273–280.

MATSUNAGA, T., TOMODA, R., MATSUDA, H. (1984), Photomicrobial electrode for selective determination of sulphide, *Appl. Microbiol. Biotechnol.* **19**, 404–408.

MATTIASSON, B. (1991), Biosensor technology for environmental analysis, *Biotech Forum Europe* **8**, 740–745.

MATTIASSON, B., MOSBACH, K., SVENSON, A. (1977), Application of cyanide-metabolizing enzymes to environmental control; enzyme thermistor assay of cyanide using immobilized rhodanase and injectase, *Biotechnol. Bioeng.* **19**, 1643–1651.

MATTIASSON, B., DANIELSSON, B., HERMANSSON, C., MOSBACH, K. (1978), Enzyme thermistor analysis of heavy metal ions with use of immobilized urease, *FEBS Lett.* **85**, 203–206.

MIONETTO, N., MARTY, J. L, KARUBE, I. (1994), Acetylcholinesterase in organic solvents for the detection of pesticides: biosensor application, *Biosens. Bioelectron.* **9**, 463–470.

NEUDÖRFER, F., MEYER-REIL, L.-A. (1997), A microbial biosensor for the microscale measurement of bioavailable organic carbon in oxic sediments, *Mar. Ecol. Prog. Ser.* **147**, 295–300.

NEUJAHR, H. Y. (1984), Biosensors for environmental control, *Biotechnol. Gen. Eng. Rev.* **1**, 167–186.

NEUJAHR, H. Y., GAAL, A. (1973), Phenol hydroxylase from yeast, *Eur. J. Biochem.* **35**, 386–400.

NEUJAHR, H. Y., KJELLEN, K. G. (1979), Bioprobe electrode for phenol, *Biotechnol. Bioeng.* **21**, 671–678.

ÖGREN, L., JOHANSSON, G. (1978), Determination of traces of mercury (II) by inhibition of an enzyme electrode loaded with immobilized urease, *Anal. Chim. Acta* **96**, 1–11.

ORUPOLD, K., TENNO, T., HENRYSSON, T., MATTIASSON, B. (1996), The application of genetically engineered *Pseudomonas putida* strains and activated sludge in biosensors for determination of phenol(s), *Res. Environ. Biotechnol.* **1**, 179–191.

PAPKOVSKY, D. B., GHINDILIS, A. L., KUROCHKIN, I. N. (1993), Flow-cell fiber-optic enzyme sensor for phenols, *Anal. Lett.* **26**, 1505–1518.

PETER, M. G., WOLLENBERGER, U. (1997), Phenol-oxidizing enzymes: mechanisms and applications in biosensors, in: *Frontiers in Biosensorics I*, Fundamental Aspects (SCHELLER, F. W., SCHUBERT, F., FEDROWITZ, J., Eds.), pp. 63–82. Basel: Birkhäuser.

PETER, J., HUTTER, W., STÖLLNBERGER, W., HAMPEL, W. (1996), Detection of chlorinated and brominated hydrocarbons by an ion sensitive whole cell biosensor, *Biosens. Bioelectron.* **11**, 1215–1219.

PILETSKY, S. A., PILETSKAYA, E. V., ELGERSMA, A. V., YANO, K., KARUBE, I. (1995), Atrazine sensing by molecularly imprinted membranes, *Biosens. Bioelectron.* **10**, 959–964.

REISS, M., SOKOLLEK, S., HARTMEIER, W. (1993), Quecksilberbestimmung durch Fließinjektionsanalyse über immobilisierte β-Glucosidase, *Bioforum* **16**, 344–346.

REISS, M., METZGER, J., HARTMEIER, W. (1995), An amperometric microbial sensor based on *Azotobacter* species for phenolic compounds, *Med. Fac. Landbouww. Univ. Gent* **60**, 2227–2230.

RIEDEL, K., ALEXIEV, U., NEUMANN, B., KUEHN, M., RENNEBERG, R., SCHELLER, F. (1989), A microbial sensor for BOD, in: *Biosensors-Application in Medicine, Environmental Protection and Process Control* (SCHMID, R. D., SCHELLER, F., Eds.), pp. 71–74. Weinheim: VCH.

RIEDEL, K., LANGE, K.-P., STEIN, H.-J., KÜHN, M., OTT, P., SCHELLER, F. (1990), A microbial sensor for BOD, *Water Res.* **24**, 883–887

RIEDEL, K., NAUMOV, A. V., BORONIN, A. M., GOLOVLEVA, L. A., STEIN, H. J., SCHELLER, F. (1991), Microbial sensors for determination of aromatics and their chloro derivatives. I: Determination of 3-chloro-benzoate using a *Pseudomonas*-containing biosensor, *Appl. Microbiol. Biotechnol.* **35**, 559–562.

RIEDEL, K., NEUMANN, B., SCHELLER, F. (1992), Mikrobielle Sensoren auf Basis von Respirationsmessungen, *Chem. Ing. Tech.* **64**, 518–528.

RIEDEL, K., BEYERSDORF-RADECK, B., NEUMANN, B., SCHELLER, F. (1995), Microbial sensors for determination of aromatics and their chloro derivatives. III: Determination of chlorinated phenols using a biosensor containing *Trichosporon beigelii* (*cutaneum*), *Appl. Microbiol. Biotechnol.* **43**, 7–9.

ROGERS, K. (1995), Biosensors for environmental applications, *Biosens. Bioelectron.* **10**, 533–541.

SANGEETHA, S., SUGANDHI, G., MURUGESAN, M., MADHAV, V. M., BERCHMANS, S. et al. (1996), *Torulopsis candida* based sensor for the estimation of biochemical oxygen demand and its evaluation, *Electroanalysis*, **8**, 689–701.

SASAKI, S., KARUBE, I., HIROTA, N., ARIKAWA, Y., NISHIYAMA, M. et al. (1998), Application of nitrite reductase from *Alcaligenes faecalis* S-6 for nitrite measurement, *Biosens. Bioelectron.* **13**, 1–5.

SCHELLER, F., SCHUBERT, F. (1989), *Biosensoren*. Berlin: Akademie-Verlag.

SCHELLER, F. W., SCHUBERT, F., FEDROWITZ, J. (1997), *Frontiers in Biosensorics I*. Basel: Birkhäuser.

SCHILLER, J. G., CHEN, A. K., LIU, C. C. (1978), Determination of phenol concentrations by an electrochemical system with immobilized tyrosinase, *Anal. Biochem.* **85**, 25–33.

SCHOWANEK, D., WEYME, M., VANCAYSEELE, C., DOMS, F., VANDEBROEK, R., VERSTRAETE, W. (1987), The rodtox biosensor for rapid monitoring of biochemical oxygen demand and toxicity of wastewaters, *Med. Fac. Landbouww. Rijksuniv. Gent* **52**, 1757-1779.

SCHÜGERL, K., ULBER, R., SCHEPER, T. (1996), Development of biosensors for enantiomeric analysis, *Trends Anal. Chem.* **15**, 56–62.

SCHWEDT, G., HAUCK, M. (1988), Reaktivierbare Enzymelektrode mit Acetylcholinesterase zur differenzierten Erfassung von Insektiziden im Spurenbereich, *Fresenius Z. Anal. Chem.* **331**, 316–320.

SKLADAL, P., MASCINI, M. (1992), Sensitive detection of pesticides using amperometric sensors based on cobalt phthalocyanine-modified composite electrodes and immobilized cholinesterases, *Biosens. Bioelectron.* **7**, 335–343.

SUN, Y., LIU, X. (1992), Study on a long life BOD microbial sensor, *Chin. J. Environ. Sci.* **13**, 59–62.

SVENSSON, A., HYNNING, P.-A., MATTIASSON, B. (1979), Application of enzymatic processes by monitoring effluents, *J. Appl. Biochem.* **1**, 318–324.

SZWEDA, R., RENNEBERG, R. (1994), Rapid BOD measurement with the Medingen BOD-module, *Biosens. Bioelectron.* **9**, IX–X.

TESCIONE, L., BELFORT, G. (1993), Construction and evaluation of a metal ion biosensor, *Biotechnol. Bioeng.* **42**, 945–952.

TRAN-MINH, C. (1993), Biosensors for the analysis of pesticide residues, *Anal. Proc.* **30**, 73–74.

VAN HOOF, F. M., DE JONGHE, E. G., BRIERS, M. G., HANSEN, P. D., PLUTA, H. J. et al. (1992), The evaluation of bacterial biosensors for screening water pollutants, *Environ. Toxicol. Water Qual.* **7**, 19–33.

VIANELLO, F., SIGOR, L., PIZZARIELLO, A., DI PAOLO, M. L., SCARPA, M. et al. (1998), Continuous flow immunosensor for atrazine detection, *Biosens. Bioelectron.* **13**, 45–53.

WANG, J., NASER, N., KWON, H.-S., CHO, M. Y. (1992), Tissue bioelectrode for organic-phase enzymatic assays, *Anal. Chim. Acta* **264**, 7–12.

WENDZINSKI, F., GRÜNDIG, B., RENNEBERG, R., SPENER, F. (1997), Highly sensitive determination of hydrogen peroxide and peroxidase with tetrathiafulvalene-based electrodes and the application in immunosensing, *Biosens. Bioelectron.* **12**, 43–52.

WIJESURIYA, D. C., RECHNITZ, G. A. (1993), Biosensors based on plant and animal tissues, *Biosens. Bioelectron.* **8**, 155–160.

WOLLENBERGER, U., NEUMANN, B., RIEDEL, K., SCHELLER, F. W. (1994), Enzyme and microbial sensors for phosphate, phenols, pesticides and peroxides, *Fresenius J. Anal. Chem.* **348**, 563–566.

WONG, K.-Y., ZHANG, M.-Q., LI, X.-M., LO, W. (1997), A Luminescence-based scanning respirometer for heavy metal toxicity monitoring, *Biosens. Bioelectron.* **12**, 125–133.

YANG, Z., SUZUKI, H., SASAKI, S., KARUBE, I. (1996), Disposable sensor for biochemical oxygen demand, *Appl. Microbiol. Biotechnol.* **46**, 10–14.

ZHAO, C. Q., ANIS, N. A., ROGERS, K. R., KLINE, R. H., WRIGHT, J. et al. (1995), Fiber optic immunosensor for polychlorinated biphenyls, *J. Agric. Food Chem.* **43**, 2308–2315.

8 Laws, Statutory Orders and Directives on Waste and Wastewater Treatment

PETER NISIPEANU

Essen, Germany

List of Abbreviations

AbwAG	Abwasserabgabengesetz (wastewater charges act)
AOX	organic halogenated compounds
ATV	Abwassertechnische Vereinigung e.V. (wastewater technical association)
BAT	best available technology
BImSchG	Bundesimmissionsschutzgesetz (act on the prevention of harmful effects on the environment caused by air pollution, noise, vibrations, and similar phenomena – federal immission control act)
BOD	biological oxygen demand
BOD_5	biological oxygen demand in 5 d
CEN	European standards
COD	chemical oxygen demand
DIN	Deutscher Industrie- und Normungsausschuß (German standard association)
EC	European Community
EIA	act on environmental impact assessment (UVPG)
EU	European Union
GG	Grundgesetz (basic law, constitution)
KrW-/AbfG	Kreislaufwirtschafts- und Abfallgesetz (recycling and waste management law)
LAWA	Länderarbeitsgemeinschaft Wasser (working group for environmental issues on the sphere of water)
LWG	Landeswassergesetz (water law of Länder)
UVPG	Umweltverträglichkeitsprüfung (act on environmental impact assessment)
VwVen	Verwaltungsvorschriften (administrative directives or rules)
WHG	Wasserhaushaltsgesetz (water supply act)

1 Introduction

National water politics and national water legislation have to pay more and more attention to international – and that means in Europe especially: European – politics of protection of the water resources by law. Because of this correlation there are requirements to the material and formal adjustment of national law or the requests for common handicaps of superior legislation instances.

In the European Union this handicaps are set by special guidelines, in the field of water and wastewater especially by the Directive of May 21, 1991 of disposal of municipal wastewater (91/271/EWG) and the amended proposal for a Council Directive establishing a framework for Community action in the field of water policy. In spite of this handicaps, which are valid for all members (member countries) of the European Union the results of national activties on putting this into action are very different. In Tabs. 1 and 2 this is exemplified at some selected countries in the European Union with data with regard to water and wastewater management selected from the European Wastewater Catalog 1997.

The grade of clarification at the sewage treatment plants is very different. Tab. 3 shows, e.g., grades of purification, of the plant nutrients (phosphorus and nitrogen).

The reasons for the different national results on the same European handicaps have great variety: national traditions, settlement structure, tradited looks of lawfulness, various problems with regard to water management, various commercial strength and different readiness to strict and fast putting into action of European handicaps.

Let me show this at the example of the Bundesrepublik Deutschland (Federal Republik of Germany), where it is tried to put the European and the own national ideas of protection of waters into action in the former Deutsche Demokratische Republik (German Democrat-

Tab. 1. Connecting Grade of Population to Communal Sewage Systems

Country	Grade (%)
Germany	92
Finland	78
France	51
Greece	58
Great Britain	97
Italy	82
Luxemburg	99
Netherlands	97
Portugal	57
Spain	82

Tab. 2. Connecting Grade of Population to Communal Sewage Treatment Plants

Country	Connecting Grade to Sewage Treatment Plants (%)	Connecting Grade of Communal Sewage Systems to Sewage Treatment Plants (%)
Germany	90	98
Finland	78	100
France	49	96
Greece	54	93
Great Britain	81	83
Italy	57	70
Luxemburg	90	91
Netherlands	97	100
Portugal	18	32
Spain	36	44

Tab. 3. Grades of Purification Shown at Plant Nutrients

Country	Grade (%)
Germany	87
Finland	88
France	2
Greece	36
Great Britain	14
Italy	9
Luxemburg	9
Netherlands	75
Portugal	6
Spain	2

ic Republic) [the so-called Neue Länder (new German states)]. There are similar problems as in other member countries of the European Union (EU) with a great need to catch up on the niveau of the European law.

2 Water Legislation

Constitutional law does not embody any compulsory obligation to establish national or international water regulations. Ultimately it is society which determines what it expects from the legislative and regulatory bodies. So there may well develop some sociologically founded actual legal obligation to regulate exploitation and protection of water resources by law. The decision *whether* such an actual requirement does exist and *how* it should be met – by issuing appropriate regulations – depends on a variety of factors, typically relating to the given hydrological, meteorological and geographic conditions, population density, economic prosperity, political attitudes, and efficiency of administrative and legislative authorities. Even religious and anthropocentric-philosophical considerations may influence the juridical approach to the use of aquatic systems.

Against this rather empirical than normative background, three or even four different development stages of juridical regulation can be identified in a general overview, leaving national particularities out of account.

(1) In times when availability of water resources appears to be unlimited or when resources are only tapped and exploited to a minor degree there will be no distribution conflicts, and if any, these will mainly affect direct neighborhood relationships. Such rather casual conflicts will be settled in accordance with the general rules for legal protection of ownership or possession of rights to achieve an adequate agreement of interests. So it is primarily the use of water resources that is to be regulated – drainage and irrigation (with storage), water withdrawal and wastewater discharge. The colliding interests

may be subsumed under the headlines of use by proprietor, use by adjacent owner, and general use and their eminence ranked in this order: As a matter of fact, water use interests of the owner of the land on which the waterbody is located or of the adjacent estate is given priority *vis-à-vis* the interests of third parties. There is neither any state initiated "public law" protection of water resources, nor is there any state initiated preferential treatment of specific water uses.

(2) In as much as awareness of the limited availability of water resources due to scarcity or pollution is becoming evident – causing a steadily rising number of conflicts of regional or national impact – public interest in and actual need for state regulation will grow. This scenario generally leads to water quantity and quality control, which may, e.g., be put into practice by the instrument of unilateral rights of reservation by sovereign powers to grant permissions for water use so as to establish a state governed distribution and allocation system in due pursuance of the priorities fixed by the state. Another instrument is a contractual agreement to be concluded between the state and the user of the water body. Sometimes such reservations, finding their expression in permissions and contracts, may be strengthened by otherwise existing prohibitions of any use of the water body whatsoever. As a rule, there will always be an upsurge of environmental consciousness and growing pressure for juridical regulation and political action in the field of water management in response to incidents or disasters endangering public life (epidemics, water eutrophication, floods, etc.).

(3) The present statuses of technology, environmental science and financial capacity have enhanced the chances for implementation of ecopolitical, administrative and legislative measures in water management, whereby the focus has clearly shifted away from a purely water supply oriented management towards a more ecological approach: The entire aquatic system becomes the object in need of protection by juridical regulation, so that any harm it may suffer must be reasonably and objectively justified.
For this purpose, immission oriented critical values for water use are to be defined. To facilitate administration, a blanket immission protection may be implemented by establishing an obligatory limitation of immissions which does not need single-case justification, but may also be subject to more stringent requirements in case of particularities of the given receiving water (already existing pollution, inefficient hydraulic or qualitative capacity).
Some economic regulatory systems follow this approach by, e.g., charging specific fees for water abstraction or wastewater discharge, or by selling licenses for a limited use of water bodies, and thereby use the tools of market economy to internalize costs that have otherwise been externalized and "socialized" to the debit of the community (consumption of environmental resources).
Such single case considerations are often supported by an attempt to manage water on a more global scale by definitely prohibiting any water degradation whatsoever, and by providing planning instruments (management plans, clean water ordinances, etc.).

(4) In the earlier development stages the necessity to use water bodies was not principally denied, but the need emphasized to find suitable strategies to regulate utilization and minimize its impact in conformance with political objectives. In the then following development stage, the medium related environmental protection – so far restricted to water resources – is extended to a "cross-media" approach. Consequently, a mere shifting of loads from one medium (e.g., water) to another (e.g., air or soil) is phased out for the benefit of integrated environmental protection which looks already for possibilities to

reduce or even avoid loads *a priori*. Among the instruments that may help achieve this target are: integrated environmental impact assessments, alternative reports, cradle-to-grave product assessment, circulatory product management, prohibition of biologically undegradable substances, and the request for zero emissions. Another meaningful aspect is an early integration of the intending user of the waters, by convincing him to provide for internal (in-plant) control, economical use of resources and to recruit environmental personnel of his own (environmental officer).

Establishment and formulation of any specific national water legislation and its progress are not only a function of the given historic, economic and industrial framework, but also of the prevailing geological, hydrogeological and meteorological conditions. The need for regulation and the objects to be regulated are a function of varying precipitation patterns in the different regions (arid or humid), so that some countries are mainly concerned with problems relating to drainage, irrigation and storage, whereas others import and export their streams or have rivers with their entire longitudinal profile from source to mouth within their political borders. National legislation also depends on industrial development, traditions, financial resources and specific purposes for which the water bodies must serve.

Further national particularities result from the administrative structures being held available for implementation of both water management and water protection. These structures may follow either political borders (of towns, counties, districts, the Länder, federal state, or community of states) or hydrological conditions (catchment areas or partial catchment areas). Moreover, the actual efficiency of the administration strongly depends on quality and quantity of tasks and competences assigned to it, and on the number, skills and expertise of its workforce.

Additional particularities are attributable to the natural or juristic persons operating the water industry in a specific country. For example, it may well be that water protection lies in the hands of those people who are at the same time the most important water users. Persons in charge of such double functions should obviously be both personal absolutely reliable and technically competent.

Such characteristic national legal requirements for the use of water bodies (by wastewater dischargers) and for water protection, being influenced by a variety of parameters, are discussed in Sects. 2.1–2.1.2.4 – though in a simplified way – by taking a single national regulation system as an example.

2.1 Water Legislation in Germany

For a German author it suggests itself to address the subject by referring to water legislation in the Bundesrepublik Deutschland, with which he is familiar. And as German legislation is known for its tendency towards perfectionism, sometimes being on the brink of overregulation, alternating between severe administrative pressure and failing executive power, reference to this country suggests that a wide range of possible regulatory options will be covered.

National legislation in the area of aquatic systems and exploitation of waters can be traced back into the 19th century. However, the legal situation at the time did not give a uniform picture, as, e.g., all water relevant directives in the former Preußischer Staat (state of Prussia) were spread over more than 80 different acts, e.g., Privatflußgesetz (private river act) (2/18/1843), Deichgesetz (dyke act) (1/28/1848), Gesetz über die Wassergenossenschaften (act on water cooperatives) (4/1/1879), Strombauverwaltungsgesetz (act on the administration of hydraulic engineering) (8/20/1883), Hochwasserschutzgesetz (flooding protection act) (8/16/1905), Quellenschutzgesetz (act on the protection of sources) (5/14/ 1908), and Wasserstraßengesetz (waterway act) (4/1/1905). Otherwise applicable were the Allgemeines Preußisches Landrecht (general Prussian state law) of 1796, the (French) Code Civil as well as common law and not to forget the personal attitudes of the electors of Brandenburg and Saxony in the 18th century. In spite of this obvious variety, the early laws and regulations did not yet cover all segments encompassed by today's water legislation. Comprehensive codifications – such as the Preußisches Wassergesetz (Prussian water act) of 1913 – emerged as late as at the turn of the century, and were established by member states of the German state and not by the empire as an entity.

2.1.1 Distribution of Legislative Competence

Also the German Grundgesetz (GG) (basic law, constitution) confirmed at first the distribution of tasks and fundamental sole responsibility of the Länder. But in a time of rapid and stormy redevelopment after the Second World War it soon became evident that such a pluralistic and split up legislative approach would not adequately serve the tasks to be settled. This insight led to the enactment of the higher legislative competence of the Bund (federal government representing the state as a whole) confirmed by constitutional law under Art. 75 GG.

Since creation of the European Community, the national legislation of the Bundesrepublik Deutschland has increasingly been influenced by the statutory orders and directives of the European Community (EC) and the European Union (EU) respectively: statutory orders issued by the EC/EU have a direct effect on all citizens of the community. EC/EU directives still need normative incorporation into national jurisdiction. If a member state fails to carry out this transfer in due time, European citizens are entitled to refer to the EU directive straightaway – in as far as said directive pertains to regulations aimed at their personal protection. Inspite of the great number of existing directives in the sector of water management and inspite of the present emphasis on a common European water policy, a comprehensive European law has not yet been codified, so that there are still scope and need for national regulatory activities. Consequently, national particularities are still prevalent.

The same applies to the existence of international and supranational agreements relating to specific water protection issues.

2.1.1.1 Federal Legislation

According to the constitutional provisions of the Grundgesetz (Art. 75 GG), the Bund as the state as an entity is only vested with the competence to issue framework legislation in the sector of water management. This competence makes it possible that federal regulations – subject to approval by the Bundesrat ("upper house" or council of federal states) – be enforced directly, however, it shall also leave the Länder with sufficient room to formulate their own standards, by issuing, e.g., Landeswassergesetze (Länder water regulations).

On this basis, the Bund introduced the national Wasserhaushaltsgesetz (WHG) (act on the regulation of matters relating to water – federal water resources act), regulating water management, as early as in 1957, and the Abwasserabgabengesetz (AbwAG) – wastewater charges act – in 1976, setting up the criteria for charging polluters for the discharge of wastewater into aquatic systems. In the meantime, both acts have been amended repeatedly; they may be defined as substantial and formal acts having got their legitimation by the Bundestag as the lower house of parliament, and – if and in as far they affect Länder competence – also by the Bundesrat as the upper house of parliament and federal council.

A great number of statutory orders are based on the aforementioned WHG, like the ordinance on water-hazardous substances for water transport in pipelines (1973), the ordinance on the determination of criteria for wastewater discharge into water bodies (Abwasserverordnung – wastewater ordinance) of 1997 or the Grundwasserverordnung (ground water ordinance) of 1997. These ordinances are material acts expressly permitted by the Grundgesetz, and issued by the ministries concerned (as the executive), acting within the scope of their powers defined by law for the specific purpose.

Besides the WHG (water resources act) and the ordinances based on it, there are a number of other statutory regulations which though belonging to the systematics of water legislation, are governed by other law corpora for historic, authoritative or political reasons. So the ordinance on drinking water has its legal roots in the law relating to food production and distribution, the disposal of wastewater from livestock farming is based on the law on fertilizers and on waste legislation, the removal of pit water is governed by mining law, and the sewerage infrastructure for buildings and facilities is overseen by the building law. Special regulations relating to the protection of water can be found in the acts on washing and cleaning agents, oil pollution, chemicals, and in the atomic energy law. Indirect demands for protection result from application of the acts on recycling and waste management. Even the Strafgesetzbuch (criminal law) includes a specific directive for the protection of water bodies, § 324 StGB.

The federal water management act is supplemented by organizational directives (§ 18a WHG: the obligation to dispose of wastewater, irrespective of the causative agent or polluter, is in principle assigned to bodies under public law, i.e., in particular to local authorities, municipalities, and water associations); by legal provisions in the framework of corporate law relating to the foundation of water and land associations, by tax exemption for public wastewater disposal to date still being in force.

The spectrum of federal legislation is rounded off, yet not completed, by the Gesetz über die Umweltverträglichkeitsprüfung (UVPG) (act on environmental impact assessment, EIA), which regulates specific wastewater discharges and wastewater treatment plants.

For quite a long time, the so-called Verwaltungsvorschriften (VwVen) (administrative directives or rules) have played an important part on the level of federal legislation, i.e., directives issued by the administration itself, whether expressly allowed by law, or by the specific status of the authorities involved, in order to enhance standardization and simplify enforcement. They only gained some importance as dogmatic legal provisions or legal source on an internal law level within the administration ("dialog between two officers", and "order from a superior officer to an inferior") as qualifying directives in particular when used to define so-called indefinite legal terms such as "generally accepted technical rules" or "state-of-the art technology". Because of their actual legal impact, the procedures for the establishment of administrative rules were highly formalized and founded on a broad basis of understanding by integration of all quarters and parties concerned and also of the Bundesrat (federal council) (therefrom deriving "quasi-democratic legitimation"). To establish foreseeability of enforcement harmonized by these administrative orders, they were – as a rule – made publicly available (published in the relevant press organs).

Well-known in this field was primarily the general administrative directive relating to wastewater, regulating the release of wastewater into water bodies, and its annexes relating to industry-specific origins (e.g., Annex 40 on wastewater from metal working and metal processing workshops, or Annex 51 on wastewater from landfills for municipal waste). Of great practical benefit is also the general administrative directive of 1996 to the WHG (water supply act) on the classification of water polluting substances into different pollutant categories. By means of the limitations and regulations laid down in these administrative directives, the Bundesrepublik Deutschland also tried to meet the targets and standards for water protection specified by European law.

Following the decision of the European Court of Justice saying that this way of giving the directives of the EC/EU Council legal effect within national jurisdiction is not compatible with common European law, on the grounds of presumed deficient legal force, the said administrative directives are now being successively replaced by statutory orders with similar systematics and similar contents (ordinance on wastewater with annexes), to which the citizen of the EC may now refer directly. The general legal basis for such statutory orders is provided by § 6a WHG. Pending creation of said statutory orders based on article 80 GG and on ordinary written law (cf. §§ 6a and 7a, para 1 WHG), the hitherto applicable VwVens (administrative directives), according to § 7 Abwasserverordnung will remain valid in law.

2.1.1.2 Land Legislation

All 16 member states of the Bundesrepublik Deutschland have created individual water laws differing noticeably as to objects, contents, and intensity of regulation. These state laws not only fit into the framework drawn up by the (federal) WHG, but also supplement it by specific regulations relating to water legislation or other areas with a factual relationship thereto. For example, they regulate property rights in water bodies and set up a catalog of rules for charging wastewater fees. To some extent, the Länder have standardized regulations flanking the water management law in its essence by separate acts and statutory orders, as, e.g., by ordinances on competences, dyke management, indirect discharging, water quality control, compliance with EC/EU Commission rules. Some Länder have issued special acts on the organization of wastewater disposal, e.g., by water associations [e.g., the Ruhrverbandsgesetz (Ruhr river association act) in North-Rhine Westphalia] or by institutions under public law as in the city state of Hamburg. Water legislation of the Länder is supplemented by the prevalent more general rules of police and regulatory law, and by the respective law on administrative procedures.

The drifting apart of the issues covered by water related legislation of the Länder due to the federalist approach, has been delimited by the institutionalized merger of the Obersten Landeswasserbehörden (superior water administrative boards of the federal Länder) into *one* working group on water and wastewater of the Länder [Länderarbeitsgemeinschaft Wasser (LAWA)].

2.1.1.3 Standardization by Municipalities and Associations

On the municipal level there is need for standardization and authority for implementation *vis-à-vis* the users of the public water supply and wastewater disposal facilities. This purpose defined demand is met by uniform regulations for water supply and non-uniform regulations for local wastewater disposal statutes, and a catalog of rates and charges designed to refinance municipal expenditures.

2.1.1.4 Technical Standardization

Technical and environmental laws – including water and wastewater legislation – are subject to a permanent change of framework conditions and of best available technologies applied to tackle all emerging problems. This process of constant change is so dynamic and complex that legislation can hardly follow and respond to all new aspects involved. Therefore, German law includes "open" formulations – so-called indeterminate legal conceptions – not only allowing enforcement under consideration of all particularities of technological and environmental law, but also judicial control.

To fill these undeterminate legal conceptions, the professional knowledge of the concerned private organizations and their experts can be employed. Typically, this "expert knowledge" finds its expression in the so-called technical standards of which, in the wastewater sector, especially the work and instruction sheets and reports of the Abwassertechnische Vereinigung e.V. (ATV) (wastewater technical association) are of practical importance. Besides these standards, there are the German industrial standards established by the Deutsche Industrie- und Normungsausschuß (DIN), and the European standards (CEN). Application of these technical standards has proved to be of great importance in practice.

2.1.2 Structure of Water Legislation

Trying to outline the structure of federal water laws, four major areas should be addressed: water management (Sect. 2.1.2.1), water protection (Sect. 2.1.2.2), regulation of water runoff (Sect. 2.1.2.3), and information (Sect. 2.1.2.4). All these areas may gain some relevance in the disposal of wastewater.

2.1.2.1 Water Management

Management of water quantity and quality is implemented by means of economic instruments, control of water utilization through strict prohibition of any use of the water whatsoever with the reservation on the granting of consents and permissions, as well as through planning instruments.

Such economic instruments are the charges levied in all parts of the Bundesrepublik Deutschland on the release of wastewater into water bodies (AbwAG), and the charges on water withdrawal levied in some Bundesländer. Moreover, there are liability regulations (under § 22 WHG) to become effective as financial restriction for any unlawful activity endangering the water.

Permission of any water use whatsoever (§ 3 WHG) as required by § 2 WHG may be secured in different ways. These may encompass any such water utilization actually allowed by the legislator, and hence not needing permission from administrative authorities for minor water management measures in the frame of public use, typically concerning "in-house" needs, utilization by adjacent site owners and proprietors (§§ 15, 17a, 23–25, 32a, 33 WHG, LWG), and extend to any such water utilization permitted by simple license (§ 7 WHG) – which may be adapted to the requirement for special procedures according to the environmental impact assessment act (UVPG) or to § 7a WHG in connection with the wastewater ordinance – and finally to any such utilization allowed by a higher license (LWG) or even by a license being on a par with property rights (§ 8 WHG; not allowed for wastewater discharge!). Each of these water utilization licenses is subject to compliance with specific prerequisites ("for the common good") which find their expression in clear-cut water law stipulations (§ 7a WHG and wastewater ordinance). The so-called Planfeststellung (project approval procedure) is not a planning instrument in this sense, but is to be seen as a particular item of procedural law conceded by several state laws, expressly involving participation of the general public and authorities, and stringent formalization, in order to ensure correctness of the defined subject matter of the desired administrative decision. Decisions made on the basis of such project approval procedures can only be contested under extremely difficult conditions.

Planning instruments in water management are clearly defined and embodied in the framework plan (§ 36 WHG), the management plan (§ 36b WHG), the fundamental ban on degradation introduced for planning security (§ 36b para 6 WHG), and the wastewater disposal plan (§ 18a WHG). Though this package of instruments is not so important in practice because of the great expenditure of work involved, it is mandatory for any licensing procedure.

2.1.2.2 Water Protection

The protection of aquatic systems as defined under § 1a WHG is supplemented by licensing requirements for all facilities [either located upstream the projected water utilization or otherwise contributing to water pollution (LWG)], and by a range of economic and planning instruments, underpinned by the threat of sanctions.

Though under federal law (§§ 18b, 18c WHG) wastewater treatment plants do not need consent by the authorities unless they represent special facilities [designed to treat wastewater with an organic load >3,000 kg d^{-1} BOD_5 (crude) or wastewater with an inorganic load >1,500 m^3 wastewater in 2 h, with the exception of cooling water]. Nevertheless, the Länder have made provision that almost all wastewater treatment plants be subject to licensing. As regards the approval of sewerage systems, procedures have been somewhat simplified (notification or fictional approval by lapse of time), and the same is true for plants and equipment that might have been otherwise approved because of their specific construction type.

Pipelines, unless used for the purpose of water supply or wastewater discharge, fall under the special provisions of §§ 19a et seq. WHG, if, e.g., used for conveying water-hazardous substances (in particular crude oils, petrols, diesel fuels and fuel oils). Therein the legislator sees a high risk potential.

Similar regulations apply to plant and equipment in which water-hazardous substances are treated or stored, in particular acids, lyes, alkali metals, silicon alloys, metal organic compounds, halogens, acidic halides, metal carbonyls and pickling salts, mineral and tar oils as well as all products derived therefrom, liquid as well as water soluble hydrocarbons, alcohols, aldehydes, ketones, esters, halogenated, nitrogenous and sulfurous organic compounds and toxic agents.

Specific planning instruments for water protection provided for by the WHG are the identification of protected water areas (§ 19 WHG) and the determination of water quality standards (§ 27 WHG). Both prerequisites must be substantiated and defined by laws of the Länder (LWG).

The right to apply sanctions can be derived from common police and regulatory law obliging the offender to ward off and eliminate any danger or nuisance endangering the maintenance of public order, which he, as the estate owner, might have caused or induced or for which he might otherwise be responsible. In case of specific infringements, the special regulatory law "Wasserrecht" (water law) allows cancellation of a previously conceded permission or licence. Additionally, summary proceedings on account of breach of legal rules and bans may be launched (§ 41 WHG, law on irregularities, LWG), as a result of which penalties may be imposed and any financial advantage invoked by the offence be confiscated. A more rigid catalog of fines is provided by criminal law (§ 324 Strafgesetzbuch), according to which any action liable to cause harmful effects on waters may be punished with a pecunary penalty or even imprisonment, and withdrawal of any economic advantage so far realized.

2.1.2.3 Regulation of Water Runoff

A glimpse into history shows that in the beginning of water management in the densely populated Bundesrepublik Deutschland priority was given to receiving waters contributing to a rapid storm water and wastewater runoff. This trend brought about a great number of straightened rivers, built-up areas in flood plains, and ecological impoverishment, which, in turn – still aggravated by an extensive sealing of land surfaces – increased flood danger and erosion. Against this background, regulations for a low hazard and environmental compatible water runoff began to gain ground. They can be systematized as follows: directives on protection against flood, on safety of water bodies (ensuring their preservation), and on a balanced water management.

Flood protection requires identification of flood plains (§ 32 WHG), as well as sufficient flood control space in existing impounding lakes (reservoirs). These measures have recently got fresh impetus by considerations towards an improved seepage of storm water *in situ*, so as to slowdown storm water runoff by retention in the bedrock. By administrative directive it can be secured that storm water is only discharged into a less efficient receiving water when all measures concerning retention, retardation and equalization to guarantee a natural runoff will have been implemented (§ 4 WHG).

A proper runoff in the water bodies is secured by regulatory measures for all buildings and other facilities located in flood areas, at or within the water bodies (LWG). These regulations have been supplemented by specific rules for pipelines (§§ 28–30 WHG); further measures are water quantity control (LWG), and specific approval procedures for hydraulic engineering relating to water bodies (§ 31 WHG, UVPG).

2.1.2.4 Documentation and Information

All water management relevant facts are collected and passed on to the users of the water bodies, the administrative authorities concerned and the general public through the so-called "water record" (§ 37 WHG, LWG), the environmental officers for water protection to be appointed by the wastewater dischargers themselves (§§ 21a et seq. WHG) and the act on environmental information. Several Länder have added thereto statutory wastewater disposal registers or wastewater treatment plant registers; and some local authorities have established additional registers for indirect dischargers and wastewater.

2.2 Administrative Water Law Enforcement

For constitutional reasons, administrative enforcement of water law is the responsibility of the Länder with only a few exceptions (federal shipment administration). The Länder have built up a two-stage administrative hierarchy consisting of the lower and supreme water authorities: the function of the lower water authority is in the hands of the districts and autonomous towns, whereas the supreme water authority is the responsibility of the competent Land ministries for environmental protection and water protection, respectively. The Länder (except for the city states) holding an additional decentralized administration level (regional commission), delegate the function of the supreme water authority to this middle level authority. In some Länder, there are additional functional authorities directly responsible to the state offices (national agencies for water management and the environment, and some central regional authorities [as, e.g., Landesamt für Umweltschutz (county administrative office for environmental protection)] entrusted in particular with science oriented tasks. Functional authority of "water law enforcement" is as a rule exercised within the given political borders; the hierarchical allocation of tasks depends on how each specific task is weighted from a state–policy or state–law related view.

Responsibility for the enforcement of the wastewater charges act (AbwAG) is not conferred to authorities on an equal hierarchical level across all Länder, but takes state law particularities into account.

2.3 Specification of Requirements for Direct Wastewater Discharge into Water Bodies

Release of effluents into the aquatic environment is forbidden in principle. Any exception of the rule must be expressly authorized. As a precaution, the term wastewater is not to be interpreted in a narrow sense: hence wastewater is any water the properties of which have been changed due to domestic, industrial, agricultural or any other use, and the water flowing off together with it during dry weather (sewage), as well as precipitations flowing off and being collected from built-up areas or paved surfaces (storm water). Wastewater is also any seepage and effluent produced and collected in facilities treating, processing, storing or depositing wastes.

To be on the safe side, the obligation to properly dispose of the effluents is not left at the discretion of the producer (polluter), but is delegated, in principle, to bodies under public law. Any exception thereof is subject to approval. The German wastewater industry not having been privatized like, e.g., its British equivalent, is principally under public control.

Since 1996, granting of wastewater discharge licences has been put under the reservation of § 7a WHG saying that any use, avoidance or treatment measures of the water prior to its discharge must be in compliance with best available technology. Best available technology (BAT) may be defined as industrially feasible, in the relevant sector, from a technical and economic point of view, i.e., advanced facilities, processes and modes of operation suitable to prevent and reduce emissions to the environment (§ 7a para 5 WHG).

To make administration as unbureaucratic as possible, to achieve forseeability and comparability of administrative enforcement, the targets of BAT have been laid down in two different pathways: the wastewater ordinance issued according to § 7a WHG determines common characteristics of this technical standard, whereas the "daughter" annexes give precise source oriented criteria for the relevant requirements.

Major prerequisites for licensing are:

(1) Unless otherwise stipulated in the Annexes, a license for the discharge of wastewater into water bodies shall only be granted if, at the place where the wastewater is produced, the pollution load is kept "as low as possible" in as far as this can be achieved (in the given circumstances) by using water economizing processes in washing or cleaning operations, indirect cooling or low pollution operating materials and auxiliaries).

(2) It is not allowed to meet these requirements by shifting to processes liable to harm other media, like the soil or the air, in contradiction to state of the art technology.
(3) Required threshold loads shall not be met by dilution, in contradiction to state-of-the-art technology.
(4) If there are any requirements to be met prior to mixing water streams, co-treatment of the effluent mixture may be allowed, provided the pollution load per parameter can be reduced as efficiently as it would have been possible in case of separate fulfilment of the relevant requirements.
(5) If there are any requirements to be met by the location at which the wastewater occurs, a mixing will only be permitted upon fulfilment thereof.
(6) Further details can be taken from the Annexes to the wastewater ordinance and to the framework wastewater administrative regulation respectively, referring to wastewater sources.

The need to apply state-of-the-art technology may, therefore, lead to quite different requirements:

(1) to meet specific load values of wastewater (result oriented),
(2) to hold available and operate specific wastewater treatment facilities (plant oriented),
(3) to use specific processes in the production and application of hazardous substances and materials (process oriented),
(4) to prevent specific substances from entering the wastewater (substance oriented), possibly going hand in hand with the demand,
(5) to operate specific production or application processes without producing any wastewater whatsoever (zero emission oriented),
(6) to determine the values to be achieved at the location where the wastewater occurs or for wastewater streams diluted prior to direct discharge (partial-stream-oriented).

The result of these requirements generally manifests itself in the establishment of parameters – single, collective or efficiency related parameters – for undesired wastewater constituents, specifying concentration and load limits, etc.

(1) The biological oxygen demand of the substances admitted for discharge is principally limited, i.e., for the easily degradable C compounds as 5 d biological oxygen demand (BOD_5), otherwise as chemical oxygen demand (COD). Other major effluent standards refer to specific metals (copper, nickel, lead, cadmium, mercury and chromium), plant nutrients (nitrogen and phosphorus compounds), organic halogenated compounds (AOX) and toxicity to fish (T_f).
(2) These limits generally refer to pollutant concentrations in 1 L water (mg L^{-1} or µg L^{-1}) or, in case of toxicity to fish, to the dilution factor.
(3) Discharge quantities are defined as maximum annual values, and as admissible values for shorter periods (e.g., 0.5 or 2.0 h).
(4) For wastewater law related reasons the annual effluent quantity is determined without consideration of rainfalls, as absolute wastewater quantity.
(5) For the purpose of reproducibility of state control, further principles have been laid down:
   (•) definition of exact sampling point (in most cases in the effluent of the last treatment stage),
   (•) definition of sampling technique (e.g., 2 h composite sample or qualified random sampling),
   (•) definition of relevant sample (settled or homogenated sample),
   (•) definition of sampling periods (e.g., for nitrogen only during the summer half year),
   (•) definition of chemical analytical technique (mostly according to DIN standards),
   (•) definition of presumed compliance with limit values (e.g., arithmetic average, rule four out of five meas-

ures with limitation of maximum allowable concentration or percentile method).

(6) Typical subordinate provisions relate to voluntary internal control and documentation as well as to appointment of an environmental officer for water protection, further to notification of any noticeable or foreseeable operating trouble. And there is the obligation to assist the water authorities in the enforcement of regulations.

Current dischargers are granted transition periods and special regulations in case the updated requirements should involve economically and technically unreasonable expenditures (§ 7a paras. 2 and 3, § 5 para. 1 sentence 2 WHG).

2.4 General Requirements for Indirect Discharges

When wastewater from industry and trade is not discharged directly into the aquatic environment, but into the municipal sewerage system, it may contain substances defying treatment in the municipal sewage plant, so that it is in the end released into the aquatic system. Such a case is referred to as an indirect discharge.

The Länder are bound by federal water laws to make all necessary provisions to avoid such undesirable harmful effects on the water bodies by a partial stream approach (§ 7a para 4 WHG). And the Länder have responded by issuing so-called ordinances for indirect dischargers and appropriate acts saying that in principal indirect discharges from specific places of origin shall have to meet the same requirements as direct discharges; however, the Länder concede that part of the degradation and elimination of pollutant loads or pollution impact can be and is allowed to be carried out in the municipal sewage treatment plant.

Since such indirect discharges may have a harmful impact on the operation of the sewerage system and sewage treatment plant, as well as on sludge treatment and sludge disposal, additional regulations and limitations have been laid down in a municipal charter for local sewage disposal. Further criteria for regulations based on municipal law relate to the safety of operating personnel and wastewater facilities, as well as to the municipality's capacity to fulfil any wastewater relevant obligation required by law and which is its responsibility.

Any deviation from the targets of the municipal charter must be approved by the local authorities.

2.5 Requirements for Wastewater Treatment

Under legal aspects, the problem of how wastewater should be treated is the implication of the underlying question whether there is need for such a treatment: In as far as granting of consent for wastewater direct or indirect discharge is subject to limitations as a function of wastewater treatment according to state-of-the-art technology, this standard is to be applied for the treatment of wastewater prior to its discharge. Hence, it is of no importance which process is applied to achieve the target values: mechanical, biological, physical or chemical processes, if only they help to reach the required minimization of pollutants or pollutant effects. § 18b WHG takes this aspect into account by requiring only a result oriented operatability, without constituting therefrom a demand for approval. Such a water law related demand is only applicable to large scale sewage treatment plants (§ 18c WHG), which must also pass an environmental impact assessment test according to the relevant EIA act. Whereas Land law (LWG) stipulates that in principle all wastewater facilities – and in particular all wastewater treatment plants – be subject to licensing procedures to examine the plants' viability.

2.6 Structure of Wastewater Charges

In the Bundesrepublik Deutschland, charges have been imposed on any discharge of wastewater whatsoever since 1981, separately for sewage and rain water. Sewage charges for large quantities are calculated on the basis of given pollutant loads – as the arithmetical product of pollutant concentration and annual wastewater quantities – for which specific units of noxiousness are defined in accordance with the given pollutant category (COD, Cu, Pb, Cr, Cd, Hg, P, N_{total}, AOX, T_F) (e.g., 50 kg COD), these unit equivalents are multiplied with the collection rate of the year (e.g., for 1997: DM 70 per noxious unit) to give the fee to be paid [e.g., for COD: (annual COD-load): (50 kg·70 DM)]. The required figures must principally be taken from the individual wastewater discharge permit, or be otherwise measured or estimated.

The fees for less important sewage discharges as well as for storm water are calculated on a flat rate basis.

The given diversity of special regulations, exemptions and charging variants, and multiple law amendments do not contribute to a smooth and rapid enforcement of this law.

2.7 Charging of Contributions and Fees

In as far as municipalities, towns and water associations undertake wastewater disposal on behalf of third parties, they are not only entitled to control and manage wastewater effluents in terms of time, origin, quality and quantity (via the obligatory approval for all new construction projects), but they are also entitled to charge the citizens or facility operators benefitting from effluent disposal with the costs incurred. This passing along of costs may be effected through payments under private law. Practically, however, it is done in compliance with the municipal charges act applicable in the different Länder, by act of state (administrative act) with title function (i.e., it has the same juristic meaning as a sentence). By paying a single contribution, the party entitled to use the land reimburses the service provider for the advantage gained through the estate's being connected to the public sewage disposal plant and the inherent possibility to use the estate properly (e.g., as industrial site or as development area). Generally, this fee is seen as a contribution to the capital expenditures for construction of the sewage plant being held available. The charges are to be seen as an equivalent for the factual use of the offered dewatering services; they cover the operating costs and write-offs. In Germany, wastewater charges differ widely and range from 3.50 to 16.00 DM m^{-3} of wastewater. The funds collected may also be considered in the calculation of charges. Factors playing a role in these calculations are so complex that to address them properly is beyond the scope of this chapter.

3 Waste Legislation

As has been stated for the area of water legislation, there is no constitutional obligation to create specific waste relevant rights, but rather an actual demand or a demand formulated by environmental policy, which may be determined by complex factors, to take action. And as has been set out for water legislation, also waste legislation has developed in several dogmatic stages:

(1) In times and regions when and where only a few people led a nomadic existence as hunter-gatherers, there was no factual demand for judicial regulation of what should happen with the negligible bulk of dispensable objects. And moreover, the waste then produced was easily organically degradable and inert at that, so that it would only involve hygienic problems, if any.

(2) With increasing populations, growing wealth and more advanced products, the number and quantity of objects that could no longer be used for their original purposes and the bulk of residues and leftover materials resulting from production of consumer goods, rose and still rises. Such residuals had so far been disposed of by the individual producer, which, in the face of growing urbanization, led to a lot of problems, and, in the end, to a regulated waste disposal on the municipal level: The

"producer" had to carry the waste to central places or the waste was collected by people put in charge with this task to be then centrally disposed of. Judicial regulations set up on the municipal level defined the details of this pathway in waste disposal.

(3) In a further development stage – not progressing concurrently with wealth and the resulting "throwaway mentality" – waste disposal turns into waste management. The State recognizes that waste disposal should be performed on an orderly and regulated basis to meet the challenge of preserving the natural basis of life for the community of man and of maintaining a well-functioning economy. This supraregional target cannot readily be achieved on a municipal level. Hence, responsibility to hold available the necessary disposal structures falls to the State: It defines what is to be understood by "waste" by issuing statutory regulations (e.g., whether there is a subjective demand and/or an objective need for disposal), and simultaneously separates it from other phenomena (wastewater, gas and fume immissions). The State differentiates between domestic and industrial, harmless and harmful wastes. It claims that all wastes be collected and – if unrecyclable – be disposed of separately. The State says who shall perform these tasks and defines waste disposal standards (e.g., for waste technology) and control mechanisms (e.g., transport permits, obligatory documentation, appointment of an in-plant environmental officer for wastes). And if the economy (as a whole) is strong enough, standards can be established claiming inertization of waste before its deposition ("thermal treatment") in order to eliminate any future environmental hazards and subsequent costs.

(4) The next step is a holistic view not only stressing awareness of limited availability of resources and space for waste depositing, but also responsibility for the generations to come. Such a holistic approach toward a sustainable development is targeted at a circulatory product management "from cradle-to-grave-to-cradle": Focus is primarily on waste avoidance and recycling, and – if technically or economically absolutely unfeasible – on disposal. And the definitely undesired disposal must be performed in a cross media approach, i.e., all environmentally relevant media must be considered. These objectives are supported by governmental act on overall product responsibility which may even include the manufacturers obligation to take back his products after use by the consumer. Waste substances finally removed from the economic cycle must be disposed of in an environmentally compatible way so as not to harm the ecosystem. This requires inertization in as far as possible and immobilization of waste constituents (e.g., by vitrification).

(5) An even more advanced approach is the juristic claim to provide against any form of waste that cannot be recycled, but so far only a few cases have been pursued and tested for feasibility. For this vision of absolute waste minimization it is essential to redefine and reengineer economic cycles: The course should be changed from end-of-pipe solutions for single products to synergetic thinking and acting.

As it has been described in Sect. 2 on water law and wastewater disposal, implementation of national waste laws also depends on manifold framework conditions and factors.

3.1 Waste Legislation in Germany

In Germany, waste legislation does not have a long history; first regulation was carried out on a municipal level and mainly dealt with epidemic control and urban development. With emerging disposal bottlenecks and severe incidents in the elimination of wastes from industry and trade it became more and more evident that there was an urgent need for governmental action. So the first waste related laws were initiated.

3.1.1 Distribution of Legislative Responsibilities

In accordance with basic law, the Bundesregierung has so-called concurring legislative responsibility for waste management (Art. 74 No. 24 GG). This legislative competence is complex and leaves the Länder with only little room for amendments. Since 1972, the Bundestag has used its competence and passed a first waste disposal act, which was abolished in 1986 and replaced by the act on avoidance and disposal of wastes. This act was followed by the new recycling and waste management law [Kreislaufwirtschafts- und Abfallgesetz (KrW-/AbfG)] which came into force on October 7, 1996, as a constituent part of the act on the avoidance, utilization and disposal of wastes of 1994, to promote the recycling economy and safety of environmentally compatible disposal of wastes. To ensure its applicability, a number of relevant provisions from the Bundesimmissionsschutzgesetz (BImSchG) (act on the prevention of harmful effects on the environment caused by air pollution, noise, vibrations and similiar phenomena – federal immission control act) are used.

This act – which follows the recommendations of the European waste laws – is only enforceable by implementation of several statutory orders (e.g., the ordinance on the determination of wastes with particular need for control, ordinance on the determination of wastes in need of control for reuse, ordinance on the methods of determination, ordinance on waste management concepts and accounting principles, ordinance on transport permits) as well as of administrative regulations [e.g., Technische Anleitung Siedlungsabfall (technical directive on residential waste), general administrative directives on the protection of ground water in case of storage or deposition of wastes].

The exercise of responsibility by the Länder as regards establishment of federal statutory orders and federal administrative regulations is restricted to their role in the Bundesrat (federal council) and the creation of rules for procedures and authority lines.

3.1.2 Structure of Waste Legislation

More recent German waste legislation defines "waste" in a rather broad sense: as a matter of fact any new product does embody specific waste related properties, so that when planning a new product a critical eco-audit should be performed. This eco-audit concerns not only the question of whether and what type of residual materials are likely to be generated during production, and how these might be avoided, re-used or disposed of without environmental adverse effects, but also whether and how the product itself might be recycled or regularly disposed.

This demonstrates that the new waste legislation stresses responsibility of the producing and manufacturing industries for the residuals they produce. Based on the "polluter pays" principle, focus is primarily on waste avoidance. Wastes that can be reused must be recycled as material or energy. Unavoidable or unrecyclable wastes must be disposed of in an environmentally compatible way. This shows to what an extent legislation has moved away from a more remedy oriented attitude towards multiple use and technical longevity of business goods.

Producers and owners of wastes are held responsible for compliance with legal avoidance, utilization and disposal regulations. However, some specified producers or owners of wastes may transfer their disposal duties to private third parties, existing associations or other bodies which might have been set up by the self-governing (quasi) corporations of the industry (§§ 16, 17 KrW-/AbfG). Besides, the overall responsibility of municipal bodies (§ 15 KrW-/AbfG) for the disposal of waste from private households (utilization and disposal) and other sources remains unchanged. This demonstrates the complexity of a circulatory management of products, comprising such activities as the holding available, assigning for beneficial use, collecting by pickup and delivery systems, transporting, storing and treating of wastes.

3.2 Administrative Enforcement of Waste Legislation

For constitutional reasons, the administrative implementation of waste legislation in the Bundesrepublik Deutschland is the sole responsibility of the Länder. These hold available at least a two-stage hierarchical administrative structure (lower waste management authorities with autonomous towns and districts, and superior waste management authorities in the ministries concerned); in the Länder except for the small city states, there is an additional middle echelon between the two authorities described above (higher waste management authority).

In practice, implementation of tasks is a largely privatized matter: On account of the favored and settled approach of product responsibility, the waste producer in trade and industry is responsible himself for an orderly avoidance strategy, utilization, or disposal of all wastes he generates or holds. Only for residential wastes there are special obligations on the part of local authorities to collect and dispose of them. However, they may delegate these tasks to private companies. To channel specific waste streams, municipalities have issued ordinances relating to separation and delivery of waste [e.g., separate collection and disposal (recycling) of glass, paper, compostable materials, metals, and plastics] within the framework of the "dual system" of waste collection and recycling (with the "green point" as its symbol for recyclable waste).

4 Identification of Targets for Waste Avoidance, Waste Recycling, and Waste Disposal

Waste legislation leaves the definition of what is to be understood by waste avoidance at the discretion of the producer. Art. 4 para. 2 KrW-/AbfG suggests the following: In-plant circulatory management of materials and substances, low waste product design, and a consumer attitude geared to the purchase of low waste and low pollution products. However, as the general rules of, e.g., trade and industrial safety laws continue to be valid, waste avoidance does not have absolute priority over other objects to be protected.

According to § 4 para. 1 KrW-/AbfG waste shall be recycled as material or as energy. Utilization of materials means recovery of raw materials from waste (secondary raw materials) or utilization of the inherent material properties for their original purpose or any other purpose with the exception of direct energy recovery. To sum it up: recovery of materials implicates that the main objective of the recycling measure is – from an economic point of view and under consideration of the pollutants present in the specific waste – to utilize the waste and not to remove its pollution potential.

Energetic utilization means the utilization of fuels as substitute fuel; the impetus on thermal use does not affect the thermal treatment of wastes for their disposal, in particular of domestic waste (incineration).

The requirements for waste disposal are defined in §§ 10 et seq. KrW-/AbfG and in the statutory orders and other directives based on this act, respectively.

Here, the following acts should be mentioned in particular:

(1) Siebzehnte Verordnung zur Durchführung des Bundes-Immissionsschutzgesetzes – Verordnung über Verbrennungsanlagen für Abfälle und ähnliche brennbare Stoffe (17th ordinance to the act on protection from the harmful effects on air pollution, noise, vibrations and other nuisances – ordinance on incineration plants for wastes and other combustible materials)
(2) Erste Allgemeine Verwaltungsvorschrift über Anforderungen zum Schutz des Grundwassers bei der Lagerung und Ablagerung von Abfällen (first general administrative instruction on the requirements for the protection of ground water in case of storage and deposition of wastes)
(3) Zweite Allgemeine Verwaltungsvorschrift zum Abfallgesetz; Teil 1: Technische Anleitung zur Lagerung, chemisch/physikalischen, biologischen Behandlung, Verbrennung und Ablagerung von besonders

überwachungsbedürftigen Abfällen – TA Abfall (second general administrative instruction to amend the waste act; part 1: technical directive on storage, chemical/physical, biological treatment, incineration and deposition of special wastes in particular need of control – TA Abfall, technical directive on waste disposal),

(4) Dritte Allgemeine Verwaltungsvorschrift zum Abfallgesetz; Technische Anleitung zur Verwertung, Behandlung und sonstigen Entsorgung von Siedlungsabfällen – TA Siedlungsabfall (third general administrative instruction to amend the waste act – technical directive on the utilization, treatment and otherwise disposal of domestic wastes – TA Siedlungsabfall; technical directive on residential waste disposal).

5 References

Abwasserabgabengesetz (wastewater charges act).

Abwasserverordnung (wastewater ordinance).

Allgemeines Preußisches Landrecht (general Prussian state law) of 1796.

Anhang 40 zur Abwasserverordnung: Metallverarbeitende und metallbearbeitende Betriebe (Annex 40 to the wastewater ordinance on wastewater from metalworking and metalprocessing workshops).

Bundes-Immissionsschutzgesetz, 1990 (act on the prevention on harmful effects on the environment caused by air pollution, noise, vibrations and similiar phenomena).

Deichgesetz (dyke act) (1/28/1848).

Dritte Allgemeine Verwaltungsvorschrift zum Abfallgesetz, Technische Anleitung zur Verwertung, Behandlung und sonstigen Entsorgung von Siedlungsabfällen – TA Siedlungsabfall (third general administrative instruction to amend the waste act – technical directive on the utilization, treatment and otherwise disposal of domestic wastes – TA Siedlungsabfall, technical directive on residential waste disposal).

Erste Allgemeine Verwaltungsvorschrift zum Abfallgesetz über Anforderungen zum Schutz des Grundwassers bei der Lagerung und Ablagerung von Abfällen (first general administrative instruction on the requirements for the protection of ground water in case of storage and deposition of wastes).

Europäische Wasserrahmenrichtlinie (amended Council Directive establishing a framework for Community action in the field of water policy) (Entwurf).

European Directive of disposal of municipal wastewater (5/21/1991).

Französischer Code Civil (French Code Civil).

Gesetz über die Wassergenossenschaften (act on water cooperatives) (4/1/1879).

Gesetz zur Vermeidung, Verwertung und Beseitigung von Abfällen, 1994 (act on the avoidance, utilization and disposal of wastes).

Grundgesetz (basic law, constitution).

Grundwasserverordnung (groundwater ordinance).

Hochwasserschutzgesetz (flooding protection act) (8/16/1905).

Kreislaufwirtschafts- und Abfallgesetz, 1996 (recycling and waste management law).

Landeswassergesetze (Länder water regulations).

Privatflußgesetz (private river act) (2/18/1843).

Quellenschutzgesetz (act on the protection of sources) (5/14/1908).

Ruhrverbandsgesetz (Ruhr river association act).

Siebzehnte Verordnung zur Durchführung des Bundes-Immissionsschutzgesetzes – Verordnung über Verbrennungsanlagen für Abfälle und ähnliche brennbare Stoffe – 17. BImSchV – (17th ordinance to the act on protection from the harmful effects on air pollution, noise, vibrations and other nuisances – ordinance on incineration plants for wastes and other combustible materials).

Strafgesetzbuch (criminal law).

Strombauverwaltungsgesetz (act on the administration of hydraulic engineering) (8/20/1883).

Technische Anleitung (TA) Siedlungsabfall (technical directive on residential waste disposal).

Verordnung über wassergefährdende Stoffe bei der Beförderung in Rohrleitungen (ordinance on water hazardous substances for water transport in pipelines, 1973).

Wasserhaushaltsgesetz (act on the regulation of matters relating to water – federal water resources act).

Wasserstraßengesetz (waterway act) (4/1/1905).

Zweite Allgemeine Verwaltungsvorschrift zum Abfallgesetz; Teil 1: Technische Anleitung zur Lagerung, chemisch/physikalischen, biologischen Behandlung, Verbrennung und Ablagerung von besonders überwachungsbedürftigen Abfällen – TA Abfall, (second general administrative instruction to amend the waste act; part 1: technical directive on storage, chemical/physical, biological treatment, incineration and deposition of special wastes in particular need of control – TA Abfall, technical directive on waste disposal).

II Processes of Wastewater Treatment

Wastewater Sources and Composition

9 Municipal Wastewater and Sewage Sludge

PAUL KOPPE
ALFRED STOZEK
VOLKMAR NEITZEL
Mülheim/Ruhr, Germany

List of Abbreviations

AOX	adsorbable organically bound halogens
BOD	biological oxygen demand
COD	chemical oxygen demand
DHEP	di-(2-ethylhexyl)-phthalate
DOC	dissolved organic carbon
EN	European Norms
FIAS	Flame Ionization Absorption Spectrometry
GC	gas chromatography
HC	hydrocarbons
IR	infrared spectrometry
ISO	International Organization for Standardization
m_d	mass of dry substance
MPN	most probable number
PAH	polycyclic aromatic hydrocarbons
PCDD	polychlorinated dibenzodioxins
PCDF	polychlorinated dibenzofurans
TIC	total inorganic carbon
TOC	total organic carbon

1 Definition of Municipal Wastewater

1.1 Introduction

Municipal wastewater and sewage sludge represent a suitable object for biotechnology because of their special composition. They contain all substances which are needed for the growth of microorganisms.

The great amount of wastewater and sludge which come up daily make biological treatment also attractive from a technical and economic point of view. Only few products of human activities reach the quantity of wastewater, which has to be treated and disposed of continuously.

Hygienic considerations forced the treatment of wastewater during the last third of the 19th century. But all experimentation was done with physical or chemical processes. This is surprising because nature shows scientists and engineers biological methods with the self-purification processes of waters.

Not until the beginning of this century treating methods on a microbiological basis were used on a larger scale such as Emscher tanks, trickling filters and activated sludge tanks (IM-

HOFF, 1979). The microbiological processes were developed further on the basis known. In the meantime planning and operation of biological wastewater treatment plants are supported by sophisticated mathematical models (DOLD and MARAIS, 1986a, 1995).

1.2 Definition According to Origin and Possibilities of Treatment

Municipal wastewater is a mixture of many substances. It contains an immense number of single substances and is defined according to its origin. Usually it is cleaned in a public treatment plant and then discharged into the next recipient (lake, sea, river).

During dry weather the lion's share of municipal sewage is domestic wastewater. It arises daily from toilet flushing, showers, bathing, and hand washing, from cleaning, cooking, and dishwashers in private households. According to the structure of the settlement in the catchment area excrement sludge, e.g., from 3-room systems, is collected and delivered with tankers to a municipal treatment plant. The contents of chemical toilets, which are installed on camping sites, building sites and vehicles like coaches, is mixed into the wastewater of larger treatment plants.

The wastewater of hotels, restaurants, barracks and hospitals, which has a similar composition to domestic sewage and is treated in the same way, is also discharged into the public sewer. For restaurants and canteen kitchens normally a grease separator has to be installed and for barracks with their great number of vehicles an oil separator. The wastewater from hospitals contains higher concentrations of disinfectants and often iodine containing contrast agents. If there is highly infectious wastewater, e.g., from infectious disease departments or pathological institutes, it must immediately be disinfected chemically or by heating at the place of its origin.

In addition to private households and similar institutions a large number of trade companies, sometimes also large factories (e.g., food and beverage industry), discharge their sewage into the public sewer (ATV-Regelwerk Abwasser Abfall, 1992, 1994). This so-called "indirect discharger" produces very different types of wastewater. Toxic substances lead to trouble at the treatment plants and in some cases to fish mortality in the recipient. Today it is required in Germany that indirect dischargers with dangerous substances in the sewage have to minimize the concentrations of the substances just like direct dischargers (Bundesgesetzblatt, 1997). If wastewater tributary currents are mixed, it is a general rule that the decrease in the quantity of waste must at least be as great as in the case of separate treatment. Some of the dangerous substances are specified in German regulations (Gesetz- und Verordnungsblatt Nordrhein-Westfalen, 1989):

(1) aldrin, dieldrin, endrin, isodrine
(2) asbestos
(3) cadmium
(4) chlorine
(5) DDT
(6) 1,2-dichloroethane
(7) hexachlorobenzene
(8) hexachlorobutadiene
(9) hexachlorocylohexane
(10) pentachlorophenol
(11) mercury
(12) tetrachloroethane
(13) tetrachloromethane
(14) trichlorobenzene
(15) trichloroethane
(16) trichloromethane

Not only the companies, which possibly have some of the substances listed above in their wastewater have to treat their sewage before discharging in a public sewer, but also those with a toxic effect in biological tests. The origin of the branch of the polluter can be listed as follows (Bundesgesetzblatt, 1991):

(1) heat generator, energy, mining
  - treatment of exhaust gas and polluted air, slag, condensates from firing
  - cooling systems
  - coal and ore preparation
  - coal refining and recycling of recyclable material, briquetting process
  - production of coal, activated carbon, soot
(2) quarry, building material, glass, ceramics
  - production of fiber cement and fi-

ber cement products
- production of processing of glass, fiber glass and mineral fiber
- production of ceramic products

(3) metals
- metal working industry (e.g., galvanizing industry, pickling installation, production of circuit boards, battery production)
- production of iron and steel including foundries
- production on non-iron metals including foundries
- production of ferric alloys

(4) inorganic chemistry
- production of basic chemicals
- production of mineral acids, bases and salts
- production of alkali and chlorine by electrolysis
- production of mineral fertilizer
- production of soda and corundum
- production of inorganic pigments and mineral dyes
- production of semiconductors and photocells
- production of explosives
- production of highly dispersed oxides
- production of barium compounds

(5) organic chemistry
- production of basic chemicals
- production of dyes
- production and treatment of synthetic fibers
- production and treatment of synthetic materials
- production of halogenated compounds
- production of organic explosives
- production of paper and leather
- production of pharmaceuticals
- production of pesticides
- production of raw material for detergents
- production of cosmetics
- production of gelatin and glue

(6) mineral oil, synthetic oil
- processing of mineral oil
- recycling of oil from oil–water mixtures
- production of synthetic oil

(7) printing office, reproduction companies, processing of resin and synthetic products
- production of print material and graphical products
- copy shops and developing institutions
- production of foils, picture and sound carriers
- production of coated and soaked materials

(8) wood, cellulose and paper
- production of pulp, paper and paste board
- production and coating of fiber boards

(9) textiles, leather and fur
- production of textiles and textile finishing
- production of leather and leather finishing
- chemical cleaning and laundry

(10) others
- utilization, treatment, storage and handling of waste and chemicals
- medical and scientific research and development, hospitals, laboratories and institutes
- technical cleaning companies
- vehicle garage and car
- water recycling
- painting and lacquer firms
- production and processing of plant and animal extracts
- production and use of microorganisms and viruses
- plants for soil washing

The substances listed above are part of the Sewage's Origin Ordinance which is changed by Bundesgesetzblatt (1997), but it has kept its importance as source of information.

Appendix 40 of Bundesgesetzblatt (1997), which is applied to metal working factories, shows the expenditure of the pretreatment of the sewage before it is discharged into a public sewer. For the following 18 substances concentration limits are established: arsenic, barium, lead, cadmium, free chlorine, chromium, chromium (VI), cobalt, easily emittable cyanide, copper, nickel, mercury, selenium, silver, sulfide, tin, zinc, and adsorbable organic halogenated compounds.

At some smaller indirect dischargers not the concentrations are supervised, but the correct operations. So photo shops have to dispose of developing and fixing baths by special firms and to record and prove it. In dental practice the rinsing water contains particles of amalgam, which originates from removal of old fillings. These practices have to prove the installation of small centrifuges.

A further source of sewage in municipal wastewater is the drainage water from dumps of waste. It could be waste from domestic refuse, trade refuse and thickened sludge of drinking water or wastewater treatment. The regulations of the ordinance (Bundesgesetzblatt, 1997) are valid also for this drainage water, not only for direct discharge into bodies of water, but also for indirect discharge into a public sewage treatment plant.

In Germany it is specified by acts, ordinances and local statutes, that the sewage in tributary currents of indirect dischargers do not have an inhibiting effect on the microbiological sewage treatment (especially on the metabolism and the reproduction of bacteria). In other European countries and in the United States of America the situation is similar (see Chapter 1, this volume). In many cities the precipitation from roofs and streets is discharged into municipal treatment plants by mixing in the sewer. In the case of rain and thaw it forms a considerable quantity of the municipal sewage (KOPPE and STOZEK, 1999). In Germany only a few communities own a separate sewer in which the precipitation flow is collected, sometimes mechanically cleaned in storm water settling tanks and discharged directly into the next recipient. To an increasing extent, it is tried to percolate rain water into subsoil and raise the amount of ground water production (ATV-Information, 1996b).

Cooling water from larger thermal power stations is only exceptionally discharged into a municipal sewage disposal plant. Beyond an increase in temperature it mainly results in an undesirable dilution of the contained substances.

Principally it is not allowed to discharge radioactive cooling water and wastewater from nuclear power stations and reprocessing plants for fuel rods into a municipal sewer disposal plant. Also the wastewaters from firms working with radioactive nuclides (e.g., in nuclear medicine), have to be disposed in another way. Nevertheless, it could not be excluded that radioactive nuclides get into the municipal sewage by rain (e.g., in the case of an accident like that in Tschernobyl 1986) and causes trouble, because of accumulation in the sewage sludge (ATV-Merkblatt M 267, 1995).

Another undesirable compound of municipal sewage is the so-called "imported water". Because of local conditions often great amounts of water from springs and brooks drain into the subterranean sewer. Water from brooks, which had flown through towns earlier and impeded the increasing motor traffic were directed into the sewer in the first half of this century and thus mixed with the sewage. Groundwater also flows into the sewage by an unwanted drainage of the sewer. It is discernible by a dilution effect and a decrease of the temperature in summer.

2 Amount of Municipal Wastewater

According to the different tributary currents which build up the influent of a municipal sewage treatment plant, the amounts vary to a great extent, locally depending on countries and regions as well as depending on time of the day and the season.

Mostly the influent of a treatment plant is measured in L s^{-1} or as an annual average in L a^{-1}. From the daily average of the influent and the number of connected inhabitants (in Germany about 90% of the whole population) the inhabitant-specific influent can be calculated in L n^{-1} d^{-1}. It has to be considered that the number of inhabitants (n) living in the catchment area of the treatment plant, whose apartments and houses are connected to the plant, differ from that of the number of inhabitants who stay there for a short time and use the sanitary installations. The number of the last category can be about 3 times that of residents (e.g., in a holiday resort with seasonal tourism).

The domestic sewage consists of the following 5 main components, which occur in different amounts depending on the consuming behavior. Tab. 1 shows the usual variety. In Germany it is calculated with a value of 150 L n^{-1} d^{-1} according to the consumption of drinking water.

During a day the influent of municipal sewage shows a typical course, which is affected by the different time of flow in the sewer. So the minimal influent of a treatment plant with a daily average of 100 L min^{-1} amounts to 30 L m^{-1} at 5 a.m. and the maximal influent amounts to 300 L min^{-1} at 2 p.m.

The indirect dischargers enhance the amount of sewage in the inflow of a municipal sewage treatment plant depending on their kind and number. In the catchment area of the Ruhr river there is a relation of 200 inhabitants to one firm with production-related sewage greater than 1 m^3 d^{-1}. In Germany there are municipal sewage treatment plants with only 10 L n^{-1} d^{-1} of trade sewage (e.g., from butchers or bakers) to 100 L n^{-1} d^{-1} (e.g., from a brewery or paper mill).

In the case of shift work the trade sewage flows to the treatment plant equally distributed throughout the day. Some firms produce wastewater only during daytime, not on Sundays or on public holidays, and they have work holidays. This can be seen by the decreasing amount of sewage in the influent of a municipal sewage treatment plant.

Without "imported water" the quantity of domestic and trade sewage, which ordinarily flows into a municipal sewage treatment plant, amounts on average to 200 L n^{-1} d^{-1}. The quantity of "imported water", which comes from ground water and brooks should in fact be unimportant. In some catchment areas it could amount to the sum of sewage from households and indirect dischargers. In the Ruhr area the average amounts to about 100 L n^{-1} d^{-1} "imported water".

In the case of combined sewage the municipal sewage contains large amounts of rain water during or after rain or melting of snow. While falling down this precipitation is polluted by the atmosphere, and it picks up more mud from roofs, streets, and fastened squares (XANTHOPULOS and HAHN, 1993; BRINKMANN, 1985).

In Germany, which has a humid climate, the amount of precipitation which flows into the public sewer, equals the quantity of municipal sewage in dry weather at the annual average (about 300 L n^{-1} d^{-1}). By calculation the whole amount of domestic, trade, and precipitation water could theoretically be treated, if the maximal effluent was held back in storm water retention tanks and if the sewage treatment plants were laid out at twice their size. In practice this could not be done, because the maximum of effluent from heavy rain amounts to 100 times that in dry weather. Normally the lion's share of municipal sewage is discharged into the recipient when such high effluents occur (IMHOFF and IMHOFF, 1993).

The whole flow off of precipitation water out of the sewer with its storage capacity and storm water retention tanks may require several days. For a determination of the effluent in dry weather, a longer precipitation-free period is needed (ANNEN, 1980).

The great variation of influent of sewage and the variation of concentrations of contained compounds make the biological treatment of municipal sewage difficult. This can only partly be corrected by regulation tanks with a great volume.

Tab. 1. Main Components of Municipal Sewage

Component of Municipal Sewage	Specific Daily Amount of Sewage per Person [L n^{-1} d^{-1}]
Drinking, cooking and dishwasher water	3– 10
Urine and feces	1– 3
Toilet water, WC	10– 30
Cleaning and washing water	5– 50
Washing, bathing and shower water	5–500

3 Effect of the Sewer on the Municipal Sewage

Beyond the kind of an urban drainage (combined or separate sewers or a combination of both) the whole length, the degree of ramification, and the inclination have great influence on the composition of municipal sewage which flows into a wastewater treatment plant. The velocity of flow must be 1 m s^{-1} at least so that urban sewage can be treated as soon as possible. Especially in older sewage systems this is not always the case. At a channel length of 10 km the average flow time of the sewage is about 3 h, if no hydraulic tailback occurs. During flow the coarse particles (rests of food, paper, oil drops, feces) are crushed and stabilized by detergents which are present in surplus. So the dispersion is enhanced noticeably at the inflow of a treatment plant. If the turbulence is sufficient, highly volatile substances like gasoline are emitted into the air of a channel. On the other hand oxygen can be picked up.

Some substances of trade sewage such as heavy metals are adsorbed at the suspended solids of the municipal sewage. Tributary currents of different indirect dischargers can neutralize each other. In doing so precipitation reactions can occur (e.g., iron hydroxide). Especially the biochemical reactions in the sewer have to be pointed out. Free enzymes and suspended microorganisms, which are plentiful in municipal sewage, bring about a fission of some organic compounds.

Most urea which is available in every fresh sewage is hydrolyzed into ammonia and carbonic acid by urease. At higher temperatures the degree of hydrolysis is higher (normally 5–25 °C). The sewage slime developing at locations of the sewer with continuous wetting is created by aerobic and anaerobic microorganisms (SAUER et al., 1997). According to the construction (partly run through ducts with free fall, fall structure and overfall in the systems, full run through ducts with overpressure or underpressure) aerobic or anaerobic decomposing processes prevail.

The fast oxidizable compounds of municipal sewage are mineralized with the help of nitrate (mainly from drinking and trade water), if there is a lack of oxygen. Therefore, nitrate is only present in low concentrations or is not analytically detectable in the influent of a municipal sewage treatment plant.

Analogously the load of detergents in the influent to the treatment plant is smaller than can be calculated from the consumption in households.

If the nitrate is consumed and there is too little oxygen from the air of the channel, sulfate is reduced to hydrogen sulfide by anaerobic bacteria of the sewage slime. This can create a very bad odor. With iron(II) ions iron sulfide is created, which gives sewage a grey to black color.

Generally it can be stated, that the amount and composition of domestic and trade sewage is not the same at the location of emission and at the influent of wastewater which is treated in a municipal sewage treatment plant.

4 Composition of Municipal Wastewater

4.1 Classification of Compounds According to Different Characteristics

The nearly constant eating and living habits of people in European countries determine the composition of municipal wastewater. Municipal wastewater is a complex mixture of many different substances. The main ingredient is water with about 99%. The multitude of other components makes it necessary to divide them into several classes:

(1) oxygen consuming substances,
(2) nutrients,
(3) trouble causing substances,
(4) toxic substances.

Tab. 2 gives some examples of components of wastewater and their effects.

From the point of view of wastewater treatment it seems to be suitable to classify the compounds in the sewage according to physical, chemical, and biochemical aspects.

Tab. 2. Examples of Components of Wastewater and their Effects

Main Class	Oxygen-Consuming Substances	Nutrient	Substances Causing Problems	Toxic Substances
Subclass	spontaneous oxygen-consuming	nitrogen compounds	technically	somatically
	slowly oxygen-consuming	phosphorous compounds	ecologically	genetically
Damaging effect	load of the oxygen balance of waters	the consequence of overfertilization is eutrophication of waters	corrosive, odor trouble and discoloration of waters, decreasing the exchange of compounds with the atmosphere	acute and chronic toxicity, bioaccumulation, mutagenicity, carcinogenicity, reproduction endangering

4.1.1 Compounds Classified According to Physical Aspects

The following parameters belong to the typical physical characteristics:

(1) state of aggregation,
(2) density, and
(3) dispersion.

According to the state of aggregation the compounds of sewage can be solid, liquid, and gaseous. Solid and liquid compounds can float on the surface, be suspended in water or settle down, depending on the density. Some examples are given in Tab. 3. According to the dispersion it can be distinguished between

(1) dissolved,
(2) partially dissolved (colloidal), and
(3) suspended

substances in a semiquantitative way. A more precise classification can be done according to the average diameter. The diameter is between 10^2 mm and 10^{-7} mm. The separation of suspended solids can be effected by several operations, e.g., screening, filtration, or sedimentation. Tab. 4 contains some typical examples in relation to practice (KOPPE and STOZEK, 1999).

4.1.2 Compounds Classified According to Chemical Aspects

In municipal sewage nearly all stable elements of the periodic system can be present. But this does not mean that an analysis of all substances is possible and convenient. In a first step the compounds of sewage can be classified into inorganic and organic substances.

The inorganic substances consist of

(1) non-metal compounds (e.g., nitrate, phosphate, sulfate)
(2) semimetal compounds (e.g., borate, silicate), and metal compounds (e.g., Na, Ca, Cu salts).

Tab. 3. Examples of Components with Different Density

Density [g cm^{-3}]	Aggregate State		
	Solids	Liquids	Gases
>1	starch, clay, sand	bitumen, trichlorethene, chloroform	
~1	celluloseethyl ether	chlorinated paraffin	
<1	polyethylene, wax	gasoline, vegetable oil	oxygen gas bubbles, nitrogen foam

Tab. 4. Dispersity of Sewage Compounds

Particle Diameter [mm]	Kind of Particle (examples)	Run-Through	Retention by
10^2	tins, pieces of wood, corks, rats		coarse screen
10^1	matchsticks, toilet paper	coarse screen	fine screen
10^0	coarse sand, tomato pips, worm eggs	fine screen	1 mm sieve
10^{-1}	fine sand, rubbish, starch, textile fiber	1 mm sieve	0.1 mm sieve
10^{-2}	large bacteria, protozoa	0.1 mm sieve	paper filter
10^{-3}	smaller bacteria, large viruses	paper filter	membrane filter
10^{-4}	small viruses, large colloids	membrane filter	ultrafilter
10^{-5}	medium-sized colloids	membrane filter	ultrafilter
10^{-6}	small colloids	membrane filter	ultrafilter
10^{-7}	dissolved molecules (sugar), ions	ultrafilter	reverse osmosis

The organic compounds can be classified according to the structure used in textbooks of organic chemistry, but this is difficult and not practicable because of the great number of substances. A good way is to distinguish between compounds of natural provenience and synthetic compounds. Examples of both groups are given in Tabs. 5, 6 (KOPPE and STOZEK, 1999).

Tab. 5. Sewage Compounds of Natural Origin

Compound	Origin
Fats	food, excrement
Carbohydrates	food, excrement
Proteins	food, excrement
Urea	urine

Tab. 6. Synthetic Sewage Compounds

Compound	Origin
Detergents	washing powder, dishwashing powder, personal hygiene shampoo, soap
Chlorinated hydrocarbons (AOX)	lacquer and dye industry, motor vehicle garage

4.1.3 Compounds Classified According to Biochemical Aspects

This category of compounds in sewage is characterized by the rate of degradation by microorganisms in the biological treatment tank as follows:

(1) fast degradation (e.g., ethanol),
(2) slow degradation (e.g., ethanediol),
(3) non-degradable (e.g., chlorinated hydrocarbons), and
(4) toxic (e.g., sulfonamides).

After this more qualitative survey of compounds in sewage, in a second step something has to be said about the average values of basic parameters of municipal wastewater. But above all some remarks are needed about the analysis of sewage, the parameters, and sampling (Sect. 4.2).

4.2 Remarks on the Analysis of Waters, the Parameters, and Sampling

From the scientific point of view it would be interesting and sometimes reasonable to analyze as many single substances as possible. But for the routine of sewage investigations it is too expensive, it takes too much time and is not realistic. Therefore, it is necessary to limit the analyses, reduce the effort, and restrict the number of parameters in a treatment-oriented, but sufficient way. Some different purposes have to be distinguished, e.g.,

(1) planning of treatment plants,
(2) operating treatment plants,
(3) controlling the function of treatment plants,
(4) controlling the observance of limits,
(5) fixing sewage taxes,
(6) investigating the behavior of harmful substances.

Practically the control of municipal treatment plants is not based on measuring single substances, but on sum or group parameters. Tab. 7 gives an overview of these 3 categories of parameters with a short characterization and some examples. The most important summarizing parameters of organic and inorganic compounds in sewage are shown in Tab. 8 and for physical and physical-chemical parameters in Tab. 9.

For the analysis of the above parameters there are about 150 national and international standards. In Germany there are the German standards for water, wastewater and sludge analysis, of which the biggest part consists of

Tab. 7. Survey of Parameters

Parameter	Characterization	Example
Summary parameter	estimation of all compounds which have a common component in structure or the same chemical feature	• chemical oxygen demand (COD) • biochemical oxygen demand (BOD)
Group parameter	selective estimation of a group of compounds with a characteristic structure or specific feature	• organic acids • fat • phenols • detergents
Single compounds	estimation of single compounds with special analyzing methods	• Fe^{3+} • Cl^- • NH_4^+

Tab. 8. Commonly Used Summarizing Parameters

Parameter	Unit
Dissolved volatile solids	mg L^{-1}
Ignition residue	mg L^{-1}
Dry residue	mg L^{-1}
Concentration of filterable substances	mg L^{-1}
Volume concentration of settleable substances	ml L^{-1}
Total organic carbon (TOC)	mg L^{-1} C
Dissolved organic carbon (DOC)	mg L^{-1} C
Total inorganic carbon (TIC)	mg L^{-1} C
Chemical oxygen demand (COD)	mg L^{-1} O_2
Biochemical oxygen demand (BOD)	mg L^{-1} O_2
Organic nitrogen	mg L^{-1} N
Adsorbable organically bound halogens (AOX)	mg L^{-1}
Acid capacity	mmol L^{-1}
Base capacity	mmol L^{-1}

Tab. 9. Commonly Used Physical and Physical-Chemical Parameters

Parameter	Unit
Coloring	m^{-1}
Turbidity	FNU
pH value	
Conductivity of electricity	$\mu S\ cm^{-1}$
Spectral absorption at 254 nm	m^{-1}
Redox voltage	mV
Density	$g\ cm^{-3}$

DIN, EN, and ISO norms (SCHMIDT, 1993; American Health Association, 1985). Because of the great number of standards only the groups of methods are mentioned. The German standards comprise

A general declarations
B smell and taste
C physical and physical-chemical parameters
D anions
E cations
F common measurable groups of parameters
G gaseous compounds
H summary effect and component parameters
K microbiological methods
L tests with water organisms
M methods for biological-ecological investigations
P single substances
S sludge and sediments

It is desirable to harmonize all methods so that the results of the measurements are comparable. A step in this direction is the standardization by ISO.

The compounds in sewage have very different concentrations (from mg L^{-1} over µg L^{-1} to pg L^{-1}). Hence different analytical methods with different measuring areas are needed. Often substances have to be concentrated before measuring. Tab. 10 gives an overview of usual methods with their measuring areas and some examples for substances.

Tab. 10. Commonly Used Analyzing Methods with their Working Areas

Method	Example	Working Area
Gravimetry	filterable substances, SO_4^{2-}	mg L^{-1}–g L^{-1}
Titrimetry	organic acids, COD, Cl^-, NH_4^+	mg L^{-1}
Photometry (combined with FIAS)	CN^-, Fe^{3+}, NO_2^-, NO_3^-, Mn^{2+}	µg L^{-1}–mg L^{-1}
Conductometry	electric conductivity	S m^{-1}
Colorimetry	AOX	µg L^{-1}–mg L^{-1}
Potentiometry	COD, pH, oxygen, F^-	µV–mV µg L^{-1}–mg L^{-1}
Polarography, Voltametry	NTA, EDTA, heavy metals	µg L^{-1}–mg L^{-1}
Atomic absorption spectrometry	heavy metals, special elements	µg L^{-1}–mg L^{-1}
Atomic emission spectrometry	heavy metals, special elements	mg L^{-1}
Infrared spectrometry (IR)	hydrocarbons	mg L^{-1}
Gas chromatography (GC)	NTA, EDTA, chlorinated hydrocarbons, pesticides	µg L^{-1}
HPLC	pesticides, polyaromatic hydrocarbons	ng L^{-1}–µg L^{-1}
Ion chromatography	anions, cations	mg L^{-1}
Enzymatic tests	acetylcholinesterase inhibition test	
Biotests	fish test, algae test	

In wastewater technique concentrations are often measured in g m^{-3}. For the calculation of load the discharge has to be considered.

$$Q\,[m^3\,d^{-1}] \cdot c\,[g\,m^{-3}] = load\,[g\,d^{-1}] \quad (1)$$

In practice inhabitant-specific loads are often used. With the help of inhabitant-specific volume of water usual concentrations can be calculated.

Example:

Urea (load): 30 g n^{-1} d^{-1}
Q (inhabitant-specific): 200 L n^{-1} d^{-1}
Concentration: 30 g n^{-1} d^{-1}/200 L n^{-1} d^{-1} = 0.15 g L^{-1} = 150 mg L^{-1}

In Germany there are legal preconditions for self-monitoring sewage treatment plants. For this internal control (function control) which is done by operating personnel simple, secure, and quick measuring methods are used in the plant. Several producers have developed short-time tests for the required parameters. For practical use there are short-time tests, simple operating methods, and standardized methods. Simple operating methods are based on standardized methods, but their handling is easier. Their results are not as precise as those from standardized methods. Therefore, the use of operating methods has a natural limit, above all if the results exceed a limit. There exists a catalog of requests for the producers of the tests, for the operating personnel, and for the principal purpose of use. In Tab. 11 standard and operating methods are compared based on some features (KOPPE and STOZEK, 1999).

For supervision and control of clarification processes continuous measurements with online instruments are used. For several years some of the parameters have been relevant for measurements according to self-monitoring. But the use of process analyzers has not only advantages. Some important aspects are:

(1) the operator of a treatment plant gets quick information on the state of operation;
(2) online instruments are conceived in the way that they have a high availability, sensitivity, and reproducibility for steering the treatment process;
(3) the expenditure is considerable in relation to costs of investments and operation as well as for quality assurance, maintenance, and deployment.

Fig. 1 shows possibilities of continuous measurements of parameters for physical, physical-chemical methods, and process analyzers. The most important requirement for a correct result of water analysis is flawless sampling. The best analysis is useless, if sampling provides elementary faults. Samples must be representative for the relevant investigation. In several standardized methods there are guidelines for the technique and operation of sampling as well as the development of a sampling program for drinking water, wastewater, and surface water. Fig. 2 shows kinds of continuous and discontinuous sampling applied in practice.

4.3 Typical Values for Summary Parameters

The knowledge of typical values for parameters of municipal wastewater is not only important for planing and dimensioning of wastewater treatment plants. The values and some of their calculable relations inform about the influence of industrial sewage and "imported water". Tab. 12 gives an overview of typical values and their variance of inhabitant-specific parameters as well as the total pollution with inorganic and organic parts. The basis for the calculation of concentrations is the inhabitant-specific volume of water of 200 L n^{-1} d^{-1}.

The dry residue of an average sewage is about 1 g L^{-1}. Wastewater with values $\leq$0.5 g L^{-1} is described as relatively less concentrated (weak), wastewater with >2 g L^{-1} as relative high concentrated (strong wastewater). The parts of inorganic and organic compounds are nearly 50% of the total pollution. The ratio of dissolved to suspended substances depends on the filtering material. With filter

Tab. 11. Typifying Physical-Chemical Methods for Water and Sludge Investigation (KOPPE and STOZEK, 1999)

	Standardized Methods		Non Standardized Methods	
	Reference Methods	Standardizeable Laboratory Methods	Operating Methods	Quick Tests
Aim	(1) determination of tax (2) supervision of limited values (3) determination of loads (4) balance of material (5) reference assay (6) trend analyses (7) operation of plants		(1) presorting of samples for supervision (2) predecision whether samples had to be analyzed with reference methods (3) early information about great pollution to find the occasion (4) early information about great pollution to repulse danger (5) decision, whether samples had to be taken (6) decision whether quick measures are necessary (7) operation of plants (8) analyzing some quick altering substances (9) self-monitoring of sewage treatment	rough orientation about the presence of substances and groups of substances
Precision	mostly very precise (±10%)		quantitative to semiquantitative (10–50%)	qualitative to semiqualitative (>50%)
Selectivity	mostly good		good to moderate	mostly less selective
Preparation	often complex and large scale		mostly less complex	none
Operation	nearly always in the laboratory		in the laboratory and in the plants	in the plant (position of sampling)
Use for	all waters and sludge		limited to special kinds of waters and sludge	limited to special waters
Personnel	experts in the laboratory		experts and service personnel	introduced non-experts
Time needed	often long	often less than for reference methods	mostly less	much less
Costs per measurement	often expensive	often less than for reference methods	low	very low

paper the ratio amounts to one third, with membrane filters it amounts to two thirds.

Tab. 13 shows the average composition of municipal sewage on the basis of summarizing parameters in different ranges of concentration. COD is the "chemical oxygen demand" (oxidation by dichromate with silver ions as a catalyst); BOD is the "biological oxygen demand" and TOC the "total organic carbon" (KOPPE and STOZEK, 1999).

For the biological treatment of wastewater it is important to know the load of organic compounds, especially the amount of biodegradable substances. Above all the summarizing parameter BOD has a key function for the biotechnology and for the oxygen balance of a

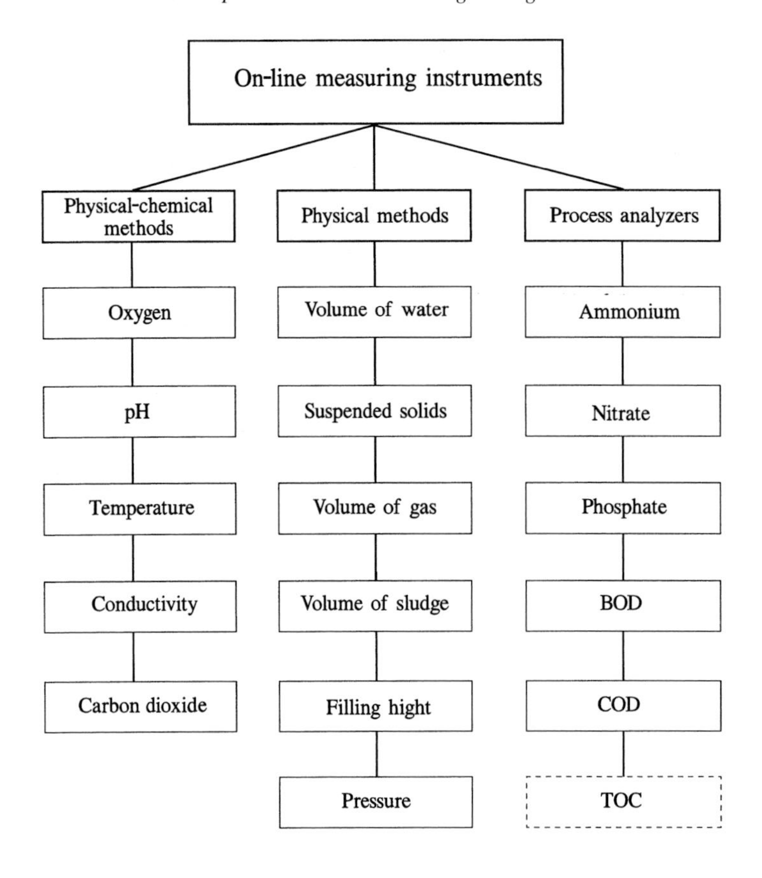

Fig. 1. On-line measurements.

discharge into a surface water (self-purification). BOD is defined as the consumption of oxygen in water caused by the activity of bacteria. As the BOD depends on the time of exposure this must be limited. BOD_5 means an exposure of 5 d, according to the average flow time of recipient rivers in Great Britain.

The "total BOD" comprises the oxygen demand for the biological oxidation of carbon compounds and for nitrification. With the addition of a nitrification inhibitor (e.g., 0.5 mg L^{-1} allylthiourea) only the "carbon BOD" is determined.

At that time the method was set down to 5 d. The degradation process is like that in the biological treatment tank, but with a lower density of microorganisms.

Toxic effects are indicated if the BOD does not decrease proportionally to a stepwise dilution of a water sample. To shorten the measuring time for operational purposes the so-called "short-time-BOD" has been developed. The measured values are proportional to the BOD_5 but not very precise.

Tab. 14 shows inhabitant-specific pollution loads. On the average the load of BOD_5 amounts to about 60 g n^{-1} d. Tab. 15 shows proportional values for COD, BOD_5 and TOC typical for domestic wastewater. If there are essential deviations, the reason would be the influence of other kinds of wastewater, especially of industrial origin. A quotient COD/ BOD_5 >3 indicates a high share of persistent (refractory; recalcitrant) compounds, if there are no toxic effects.

4.4 Group Parameters for Organic Compounds

Three of the 10 sewage-relevant group parameters are very important because of their

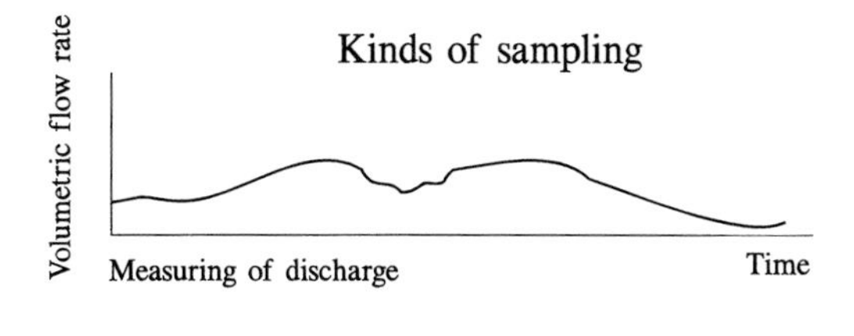

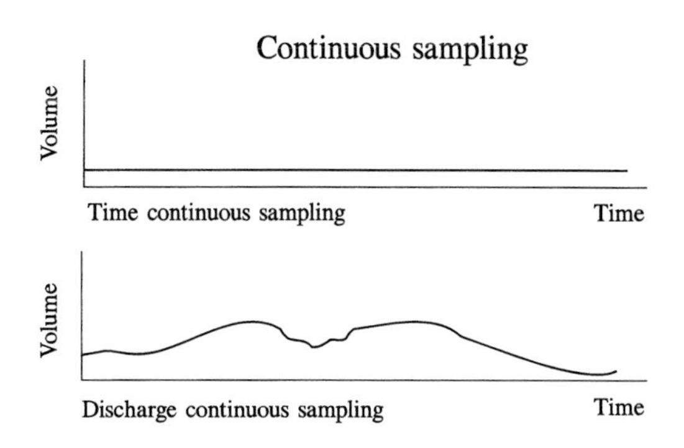

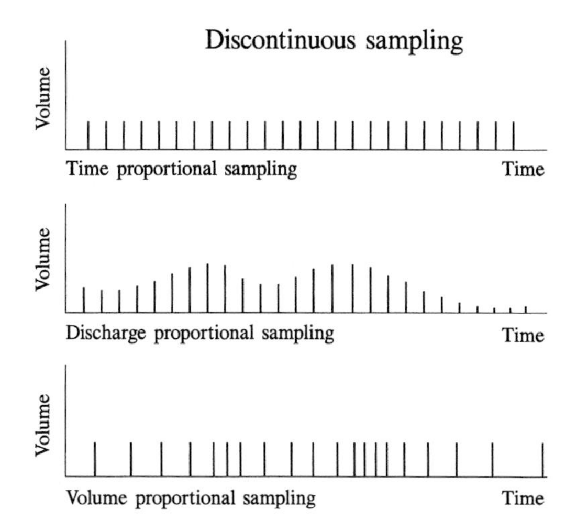

Fig. 2. Kinds of sampling.

origin. They are natural compounds and food at the same time:

(1) fat/oil,
(2) carbohydrate, and
(3) protein.

In sewage these compounds are partly dissolved (e.g., sugar), dispersed and suspended. They are the basic substrates for the microorganisms in the biological treatment plant. Their concentrations are fairly constant with only a small variation (Tab. 16).

As the average concentration of organic compounds is about 500 mg L^{-1} (see Tab. 12) in municipal sewage, the 3 groups of compounds are about 66%. An unhindered biochemical degradation can be expected, if there is no toxic influence.

Mineral oil and its distillation products are completely different chemicals. They mainly consist of hydrocarbons (see below).

The largest group of synthetic compounds in municipal sewage are detergents. All washing powders and detergents in households, trade, and industry contain them. They are also used in wetting, dispersing, and flotation agents. According to their chemical structure they are classified into

(1) anionic detergents,
(2) cationic detergents,
(3) amphoteric detergent, and
(4) non-ionic detergents.

Their adsorbability at the surface liquid–solid, liquid–liquid and liquid–gaseous is the most important characteristic of detergents. This process results in a decrease of surface tension.

On the basis of the actual consumption of the single groups of detergents concentrations for municipal sewage can be calculated (Tab. 17).

In the inflow of sewage treatment plants the concentration is usually lower as a part of the detergents is adsorbed on suspended solids and the determination is done in the filtered sample. Detergents of nearly 50 mg L^{-1} enhance the amount of the organic load from 66–76%.

The following sewage compounds are mentioned as to their origins. The concentrations are shown in Tab. 18 in decreasing order.

4.4.1 Hydrocarbons

Hydrocarbons are found in nature and are not only produced by geochemical processes (e.g., mineral oil). Biogenic compounds also belong to this group (e.g., special waxes on the surface of fruits and leaves). Undigested food-

Tab. 12. Characterization of Compounds in Sewage

A	(1)	Total pollution	190–210 g n^{-1} d^{-1} ≃ 950–1,050 mg L^{-1}
	(2)	Total dry residue	950–1,050 mg L^{-1}
	(3)	Total inorganic part (ash of A2)	100 n^{-1} d^{-1} ≃ 500 mg L^{-1}
	(4)	Total organic part (loss on ashing of A2)	90–110 g n^{-1} d^{-1} ≃ 450–550 mg L^{-1}
B	(1)	Quantity of soluble substances (organic and inorganic)	125–140 g n^{-1} d^{-1} ≃ 625–700 mg L^{-1}
	(2)	Dry residue of filtrate	650–700 mg L^{-1}
	(3)	Inorganic part (ash of B2)	75 g n^{-1} d^{-1} ≃ 375 mg L^{-1}
	(4)	Organic part (loss on ashing of B2)	50–65 g n^{-1} d^{-1} ≃ 250–325 mg L^{-1}
C	(1)	Quantity of suspended solids	65–70 g n^{-1} d^{-1} ≃ 325–350 mg L^{-1}
	(2)	Dry substance	325–350 mg L^{-1}
	(3)	Inorganic part (ash of C2)	25 g n^{-1} d^{-1} ≃ 125 mg L^{-1}
	(4)	Organic part (loss on ashing of C2)	40–45 g n^{-1} d^{-1} ≃ 200–225 mg L^{-1}

Tab. 13. Typical Values of some Parameters in Different Concentrated Sewage (KOPPE and STOZEK, 1999)

Concentration	pH Value	Sedimentable Substances [ml L^{-1}]	Filterable Substances [mg L^{-1}]	Total Dry Residue [mg L^{-1}]	COD [mg L^{-1}]	BOD_5 [mg L^{-1}]	TOC [mg L^{-1}]
Low	6.6	2	170	500	300	150	90
Medium	7.6	6	350	1,000	600	300	200
High	8.6	12	700	2,000	1,000	500	300

Tab. 14. Specific Pollution per Person According to IMHOFF and IMHOFF (1993) as C-BOD_5

Pollution	Load of BOD_5 [g n^{-1} d^{-1}]
Total pollution	60
Soluble substances (in the filtrate)	30
Filtrable substances (filtration residue)	30
Sedimentable substances	20
Non-sedimentable substances	10

Tab. 15. Ratios for Parameters of Organic Compounds (A/B) (KOPPE and STOZEK, 1999)

A/B	COD	BOD_5	TOC
COD	1.0	0.5	0.3
BOD_5	2.2	1.0	0.7
TOC	3.1	1.5	1.0

Tab. 16. Concentrations of the Three Basic Substrates in Sewage

Substrate	Concentration [mg L^{-1}]
Fat/oil	100–120
Carbohydrate	100–120
Protein	100–120
Sum	300–360

Tab. 17. Concentrations of Detergents in Municipal Sewage

Detergent	Concentration [g n^{-1} d^{-1}]	[mg L^{-1}]
Anionic detergents	4.5	23
Non-ionic detergents	4.5	23
Cations detergents	1	5
Amphoteric detergents	not relevant	
Total		51

stuffs and lipids of the feces contribute to this group of substances among sewage from kitchens and baths.

4.4.2 Phenols

The phenolic compounds in municipal sewage (e.g., phenol, cresol, brenzcatechin) are derived from excrements of humans and animals as well as from decomposition processes of organic material.

The greatest amounts of phenolic compounds enter the municipal sewage from industrial products (e.g., disinfecting and conserving agents). Phenols can be perceived by sensoring organs (smell, taste). The odor limiting concentration lies between 10 and 10^4 µg L^{-1}. A chlorination process creates chlorinated phenols which belong to the most intensively smelling substances.

4.4.3 Organic Nitrogen Compounds

Tab. 19 shows the great variety of nitrogen containing compounds in municipal sewage which mostly come from sanitary and kitchen sewage and washing liquids. The substances can be analyzed as Kjeldahl nitrogen. The amount of organic nitrogen compounds varies between 10 and 20 mg L^{-1}. Including the nitrogen of non-decomposed urea, the total concentration lies between 50 and 95 mg L^{-1}.

Tab. 18. Concentrations of Group Parameters in Municipal Wastewater (KOPPE and STOZEK, 1999)

Parameter	Mass Concentration
Hydrocarbons	3–30 mg L^{-1}
Organic nitrogen compounds	10–20 mg L^{-1}
Phenols	≈1 mg L^{-1}
Adsorbable organic chlorinated compounds	100–300 µg L^{-1}

4.4.4 Adsorbable Organically Bound Halogens (AOX)

A small amount of AOX is created by chlorination of drinking water (5–25 µg L^{-1}). About 10 µg L^{-1} come from urine. The quantity from toilet paper amounts to 15 µg L^{-1}. About 100 µg L^{-1} come from rinsing, washing, and cleaning water. Further origins are sewage from hospitals (disinfecting agents) and motor vehicle garages. The total amount lies between 100 and 300 µg L^{-1}. Further information is given by KOPPE and STOZEK (1999) and NEITZEL and ISKE (1998).

5 Single Compounds

5.1 Non-Metal Compounds

This section deals with the group of inorganic compounds, especially the non-metallic compounds. The nitrogen and phosphorous compounds belong to this group of substances which are well known as nutrients. Furthermore, sulfur and halogen compounds must be considered. Nitrogen in its elemental state plays only a secondary role for crude sewage.

It has to be pointed out that some compounds of the group of non-metals occur in drinking water in different concentrations as initial loading. Tab. 20 gives examples of some of these compounds.

The origin of ammonia or ammonium in municipal sewage results from the nearly complete enzymatic decay of urea from human urine during the flow from the household to the sewage treatment plant:

$$NH_2-CO-NH_2+H_2O \rightarrow 2NH_3+CO_2 \quad (2)$$

The so created NH_3/NH_4^+ is a sign of human and animal influence on the composition of waters.

Drinking water has hardly any initial loading with ammonia. In comparison to ammonia nitrate occurs in higher concentrations according to the region of production. Therefore, in

Tab. 19. Nitrogen Containing Compounds in Municipal Wastewater

Compound	Origin
Protein	urine, feces
Amino acid	urine
Urea	urine
Creatine	urine, feces
Creatinine	urine, feces
Hippuric acid	urine
Amine	urine
Indol	urine, feces
Scatol	feces
Imidazole	urine
Porphyrine	urine, feces
Purine (e.g., uric acid)	urine, feces
Lecithin	feces
Cationic detergents	soft rinser
Nitrilo-triacetic acid (NTA)	detergent
Natural dyes	urine, feces, blood, vegetables
Synthetic dyes	washing liquor, coloring agent

Germany the concentrations had to be limited (drinking water ordinance). The concentration may not exceed 50 mg L^{-1}. In the influent of municipal sewage treatment plants there are only low concentrations of nitrate – not more than 1–2 mg L^{-1} – which can be explained by a fast reduction of nitrate over nitrite to elemental nitrogen as a result of partly anoxic conditions in the sewer.

Under present circumstances the compounds of phosphorus in sewage stem mainly from sanitary and kitchen sectors (urine, feces, and nutrients). In former times considerable quantities came from washing and cleaning agents (triphosphate). Drinking water is only moderately loaded with phosphorus compounds.

The situation of sulfur compounds is absolutely different, especially of sulfate, which are not only important in drinking water but also in wastewater. Concentrations >600 mg L^{-1} have corrosive effects on components of the sewer. Furthermore, sulfate can be reduced to sulfite, hydrogen sulfide and sulfide under anaerobic conditions along the way of flow. Under these conditions proteins also begin to rot. At pH values <6 hydrogen sulfide can escape into the atmosphere of the sewer. Because of its high toxicity hydrogen sulfide is a direct danger for the people working in or at the sewer. In low concentrations the odor of H_2S is objectionable.

In the atmosphere of the sewer H_2S can be oxidized to sulfate by biogenic processes:

$$H_2S + 2\,O_2 \rightarrow H_2SO_4 \tag{3}$$

This process also leads to corrosion of the concrete tubes which can lead to a breakdown of the tubes.

From the group of halogens only chlorides are of central importance. They originate as preload from drinking water as well as from sanitary and kitchen sectors.

It has to be pointed out, that the inorganic compounds of phosphor and nitrogen are an essential part of the nutrients for the microorganisms in the biological treatment tank. The ratio of nutrients should be in the order of

$$BOD_5 : N : P = 100 : 5 : 1$$

The conditions in municipal sewage are actually

$$BOD_5 : N : P = 100 : 25 : 3.5$$

Tab. 21 gives an overview over non-metal compounds under quantitative aspects.

Tab. 20. Non-Metal Compounds in Real Municipal Sewage According to Kind and Origin (Dissolved/Suspended)

Compound	Formula	Origin
(1) **Nitrogen**		
• Ammonium/ammonia	NH_3/NH_4^+	feces and urine (after decay of urea)
• Nitrite	NO_2^-	pickle salt
• Nitrate	NO_3^-	food, drinking water
(2) **Phosphorus**		
• Dihydrogen phosphate	$H_2PO_4^-$	} urea, drinking water, food
• Hydrogen phosphate	HPO_4^{2-}	
• Phosphate	PO_4^{3-}	
• Phosphoric acid	H_3PO_4	beverages
• Magnesium ammonium phosphate	$MgNH_4PO_4$	urea
• Calcium phosphate	$Ca_3(PO_4)_2$	feces
• Apatite	$Ca_5(PO_4)_3OH$	cleaning water, household
(3) **Sulfur**		
• Sulfates	SO_4^{2-}	urea, washing agents, drinking water
• Sulfites	SO_3^{2-}	created during anaerobic processes
• Hydrogen sulfide, sulfide	H_2S/S^{2-}	reduction of sulfate and proteins under anaerobic conditions
(4) **Halogens**		
• Chlorides	Cl^-	urea, drinking water

5.2 Semimetal Compounds

Salts of boron from perborates, which are contained in washing powder belong to this group of compounds. Similar to halogen compounds in wastewater borates are not eliminated during the treatment of sewage.

Silicon compounds belong to the main inorganic substances in municipal sewage. They are derived from washing, cleaning, and scrubbing agents, furthermore, from clay and sand which is washed out by precipitation and enters the sewer.

5.3 Metal Compounds

Alkali and alkaline earth compounds dominate as components of wastewater. Their origin is partly the drinking water (e.g., the hardness of water due to Ca^{2+} and Mg^{2+}). Other sources are several uses in human households (sanitary and kitchen water).

From the physiological point of view calcium and magnesium are essential elements for plants, animals, and humans. The daily supply must be regulated by the food intake.

Tab. 22 shows a survey of average concentrations of semimetals and metals in municipal sewage under actual and realistic relations.

5.4 Single Organic Substances

Today the scientific investigations of sewage pay increasing attention to single organic substances. This is necessary and justified, if special questions cannot be answered by summary specifications. Of course, it is possible to get redundant information. This is no problem provided the analytic concept is coordinated with the cost–profit relation.

It is helpful and practicable at the same time to divide the great number of organic compounds in wastewater according to

(1) primary substances with concentrations >1 mg L^{-1}, which have a natural or synthetic origin,
(2) secondary substances with concentrations <1 mg L^{-1}, which also have a natural or synthetic origin.

Tab. 21. Non-metal Compounds in Raw Municipal Sewage (without Initial Loading of Drinking Water) (KOPPE and STOZEK, 1999)

Compound	Concentration $[g\,n^{-1}\,d^{-1}]$	$[mg\,L^{-1}]$
Nitrogen compounds (N reduced)	11–13	55–65
Phosphorous compounds (total P)	2–2.5	10–13
Sulfur compounds	9–10	45–50
Halogen compounds (mainly Cl^-)	≈8	40

Reference value 200 $n^{-1}\,d^{-1}$.

Examples for such substances in the different concentration ranges are shown in Tabs. 23–25. The values are self-explanatory.

Tab. 22. Average Concentrations of Metals and Semimetals in Domestic Sewage

	Element	Concentration $[\mu g\,L^{-1}]$
Sodium	Na	80,000
Calcium	Ca	50,000
Silicon	Si	30,000
Potassium	K	20,000
Magnesium	Mg	12,000
Iron	Fe	3,000
Boron	B	2,000
Aluminum	Al	2,000
Strontium	Sr	400
Zinc	Zn	200
Manganese	Mn	150
Copper	Cu	150
Lead	Pb	100
Barium	Ba	40
Nickel	Ni	40
Chromium	Cr	30
Lithium	Li	20
Tin	Sn	15
Silver	Ag	10
Molybdenum	Mo	5
Cobalt	Co	2
Arsenic	As	2
Cadmium	Cd	1
Cerium	Ce	1
Vanadium	V	1
Lanthanum	La	1
Tungsten	W	1
Selenium	Se	<1
Uranium	U	0.5
Mercury	Hg	0.1
Gold	Au	0.1

Tab. 23. Principal Organic Substances; Dissolved Single Substances in Raw Municipal Sewage (KOPPE and STOZEK, 1999)

Designation	Structure	Concentration Range $[mg\,L^{-1}]$ from	to
Urea	CH_4ON_2	50	300
Creatinine	$C_4H_7ON_3$	5	30
Hippuric acid	$C_9H_{10}O_3N$	4	16
Uric acid	$C_5H_4O_3N_4$	2	10
Cane sugar	$C_{12}H_{22}O_{11}$	2	10
Lactose	$C_{12}H_{22}O_{11}$	1	2
Glucose	$C_6H_{12}O_6$	1	2
Glycine	$C_2H_5O_2N$	2	10
Leucine	$C_6H_{13}O_2N$	1	2
Glutamic acid	$C_5H_9O_4N$	1	2
Ethanol	C_2H_6O	1	10
Glycerin	$C_3H_8O_3$	2	20
Formic acid	CH_2O_2	1	10
Acetic acid	$C_2H_4O_2$	5	50
Propionic acid	$C_3H_6O_2$	1	10
Butyric acid	$C_4H_8O_2$	1	10
Lactic acid	$C_3H_6O_3$	1	10
Citric acid	$C_6H_8O_7$	1	10
Coprosterine	$C_{27}H_{48}O$	3	10
Coprostanon	$C_{27}H_{47}O$	3	10

5.5 Persistent, Accumulative and Toxic Substances

Compounds with low degradability belong to the large number of organic substances in raw municipal sewage. They may accumulate in organisms and exert toxic effects.

Tab. 24. Natural Organic Substances Present in Raw Municipal Sewage in Low Concentrations

Designation	Structure	Concentration Range [μg L⁻¹] from	to
Androsterone	$C_{19}H_{30}O_2$	1	30
Testosterone	$C_{19}H_{28}O_2$	0.1	0.8
Estradiol	$C_{18}H_{24}O_2$	0.01	0.1
Progesterone	$C_{21}H_{30}O_2$	10	50
Scatol	C_9H_9N	1	10
Indol	C_8H_7N	10	1,000
Thiamine	$C_{12}H_{17}ON_4SCl$	0.2	3.4
Toluene	C_7H_8	1	10
Xylene	C_8H_{10}	1	10
Trimethylbenzene	C_9H_{12}	1	10
Benzpyrene	$C_{20}H_{12}$	0.1	10

Tab. 25. Synthetic Organic Substances Present in Raw Municipal Sewage in Low Concentrations (KOPPE and STOZEK, 1999)

Designation	Structure	Concentration Range [μg L⁻¹] from	to
Saccharin	$C_7H_5O_3NS$	2	20
Acetyl salicylic acid	$C_9H_8O_4$	1	10
Oxybenzoic acid	$C_7H_6O_3$	5	50
Sorbic acid	$C_6H_8O_4$	5	50
Diethylhexylphthalate	$C_{24}H_{38}O_4$	10	300
Dibytylphthalate	$C_{16}H_{22}O_4$	2	20
Chloroform	$CHCl_3$	5	50
Dichloromethane	CH_2Cl_2	10	100
Trichloroethene	C_2HCl_3	50	500
Tetrachloroethene	C_2Cl_4	50	500
Hexachlorocyclohexane	$C_6H_6Cl_6$	0.01	0.1
Hexachlorobenzene	C_6Cl_6	0.01	0.1
Decachlorodiphenyl	$C_{12}Cl_{10}$	0.01	0.1
Dichlorobenzene	$C_6H_4Cl_2$	100	1,000
Nitrilotriacetic acid	$C_4H_9O_6N$	500	1,000
Chlofibrinic acid (pharmaceutical residue)	$C_{10}H_{11}O_3Cl$	1	10

For example, there are products in human excrements which are more or less degradable. They are organic indicators for the pollution with domestic sewage. Lipophilic steroids as well as some sexual hormones (e.g., estrogen) and coprosterine belong to these compounds (for concentration ranges see Tabs. 23, 24).

Sewage can contain polycyclic aromatic hydrocarbons, for which concentrations are limited in drinking water, as well as chlorinated hydrocarbons (e.g., trichloromethane), polychlorinated biphenyls, dibenzodioxins and dibenzofurans of different origins. The ranges of concentrations are very low (ng L^{-1} range). One of their special characteristics is the accumulation in sewage sludge (see Sect. 6).

It seems to be meaningful to investigate municipal sewage systematically regarding water-endangering organic substances, even though it mostly contains harmless natural compounds. Water-endangering organic substances originate (among others) from disinfection and conservation agents.

It is not possible to completely eliminate all organic compounds in sewage by biological treating processes. Organic substances always remain in the sewage if they are not degraded or adsorbed on activated carbon. During the treating process the microorganisms of the activated sludge produce metabolites from the contained substances, which sometimes have the same negative characteristics as described above.

For details, the reader is referred to the scientific literature (KOPPE and STOZEK, 1999; IMHOFF and IMHOFF, 1993; NEITZEL and ISKE, 1998).

6 Sewage Sludge

6.1 Definition According to Sewage Sludge Origin

Sewage sludge is a mixture of water and solid material of varied dispersion and shape, with a concentration of dry matter (m_d) of about 1–10% and a marked structural viscosity (PROFF and LOHMANN, 1997).

The sludges to be discussed come from collection, drainage, and treatment of domestic and municipal sewage. Sewage sludge from industrial production and sewage treatment (e.g., sugar factory or galvanizing mill) are not considered here.

In detail there are: excrement sludge (privy pit, 3-compartment septic tank), sludge from storm water overflow tanks of mixed sewers (flow through tanks), primary sludge from the primary purification as well as secondary sludge (surplus sludge) from final purification of a mechanical-biological sewage treatment plant. The sludge which is separated during the chemical removal of phosphate with iron and aluminium salts belongs to the sewage sludge.

Before further treatment the single sorts of sludge are mostly mixed (so-called "raw sludge").

6.2 Formation of Sludge

The amounts of raw sludge can be specified as volume or as to the content of water, and so the volume of sludge is very different according to the treatment of wastewater and the intermediate storage (e.g., in thickeners). The specification as mass of dry substance (m_d) is generally more precise. According to an estimation of the Environmental Ministry of Germany in 1995 it was predicted that the amounts of sludge will increase to about 4 million t a^{-1} m_d in Germany (ATV-Information, 1996c). At an average concentration of dry matter of 5% an average increase of 1 m^3 n^{-1} a^{-1} can be evaluated, which corresponds well to the experience of several sewage treatment plants. The predominant part of dry matter results from the primary sludge of precleaning of the municipal treatment plant. With screens and grit chambers coarse and heavy solid materials are separated before precleaning and disposed sepa-rately.

6.3 Chemical Constitution of Sewage Sludge

The main component of sewage sludge is water which exists in three different spatial distribution forms: cavity (or void) water, external water (adhesion, adsorption, capillary water), and internal water (cell liquid, water of hydration) (ATV-Arbeitsbericht, 1997).

By disintegration (e.g., by ultrasound), by sludge conditioning (e.g., by hydrolysis with sodium hydroxide), and by stabilization (e.g., by aerobic-thermophilic bacteria) the amount of external water increases, whereby the ability to dewater the sludge is enhanced (DICHTL et al., 1997; DICHTL, 1997; FUCHS and SCHWINNING, 1997).

The dry residue (also called dry matter m_d) is estimated by drying of a weighed out quantity of sludge at 105 °C until the weight remains constant. The result is specified in weight %. The difference between 100% and dry residue in % gives the approximate quantity of water in wet sludge. If very precise results are required, it has to be considered that some substances evaporate at 105 °C, e.g., ammonium carbonate, which may not be added to the water. There is a difference between the dry residue and the dry matter. The latter is estimated after separation of the greatest part of water with the solved compounds by filtration. Normally this is done with activated sludge for which the difference of both results could amount to about 10%.

When specifying the results of quantitative chemical analyses, it has to be considered, whether the values relate to the wet sludge, the dry residue or the dry matter.

The chemical oxygen demand and the biochemical oxygen demand are estimated in original wet sewage sludge and then are calculated for the dry matter if necessary. An example is the sludge from a secondary clarification tank of a mechanical-biological sewage treatment plant:

The dry residue amounts to 3%. The COD is estimated at 20,000 mg^{-1} (660 g kg^{-1}) and the BOD_5 to 10,000 mg L^{-1} (330 g kg^{-1}) (with the Warburg method). The loss of the dry residue by incineration at 550 °C amounts to 65%, the ash to 35%.

After a mesophilic digestion of 4 weeks in which no turbid water is removed the dry matter diminishes to 2%, whereby the concentration of heavy metals increases as a percentage of the dry residue, for instance. The COD amounts now to 10,000 mg L^{-1} (500 g kg^{-1}), the BOD_5 to 1,000 mg L^{-1} (50 g kg^{-1}). An ashing loss of about 40% and ash of 60% can be measured based on the dry residue.

For converting the volume to mass the density of 1 kg L^{-1} is assumed, for more precise investigations the density (between 1.01 and 1.2 kg L^{-1}) has to be estimated separately.

The loss on ashing is usually equal to the organic part of sludge, which is sufficient for practical purposes. The organic mass of primary sludge from the preliminary sedimentation tank consists approximately of 50% carbohydrates (polysaccharides, cellulose), 30% proteins and 10% oil and fat from plants and animals. The residual 10% consist of many different natural and synthetic organic compounds (e. g., lignin, adsorbable detergents). In digested sludge an average concentration of 3 g kg^{-1} dry residue of anionic detergents can be measured (Ruhrwassergütebericht, 1994, 1995).

The concentrations of compounds of nitrogen and phosphorus are important for the utilization of digested sludge in agriculture. The content of the total nitrogen in sludge after clarification in a mechanical-biological sewage treatment plant is considerable. The average concentration of nitrogen amounts to 40,000 mg kg^{-1} dry residue and causes the high concentrations in turbid water, filtrate, and concentrate, which result from conditioning and concentrating the sludge.

The average content of phosphorus of 10,000 mg kg^{-1} dry residue is remarkably high. Because of further sewage treatment by chemical and biochemical elimination of phosphorous it is clearly enhanced (ROSCHKE et al., 1997).

Sulfur and its compounds are important for the formation of hydrogen sulfide during the anaerobic stabilization. The concentration of hydrogen sulfide in digester gas is between 10 and 10,000 mg L^{-1} gas, according to the S–C-relation and depends on the pH value and the content of iron. The content of total sulfur in sewage sludge of the post-clarification amounts to about 1% S based on the dry residue.

Organic trace substances which are persistent and toxic accumulate in sewage sludge. For utilization in agriculture the concentrations of special compounds may not exceed fixed limits. The following adsorbable organically bound halogens, polycyclic aromatic hydrocarbons (PAH), polychlorinated dibenzodioxins and -furans (PCDD, PCDF) as well as mineral hydrocarbons (HC) belong to this group (see Tab. 26) (ROSCHKE et al., 1997; Bundesgesetzblatt, 1992).

Phthalates are regularly found in sewage sludge, e.g., di-(2-ethylhexyl)-phthalate (DHEP) 4–103 mg kg^{-1} in the dry residue (MERKEL and APPUHN, 1996). These organic compounds belong to the dangerous substances for which limiting values are fixed not only for sludge, but also for wastewater when discharged into a public sewer.

Beyond phosphorous and nitrogen sewage sludge contains further elements like potassium, magnesium, and calcium which are regarded as nutrients for plants. For a fertilization compound the concentrations are too small, as the following example shows (ROSCHKE et al., 1997). A digested sludge has a dry residue of 3.9% and a loss of 590 g kg^{-1} on ashing. Tab. 27 shows the content of some elements in the dry residue.

Analogous to toxic organic compounds the concentrations of heavy metals are limited according to the German Ordinance on Sewage Sludge, when it is used in agriculture (see Tab. 28) (Bundesgesetzblatt, 1992).

Tab. 26. Concentration Ranges of Organic Compounds in Sewage Sludge

Group Parameter	Concentration in Dry Residue
AOX	120–220 mg kg^{-1}
PAH	1.8–4.7 mg kg^{-1}
HC	1,680–2,420 mg kg^{-1}
PCDD, PCDF	12–40 ng kg^{-1} toxicity equivalents

Tab. 27. Concentrations of some Elements in Digested Sludge

Element	Concentration [g kg^{-1} Dry Residue]
Mg	2.2
K	5.1
Ca	32.2
N	53.8
P	21.3

Tab. 28. Limits of Heavy Metals in Sewage Sludge for Utilization in Agriculture According to German Ordinance on Sewage Sludge

Heavy Metal	Limit [mg kg^{-1} Dry Residue]
Lead	900
Cadmium	10
Chromium	900
Copper	800
Nickel	200
Mercury	8
Zinc	2,500

Tab. 29. Accumulation Factors of Heavy Metals in Sewage Sludge

Heavy Metal	Accumulation Factor [L kg^{-1}]
Nickel	1,000
Copper	2,000
Zinc	3,000
Chromium	3,000
Cadmium	4,000
Mercury	5,000
Lead	5,400

By the way sewage sludge contains all essential trace elements down to low concentration ranges of 1 mg kg^{-1} of dry residue. The heavy metals have a strong tendency to accumulate on suspended solids of sewage sludge, partly by sorption, partly by coprecipitation. The ratio of concentration c_1, in the dry residue to concentration c_2 in untreated wastewater is defined as accumulation factor AF [L kg^{-1}].

$$AF = c_1/c_2 \tag{4}$$

Tab. 29 shows accumulation factors arranged in an increasing order.

In digester gas from anaerobic stabilization of sludge heavy metals occur in the form of gaseous compounds such as $Cd(CH_3)_2$, $Hg(CH_3)_2$, $Pb(CH_3)_4$, $Pb(C_2H_5)_2$, $Bi(CH_3)_3$, $Sn(CH_3)_4$, and $Sb(CH_3)_3$ (FELDMANN and KLEIMANN, 1997).

The elements and their oxides which are the main components of the ash of municipal sewage sludge can be seen from the many analyses of ashes from incineration plants. According to the type of the treatment plant one can be distinguish between iron-rich and calcium-rich ashes (WIEBUSCH et al., 1997). Tab. 30 shows two typical examples. The remaining approximately 10% are distributed over the other elements like Mg, Mn, Na, Cl, etc. A part of calcium exists as $CaSO_4$.

At the end of this chapter something must be said about the control and decontamination of stabilized sewage sludge in a treatment plant. In practice, the number of fecal coliform bacteria of a sludge sample with a dry residue of about 1 g are estimated. The most probable number (MPN) should be smaller than 1,000 per g dry residue (FUCHS and SCHWINNING, 1997).

7 Overview on Possible Treating Methods for Wastewater and Sewage Sludge

7.1 Aim of the Treatment

In Europe municipal sewage is not regularly used in agriculture (e.g., spread together with fertilizer from animals on farm land). It is discharged into the next brook, river, lake, or ocean after treatment in a municipal sewage treatment plant.

Tab. 30. Content of some Components in the Ash of Sewage Sludge (KOLISCH, 1996)

Component	Weight %	
	Fe Ashes	Ca Ashes
SiO_2	36	33
Al_2O_3	14	17
Fe_2O_3	16	4
CaO	10	18
P_2O_5	12	16
S	0.8	0.7
Total	88.8	88.7

The treatment of sewage has to protect the recipients, therefore, in a wide sense to consider their utilization. In the case of flowing water bodies the sea into which it flows has also to be considered.

Today increasing requests concern such uses as navigation (no hindrance by sludge banks), holiday, and water sports (no nuisance by smell and ugly moving material), supply with drinking water (no acute or chronic active toxic substances, only small amounts of precursors of haloforms, only small amounts of organic compounds which promote the regermination in the pipe system of drinking water, only small amounts of inorganic compounds which promote the corrosion of materials). Also a stable water biocoenosis according to the location from the point of view of fisheries should be permitted (no overload of the oxygen balance, no eutrophication by nutrients like phosphor and nitrogen, no critical concentrations of toxic substances for water organisms owing to bioaccumulation and biomagnification).

In this field the oxygen balance which is influenced by the self-purification processes of waters has a key function. This is the reason why the BOD of wastewater and its decrease stand in the center of wastewater technique (see Sect. 4.2).

Secondarily, the nutrients phosphorus and nitrogen as well as their compounds must be considered as they result in a rapid reproduction of phytoplankton and other aquatic plants. The danger of eutrophication is also valid for the bays of the ocean into which the recipients flow.

Municipal sewage substances like salts, heavy metals, and toxic compounds have such low concentrations, if the wastewater of the indirect dischargers is treated correctly so that the utilization is not affected by enrichment in sediments and organisms. The persistent organic compounds in municipal sewage cause special problems by interfering with the drinking water treatment from river water or bank filtrate. They are called "waterworks relevant" and "drinking water relevant" according to VÖLKER and SONTHEIMER (1988). Partly they are caused by harmless substances during the biological sewage treatment (KLOPP and KOPPE, 1990; MÜLLER et al., 1993).

Nowadays in Germany the legal situation is that the goal of cleaning the sewage depends on the size of the treatment plant. According to appendix 1 of the Wastewater Ordinance for Domestic and Municipal Sewage 5 size classes with increasing standards for the location of discharge are specified. The classification is focused on the daily load of BOD and extends from 60 kg d^{-1} to 6,000 kg d^{-1} raw BOD_5. The last class has the following limits: 75 mg L^{-1} COD, 15 mg L^{-1} BOD_5, 10 mg L^{-1} NH_4-N, 18 mg L^{-1} total nitrogen and 1 mg L^{-1} total phosphorus. These limits are valid for a qualified random sample or a 2 h composite sample.

For NH_4-N and total nitrogen the standards are valid for wastewater temperatures $>12\,°C$ in the effluent of the biological reactor or instead, for the time between 5/1 and 10/31. The concentration limit for nitrogen can be raised to 25 mg L^{-1}, if the decrease of the total nitrogen load exceeds 70%. The decrease is measured from the relation of nitrogen load in the influent to that in the effluent in a representative time period, which may not exceed 24 h. The load in the influent is based on the sum of inorganic and organic nitrogen (Bundesgesetzblatt, 1997).

In the calculation of the total costs of sewage disposal in Germany the sewage charge is taken into account (Bundesgesetzblatt, 1990). For the discharge of a "damage unit" of organic substances, e.g., 50 kg COD, 80 DM have to be paid at present. The same amount is valid for a "damage unit" of phosphorus (3 kg total P) or a "damage unit" of nitrogen (25 kg total N).

7.2 Fundamental Treating Methods

For processes or combinations of processes for treating municipal sewage the following conditions have to be fulfilled: flexibility to meet quickly changing quantities and concentrations, not prone to trouble with variations of temperature and disturbing compounds of unknown combination, low specific costs due to the great amounts of water. The same conditions are also valid for the treatment of sewage sludge, but less so, as the variation in composition and quantity is not as great as in the case of municipal sewage.

The physical unit operations comprise: mechanical processes for separation of solid material by filtration and sieving, by sedimentation and flotation as well as centrifugation; blow out of gases and readily volatile substances (e.g., ammonia after an increase of the pH value); processes of adsorption on organic and inorganic adsorbing substances, radiation with UV, X-rays and γ-rays. The chemical processes comprise the oxidation with elementary oxygen (with an increase in temperature and pressure), the oxidation with other oxidants (e.g., ozone, hydrogen peroxide, permanganate) (WIMMER et al., 1995), precipitation and flocculation (with calcium hydrate, salts of iron and aluminum), which is always followed by a mechanical treatment. A special case is the chlorination, which is regularly used for disinfection of cleaned wastewater in other countries. In Germany pathogenic germs are killed by chorination only in the case of epidemics (e.g., *Mycobacterium tuberculosis*, *Salmonella typhi*, *Salmonella enteritidis*, *Vibrio cholerae*).

The biological processes can be classified into natural and artificial processes. Natural processes need large areas and longer times (weeks to years). In artificial processes the requirements of room and time are much smaller because of the extreme density of the microorganisms.

The natural processes comprise, e.g., wastewater lagoons, irrigation fields with or without utilization in agriculture, plant sewage treatment works and root space plants, in which besides microorganisms plants and animals also participate in the work of purification. In practice mostly artificial biological processes are used (in Germany to about 95%). These processes can be classified as aerobic (which need an intensive supply of oxygen) and anaerobic treatments.

A further classification can be done according to the technical realization, as the continuous flow and batch processes. The microorganisms can be freely suspended in the wastewater or immobilized on solid material (e.g., the sludge activation process on one hand and the percolation filter on the other).

7.3 Combinations of the Methods

In most cases the clarification processes for wastewater are combinations of physical, chemical, and biological treatments. Usually undissolved material is separated with screens, sand catchers, and sedimentation tanks before further treatment. During the aerobic biological treatment excess sludge is produced and has to be separated mechanically.

According to the sensitivity of the recipient or to the strictness of the governmental conditions advanced clarification (more than is possible with mechanical-biological methods) of the municipal sewage has to be done.

In Germany for wastewater treatment plants of size class 4 and 5 phosphorus must be reduced to residual concentrations of 2 and 1 mg L^{-1}, respectively (Bundesgesetzblatt, 1997). It is possible to meet these standards by chemical treatment (precipitation with iron or aluminum salts) in combination with mechanical-biological treatment (pre-precipitation, simultaneous precipitation or post-precipitation) (ATV-Merkblatt M 206, 1994).

The elimination of nitrogen compounds by a combination of microbiological processes under aerobic (nitrification) and anoxic conditions (denitrification) also belongs to the advanced treatment (ATV-Arbeitsbericht, 1994a; KROISS, 1994). A consequence of this change in concentrations of oxygen could be an increase in the elimination of phosphorus (ATV-Merkblatt M 208, 1994).

If the amount of persistent organic compounds is tremendous as a result of large quantities of wastewater of a special indirect discharger (e.g., pulp producing plants) it is sometimes possible to decrease the remaining concentrations of refractive substances by a combination of aerobic and anaerobic biological clarification (including the necessary mechanical intermediary stages). Using such combinations it is possible to save energy costs by the treatment of municipal sewage with a great part of wastewater from a brewery.

The treatment of wastewater with chlorinated organic compounds should be mentioned as a typical example for the combination of physical, chemical, and biological methods (BACHMANN et al., 1997). In some cases the mechanically-biologically cleaned sewage is treated in filter columns or polishing ponds and so cleansed of remaining suspended solids (ATV-Arbeitsblatt A 203, 1985; IMHOFF, 1979). In the polishing ponds the number of heterotrophic bacteria decreases as the number of autotrophic alga increases, while the character of the water body becomes similar to that of the recipient.

It would be beyond the scope of this chapter to describe all combinations of treating methods for wastewater. From the mathematical point of view the number of all combinations of two biological (aerobic and anaerobic), two chemical (oxidation and precipitation) and a physical method (filtration) can be calculated to $5! = 120$.

The number of possible wastewater treating methods becomes much greater, if the methods are taken into consideration, with which water of drinking water quality is produced from wastewater. For the elimination of organic compounds adsorption processes are mostly used (DREWES and JEKEL, 1997), and for inorganic compounds those methods, which are tested for the desalination of sea water (evaporation, reverse osmosis, electrical osmosis, etc.).

7.4 Treatment and Use or Disposal of Residual Sewage Sludge

The goal to treat sewage sludge is its biochemical stabilization in order to avoid the development of troublesome gases at a later use or deposition of residual material. Further goals which are striven for simultaneously, are a decrease of volume, an increase in shear stress and a decrease of pathogenic germs and worm eggs.

The sewage sludge and the resulting residual material have to be disposed of in a manner that stresses the environment as little as possible. For this purpose there are some fundamental methods: the utilization in agriculture and for land restoration, the deposition as a mono dump or together with domestic waste, the industrial utilization of ashes and the deposition of ashes after combustion of the treated sludge. An overview over the most important types of processes for treating, material utilization and other deposition of sewage sludge is provided by MÖLLER (1994) under special consideration of the "Technical Instruction of Treatment and of Removal of Domestic Waste" in Germany (TA Siedlungsabfall, 1993).

It is necessary to stabilize raw sludge by biological treatment for utilization in agriculture, which can be realized aerobically or anaerobically. Today, in most municipal sewage treatment plants the sewage sludge is digested by mesophilic or thermophilic bacteria in big digestion tanks during time periods of some weeks. As a result the so-called "digester gas" or "biogas" is created (on the average $24 \text{ L } n^{-1} \text{ d}^{-1}$), which can be used for heating or turbine propulsion because of its high heat value (ATV-Arbeitsbericht, 1994b).

After disinfection by pasteurization or addition of lime the biologically stabilized sludge can (if necessary) be spread as fertilizer in liquid form. It is also useful to partly dewater the sludge to save storage space and cost for transportation. In Germany the rather restricting regulations for quality, spreading quantity, and control have to be followed (Bundesgesetzblatt, 1992; ATV-Information, 1996a).

For dumps and composting (e.g., together with waste from the garden) biological stabilization is required. For the following dewatering the digested sludge is conditioned which can be done thermically, chemically or mechanically. A certain conditioning (promoted by freezing and defrosting in winter time) also happens during interim storage in sludge beds. The conditioned sludge is dewatered by centrifuges, band filter presses or filter presses to dry residues of about 28–45%. The separated sludge water is highly polluted and must be treated biologically (ATV-Arbeitsbericht, 1995b; KOLISCH, 1996).

For combustion of sewage sludge biological stabilization is not required by law, as the heat value of the sludge decreases to one half, but it may facilitate the dewatering process and the combustion and save operating costs. According to the German Emission Control Act not only the classical combustion components like CO, organic carbon and NO_x are limited for combustion, but also further substances like dust, HCl, HF, SO_2, heavy metals and the polychlorinated dioxins and furans (ATV-Arbeitsbericht, 1995a).

8 References

ABDULLAEV, K. M., MALAKHOV, I. A., POLETAEV, L. N., SOBOL, A. S. (1992), *Urban Waste Waters*. Chichester: Ellis Horwood.

American Health Association (1985), *Standardized Methods for Examination of Water and Wastewater*, 16th Edn. Washington, DC: American Health Association.

ANNEN, G. W. (1980), Trockenwetterabfluß und Jahresschmutzwassermenge, *Korrespondenz Abwasser* **27**, 411–418.

ATV-Arbeitsbericht (1994a), *Kenn-Nr. 2.6 – Aerobe biologische Abwasserreinigungsverfahren – Umgestaltung zweistufiger biologischer Kläranlagen zur Stickstoffelimination*. Hennef: Gesellschaft zur Förderung der Abwassertechnik e.V.

ATV-Arbeitsbericht (1994b), *Gewinnung, Aufbereitung und Verwertung von Biogas*. Hennef: Gesellschaft zur Förderung der Abwassertechnik e.V.

ATV-Arbeitsbericht (1995a), *Klärschlammverbrennung, Emissionen*. Hennef: Abwassertechnische Vereinigung e.V.

ATV-Arbeitsbericht (1995b), *Maschinelle Schlammentwässerung*. Hennef: Gesellschaft zur Förderung der Abwassertechnik e.V.

ATV-Arbeitsbericht (1997), Trocknung kommunaler Klärschlämme in Deutschland. *Korrespondenz Abwasser* **44**, 1869–1884.

ATV-Arbeitsblatt A 203 (1985), *Abwasserfiltration durch Raumfilter nach biologischer Reinigung*. Hennef: Gesellschaft zur Förderung der Abwassertechnik e.V.

ATV-Information (1996a), *Landwirtschaftliche Klärschlammverwertung*. Hennef: Gesellschaft zur Förderung der Abwassertechnik e.V.

ATV-Information (1996b), *Regenwasserversickerung*. Hennef: Abwassertechnische Vereinigung e.V.

ATV-Information (1996c), *Zahlen zur Abwasser- und Abfallwirtschaft*. Hennef: Abwassertechnische Vereinigung e.V.

ATV-Merkblatt M 206 (1994), *Automatisierung der chemischen Phosphorelimination*. Hennef: Gesellschaft zur Förderung der Abwassertechnik e.V.

ATV-Merkblatt M 208 (1994), *Biologische Phosphatentfernung bei Belebungsanlagen*. Hennef: Gesellschaft zur Förderung der Abwassertechnik e.V.

ATV-Merkblatt M 267 (1995), *Radioaktivität in Abwasser und Klärschlamm*. Hennef: Gesellschaft zur Förderung der Abwassertechnik e.V.

ATV-Regelwerk Abwasser Abfall (1992), *Arbeitsblatt A 163: Indirekteinleiter, Teil 1: Erfassung; Teil 2: Bewertung und Überwachung* (draft). St. Augustin: Abwassertechnische Vereinigung.

ATV-Regelwerk Abwasser Abfall (1994), *Arbeitsblatt A 115: Einleiten von nicht häuslichem Abwasser in eine öffentliche Abwasseranlage*. St. Augustin: Abwassertechnische Vereinigung.

BACHMANN, J., FISCHWASSER, K., REICHERT, K. J. (1997), Untersuchung zur Schnittstelle zwischen UV/Oxidationsbehandlung beim Abbau chlorierter organischer Verbindungen, *Vom Wasser* **89**, 1–11.

BRINKMANN, W. L. F. (1985), Urban storm water pollutants, sources and loadings, *Geo. J.* **11**, 277–283.

Bundesgesetzblatt (1990), Gesetz über Angaben für das Einleiten von Abwasser in Gewässer (Abwasserabgebengesetz – AbwAG), Neufassung in der Bekanntmachung vom 6. 11. 1990, *Bundesgesetzblatt* **Teil 1**, 2432.

Bundesgesetzblatt (1991), Verordnung über die Herkunftsbereiche von Abwasser (Abwasserherkunftsverordnung – AbwHerkV) vom 3. Juli 1987, *Bundesgesetzblatt* **Teil 1**, 1578–1579; zuletzt geändert vom 27. 5. 1991, *Bundesgesetzblatt* **Teil 1**, 1197.

Bundesgesetzblatt (1992), Klärschlammverordnung (AbfKlärV) vom 14. 4. 1992, *Bundesgesetzblatt* **Teil 1**, 912.

Bundesgesetzblatt (1997), Verordnung über Anforderungen an das Einleiten von Abwasser in Gewässer und zur Anpassung der Anlage des Abwasserabgabengesetzes vom 21. März 1997, *Bundesgesetzblatt* **Teil 1 Nr. 19**, 567–583.

DICHTL, N. (1997), Stellenwert und Kosten der Stabilisierung, *Korrespondenz Abwasser* **44**, 1740–1750.

DICHTL, N., MÜLLER, J., ENGLMANN, E., GÜNTHERT, F. W., OSSWALD, M. (1997), Desintegration von Klärschlamm – ein aktueller Überblick, *Korrespondenz Abwasser* **44**, 1726–1736.

DOLD, P. L., MARAIS, G. v. R. (1986a), Evaluation of the general activated sludge model proposed by the IHWPR Task Group, *Water Sci. Technol.* **18**, 63–89.

DOLD, P. L., MARAIS, G. v. R., WINKLER, U., VOIGTLÄNDER, G. (1995), Anwendung neuronaler Netze für die Simulation von Prozeßabläufen auf vorhandenen Kläranlagen, *Korrespondenz Abwasser* **42**, 1784–1792.

DREWES, I. E., JEKEL, M. (1997), Untersuchungen zur konkurrierenden Adsorption von DOC und AOX in Kommunalabwässern, *Vom Wasser* **89**, 97–114.

FELDMANN, J., KLEIMANN, J. (1997), Flüchtige Metallverbindungen im Faulgas, *Korrespondenz Abwasser* **44**, 99–104.

FUCHS, L., SCHWINNING, H.-G. (1997), Zum Stand der aerob-thermophilen Stabilisierung und Entseuchung von Klärschlamm, *Korrespondenz Abwasser* **44**, 1834–1842.

Gesetz- und Verordnungsblatt Nordrhein-Westfalen (1989), Ordnungsbehördliche Verordnung über die Genehmigungspflicht für die Einleitung von Abwasser mit gefährlichen Stoffen in öffentliche Abwasseranlagen (VGS) vom 25. September 1989, *Gesetz- und Verordnungsblatt Nordrhein-Westfalen,* **564**.

IMHOFF, K. R. (1979), Die Entwicklung der Abwasserreinigung und des Gewässerschutzes seit 1868, *Gas Wasserfach Wasser Abwasser* **120**, 563–576.

IMHOFF, K., IMHOFF, K. R. (1993), Taschenbuch der Stadtentwässerung, 28th Edn. München, Wien: R. Oldenbourg Verlag.

KLOPP, R., KOPPE, P. (1990), Quantitative Charakterisierung von Abwässern hinsichtlich ihrer Dispersität und Abbaubarkeit, *Vom Wasser* **75**, 307–329.

KOLISCH, G. (1996), Zweistufige biologische Stickstoffelimination aus Filtraten oder Zentraten der Schlammentwässerung, *Korrespondenz Abwasser* **43**, 1040–1045.

KOPPE, P., STOZEK, A. (1999), *Kommunales Abwasser*, 4th Edn. Essen: Vulkan-Verlag.

KROISS, H. (1994), Die Bedeutung der Massenbilanz für den Betrieb und die Bemessung von Kläranlagen zur Stickstoffentfernung, *Korrespondenz Abwasser* **41**, 416–424.

MERKEL, D., APPUHN, H. (1996), Untersuchung von Klärschlamm und Böden auf 2-Di-(2-ethylhexyl)-phthalat (DEHP), *Korrespondenz Abwasser* **43**, 578–582.

MÖLLER, U. (1994), Biologische Voll-Stabilisierung, *Korrespondenz Abwasser* **41**, 1290–1300.

MÜLLER, U., BURKHARD, W., SONTHEIMER, H. (1993), Wasserwerks- und trinkwasserrelevante Substanzen in der Elbe, *Vom Wasser* **81**, 371–386.

NEITZEL, V., ISKE, U. (1998), *Abwasser – Technik und Kontrolle*. Weinheim: Wiley-VCH.

PROFF, E. A., LOHMANN, J. H. (1997), Rheologische Charakterisierung flüssiger Klärschlämme, *Korrespondenz Abwasser* **44**, 1615–1621.

ROSCHKE, M. et al. (1997), Klärschlamm als Dünger im Land Brandenburg, *Korrespondenz Abwasser* **44**, 1795–1805.

Ruhrwassergütebericht 1994 (1995), *Anionische Tenside in der Ruhr, in Abwässern und in Klärschlämmen*, pp. 56–58. Essen: Eigenverlag Ruhrverband.

SAUER, J., AUTUSCH, E., RIPP, CHR. (1997), Monitoring lipophiler organischer Schadstoffe im Kanalnetz mittels Sielhautuntersuchungen, *Vom Wasser* **88**, 49–69.

SCHMIDT, S. (1993), Genormte Verfahren in der Wasseranalytik, *Vom Wasser* **80**, 1–5.

TA Siedlungsabfall (1993) Technische Anleitung zur Verwertung, Behandlung und sonstigen Entsorgung von Siedlungsabfällen – Dritte allgemeine Verwaltungsvorschrift zum Abfallgesetz vom 14. 5. 1993, *Bundesanzeiger* **45. Jahrgang**, Nr. 99a vom 29. 5. 1993.

VÖLKER, E., SONTHEIMER, H. (1988), Charakterisierung und Beurteilung von Kläranlagenabläufen aus der Sicht der Trinkwasserversorgung, *Gas Wasserfach Wasser Abwasser* **129**, 216–230.

WIEBUSCH, B., SEYFRIED, C. F., HAUCK, D. (1997), Einsatzmöglichkeiten von Aschen aus der Mono-Klärschlammverbrennung in der Ziegelindustrie, *Korrespondenz Abwasser* **44**, 1762–1777.

WIMMER, B., BISCHOF, H., SCHEUER, C., WABNER, D. (1995), Oxidativer Abbau von organischen Wasserinhaltsstoffen – Abbauprodukte und Reaktionswege, *Vom Wasser* **85**, 421–432.

XANTHOPULOS, C., HAHN, H. H. (1993), *Anthropogene Schadstoffe auf Straßenoberflächen und ihr Transport mit dem Niederschlagsabfluß*. Karlsruhe: Institut der Siedlungswasserwirtschaft der Universität Karlsruhe.

10 Industrial Wastewater Sources and Treatment Strategies

Karl-Heinz Rosenwinkel
Ute Austermann-Haun
Hartmut Meyer
Hannover, Germany

1 Introduction and Targets

This chapter deals with the wastewater flows discharged from industrial plants and offers a synopsis of the applicable treatment methods.

First, the different industrial branches are introduced and classified, differentiating between industries emitting wastewater with mainly organic pollutants and industries emitting wastewater with mainly inorganic pollutants. The central topic of this book being on biotechnology, the main emphasis will be on industries emitting wastewater with organic pollutants as these can be treated biologically. Following that, the possible wastewater flow fractions occurring in industrial plants will be listed. According to each type of industry and individual local conditions, certain wastewater flow fractions will not occur, will be disposed of, or treated in a different manner. Then, various wastewater pollutants will be investigated with regard to their direct importance for certain industries. This is followed by a typical treatment sequence for wastewater. In the main section, wastewater composition and possible treatment strategies for the individual industries will be examined, with particular emphasis on the most important branches of the food processing industry.

It should be noted, that there are not only substantial differences between the various industrial branches, but that even within one branch wastewater composition and respective treatment methods are determined by the following factors:

(1) production methods,
(2) water supply and water processing,
(3) technical condition and age of the production site,
(4) training and motivation of employees,
(5) application of certain additives and cleaning agents, etc.,
(6) number of shifts, seasonal differences (campaign operation),
(7) effluent requirements (direct or indirect discharge),
(8) extent of production-integrated environmental protection,
(9) number of wastewater treatment facilities.

2 Classification of Industries

Industries can be subdivided into the following main groups:

(1) primary industry and process manufacturing,
(2) manufacturing equipment industry,
(3) consumer goods industry,
(4) food and beverage industry,
(5) utilities,
(6) miscellaneous industries.

In the following classification, the first level of differentiation is made between industries with mainly organic or inorganic pollutants, since this criterion has the greatest impact on the wastewater treatment strategy.

2.1 Industries with Mainly Organic Pollutants

The following industrial groups produce wastewater with mainly organic pollutants:

(1) Food production:
  (–) animal production,
  (–) fruit and vegetable production,
  (–) plant growing.
(2) Food processing industry:
  (–) sugar factories,
  (–) starch factories,
  (–) vegetable oil and fats production,
  (–) potato processing industry,
  (–) vegetable and fruit processing industry,
  (–) abattoirs, meat, and fish processing,
  (–) sweets and candy industry,
  (–) dairy industry,
  (–) fruit juice and beverage production,
  (–) breweries,
  (–) wine production,
  (–) distilleries and yeast industry.
(3) Animal by-product industry:
  (–) animal cadaver utilization plants,
  (–) gut processing,

(–) tanneries,
(–) glue and gelatine production,
(–) animal and fish meal factories.
(4) Paper and pulp mills:
(–) wood pulp manufacturing,
(–) paper factories (including de-inking plants).
(5) Textile industry:
(–) wool washing factory,
(–) textile finishing industry (including dyeing).
(6) Pharmaceutical and chemical industry:
(–) pharmaceuticals,
(–) organics,
(–) soaps, washing powders,
(–) synthetics,
(–) miscellaneous.
(7) Miscellaneous:
(–) wood processing industry,
(–) textile cleaning,
(–) transport container cleaning,
(–) hospital wastewater.

2.2 Industries with Mainly Inorganic Pollutants

The following industrial groups produce wastewater with mainly inorganic pollutants:

(1) Heavy industry:
(–) mining,
(–) steel mills,
(–) power plants,
(–) quarries,
(–) ceramic industry,
(–) glass industry.
(2) Metal industry:
(–) metallic mordant plants,
(–) electroplating plants,
(–) machine manufacturing,
(–) vehicle manufacturing.
(3) Mineral oil industry:
(–) refineries,
(–) storage tanks,
(–) oil recycling.
(4) Chemical industry:
(–) primary inorganics,
(–) paint and lacquer production,
(–) detergents,
(–) adhesives,
(–) synthetics,
(–) pesticides,
(–) fertilizers,
(–) miscellaneous.

3 Wastewater Flow Fractions from Industrial Plants

3.1 Synopsis

Due to the multitude of products and production methods in industrial plants, there is also a wide range of different wastewater flow fractions for which the respective industrial branch has coined their own particular terms (e.g., "singlings" is a term used only in the distillery business). The wastewater flow fractions listed below represent the most important wastewater sources, without each fraction being necessarily produced by each of the single branches:

(1) rain water,
(2) wastewater from the sanitary and social facilities,
(3) cooling water,
(4) wastewater from in-plant water preparation,
(5) production wastewater (the following flow fractions are in some branches part of the production water):
(–) wash and flume water,
(–) fruit water,
(–) condensates,
(–) cleaning water.

3.2 Rain Water

Rain water can periodically constitute a substantial proportion of wastewater, depending on the amount of sealed surfaces. It is important to differentiate between non-polluted to slightly polluted rain water, such as that coming from roofs, which should either be percolated or can be discharged directly into the rain water sewer or waterways, and rain water that is collected from sealed areas where product handling or vehicles contribute to the contamination of rain water, making treatment of rain water necessary.

The amount of rain water Q_r [L s^{-1}] can be calculated by the following formula:

$$Q_r = r_{T,n} \cdot \psi_S \cdot A_E$$

A_E [ha] (1 ha = 10,000 m^2) is the catchment area, i.e., all areas that create run-off into the rain water canal in the event of rain fall.

ψ_S [–] is the run-off coefficient, which indicates the percentage of fallen rain water that actually increases the run-off flow (i.e., the portion which does not percolate, evaporate, or is held back by low lying areas). This amount varies greatly, depending on the surface conditions of the catchment area. For sealed industrial areas it is safe to use the run-off coefficient $\psi_S = 1$.

$r_{T,n}$ represents the rain fall intensity as [L s^{-1} ha^{-1}]. For a 10 min rain storm (with a sewer flow time of $\leq$10 min, the standard design rain fall intensity), which in this intensity only occurs once a year, this value amounts in Germany to 100–200 [L s^{-1} ha^{-1}]. If a sewer overflow is not permissible once a year, but only once every 5 years, the given amounts need to be increased by the factor 1.8.

A detailed design capacity for the rain fall amount can be found in ATV (1992).

The nature of the pollutants and the resulting pollutant load in the rain water depend mainly on the prevailing degree of air and surface pollution. According to ATV (1992), the average COD-value in rain water run-off is 107 mg L^{-1}. Peak values which occur especially after long dry periods at the onset of the rain fall are substantially higher.

3.3 Wastewater from Sanitary and Social Facilities

Wastewater from sanitary and social facilities consists of the water used by the employees for washing, the flushing of toilets and of the wastewater from social facilities. If there is a large canteen the wastewater from this source should be evaluated separately.

Wastewater from sanitary and social facilities has the same basic composition as domestic wastewater and should, therefore, be kept separate from the effluent emerging from the production line. This flow fraction should be discharged into the municipal sewage system, but treatment in the factories own treatment plant is also possible.

The amount of wastewater per employee and shift is approx. 40–80 L, while the condition of the plumbing fixtures and the degree of dirt involved in the operation (use of showers at the end of the shift), can have a major impact on the amount of wastewater produced.

The load per employee and shift can be assumed to be 20–40 g BOD_5, 3–8 g N and 0.8–1.6 g P.

3.4 Cooling Water

If cooling is necessary and done using water, one must differentiate between continuous flow (high water consumption) and recirculation (increase of salt concentrations).

For the quantity of cooling water being necessary, no general amounts can be given as the need for cooling water varies greatly from one industry to the other. As a first approach one can say, that for industries using recirculation cooling towers approx. 3–5% of the water capacity (depending on the salt concentration in the cooling water, this ratio may even be higher) need to be extracted to prevent the excessive increase of pollutant concentration.

If there are no leaks in the a continuous-flow cooling tower and no chemical additives are used in the cooling water, the composition of the water does not change, except for an increase of temperature. In recirculation cooling towers the concentration of pollutants increases with the evaporation of the cooling water.

3.5 Wastewater from In-Plant Water Preparation

Wastewater from in-plant water preparation can result from the preparation of drinking water from well water or from water softening, decarbonation, or desalination systems. The contents of this water depends on the characteristics of the raw water, of the respective treatment method, and of the chemicals used.

3.6 Production Wastewater

Production wastewater is a generic term, which is comprised of a variety of flow fractions. Exact specifications can only be made in connection with a particular industry. Certain flow fractions arising in specific industries have been designated with a particular name (e.g., "singlings") and are not listed here. The following wastewater flow fractions will either be investigated as a separate flow fraction or as part of the production wastewater.

- *Produce washing water* is in many food processing industries also used as *flume water* for the transport of the produce.
- *Fruit water* is the term used in the food processing industry for that water which is extracted from the processed fruit, such as the water extracted from potatoes during starch production.
- *Condensates* are the exhaust vapors that have been condensed after having been removed during an evaporation or drying process.
- *Cleaning water* results from the washing of production lines (pipes, containers, etc.), the cleaning of the production facilities, and the transportation containers (bottle washing).

4 Kinds and Impacts of Wastewater Components

Below, the most important wastewater components are listed, with their influence on the value retention of the production facilities and their impact on the necessary wastewater treatment measures.

4.1 Temperature

The temperature has a major influence on the construction material. High temperatures are not suitable for synthetic materials; e.g., they may cause damage to gaskets. Thus, the temperature of wastewater which is discharged into municipal sewage systems is restricted (in Germany the standard discharge temperature is $<35\,°C$). With regard to the corrosion of metallic materials, the temperature is a major factor, next to chloride and oxygen contents. Another aspect is that volatile components are easily removed from the wastewater and volatilized at higher temperatures.

An increased wastewater temperature has a positive influence on biological wastewater treatment methods, since an increase in temperature also increases the activity of microorganisms. On the other hand, any temperature increase results in a lower oxygen input capacity of the aeration system.

4.2 pH Value

Similar to the temperature, the pH value has an impact on determining the suitable construction materials and the activity of microorganisms. In order to avoid damage to the sewage systems and the connected treatment plants, in Germany the pH value of the discharged effluent should be between 6.5 and 10.

For many microorganisms the ideal living conditions are given at a relatively neutral pH value. Anaerobic wastewater treatment processes are especially sensitive to pH value fluctuations. It is important to keep in mind that organic acids can at suitable organic loads be degraded by biological plants without having to be neutralized. This, however, does not apply to mineral acids.

4.3 Obstructing Components

With regard to wastewater treatment obstructing components are larger inorganic particles, such as glass shards, plastic parts, cigarette butts, sand, etc., which are biologically inert and need to be removed mechanically in order to avoid damage, clogging, or caking in the further treatment process.

4.4 Total Solids, Suspended Solids, Filterable Solids, Settleable Solids

The content of solid particles has a substantial effect on the amount of organic matter. One differentiates between: total solids (TS), which consist of the suspended solids (SS; all particles that do not pass through a membrane filter with a pore size of 0.45 μm) and the filterable solids (FS). Settleable solids (unit: $mL\ L^{-1}$) are the solids that will settle to the bottom of a cone-shaped container, therefore, it does not include all of the suspended and floating solids.

4.5 Organic Substances

Organic substances constitute the main pollutant fraction in most industrial plants. In order to prevent direct oxygen consumption in the waterways into which the effluent is discharged, organic substances need to be eliminated as far as possible. There is a variety of parameters which can be used to determine the content of organic matter in a given wastewater sample: BOD, COD, TOC, DOC, etc. These sum parameters, however, do not allow any indication about the kind of organic substances measured.

The COD has developed into a major parameter, as the results of COD analysis are available much quicker than those of the BOD analysis. The use of cuvette tests gives an excellent cost–benefit ratio and it takes less efforts, space, and time to get results.

The COD–BOD ratio is an important value in determining the biodegradability of the pollutants in a particular wastewater. If the ratio is <2, the load is considered easily biodegradable.

4.6 Nutrient Salts (Nitrogen, Phosphorus, Sulfur)

Nutrient salts are inorganic salts such as: NH_4, PO_4, SO_4, which are considered vital for the growth of plants and microorganisms. Since nitrogen and phosphorus can cause massive growth of biomass and may, therefore, lead to eutrophication problems in waterways into which the effluent is discharged, in the European Union nitrogen and phosphorus must be eliminated before being discharged into sensitive waterways.

In order to biologically eliminate nitrogen and phosphorus sufficient organic pollutants must be present. Therefore, not only the concentration of these substances in the industrial wastewater is important, but also the ratio of these concentrations to the COD or BOD.

Some kinds of industrial wastewater have such low nitrogen and/or phosphorus concentrations that nitrogen (in the form of urea) and/or phosphorus (as phosphoric acid) must be added to obtain the necessary minimum nutrient ratio for the growth of the microorganisms needed for biodegradation.

4.7 Hazardous Substances

Hazardous substances is a generic term used for those substances or substance groups contained in the wastewater which must be regarded as dangerous because they are toxic, long-lived, bioaccumulative, or which have a carcinogenic, teratogenic, or mutagenic impact. In industrial wastewater, the following substances are of major importance:

- absorbable organic halogen compounds (AOX),
- chlorinated hydrocarbons and halogenated hydrocarbons,
- hydrocarbons (benzene, phenol, and other derivatives),
- heavy metals, in particular mercury, cadmium, chromium, copper, nickel, and zinc,
- cyanides.

Whether a substance is to be regarded as toxic or hazardous (according to the definition given above) is primarily a matter of concentration. Many heavy metals are vital as trace elements for the growth of microorganisms, but toxic in higher concentrations.

In Germany, the regulations for hazardous substances are so extensive that certain industrial branches not only have to comply with the required discharge quality of the effluent of the entire plant, but also must meet further requirements for the locations where certain flow fractions emerge before they are mixed with the eventual effluent (fractional flow treatment). Some substances have been banned entirely, which requires a specific design and schedule of particular production steps.

4.8 Corrosion Inducing Substances

The evaluation of which substances are corrosion-inducing in which concentrations depends not only on the substances themselves, but also on the choice of material.

For cement-bound materials three kinds of corrosion are distinguished, swelling impact through sulfates, dissolving impact through acids, and dissolving impact through exchange reactions, e.g., with chloride, ammonium, magnesium. A sufficient concrete resistance is given when the following limits are not exceeded with quality concrete and long-term strain: SO_4 <600 mg L^{-1} (for HS-concrete <3,000 mg L^{-1}; inorganic and organic acids pH >6,5; lime-dissolving carbonic acid <15 mg L^{-1}; magnesium <1,000 mg L^{-1}; ammonium nitrogen <300 mg L^{-1}.

Corrosion of metals is an electrochemical process for which a conductive liquid must be present, e.g., water. For metallic materials problems arise particularly with high concentrations of chloride, which can lead to corrosion even when high-grade steel is used. It is not possible to determine any general chloride concentrations, as the corrosion potential depends on a number of parameters such as material quality, redox potential, gap width, flowing velocity, temperature, or manufacturing quality.

Synthetic materials are commonly regarded mainly corrosion-proof. There are some synthetics, however, which are not stable towards acids, lyes, solvents, or oils, so that when dealing with aggressive media one has to consult resistance tables and check the producers' advice.

4.9 Cleaning Agents, Disinfectants, and Lubricants

In industrial factories a great number of cleaning agents, disinfectants, lubricants, dyes, etc., are used. If they occur in higher concentrations, many of these materials have an inhibiting or even toxic impact on biological treatment methods. It is also possible that they may contain components which cannot be eliminated in the wastewater treatment plants, e.g., AOX.

Although the unrestricted production flow is the superior maxim for industrial companies (e.g., in food industry factories getting unsterile must be entirely ruled out), examples tested in practice have evinced that consumption control systems or substitution of particularly polluting substances have allowed for useful and cost-effective improvements.

5 General Processes in Industrial Wastewater Treatment Concepts

5.1 General Information

To an increasing extent, wastewater treatment plants have changed from being pure "end-of-pipe" units towards being modules which are fully integrated into the production process. In such cases, this is referred to as production-integrated environmental protection. The technological basis of this tendency is that production residues can often be used in other ways than disposal, that the wastewater flow often presents some "product at the wrong place", and that it is simpler and more cost-effective to clean the single flow fractions individually and at higher concentrations. Thus, the general process of wastewater treatment by industrial companies can be roughly divided into the sections of production integrated environmental protection and post-positioned wastewater purification. The borderline between these sections is not clear-cut.

The first step in any procedure of developing wastewater treatment concepts is a detailed stock-taking of the situation of the company with regard to production methods, water supply, and wastewater production. For instance, the production-specific amounts of water, wastewater, and pollutant loads should be ascertained as well as the characteristics of the different flow fractions of the company and the places where lyes, acids, detergents, etc. are used.

The next step would comprise the different propositions for production integrated measures and the examination of the different options for the post-positioned wastewater purification unit. The following sections will provide further in-depth information on this point.

The last step would be the comparison and evaluation of the different propositions and options, the most important criteria for the evaluation being operational safety, economic viability, possibilities for a sustainable way of production, and the consideration of the overall concept (combination of industrial pretreatment and post-positioned municipal wastewater treatment plant).

5.2 Production Integrated Environmental Protection

The following methods should be examined with regard to their suitability as production integrated measures. The main principle should be that avoidance should take precedence over utilization and utilization over disposal.

(1) Careful treatment of raw materials (short storage periods, careful handling),
(2) changes of the transport facilities (dry conveyance, establishment of conveyance circuits),
(3) changes of the production methods (reduction of water demand, substitution of water, improvements of the production organisation),
(4) avoidance of surplus batches and production losses,
(5) product recovery,
(6) production circuits and multiple use of water (flume and washing water circuits, reverse flow cleaning, lye recirculation in CIP (cleaning in place) plants, cooling water recirculation),
(7) utilization of raw materials from production residues (protein coagulation, valuable substance recycling, by-product yield, forage production),
(8) separate collection of residues,
(9) extensive retention of production losses in collection containers and utilization separate from the wastewater flow,
(10) cleaning in different steps (1. dry cleaning, e.g., with high-pressure air; 2. washing; 3. rinsing),
(11) useful treatment of the flow fractions,
(12) general operation and organization (training of staff, control of water and wastewater amounts, installation of water-saving armatures, use of high-pressure cleaning tools, etc.).

5.3 Typical Treatment Sequence of a Wastewater Treatment Plant

A typical treatment sequence of a wastewater treatment plant consists of the stages listed below. Most industrial factories, however, have only a few of these operation stages, as they either do not have to cope with the respective pollutants in their wastewater, or because they are exempted from particular treatment steps due to specific regulations, which is the case, e.g., with indirect dischargers.

(1) removal of obstructing substances (screens, grit chamber),
(2) solids removal (strainers, settling tank, flotation),
(3) storing equalization cooling,
(4) neutralization or adjustment of the pH value,
(5) special treatment (detoxification, precipitation/flocculation, emulsion cracking, ion exchange),
(6) biological treatment or concentration increase (evaporation) or separation (membrane methods).

6 Wastewater Composition and Treatment Strategies in the Food Processing Industry

6.1 General Information

In view of the great variety of industrial branches, this section is by no means comprehensive in examining all areas of industry, but concentrates on the most important branches of the food processing industry.

When examining the specific wastewater amounts and pollutant concentrations, it is important to consider the points already mentioned in Sect. 1 (e.g., age and technical state of the equipment used at the plants). One crucial aspect mentioned in the introduction as well is training and motivation of the employees, which can have a substantial impact on the amount of product loss (spillage, etc.) during the production process, as these losses have a considerable impact on the concentration of pollutants in the wastewater. Apart from the hard statistical facts about production processes, a conscientious plant designer should, when planning the layout for the wastewater treatment units, also try to consider all operational factors of the respective plant.

The general pressure to remain cost-effective is prodding the industry to find ever new solutions, which makes for an ongoing optimization process. This has a considerable impact on the amount and composition of the wastewater produced and thus directly on the treatment methods. Two basic tendencies have resulted from this: the first is the attempt to reduce the amount of water used, e.g., by recycling or reusing water in other processes. The result of this is a decreased amount of wastewater, but with a higher concentration of pollutants. The second tendency is to separate the fractions and to reduce the specific load (kg COD t^{-1} product), e.g., by disposing of the dried solid matter (in some cases as nutrients for agricultural use). A detailed description of what is generally referred to as *production-integrated environmental protection* will be presented in the following section.

In order to make precise statements about wastewater amount and composition of any particular industry, it is recommended to obtain the latest data pertaining to the respective plant. Apart from the relevant periodicals dealing with this topic, the most important data source are the *Handbücher zur Industrieabwasserreinigung* (manuals on industrial wastewater treatment) published by the "Abwassertechnische Vereinigung ATV" (association for wastewater technology, Germany) (ATV, 1985, 1986). Furthermore, relevant data can be found in the reports, codes of practice, and leaflets developed and issued by the single ATV commissions and ATV working groups.

6.2 Sugar Factories

Sugar consumption in Germany is approx. 36 kg per capita and year. Most of the sugar is extracted from sugar beets. The average sugar content of sugar beets is 18%, so that the sugar beets processed per capita are approx. 220 kg per year. The processing of sugar beets is seasonal (campaign operation) and generally takes place from September to mid-December.

In recent years, the amount of wastewater has been greatly reduced from 15 m^3 to approx. 0.8 m^3 t^{-1} of sugar beets by water saving methods and by extensive sealing of the water circuits. In modern plants the water derives almost exclusively from the sugar beets themselves (sugar beets have approx. 0.78 m^3 fruit water per ton).

According to ATV (1990), a typical sugar factory produces the following wastewater flow fractions (Tab. 1).

Regarding the wastewater components, most important are the high organic loads derived from the sugar. In contrast, the amounts of nitrogen and phosphorus are comparatively low; due to the fact that growing ecological awareness has led to a more controlled fertilization of the sugar beet fields, the amounts of these pollutants are everdecreasing. In Germany, lime milk is added to the flume water and at the juice purification step to prevent getting unsterile. Therefore, the wastewater has an increased calcium content. Tab. 2 shows the concentrations of the most important parameters. Due to the recirculation of water in

Tab. 1. Wastewater Amounts in $m^3\,t^{-1}$ of Sugar Beets (ATV, 1990)

Wastewater Flow Fraction	Amount [$m^3\,t^{-1}$]
Pump sealing water and cooling water	0.4–0.7
Surplus condensates	0.4–0.6
Soil sludge transport water	0.3–0.4
Cleaning water	0.02
Water from the ion exchange unit	0.05–0.13
Surplus water from the wet dust collection of the pulp drying vapors	0.02

Tab. 2. Concentrations of Pollutants from Sugar Factory Wastewater (JÖRDENING, 1997)

COD [mg L^{-1}]	BOD_5 [mg L^{-1}]	N [mg L^{-1}]
6,000–30,000	4,000–18,000	50–180

partially closed cycles, the concentrations increase steadily during the course of the campaign operation.

For the treatment of sugar factory wastewater, the following processes are common:

- soil treatment,
- long-term batch processing,
- small-scale technical processes, in most cases consisting of anaerobic pretreatment and aerobic secondary treatment.

The agricultural soil treatment is primarily geared towards the utilization of the contained nutrients and water. The simultaneous degradation of the organic substances is of less importance; it does not form the basis of dimensioning. If large quantities of ion exchange waters occur, the salt compatibility of the irrigated plants has to be considered. 90–120 mm of total annual wastewater irrigation can be regarded as standard value, while the maximum single occurrence should not exceed 30 mm. In Germany, the total amount of wastewater irrigation is partly restricted to <50 mm per year, so that because of the high area demand these methods are no longer economically viable.

With long-term batch processing, the wastewater is discharged into sealed ponds. The degradation period lasts until the summer of the following year and consists of an anaerobic phase succeeded by an aerobic phase (if necessary, aeration equipment is used). If the batch depth does not exceed 1.2 m the COD concentration can be reduced to less than 300 mg L^{-1}. The disadvantages of this process are the great area demand and the emergence of noxious odors in springtime.

In recent years, mainly anaerobic wastewater treatment processes have become established in the sugar industry. In comparison to the processes described above and to the aerobic biological treatment, anaerobic methods have a number of distinct advantages: less area demand, lower sludge production, low energy consumption, energy generation from biogas, no emergence of noxious odors. In Germany, mainly anaerobic contact sludge processes are used due to the high calcium contents, whereas in the Netherlands, e.g., where caustic soda is used instead of lime milk, many UASB reactors are used. Anaerobic fixed film reactors have not been tried yet. The successful operation of a fluidized bed reactor has been described by JÖRDENING (1996a, b). In properly working reactors, the desired pH range of 6.8–7.5 automatically remains stable, the operating temperature normally is 37 °C. With a volumetric load of 5–10 kg COD $m^{-3}\,d^{-1}$ (for UASB up to 15 kg), a COD reduction efficiency of approx. 90% is achieved. The nitrogen and phosphorus elimination is only achieved to a small degree by integration into the excess sludge; thus, further treatment is necessary before a direct discharge of the effluent is possible. As a rule of thumb nutrient salts are eliminated in anaerobic reactors in the ratio COD:N:P=800:5:1. In order to avoid operational difficulties with wastewater with high calcium concentrations, for contact sludge processes decanter centrifuges have proved to be most suitable to separate the calcium carbonate particles from the bacterial sludge.

If direct discharge into a waterway is planned, aerobic secondary purification is necessary, ideally by an activated-sludge system. The condensates may then be directly fed into the aerobic stage.

6.3 Starch Factories

In 1995, 1.4 million tons of starch were produced in Germany, consisting of: 0.5 million tons of potato starch, 0.5 million tons of corn starch, and 0.4 million tons of wheat starch. Depending on the original raw material the wastewater fractions, amounts, and loads emerging during starch production vary considerably.

Corn starch: fractional flow sources: maceration station, germ washing, starch milk dewatering, gluten thickener, glue of gluten dewatering, and chaff dehydration. The various fractions of wastewater are mainly recycled and used as processing water. The wastewater fractions which have to be treated are

(1) processing water and
(2) the condensates resulting from the evaporation of the maceration water.

Wheat starch: during wheat starch production, wastewater is derived from the separation step and in the thickening of the secondary starch. All other flow fractions are returned into the process water. The wastewater flow fraction which has to be treated is referred to as process water.

Potato starch: during potato starch production, washing water (flume and transport water), fruit water (in some cases condensates from a fruit water evaporation plant), and production wastewater are produced. By recycling it has been possible to reduce the washing water demand from 5–9 $m^3 t^{-1}$ to 0.25–0.6 $m^3 t^{-1}$ (status: 1996). Fruit water separated from the potatoes is usually fed into the protein recovery unit. Approx. 50% of the protein compounds contained in the fruit water are coagulated and separated subsequently. Protein production from fruit water is always economically viable, since the price for the protein produced is approx. 1 DM kg^{-1}, whereas the biological treatment, particularly for the elimination of nitrogen (10 g protein contain 1.7 g N), is quite expensive (approx. 4–10 DM kg^{-1} N).

According to ATV (1994), the specific wastewater amounts of starch production are as follows (Tab. 3).

Wastewater from starch factories has a high organic load and usually consists of easily degradable matter. The undissolved organic contents are mainly carbohydrates and proteins. The fat content is normally $<10\%$. In most cases, this kind of wastewater does not contain any toxic substances. The substrate ratio COD:N:P is satisfactory; actually, there may even occur an excess of N and P, so that it is not necessary to add nutrient salts. Tab. 4 shows standard wastewater values for corn, wheat, and potato starch wastewater, which show distinct differences.

Tab. 3. Specific Wastewater Amounts of the Starch Industry (ATV, 1994)

	Specific Total Wastewater Amount [$m^3 t^{-1}$]	Specific Amount of Certain Flow Fractions [$m^3 t^{-1}$]
Corn starch production	1.5–3.0	condensates: 0.4–0.7
Wheat starch production	1.5–2.5	
Potato starch production	1.5–2.3 without fruit water	– flume and washing water: 0.25–0.6 – process water (refinement + fruit water separation): 0.6–1.4 – condensates: 0.5–0.8 – potato fruit water: approx. 0.8

Tab. 4. Wastewater Concentrations in the Starch Industry (AUSTERMANN-HAUN, 1997)

	pH	COD [mg L^{-1}]	BOD$_5$ [mg L^{-1}]	N [mg L^{-1}]	P [mg L^{-1}]	S [mg L^{-1}]
Corn	–	2,500– 3,500	1,700– 2,500	100	60	110
Wheat	3.5–4.6	18,750–48,280	11,600–24,900	610–1,440	115– 240	120–410
Potato						
– washing water	6–7	2,000– 4,000		200– 300	20– 40	
– process water	6–7	4,000– 8,000		300– 600	20– 40	
– potato fruit water	5–6	50,000–60,000		3,600–4,400	800–1,000	

The following processes are used for the treatment of starch factory wastewater:

- soil treatment,
- pond processing,
- small-scale technical processes (anaerobic, aerobic evaporation).

Until recently, the wastewater from most starch factories in Germany was used to irrigate fields. A summary of the results of this practice can be found in SEYFRIED and SAAKE (1985). A lack of sufficient agriculturally used areas and the low amounts permitted to be used thus put an end to this practice.

Pond processing in batch operation is only possible for the production of potato starch, since this is a campaign operation. Continuously operating ponds are rare, since they require additional aeration, due to the high load concentration. This can only be recommended as a secondary treatment method, as the demands for area and aeration are too high.

Anaerobic wastewater treatment has become established alongside conventional activated sludge processes. Fixed film reactors and UASB reactors are mainly used, which at loads between 7–30 kg COD m^{-3} d^{-1} can achieve degradation rates of 70–90%. In order to maintain operational stability, a separator for solids is usually placed ahead of the reactor. Furthermore, a separate pre-acidification stage with a sludge removal system is recommended. The amounts of available phosphorus and nitrogen are sufficient for the anaerobic treatment. Sometimes, trace amounts of cobalt and nickel need to be added. The pH value should not exceed 7.0, since at higher pH values the precipitation of MAP (magnesium ammonium phosphate) might take place.

In the potato starch industry a further possibility for fruit water treatment is the reduction of the wastewater by evaporation to a solids content of 70%. The dried solids can then be utilized, e.g., as fertilizer. In order to save evaporation energy, a membrane process can be used to increase the concentration prior to evaporation, which would be done as a cascade operation under vacuum conditions.

Aerobic secondary treatment is required in any case, if the respective company intents to discharge its wastewater directly into waterways.

6.4 Vegetable Oil and Shortening Production

The nutrient fat consumption in Germany is approx. 29 kg per capita and year. It consists of the following categories: margarine 8.4 kg, butter 7.1 kg, lard tallow 6.4 kg, shortening 3.6 kg, and salad oil 3.5 kg.

In the vegetable oil industry, one differentiates between the following branches: extraction plants for the extraction of raw fat and oil, refining plants for the refining of raw fat and oil, plants for further processing of the nutrient fats, e.g., into margarine, which consists of approx. 80% fat, 20% water, plus some other ingredients.

The process technology for nutrient fat and oil production largely depends on the respective raw materials. The differentiation is based on whether the oil is extracted from seeds (soy, sunflower, coconut, rapeseeds, etc.), animal fats from tallow and fat liquefiers (tallow or lard), and fruit flesh fats (palm oil, olive oil, which due to its poor storage stability is usually extracted at the site of production), and fats from marine animals.

The processing of seed oils consists of the following stages: cleaning, in some cases peeling/shelling, masticating, conditioning (heating and moistening), pre-pressing with ensuing extraction or direct extraction without pre-pressing. Extraction is accomplished by means of a solvent–oil mixture (Miscella), followed by the removal of the solvent (usually hexane) by steam. The vapors are condensed and the hexane is separated from the water in static separators. In a further distillation step, the remaining hexane is removed from the wastewater. The wastewater amount produced in seed oil extraction processes is less than 10 $m^3\ t^{-1}$ of seeds. Data for the wastewater pollutant concentrations are only available for COD which is in the range of 700–2,000 mg L^{-1}.

Extraction of animal fats is accomplished by masticating fatty tissues and heating the macerate, followed by a separation process. A wet and a dry melting process are distinguished. The dry melting process does not produce any wastewater, whereas the wet process produces approx. 0.35 $m^3\ t^{-1}$ raw material, which mainly consists of glue water and pump sealing water from the separators. Depending on the extraction method and on the cleaning procedures applied in the production plant, the concentration of pollutants in the wastewater from tallow melting plants can vary considerably. For wastewater which has passed the fat separators with topped sludge catcher implements one can assume the following values (the ratio of emulsified fats, however, may be dramatically higher than given here) (Tab. 5).

In refining plants, the raw oil is purified (refined) by removing undesired ingredients. The basic process steps consist of desliming, neutralizing, bleaching, and deodorizing (steaming). Depending on the raw material and the desired end product, further processes such as winterizing, fractionating, transesterification, or hydrogenation may be added. In addition to this, other waste products such as saponification water (soapstock) are being processed. Wastewater fractions occur during desliming/neutralizing, oil drying and cooling, steaming (condensates) as well as during soap cracking (acidic water). The amount and composition of the wastewater depends mainly on the characteristics of the fresh water, with regard to temperature and degree of hardness, the mode of process operation (continuous or discontinuous methods, process temperature and pressure), the type and quality of raw materials, and the frequency with which the raw materials are changed. The data given in the reference literature vary greatly, due to the multitude of factors involved. A standard value for wastewater of a refining plant which is recirculating the falling water is <10 $m^3\ t^{-1}$ of raw material.

Provided that no falling water recirculation or static fat separators are used in the wet chemical refining process, the total wastewater of a refinery has the following characteristics (Tab. 6).

Tab. 5. Wastewater Concentrations in Animal Fats Processing (RÜFFER and ROSENWINKEL, 1991)

Settleable Solids [mL L^{-1}]	COD [mg L^{-1}]	Temperature (after cooling) [°C]	pH	Lipophilic Substances [mg L^{-1}]
<1	5,000–10,000	30–35	6–7	<100

Tab. 6. Wastewater Concentrations in Refinement Plants (RÜFFER and ROSENWINKEL, 1991; ATV, 1981)

Settleable Solids [mL L^{-1}]	Lipophilic Substances [mg L^{-1}]	COD [mg L^{-1}]	COD/ BOD_5	Temperature [°C]	pH	Sulfate (when sulphuric acid is used to crack soapstock) [mg L^{-1}]
<1	<150	<600	1.5–2.0	<35	5–9	500–1,000

For margarine production, wastewater results only from the cleaning circuits of the CIP plants amounting to approx. 1–3 $m^3 t^{-1}$ margarine. After the rinsing and cleaning water has passed the fat separator, the wastewater has the following characteristic parameters (Tab. 7).

Production-integrated environmental protection must be the central part of the wastewater treatment concept, with the major factors being the efficiency of the production process and the avoidance of product losses during the process.

Wastewater treatment processes usually consist of a physical-chemical pre-treatment by fat separators or flotation systems for the reduction of undissolved solids and lipophilic substances. Whereas the installation of fat separators is only a minimal pre-treatment concept, because the efficiency of these systems may be completely reduced by a hot water surge, a more expensive flotation system using additives also allows the elimination of emulsified, mainly lipophilic, substances. Particularly effective are pressure flotation systems designed for a surface load of 4–5 $m^3 m^{-2} h^{-1}$ with a recirculation flow of 10–40% and operating at 4–7 bar.

Often a mixing and equalizing (M+E) tank is installed behind the flotation system in order to equalize pH value and temperature peaks and to allow a more constant loading of the subsequent biological treatment stage.

In biological wastewater treatment, aerobic activated sludge treatment with a sludge load of approx. 0.1 kg BOD kg $MLSS^{-1} d^{-1}$ has proved to be effective. In order to prevent the flotation of sludge and the development of scum layers the fat contents of the respective wastewater should be <200 mg L^{-1}.

6.5 Potato Processing Industry

Potatoes are used as fodder or feed potatoes, as industry potatoes (starch production, distilleries), or as market potatoes. In Germany, the annual market potato consumption is approx. 75 kg per capita, approx. 30% of which is processed potato products, compared to over 50% in the USA.

Potato processing can be divided into 3 main production ways: fresh products (pre-peeled potatoes, pre-cooked potatoes, potato salads, sterilized potatoes, potato products), dried products (dehydrated potatoes, instant mashed potatoes, raw dehydrated potatoes), and fried products (French fries, chips, sticks, etc.).

Prior to the final processing the potatoes must be dry cleaned and sorted, in some cases stored for some time and then transported to the processing plant. The conveyance to the processing plant can be either wet or dry, depending on the mode of conveyance. It is followed by washing and peeling (mechanically and/or with steam). In the fried production process, peeling is followed by cutting, sorting/washing, blanching, steaming (drying), dewatering, frying and, if necessary, cooling.

For each of the different production methods, wastewater occurs during washing, peeling, cutting, sorting, blanching and steam drying. The specific wastewater amounts and the contents of the fractions of each process step are shown in Tab. 8 (SCHEFFEL, 1994). The most significant wastewater fraction results from the potato fruit water due to the contained starch. 1 g L^{-1} starch has approx. 1230 mg BOD L^{-1}. The wastewater load depends mainly on the type of potato (cell size), the growth and storage conditions as well as on the careful handling (cut and bruised potato cells increase the loads). Usually, the COD–BOD ratio is between 1.6 and 2, which makes for a

Tab. 7. Wastewater Concentrations in Margarine Production (RÜFFER and ROSENWINKEL, 1991)

COD [mg L^{-1}]	COD/BOD_5	Lipophilic Substances [mg L^{-1}]	Temperature [°C]	pH
1,000–2,000	1.5–2.0	<250	<35	5–9

Tab. 8. Wastewater Volume and Loads in Various Potato Processing Steps (SCHEFFEL, 1994)

Process Step	Spec. WW Volume [$m^3 t^{-1}$]	Settleable Solids [$ml L^{-1}$]	COD [$g L^{-1}$]	BOD_5 [$g L^{-1}$]	N [$mg L^{-1}$]	P [$mg L^{-1}$]
Washing	0.3–0.5					
Peeling						
– mechanical process	0.8–1.5	peel loss: 15–25%; settleable solids: 80–90% of peel loss				
– steam process	0.2–0.3	peel loss: 8–15%; settleable solids: 75–80% of peel loss				
Cutting/sorting/washing	<0.1					
Blanching	0.05–0.15			12.0–18.0		
Steaming (drying)	0.15–0.25			6.0–8.0		
Wet potato dough products	2–3	4.5–8.0	7.0–12.0	3.5–6.0	300–400	30–50
Dehydration products						
– dried potatoes	5–7		6.0–8.0	3.0–4.0	150–300	
– instant mashed potatoes	3–6			2.0–4.0	100–140	15–30
Fried products			3.6–7.5	2.0–5.0	120–600	25–250
Potato chips	3.9		2.0–6.0	1.0–3.5	90–500	6–50

favorable biodegradation. The C–N ratio is also advantageous for denitrification. Wastewater from the potato processing industry contains comparatively much nitrogen and phosphorus. Nutrient salt limitations only occur in exceptional cases.

Wastewater treatment systems often consist of the following sections:

(1) grit chamber, screening system, settling tanks for purification of the flume and washing water recirculation;
(2) production-integrated screening systems and separators to recover organic solids which have been separated from the production wastewater and dewatered, for the recycling of valuable substances (e.g., cattle forage);
(3) fat separators for wastewater with fat contents, as is the case when deep fat fryers are used in the production process;
(4) biological treatment for pretreatment and full treatment.

Although there are large scale applications for the entire range of aerobic and anaerobic treatment methods, good results have been achieved using activated sludge systems, which are either designed as one-stage or cascade units and which can cope with a sludge load of 0.1 kg BOD kg^{-1} MLSS d^{-1}. A problem that may arise is bulking sludge, which can be reduced by using the cascade design. In recent years, more anaerobic plants have been built, with the UASB reactor being the most common type in the potato processing industry. In order to guarantee stable operation of these reactors, it is imperative to limit the amount of filterable solids in the wastewater by installing separators or settling tanks preceding the anaerobic treatment. A relatively stable and neutral pH value is also essential. Due to the large amount of starch particles, the reactors should be preceded by an acidification unit with a retention period of more than 6 h for hydrolysis.

6.6 Abattoirs

Germany has an annual per capita meat consumption of approx. 90 kg, which makes it one of the foremost meat consumer nations in the world. Approximately 60% is consumed as fresh meat while the remaining 40% are processed. The abattoirs and meat processing plants can be divided into 4 categories according to their different production processes.

(1) slaughter houses,
(2) slaughter houses for poultry,
(3) meat cutting plants,
(4) meat processing industry.

Local butcher shops are not taken into account, due to their small turnover volume.

The following statements refer only to wholesale slaughter houses which process up to 8,000 hogs per day. In the slaughtering process not only meat from muscle tissue, but also by-products and residues are produced. The meat and the by-products (such as liver, etc.) are fit for human consumption and can be traded freely. The residues produced in the slaughtering process are divided into two categories:

(1) residues that have commercial value and are tradable, such as fats and bones which find use as raw materials in feed plants or in the pharmaceutical industry and
(2) non-tradeable residues such as meat unfit for consumption and other by-products which need to be treated in animal cadaver disposal plants. Additional waste products are primary waste (stomach and intestine contents) and secondary waste (solids from wastewater screening and flotation sludge).

The primary wastewater and waste product sources are divided into 3 production areas:

(1) truck washing and stables (green line),
(2) slaughter and cutting (red line), and
(3) stomach, intestine, and entrails cleaning (yellow line) – the latter, however, often being not done on the abattoir grounds.

The waste occurring in trucks and stables consists of bedding material, faeces, and urine which amount to 2.0 kg per hog and 10 kg per cow, if the stable time is kept short. These materials should be removed without the use of water and spread on agricultural land. Sparing water consumption during truck cleaning can keep the washing water down to 100 L per truck. This wastewater should be screened in order to remove large particles and should then be fed into the main wastewater stream.

The specific amounts and contents of wastewater (red line) are presented in Tab. 9. The presented data can only be achieved, however, as long as the blood retention is approx. 90% (the COD of blood is 375,000 mg L^{-1}. The total blood volume is approx. 5 L per hog and approx. 30 L per cow).

In abattoirs, wastewater pretreatment is mainly done with mechanical procedures. Up to now, the number of cases where physicochemical or biological operational steps are added is comparatively small. The main cleaning effect of mechanical and physical procedures consists of the retention and separation of solids with the help of stationary strainers, rotating screening drums, separators, fine rakes, screening catchers, or fat separators with preceding sludge catcher.

In wastewater from abattoirs, one part of the organic matter consists of oils and fats in emulsified form. Thus, it may be wise to add a chemical/physical stage to the mechanical pretreatment, which would consist of a precipitation/flocculation stage and a flotation unit.

Although most of the abattoirs operating in the Federal Republic of Germany do not have their own biological stages, the wastewater from the abattoirs is suitable for biological methods, due to its composition. The following biological methods have up to now been used successfully on an industrial scale:

Tab. 9. Specific Wastewater Amounts and Concentrations of Abattoir Wastewater (red line) (ATV, 1995)

	Amount [L per animal]	COD [g L^{-1}]	BOD_5 [g L^{-1}]	N [mg L^{-1}]	P [mg L^{-1}]	fat [mg L^{-1}]	AOX [µg L^{-1}]	Settleable Solids [mL L^{-1}]
Hogs	100– 300	2.0–8.0	1.0–4.0	150–500	15–50	500–2,500	20–100	10–60
Cattle	500–1,000		1.5–3.0					
Chicken	approx. 25	2.2–4.0	1.0–2.5	150–350	5–30	300–1,200		

- large space biological methods (oxidation ponds, frequent formerly),
- activated sludge systems in different variations (single-stage, cascade, two-stage),
- anaerobic biological methods.

For activated sludge systems the sludge load should not exceed 0.15 kg BOD_5 kg^{-1} MLSS d^{-1}. Direct anaerobic treatment in contact sludge reactors or joint treatment in municipal digestion tanks is particularly suitable for fats, floating materials, stomach and gut contents, and the liquid phase from the dewatering of rumen contents.

6.7 Dairy Industry

In 1995, 26.74 mio. t of milk were delivered to the dairies of the Federal Republic of Germany, which makes for an annual per capita consumption of approx. 330 L of milk. The products are classified into drinking milk 5.8 mio. t, cream products 0.52 mio. t, sour milk and milk mix drinks 2.13 mio. t, butter 0.49 mio. t, curd products 0.72 mio. t, hard and soft cheese 0.72 mio. t, condensed milk 0.55 mio. t, and dried milk 0.95 mio. t. The difference between the listed products and the entire delivery amount is mainly due to the return of skimmed milk and whey, and to the extraction of exhaust vapors during coagulation and drying of milk and whey.

For milk processing, the delivered raw milk is at first – regardless of the production schedule – cleaned with separators, separated into cream and skimmed milk, and pasteurized. Cream is mainly used for the production of butter. Skimmed milk is processed into drinking milk, fresh milk products, and cheese, partly by adding cream or bacterial cultures. Skimmed milk and whey are the basic materials for the production of dried milk, lactose, and casein.

The wastewater produced in milk processing plants consists of cooling water, condensation water, sanitation water, and process water. The process water consists of the wastewater from the pretreatment, water losses during production, those residues which can no longer be used economically, washing water, detergents, rinsing and cleansing water, and water processing. In dairies the wastewater derives almost exclusively from the cleaning of the conveyance and production implements. More than 90% of the organic solids in the wastewater result from milk and production residues. For the milk processing industry, wastewater discharge is almost always identical with the loss of products which could otherwise be utilized or sold. This is a great incentive to cut down on the production of wastewater by production-integrated measures. For untreated dairy wastewater, the following data are valid; peak values can even exceed these values (Tab. 10).

Dairy wastewater has only a small ratio of settleable solids. Thus, conventional mechanical procedures, such as settling tanks, are ineffective. Inevitable losses of fat can be retained in fat separators, which, however, have only a limited efficiency, if the wastewater temperatures are comparatively high.

One major problem with dairies are the considerable variations in wastewater volume and concentrations. Thus, as a first step after straining and a sand trap it is recommended to install a mixing and equalizing tank (M+E tank). In the M+E tank the different concentrations are mixed (including pH value), the wastewater flow is equalized, and there is a partial biological degradation, which can also result in a biological neutralization. As a minimum volume, 25% (approx.) of the daily water flow has proved to be a favorable value, but it is also possible to adjust the facilities to a daily or weekly equalization. With unaerated M+E

Tab. 10. Amounts and Concentrations of Dairy Wastewater (BERTSCH, 1997)

Wastewater Amount [m^3 t^{-1} of milk]	BOD_5 [g L^{-1}]	COD [g L^{-1}]	NO_3–N [mg L^{-1}]	N [mg L^{-1}]	P [mg L^{-1}]	Settleable Solids [mL L^{-1}]	pH	Lipophilic Substances [mg L^{-1}]
1–2	0.5–2.0	0.5–4.5	10–100	30–250	10–100	1–2	6–11	20–250

tanks there is no biological degradation worth mentioning; instead, a substance conversion (acidification) happens, which leads to the emergence of obnoxious odors. Aerated M+E tanks are mostly operated as washing-off reactors. They are able to reach BOD_5 efficiency rates of 20–60% for homogenized samples.

As a further pretreatment stage, a flotation implement is recommended, which can either replace the M+E tank or be post-positioned to it; this facility would allow for the removal of fats and proteins, i.e., the major part of organic pollutants.

Although more than 90% of the milk processing companies in Germany discharge their wastewater indirectly, direct discharge could under certain conditions be economically viable. Direct dischargers mostly have an activated sludge system, which should, if strong variations occur, be preceded by an M+E tank. As dairy wastewater has a tendency to develop bulking sludge, it is recommended to design the plant in such a way that the microorganisms of the activated sludge are intermittently subjected to high loads, which could be achieved by installing an activated sludge system in plug flow design, by a preceding contact tank (selector), or by a SBR method (sequencing batch reactor: the process steps filling, denitrification, aeration, sedimentation happen one after the other, but in the same tank).

6.8 Fruit Juice and Beverage Industry

In 1996, in Germany the per capita consumption of mineral water amounted to approx. 91 L (natural mineral water, spring water, and table water), of soft drinks to approx. 92 L, and of fruit juice and fruit syrup to approx. 41 L.

Natural mineral water and so-called spring water are gathered and bottled at the location of the respective spring. Table water consists of drinking water or natural mineral water to which salts are added. Refreshment beverages are produced from water, flavoring substances, sugar or sweetener, and carbon dioxide. The technology of fruit juice production can, in a simplified manner, be divided into the production stages of washing, grinding, refining, filtering, heating, recooling, and bottling.

The wastewater produced in these three industrial branches consists of the following streams (some may not occur in each branch): wastewater from the cleaning of bottles and containers and from bottling, rinsing and washing water, exhaust vapor condensate, wastewater from the production facilities, wastewater from surface cleaning (the floors of the production sheds and the parts of the yards where the production takes place), and wastewater from cleaning the conveyor facilities.

Wastewater produced in the mineral water industries contains the following components: adhesive materials and fibrous substances, cleaning lyes and acids, and soiling from the deposit bottles. For soft drinks, one has to consider the fact that the wastewater additionally contains organic pollutants (with a high ratio of carbohydrates, a big part of it is sugar), which derive from residues and product losses. For the fruit juice industry, product losses – in particular the loss of fruit concentrates – and the sugar, which in many cases is added, are a considerable part of the wastewater pollution. The COD of fruit juices ranges from about 50 g L^{-1} (tomato juice) to about 200 g L^{-1} (apricot juice). 1 kg of glucose (fructose) is equivalent to 1066 g COD. Apart from the wastewater, the fruit juice industry also produces cooler sludge, filtration residues, sludge from clarifying agents, pomace, and kieselguhr. Because of their high pollution potential, these substances should be in the focus of production integrated measures. If possible, they should separately be utilized or disposed of.

The specific wastewater amounts and pollutant concentrations of the three industry branches are presented in Tab. 11. For the companies of the fruit juice industry, one has to differentiate between those which only do the bottling, those which only do the processing, and those who do both, processing and bottling. It is apparent that the wastewater from the beverage industry has low nitrogen and phosphorous values in relation to the BOD. The pH value can range from 3.5–11.5. At the time of contact with the product the pH value is mostly within the acidic range.

Tab. 11. Specific Wastewater Amounts and Concentrations in Mineral Water, Refreshment Beverage, and Fruit Juice Industry (ATV, 1998)

	Specific Demand [m^3 1000 L^{-1} drink]	BOD_5 [mg L^{-1}]	COD [mg L^{-1}]	N [mg L^{-1}]	P [mg L^{-1}]
Mineral water and beverages	0.9–1.3 (mineral) 1.1–3.3 (beverages)	110– 800	200– 1,600	2–35	0–18
Fruit juice (bottling only)		250–1,000	1,500– 3,000	1.2–10	1.5–12
Fruit juice (production only)		1,700–4,000	2,500–45,000	5–30	3–15
Fruit juice (production + bottling)		400–2,000	400– 3,000	9–25	2–14

In the non-alcoholic beverage industry, the major part of the companies are indirect dischargers. In order to meet the discharge limits of the municipal sewer, it is often necessary to equalize the pH value peaks and sometimes to reduce the temperature (discharge limit in most cases: <35 °C). Thus, a wastewater pretreatment plant could – apart from a straining station – consist only of a neutralization stage. Because of the high cost of chemicals, in most cases biological neutralization is recommended, e.g., in an aerated mixing and equalizing tank, which as a washing-off reactor reaches BOD_5 elimination rates between approx. 35% (daily equalization) and >50% (weekly equalization). Particular heed, however, should be paid to the alkaline water from the bottle washing machines, as it might be necessary to collect this water in a separate container and to discharge it in controlled doses.

Extensive wastewater pretreatment can successfully be done with anaerobic reactors. Two-stage implements (first stage: acidification reactor with mixing and equalization function, second stage: methane reactor) have proved to be advantageous. At volumetric loads in the methane reactor up to >10 kg COD m^{-3} d^{-1} the COD elimination rate amounts to about 80%.

For direct discharge into waterways the activated sludge system with cascade design has proved to be viable, which is operated at sludge loads <0.1 kg BOD_5 kg^{-1} MLSS d^{-1}. Because of the low nitrogen and phosphorus amounts in the raw wastewater it is generally necessary to additionally dose these substances. For the bottling of deposit bottles, denitrification may be necessary, because as part of the label glue nitrogen is added to the water. Moreover, one has to deal with the danger of bulking sludge.

6.9 Breweries

Beer is a beverage containing alcohol and carbonic acid. In Germany, legal regulations prescribe that it may be produced only from malt (germinated and oasted barley), hop, yeast, and water. The annual per capita consumption in Germany is approx. 120 L of beer which is produced in about 1,200 breweries. The different kinds of beer (lager, stout, top-fermented, bottom-fermented) are produced mainly by varying the original wort concentrations and by using different kinds of malt and yeast.

After malting, the main operation steps of beer production are wort production, fermentation, storing, filtration, and bottling. Prior to the fermentation into alcohol, the starch contained in the malt (which in Germany is obtained from barley) has to be converted into fermentable sugar.

Residues and wastewater flow fractions occur in the brewing room, in the fermentation and storage cellars, in the filter and pressure tank cellar, in the de-alcoholization, and during bottling (bottle, barrel, other containers). Brewery wastewater is prone to heavy variations in regard to volume and concentration in the single flow fractions. In cases where production-integrated measures have already been applied, one can assume the following characteristics for the entire wastewater flow of an average brewery (Tab. 12).

Brewery wastewater has comparatively high temperatures (25–35 °C). With decreasing wastewater volume the temperature tends to rise to up to 40 °C. It is likely that the pH values will vary very strongly. In companies with bottling of deposit bottles, the wastewater from the bottle washing will generally cause alkaline pH values. Acidic wastewater may at times result from the cleaning processes and from the regeneration by ion exchange implements (for the water processing). The nitrogen consists mainly of organic nitrogen (albumen, yeast), and to some extent of nitrate (nitric acid). Furthermore, one has to expect the wastewater to be impaired by cleaning and detergent agents as well as by kieselguhr and by abrasion from bottles and shards.

Through production-integrated measures, considerable contributions to the reduction of amounts and loads, temperature, solids contents, and also of the pH value can be achieved. Some very important measures are the retention and separate disposal of cooler sludge, kieselguhr, yeast and the addition of lye.

The majority of breweries in the Federal Republic of Germany discharge their wastewater indirectly into the municipal wastewater treatment plants. For wastewater pretreatment, the first step should be the removal of settleable solids, such as shards, labels, spent hops, bottle caps, etc. by suitable screen and strainer implements. In order to neutralize the mainly alkaline wastewater it is common to use carbonic acid from the fermentation or flue gas. It is also possible to biologically neutralize the lyes with the carbon dioxide which is produced during the BOD degradation. Because of the heavy variations, it is always recommended to use an equalization tank, which can either be run as an aerated mixing and equalizing tank with a biological partial purification (the elimination rates of these washing-off reactors range from approx. 35% with daily equalization and >50% with weekly equalization), or which may serve as an unaerated pretreatment tank or acidification reactor for the anaerobic plant. For the pretreatment of brewery wastewater the most common implements are UASB and fixed-film anaerobic plants, which in most cases are run without heating of the wastewater (reactor temperatures are in the range of 24–36 °C).

For the full-scale purification of brewery wastewater to direct discharge quality, aerobic activated sludge systems have proved to be best, due to the need to eliminate nitrogen and phosphorus. The activated sludge system can either be the sole treatment stage or it can be post-positioned to an anaerobic plant or an aerobic trickling filter unit. Another sensible solution is the use of SBR methods (sequencing batch reactor), where all treatment steps are run one after the other, but in the same tank.

6.10 Distilleries

Distilleries are factories which produce alcohol for human consumption by fermentation and distillation of agricultural products which contain sugar or starch. Some of this alcohol is also used for vinegar production and in the pharmaceutical and cosmetics industries. In companies which produce spirits, the alcohol is diluted to make it potable and it is enhanced with flavoring additives. In Germany, the per capita consumption in 1996 of pure alcohol (100 vol.%) produced from grain, potatoes, molasses, fruit, or wine amounted to

Tab. 12. Brewery Wastewater Amounts and Concentrations (RÜFFER and ROSENWINKEL, 1991)

Specific Wastewater Amount [m^3 100 L^{-1} beer]	BOD_5 [mg L^{-1}]	COD [mg L^{-1}]	N [mg L^{-1}]	P [mg L^{-1}]	Settleable Solids [mL L^{-1}]
0.25–0.60	1,100–1,500	1,800–3,000	30–100	10–30	10–60

1.83 L. There is quite a large number of rather small fruit schnapps distilleries. Distilleries are subjected to a number of legal regulations, in Germany in particular to the law on the spirits monopoly (BranntwMonG).

For the production process, it is important that raw materials containing starch are turned into sugar by enzymes, fermented into ethanol, and then distilled, whereas raw materials containing sugar are only fermented and then distilled. Wine is only distilled. Normally, the first steps are mechanical disintegration of the fruit and mashing with water. In some cases, the saccharification must artificially be boosted. After the fermentation is finished, the raw spirit (approx. 80 vol.% alcohol) is cleaned of its distillation residues (slops) by a first distillation. In a further refining step, either a so-called fine spirit (approx. 86 vol.% alcohol) is produced by a second discontinuous distillation, or a fine spirit or neutral alcohol (approx. 96 vol.% alcohol) is produced by continuous rectification. The residues of this second refinement step are called singlings.

Depending on the respective raw product and production methods, distilleries produce washing water, steaming water or fruit water, slops, singlings, and cleaning water. The specific wastewater amounts as well as the concentrations of the major components are listed in Tab. 13.

Washing water is only produced during the cleaning of potatoes or roots, which is mostly done already at the supplier's. During steaming of the potatoes a mixture of condensate and fruit water emerges, which as a rule is added to the mash. Modern methods of unpressurized breaking down of the starch (DSA methods), however, do no longer produce any steaming water. As singlings derive from the distillate of the first distillation stage, it does contain absolutely no solids and is hardly polluted. For the bottling of spirits, almost exclusively new bottles, are used so that cleaning water derives only from washing the implements, containers, and factory sheds and is not highly polluted.

Slops contain a high amount of organic acids, proteins, minerals, trace elements, unfermentable carbohydrates, or – in particular with fruit slops – high solids ratios (cores, stalks, stones, skins), which result in very high COD and BOD_5 values as well as in low pH values.

In regard to the wastewater treatment the slops are of particular importance, as the other fractions of the wastewater can normally be discharged into the wastewater sewage without further treatment. Thus, slops should, if possible, be collected and utilized separately from the wastewater flow. The most common utilization method for slops from grain, potatoes, or fruit is direct feeding in cattle farming (if necessary, after thickening with decanters). If direct feeding is not possible, one should consider the spreading on agricultural areas (if necessary, after anaerobic treatment of the slops in a factory owned biogas plant or after injection as co-substrate into a municipal digestion tank). Only if these two utilization methods are not possible, may the slops wastewater be mixed with the other production wastewater flow. Direct discharge, however, is then only possible with sufficiently powerful municipal wastewater treatment plants. In any case, the wastewater should be neutralized as a major pretreatment step. Another point one has to consider is the danger of bulking sludge development and hydrogen sulfide emission (corrosion, odors).

Tab. 13. Specific Wastewater and Slops Amounts and Concentrations in Distilleries (ATV, 1997a, b)

	Amount $[m^3\ 100\ L^{-1}]$	COD $[g\ L^{-1}]$	BOD_5 $[g\ L^{-1}]$	TKN $[g\ L^{-1}]$	P $[mg\ L^{-1}]$
Washing water (potatoes)	0.2–0.5 m^3/t		0.3–1.7		
Slops					
– wine	0.06–0.08	10–39	6–25	0.24–0.45	0.044–0.092
– potatoes	0.5–1.0	72	44	2.5	
– grain (wheat)	0.79–0.97	71	32	2.8	0.19
Singlings (grain)	0.014–0.14	<0.05			

Slops from molasses distilleries often retain very high residual COD ratios after the biological treatment. There, evaporation of the slops down to a dry solids content of approx. 75% with ensuing separation of potassium sulfate (fertilizer) and utilization of the evaporated slops as an additive for cattle feed has proved to be a suitable utilization method. The condensed exhaust vapors from the evaporation unit are often subjected to anaerobic secondary purification.

7 References

7.1 General Literature

ANDREADAKIS, A. (Ed.) (1997), Pretreatment of industrial wastewaters II, *Water Sci. Technol.* **36**.

ANDREADAKIS, A., CHRISTOULAS, D. G. (Eds.) (1994), Pretreatment of industrial wastewaters, *Water Sci. Technol.* **29**.

ATV (1985), *Organisch verschmutzte Abwässer der Lebensmittelindustrie*, 3rd Edn., Vol. V. Berlin: Ernst & Sohn.

ATV (1986), *Organisch verschmutzte Abwässer sonstiger Industriegruppen*, 3rd Edn., Vol. VI. Berlin: Ernst & Sohn.

ATV (1991), Empfehlungen zum Korrosionsschutz von Stahlteilen in Abwasserbehandlungsanlagen durch Beschichtungen und Überzüge, *ATV-Merkblatt* **M 263**.

ATV (1992), Richtlinien für die Bemessung und Gestaltung von Regenentlastungsanlagen in Mischwasserkanälen, *ATV-Arbeitsblatt* **A 128**.

BALLAY, D., IAWQ Programme Committee (Ed.) (1994), Water Quality International '94, Part 8: Anaerobic digestion, sludge management, appropriate technologies, *Water Sci. Technol.* **30**.

BALLAY, D., IAWQ Programme Committee (Ed.) (1996), Water Quality International '96, Part 7: Agro-industries waste management, appropriate technologies, *Water Sci. Technol.* **34**.

BRITZ, T. J., POHLAND, F. G. (Eds.) (1994), Anaerobic digestion VII, *Water Sci. Technol.* **30**.

BRAUER, H. (1996), Produktions- und produktintegrierter Umweltschutz, in: *Handbuch des Umweltschutzes und der Umweltschutztechnik* (BRAUER, H., Ed.) Vol. 2. Heidelberg: Springer-Verlag.

CECCI, F., MATA-ALVAREZ, J., POHLAND, F. G. (Eds.) (1993), Anaerobic digestion of solid waste, *Water Sci. Technol.* **27**.

MALINA, J. F., POHLAND, F. G. (1992), *Design of Anaerobic Processes for the Treatment of Industrial and Municipal Wastes.* Lancaster, PA: Technomic Publishing.

METCALF, E. (Ed.) (1991), *Wastewater Engineering.* New York: Mc Graw-Hill.

NOIKE, T., TILCHE, A., HANAKI, K. (1997), Anaerobic digestion VIII, *Water Sci. Technol.* **36**.

NYNS, E.-J. (1994), *A Guide to Successful Industrial Implementation of Biomethanisation Technologies in the European Union, Report Prepared for the European Commission Directorated General for Energy Thermie Programme.* Namur, Belgium: Institut Wallon.

RÜFFER, H., ROSENWINKEL, K.-H. (Eds.) (1991), *Taschenbuch der Industrieabwasserreinigung.* München, Wien: R. Oldenbourg Verlag.

SPEECE, R. E. (1996), *Anaerobic Biotechnology for Industrial Wastewater.* Nashville, TN: Archae Press.

7.2 Special Literature

7.2.1 Sugar Factories

ATV (1979), Reinigung organisch verschmutzten Abwassers, *Korrespondenz Abwasser* **3**, 156–161.

ATV (1990), Abwasser aus Zuckerfabriken, *Korrespondenz Abwasser* **3**, 285–290.

JÖRDENING, H.-J. (1996a), Produktionsintegrierter Umweltschutz in der Zuckerindustrie, *Handbuch des Umweltschutzes und der Umweltschutztechnik* (BRAUER, H., Ed.) Vol. 2, pp. 616–635. Heidelberg: Springer-Verlag.

JÖRDENING, H.-J. (1996b), *Abwasserreinigung in Zuckerfabriken, ATV-Seminar* – Abwasserbehandlung in der Ernährungs- und Getränkeindustrie, Essen.

JÖRDENING, H.-J. (1997), *Abwasserreinigung in Zuckerfabriken, ATV-Seminar* – Abwasserbehandlung in der Ernährungs- und Getränkeindustrie, Essen.

NYNS, E.-J. (1994), *The Anaerobic Treatment of the Wastewater of the Sugar Refinery at Tienen, Report* prepared for the European Commission Directorated General for Energy Thermic Programme. Namur, Belgium: Institut Wallon.

7.2.2 Starch Factories

ALTHOFF, F. (1995), *Betriebserfahrungen mit einer anaerob/aerob-Betriebskläranlage in der Weizenstärkeindustrie, ATV-Seminar* – Anaerobtechnik in der Abwasserbehandlung, Magdeburg.

ATV (1992), Abwasser der Stärkeindustrie, *Korrespondenz Abwasser* **8**, 1177–1203.

ATV (1994), Abwasser der Stärkeindustrie, Gewinnung nativer Stärke, Herstellung von Stärkeprodukten durch Hydrolyse und Modifikation, *Korrespondenz Abwasser* **7**, 1147–1174.

AUSTERMANN-HAUN, U. (1996), *Stärkefabriken, ATV-Seminar* – Abwasserbehandlung in der Ernährungs- und Getränkeindustrie, Essen.

AUSTERMANN-HAUN, U. (1997), *Stärkefabriken, ATV-Seminar* – Abwasserbehandlung in der Ernährungs- und Getränkeindustrie, Essen.

SEYFRIED, C. F., SAAKE, M. (1985), Stärkefabriken, Stärkezucker- und Stärkesirupherstellung, in: *Lehr- und Handbuch der Abwassertechnik* (ATV, Ed.), Vol. V, pp. 182–219. Berlin: Ernst & Sohn.

7.2.3 Vegetable Oil and Shortening Production

ATV (1979), Organisch verschmutzte Industrieabwässer, *Korrespondenz Abwasser* **11**, 664–667.

ATV (1981), Organisch verschmutzte Industrieabwässer, *Korrespondenz Abwasser* **9**, 651–656.

HEINRICH, D. (1996), *Speisefett-/Speiseölfabriken*, Hamburg: Heinrich Umweltschutztechnik.

HEINRICH, D. (1997), *Speisefett-/Speiseölfabriken, ATV-Seminar* – Abwasserbehandlung in der Ernährungs- und Getränkeindustrie, Essen.

KRAUSE, A. (1985), Fabriken zur Gewinnung und Verarbeitung von Speisefetten und -ölen, *Lehr- und Handbuch der Abwassertechnik* (ATV, Ed.), Vol. V, pp. 219–252. Berlin: Ernst & Sohn.

7.2.4 Potato Processing Industry

ATV (1985), Abwässer der Kartoffelindustrie, *ATV-Merkblatt* **M 753**.

NEUMANN, H. (1985), Kartoffelveredelungsindustrie, in: *Lehr- und Handbuch der Abwassertechnik* (ATV, Ed.), Vol. V, pp. 252–276. Berlin: Ernst & Sohn.

NYNS, E.-J. (1994), *Anaerobic wastewater treatment of the potato chips factory convention at Frankenthal*, Germany. Namur, Belgium: Institut Wallon.

ROSENWINKEL, K.-H., AUSTERMANN-HAUN, U. (1997), *Abwasserreinigung in der Gemüse- und Kartoffelindustrie, ATV-Seminar* in der Ernährungs- und Getränkeindustrie, Essen.

SCHEFFEL (1994), Kartoffelverarbeitung, *ATV-Seminar* in der Ernährungs- und Getränkeindustrie, Essen.

7.2.5 Abattoirs

ATV (1992), Abwasser aus Schlacht- und Fleischverarbeitungsbetrieben, in: *ATV-Regelwerk Abwasser – Abfall. ATV Merkblatt* **M 767**.

ATV (1995), Behandlung und Verwertung von Reststoffen aus Schlacht- und Fleischverarbeitungsbetrieben, *ATV-Regelwerk Abwasser – Abfall, ATV Merkblatt* **M 770**.

BLAHA, M.-L. (1995), Schlachthofabfälle; Mengenanfall und Inhaltsstoffe, *Die Fleischwirtschaft* **75**, 648–654.

JÄPPELT, W., NEUMANN, H. (1985), Verarbeitung tierischer Produkte; Schlacht- und Fleischverarbeitungsbetriebe, in: *Lehr- und Handbuch der Abwassertechnik* (ATV, Ed.), Vol. V, pp. 320–382. Berlin: Ernst & Sohn.

Oswald-Schulze-Stiftung (1986), Stand und Entwicklungspotentiale der anaeroben Abwasserreinigung unter besonderer Berücksichtigung der Verhältnisse in der Bundesrepublik Deutschland; Abwässer der Fleischwirtschaft, *Mitteilungen der Oswald-Schulze-Stiftung*, **7**, pp. 563–610. Gladbeck.

SAYED, S. K. I. (1987), *Anaerobic treatment of slaughterhouse wastewater using the UASB process, Thesis*. Wageningen, NL: Eigenverlag.

TRITT, W. P. (1992), *Anaerobe Behandlung von flüssigen und festen Abfällen aus Schlacht- und Fleischverarbeitungsbetrieben, Thesis*, Vol. 83, Hannover: Institut für Siedlungswasserwirtschaft und Abfalltechnik der Universität Hannover.

7.2.6 Dairy Industry

ATV (1978), Reinigung organisch verschmutzten Abwassers, 1.5 Molkereien und Milchindustriebetriebe, *Korrespondenz Abwasser* **4**, 114–117.

ATV (1994), Abwasser bei der Milchverarbeitung, *ATV-Merkblatt* **M 708**.

BERTSCH, R. (1997), *Molkereien, ATV-Seminar* – Abwasserbehandlung in der Ernährungs- und Getränkeindustrie, Essen.

DOEDENS, H. (1985), Molkereien (Verarbeitung von Milch und Milchprodukten), in: *Lehr- und Handbuch der Abwassertechnik* (ATV, Ed.), Vol. V, pp. 410–456. Berlin: Ernst & Sohn.

DOEDENS, H. (1996), *Molkereien, ATV-Seminar* – Abwasserbehandlung in der Ernährungs- und Getränkeindustrie, Essen.

7.2.7 Fruit Juice and Beverage Industry

ATV (1998), Abwasser aus Erfrischungsgetränke- und Fruchtsaft-Industrie und der Mineralbrunnen, *ATV-Regelwerk Abwasser – Abfall, ATV Merkblatt* **M 766**.

ROSENWINKEL, K.-H. (1997), *Getränkeindustrie, ATV-Seminar* – Abwasserbehandlung in der Ernährungs- und Getränkeindustrie, Essen.

RÜFFER, H. M., ROSENWINEKL, K.-H. (1984), *The Treatment of Wastewater from the Beverage Industry*. Pitmens Publishing "Food and Allied Industries".

RÜFFER, H. M., ROSENWINKEL, K.-H. (1985), Getränkeindustrie und Gärungsgewerbe, in: *Lehr- und Handbuch der Abwassertechnik* (ATV, Ed.), 3rd Edn., Vol. V, pp. 457–509. Berlin: Ernst & Sohn.

7.2.8 Breweries

ATV (1977), Reinigung hochverschmutzten Abwassers, 1.18 Brauereien, *Korrespondenz Abwasser* **6**, 182–186.

Oswald-Schulze-Stiftung (1986), Stand und Entwicklungspotentiale der anaeroben Abwasserreinigung unter besonderer Berücksichtigung der Verhältnisse in der Bundesrepublik Deutschland; Abwässer der Brauereien, *Mitteilungen der Oswald-Schulze-Stiftung* **7**, pp. 367–402. Gladbeck.

ROSENWINKEL, K.-H. (1996), *Entwicklungstendenzen bei der Brauereiabwasserbehandlung, Brauereiabwasser-Seminar*, Institut für Siedlungswasserwirtschaft und Abfalltechnik der Universität Hannover (26. 03. 1996).

SEYFRIED, C. F. (1996), *Verfahrenstechnische Grundlagen der Brauereiabwasserbehandlung, Brauereiabwasser-Seminar*, Institut für Siedlungswasserwirtschaft und Abfalltechnik der Universität Hannover (26. 03. 1996).

SEYFRIED, C. F., ROSENWINKEL, K.-H. (1985), Brauereien und Malzfabriken, in: *Lehr- und Handbuch der Abwassertechnik* (ATV, Ed.), 3rd Edn., Vol. V, pp. 509–549. Berlin: Ernst & Sohn.

7.2.9 Distilleries

ATV (1997a), Abwässer aus Brennereien und der Spirituosenherstellung, *ATV-Merkblatt* **M 772**.

ATV (1997b), Abwasser aus Hefefabriken und Melassebrennereien, *ATV-Merkblatt*.

CRAVEIRO, A. M., SOARES, H. M., SCHMIDELL, W. (1996), Technical aspects and cost estimation for anaerobic systems treating vinasse and brewery/ soft drink wastewaters, *Water Sci. Technol.* **18**, 123–134.

EGGERT, W., BORGHANS, JR., A. J. M. L. et al. (1990), Upflow Fluidized-Bed-Reaktor: In der Praxis bewährt, *umwelt & technik* **7/8**, 10–16.

HABERL, R., ATANASOFF, K., BRAUN, R. (1991), Anaerobic-aerobic treatment of organic high strength industrial wastewaters, *Water Sci. Technol.* **23**, 1909–1918.

NYNS, E.-J. (1994), *Anaerobic Treatment of Distillery Wastewater: Distercoop at Faenza, Report* prepared for the European Commission Directed General for Energy Thermic Programme. Namur, Belgium: Institut Wallon.

RECAULT, Y. (1990), Treatment of distillery waste waters using an anaerobic downflow stationary fixed film reactor, *Water Sci. Technol.* **22**, 361–372.

VIAL, D., MALNOU, D., HEYARD, A., FAUP, G. M. (1987), Use of yeast in polutant removal from concentrated agricultural food effluents. Application to the treatment of vinasse from molasses, *LEBEDEAU* **40**, 27–36.

WELLER, G. (1985), Brennereien und Spirituosenbereitung, in: *Lehr- und Handbuch der Abwassertechnik* (ATV, Ed.), 3rd Edn., Vol. V, pp. 591–616. Berlin: Ernst & Sohn.

11 Agricultural Waste and Wastewater Sources and Management

PETER WEILAND
Braunschweig, Germany

1 Introduction

The generation of animal and plant products such as meat, poultry, milk, fish, grain, fruits, and vegetables has been growing tremendously within the last few decades due to the increasing world population and the rising living standard in all industrialized countries. As a result the amounts of wastes and residues from growing and processing of agricultural products increased considerably, which has caused a variety of environmental problems in countries of high agricultural productivity. Environmental pollution problems result from handling, storing, and disposing of agricultural wastes which can affect air, water, and soil quality and may be a nuisance to those who live nearby. These problems are forced by the change from small, individual farms into large-scale enterprises which often cause regional surpluses of wastes.

In many industrial countries about two thirds of the whole amount of organic wastes are generated in agriculture and agro-industry. Plant materials become wastes after harvesting the main crop (straw in cereal production), after processing the product in industrial operations (pulps, oil cakes, bagasse), if the products do not fulfil the quality standards of the market (refuse fruits and vegetables), and if the plant products cannot be sold at the right time (overlayed products). Wastes from animal production are mainly formed during livestock breeding (manure, bedding material), slaughtering, and meat processing (paunch manure, flotation fat).

Typical environmental problems caused by disposing of these wastes are gaseous emissions of ammonia, methane, and laughing gas, which have a harmful effect on atmosphere and climate. Nutrient losses by runoff result in an eutrophication of surface water by nitrogen, phosphate, and potassium, which unbalance the natural ecological system by the formation of bacteria and algae. Leaching of nitrogen into groundwater impairs the use as potable water and complicates water treatment considerably. Hence, a treatment of these wastes is necessary for environmental protection, but biotechnological processes can also be utilized for upgrading these wastes into higher value products, such as energy, animal feed, soil conditioners, fertilizer, or chemical feedstocks. Therefore, some agricultural and agro-industrial wastes can also be considered as a renewable resource.

Waste generation is dynamic, reflecting the requirements of laws, regulations, and prices of the inputs and outputs. As these factors change, the extent of waste recovery and utilization may increase or decrease. In this overview the amounts and characteristics of agricultural wastes and wastewater are discussed and the most important biological treatment technologies for environmental protection and waste utilization and strategies for waste management are presented.

2 Agricultural Wastes and Wastewater

2.1 Types and Sources

Wastes from agriculture are mainly organic and can be in solid, slurry, or liquid form. A part of the agricultural residues results from harvesting and processing of plants and another part is generated in livestock breeding, by slaughtering and meat processing. The origin of some typical agricultural and agro-industrial wastes is shown in Fig. 1.

Typical harvesting residues are mainly straw, leaves, and roots. Generally, the level of pollution resulting from these residues is low and the majority of the harvesting residues is returned directly to the soil. Straw, the major by-product of cereals, is produced in the same or higher quantities than edible grain. When disposed on the soil surface, straw can facilitate the production of phytotoxic metabolites when pathogenic microorganisms and particularly saprophytes colonize straw. Straw can also be utilized for animal feed and for the production of sugar hydrolysates, but often more than 50% is wasted. Leaves, e.g., from sugar beets, can be used as animal feed. They usually appear in large quantities in short periods of

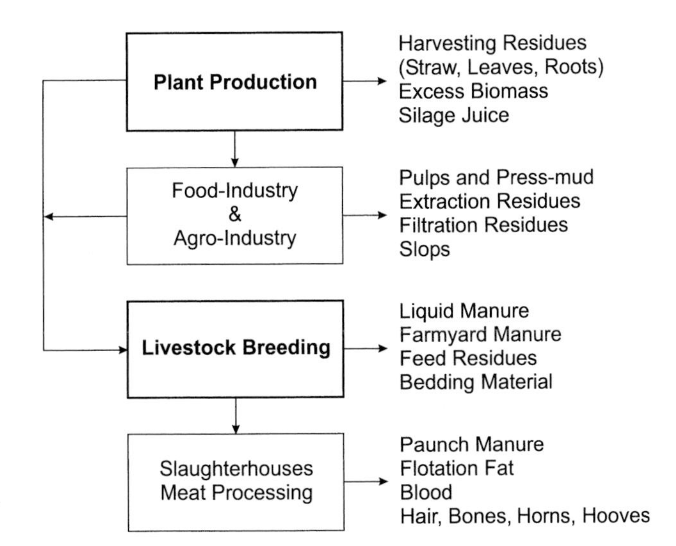

Fig. 1. Origin of agricultural and agro-industrial wastes.

time and have to be ensiled for use as animal feed. Large quantities of green biomass are also formed on set-aside areas which have to be harvested for maintenance. Ensiling of fresh biomass, often used in agriculture for the conservation of fodder crops, results in the formation of silage juice which is heavily polluted with organics and nutrients. If the silage juice cannot be added to manure and spread on fields, it has to be treated in order to reduce the high organic load.

The processing of plant materials in the food and agro-industry results in a large number and diversity of solid and liquid by-products and wastes. Processing wastes result from peeling, juicing, extraction, and filtration, from washing, blanching, and pasteurizing, and from fermentation of cereals, fruits, and vegetables. The processing wastes consist mainly of carbohydrates, proteins, oils, or greases. If these wet wastes and highly loaded wastewater cannot be upgraded into animal feed or other useful products they have to be treated in order to prevent environmental pollution.

Livestock farming is the main source of organic wastes which are formed in agriculture. The high amount of animal wastes results from the inefficient conversion of fodder plants into beef, because it takes about 7–8 kg of feed to produce 1 kg of beef. Therefore, livestock wastes consist of undigested feed, mostly cellulose or fibers, undigested protein which is excreted in the feces, and digested protein mainly excreted in the urine as uric acid and urea. A portion of the other nutrients also escapes digestion and causes the high nutrient content of manure. The main part of livestock wastes remains in the pasture or is recycled to the farmland, but large volumes have to be treated in order to avoid methane, nitrous oxide, and odor emissions, nitrate accumulation in groundwater, and potassium accumulation in soils. Treatment of manure is also necessary to avoid the runoff of nutrients and oxygen consuming organics into rivers and seas.

The slaughtering and processing of animals for the preparation of meat products result in wastes like paunch manure, fat and grease, blood, hair, bones, horns, and hooves. Many of these wastes can be converted by biological and chemical treatment processes into useful products. Large amounts of highly loaded wastes are also formed in dairy food plants. Whey from cheese production is the most important waste. It contains about 5% lactose and can be utilized as a substrate for many fermentation processes.

2.2 Quantities and Characteristics

2.2.1 Wastes from Animal Production and Processing

Typical wastes from animal production and processing which occur in large quantities are feces of livestock, wastes from milking, and by-products from dairy food production and wastes from slaughtering and meat processing. The quantities and characteristics of feed wastes which are formed during forage plant processing are described in Sect. 2.2.2.

2.2.1.1 Wastes from Livestock Farming

Animal manure is the most important organic waste because it is generated in very high quantities. An estimate of the organic waste quantity from agriculture shows that in Germany and other industrialized countries about 45–55% of the waste organic dry matter or nearly 80% of the raw organic agricultural wastes are formed in livestock farming. The daily excretion is about 8–9% of the body weight for cattle and about 6% for pigs. The natural dry matter content of manure from cows and pigs is between 9 and 11%, but dependent on the housing systems manure with different dry matter content (DM) is produced: liquid manure with 6–14% DM, farmyard manure with 20–25% DM, and urine with only 1–2% DM.

Housing systems which generate liquid manure dominate in developed countries. The total amount of livestock manure per year was estimated for Germany with 230 Mt in 1997 (DÖHLER et al., 1997). About 65% is liquid manure, 30% solid manure, and 5% urine.

Tab. 1 shows the typical quantities and nutrient contents of manure from different animals. All dates are related on a dry matter content of 10%. The manure quantities are shown for one live weight unit (LU), which is equivalent to a live weight of 500 kg.

Manure contains considerable amounts of nutrients, fatty acids, and solids. The nutrient content of manure depends on the feed type and the formulation of the feed ratios. According to the "Code of Good Agricultural Practice" the quantities of manure which can be spread on farmland is limited by the nutrient content. In Germany it is permitted to spread 210 kg N ha^{-1} a^{-1} on grassland and 170 kg N ha^{-1} a^{-1} on farmland (Deutscher Bundestag, 1996). In order to reduce the phosphate and nitrogen content of manure, feed supplements can be applied. The addition of microbial phytase enhances the P-digestibility and reduces the P-content in manure up to 35–40% (HOPPE et al., 1993). By more-phase feeding and the addition of essential amino acids, e.g., lysine, methionine, and threonine to the feed the excretion of nitrogen can also be reduced by 50–60% (KIRCHGESSNER et al., 1991). For anaerobic and aerobic treatment of manure the ratio of carbon, nitrogen, and phosphorus plays a key role. For both processes nitrogen, phosphate, and potassium occur in higher proportions than necessary, which results in high residual concentrations in the treated liquid.

Volatile fatty acids occur in a concentration of 3,000–10,000 mg L^{-1} in cow manure and 4,000–18,000 mg L^{-1} in pig manure with acetic

Tab. 1. Manure Quantities and Typical Nutrient Contents

Animal Species	LU[a]	Manure/Year [m^3 LU^{-1}]	N [g kg^{-1}]	NH_4-N [g kg^{-1}]	P_2O_5 [g kg^{-1}]	K_2O [g kg^{-1}]	MgO [g kg^{-1}]	CaO [g kg^{-1}]
Dairy cow	1.2	17	5.3	2.7	2.0	8.0	1.1	2.7
Beef cow	0.7	12	6.0	3.0	2.0	4.7	1.1	1.7
Calf	0.2	7	7.9	3.9	2.7	6.8	1.0	2.6
Fattened pig	0.12	13	8.0	5.3	4.0	4.0	1.3	4.0
Layer/poultry	0.004	146	6.5	4.6	5.0	3.0	1.0	11.0

[a] 1 LU (live weight unit) = 500 kg

Tab. 2. Typical Distribution of Fatty Acids of Stored Pig Manure

Fatty Acid	% in Total Quantity[a]
Acetic acid	50–60
Propionic acid	10–15
n-Butyric acid	6–10
n-Valeric acid	3–6
n-Caproic acid	1–3
iso-Caproic acid	1–3

[a] Total fatty acid content: 4–18 g kg^{-1}

Tab. 3. Average Composition of Whey

		Sweet Whey	Acid Whey
Water content	[%]	93–94	94–95
Lactose	[%]	4.5–5.0	3.8–4 5
Total protein	[%]	0.5–1.0	0.5–1.0
Fat	[%]	0.3–0.8	<0.3
Mineral components	[%]	0.5–0.7	0.7–0.8
pH value		5–7	4–5

acid as the main component. The acid concentration but also the ammonia nitrogen content increase during storage due to the activity of aerobic and facultative anaerobic bacteria. The typical spectrum of volatile fatty acids in stored pig manure is shown in Tab. 2.

Manure contains components of intensive odor, like indole, skatole, and cresols, which can be degraded by aerobic or anaerobic treatment. Depending on the farm management manure may contain also antibiotics, sulfonamides, and hormones from therapeutic measures and chemicals from disinfecting procedures which can influence the biological degradability.

2.2.1.2 Wastes from Milk Production and Processing

Wastewater from dairy plants results from cleaning the milking machines, the milking place, and the cow udder. The wastewater is loaded with milk, excrements, cleaning and disinfecting agents. The typical chemical oxygen demand (COD) is between 2,200 and 3,700 mg L^{-1} and the biological oxygen demand (BOD) between 350 and 650 mg L^{-1} (HÖRNIG and SCHERPING, 1993). The wastewater quantity per cow is between 17 and 46 L d^{-1} and can be fed to the manure storage tank, if a separate aerobic or anaerobic treatment is not economic.

Much higher quantities of very highly loaded wastewater are formed during dairy processing. The largest quantities are whey from cheese production followed by wash water and pasteurization water. 5–10 kg whey is produced per kg of cheese. Manufacture of cheese from the whole milk results in sweet whey of pH 5–7 and the use of skim milk to produce cottage cheese results in acid whey of pH 4–5. The BOD of whey ranges from 32,000 to 60,000 mg L^{-1} depending on the specific cheese making process. Whey contains an average of 6–7% dry matter, of which approximately 70% is lactose (KESSLER, 1988). The average composition is shown in Tab. 3.

Whey is a good medium for direct use in fermentation processes to produce microbial biomass and metabolic products like lactic acid, alcohols, and vitamin B_{12} (MOULIN and GALZY, 1984).

2.2.1.3 Wastes from Slaughtering and Meat Processing

Slaughtering of animals is allowed in slaughterhouses only because most of the slaughtering and meat residues are classified as hazardous wastes which have to be treated in rendering plants or special disposal plants. Slaughtering on farms is allowed only under special circumstances.

The non-meat portion of animals results in significant waste loads, because about 25% of the total weight of slaughtered animals are residues and wastes. The main wastes originate from killing, hide removal, dehairing, paunch handling, processing and cleanup operations. The wastewater contains blood, fat, grease, bowel contents, cleaning and disinfection agents, and solid wastes. Typical solids are

Tab. 4. Amount of Solid and Liquid Wastes per Live Weight Unit (LU) from Slaughtering

Residues and Wastes	Pig[a] [kg LU^{-1}]	Cattle[b] [kg LU^{-1}]	Composition			
Solid wastes						
Bones	70–75	60–65	DM:	57%	ODM:	63% of DM
Bristles/claws	3–5	1.5–2.5	DM:	76%	ODM:	98% of DM
Stomach contents	3–12	40–80	DM:	11–15%	ODM:	80–87% of DM
Fat	2–4	2–3	DM:	30–70%	ODM:	90–95% of DM
Liquid wastes						
Blood	30–45	20–35	COD:	375 g L^{-1}	BOD:	150–200 g L^{-1}
Fat flotate	5–35	4–25	COD:	95–400 g L^{-1}	DM:	5–24%
Total wastewater	1,500–4,000	1,000–1,500	COD:	3–17 g L^{-1}	BOD:	2–5 g L^{-1}

[a] 1 LU = 7.7 pigs
[b] 1 LU = ~1 cattle

paunch manure, hides, bristles, hair, hornes, bones, hooves, and feathers. All residues can be contaminated with pathogenic microorganisms or worm eggs, which necessitates careful treatment. The average quantities of solid and liquid wastes produced per live weight unit (LU) are shown in Tab. 4 (TRITT and SCHUCHARDT, 1992).

The quantities and the composition can vary considerably between different plants. Bones can be cleaned by enzymatic extraction for gelatin production (OLSEN, 1988) and blood can be utilized for animal feed production or for the production of blood cell hydrolysates (OLSEN, 1995). The stomach content of cattle can be composted (SCHUCHARDT and FERCH, 1994) and the highly loaded wastewater and flotation fats can be treated by anaerobic digestion (TRITT, 1992).

2.2.2 Wastes from Plant Production and Processing

Large quantities of plant residues result from harvesting of crops and vegetables. During their processing to foodstuffs, animal feed, or industrial raw materials a large variety of processing wastes with strongly different physicochemical properties are formed. The quantity and quality of these residues depend upon the type of raw materials and the process type used for plant processing. The residues may be liquid or solid or a combination of both forms (WINTZER, 1996).

2.2.2.1 Harvesting Residues

Residues from harvesting are mainly leaves and tops from beets and sugar cane, greenstuff from potatoes, and straw from cereals, maize, and legumes. For several important cultivated plants the typical harvesting residues, their quantities generated per hectare and year, and their main characteristics are shown in Tab. 5 (FRÜCHTENICHT et al., 1993).

Large quantities of harvesting residues result also from the production of sugar cane, cassava, manioc, rice, and cotton and other tropical crops. The quantities of harvesting residues are usually high compared to their product formation. The ratio of product to residues is about 1:1 for cereal straw, 1:1.7 for rape straw, and 1:0.8 for sugar beets.

A part of the residues is returned directly to the soil for fertilization and a part can be utilized for animal feed. Lignocellulosic harvesting residues (straw, cobs) can also be utilized for the production of compost or sugar hydrolysates whereas residues with digestible carbohydrates (beet leaves) can be converted to biogas.

Leaves and tops from sugar and fodder beets are mainly preserved by fermentation

Tab. 5. Quantities and Characteristics of Some Typical Harvesting Residues

Residue	Yield [t ha^{-1}]	Dry Matter [%]	N [%]	P_2O_5 [%]	K_2O [%]	MgO [%]	CaO [%]
Sugar beet leaves	40	16	0.25–0.33	0.08–0.11	0.4 –0.7	0.06–0.10	0.7 –1.5
Fodder beet leaves	30	10	0.23–0.25	0.06–0.08	0.41–0.43	0.09–0.11	0.20–0.30
Potato haulm	13	25	0.3 –0.5	0.1 –0.2	0.5 –0.7	0.15–0.25	—
Pea and bean straw	6	86	1.3 –1.7	0.1 –0.3	0.8 –1.8	0.1 –0.3	0.2 –0.3
Cereal straw	6	86	0.3 –0.8	0.2 –0.4	1.5 –2.0	0.1 –0.3	0.3 –0.6
Maize straw	10	86	0.5 –0.9	0.5 –0.7	1.5 –2.5	0.2 –0.4	0.5 –0.7
Rape straw	5	86	0.6 –0.8	0.2 –0.4	2.0 –3.0	0.2 –0.3	—

and stored as silage in order to utilize these residues for animal feeding. Ensiling itself results in the formation of a strongly polluted fermentation juice. High quantities of fermentation juice are also formed during the conservation of forage crops.

2.2.2.2 Ensiling Juice

Ensilage, a process similar to the lactic acid fermentation of cabbage to sauerkraut, is widely used for conserving forage crops as a succulent feed. Silage is made from beets, beet leaves, maize, grain sorghum, forage grasses, alfalfa, and similar crops. The microbial conversion of the crop material into silage in airtight pits or silos results in the formation of fermentation juice, because the plant cell walls are degraded during the acidification process.

The amount of fermentation juice which is formed during ensiling depends on the crop type, the degree of compression, and the mean particle size of the harvested crops. Tab. 6 shows the total amount of fermentation juice which is formed per ton of fresh harvested forage (ATV, 1995).

About 50% of the juice is formed during the first 10 d and about 75% is run off after 20 d fermentation time. Fermentation juice is characterized by a low pH value between 3.8–4.5 and an extremely high BOD of 45–120 g L^{-1} which is about the 150- to 400-fold value of domestic sewage. The composition of ensiling juice from different forage crops is similar and their concentration increases with increasing ensiling time. The typical properties reported by ZIMMER et al. (1993) are shown in Tab. 7.

The fermentation juice can be stored and spread on fields together with liquid manure. A treatment of the juice in anaerobic digesters results in high biogas yields, because about 50% of the organic matter exists in the form of organic acids which can be degraded completely. Due to the low pH value the juice has to be treated together with substrates of high buffer capacity, e.g., manure from pigs or cows.

Tab. 6. Total Yield of Fermentation Juice from Ensiling of Fresh Crops

Forage Crop	Dry Matter [%]	Fermentation Juice [L t^{-1}]
Fodder beets	10	350–500
Sugar beet leaves	16	250–500
Rape (interfruit)	12	350–500
Grass, clover	20	220–250
Maize	25	100–120

Tab. 7. Composition of Fermentation Juice from Ensiling of Forage Crops

Characteristics	Concentration [g L^{-1}]
Dry matter	40–120
Organic matter	30–90
Organic acids	10–35
N	1–2
P_2O_5	0.5–1.5
K_2O	3.5–6.0

2.2.2.3 Wastes from Crop Processing

The processing of agricultural crops such as grain, fruits, and vegetables into foodstuffs, agro-industrial raw materials, energy sources, or chemicals results in the formation of a wide range of different solid wastes and highly loaded wastewater whose characteristics may be different for each processing plant. Crop processing wastes result from washing, trimming, blanching, extraction, juicing, pasteurization, and biological conversion processes. The wastes consist mainly of carbohydrates such as starch and sugar, pectins, cell wall components such as cellulose and hemicellulose, nutrients, vitamins, and other crop components. Sometimes they contain also pollutants such as soil, insecticides, or lye from caustic peeling processes. Tab. 8 indicates some typical processing wastes which are formed in large quantities (THOMÉ-KOZMIENSKY, 1995; KOBALD and HOLLEY, 1990; BÖHM et al., 1996).

Most of the solid wastes show a dry matter content (DM) between 15 and 30%. About 90–95% of the dry matter is organic and biodegradable by anaerobic or aerobic processes. These wet wastes undergo a rapid decomposition resulting in the formation of fatty acids and odours compounds, which makes their storage and utilization more difficult. Only residues from milling of grains or deoiling of oil seeds result in stable products with a dry matter content of 85% and more. Solid process residues can often be utilized as animal feed, for biogas formation, and sometimes for the production of higher value products such as enzymes, organic acids, and edible fungi.

The processing of agricultural plant materials often results in the formation of highly loaded wastewater with a typical COD between 2 and 10 g L^{-1}. The effluents from crop processing plants generally contain high amounts of nutrients and some types of wastewater, such as those from beet processing, are also highly colored, which complicates their biological treatment. Extremely highly loaded wastewater results from fermentation processes for beverage alcohol production. The high polluting capacity (COD: 20–90 g L^{-1}) of distillery stillage from beet, potato, or grain associated with the high mineral content makes further utilization difficult. An aerobic treatment of these process wastewaters by activated sludge or trickling filters is rarely cost-effective and reliable because of the extremely high COD content. Generally, above a COD level of about 3 g L^{-1} anaerobic digestion processes are preferred for wastewater pretreatment and aerobic processes are utilized only for polishing the anaerobic digester effluent and for removal of the nutrients.

Tab. 8. Typical Wastes and Residues from Grain, Fruit, and Vegetable Processing

Processing Category	Solid Wastes			Wastewater	
	Type	DM [%]	ODM [%]	Type	COD [g L^{-1}]
Breweries	spent grain	22–24	19–21	total wastewater	1.5–3.0
Canneries (fruits, vegetables)	sorting residues, fragments	18–25	17–24	blanching water	2.0–5.0
Distilleries	thick stillage	16–27	15–26	stillage	20–90
Fruit juices	sorting residues, pulps	21–23	20–22	process water	4.0–6.0
Grain mills	chaff	84–86	79–81	—	—
Oil seed mills	oil seed cake	88–90	79–81	—	—
Potato processing	sorting residues, peels	20–25	19–24	process water	2.0–8.0
Potato starch production	potato pulp	18–20	15–17	potato fruitwater	7.0–10
Sugar production	beet chips	20–30	18–27	washing water	5.0–9.0

3 Waste Management and Waste Utilization

3.1 Anaerobic Treatment

Anaerobic digestion processes are most widely used for the treatment of wastewater and wastes from agriculture, because this process simultaneously solves the pollution problem and recovers valuable products such as biogas and compost (WEILAND, 1997b). Furthermore, the process reduces the load of pathogenic germs or weed seeds, degrades malodorous compounds, increases the fertilizing effect of wastes due to the conversion of organic nitrogen into ammonia nitrogen, and improves the handling properties of manure and other sludges. The microbial process consists of three interdependent steps which occur simultaneously or in separated digesters:

(1) hydrolysis and fermentation of polymers and complex organic compounds;
(2) production of organic acids, mainly acetic acid, from smaller soluble organic molecules;
(3) formation of methane and carbon dioxide (biogas).

At least three groups of microoganisms are involved in the degradation process which allows the conversion of the dominating soluble and solid compounds of agricultural wastes such as carbohydrates, lipids, and proteins, into biogas. The process needs for a complete and reliable degradation a well balanced population of the different groups of microorganisms that are necessary for the individual degradation steps, a well balanced C:N ratio between 15:1 and 35:1, and some essential trace elements such as Fe, Ni, Co, and Mo. Therefore, the process conditions and the reactor configurations have to be well adapted to the individual properties of each waste type in order to avoid process failure and to achieve a high process efficiency. The different anaerobic treatment processes which dominate in agricultural waste management are shown in Fig. 2.

Wastewater and slurries are treated in one-stage or two-stage processes using digesters with or without biomass accumulation. In the two-stage mode of operation, hydrolysis and acidification take place in the first stage and methanation in the second stage whereas in the one-stage operation all process steps occur in the same reactor simultaneously. Solid wastes are treated in wet or dry fermentation processes, which differ in the solids content of the digester influent. Another process type, called co-fermentation, treats a mixture of liquid manure and solid wastes, which can make waste management more economical and efficient.

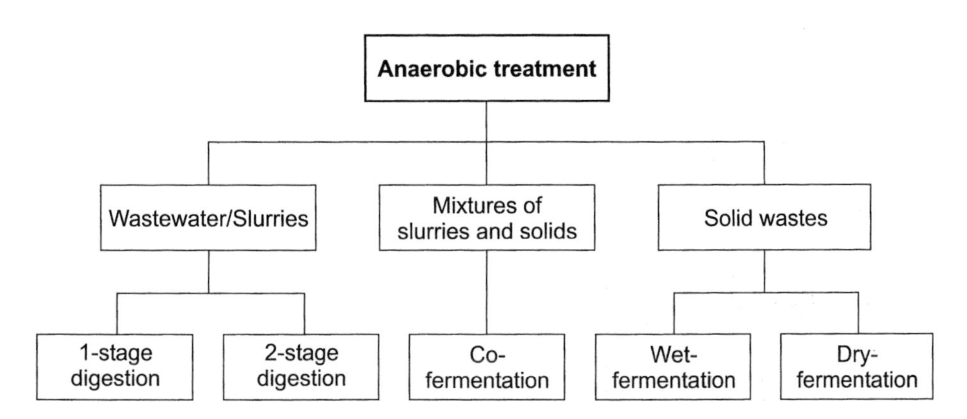

Fig. 2. Classification of anaerobic treatment processes.

3.1.1 Wastewater and Slurry Treatment

For the treatment of wastewater and slurries the one-stage process is the most frequently used variant, because the hydrolysis and acidification process has taken place to a great extent within the storage tank which is usually necessary for a proper waste management. Two-stage processes have to be used only for effluents with a fast acidification rate or for wastewater with high contents of biodegradable solids (SCHULZ, 1996). Stillage from alcohol distilleries is a typical wastewater which should be treated in a two-stage operation because the residual carbohydrates are acidified very quickly, which can lead to process inhibition or complete process failure by pH drop and accumulation of organic acids. Furthermore, an efficient liquefaction of the suspended yeast cells can be obtained only in a separate hydrolysis reactor which is operated under controlled conditions (WEILAND, 1988a). Optimum results are achieved, if the first reactor is kept at a pH of 5.0–6.0 and if the methanogenic stage is operated at a pH of 7.2–8.0. The two-stage operation allows a better adaptation of the process conditions to the demand of the individual anaerobic degradation steps, which results in a better process stability, higher degradation rates, and higher gas yields. On the other hand, the treatment of liquid manure in a two-stage process shows no essential advantages because hydrolysis and acidification have taken place in the animal houses during the storage period. Furthermore, the buffer capacity of manure is strong enough to avoid a pH drop by accumulation of fatty acids. One-stage operation entails lower investment costs, which is the most important reason for the preferred application of this process type.

For wastewater treatment completely mixed reactors (CSTR), contact reactors with biomass recirculation, fixed bed and fluidized bed reactors with fixed films, or sludge blanket reactors with granulated biomass (UASB reactors) are used (see Chapters 17, 18, 23–25, this volume). Sometimes also a combination of

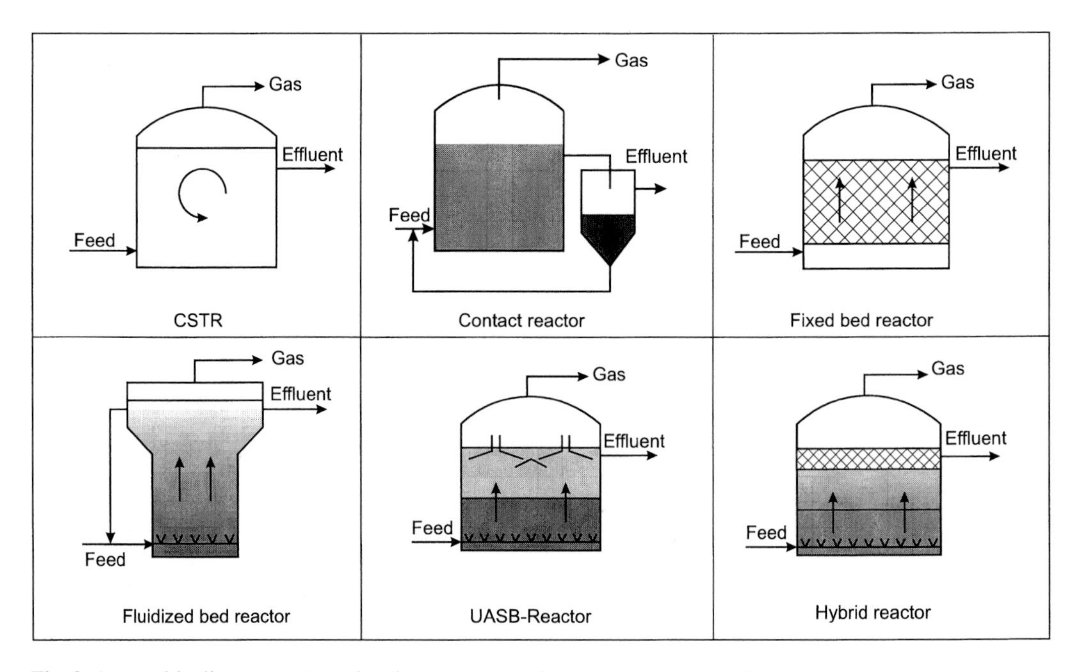

Fig. 3. Anaerobic digester systems for the treatment of wastewater and slurries.

fixed bed and UASB is used, which is called hybrid reactor. The different reactor configurations are shown schematically in Fig. 3. For wastewater with low suspended solids content, e.g., washing water or ensiling juice, reactors with biomass accumulation should be used because the loading capacity is much higher and the process is more stable. Most sensitive against solids are fixed bed reactors and sludge bed reactors. Only CSTR reactors are suitable for liquid manure and other slurries if the solids are not separated before treatment. Good mixing is necessary to prevent scum layers or sediment layers within the reactor. Tank digesters are equipped with mechanical stirrers, gas injectors, or external pumps for hydraulic mixing. Beside the vertical-type digesters horizontal channel-type digesters with slowly moved paddle stirrers are also used for slurries rich in coarse solids (WEILAND, 1995).

On farms the treated effluent is usually spread on fields. Particular attention has to be payed to the application techniques because the anaerobic process transforms organic nitrogen into ammonia nitrogen. For avoiding ammonia emissions ground-near distribution systems have to be used. Generally, a direct discharge of the treated wastewater to a receiving water is not possible because the anaerobic process results only in a partial purification without nutrient elimination. For further purification and nutrient recovery various new technologies have recently been tested in full-scale plants for manure treatment (HÜTTNER and WEILAND, 1996). Ammonia can be recovered by air or steam stripping in the form of $(NH_4)_2SO_4$ or $(NH_4)HCO_3$ which can be utilized as fertilizer. Phosphorous compounds are mainly present in the form of suspended solids and can be separated mechanically. Dissolved orthophosphates can be eliminated by addition of lime with subsequent filtration of the precipitated phosphates. The schematic diagram of a typical manure treatment plant is shown in Fig. 4.

The process results in a partial purification of manure and reduces the nutrient content to such an extent that the treated manure can be disposed on the available farmland without environmental pollution. The properties of the fresh and treated manure are shown in Tab. 9.

Membrane techniques or evaporation processes can be used after anaerobic digestion for a total purification to river quality. Reverse osmosis (RO) needs a pretreatment of the digester effluent by centrifugation and ultrafil-

Tab. 9. Properties of Fresh and Treated Cow Manure

Parameter	Fresh Manure	Treated Manure	Removal Efficiency
COD [mg L^{-1}]	90,100	40,000	55.6
N [mg L^{-1}]	4,900	1,900	61.2
P [mg L^{-1}]	740	385	48.0

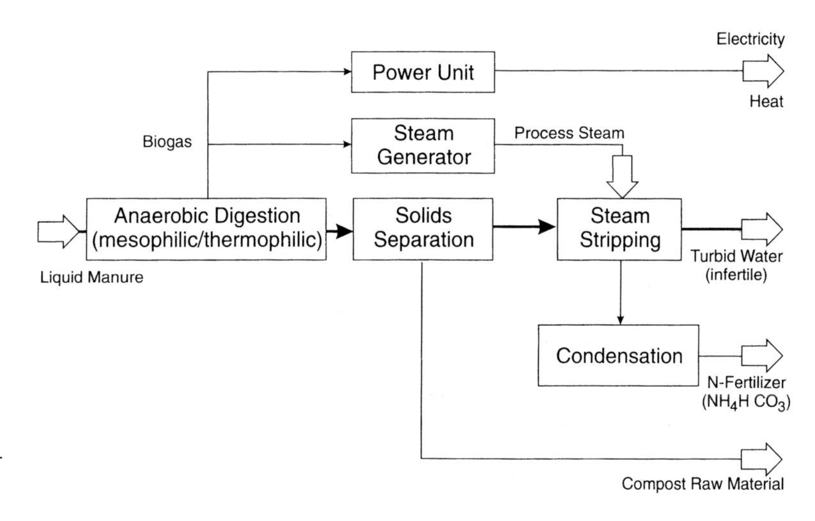

Fig. 4. Schematic diagram of the Schwarting-Uhde plant in Göritz.

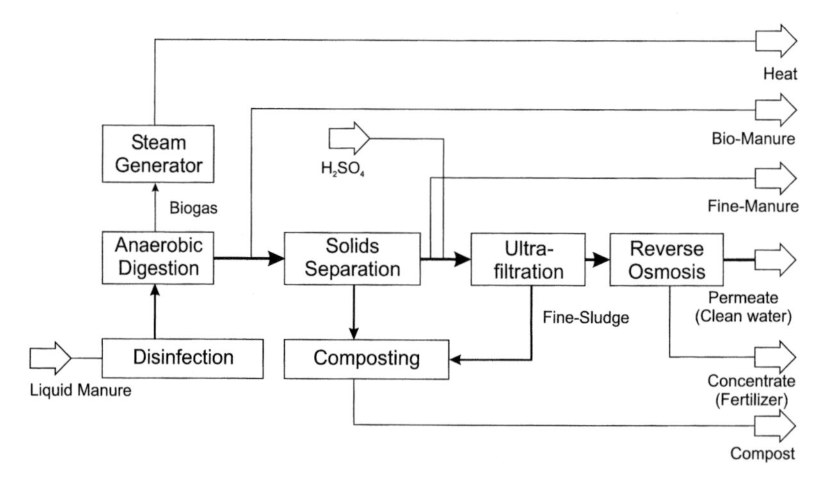

Fig. 5. Schematic diagram of the manure treatment plant in Surwold-Börgermoor.

tration for achieving reliable operation conditions. The RO-permeate is of high quality (COD <15 mg L^{-1}, N <10 mg L^{-1}, P <1 mg L^{-1}) and can be discharged to a river or reused for irrigation. The concentrate contains all minerals and can be utilized as fertilizer (SCHIEDING, 1997). The schematic diagram of the manure treatment plant in Surwold-Börgermoor, which is operated with a mixture of pig and cow manure, is shown in Fig. 5. Evaporation processes produce a low loaded condensate which needs an aerobic posttreatment before the water can be discharged to a river (KARLE and WEILAND, 1996). In order to avoid a transfer of ammonia into the condensate or permeate the pH of the anaerobic treated waste has to be lowered by addition of acids.

3.1.2 Co-Fermentation

The combined conversion of liquid manure together with semisolid or solid wastes mainly from harvesting, crop or vegetable processing has several technical and economical advantages. Co-fermentation makes manure treatment more economical because the addition of co-substrates increases the specific biogas productivity per reactor volume considerably. The maximum biogas yield per ton of cow or pig manure is less than 25 m^3 and 35 m^3, respectively, whereas co-substrates usually show a two- or four-fold gas yield due to the higher volatile solids (VS) content (WEILAND, 1997a). Furthermore, co-fermentation allows treatment of solid wastes of unbalanced composition because a lack in nutrients, trace elements, and an insufficient buffer capacity of the waste can be adjusted by the addition of manure. Fat residues from slaughtering or oil seed processing are typical wastes which are difficult to treat as monocharges because the anaerobic metabolism is inhibited due to a lack in nitrogen and trace elements. The combined digestion of fats and manure results in well balanced environmental conditions and improved material flow properties (KTBL, 1998).

For avoiding process instabilities and insufficient mixing conditions the portion of co-substrates should not exceed about 40% of the reactor input. At higher co-substrate portions the buffer capacity of manure is quite often not high enough to avoid process failure at higher loadings. The flow diagram of Fig. 6 shows the typical process steps of co-fermentation plants.

The co-substrates have to be pretreated by milling, impurity separation, and pasteurization if they contain coarse solids, impurities, or pathogenic germs. Impurities are separated by dry or wet preparation and the disinfection is carried out normally at 70°C for 1 h. Disinfection is necessary if residues from slaughterhouses, food residues from restaurants, or non-

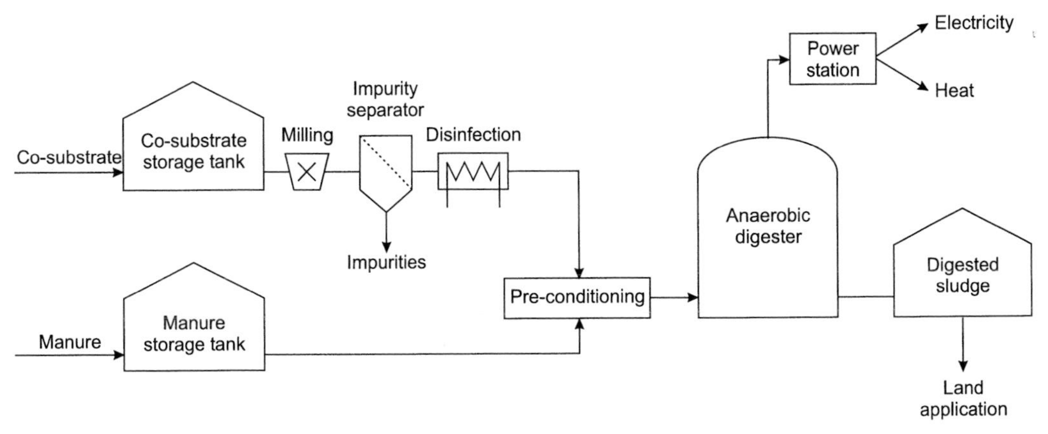

Fig. 6. Schematic diagram of a co-fermentation treatment plant.

agricultural wastes, e.g., household biowastes are used for co-fermentation (KTBL, 1998). Manure and co-substrates have to be mixed intensively before they are fed to the digester to achieve a constant load and to avoid scum or bottom layers. The total solids content of the reactor influent should not exceed 12% DM, otherwise special pumps and mixers are necessary. In most cases stirred tank digesters with residence times between 15–25 d are used for co-fermentation (Kuhn, 1995).

For the land application of the digested residues, it has to be noticed that the treated mixture usually contains more nutrients than pure manure. In order to apply the mixture according to the "Code of Good Agricultural Practice" an additional land area is necessary for the controlled land application. The German "Fertilizer Ordinance" limits the amount of nitrogen which can be disposed of per year on land to 210 kg ha^{-1} on grassland and 170 kg ha^{-1} on arable land (Deutscher Bundestag, 1996). Therefore, co-fermentation processes should be used only on farms with a specific livestock size of <2 LU per hectare. The land application is also limited by the content of inorganic contaminants. Tab. 10 shows the limiting values for heavy metals in Germany according to the regulations of the planned Biowaste Ordinance (Deutscher Bundestag, 1998).

Constant quality control of the treated co-substrates is necessary in order to fulfill these criteria.

Tab. 10. Limits for Land Application of Digested Residues

Component	Heavy Metal Content [mg/kg DM]	
	20 t DM/ha within 3 years	30 t DM/ha within 3 years
Pb	150	100
Cd	1.5	1.0
Cr	100	70
Cu	100	70
Ni	50	35
Hg	1	0.7
Zn	400	300

3.1.3 Solids Treatment

Solid organic wastes from harvesting or crop processing can also be treated by wet or dry fermentation processes which need no addition of other substrates. The typical reactor configurations are shown in Fig. 7. Wet fermentation takes place in a slurry phase at a total solids content of 8–12%. This method can be used for different types of degradable solid

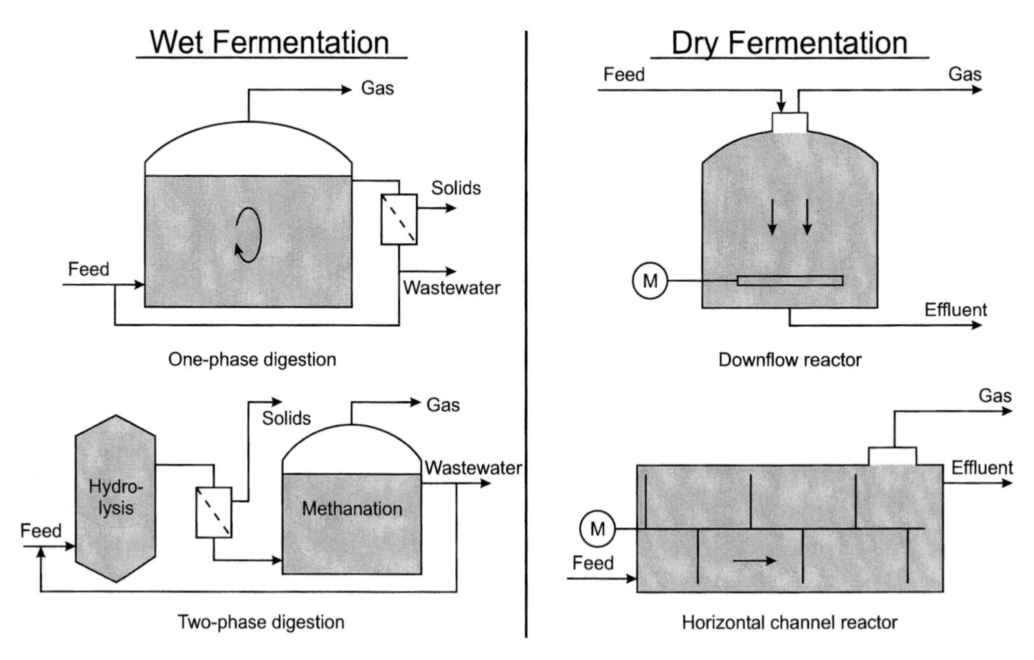

Fig. 7. Anaerobic digester systems for the treatment of solid organic wastes.

wastes because the total solids content of the reactor input can be adapted by recirculation of process water. Dry fermentation processes are operated at a total solids content over 25%, which reduces the mass flow, the reactor size, and the technical effort for dewatering the digested residues. Its application is restricted to wastes with more than 25% DM.

Both process types involve some form of pretreatment for size reduction, impurity separation, and homogenization depending on the properties of the waste. Wet fermentation is performed in its simplest form in a one-stage completely mixed digester or in a two-stage mode of operation with a separate hydrolysis and acidification reactor. The two-stage processes can be operated with or without solids separation after the first process stage. Solids separation after hydrolysis and acidification allows the application of high-rate reactors, e.g., fixed film or UASB reactors for biomethanation, but makes an aerobic posttreatment of the separated solids necessary due to the high concentration of malodorous fatty acids (KLEMPS et al., 1997).

Dry fermentation processes are usually performed in downflow operated tower reactors or in horizontal channel reactors. Vertical tower-type reactors are operated without mixing equipment but need special discharging devices at the reactor bottom. Channel reactors need slowly rotating horizontal paddles for smooth stirring (SCHÖN, 1994). The typical mass balance for the anaerobic treatment of solid agricultural wastes with about 35% TS of the input matter is shown in Fig. 8.

The amount of biogas and compost that will be produced by anaerobic solids digestion depends on the composition and pretreatment of the input material, and on the process conditions. Soft plant wastes, such as leaves or green crop residues, result in higher gas yields whereas wastes of lignocellulosic structure such as straw, form less gas and more compost as shown in Fig. 8.

The undigested solids are removed after biogasification by a screw or sieve drum press. The compost-like solids can be used directly in agriculture for soil conditioning or have to be posttreated by composting to make a marketable product. The liquid phase can be used for

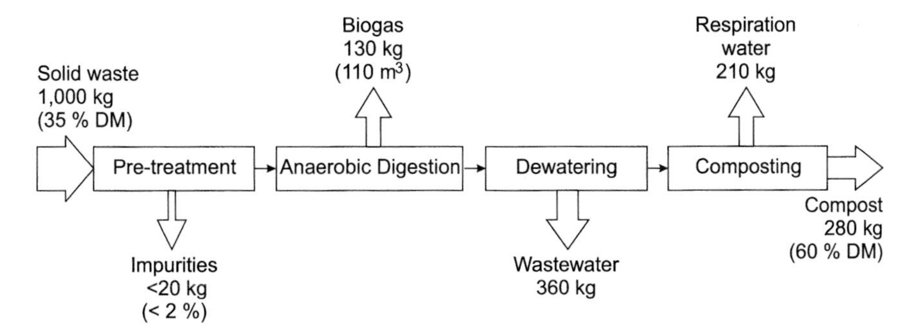

Fig. 8. Typical mass balance for biogasification of solid agricultural wastes.

land application, if the heavy metal content is below the accepted limit or has to be purified in a sewage treatment plant before it can be discharged. The majority of wet and dry fermentation processes is used for the treatment of source-separated biowaste from households but several plants are also used for the treatment of solid wastes from agriculture, food-, and agro-industry (ANS, 1995; THOMÉ-KOZMIENSKY, 1995).

3.1.4 Process Conditions

Anaerobic treatment processes are operated under mesophilic (30–38°C) or thermophilic (55–65°C) temperature conditions. Under thermophilic conditions the degradation will occur at faster rates which allow smaller reactor sizes due to shorter necessary residence times and sometimes also higher biogas yields. The higher process temperature results in a complete destruction of pathogenic germs and weed seeds. This makes the treatment of unsanitary wastes, e.g., residues from slaughtering processes, and the land application of the treated wastes possible. Experiments with salmonella, worm eggs, and different types of viruses have shown that most pathogenic germs are inactivated under thermophilic process conditions within 2 h, whereas under mesophilic conditions the period of survival is often more than 6 d (STRAUCH, 1996). Nevertheless, most of the anaerobic processes are operated under mesophilic temperature conditions, because lower quantities of heat have to be supplied to maintain the process temperature. Several times a better process stability has been observed under mesophilic conditions, because the inhibiting effect of some components, such as ammonia, is lower under mesophilic temperature conditions (WELLINGER, 1991).

3.1.5 Biogas Formation

The amount of methane which is generated during anaerobic digestion is a function of the biodegradable fraction of the total waste that is available to the anaerobic bacteria and the operating and environmental conditions of the process. The anaerobic degradability depends not only on the waste type but also on its pretreatment and handling prior to the digestion. Soluble organic compounds, e.g., sugars or volatile fatty acids, are almost completely biodegradable whereas the degradability of solid wastes from crops decreases at increasing maturity and senescence due to an increase of the lignin and lignocellulose content. The specific gas yield which can be obtained within a realistic retention time of at most 20 d is between 150 and 800 L biogas per kg of volatile solids (VS). The generated biogas contains 55–75% methane, 24–44% carbon dioxide, and 100–10,000 ppm hydrogen sulfide. Tab. 11 shows the typical gas yield and VS degradation efficiency for different agricultural wastes.

Depending on the methane content the caloric value is between 5.5 and 7.5 kWh m^{-3}. Although hydrogen sulfide is a minor component, its effect is important because H_2S can inhibit the degradation process and forms

Tab. 11. Biogas Yield and Degradation Efficiency of Different Agricultural Wastes at Mesophilic Digestion and 20 d Retention Time

Waste Type	Biogas Yield [L/kg VS_{in}]	VS-Degradation [%]
Wastewater		
Blanching water	600–700	70–85
Distillery slops	400–700	50–85
Molasse	300–350	35–45
Silage juice	740–780	90–95
Whey	750–800	94–98
Slurries		
Beef cattle manure	300–350	35–45
Dairy cow manure	200–220	25–28
Pig manure	300–450	35–55
Poultry manure	200–400	25–48
Flotation fat	600–800	65–90
Solids		
Farmyard manure (beef)	200–280	25–28
Paunch manure	250–400	30–50
Brewery spent grain	380–400	45–55
Cereal straw	150–200	20–25
Potato haulm	550–570	65–70
Potato pulp	400–450	75–80
Sugar beet leaves	500–600	60–70
Sugar beet pulp	740–760	80–85
Vegetable pomace	500–800	70–90

corrosive sulfuric acid during combustion (KROISS, 1985). Especially the degradation of high protein containing wastes results in the formation of biogas with high H_2S concentrations, which makes purification of biogas necessary. The typical limit for the operation of gas engines is <500 ppm H_2S. Biogas is mainly utilized in block-type power stations for electricity and heat production that achieve a high electrical efficiency factor of 30–35%.

3.2 Aerobic Treatment

3.2.1 Wastewater and Slurry Treatment

Aerobic processes are used relatively seldom for the treatment of slurries and highly loaded wastewater, because the aeration is difficult, a high energy input is necessary, and the process is often disturbed by the formation of foam or scum layers. In general, aerobic processes are useful for COD and nitrogen removal and odor control. Some full-scale plants and pilot experiments have shown that animal wastes can be purified according to the principles of industrial sewage treatment with a COD elimination of 70–95% and about 50% N-reduction, but the aeration time of about 10–20 d is too long in order to achieve economical process conditions (KÜHL, 1988). Higher efficiencies can be obtained if rotating biological discs contactors (RUDOLPH and BÜSCHER, 1994) or trickling filters (ZÖTTL et al., 1994) are used, but all aeration processes result in emissions of ammonia, malodorous gases, and laughing gas.

Aerobic thermophilic processes, well known from sewage sludge treatment, will become more important for the agricultural waste treatment in the future, because the process allows an efficient disinfection, a biological stabilization of liquids and slurries, and a separation of ammonia nitrogen. If the process is applied to separated liquid manure in a well insulated sequencing batch reactor the temperature increases by biological oxidation up to 60°C and the pH rises to about 9.5. Consequently, ammonia is stripped out and can be recovered in the form of ammonium sulfate by washing the exhaust air with sulfuric acid. Within aeration times of 3–4 d about 40% of the COD, 90% of the BOD_5, and 50–60% of total nitrogen are removed. Coarse suspended solids have to be removed before aeration in order to avoid process failure by scum formation. For cow and pig manure specific air flow rates up to 20 m^3 m^{-3} h^{-1} are necessary (HAHNE and SCHUCHARDT, 1996). The schematic diagram of the thermophilic FAL process is shown in Fig. 9.

Enterobacteria are completely inactivated within the typical retention time of 3–4 d. A higher nutrient removal can be achieved if the process is combined with a centrifuge for the separation of the fine-size solids. The treated liquid manure can be used for land application without odor emissions and ecological risks. Wastewater from fruit and vegetable processing with COD loadings below 3,000 mg L^{-1} is

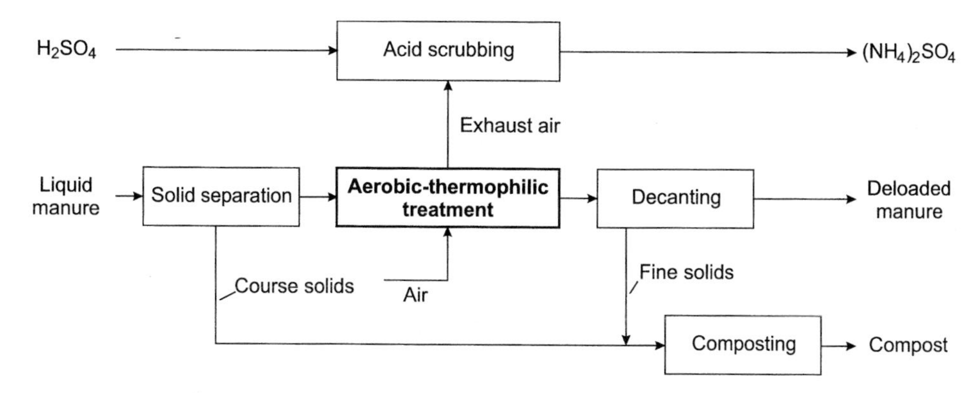

Fig. 9. Schematic diagram of the thermophilic process for manure treatment.

normally treated in conventional activated sludge processes with biological nitrogen and phosphate removal.

3.2.2 Solids Treatment

Aerobic treatment processes are well established in agriculture for composting of solid organic wastes in simple heap, forced aerated windrow systems, or closed composting reactors to produce a stable humified end product which can be used as fertilizer or soil conditioner. Fig. 10 shows some typical composting configurations for agricultural wastes. The complex microbial population of composting processes is able to degrade sugars, fats, proteins, hemicellulose, and cellulose to carbon dioxide and water. The process needs a well balanced C:N ratio of 25:1–30:1, a moisture

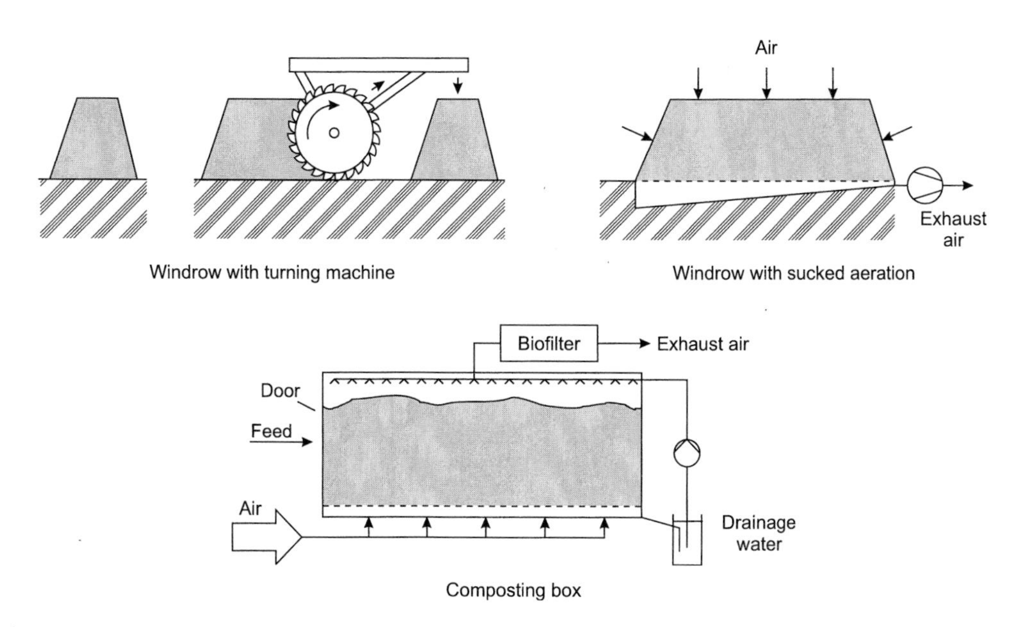

Fig. 10. Typical composting processes.

content of 50–60%, and a free air space of about 30%. Many agricultural wastes, e.g., manures and vegetable wastes, have a lower C:N-ratio, a higher moisture content, and are difficult to aerate. They need to be blended with an adsorbent solid matrix, e.g., straw or bulking agents like wood chips, which provides extra carbon and an appropriate bulk structure (BIDDLESTONE et al., 1987).

The amount of heat produced during the biological oxidation is high enough that in a large composting mass temperatures of 60–70°C can be reached for several weeks. The majority of disease and parasitic organisms but also weed seeds are killed at these temperatures within a few days and only few aerobic spore forming *Bacillus* species can survive. Agricultural wastes from livestock farming are always contaminated with pathogens, but also many harvesting residues are polluted with nematodes, worms, fungi, and viruses. Consequently, these contaminated solid wastes have to be composted before recycling if the material properties allow a natural or forced aeration. Dependent on its nutrient content, quality, and structure the compost can be used as soil conditioner, fertilizer, mulch, or for peat replacement.

3.3 Special Upgrading Processes

Several agricultural wastes from harvesting and plant materials processing as well as wastes from milk and meat processing can be utilized in larger quantities also as substrate resources for biotechnological processes to produce higher value products. By-products which contain low-molecular carbohydrates, starch, proteins, or fats can be utilized sometimes directly for fermentation processes without pretreatment, whereas cellulosic and hemicellulosic wastes generally need an enzymatic or physicochemical pretreatment to become utilizable substrates. Fig. 11 shows the typical process routes which are used for upgrading agricultural wastes to higher value products (BAADER and WEILAND, 1991).

Submerged fermentation processes are used for converting liquid and solid agricultural wastes to organic acids, alcohols, enzymes, single cell protein (SCP), etc., which can be

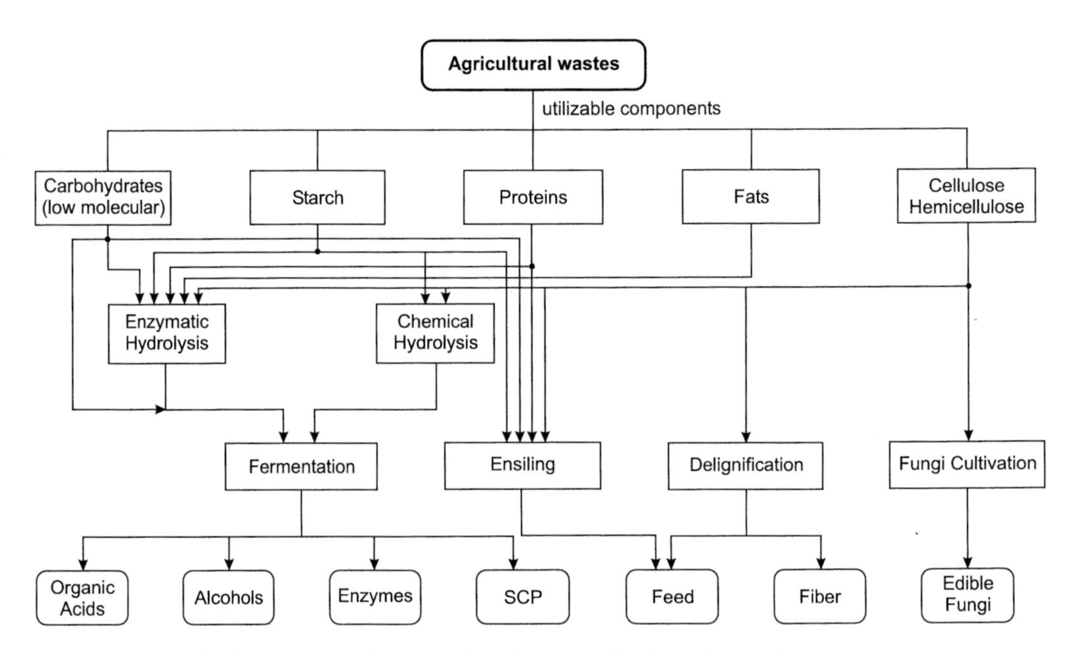

Fig. 11. Biotechnological processes for upgrading of wastes to higher value products.

used in food, feed, and chemical industry in a wide range of applications. Ensiling, used for animal fodder production, is a solid-state fermentation process in which the preservation of wastes is aimed at under not strictly controlled conditions. Under strictly controlled conditions operated solid-state fermentation processes are used to improve the feed value of crop residues and to produce several metabolic products (SCHUCHARDT and ZADRAZIL, 1988; WEILAND, 1991). Delignification of cellulosic wastes (straw) is a controlled solid-state fermentation process that is used for an improvement of the digestibility and protein content of solid wastes. The cultivation of fungi represents nothing more than a solid-state fermentation process using a large variety of agricultural wastes, such as manure, straw, plant stems, leaves, and other residues from plant processing for food production (ZADRAZIL, 1978).

Some typical higher value products that can be obtained through biotechnological upgrading of agricultural wastes are summarized in Tab. 12.

Tab. 12 shows that solid-state fermentation (SSF) can be used not only for animal feed and fungi production but also for the formation of various metabolites, e.g., enzymes and organic acids. SSF offers numerous advantages over conventional submerged fermentation since the quantities of process wastewater are lower and sometimes no further downstream processing is necessary for product utilization. Solid-state fermentation needs special reactor types and makes process control more difficult (WEILAND, 1988b). For submerged fermentation of solid substrates an enzymatic or chemical hydrolysis is necessary. Depending on the process type both process steps can be carried out in separate reactors or together in one stage. Simultaneous saccharification and fermentation is a typical method for converting cellulosic wastes into alcohols, acids, and other chemicals. By carrying out both processes together, the glucose inhibition of hydrolysis can be minimized resulting in faster saccharification rates and reduced reactor volumes (SOUTH et al., 1995).

Factors which often limit the use of agricultural wastes for biotechnological upgrading processes are the short harvesting and processing periods that make the availability of wastes

Tab. 12. Higher Value Products from Agricultural Wastes

Product Class	Product	Substrate	Process Type	Reference
Alcohols	ethanol	whey	submerged fermentation	SISO (1966)
	ethanol	cellulosic wastes	saccharification and fermentation	WYMAN (1994)
	2,3-butanediol	straw, corn cobs	hydrolysis and fermentation	GARG and JAI (1995)
Organic acids	amino acids	whey	submerged fermentation	MACRIS (1989)
	acetic acid	banana peelings, rice hulls	hydrolysis and fermentation	FAO (1982)
	citric acid	brewery grains, fruit peels	solid-state fermentation	LU et al. (1995)
	lactic acid	sugar cane bagasse	saccharification and fermentation	VENKATESH (1997)
Enzymes	cellulases	cereal straw	solid-state fermentation	SERMANNI et al. (1994)
	lactase	whey	submerged fermentation	FAO (1982)
	pectinases	bean hulls	submerged fermentation	MACRIS (1989)
	xylanases	beet pulp, apple pomace	solid-state fermentation	WIACEK-ZYCHLINSKA et al. (1994)
Biopolymers	PHB	potato waste	hydrolysis and fermentation	RUSENDI and SHEPPARD (1995)
	xanthan	citrus waste	submerged fermentation	BILANOVIC et al. (1994)
Animal feed SCP	feedstuff	rice/wheat straw	solid-state fermentation	SCHUCHARDT and ZADRAZIL (1988), BISARIA et al. (1997)
	feedstuff	cassava peel	solid-state fermentation	OFUYA and OBILOR (1993)
	feedstuff	sugar beet pulp	semisolid fermentation	ISRAILIDES et al. (1994)
Edible fungi	mushroom	cereal straw	solid-state fermentation	ZADRAZIL (1978)

difficult. Other factors which limit the upgrading of wastes are their low concentration of relevant substrates and their high contents of water that increase the transportation costs. Sometimes the utilization of wastes is also limited due to the rapid microbial decomposition of the valuable substances. Many wastes also contain substrates which are difficult to convert by conventional microorganisms. Xylose, the main sugar derived from hemicellulosic and cellulosic wastes, is a typical substrate that cannot be converted to ethanol by conventional yeasts. The application of recombinant microorganisms may open new ways for converting such substrates to higher value products and may expand the range of agricultural wastes that can be used as substrates for microbial processes. Today, the economic feasibility for the application of genetically modified microorganisms for waste treatment and waste utilization is still poor.

4 Closing Remarks

Within the last decades the quantities and the diversity of agricultural and agro-industrial wastes have considerably increased due to the dramatic improvement in agricultural productivity and the development of new processes for finishing agricultural raw materials. Consequently, a number of environmental problems have occurred because the usual approach for agricultural waste management was disposal.

The need for a better protection of the environment and the need for a more efficient management of the increasing waste streams has focused attention on biological processes for a better use of agricultural wastes. Today, wastes from agricultural production and processing should be considered as a potential resource for the production of energy, animal feeds, chemicals, soil conditioner, and biopolymers, etc., and not as an undesired waste with a negative economic value. Many biotechnological processes are available for the microbial or enzymatic conversion of agricultural wastes, but usually they are used only if conventional disposal techniques are more expensive. The use of wastes as renewable substrate resource depends on the regulations and laws for environmental protection and the prices for conventional substrates and products. Generally, there is no one best approach for a successful utilization of waste, because the situation in each country is different and depends on local and regional conditions.

Anaerobic digestion and composting are the most important technologies applied today for an environmentally compatible and resource saving management of agricultural wastes. Microbial upgrading processes to form higher value products are only well established for few residues and are used relatively seldom compared to the large number and diversity of agricultural wastes. The development of genetically modified microorganisms may help to increase the number of utilizable wastes and formed products. In the near future waste utilization may be promoted by the development of integrated production and processing chains that can be used for a wide variety of purposes, e.g., in food, feed, and energy production to achieve a sustainable development in agriculture.

5 References

ANS (1995), Anaerobe Bioabfallbehandlung in der Praxis, *ANS-Schriftenreihe*, Heft 30. Mettmann: ANS.

ATV (1995), Wirtschaftsdünger, Abfälle und Abwässer aus landwirtschaftlichen Betrieben, *Merkblatt ATV-M702*. Hennef: GFA.

BAADER, W., WEILAND, P. (1991), Biotechnological methods for the utilization of residues and by-products of agricultural production, in: *Natural Resources and Development*, Vol. 34 (Institute for Scientific Co-operation in Conjunction with the Federal Institute for Geosciences and Natural Resources and Numerous Members of German Universities, Ed.), pp. 48–65. Metzingen: Georg Hauser.

BIDDLESTONE, A. J., GRAY, K. R., DAY, C. A. (1987), Composting and straw decomposition, in: *Environmental Biotechnology* (FORSTER, C. F., WASE, D. A. J., Eds.), pp. 135–175. Chichester: Ellis Horwood.

BILANOVIC, D., SHELEF, G., GREEN, M. (1994), Xanthan fermentation of citrus waste, *Biores. Technol.* **48**, 169–172.

BISARIA, R., MADAN, M., VASUDEVAN, P. (1997), Utilization of agro-residues as animal feed through bioconversion, *Biores. Technol.* **59**, 5–8.

BÖHM, E., KUNZ, P., MANNSBART, W., MIELICKE, M. U. (1995), *BMFT-Forschungsbericht* T85-027, Karlsruhe: FhG Systemtechnik und Innovationsforschung.

Deutscher Bundestag (1996), *Dünge-Verordnung (Fertilizer Ordinance)*, BGBl. Teil I vom 6. 2. 1996, 118.

Deutscher Bundestag (1998), *Bioabfall-Verordnung (Biowaste Ordinance)*, Bonn (BGBL, Teil 1 vom 21. 98. 98, 2955).

DÖHLER, H., KUHN, E., SCHWAB, M., METZGER, H. J. (1997), Umweltprobleme der Güllewirtschaft und Optionen zur Lösung, in: *KTBL-Arbeitspapier* 242, pp. 9–16. Münster-Hiltrup: Landwirtschaftsverlag.

FAO (1982), *Agricultural Residues – Compendium of Technologies*, Agricultural Services Bulletin 33, Rev. 1. Rome: Food and Agricultural Organization.

FRÜCHTENICHT, K., HEYN, J., KUHLMANN, H. (1993), Pflanzenernährung und Düngung, in: *Faustzahlen für Landwirtschaft und Gartenbau*, 12nd Edn. (Hydro Agri Dülmen GmbH, Ed.), pp. 254–295. Münster-Hiltrup: Landwirtschaftsverlag.

GARG, S. K., JAI, A. (1995), Fermentative production of 2,3-butanediol: A review, *Biores. Technol.* **51**, 103–109.

HAHNE, J., SCHUCHARDT, F. (1996), Aerob-thermophile Güllebehandlung zur N-Elimination und Entseuchung, *Korrespondenz Abwasser* **43**, 1256–1263.

HOPPE, P. P., SCHÖNER, F.-J., WIESCHE, H., SCHWARZ, G., SAFER, S. (1993), Phosphor-Äquivalenz von *Aspergillus niger*-Phytase für Ferkel bei Fütterung von Getreide-Soja-Diät, *J. Anim. Physiol. Anim. Nutr.* **69**, 225–234.

HÖRNIG, G., SCHERPING, E. (1993), Water consumption and wastewater volume in milking plants and milk houses of large size, in: *Technology in Animal Production, Bornimer Agrartechnische Berichte*, Heft 3, pp. 149–171, Potsdam-Bornim.

HÜTTNER, A., WEILAND, P. (1996), Verfahren zur umweltverträglichen Gülleaufbereitung mit Nährstoffrückgewinnung, in: *Verfahrenstechnik der Abwasser- und Schlammbehandlung*, Vol. 2 (GVC-VDI, Eds.), pp. 1019–1035. Düsseldorf: VDI.

ISRAILIDES, C. J., ICONOMOU, D., KANDYLIS, K., NIKOKYRIS, P. (1994), Fermentability of sugar beet pulp and its acceptability in mice, *Biores. Technol.* **47**, 97–101

KARLE, G., WEILAND, P. (1996), Technische Gülleaufbereitung – FuE-Anlage Bakum/Landkreis Vechta, *NLM-Abschlußbericht*. Braunschweig: Institut für Technologie, Bundesforschungsanstalt für Landwirtschaft.

KESSLER, H. G. (1988), *Lebensmittel- und Bioverfahrenstechnik – Molkereitechnologie*. Freising-Weihenstephan: A. Kessler.

KIRCHGESSNER, M., ROTH, F. X., KREUZER, M. (1991), Bestimmungsfaktoren der Güllecharakteristik beim Schwein – Einfluß von N-Zufuhr und Aminosäurezusammensetzung des Futters, *Agribiol. Res.* **44**, 345–356.

KLEMPS, R., HARMSSEN, H., WEILAND, P., GROTE, J. (1997), Biologisches Verfahren zur anaeroben Behandlung organischer Abfälle unter Gewinnung von Biogas, *Forschung, Planung und Betrieb* (Preussag) **21**, 1–5.

KOBALD, M., HOLLEY, W. (1990), *BMBF-Studie zur Emissionssituation in der Nahrungsmittelindustrie*. München: FhG für Lebensmitteltechnologie und Verpackung.

KROISS, H. (1985), Anaerobe Abwasserreinigung, *Wiener Mitt. Wasser–Abwasser–Gewässer*, Vol. 62.

KTBL (1998), Kofermentation, *KTBL-Arbeitspapier* 249. Münster-Hiltrup: Landwirtschaftsverlag.

KÜHL, H. (1988), Erfahrungen bei der Gülleaufbereitung mit dem Belebtschlammverfahren, *Agrartechnik* **38**, 178–180.

KUHN, E. (1995), Kofermentation, *KTBL-Arbeitspapier* 219. Münster-Hiltrup: Landwirtschaftsverlag.

LU, M. Y., MADDOX, I. S., BROOKS, J. D. (1995), Citric acid production by *Aspergillus niger* in solid-substrate fermentation, *Biores. Technol.* **54**, 235–239.

MACRIS, B. J. (1989), Upgrading of agricultural and agroindustrial by-products to food and feed by biotechnology, in: *International Biosystems*, Vol. 2 (WISE, D. L., Ed.), pp. 169–205. Boca Raton, FL: CRC Press.

MOULIN, G., GALZY, P. (1984), Whey, a potential substrate for biotechnology, in: *Biotechnology and Genetic Engineering Reviews*, Vol. 1 (RUSSEL, G. E., Ed.), pp. 347–374. Newcastle-upon-Tyne: Intercept.

OFUYA, C. O., OBILOR, S. N. (1993), The suitability of fermented cassava peel as a poultry feedstuff, *Biores. Technol.* **44**, 101–104.

OLSEN, H. (1988), Enzymes and food proteins, in: *Food Technology International Europe*, pp. 245–250. Northampton, MA: Sterling Publications.

OLSEN, H. (1995), Enzymes in food processing, in: *Biotechnology*, Vol. 9 (REHM, H.-J. et al., Eds.), pp. 663–736. Weinheim: VCH.

RUDOLPH, K. U., BÜSCHER, E. (1994), Einsatz von schwimmenden Scheibentauchkörpern zur Güllebehandlung in Standardsilos, in: *Umweltverträgliche Gülleaufbereitung* (BMBF and KTBL, Eds.), pp. 33–42. Darmstadt: KTBL.

RUSENDI, D., SHEPPARD, J. D. (1995), Hydrolysis of potato processing waste for the production of poly-β-hydroxybutyrate, *Biores. Technol.* **54**, 191–196.

SCHIEDING, D. (1997), Demonstration einer regional zentralisierten Güllebehandlung durch ein mehrstufiges Verfahren mit einer Umkehrosmosestufe zur Erzeugung vorflutfähigen Wassers, *KTBL-Arbeitspapier* 242, pp. 50–58. Darmstadt: KTBL.

SCHÖN, M. (1994), *Verfahren zur Vergärung organischer Rückstände in der Abfallwirtschaft*. Berlin: Erich Schmidt Verlag.

SCHUCHARDT, F., FERCH, A. (1994), Untersuchung zur Kompostierung von Panseninhalt, *Korrespondenz Abwasser* **41**, 1812–1818.

SCHUCHARDT, F., ZADRAZIL, F. (1988), A 352-liter fermenter for solid state fermentation of straw by white-rot fungi, in: *Treatment of Lignocellulosics with White-Rot Fungi* (ZADRAZIL, F., REINIGER, P., Eds.), pp. 77–89. London: Elsevier Applied Science.

SCHULZ, H. (1996), *Biogas-Praxis*. Staufen: Ökobuch-Verlag.

SERMANNI, G. G., ANNIBALE, A., DI LENA, G., VITALE, N. S., DI MATTIA, E., MINELLI, V. (1994), The production of exo-enzymes by *Lentinus edodes* and *Pleurotus ostreatus* and their use for upgrading corn straw, *Biores. Technol.* **48**, 173–178.

SISO, M. I. G. (1996), The biotechnological utilization of cheese whey: A review, *Biores. Technol.* **57**, 1–11.

SOUTH, C. R., HOGSETT, D. A. L., LYND, L. R. (1995), Modeling simultaneous saccharification and fermentation of lignocellulose for production of fuel ethanol in batch and continuous reactors, *Enzyme Microb. Technol.* **17**, 797–803.

STRAUCH, D. (1996), Hygieneaspekte bei der Cofermentation, in: *Internationale Erfahrungen mit der Verwertung biogener Abfälle zur Biogasproduktion* (Federal Environment Agency Austria, Ed.), pp. 53–92. Vienna.

THOMÉ-KOZMIENSKY, K. J. (1995), *Biologische Abfallbehandlung*. Berlin: EF-Verlag für Energie- und Umwelttechnik.

TRITT, W. P. (1992), The anaerobic treatment of slaughterhouse wastewater in fixed-bed reactors, *Biores. Technol.* **41**, 201–207.

TRITT, W. P., SCHUCHARDT, F. (1992), Materials flow and possibilities of treating liquid and solid wastes from slaughterhouses in Germany, *Biores. Technol.* **41**, 235–245.

VENKATESH, K. V. (1997), Simultaneous saccharification and fermentation of cellulose to lactic acid, *Biores. Technol.* **62**, 91–98.

WEILAND, P. (1988a), Development of an anaerobic fixed bed reactor for agro-industrial wastes, *Med. Fac. Landbouw. Rijksuniv. Gent* **53**, 1971–1979.

WEILAND, P. (1988b), Principles of solid state fermentation, in: *Treatment of Lignocellulosics with White-Rot Fungi* (ZADRAZIL, F., REINIGER, P., Eds.), pp. 64–76. London: Elsevier Applied Science.

WEILAND, P. (1991), Entwicklung eines Feststoff-Fermentationsprozesses zur Herstellung von Ethanol aus Zuckerrüben und anderen zuckerhaltigen Früchten, in: *Schriftenreihe des BML, Reihe A, Produktions- und Verwendungsalternativen für die Land- und Forstwirtschaft*, pp. 32–38. Münster-Hiltrup: Landwirtschaftsverlag.

WEILAND, P. (1995), Biogastechnik – Verfahrenstechnische Grundlagen, Anwendungsmöglichkeiten und Entwicklungstendenzen, *Der Tropenlandwirt*, Beiheft **53**, 153–170.

WEILAND, P. (1997a), Co-digestion of agricultural, industrial and municipal wastes, *REUR Technological Series* **52**, pp. 42–52, Rome: FAO.

WEILAND, P. (1997b), Recent German technology in anaerobic waste treatment, *Environ. Eng. Res.* **2**, 217–223.

WELLINGER, A. (1991), *Biogas-Handbuch*. Aarau: Verlag Wirz.

WIACEK-ZYCHLINSKA, A., CZAKAJ, J., SAWICKA-ZUKOWSKA, R. (1994), Xylanase production by fungal strains in solid-state fermentations, *Biores. Technol.* **49**, 13–16.

WINTZER, D. (1996), Wege zur umweltverträglichen Verwertung organischer Abfälle, *Abfallwirtschaft in Forschung und Praxis*, Vol. 97, Berlin: Erich Schmidt Verlag.

WYMAN, C. E. (1994), Ethanol from lignocellulosic biomass: Technology, economics, and opportunities, *Biores. Technol.* **50**, 3–16.

ZADRAZIL, F. (1978), Cultivation of Pleurotus, in: *The Biology and Cultivation of Edible Mushrooms* (CHANG, S. T., HAYES, W. A., Eds.), pp. 521– 557. New York: Academic Press.

ZIMMER, E., HONIG, H., KÜNTZEL, U. (1993), Futterkonservierung, in: *Faustzahlen für Landwirtschaft und Gartenbau*, 12nd Edn., pp. 497–511. Münster-Hiltrup: Landwirtschaftsverlag.

ZÖTTL, H. W., WAGNER, F., SCHALK, P. (1994), Entwicklung eines Verfahrens zur biologischen Güllebehandlung mittels Rindenfilterkörpern, in: *Umweltverträgliche Gülleaufbereitung* (BMBF and KTBL, Eds.), pp. 43–53. Darmstadt: KTBL.

Aerobic Carbon, Nitrogen, and Phosphate Removal

12 Biological Processes in Wetland Systems for Wastewater Treatment

PETER KUSCHK
ARNDT WIEßNER
ULRICH STOTTMEISTER
Leipzig, Germany

1 Introduction

Wetland systems have now gained international acceptance aside conventional techniques for the treatment of wastewater.

They comprise natural wetlands such as swamps, bogs, wet meadows, as well as natural and constructed wastewater ponds, and lagoons and artificially installed constructed wetlands with different design and flow characteristics such as surface horizontal flow systems and subsurface horizontal or vertical flow systems (KADLEC, 1987; WISSING, 1995). Generally speaking, constructed wetlands are defined as wetlands which have been deliberately created from non-wetland sites in order to treat polluted water (BRIX, 1995). In all such systems the treatment of mainly domestic wastewater (but also industrial effluents, agricultural wastewater, landfill leachate, groundwater, etc.) is in practice often implemented in conjunction with other methods of treatment (WISSING, 1995).

Generally speaking, the active reaction matter (the rhizosphere, Fig. 1, with its physicochemical and mainly specific biological process forces of varying importance induced by the interaction of plants, microorganisms, soil, and pollutants) is treated like a "black box" and knowledge thereof is largely inadequate. More detailed knowledge about the role of various special plants, the settlement of different microbiological consortia, and the interactions of the biological compounds in addition to the specific contaminants and the given filter bed material can in the end help to make treatment more effective by permitting the installation of defined systems.

Therefore, the goal must be to optimize the technology required by maximizing the biological interaction forces in the rhizosphere – an approach which is also essential in order to develop cheap, simple systems to tackle the wastewater problems of third-world countries. To this end and in view of the prominent application of subsurface flow constructed wetlands compared to other systems, the key aspects of the biological processes in the rhizosphere of these systems are described and discussed in this chapter.

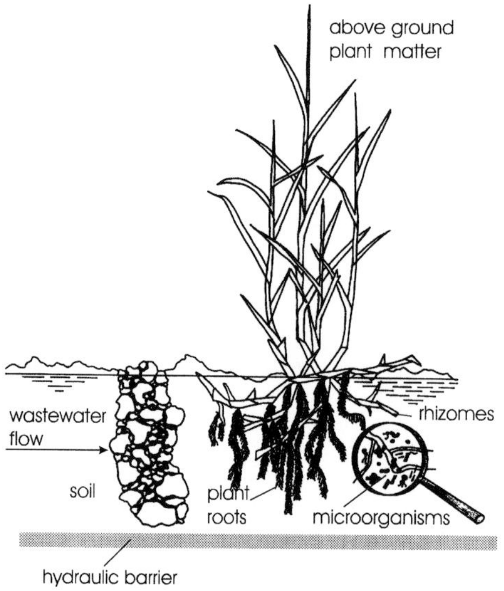

Fig. 1. Simplified model of a horizontal subsurface flow constructed wetland.

2 The Role of Plants

2.1 Wetland Species

There are hundreds of wetland species (MÜHLBERG, 1980). The applicability of certain macrophytes for planting in wetland systems for wastewater treatment depends on a number of their characteristics, chiefly whether they are able to grow in waterlogged soils and withstand extremes of various chemical soil and water (or wastewater) parameters, such as:

- oxygen content (often anaerobic conditions with the presence of H_2S);
- pH;
- toxic wastewater constituents (such as phenols, surfactants, biocides, heavy metals, etc.);
- salinity.

Mainly helophytes can grow under these conditions and can cope with more or less anaerobic soil conditions by effecting the passage of air into the rhizosphere (see Sect. 2.4).

Over the past decades many species have been tested for their ability to grow under such extreme conditions. Practical experiments have found about a dozen different species to be suitable (Tab. 1).

Recently, rapidly growing trees have also been planted in constructed wetlands (GREENWAY and BOLTON, 1996).

2.2 Storage Potential of Nutrients

The main removal mechanisms of nutrients are microbial processes such as nitrification, denitrification, and physicochemical processes like the fixation of phosphate by iron and aluminum in the soil filter. Nevertheless, plants are able to tolerate high concentrations of nutrients and heavy metals, while some can also accumulate them in their tissues.

The phosphorus incorporated in the surface biomass of bulrushes amounts to about 6.7 g m^{-2} a^{-1} (SEIDEL, 1966). More recently a great variety (41) of different wetland plants have been investigated with respect to their mean tissue P concentrations, values being about 0.15–1.05% of dry weight (MCJANNET et al., 1995).

From a technical viewpoint, nitrogen incorporation into plant biomass is also of minor significance. Thus only about 5–10% nitrogen is removed when the plant biomass is harvested (THABLE, 1984). TANNER (1996) estimated in wetland plants surface tissue concentrations of nitrogen in the range of 15–32 mg N g^{-1} dry weight.

As a result of this relatively low significance of nutrient accumulation, the plant biomass generally tends not to be harvested in Europe.

Metal removal seems to be realized by different mechanisms such as the precipitation of iron (and other transition metals) as complex ferric hydroxides in oxic zones and probably as sulfides precipitated by H_2S evolved during the process of dissimilatory sulfate reduction.

At present intensive research is focused on plant selection by conventional methods and the creation of new transgenic plants with a high capacity to tolerate and accumulate heavy metals (MACEK et al., 1997). The aim is

Tab. 1. Plant Species Commonly Used for Wastewater Treatment in Constructed Wetlands

Species	Some Characteristics
Phragmites australis (common reed)	worldwide distribution, optimum pH: 2–8, salinity tolerance <45 g L^{-1} (REED et al., 1995)
Typha latifolia (broad-leaf cattail)	worldwide distribution, optimum pH: 4–10, salinity tolerance: <1 g L^{-1} (REED et al., 1995)
Typha angustifolia (narrow-leaf cattail)	worldwide distribution, optimum pH: 4–10, salinity tolerance: 15–30 g L^{-1} (REED et al., 1995)
Scirpus sp. (bulrushes)	worldwide distribution, optimum pH: 4–9 (REED et al., 1995)
Juncus sp. (rushes)	worldwide distribution, optimum pH: 5–7.5, salinity tolerance: 0–25 g L^{-1} depending on type (REED et al., 1995)
Iris pseudacorus (yellow flag)	
Carex sp. (sedges)	worldwide distribution, optimum pH: 5–7.5, salinity tolerance: <0.5 g L^{-1} (REED et al., 1995)
Glyceria maxima (reedgrass)	
Acorus calamus (sweet flag)	

to develop economical methods for "rhizofiltration" to remove heavy metals and radionuclides, etc., from aqueous streams.

2.3 Rhizodeposition

Plants cause the passage of carbon from their roots into the soil system (rhizodeposition). Such rhizodeposition products include exudates, mucilage, and dead cell material, etc., which affect the biological processes in the rhizosphere. The quantity of organic carbon exuded by the roots has been estimated to be 10–40% of the photosynthetic net production of agricultural plants investigated (HELAL and SAUERBECK, 1989). The nature of the exudates is quite different.

Generally speaking, compounds which are found in plant tissues are also exuded. For instance, the compounds which have been found to be root exudates include sugars and vitamins such as thiamine, riboflavin, pyridoxin, etc., organic acids like malate and citrate, amino acids, benzoic acids, phenolic compounds, and other compounds (MIERSCH et al., 1989). The range of compounds is specific in terms of both species and subspecies.

The root deposition products may play the following roles in the rhizosphere:

- Mobilization of nutrients:

Nutrient limitation can induce the excretion of organic acids and other compounds which improve the solubility of, e.g., iron and phosphate and thus improve plant nutrient acquisition (HOFFLAND et al., 1992).

- Allelopathic effects:

Certain species excrete special compounds into the rhizosphere which retard the growth of other plant species (MIERSCH et al., 1989). This effect has been intensively investigated in plants used for agriculture. Concerning helophytes, it is difficult to find any clear indications of allelopathy in the literature (GOPAL and GOEL, 1993).

- Rhizosphere effects:

Compounds such as sugars and amino acids, etc. may function as substrates for microorganisms. Microbial growth may be stimulated by the vitamins released. HELAL and SAUERBECK (1989) reported that the bulk (almost 80%) of the exuded organic compounds from *Zea mays* is converted by the rhizosphere microorganisms into carbon dioxide, causing the microbial biomass of the rhizosphere to increase. Furthermore, it has been shown that plant residues affect the microbial degradation of xenobiotics (HORSWELL et al., 1997).

Knowledge of the root exudation of helophytes is very limited. Indeed, there are not yet any results at all for adult plants. KAITZIS (1970) investigated rhizome extracts of *Scirpus lacustris* and found several benzene derivatives with hydroxyl-, methoxyl-, aldehyde, and carboxyl groups. These extracted compounds showed bactericidal effects, and so it was assumed that they are responsible for a "negative" rhizosphere effect. For example, it was reported that as a result of the root exudation of plants such as *Alisma plantago*, *Mentha aquatica*, *Juncus effusus*, *Scirpus lacustris*, and *Alnus glutinosa*, etc. the cell numbers of *Escherichia coli*, enterococci, and salmonellae were drastically reduced in model suspensions and a hospital wastewater (SEIDEL, 1973). Similar effects were observed by BURGER and WEISE (1984), who found an accelerated decrease of the *E. coli* cell number in a planted sand bed compared to the unplanted control when the helophytes *Glyceria maxima*, *Schoenoplectus lacustris*, *Alisma plantago-aquatica*, and *Mentha aquatica* were used. However, it is hard to draw any further conclusions because of the poor description of experimental methods in the cited references.

Later VINCENT et al. (1994) investigated the action of *Mentha aquatica*, *Phragmites australis*, and *Scirpus lacustris* in aseptic *in vitro* culture on *E. coli* suspensions. All three helophytes inhibited the growth of *E. coli* in the test system, bulrush being most effective. After removing the plants from the nutrient solutions, only the growth medium of *Mentha aquatica* displayed a feeble bactericidal effect. Thus, the authors concluded that the process depends on the direct presence of or even contact with the plant roots. Furthermore, during microcosm investigations RIVERA et al. (1995) observed the significantly better cell removal of *E. coli* in the rhizosphere of reed and cattail (35–91%) compared to the unplanted control (0–35%). In these experiments no differences were observed regarding cell removal efficiency depending on the plant species (cattail and reed).

Despite the findings described above, it is hard to understand that the root exudates both stimulate and inhibit bacterial growth. Additional mechanisms such as other indirect effects from the plant, adsorption, aggregation, filtration, the action of protozoa, etc. could play a role (KADLEC and KNIGHT, 1996).

2.4 Oxygen Release via the Roots

Higher plants vary considerably in their response to oxygen deprivation in their rhizosphere. Some are very sensitive and die, whereas others adapt well to anoxic conditions by means of various mechanisms (VARTAPETIAN and JACKSON, 1997). It is well known that helophytes are able to supply their roots with atmospheric oxygen. The supply of the air from the leaves to the roots is effected by a special tissue, namely the aerenchyma – long open channels for gas flow. This special tissue accounts for up to 60% of the total tissue volume of the plant (GROSSE and SCHRÖDER, 1986).

Fig. 2 provides a simplified illustration of the processes effecting the entry, transport, consumption, and emission of the gas molecules.

The different temperatures of young leaves inside T_1 and the surrounding atmosphere T_0 (warming-up caused by light absorption, thermal energy from water) cause a flow of gas (air) from the atmosphere into the young leaf (Knudsen effect). The higher temperature T_1, inside the leaf increases the average free path λ of the gas molecules to prevent the gas from flowing back out of the leaf and consequently the gas pressure P_1 inside the leaf increases. This excess pressure P_1 in the young leaves and in addition a subatmospheric pressure inside the oxygen consuming tissues (roots) caused by the higher solubility of the generated CO_2 compared to the consumed O_2 cause a convective gas flow inside the plant via the

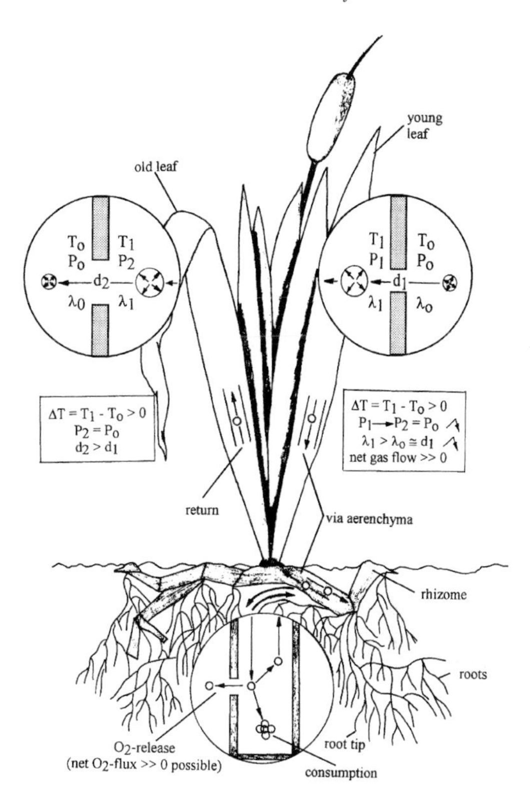

Fig. 2. Model of thermoosmotic gas transport by helophytes (GROSSE and SCHRÖDER, 1986, modified).

aerenchyma system. The resulting convective gas flow is much more effective than just the diffusive gas flow and also guarantees the O_2 supply over longer distances up to the roots tips of deeply rooted plants such as *Phragmites* species. Finally the gas flows out of older leaves owing to an equalization of pressure realized by the excess pressure P_1 in the young leaves and the atmospheric pressure P_0 outside the old leaves.

The flow of oxygen through the plant into the root serves more than just root respiration. It is also possible that young roots, especially the top section, release oxygen through the root wall into the rhizosphere (ARMSTRONG et al., 1991).

This oxygenation of the rhizosphere changes the redox conditions of anoxic zones and makes biological and abiotic oxidation processes possible. Depending on the helophyte species different amounts of oxygen released into the rhizosphere were practically measured in different technical systems. Several laboratory-scale investigations have been carried out, partly with the intention of calculating *in situ* values. Owing to methodological problems and the heterogeneity of the process parameters, calculating specific oxygen release rates quantitatively proves to be difficult (BRIX, 1993). Most studies have focused on evaluating radial oxygen losses in oxygen-depleted solutions using microelectrodes mainly via short-term experiments on entire root systems or parts of roots (ARMSTRONG et al., 1994; BRIX and SCHIERUP, 1990). Oxygen release rates of 10–160 ng min^{-1} cm^{-2} root surface have been measured.

Other experiments using root chambers and entire plants have been performed to investigate oxygen release via the roots into oxygen-depleted solutions. Long-term investigations have been executed under continuous and discontinuous flow conditions. The results vary from net uptake by the roots via slight oxygen release to relatively high rates of release (up to 1.52 μmol O_2 h^{-1} g^{-1} root dry weight) (BEDFORD et al., 1991; GRIES et al., 1990). The investigations and the contradictory results of the studies mainly highlight the dependence of oxygen release on a variety of factors affecting the plants: light intensity, photosynthetic intensity, phyllosphere temperature, the temperature of the rhizosphere, aerodynamics, the subterranean and above-ground state of plant development, the permeability of root walls, the level of oxygen inside the root tissues, and the redox status of the rhizosphere all affect the uptake of atmospheric oxygen and its release into the rhizosphere. SORELL and ARMSTRONG (1994), e.g., found that compared to the oxygenation of an oxygen-depleted solution, the oxygen release rates into a solution with a lower redox potential (in the range of actual rhizosphere conditions, $E_h \cong -200$ mV) are more than 80 times higher. WIEßNER and KUSCHK (unpublished data) corroborated the importance of the redox state for oxygen release into the rhizosphere. Under defined laboratory conditions, an oxygen release rate for a *Scirpus* species of 170 μmol O_2 h^{-1} g^{-1} root dry weight, e.g., was measured for $E_h \cong -240$ mV. The oxygen flux was found to be closely related to the redox potential, the plant species, and the light intensity. Initial investigations with young *Typha latifolia*, e.g., even show a much higher maximum release rate (WIEßNER and KUSCHK, unpublished data).

Because of the immense practical importance of this process for the treatment of wastes in natural or near-nature systems, the quantitative magnitude of *in situ* oxygen release has to be systematically investigated taking into account all the essential parameters and the different plants.

2.5 Metabolism of Xenobiotics

Pioneering work on the application of helophytes for wastewater treatment was originally performed by SEIDEL (1968). It was she who tested the removal of various phenolic compounds by helophytes (planted in hydroponic vessels) and the tolerance of the plants to these compounds. As the investigations were carried out in the batch mode under septic conditions, it may be assumed that removal was caused by the action of both microorganisms and plants. Furthermore, constant test concentrations were not established concerning the characterization of plant tolerance.

As a result of these test conditions and further experiments, FELGNER and MEISSNER (1968) came to the conclusion that the activities of plants for phenol removal are negligible under these conditions and that the main removal mechanism is bacterial degradation. KICKUTH (1970) found an uptake of about 0.08 mg phenol g^{-1} fresh weight d^{-1} in sterile infusion experiments into plant tissues (*Scirpus lacustris*). As the main metabolite picolinic acid was detected, it was concluded that the major part of phenol was metabolized via catechol and further *meta* ring fission. In *Lemna gibba* phenyl-β-D-glucopyranoside was identified as a metabolite of phenol conversion (BARBER et al., 1995).

Many investigations have been carried out with the water hyacinth *Eichhornia crassipes*. WOLVERTON and MCKOWN (1976) estimated an uptake of approximately 100 mg phenol per 2.75 g dry weight of this aquatic plant within 72 h. O'KEEFFE et al. (1987a) investigated the removal of several substituted phenols. The relative removal rate of the isomers was found to be para ≫ meta > ortho. Toxicity increased with increasing removal rate. The acutely toxic phenol concentration was determined to be about 400 mg L^{-1} (O'KEEFFE et al., 1987b), with catechol being detected as an early transformation product.

Much literature deals with the metabolism of agro- and other chemicals in non-aquatic plants. SANDERMANN (1992) generally divides the metabolism of xenobiotics by plants into three phases: transformation, conjugation, and compartmentation. The following enzymes were mainly discussed: cytochrome P450, glutathione transferase, carboxyl esterase, O- and N-glucosyl transferase, and O- and N-malonyl transferase. The final step of detoxification comprises three major pathways: (1) export into cell vacuoles; (2) export into the extracellular space; (3) deposition into lignin or other cell wall components.

Despite the above summarized ability of detoxification of xenobiotics, the plants directly play a minor role in degrading organic chemicals in soils and constructed wetlands in comparison to microbes.

2.6 Transpiration

The transpiration of plants is not only of ecological importance – it also influences their technological application for water and soil/sludge treatment. Evapotranspiration rates vary widely. The values for tropical rain forests amount to about 1.5–2.0 m a^{-1}, compared to about 0.4–0.5 m a^{-1} in Central Europe for cereal fields and forests and about 1.3–1.6 m a^{-1} for wetlands with helophytes (LARCHER, 1994). The rate is subject to a variety of parameters determining the micro climate of the ecosystem. These parameters have already been well summarized by KADLEC and KNIGHT (1996). In constructed wetlands for wastewater treatment, water loss by evapotranspiration amounts in Central Europe in the summer to 5–15 mm d^{-1}, i.e., about 20–50% of the total flow. This must, therefore, be taken into account in warmer seasons (SCHÜTTE and FEHR, 1992).

2.7 Influence on the Soil Matrix

One aspect of the complex rhizosphere processes is the interactions of roots/rhizomes with the soil matrix. Both chemical soil composition and physical parameters such as grain size distributions, interstitial pore spaces, effective grain sizes, degrees of irregularity, and the coefficient of permeability are important factors affecting the biotreatment system. These physical parameters cause particular hydraulic states in the soil and affect nutrient supply inside the rhizosphere. They also considerably influence the retention time of wastewater in specially constructed wetlands and the removal of contaminants. Root growth causes changes in the physical (hydraulic) quality of soils (KICKUTH, 1984; WISSING, 1995). On the one hand, roots and microbial biomass cause soil pores to be clogged; on the other, root growth and the microbial degradation of dead roots give rise to the formation of new secondary soil pores.

In constructed wetlands, it seems that the main parameter influencing soil hydraulics is the primary state of grain size distribution. Experiments and intensive investigations over a number of years in Germany in the field of hydraulics on constructed wetlands with various soil parameters have produced the best results regarding hydraulic conditions and also the removal of contaminants when a mixture of sand and gravel was used (BÖRNER, 1990; NETTER, 1990; WISSING, 1995). For constructed wetlands with vertical flow, a relatively small range of effective grain size d10 (a grain size which falls short of 10% (w/w) of the soil is the leading characteristic of soils) of 0.06–0.1 mm. For those with horizontal flow a higher range of >0.1 mm (because of higher susceptibility to obstruction) was evaluated (WISSING, 1995). The grain size distribution of >0.06 mm (up to 10 mm) with various distribution characteristics evidently realizes effective coefficients of permeability in the range of $>10^{-5}$ m s^{-1} (BAHLO and WACH, 1993; WISSING, 1995), a sufficient immobilization surface area for biofilm growth, positive impacts of root growth and on the hydraulic conductivity of the soil, and overall the generally good removal of contaminants.

Scientific research has been conducted for years on a whole series of constructed wetlands which were exclusively installed by using soil material with $K_f < 10^{-8}$ m s^{-1} to maximize the area for biofilm growth and the adsorption of wastewater chemicals. The systems suffered hydraulic problems mainly due to short-circuit flow on the wetland surface, and the predicted improvements in hydraulic conditions due to root growth over the course of time were not observed (Börnert, 1990; Netter, 1990). Despite small increases in the K_f-values (up to about 10^{-7} m s^{-1}), the hydraulic conditions did not change sufficiently. The main root growth of reed was only recorded in the top soil zone at a depth of 20–30 cm (Börnert, 1990); the main share of the bed volume of the soil filter was thus more or less ineffective.

3 The Role of Microorganisms

As already mentioned in Sect. 2.5, it is not the plants but rather the microorganisms which play the major role in the transformation/mineralization of nutrients/organic load. Depending on the oxygen input by helophytes or other technical measures as well as the other electron acceptors available, the wastewater ingredients are metabolized by different pathways as described in Chapters 2 and 22, this volume. In subsurface flow systems near the roots and on the rhizoplane, aerobic processes predominate. In the oxygen-free zones anaerobic processes such as denitrification, sulfate reduction, methanogenic digestion, etc. take place. Owing to the different redox conditions, this system represents a metabolically multipotent "technical ecosystem". The immobilization of the microorganisms in biofilms attached to soil particles and roots doubtless also plays an important role.

Little is currently known about the effects plant species have on the composition of the microbial cenosis in the rhizosphere.

Initial results from wheat, rye grass, bent grass, and clover grown in two different soil types (with the moisture content adjusted to 70% field capacity) are described by Grayston et al. (1998). Under these conditions, the community was subject to greater influence from the different plant species than the two soil types. It was suggested that plants differ in their exudation into the rhizosphere and by this exert a selective impact on the microbial cenosis. Under the conditions of wetlands, many other factors also influence the microbial community of the rhizosphere. Especially in the case of high hydraulic and contaminant loading rates, the amount of root exudates is negligible in comparison with the carbon load. Hence Calhoun and King (1998) report that the root-methanotrophic association of three common aquatic macrophytes under distinct laboratory conditions did not appear to be highly specific. Nevertheless, it is conceivable that in low loading areas of constructed wetlands the significance increases and root deposits (exudates and dead cell matter) play a role in the microbial cometabolic degradation of recalcitrant organic compounds.

4 Technological Aspects

The knowledge accumulated over time regarding the possibility of removing contaminants by the simple natural combination of water, plants, and soils has led to the deliberate application of such systems in nature and finally to the creation of artificial systems with different states of naturalness. Wissing (1995) divides the systems into the following three main groups:

- Aquaculture systems (Fig. 3a, b): Installations without active soil filters like ponds and ditches with intensive growth of submerged aquatic and/or free floating plants.
- Hydrobotanical systems (Fig. 3c): Installations with a few active soil filters with removal being mainly affected by aquatic plants, helophytes, and microorganisms – ponds and ditches with intensive growth of mainly helophytes.
- Soil systems (Fig. 3d, e).

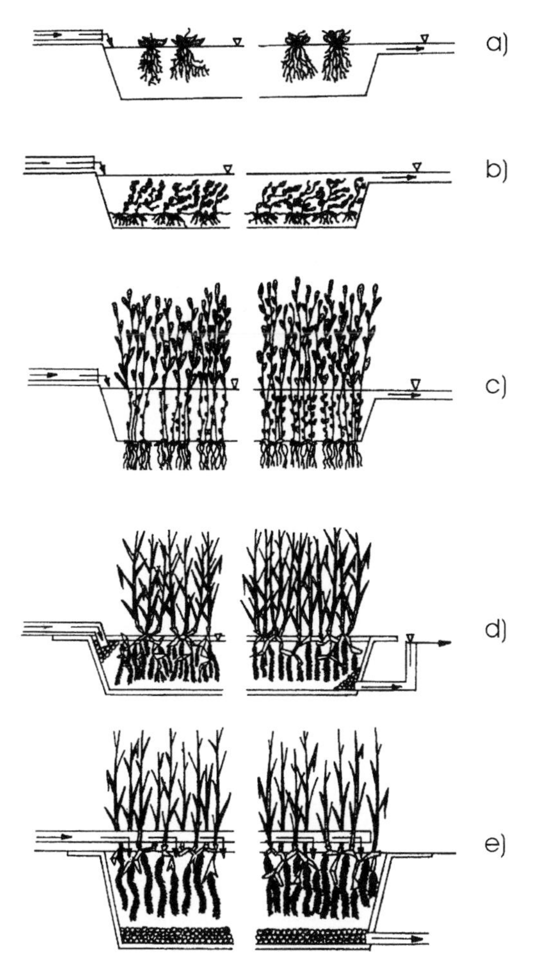

Fig. 3. Hydrobotanical systems for wastewater treatment.
a Pond with flotating plants; **b** pond with submerged plants; **c** pond with emerged plants; **d** planted soil filter with horizontal subsurface flow; **e** planted soil filter with vertical flow.

The basic types of soil-based constructed wetlands are horizontal surface flow systems (with the wastewater level above the soil surface), horizontal subsurface flow systems (with the wastewater level below the soil surface), and vertical subsurface flow systems with upstream or downstream characteristics and continuous or periodical loading. The soil systems are the constructed wetlands most widely used and there exist numerous different technological variants (design, peripheral equipment, etc.). They are mainly distinguishable by the grain sizes of the soil bed. In each case, the most successful system can be adapted to specific waste problems and local conditions. In addition, combination with other common methods of waste pre- or posttreatment increases the available possibilities.

Constructed wetlands represent the state of the art and are now well accepted by state authorities and professional associations in many countries.

5 References

Armstrong, W., Beckett, P. M., Justin, S. H. F. W., Lythe, S. (1991), Modeling and other aspects of roots aeration by diffusion, in: *Plant Life under Oxgen Deprivation* (Jackson, M. B., Davies, D. D., Lambers, H., Eds.), p. 267. The Hague, The Netherlands: SPB Academic Publishing.

Armstrong, W., Strange, M. E., Cringle, S., Beckett, P. M. (1994), Microelectrode and modeling study of oxygen distribution in roots, *Ann. Bot.* **74**, 287–299.

Bahlo, K., Wach, G. (1993), *Naturnahe Abwasserreinigung*. Staufen bei Freiburg: Ökobuch Verlag.

Barber, J. T., Sharma, H. A., Ensley, H. E., Polito, M. A., Thomas, D. A. (1995), Detoxification of phenol by the aquatic angiosperm, *Lemna gibba*, *Chemosphere* **31**, 3567–3574.

Bedford, B. L., Bouldin, D. R., Beliveau, B. D. (1991), Net oxygen and carbon dioxide balances in solutions bathing roots of wetland plants, *J. Ecol.* **79**, 943–959.

Börner, T. (1990), Einflußfaktoren für die Leistungsfähigkeit verschiedener Konstruktionsvarianten von Pflanzenkläranlagen, in: *Pflanzenkläranlagen – besser als ihr Ruf?* 21. Wassertechnisches Seminar am 18. 09. 1990 an der TH Darmstadt (Eigenverlag), pp. 239–256.

Börnert, W. (1990), Wissenschaftliche Begleituntersuchungen an der Pflanzenkläranlage Hofgeismar-Beberbeck, in: *Pflanzenkläranlagen – besser als ihr Ruf?* 21. Wassertechnisches Seminar am 18. 09. 1990 an der TH Darmstadt (Eigenverlag), pp. 69–90.

Brix, H. (1993), Macrophyte-mediated oxygen transfer in wetlands: Transport mechanisms and rates, in: *Constructed Wetlands for Water Quality Improvement* (Moshiri, G. A., Ed.), pp. 391–398. Boca Raton, FL: Lewis Publishers, CRC Press.

Brix, H. (1995), Use of constructed wetlands in water pollution control: Historical development, present status, and future perspectives, *Water Sci. Technol.* **30**, 209–223.

BRIX, H., SCHIERUP, H.-H. (1990), Soil oxygenation in constructed reed beds: The role of macrophyte and soil atmosphere interface oxygen transport, in: *Constructed Wetlands in Water Control* (COOPER, P. F., FINDLATER, B. C., Eds.), *Proc. Int. Conf. Use of Constructed Wetlands in Water Pollution Control*, 24–28 September 1990, Cambridge, UK, pp. 53–66. Oxford, New York: Pergamon Press.

BURGER, G., WEISE, G. (1984), Untersuchungen zum Einfluß limnischer Makrophyten auf die Absterbegeschwindigkeit von *Escherichia coli* im Wasser, *Acta Hydrochim. Hydrobiol.* **12**, 301–309.

CALHOUN, A., KING, G. M. (1998), Characterization of root-associated methanotrophs from three freshwater macrophytes: *Pontederia cordata*, *Sparganium eurycarpum*, and *Sagittaria latifolia*, *Appl. Environ. Microbiol.* **64**, 1099–1105.

FELGNER, G., MEISSNER, B. (1968), Untersuchungen zur Reinigung phenolhaltiger Abwässer durch die Flechtbinse (*Scirpus lacustris* und *S. tabernaemontani*) 2. Mitteilung, *Fortschritte der Wasserchemie* **9**, 199–214.

GOPAL, B., GOEL, U. (1993), Competition and allelopathy in aquatic plant communities, *Bot. Rev.* **59**, 155–210.

GRAYSTON, S. J., WANG, S., CAMPBELL, C. D., EDWARDS, A. C. (1998), Selective influence of plant species on microbial diversity in the rhizosphere, *Soil Biol. Biochem.* **30**, 369–378.

GREENWAY, M., BOLTON, K. G. E. (1996), From wastes to resources – turning over a new leaf: *Melaleuca* trees for wastewater treatment, *Environ. Res. Forum* **5–6**, 363–366.

GRIES, C., KAPPEN, L., LOSCH, R. (1990), Mechanism of flood tolerance in reed, *Phragmites australis* (Cav.) Trin. ex Steudel, *New Phytol.* **114**, 589–593.

GROSSE, W., SCHRÖDER, P. (1986), Pflanzenleben unter anaeroben Umweltbedingungen, die physikalischen Grundlagen und anatomischen Voraussetzungen, *Ber. Dtsch. Bot. Ges.* **99**, 367–381.

HELAL, H. M., SAUERBECK, D. (1989), Carbon turnover in the rhizosphere, *Z. Pflanzenernähr. Bodenk.* **152**, 211–216.

HOFFLAND, E., VAN DEN BOOGAARD, R., NELEMANS, J., FINDENEGG, G. (1992), Biosynthesis and root exudation of citric and malic acids in phosphate-starved rape plants, *New Phytol.* **122**, 675–680.

HORSWELL, J., HODGE, A., KILLHAM, K. (1997), Influence of plant carbon on the mineralisation of atrazine residues in soils, *Chemosphere* **34**, 1739–1751.

KADLEC, R. H. (1987), Northern natural wetland water treatment systems, in: *Aquatic Plants for Water Treatment and Resource Recovery* (REDDY, K. R., SMITH, W. H., Eds.), pp. 83–98. Orlando, FL: Magnolia Publishing.

KADLEC, R. H., KNIGHT, R. L. (1996), *Treatment wetlands*. Boca Raton, FL: Lewis Publishers.

KAITZIS, G. (1970), Mikrobiozide Verbindungen aus *Scirpus lacustris* L. (Ein Beitrag zur Ökochemie des Wurzelraumes), *Thesis*, Universität Göttingen, Germany.

KICKUTH, R. (1970), Ökochemische Leistungen höherer Pflanzen, *Naturwissenschaften* **57**, 55–61.

KICKUTH, R. (1984), Das Wurzelraumverfahren in der Praxis, *Landschaft und Stadt* **16**, 145–153.

LARCHER, W. (1994), *Ökophysiologie der Pflanzen*, pp. 216–217. Stuttgart: Verlag Eugen Ulmer.

MACEK, T., MACKOVA, M., KOTRBA, P., TRUKSA, M., CUNDY, A. S. et al. (1997), Attempts to prepare transgenic tobacco with higher capacity to accumulate heavy metals, containing yeast metallothionein combined with a polyhistidine, in: *Int. Symp. Environ. Biotechnol.* (VERACHTERT, H., VERSTRAETE, W., Eds.), pp. 263–266. Oostende, Belgium: Technologisch Instituut.

MCJANNET, C. L., KEDDY, P. A., PICK, F. R. (1995), Nitrogen and phosphorus tissue concentrations in 41 wetland plants: a comparison across habitats and functional groups, *Funct. Ecol.* **9**, 231–238.

MIERSCH, J., KRAUSS, G.-J., SCHLEE, D. (1989), Allelochemische Wechselbeziehungen zwischen Pflanzen – eine kritische Wertung, *Wiss. Z. Univ. Halle* **38**, 59–74.

MÜHLBERG, H. (1980), *Das große Buch der Wasserpflanzen.* Hanau/Main: Verlag Werner Dausien.

NETTER, R. (1990), Leistungsfähigkeit von bewachsenen Bodenfiltern am Beispiel der Anlagen Germerswang am See, in: *Pflanzenkläranlagen – besser als ihr Ruf?* 21. Wassertechnisches Seminar am 18. 09. 1990 an der TH Darmstadt (Eigenverlag), pp. 135–156.

O'KEEFFE, D. H., WIESE, T. E., BENJAMIN, M. R. (1987a), Effects of positional isomerism on the uptake of monosubstituted phenols by the water hyacinth, in: *Aquatic Plants for Water Treatment and Resource Recovery* (REDDY, K. R., SMITH, W. H., Eds.), pp. 505–512. Orlando, FL: Magnolia Publishing.

O'KEEFFE, D. H., WIESE, T. E., BRUMMET, S. R., MILLER, T. W. (1987b), Uptake and metabolism of phenolic compounds by the water hyacinth (*Eichhornia crassipes*), *Recent Adv. Phytochem.* **21**, 101–121.

REED, S. C., CRITES, R. W., MIDDLEBROOKS, E. J. (1995), Natural Systems for Waste Management and Treatment, pp. 178–181. New York: McGraw-Hill.

RIVERA, F., WARREN, A., RAMIREZ, E., DECAMP, O., BONILLA, P. et al. (1995), Removal of pathogens from wastewaters by the root zone method (RZM), *Water Sci. Technol.* **32**, 211–218.

SANDERMANN, H. (1992), Plant metabolism of xenobiotics, *Trends. Biochem. Sci.* **17**, 82–84.

SCHÜTTE, H., FEHR, G. (1992), Neue Erkenntnisse zum Bau und Betrieb von Pflanzenkläranlagen, *Korrespondenz Abwasser* **39**, 872–879.

SEIDEL, K. (1966), Reinigung von Gewässern durch höhere Pflanzen, *Naturwissenschaften* **53**, 289–297.

SEIDEL, K. (1968), Elimination von Schmutz- und Ballaststoffen aus belasteten Gewässern durch höhere Pflanzen, *Vitalstoffe – Zivilisationskrankheiten* **4**, 154–160.

SEIDEL, K. (1973), Leistungen höherer Wasserpflanzen unter heutigen extremen Umweltbedingungen, *Verh. Internat. Verein. Limnol.* **18**, 1395–1405.

SORRELL, B. K., ARMSTRONG, W. (1994), On the difficulties of measuring oxygen release by root systems of wetland plants, *J. Ecol.* **82**, 177–183.

TANNER, C. C. (1996), Plants for constructed wetland treatment systems – A comparison of the growth and nutrient uptake of eight emergent species, *Ecol. Eng.* **7**, 59–83.

THABLE, T. S. (1984), Einbau und Abbau von Stickstoffverbindungen aus Abwasser in der Wurzelraumanlage Othfresen, *Thesis*, Gesamthochschule Kassel, Universität Hessen, Germany.

VARTAPETIAN, B. B., JACKSON, M. B. (1997), Plant adaptations to anaerobic stress, *Ann. Bot.* **79**, (Suppl. A), 3–20.

VINCENT, G., DALLAIRE, S., LAUZER, D. (1994), Antimicrobial properties of roots exudate of three macrophtes: *Mentha aquatica* L., *Phragmites australis* (Cav.) Trin. and *Scirpus lacustris* L., in: *Proc. 4th Int. Conf. Wetland Systems for Water Pollution Control*, 6–10 November 1994, Guangzhou, China. pp. 290-296. ICWS '94 Secretariat: Guangzhou, China.

WISSING, F. (1995), *Wasserreinigung mit Pflanzen.* Stuttgart: Verlag Eugen Ulmer.

WOLVERTON, B. C., MCKOWN, M. M. (1976), Water hyacinths for removal of phenols from polluted waters, *Aquat. Bot.* **2**, 191–202.

13 Activated Sludge Process

ROLF KAYSER

Braunschweig, Germany

1 Process Description and Historical Development

1.1 Single-Stage Process

About 1910, investigations started to treat wastewater simply by aeration (e.g., FOWLER and MUMFORD in Manchester; VON DER EMDE, 1964). ARDERN and LOCKETT (1914) in Manchester conducted similar experiments but after a certain aeration period they stopped aeration, let the flocs settle, decanted the supernatant, filled in wastewater, and repeated the cycle again and again. After build-up of a certain amount of biomass they obtained a fully nitrified effluent at an aeration period of 6 h. The settled sludge they called "activated sludge". The first technical-scale plant was a fill and draw activated sludge plant, which today is called SBR process. Since at that time the process had to be operated manually, they had a lot of operational problems, and, therefore, decided to build the next plant in what is called today the conventional mode (Fig. 1).

An activated sludge plant is characterized by four elements:

(1) the aeration tank equipped with an appropriate aeration installation in which the biomass is mixed with wastewater and supplied with oxygen,
(2) the final clarifier in which the biomass is removed from the treated wastewater by settling (or other means),
(3) the continuous collection of return sludge and pumping back into the aeration tank,
(4) the excess sludge withdrawal to maintain the appropriate concentration of mixed liquor.

If one of the elements fails, the whole process fails.

Primary sedimentation is not required for activated sludge plants. For economical reasons, however, primary tanks are usually operated.

In the early times the activated sludge tanks were aerated with fine bubble diffused air. Due to clogging problems of the ceramic diffusers surface aerators were developed. BOLTON invented in 1921 the vertical shaft cone surface aerator at the treatment plant of Bury. Beginning about 1965 cone surface aerators were installed in numerous plants in Germany, the largest one being the Emscher river treatment plant. Another large one is the second stage of the main treatment plant of the city of Hamburg. In the Netherlands Kessener constructed in 1925 the horizontal axis brush aerator, which was installed in spiral flow aeration tanks as they were used for diffused air aeration (VON DER EMDE, 1964). Brush aerators in the 1960s were frequently used in Germany in high-rate activated sludge plants. PASVEER (1958) installed the brush aerator in the oxidation ditch for the aeration and the circulation of mixed liquor. Starting about 1965 the horizontal axis mammoth rotor, diameter 1.00 m, was installed in closed loop aeration tanks. The carousel tank with cone surface aerators represents another closed loop aeration tank (ZEPER and DE MAN, 1970). Both systems are still applied at small as well as at large plants.

After membrane diffusers were developed around 1970, diffused air aeration became

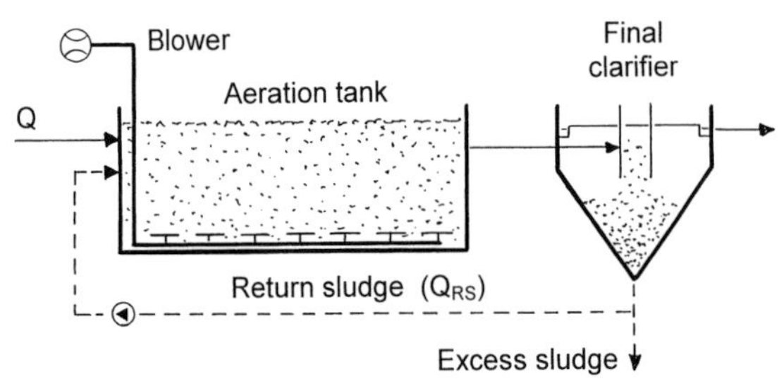

Fig. 1. Flow diagram of an activated sludge plant.

popular again. Usually it is anticipated that by aeration sufficient turbulence is created to prevent mixed liquor from settling. In order to minimize the power for aeration at plants with a low oxygen uptake rate Imhoff installed a horizontal axis paddle in a tank with diffused air aeration as early as 1924. After PASVEER and SWEERIS (1962) had postulated that oxygen transfer was considerably increased if the air bubbles rise in a horizontal flow DANJES developed a system at which the diffusers were fixed on a moving bridge (SCHERB, 1965). This system today is marketed as "Schreiber Countercurrent Aeration System". As an answer to the Schreiber system around 1970 the firm Menzel installed slow-speed propellers in circular tanks with fine bubble diffused aeration. Today the combination of membrane diffusers in circular or closed loop tanks with propellers creating a circulating flow is favored for intermittent aeration to remove nitrogen.

Industrial wastewater may contain substances which precipitate on ceramic diffusers. They may also contain grease or substances which are able to destroy membrane material. Surface aeration or coarse bubble aeration with static mixers is, therefore, a good choice. For deeper tanks the rotating turbine aerator, which is a combination of coarse bubble aeration and a mixer to break up the large bubbles into smaller ones, was successfully operated. In a jet or ejector diffuser mixed liquor is pumped through a venturi nozzle into which air is introduced (e.g., JÜBERMANN and KRAUSE, 1968). Very fine bubbles are released by the shear stress.

The early aeration tanks were rectangular with evenly distributed diffusers along the tank. Wastewater and return sludge entered the tank at one end and mixed liquor left the tank at the other end. Due to the high oxygen uptake rate the dissolved oxygen concentration in the inlet zone was almost zero. To overcome this problem the diffusers at the inlet zone were arranged with a higher density than at the outlet zone. This was called "step aeration". "Step feed" where the return sludge was introduced at one end and wastewater was distributed along the tank was a solution for tanks with evenly distributed diffusers (Fig. 2).

Since organics are removed after a short contact period with activated sludge, in the USA the "biosorption" or "contact stabilization" process was implemented in some full-scale plants (ULLRICH and SMITH, 1951). In this process return sludge is aerated for 2–4 h to oxidize adsorbed organics before it enters the aeration tank in which the detention period may be in the range of 0.5–2 h (Fig. 3).

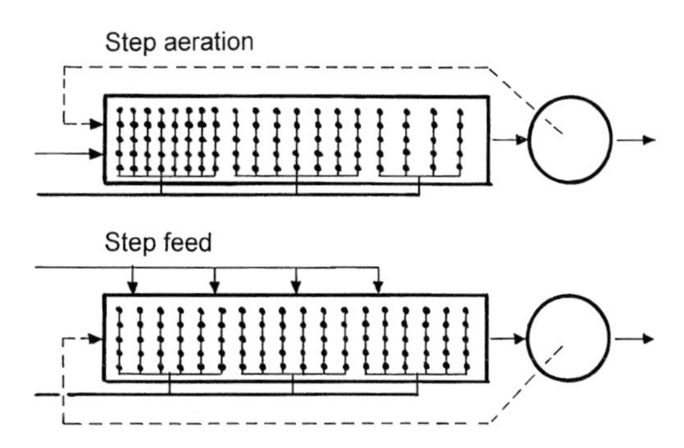

Fig. 2. Step aeration and step-feed activated sludge plants.

1.2 Two-Stage Process

IMHOFF (1955) realized the first two-stage activated sludge plant in Germany. It consists of two independent activated sludge plants in series. The first stage is characterized by a high sludge loading rate (*F*/*M*) and consequently the second stage has a rather low *F*/*M*. The excess sludge of the second stage is usually transferred to the first stage (Fig. 4). The AB process, invented (and patented) by BÖHNKE (1978), is a two-stage activated sludge plant without primary sedimentation. In this process the excess sludge of the second stage shall not be transferred to the first stage.

The two-stage process has several advantages. Harmful substances may be removed in the

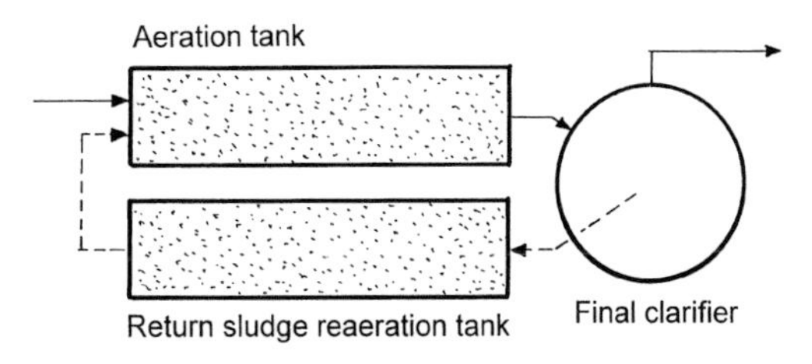

Fig. 3. Contact stabilization process.

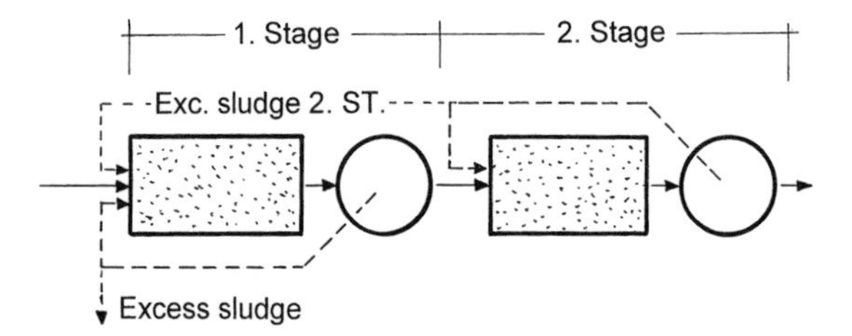

Fig. 4. Flow diagram of a two-stage activated sludge plant.

first stage, which is important for the treatment of industrial wastewater, and in the low loaded second stage, due to the high sludge age, microorganisms may be kept which are able to remove heavily biodegradable organics or to oxidize ammonia. Furthermore, bulking sludge is only rarely observed in either stage. The disadvantages are that about twice as many clarifiers are needed as for the one-stage process and that nitrogen removal as well as enhanced biological phosphate removal may be inhibited due to missing organics which are removed in the first stage.

1.3 Single Sludge Carbon, Nitrogen, and Phosphorus Removal

In the early 1960s three different methods for nitrogen removal were demonstrated (Fig. 5):

- post-denitrification (BRINGMANN, 1961; WUHRMANN, 1964),
- pre-anoxic zone denitrification (LUDZACK and ETTINGER, 1962),
- simultaneous denitrification (PASVEER, 1964).

Post-denitrification was not successful without the addition of external organic carbon. BRINGMANN (1961) tried a bypass of wastewater to enhance denitrification but then some ammonia remained in the final effluent.

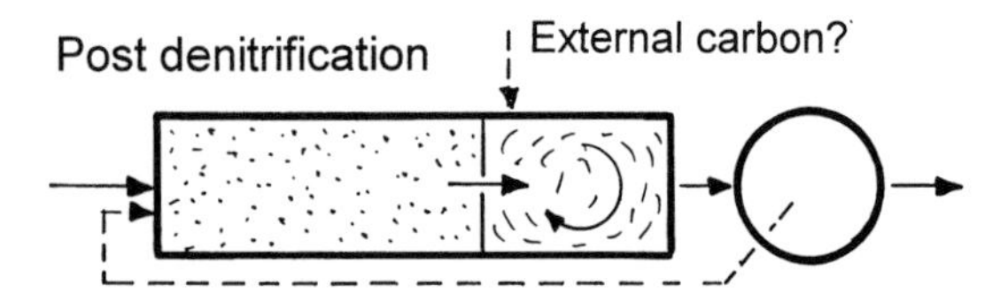

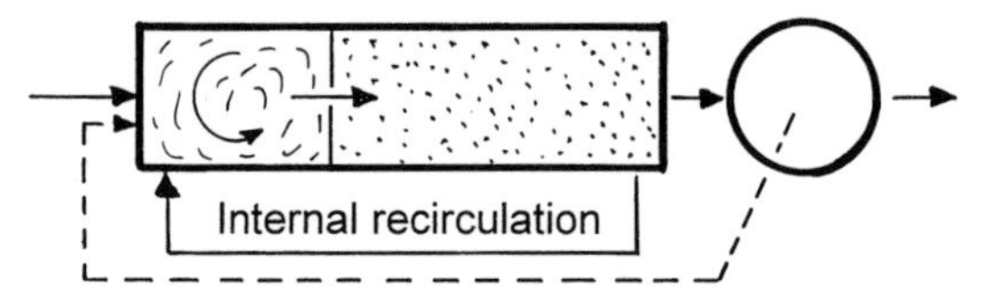

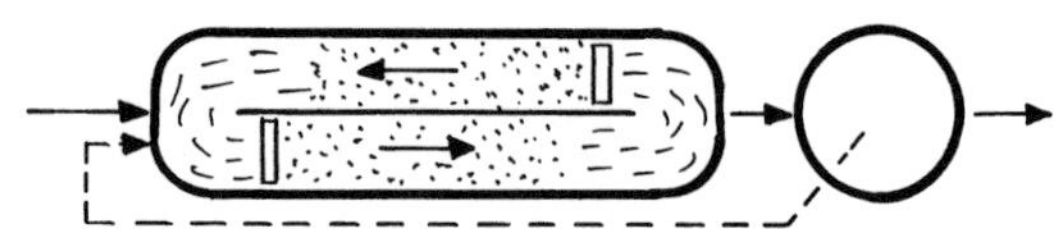

Fig. 5. Processes for nitrogen removal.

The first technical-scale pre-anoxic zone denitrification process in Germany was implemented at the research wastewater treatment plant of the University of Stuttgart (KIENZLE, 1971). After BARNARD's (1973, 1974) successful experiments the first full-scale plant with pre-anoxic zone denitrification and enhanced biological phosphate removal was built in 1974 at Klerksdorp, South Africa. In Germany the first full-scale plant with pre-anoxic zone denitrification was erected at Biet (KRAUTH and STAAB, 1988).

In 1969 the wastewater treatment plant of Vienna Blumental, designed for 300,000 p.e. was put into operation (VON DER EMDE, 1971). Two closed loop aeration tanks each 6,000 m^3 operated in series each equipped with 6 twin mammoth rotors were operated in the simultaneous denitrification mode (MATSCHE, 1972). At that time this was the largest single-stage activated sludge plant in the world for nitrogen removal without the addition of external carbon.

These and other process developments for nitrogen removal will be discussed in detail in Sect. 3.3.

Based on the work of THOMAS (1962) phosphorus is easily removed by simultaneous precipitation. BARNARD (1973) put an anaerobic tank ahead of the biological reactor for enhanced biological phosphate removal. Today most newer plants are built with means for enhanced biological phosphate removal and/or installations for simultaneous precipitation. Phosphate removal in detail is discussed in Chapter 14, this volume.

1.4 Special Developments

1.4.1 Pure Oxygen Activated Sludge Process

Due to poor aeration systems in many activated sludge plants in the USA, the plants were operated with a low mixed liquor suspended solids concentration of about 1–2 kg m^{-3} MLSS and long aeration periods of 5–8 h just for carbon removal. At the pure oxygen activated sludge process, which came up around 1970, the oxygen transfer is not limited. *MLSS* can be raised to, e.g., 5 kg m^{-3}, and the detention period could be shortened to, e.g., 2 h in order to achieve over 90% BOD removal (KULPBERGER and MATSCH, 1977). The covered aeration tanks, necessary to recycle the oxygen gas, are advantageous since the plant looks clean and no emissions are released. The disadvantage is the high carbon dioxide concentration in the gas, which may cause corrosion of the concrete and, furthermore, nitrification may be inhibited by a too low pH. In Germany only a few pure oxygen activated sludge plants were built in the 1970s. Most of them in the meantime are converted to nitrogen removal plants with diffused air aeration.

Today in some conventionally aerated activated sludge plants pure oxygen is used at periods of peak oxygen demand. The tanks are not covered and oxygen is not recycled. The oxygen gas is either introduced by special hoses with fine apertures which release very small bubbles or by jet type aerators.

At the new plant of Bremen-Farge pure oxygen in addition to diffused air is applied not only at peak loads but also in cases of power failure. This was calculated to be more economical than the installation of one more turbine blower and a much bigger emergency power station.

1.4.2 Attached Growth Material in Activated Sludge Aeration Tanks

In order to increase the biomass in an aeration tank or to enhance nitrification, elements on which biomass can grow were immersed in the mixed liquor. At first corrugated plastic sheet material as used in trickling filters, e.g., flocor, was installed (LANG, 1981; Fig. 6). It was believed that nitrifiers would grow on the material. SCHLEGEL (1986), however, demonstrated that there are almost no nitrifiers attached but a high number of protozoa. He postulated that at such plants nitrification was enhanced due to a lower sludge production caused by the high number of protozoa and the possibility to operate with higher *MLSS* because of the improved settleability (low sludge volume index) of the mixed liquor.

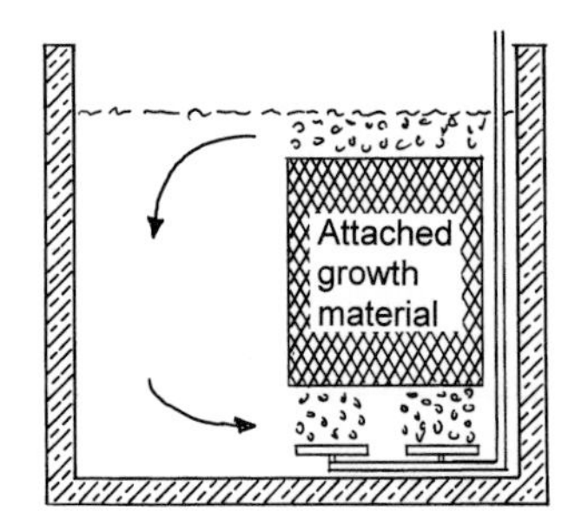

Fig. 6. Cross section of an aeration tank with a module for attached growth.

Another material installed in aeration tanks is ring-lace, vertically fixed cords with loops (LESSEL, 1991). As a nuisance of that process a massive growth of worms attracted by the attached protozoa was observed (HEINZ et al. 1996).

At the LINPOR® process porous plastic foam cubes occupy 10–30% of the aeration tank volume. Due to the attached biomass the total amount of mixed liquor solids can be increased and since the solids are kept in the aeration tanks, the clarifiers may not become overloaded (MORPER, 1994). In Japan the ANDP process was developed whereby the aerobic compartment is filled up to 40% of the volume with pellets (short polypropylene hoses of 4 mm diameter and 5 mm length) which are assumed to contain immobilized nitrifiers (TAKAHASHI and SUZUKI, 1995).

1.4.3 High-Rate Reactors

High-rate reactors are characterized by a high volumetric removal rate, e.g., 10–60 kg COD per m^3 and day and consequently by a high volumetric oxygen transfer rate. Preferably loop reactors are used. The ICI-Deep Shaft is such a reactor as well as the HCR reactor (VOGELPOHL, 1996). In the "Hubstrahl" reactor a number of plates with holes oscillates with a high frequency up and down to enhance oxygen transfer and to cause a high turbulence (BRAUER, 1996).

High-rate processes are preferably used for high-strength wastewater with a very high fraction of soluble and readily biodegradable organics. Because of the high loading rates applied for economical reasons the overall removal is restricted to 60–80% COD.

1.4.4 Membrane Separation of Mixed Liquor

After the membrane separation technique was successfully applied to separate the sample flow for continuous monitoring of, e.g., nitrate from mixed liquor, KAYSER and ERMEL (1984) and KRAUTH and STAAB (1988) installed tubular membrane units instead of a final clarifier at a pressurized aeration tank (Fig. 7). The process is marketed as "Biomembrat process". Due to improvements of the membrane technique recently numerous experiments using microfilter units are under way. The advantages are that it is possible to operate such systems with rather high *MLSS* (up to 15 kg m^{-3}) and that the permeate is almost free of suspended matter. The disadvantage is that the flux rate is only of the order of 20 L m^{-2} h^{-1}. The process, therefore, is still costly and may only be applied in case of special effluent requirements (ROSENWINKEL and WAGNER, 1998).

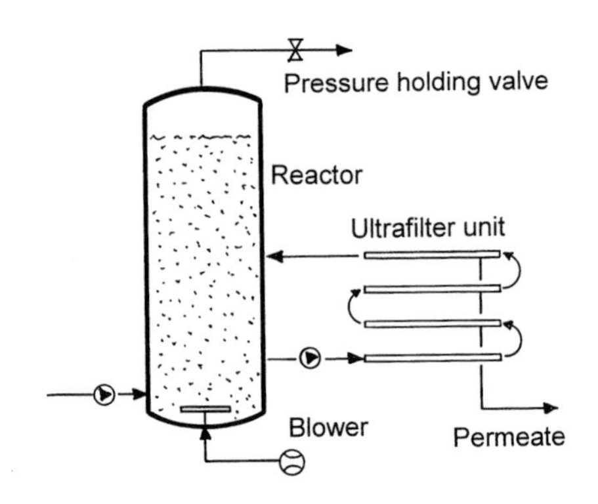

Fig. 7. Flow diagram of the Biomembrat process.

2 Technological and Microbiological Aspects

2.1 Wastewater Characteristics

Wastewater to be treated may contain organic carbon predominantly as a single soluble substance, e.g., an alcohol, as a mixture of soluble substances, or as a mixture of solids and numerous soluble organic substances.

Many industrial wastewaters are mixtures of soluble organic substances but in some cases they may contain mainly one organic substance. Municipal wastewater and wastewater from most food processing industries always is a mixture of soluble and particulate organic matter.

The microbial degradation of organic carbon requires certain amounts of nitrogen, phosphorus, calcium, sodium, magnesium, iron, and other essential trace elements to build up biomass. At industrial wastewater treatment plants missing elements have to be added. Since domestic wastewater contains all necessary elements in excess, it is favorable to treat special industrial wastewater together with municipal wastewater.

The concentration of organic matter in wastewater is measured as biochemical oxygen demand (BOD_5, or BOD_{20}, incubation period of 5 or 20 d, nitrification inhibited), chemical oxygen demand (COD), or total organic carbon (TOC). Raw municipal wastewater may be characterized by the following ratios:

COD/TOC:	3.2–3.5
COD/BOD_5:	1.7–2.0
BOD_5/TOC:	1.7–2.0

Since BOD reflects only biodegradable matter and COD and TOC also comprise nonbiodegradable components the ratios of well treated effluent (BOD_5 = 10–20 mg L^{-1}) are different (BEGERT, 1985):

COD/TOC:	3.0–3.5
COD/BOD_5:	3.0–6.0
BOD_5/TOC:	0.5–1.0

BOD_5 (or in Scandinavia BOD_7) so far was the most common parameter to characterize wastewater. The advantage of BOD is that it comprises only biodegradable organics, the disadvantages are that by BOD_5 only a fraction of the biodegradable organics is determined and that the measurement of BOD_{20} is economically not feasible. The TOC from the microbiological point of view is favorable but there are still analytical problems with solids. After the development of simplified methods to measure COD this parameter is mainly analyzed today. It allows mass balances and, therefore, is widely used for modeling of biological processes (HENZE et al., 1987, 1995).

2.2 Removal of Organic Carbon

Many attempts have been undertaken to describe the removal of organics by Michaelis–Menten or by Monod kinetics. For many single substances the k_m or k_s value is rather small, therefore, in batch experiments zero-order removal of single substances is observed. Mixed substrates like municipal wastewater, however, indicate first-order removal reactions in batch experiments. WUHRMANN and VON BEUST (1958) explained these phenomena as a series of different zero-order reactions. TISCHLER (1968) later demonstrated the removal of glucose, aniline, and phenol in batch tests. In tests with the substances separately the removal of the substances as well as of the COD follows different zero-order reactions (Fig. 8, left). It is important to notice that although the substances are completely removed a COD of about 30 mg L^{-1} remains. In a test where the three substances were mixed with adapted mixed liquor a quasi first-order reaction was observed (Fig. 8, right). Again a nonbiodegradable COD of about 60 mg L^{-1} remains. The remaining COD can be visualized as nonbiodegradable compounds produced by the bacteria; hence at least a fraction of the nonbiodegradable COD of any wastewater does not originate from the wastewater itself.

The rate constant of the first-order removal reaction of organics of municipal and many other types of wastewater depends on the wastewater characteristics as well as the load-

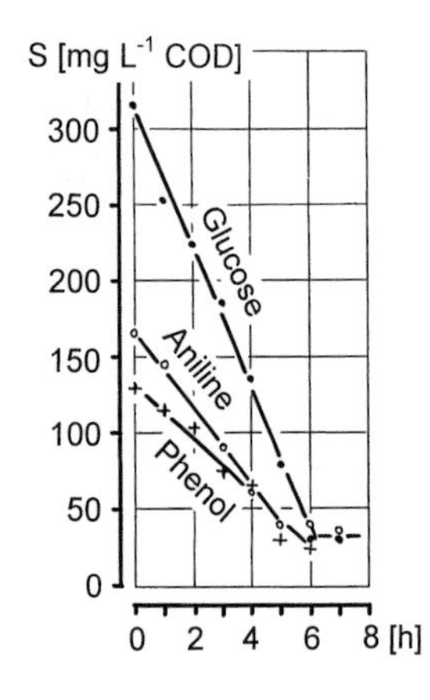

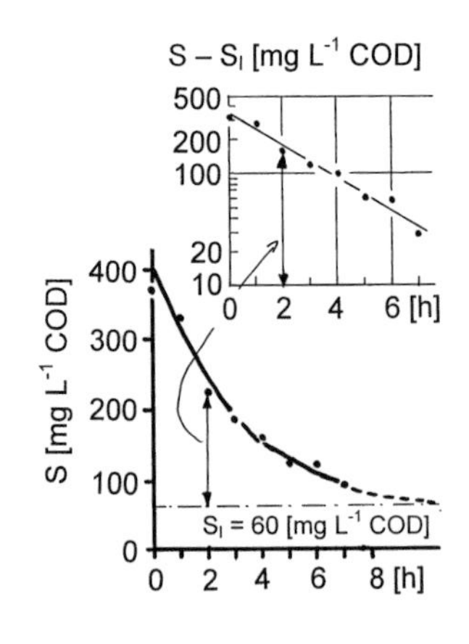

Fig. 8. Removal of single substances (TISCHLER, 1968). Left: substances separately, right: substances mixed.

ing rate of the mixed liquor. This is due to the fact that even at the same *MLVSS* (mixed liquor volatile suspended solids concentration) the number and type of microorganisms may be different for various types of wastewater as well as at different loading conditions.

Since batch experiments indicate that the removal of organics depends on the detention period and the mixed liquor volatile suspended solids (*MLVSS*), the food–microorganism ratio (*F–M* ratio) is widely used for design. It may be calculated as:

$$F/M = \frac{Q_d \cdot C}{V \cdot MLVSS \cdot 1{,}000} = \frac{C}{t^* \cdot MLVSS \cdot 1{,}000} \quad [d^{-1}] \tag{1}$$

Q_d	$m^3\ d^{-1}$	daily wastewater flow
C	$mg\ L^{-1}$	substrate concentration (BOD_5, COD, or TOC)
V	m^3	reactor (tank) volume
$MLVSS$	$kg\ m^{-3}$	mixed liquor volatile suspended solids
$MLSS$	$kg\ m^{-3}$	mixed liquor suspended solids
t^*	d	detention period ($t^* = V/Q_d$)
M_{exc}	$kg\ d^{-1}$	daily mass of excess sludge

In Germany *F–M* is expressed as BOD_5 sludge loading rate (B_{TS}) based on MLSS. The degree of the removal of organics highly depends on the *F–M* ratio. Unfortunately, similar dependencies are only observed when treating the same type of wastewater.

Today the sludge age or solids retention time (*SRT*) is applied for design more frequently.

$$SRT = \frac{V \cdot MLSS}{M_{exc}} \quad [d] \tag{2}$$

Combining Eqs. 1 and 2 leads to:

$$F/M = \frac{Q_d \cdot C}{SRT \cdot M_{exc} \cdot 1{,}000} \quad [d^{-1}] \tag{3}$$

Since the mass of excess sludge from organic carbon removal ($M_{exc,C}$) is a function of the organic load ($Q \cdot C$), the loading rate *F–M* increases with decreasing *SRT*. High *F–M* ratios require an adequate volumetric oxygen transfer rate which in the past could be visualized as the bottle neck of high-rate processes. By new reactor developments this can be overcome (see Sect. 1.4.3). The specific energy ($kW\ m^{-3}$) required for oxygen and mass transfer, however, may be considerable.

In order to estimate the excess sludge production and the oxygen consumption for organic carbon removal and to finally calculate the reactor volume, the COD balance may be used. The following calculations are partly based on the "Activated Sludge Model No. 1" (HENZE et al., 1987) and considerations of GUJER and KAYSER (1998). Some factors are restricted to almost complete biodegradation of organics which with municipal wastewater may be obtained for sludge ages of more then 5 d.

The wastewater entering the biological reactor should be analyzed for:

C_{COD}	$mg\,L^{-1}$	total COD
S	$mg\,L^{-1}$	COD of the filtrate (membrane filtration 0.45 μm pore size)
TSS	$mg\,L^{-1}$	total suspended solids
VSS	$mg\,L^{-1}$	volatile suspended solids
ISS	$mg\,L^{-1}$	inorganic suspended solids ($ISS = TSS - VSS$)

Subscripts 0 and e denote influent and effluent, I denotes inert organic matter (Fig. 9).

Since the suspended solids of the final effluent are regarded as a fraction of the excess sludge for COD balancing only S_e is of interest.

The total COD of the influent can be divided into the soluble COD (S_0) and particulate COD (X_0):

$$C_{COD,0} = S_0 + X_0 \quad (4)$$

The biodegradable COD ($C_{COD,bio}$) can be expressed as:

$$C_{COD,bio} = C_{COD,0} - S_{I,0} - X_{I,0} \quad (5)$$

The inert fraction of the particulate COD ($X_{I,0}$) may be estimated as 20% to 35% of the total particulate COD (X_0). The soluble inert influent COD ($S_{I,0}$) can be experimentally determined assuming $S_{I,e} = S_{I,0}$. Generally for municipal wastewater it is in the range of 5% to 10% of the total influent COD ($S_{I,0} = 0.05$ to $0.1\ C_{COD,0}$)

As the result of biological treatment the effluent COD (S_e) and the excess sludge (X_{exc}) remain; the gap in between represents the oxygen consumed (OU_C) for biological degradation of organics (Fig. 9).

$$C_{COD,0} = S_e + OU_C + X_{exc} \quad (6)$$

The excess sludge COD (X_{exc}) consists of the biomass (X_{BM}), the remaining inert particulate matter from endogenous decay ($X_{BM,I}$), and the inert influent particulate COD ($X_{I,0}$):

$$X_{exc} = X_{BM} + X_{BM,I} + X_{I,0} \quad (7)$$

The biomass produced is obtained as:

$$X_{BM} = C_{COD,bio} \cdot Y \cdot \frac{1}{1 + b \cdot SRT \cdot F} \quad (8)$$

The remaining inert COD of biomass decay ($X_{BM,I}$) can be assumed as 20% of the biomass lost by decay.

$$X_{BM,I} = 0.2 \cdot X_{BM} \cdot b \cdot SRT \cdot F \quad (9)$$

In which:

OU_C	$mg\,L^{-1}\,O_2$	oxygen uptake for organic carbon removal
X_{BM}	$mg\,L^{-1}$	biomass produced, measured as COD
$X_{BM,I}$	$mg\,L^{-1}$	inert biomass, measured as COD
Y	$mg\,mg^{-1}$	yield, may be assumed as $Y = 0.67\ mg\,mg^{-1}$
b	d^{-1}	decay rate, may be assumed as $b = 0.17\ d^{-1}$ at 15 °C
F	–	temperature factor $F = 1.072^{(T-15)}$

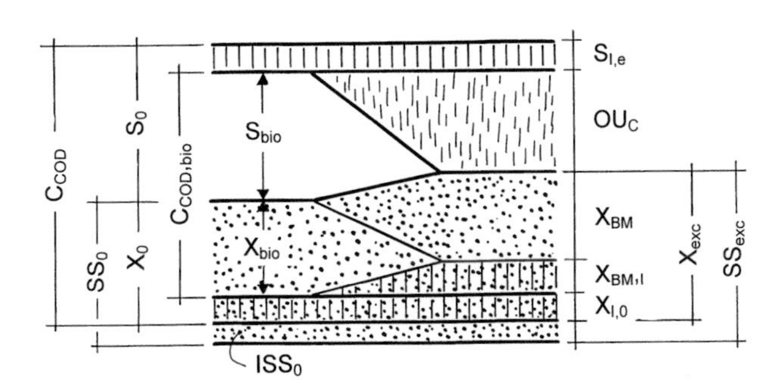

Fig. 9. Change of COD at biological treatment.

It has to be noted that X_{exc}, X_{BM}, and $X_{BM,I}$ are in mg L^{-1} COD and OU_C in mg L^{-1} oxygen (O_2) based on the daily wastewater flow Q_d [m^3 d^{-1}].

Assuming that mixed liquor is 80% organic and 1 mg organic particulate matter is equivalent to 1.45 mg COD, the excess sludge (SS_{exc}) as mg L^{-1} suspended solids is obtained ($1.45 \cdot 0.8 = 1.16$):

$$SS_{exc} = \frac{X_{exc}}{1.16} + ISS_0 \tag{10}$$

The oxygen consumption is derived from Eq. 6:

$$OU_C = C_{COD,0} - S_e - X_{exc} \tag{11}$$

The daily mass of excess sludge solids $M_{exc,C}$ [kg d^{-1}] and the daily mass of oxygen $M_{O,C}$ [kg d^{-1}] required are:

$$M_{exc,C} = Q_d \cdot SS_{exc}/1{,}000 \tag{12}$$

$$M_{O,C} = Q_d \cdot OU_C/1{,}000 \tag{13}$$

If phosphate is removed the additional excess sludge ($M_{exc,P}$) has to be taken into account. At plants with nitrogen removal the biomass of autotrophs produced ($M_{exc,N}$) may also be considered as additional excess sludge. The mass of oxygen required for nitrification and the oxygen "gained" from denitrification in any case must be determined separately.

If SRT is selected according to the desired effluent quality, the reactor volume V [m^3] can be calculated:

$$V = \frac{SRT \cdot M_{exc}}{MLSS} \tag{14}$$

In which: $M_{exc} = M_{exc,C} + M_{exc,P} + M_{exc,N}$.

2.3 Nitrification

The main step of nitrogen removal is nitrification, which is assumed to be performed in two steps. At first ammonia is converted to nitrite by *Nitrosomonas* and then nitrite is converted to nitrate by *Nitrobacter*. The overall reactions are written as:

$$NH_4^+ + 1.5\,O_2 \rightarrow NO_2^- + H_2O + 2H^+$$
$$NO_2^- + 0.5\,O_2 \rightarrow NO_3^-$$
$$NH_4^+ + 2.0\,O_2 \rightarrow NO_3^- + H_2O + 2H^+$$

The stoichiometric oxygen demand is 3.43 mg O_2 per mg of ammonia nitrogen (S_{NH_4}) converted to nitrite nitrogen (S_{NO_2}) and 1.14 mg O_2 per mg S_{NO_2} converted to nitrate nitrogen (S_{NO_3}). Thus an overall demand of 4.57 mg O_2 per mg S_{NO_3} formed is obtained. Since there is some build-up of autotrophic biomass, the total actual oxygen consumption is reported to be in the range of 4.2–4.3 mg mg^{-1} (STENSEL and BARNARD, 1992). For design calculations an overall oxygen demand of 4.3 mg O_2 per mg S_{NO_3} formed is widely used.

The 2 moles of hydrogen released per mole of ammonia converted to nitrite destroy 2 moles of alkalinity that equals 0.14 mmol mg^{-1} S_{NO_2}. If all alkalinity is destroyed the pH may drop and the flocs of the mixed liquor may disintegrate (see Sect. 2.5.2).

Compared to heterotrophic bacteria the growth rate of nitrifiers is much lower. Today the net maximum growth rates μ^*_{max} [d^{-1}] determined by KNOWLES et al. (1965) are widely adopted:

Nitrosomonas: $$\mu^*_{max} = 0.47 \cdot 1.10^{(T-15)} \tag{15}$$

Nitrobacter: $$\mu^*_{max} = 0.78 \cdot 1.06^{(T-15)} \tag{16}$$

Since at reactor temperatures below $T = 30\,°C$ the growth rate μ^*_{max} of *Nitrosomonas* is lower than μ^*_{max} of *Nitrobacter* for design calculations, Eq. 15 is generally applied. To avoid a wash out of nitrifiers the inverse net maximum growth rate must be larger than the aerobic sludge age (SRT_{aer}). In order to obtain a low ammonia effluent concentration a series of Monod terms may be applied to estimate the desired growth rate and necessary sludge age, respectively.

$$\mu_A = \mu_{max,A} \cdot \left(\frac{S_{NH_4}}{k_{NH_4} + S_{NH_4}} \cdot \frac{S_{O_2}}{k_{O_2} + S_{O_2}} \cdot \frac{S_{alk}}{k_{alk} + S_{alk}} \right) - k_{D,A} \tag{17}$$

In which:

μ_A	d^{-1}	growth rate of *Nitrosomonas*
$\mu_{max,A}$	d^{-1}	maximum growth rate of *Nitrosomonas*, may be assumed as $\mu_{max,A} = 0.52 \cdot 1.10^{(T-15)}\ d^{-1}$
$k_{D,A}$	d^{-1}	decay rate of *Nitrosomonas*, may be assumed as $k_D = 0.05 \cdot 1.072^{(T-15)}\ d^{-1}$
k_{NH_4}	$mg\ L^{-1}$	saturation coefficient for ammonium, typical value 1 $mg\ L^{-1}\ S_{NH_4}$
k_{O_2}	$mg\ L^{-1}$	saturation coefficient for dissolved oxygen, typical value 0.5 $mg\ L^{-1}$ DO
k_{alk}	$mmol\ L^{-1}$	saturation coefficient for alkalinity, typical value 0.5 $mmol\ L^{-1}\ HCO_3^-$
S_{NH_4}	$mg\ L^{-1}$	concentration of ammonia nitrogen
S_{O_2}	$mg\ L^{-1}$	concentration of dissolved oxygen (DO)
S_{alk}	$mmol\ L^{-1}$	alkalinity, concentration of HCO_3^-

The typical values are taken from the "Activated sludge model No. 2" (HENZE et al., 1995). If one assumes that each of the three terms shall not be lower than 0.8 (80% of maximum rates) the following operating parameters should be kept:

$S_{NH_4} = 4\ mg\ L^{-1}$,
$S_{O_2} = 2\ mg\ L^{-1}$ DO, and
$S_{alk} = 2\ mmol\ L^{-1}$

At 10 °C the growth rate under these conditions would become $\mu_A = 0.13\ d^{-1}$ and the aerobic sludge age required should be $SRT_{aer} > 1/0.13 = 7.7$ d. The values for DO and alkalinity reflect the usual optimum process conditions. The average concentration of 4 $mg\ L^{-1}$ ammonia nitrogen is rather high, because practical experience indicates that for an aerobic sludge age (SRT_{aer}) of 8 to 10 d almost complete nitrification is achieved at a reactor temperature of 10 °C. For design purposes a more simplified approach may be applied (KAYSER, 1997):

$$SRT_{aer} = SF_0 \cdot SF_1 \cdot SF_2 \cdot (1/0.47) \cdot 1.10^{(T-15)} \quad (18)$$

The safety factor $SF_0 = 1.5$ enables the growth of nitrifiers or, avoids the wash-out if $SF_1 = SF_2 = 1.0$. The safety factor SF_1 considers any inhibition of the growth rate; a value of $SF_1 = 1.25$ may be chosen. The safety factor SF_2 reflects the ammonia load fluctuations. Depending on the size of the plant or the load fluctuations $SF_2 = 1.3$–1.6 seems to be appropriate. For a temperature of 10 °C this again leads to $SRT_{aer} = 8$–10 d.

Nitrification is sensitive with respect to pH as shown by ANTHONISEN et al. (1976) and NYHUIS (1985). *Nitrosomonas* are inhibited by low concentrations of HNO_2. ANTHONISEN et al. suggested the beginning of inhibition at 0.8–2.8 $mg\ L^{-1}\ HNO_2$, NYHUIS found values of 0.02–0.1 $mg\ L^{-1}\ HNO_2$, the result being a build-up of ammonia nitrogen (S_{NH_4}). Free ammonia (NH_3) inhibits *Nitrobacter* starting at 0.1–1 $mg\ L^{-1}\ NH_3$ (ANTHONISEN et al.) or 1.0–10 $mg\ L^{-1}\ NH_3$ (NYHUIS) leading to a build-up of nitrite nitrogen (S_{NO_2}). Higher concentrations of free ammonia also inhibit *Nitrosomonas* [10–150 $mg\ L^{-1}\ NH_3$ (ANTHONISEN et al.) and 40–200 $mg\ L^{-1}\ NH_3$ (NYHUIS)]. Since the concentrations of HNO_2 and NH_3 depend on pH and S_{NO_2} or S_{NH_4} of the mixed liquor, ANTHONISEN et al. (1976) developed a graph indicating the interference (Fig. 10). The arrows show which concentration may accumulate at which pH. It is evident that at low values of S_{NH_4} and S_{NO_2} and pH of 6–7, as

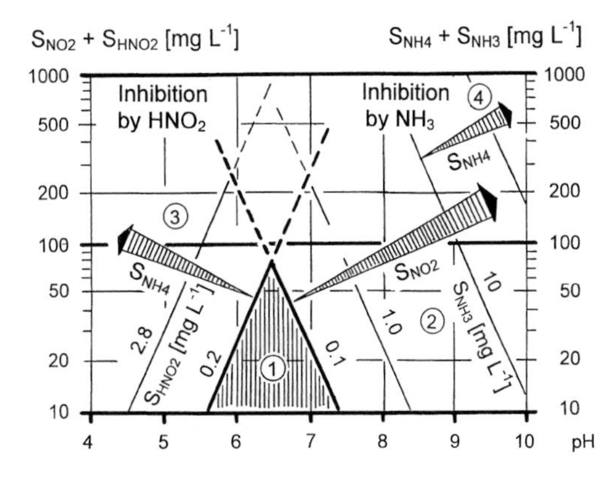

Fig. 10. Inhibition of nitrification by free ammonia and nitrous acid.

it is usual when treating municipal wastewater, inhibition will not occur (shaded area). When treating high-strength wastewater such as from rendering plants due to a build-up of nitrite (S_{NO_2}), disturbances are observed frequently.

2.4 Denitrification

From an engineering point of view denitrification can be visualized as heterotrophic respiration using nitrate instead of dissolved oxygen as electron acceptor. Nitrate by numerous reduction steps is finally converted to gaseous nitrogen (N_2):

$$NO_3^- \rightarrow NO_2^- \rightarrow NO \rightarrow N_2O \rightarrow N_2 \qquad (19)$$

Since nitrous oxide (N_2O) is a greenhouse gas some concern about biological nitrogen removal came up. Investigations in Germany indicate, however, that the contribution of N_2O from nitrogen removal of all German wastewater would be in the range of 2% of the total German N_2O emissions (WICHT and BEIER, 1995; WICHT, 1995).

A simplified overall reaction of the conversion of nitrate to molecular nitrogen may be written as follows:

$$2NO_3^- + 2H^+ \rightarrow N_2 + H_2O + 2.5\,[O_2] \qquad (20)$$

If at the left hand side a substrate (acetate) is added Eq. 20 becomes:

$$\begin{aligned}5\,CH_3COOH + 8\,NO_3^- \rightarrow 4\,N_2 + 8\,HCO_3^- \\ + 2\,CO_2 + 6\,H_2O\end{aligned} \qquad (21)$$

Both equations indicate that for one mole of nitrate denitrified, one mole of alkalinity is gained and 1.25 mole of O_2 as equivalent for heterotrophic respiration becomes available. Since bacterial synthesis is not considered in Eq. 21 the actual requirement of substrate is higher. For design calculations a ratio of 2.9 mg of oxygen equivalent gained per mg of S_{NO_3} to be denitrified may be assumed. If only nitrite has to be denitrified 1.7 mg of oxygen equivalent is gained per mg of S_{NO_2} denitrified.

Nitrate to be denitrified may be calculated as follows:

$$\begin{aligned}S_{NO_3,D} = S_{TKN,0} + S_{NO_3,0} - S_{TKN,e} \\ - S_{NO_3,e} - X_{TKN,exc}\end{aligned} \qquad (22)$$

For municipal wastewater the non-oxidized nitrogen in the effluent may be assumed to be 2 mg L^{-1} organic nitrogen plus 1 mg L^{-1} ammonia nitrogen ($S_{TKN,e} = 3$ mg L^{-1}). The effluent nitrate nitrogen $S_{NO_3,e}$ should be assumed as 2/3 of the value to be kept. Nitrogen contained in the excess sludge may be assumed as $X_{TKN,exc} = 0.02 \cdot C_{COD,0}$.

The reactor volume for denitrification (V_D) may be estimated by

- empirical denitrification rates,
- applying anoxic heterotrophic growth kinetics, or
- using the aerobic heterotrophic oxygen uptake rate in combination with empirical slowdown factors.

Empirical denitrification rates are, e.g., listed in the EPA Manual (1975). The estimation based on heterotrophic growth kinetics is outlined by, e.g., STENSEL and BARNARD (1992).

Since the driving force of nitrate uptake is similar to the driving force of the uptake of dissolved oxygen according to GUJER and KAYSER (1998) for pre-anoxic zone denitrification it can be written:

$$S_{NO_3,\,D} = \frac{0.9 \cdot (1-Y) \cdot S_{read}}{2.9} + \frac{V_D}{V} \cdot \frac{0.6 \cdot [OU_C - (1-Y) \cdot S_{read}]}{2.9} \qquad (23)$$

The first term considers that readily biodegradable COD (S_{read}) is almost immediately (factor 0.9) removed by nitrate and, therefore, is almost independent of the size of the anoxic tank volume. The remaining slowly biodegradable COD is removed at a slower rate (factor 0.6) and depending on the anoxic fraction of the reactor (V_D/V).

At the intermittent denitrification process the fraction of readily biodegradable COD which enters the reactor during aeration periods cannot be considered for denitrification:

$$S_{NO3,D} = \frac{V_D}{V} \cdot \frac{0.9 \cdot (1-Y) \cdot S_{read} + 0.6 \cdot [OU_C - (1-Y) \cdot S_{read}]}{2.9} \tag{24}$$

The readily biodegradable COD (S_{read}) can be estimated, e.g., by the procedure of KAPPELER and GUJER (1992). For municipal wastewater a fraction of 0.05–0.2 $S_{read}/S_{COD,0}$ may be assumed.

The sludge age (SRT) for the calculation of OU_C in Eqs. 23 and 24 is derived from the required aerobic sludge age for nitrification:

$$SRT = \frac{1}{1 - V_D/V} \cdot SRT_{aer} \tag{25}$$

For a design temperature of 10–12 °C the values of V_D/V may be taken from Fig. 11.

Denitrification becomes limited by the ratio V_D/V and OU_C or the COD/TKN ratio, respectively. Since by enlarging of the anoxic reactor volume (V_D) the sludge age increases and the volumetric carbonaceous oxygen uptake rate decreases, it is not economical to raise the anoxic reactor fraction beyond $V_D/V = 0.5$. The addition of external organic carbon like methanol, acetate, etc., in such cases is one possibility to achieve the desired effluent nitrate concentration. Another method to gain additional organic carbon is pre-fermentation of sludge from primary sedimentation (e.g., BARNARD, 1992).

Experience indicates that 4–6 kg COD of external substrate is required per kg of nitrate nitrogen to be removed. If external carbon is the only source of organics for denitrification, it is advisable to perform experiments in order to determine the size of the anoxic reactor. As a very rough estimate one can assume for methanol or acetate denitrification rates of 0.1–0.2 mg $S_{NO_3,D}$ per mg VSS per day at temperatures of 15 °C–20 °C. Since methanol requires the development of special bacteria, it is not recommended for temporary dosage, e.g., only during peak hours.

Dissolved oxygen inhibits any denitrification process since it removes readily biodegradable organic carbon at a higher rate than nitrate. If, at the pre-anoxic zone denitrification process, the internal recycle ratio is, e.g., $4Q$ with 3 mg L^{-1} DO, denitrification of ($4 \cdot 3/2.9$) about 4 mg L^{-1} nitrate nitrogen is prevented. On the other hand, it is possible that in an aeration tank operated with, e.g., 1 mg L^{-1} DO some nitrate disappears. This is attributed to zero DO within the flocs of mixed liquor where some nitrate is then denitrified.

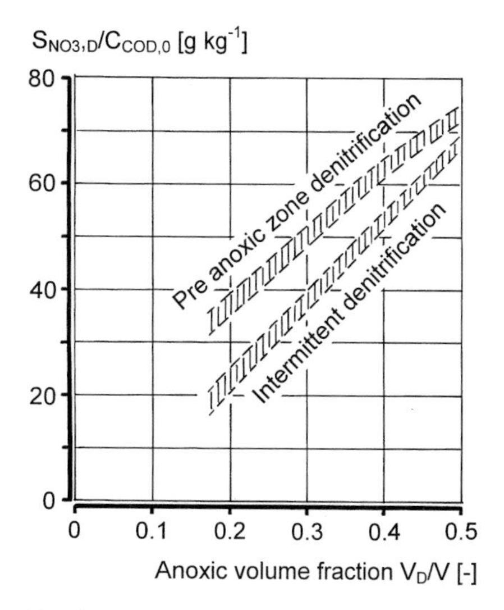

Fig. 11. Nitrate to be denitrified as a function of the influent COD and the type of denitrification process.

2.5 Environmental Factors

2.5.1 Dissolved Oxygen

The dissolved oxygen concentration (DO), the pH, and toxic substances of the wastewater are considered as environmental factors which may inhibit the biological reactions.

In plants with properly designed aeration systems process upsets may be observed only because of inadequate control of DO. The adverse effects are higher with respect to nitrification than with respect to organic carbon removal. This is due to the fact that a part of organic carbon even at low DO may be removed by adsorption while ammonia can not be adsorbed and furthermore due to the fact that the growth of nitrifiers is reduced at low DO (see Eq. 17).

2.5.2 Alkalinity and pH

The pH in the biological reactors depends only to some extent on the pH of the incoming wastewater, which may be the case if inorganic acids are contained in the wastewater. The pH is far more influenced by the remaining alkalinity and the concentration of carbon dioxide of the mixed liquor. The remaining alkalinity may be calculated as follows:

$$S_{alk,e} = S_{alk,0} - 0.07 \cdot (S_{NH4,0} - S_{NH4,e} - S_{NO3,e}) - 0.06 \cdot S_{Fe^{3+}} - 0.04 \cdot S_{Fe^{2+}} - 0.11 \cdot S_{Al^{3+}} + 0.03 \cdot (S_{P,0} - S_{P,e}) \quad (26)$$

Alkalinity is expressed in mmol L^{-1}, the nitrogen and phosphorus concentrations are in mg L^{-1}, and the possible precipitants for phosphate removal (iron and aluminum) are in mg L^{-1} of the incoming wastewater flow. If the precipitants contain free acids or bases these have to be taken into account separately.

The pH in the reactor is a function of the concentration of carbon dioxide which is assumed to be the same in the bulk of the reactor liquid as in the effluent:

$$\mathrm{pH} = 6.43 - \log \frac{S_{CO_2,e}}{S_{alk,e}} \quad (27)$$

$S_{CO_2,e}$ and alkalinity $S_{alk,e}$ are both in mmol L^{-1}. In this equation the dissociation coefficient of 6.43 for T = 15 °C is used. If reactors are operated at other temperatures the appropriate value should be inserted.

According to EPA (1975) the production of carbon dioxide can be estimated as 1.38 kg per kg of oxygen uptake for carbon and ammonia oxidation. With OU_V [mg L^{-1} h^{-1}], the volumetric oxygen uptake rate, the mass balance for production and removal of carbon dioxide can be written:

$$1.38\, OU_V = k_L a_{CO_2} \cdot (S_{CO_2,e} - S_{CO_2,S}) + Q/V \cdot S_{CO_2,e} \quad (28)$$

A second mass balance reflects the gas phase:

$$k_L a_{CO_2} \cdot (S_{CO_2,e} - S_{CO_2,S}) = Q_{Air}/V \cdot p_{CO_2,e} \cdot \rho_{CO_2} \quad (29)$$

Since OU_V is known, $k_L a_{O_2}$ for the aeration system can be calculated and $k_L a$ of carbon dioxide may be assumed as $k_L a_{CO_2} = 0.9\, k_L a_{O_2}$. The wastewater flow ($Q$) and the airflow ($Q_{Air}$) are in m^3 h^{-1}. The partial pressure of carbon dioxide in the exit gas ($p_{CO_2,e}$) may be expressed as volumetric fraction. The density of carbon dioxide is $\rho_{CO_2} = 1.98$ kg m^{-3} at $T = 0$ °C and 1,013 hPa.

The saturation value of carbon dioxide ($S_{CO_2,S}$) depends on the average partial pressure of carbon dioxide in the gas. It may be estimated as mid depth saturation:

$$S_{CO_2,S} = H \cdot \frac{p_{CO_2,e} - p_{CO_2,0}}{2} \cdot \left(1 + \frac{h_{Air}}{2 \cdot 10.35}\right) \quad (30)$$

The Henry constant (H) may be taken from tables; for $T = 15$ °C it is $H = 2{,}000$ mg L^{-1} $(1{,}013\ \mathrm{hPa})^{-1}$. The partial pressure of the incoming gas is very low ($p_{CO_2,0} = 0.0003$) and can be neglected. The diffuser distance below the water surface (h_{Air}) is in m.

Equations 28, 29, and 30 contain three variables ($S_{CO_2,S}$, $S_{CO_2,e}$, and $p_{CO_2,e}$) which are dependent on each other. The following two solutions are obtained:

$$p_{CO_2,e} = \frac{1.38 \cdot OU_V}{\frac{Q_{Air}}{V} \cdot \rho_{CO_2} + \frac{Q}{V} \cdot 0.5 \cdot H \cdot \left(1 + \frac{h_{Air}}{20.7}\right) + \frac{Q}{V} \cdot \frac{Q_{Air}}{V} \cdot \frac{\rho_{CO_2}}{k_L a_{CO_2}}} \quad (31)$$

$$S_{CO_2,e} = \frac{Q_{Air}}{V \cdot k_L a_{CO_2}} \cdot p_{CO_2,e} \cdot \rho_{CO_2} + H \cdot \frac{p_{CO_2,e}}{2} \cdot \left(1 + \frac{h_{Air}}{20.7}\right) \quad (32)$$

By inserting $S_{CO_2,e}$ after conversion to mmol L^{-1} into Eq. 27 the resulting pH is obtained. It has to be noted that with these calculations the pH is somewhat lower than shown by EPA (1975). This is due to different approaches for the estimation of carbon dioxide in the exhaust gas ($p_{CO_2,e}$, $S_{CO_2,e}$).

2.5.3 Toxic Substances

One has to distinguish between biodegradable and nonbiodegradable toxic substances. Nonbiodegradable toxic substances should be held back by on-site treatment or may be removed by pre treatment which is not possible in most cases.

Biodegradable organic toxic substances such as phenols, cyanides, etc., can be almost completely removed if

(1) they are continuously present in the wastewater so that the microorganisms can build up the appropriate enzymes and
(2) their concentration in the mixed liquor is kept as low as possible at any time; shock loads, therefore, should be avoided.

A balancing tank for the wastewater stream containing the toxics is the most appropriate measure to overcome shock load problems. The biological reactor should be completely mixed in order to avoid higher concentrations in any section. Intensive chemical analysis of the reactor preferably by on-line monitoring must be performed in order to detect anomalies like increasing concentrations of the toxic substance as early as possible. If that happens the wastewater feed must be decreased or shut down completely until the conditions in the reactor are stabilized.

Nitrifiers are far more sensitive to toxic substances than heterotrophs. If organic carbon removal is not inhibited usually denitrification also is not inhibited. But since some biodegradable organic toxic substances cannot use nitrate instead of dissolved oxygen they may accumulate in denitrification tanks and inhibit denitrification.

2.6 Properties of Mixed Liquor

The separation of mixed liquor from treated wastewater in final clarifiers is important for the whole process. The concentration of mixed liquor suspended solids (*MLSS*) as well as the sludge volume after $^1/_2$ h settling (SV_{30}) are two parameters to describe sludge properties. Since, due to wall effects and bridging in determining *SV* in 1 L measuring cylinders, erroneous results are obtained at $SV_{30} > 300$–400 mL L^{-1}, either the diluted SV_{30} (DSV_{30}) or the stirred SV_{30} (SSV_{30}) should be measured. For DSV_{30} the mixed liquor should diluted with final effluent to obtain $150 < SV < 250$ mL L^{-1}. Considering the dilution factor DSV_{30} is then calculated from the measured *SV*. It has to be noted that mixed liquor with poor settling characteristics may have $DSV_{30} > 1{,}000$ mL L^{-1}. The dilution method is common practice in Germany. In other countries the stirring method is preferred where some bars (diameter ≈ 2 mm) are rotating in the measuring cylinder at a speed of 1–2 rpm.

Combining *MLSS* and *SV* leads to the sludge volume index *SVI* [mL g^{-1}] (MOHLMANN, 1934):

$$SVI = \frac{SV_{30}}{MLSS} \tag{33}$$

If DSV_{30} or SSV_{30} are used in Eq. 33 *SVI* may be named *DSVI* or *SSVI*.

The sludge volume index is an overall parameter to characterize the sludge thickening behavior. $SVI < 100$ mL g^{-1} is regarded as "good" but if, *SVI* exceeds 150 mL g^{-1} the sludge is called "bulking". Bulking is mainly caused by the growth of filamentous organisms. It is frequently observed at low loaded plants, e.g., for nitrogen removal and at plants treating wastewater with a high fraction of readily biodegradable organics. Bulking not only depends on loading and wastewater characteristics but also on the mixing conditions of the biological reactor. In low loaded completely mixed tanks sludge tends more to bulking than in plug flow tanks.

From an engineering point of view for prevention of sludge bulking, it is important to create reactor configurations with a high concentration gradient like in plug flow tanks or cascaded tanks and to combine completely mixed tanks with a selector. With such precautions the growth of some filamentous organisms like *Microthrix parvicella* may not be prevented. Literature on causes and control of sludge and bulking or foaming is extensive (e.g., JENKINS et al., 1993).

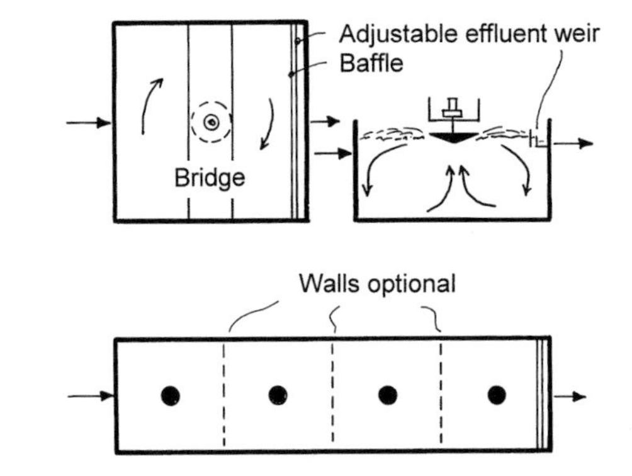

Fig. 13. Aeration tanks for vertical shaft surface aerators.

3 Plant Configurations

3.1 Typical Tanks for Mixing and Aeration

The mixing tanks for denitrification may be square, rectangular, or circular with mixers in the center or with propellers (Fig. 12). Rectangular tanks can be visualized as a series of square tanks. In closed loop tanks either propellers or vertical shaft impellers maintain the flow circulating.

Fine bubble diffused air aeration systems can be installed in almost any type of tank. Tanks for vertical shaft surface aerators are either square or rectangular of which the length is a multiple of the width (Fig. 13). Tanks for cyclic aeration preferably are circular or of the closed loop type. Aeration equipment as well as appropriate mixers must be installed. Mixing may also be performed by rotating bridges on which diffusers may be mounted (Fig. 14). Since the aeration is switched off cyclically, only non-clogging aeration systems, e.g., membrane diffusers for fine bubble aeration are appropriate. Simultaneous nitrification and denitrification in practice is performed mainly in closed loop tanks equipped with horizontal axis surface aerators, e.g., mammoth rotors or vertical shaft surface aerators as in the carousel process. Simultaneous denitrification may also be performed if air diffusers are arranged in "fields" in closed loop tanks (Fig. 15).

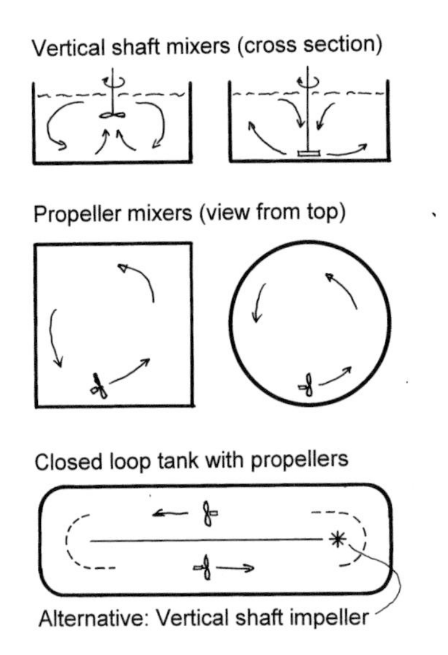

Fig. 12. Typical mixing tanks for denitrification.

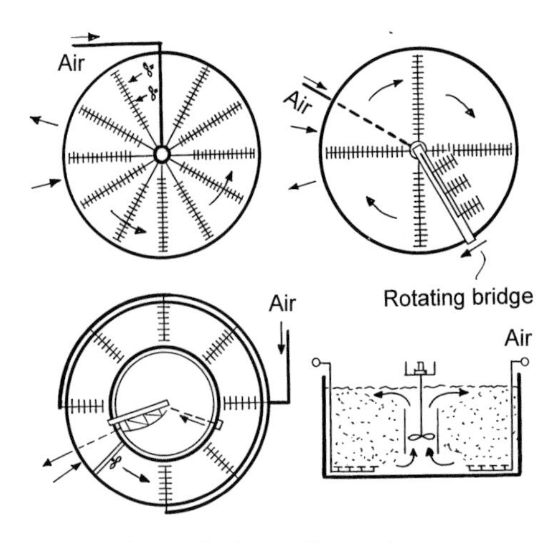

Fig. 14. Typical tanks for cyclic aeration.

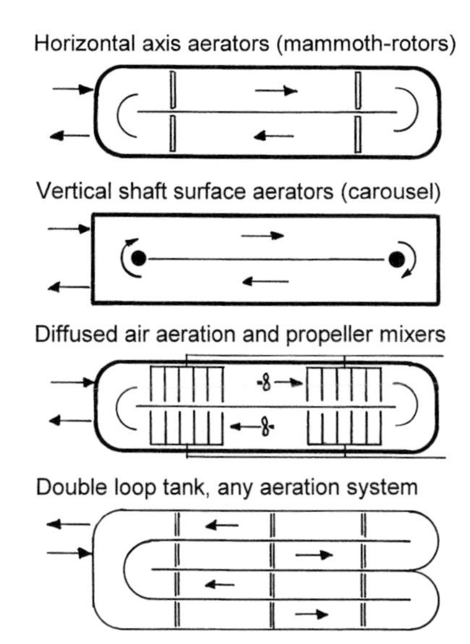

Fig. 15. Typical tanks for simultaneous nitrification and denitrification.

3.2 Carbon Removal Processes

Nowadays worldwide in industrialized countries the removal of carbon and nitrogen is more and more requested, but in developing countries and for pretreatment of industrial and trade wastewater, however, removal of only organic carbon is still important.

As a rough design value a sludge age of $SRT \approx 4$ d may be chosen to achieve a low effluent BOD_5 of 10–30 mg L^{-1}. Assuming a specific sludge production of 0.4 kg SS per kg COD the loading rate would be in the order of $F/M_{COD} \approx 0.6$ kg COD (kg $MLSS$)$^{-1}$ d^{-1}. For the treatment of specific industrial wastewater a higher sludge age may be appropriate. The mixed liquor suspended solids concentration may be as high as $MLSS \approx 3$–4 kg m^{-3}.

In designing provisions should be made for

- a well settling sludge,
- a sufficient high volumetric oxygen transfer rate, and
- a robust aeration system especially in developing countries and in industrial plants.

Sludge settling is improved by plug flow type aeration tanks, aeration tanks constructed as a cascade, and/or by implementing a selector. Two-stage plants may be a choice if the wastewater contains high loads of organic compounds which are easily biodegradable and biodegradable with difficulty. In Germany numerous plants which receive wastewater with a high fraction of readily biodegradable organics were operated with a plastic media trickling filter as a first stage followed by an activated sludge plant. In many cases intermediate settling to remove trickling filter sludge was not implemented.

Although the aeration efficiency (expressed as kg oxygen transferred per kWh at zero DO) of surface aeration systems measured in clean water may be lower than that of fine bubble diffused air systems, surface aeration in some respects is preferable. First of all no problems of diffuser clogging have to be considered and secondly under process conditions the aeration efficiency of fine bubble systems is lower than in clean water, which is not the case with surface aeration systems. The weak points of surface aerators are the bearings and the gears. If these are properly designed the maintenance is restricted to lubrication and change of gear oil.

3.3 Nitrogen Removal Processes

3.3.1 Introduction

The single-stage activated sludge process for nitrogen removal, when using the organic matter of the wastewater for denitrification, incorporates the dilution of ammonia nitrogen to a concentration equivalent to the desired effluent concentration of nitrate nitrogen. Consequently the organics are diluted by the same ratio. It depends on the wastewater characteristics, the dilution ratio, and the process configuration if conditions which are favorable for the development of filamentous bacterial growth are created. In many cases anaerobic contact tanks for enhanced biological phosphate removal are considered to suppress filamentous growth. Unfortunately, these are not successful with respect to some detrimental filamentous organisms like *Microthrix parvicella*.

The processes for nitrogen removal can be divided into three groups:

- Subdivided tanks with distinct compartments for denitrification and nitrification, respectively, e.g., pre-anoxic zone denitrification, step-feed process, post-denitrification process.
- Completely mixed or closed loop tanks in which periodically conditions for nitrification or denitrification are established, e.g., intermittent nitrification–denitrification, alternating nitrification–denitrification (Bio-Denitro process), intermittent nitrification–denitrification with intermittent wastewater feeding (JARV process).
- Closed loop tanks in which at the same time anoxic zones for denitrification and aerobic zones for nitrification are established (simultaneous nitrification–denitrification).

Activated sludge plants for nitrification only may suffer from too low a residual alkalinity. Especially if the aeration tank is completely mixed due to denitrification in the final clarifier, sludge may float and deteriorate the final effluent quality. In order to overcome such problems it is strongly recommended to implement provisions for some denitrification in cases where nitrification only is required.

3.3.2 Pre-Anoxic Zone Denitrification

The activated sludge tank is divided into two main parts, the anoxic zone and the aerobic zone. Since biodegradation of organic carbon follows first-order kinetics, the anoxic zone may be subdivided (Fig. 16). Furthermore, the last one or two anoxic compartments may also be equipped with aeration installations in order to achieve some flexibility.

The aeration tank may have any configuration as shown in Sect. 3.1. Since in many cases it is difficult to keep the dissolved oxygen concentration in the outlet zone at the desired low value, a final non-aerated zone, where the

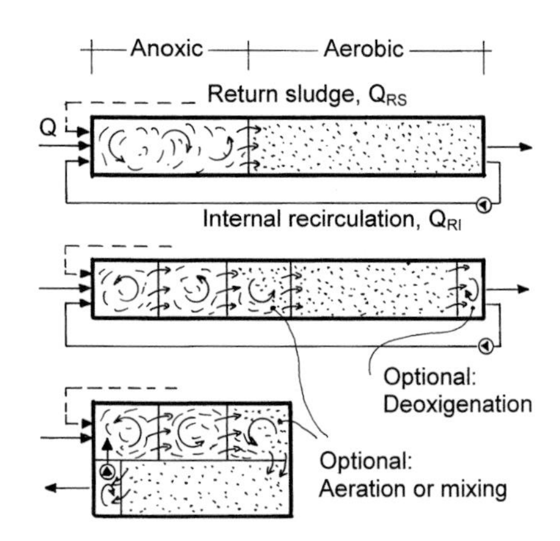

Fig. 16. Pre-anoxic zone denitrification tank configurations.

internal recirculation flow is withdrawn, may be appropriate. If the tank is arranged in a U-form the recycle pump has just to push the flow through the dividing wall.

The hydraulic headloss along such activated sludge tanks is in the range of centimeters, if the top of the dividing walls ends 0.10–0.30 m below the water level.

The internal recycle flow (Q_{RI}) at municipal treatment plants is in the range of 3–5 Q, depending on the wastewater strength and the effluent requirements. Since the headloss is small it is hard to size the recycle pump, therefore, a variable speed drive to adjust the flow is appropriate. The total recycle flow (Q_R) becomes:

$$Q_R = Q_{RS} = Q_{RI} \tag{34}$$

If no nitrification in the anoxic zone, no denitrification in the aerobic zone, and no nitrate in the effluent of the anoxic zone is assumed, by a mass balance the required flow to be recirculated can be calculated. The mass of ammonia which undergoes nitrification has to be calculated first:

$$Q \cdot S_{NH_4,N} = Q \cdot (S_{TKN,0} - S_{TKN,e}) - Q_{exc} \cdot X_{N,exc} \tag{35}$$

This mass of ammonia in the anoxic zone is diluted by the total recycle flow (Q_R) which is assumed to contain no ammonia. Ammonia in the aerobic zone is then oxidized to nitrate:

$$Q \cdot S_{NH_4,N} = (Q + Q_R) \cdot S_{NO_3,e}$$

$$\frac{S_{NO_3,e}}{S_{NH_4,N}} = \frac{1}{1 + Q_R/Q} \tag{36}$$

$$\eta N = \frac{S_{NH_4,N} - S_{NO_3,e}}{S_{NH_4,N}} = 1 - \frac{1}{1 + Q_R/Q} \tag{37}$$

Furthermore:

$$S_{NH_4,N} = S_{NO_3,e} + S_{NO_3,D} \tag{38}$$

In which:

Q_{exc}	$m^3\ d^{-1}$	excess sludge flow
S_{TKN}	$mg\ L^{-1}$	concentration of non-oxidized nitrogen
S_{NH_4}	$mg\ L^{-1}$	concentration of ammonia as N
S_{NO_3}	$mg\ L^{-1}$	concentration of nitrate as N
$X_{N,exc}$	$mg\ L^{-1}$	concentration of nitrogen of excess sludge
η_N	–	denitrification efficiency

Subscripts 0 and e denote influent and effluent, D and N denote to be denitrified and to be nitrified.

Equation 36 indicates that if a low effluent nitrate concentration is required at a high influent concentration and consequently a high concentration of nitrate is to be denitrified, the recycle ratio (Q_R/Q) must be high. Since not only nitrate but also dissolved oxygen is transferred to the anoxic zone via the recycle flow, some organic carbon will be removed aerobically and thus be lost for denitrification.

The process control may be restricted to an automatic control of the aeration intensity, in order to keep a pre-set dissolved oxygen concentration. If at the aeration tank outlet or better at the outlet of a group of aeration tanks monitors for ammonia and nitrate are installed, the signals may be used for further control measures. The purpose of this may be:

- Energy saving: If the concentration of ammonia is zero and that of nitrate is in the desired range, the aeration of the denitrification cells may be switched off or if that already is the case, the set point for aeration control may be lowered. If ammonia increases the measures have to be taken reversed.
- Keeping ammonia low in winter: If at very low temperatures of the mixed liquor the concentration of ammonia increases, one might try to improve nitrification by raising the set point for aeration control beyond the usual 1.5–2.0 mg L^{-1} of dissolved oxygen. This was observed to be a successful measure at some plants.
- Improving nitrate removal: If the concentration of ammonia is about zero but the concentration of nitrate is at the upper limit, the aeration in the denitrification cells may be stopped and/or the internal recycle flow may be increased.

Numerous larger municipalities in Germany like Berlin, Stuttgart, Bremen but also smaller cities are using single-stage activated sludge plants with the pre-anoxic zone denitrification process.

3.3.3 Step-Feed Denitrification Process

The activated sludge tank of this process consists of two to three pre-anoxic zone tanks in series, where the return sludge is diverted to the first denitrification zone and where the wastewater is distributed to each denitrification zone (Fig. 17). Although the first experiments of this process were performed in the UK (Cooper et al., 1977) and later in Japan (Miyaji et al., 1980) the first results of a full-scale plant were reported by Schlegel (1983) from Germany.

Again, considering complete nitrification in the aeration zones and complete denitrification in the anoxic zones the effluent nitrate concentration for the last denitrification zone can be calculated by a mass balance as follows:

$$(x \cdot Q) \cdot S_{NH_4,N} = (Q + Q_{RS}) \cdot S_{NO_3,e} \tag{39}$$

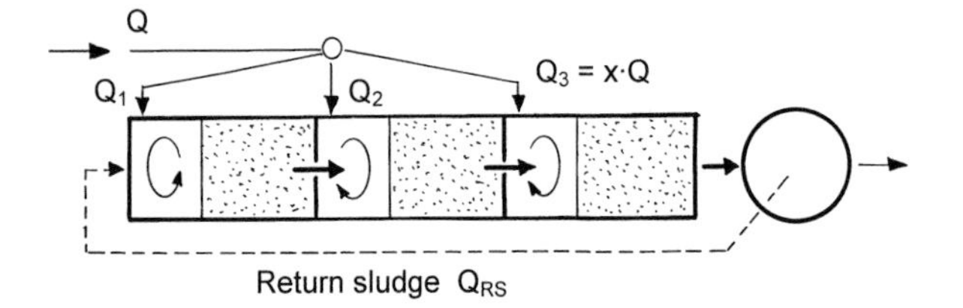

Fig. 17. Step-feed denitrification tank configuration.

The flow ($x \cdot Q$) is the fraction of the total wastewater flow entering the last denitrification zone whereby x must not be equal to the inverse number of stages.

$$\frac{S_{NO_3,e}}{S_{NH_4,N}} = \frac{x}{(1 + Q_{RS}/Q)} \tag{40}$$

$$\eta_N = 1 - \frac{x}{(1 + Q_{RS}/Q)} \tag{41}$$

In order to increase nitrate removal at the step-feed process, it is necessary to lower the fraction of wastewater diverted to the last denitrification zone (decrease x) or to increase the return sludge flow. Since by increasing the return sludge flow (Q_{RS}) disturbances of the final clarification may be caused, it seems to be more appropriate to decrease x. This is only possible if the organic carbon contained in the wastewater flow ($x \cdot Q$) is sufficient to denitrify the incoming nitrate load.

As an advantage of the step-feed process the higher average concentration of mixed liquor suspended solids (*MLSS*) as compared to the *MLSS* of the last aeration tank effluent is visualized. At a three-step process with equal distribution of wastewater ($Q/3$) the *MLSS* at $Q_{RS} = Q$ becomes, e.g., 4.5 kg m^{-3}, 3.6 kg m^{-3}, and 3.0 kg m^{-3} at zones 1, 2, and 3, respectively. In order to maintain the same loading rates at the three compartments it is possible to choose appropriate tank volumes or, what is easier, to distribute the wastewater flow accordingly. This would be the case at $Q_1 = 0.4\,Q$, $Q_2 = 0.33\,Q$, and $Q_3 = 0.27\,Q$. *MLSS* then would be, e.g., 4.3 kg m^{-3}, 3.5 kg m^{-3}, and 3.0 kg m^{-3}.

At the three-step process with $Q_{RS} = Q$ and $x = 0.27$ the denitrification efficiency would become 86%.

The similarity of the pre-anoxic zone denitrification process and the step-feed process is obvious when comparing Eqs. 37 and 41. In order to achieve 86% denitrification efficiency at the pre-anoxic zone process the recycle ratio must be $Q_R/Q = 6.4$, while at the three-step process no internal recirculation is required besides the return sludge flow of $Q_{RS}/Q = 1$.

The differences of the pre-anoxic zone process and the step-feed process are demonstrated in Fig. 18.

The advantage of the pre-anoxic zone process is that each of the three tanks is operated independently and that the tanks will have the same water level and the same water depth. In the step-feed process there will be some headloss as the water flows from one tank to the other, therefore, either the water depth differs

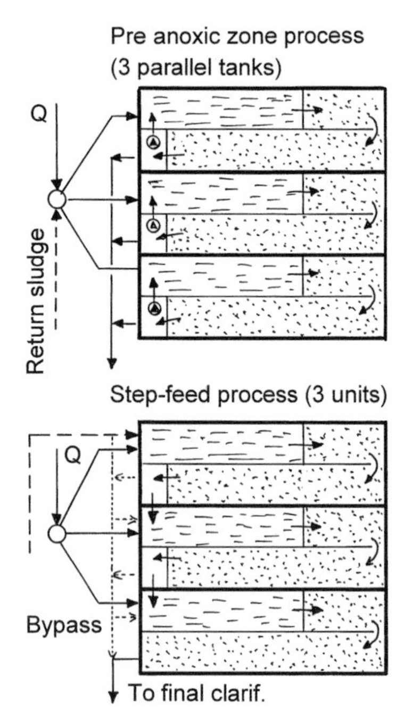

Fig. 18. Comparison of the step-feed process and the pre-anoxic zone process.

from tank to tank (same bottom level) or the water depth is kept constant (different bottom levels). For maintenance it is necessary to have a bypass for each tank of the step-feed process.

At both processes dissolved oxygen which enters the denitrification zone removes organic carbon and hence reduces denitrification. Therefore non-aerated outlet zones are shown at the pre-anoxic zone tanks. At the step-feed process only the first and second aeration tank may each be equipped with a non-aerated zone.

Several plants with the step-feed process are operated by the Emschergenossenschaft and, e.g., the cities of Wolfsburg and Bremerhaven.

3.3.4 Simultaneous Nitrification and Denitrification

The key to nitrogen removal by the simultaneous denitrification process is the appropriate setting of the aerators in order to establish simultaneously sufficiently large aerobic and anoxic zones. Since the load of any wastewater treatment plant shows a diurnal fluctuation the concentrations of nitrate and ammonia countercurrently will vary at a constant aerator setting (Fig. 19). In order to achieve the desired nitrogen removal, a process control is, therefore, required.

PASVEER (1964) was the first to report on simultaneous denitrification in an oxidation ditch. He achieved this by setting the optimal immersion depth of the surface aerator in order to create a sufficient large anoxic zone.

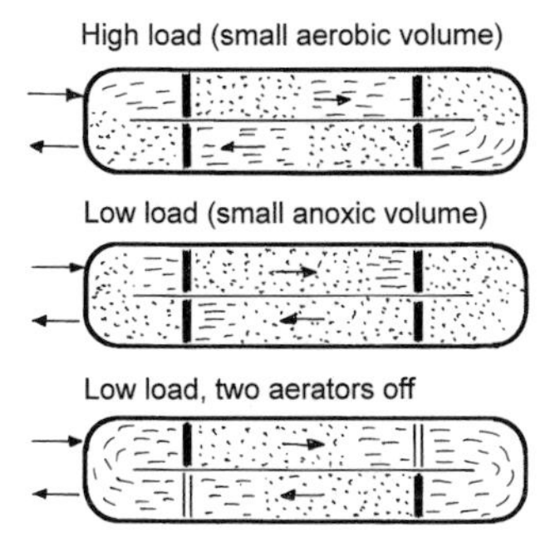

Fig. 19. Simultaneous nitrification and denitrification at different loading conditions.

MATSCHÉ (1977) showed a high and constant nitrogen removal of the plant Vienna Blumental which was designed for 300,000 p.e. The two closed loop aeration tanks of 6,000 m^3 and equipped with 6 twin mammoth rotor aerators (diameter 1 m, each 7.50 m long) were operated in series. The oxygen uptake rate was continuously measured by a special (home-made) respirometer. The respirometer output was used to activate the appropriate number of aerators in order to achieve the desired nitrogen removal.

The first real process control for simultaneous denitrification was developed by ERMEL (1983). By ultrafiltration a continuous sample flow was separated from the mixed liquor and diverted to a nitrate monitor. At the plant of Salzgitter-Bad the two closed loop aeration tanks are operated in parallel. Each tank is equipped with three mammoth rotor surface aerators. One rotor in each tank is operated continuously in order to create a sufficient circulating flow. The two other rotors of each tank were automatically switched on if the nitrate concentration of the sample flow dropped down to, e.g., $S_{NO_3}=3$ mg L^{-1}. The aerators were stopped if, e.g., S_{NO_3} reached 6 mg L^{-1} and switched on again after the set point of $S_{NO_3}=3$ mg L^{-1} was reached. Fig. 20 shows a daily course of ammonia and nitrate

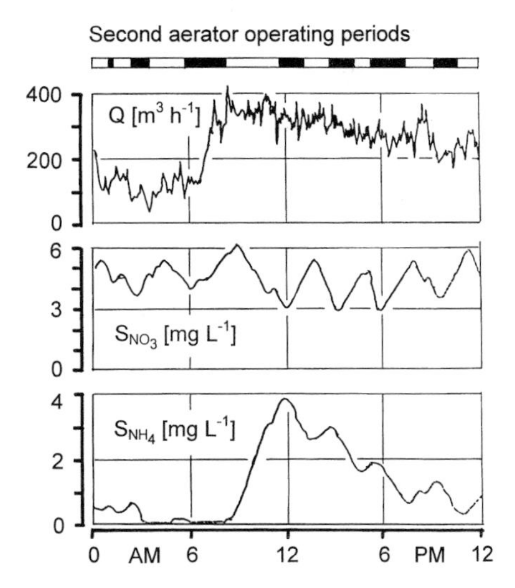

Fig. 20. Course of ammonia and nitrate in one aeration tank of the Salzgitter-Bad plant.

measured in one of the aeration tanks. Due to the stil low denitrification rate at the period of the high ammonia load (8–12 a.m.) the concentration of ammonia rises to about 4 mg L^{-1} N. As the denitrification rate increases after noon the off-periods of the two aerators become shorter and the ammonia concentration decreases.

A more sophisticated control system using nitrate monitoring and ammonia monitoring was applied at the wastewater treatment plant in Hildesheim (SEYFRIED and HARTWIG, 1991).

In Salzgitter-Bad an ORP controller was tested as a very simple and low-cost control method. Since the oxidation–reduction potential (ORP) drops sharply, if at zero DO nitrate reaches zero, the controller switches the additional aerators on at a certain slope of ORP. The additional aerators then were operated for a pre-set period of time. Fig. 21 shows the course of ORP. The effluent quality at the time, when the ORP controller was used, was as good as at times when the nitrate controller was used (KAYSER, 1990). At about the same time similar experiments were conducted in Canada by WAREHAM et al. (1993).

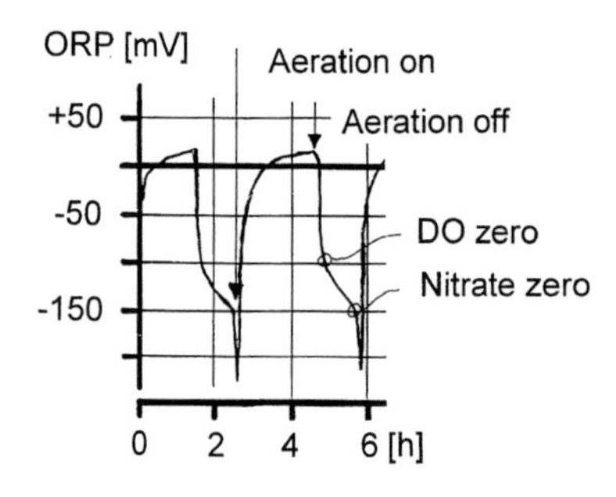

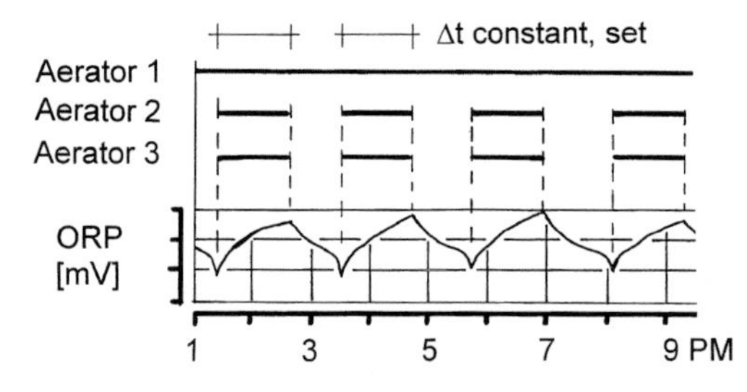

Fig. 21. ORP to control simultaneous denitrification. Top: at intermittent aeration showing the important points, bottom: denitrification control at the Salzgitter-Bad plant.

Some newer plants for simultaneous nitrification and denitrification (in addition to the surface aerators) are equipped with propellers in order to maintain a sufficient flow velocity independent of the number of aerators in operation. Such plants may also be operated in the intermittent nitrification–denitrification mode.

Besides the cities already mentioned, plants with simultaneous denitrification are also operated, e.g., at Osnabrück, Münster, Gera, Paderborn, Tel Aviv, and numerous smaller communities where oxidation ditches were converted to the simultaneous mode by means of automatic control.

It depends on the placement of the wastewater inlet and its distance to the next aerator, if and how much of the readily biodegradable organics are oxidized by dissolved oxygen and, therefore, are lost for denitrification. In any case it is advisable to introduce the flow near the tank bottom since anoxic conditions are predominant there.

The closed loop circulating flow tanks for simultaneous denitrification may be regarded as completely mixed. In addition, due to the high dilution of the incoming wastewater by the circulating flow, conditions favorable for filamentous bacterial growth are established. If the wastewater contains a higher fraction of readily biodegradable organics, a selector or an anaerobic mixing tank for enhanced biological phosphate removal is recommended.

3.3.5 Intermittent Nitrification–Denitrification Process

Aeration tanks for intermittent nitrification–denitrification must be equipped with an aeration system and mixing devices. The design of the aeration system has to take into consideration the aeration-off periods. In order to improve the aeration efficiency of plants with diffused air aeration, it may be advisable, to also operate the propellers during the aeration periods.

At intermittent nitrification–denitrification plants the fraction of readily biodegradable organics, which enters the tank while aeration is operating, may be oxidized by dissolved oxygen and, therefore, lost for denitrification. This must be considered when calculating the nitrate to be denitrified. Precautions against the growth of filamentous organisms should be taken because the tanks are completely mixed.

The duration of the aeration periods and the aeration-off periods, respectively, may be set by a timer. Any other control system as described in Sect. 3.3.4 may be applied in order to achieve a more stable effluent quality. In Germany, especially at the large number of extended aeration plants the ORP controller is frequently used. NitraReg is another low cost control system which only requires continuous measuring of the dissolved oxygen concentration (BOES, 1991). By knowing the oxygen transfer capacity the oxygen uptake rate is automatically calculated and used as control parameter.

The course of nitrate during the anoxic period (t_D) depends on the activity of the denitrifiers, and the course of ammonia and nitrate during the aerobic period (t_N) depends on the activity of the nitrifiers. Since the reactor is completely mixed the course of ammonia during the anoxic period can be calculated as follows:

$$S_{NH4,t} = S_{NH4,N} \cdot [1 - \exp.(-t/t^*)] \tag{42}$$

Neglecting the effluent loss of ammonia during the anoxic period a straight line shows how ammonia would increase:

$$S_{NH4,t} = S_{NH4,N} \cdot (t/t^*) \tag{43}$$

In which:

$S_{NH4,t}$	mg L^{-1}	concentration of ammonia nitrogen at time t
t^*	h	detention time of wastewater ($t^* = V/Q$)
t_N	h	aerobic (nitrification) time period
t_D	h	anoxic (denitrification) time period

The deviation between Eqs. 42 and 43 is lower than 5% if t_D/t^* is below $^1/_4$ which in practice is the case at activated sludge plants. In Fig. 22, therefore, only the straight line according to Eq. 43 is shown. Since ammonia is converted during the aerobic period to nitrate, the nitrate concentration at the end of a cycle will be the starting concentration at the anoxic period of the next cycle.

The average effluent concentration of nitrogen ($S_{NO3,e} + S_{NH4,e}$) theoretically becomes (see Fig. 22):

$$S_{NO3,e} + S_{NH4,e} = 0.5 \cdot S_{NH4,N} \cdot \frac{2 \cdot t_D + t_N}{t^*} \tag{44}$$

The efficiency of nitrogen removal is obtained as:

$$\eta_N = \frac{S_{NH_4,N} - (S_{NH_4,e} + S_{NO_3,e})}{S_{NH_4,N}} = 1 - \frac{t_D + 0.5 \cdot t_N}{t^*} \tag{45}$$

In order to achieve a nitrogen removal efficiency of 86% ($\eta_N = 0.86$) at, e.g., $t^* = 13$ h and $t_D/(t_N + t_D) = 0.35$ the cycle period should be $t_N + t_D = 2.75$ h.

Equation 44 indicates that technically any removal efficiency can be tuned by variation of the cycle period, but as a prerequisite the nitrate removal capacity of the process must be observed. Cycle periods of less than $t_N + t_D = 1$ h, however, are not recommended.

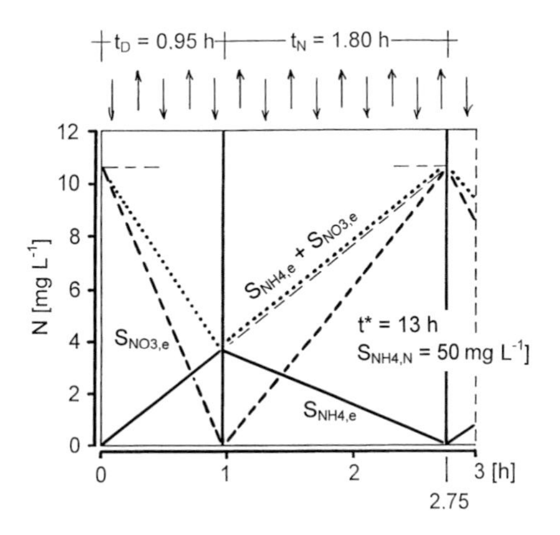

Fig. 22. Course of ammonia and nitrate at intermittent nitrification and denitrification.

The ratio $t_D/(t_N+t_D)$ can to be calculated, e.g., by Eq. 24 assuming:

$$V_D/V=t_D/(t_N+t_D) \quad (46)$$

3.3.6 Intermittent Nitrification–Denitrification Processes with Intermittent Wastewater Feeding

There are three processes which apply intermittent wastewater feeding:

- The alternating nitrification denitrification process (Bio-Denitro)
- The Tri-Cycle Process
- The Jülich wastewater treatment process (JARV)

The alternating nitrification–denitrification process consists of pairs of activated tanks, each equipped with aeration installations and mixing devices. The two parallel interconnected tanks are fed alternating with wastewater and return sludge. While one tank is fed, the mixed liquor is discharged from the other to the final clarifier (Fig. 23). This process was developed in Denmark at a plant with two parallel oxidation ditches. It was named Bio-Denitro process (THOLANDER, 1977).

The course of ammonia and nitrate during a cycle theoretically looks similar to that at intermittent nitrification–denitrification (Fig. 24). The effluent concentration of nitrogen can be calculated in a similar way as for intermittent nitrification–denitrification.

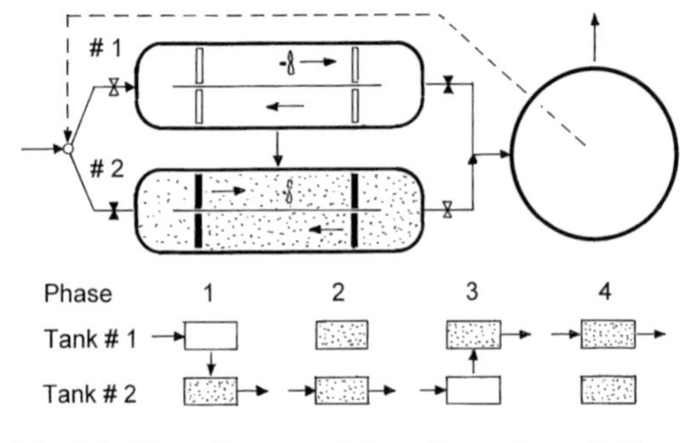

Fig. 23. Flow diagram of the alternating nitrification–denitrification process.

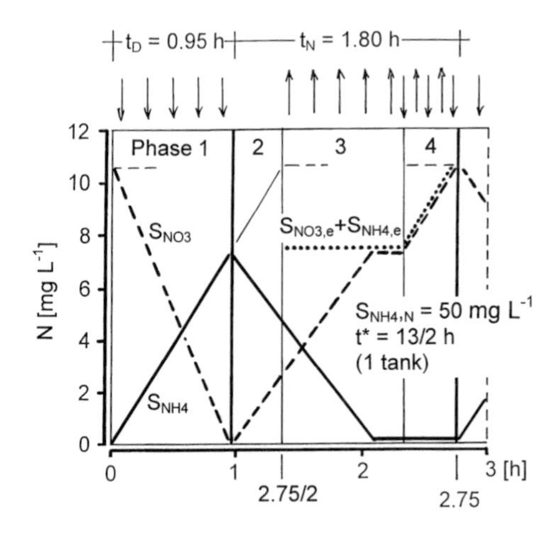

Fig. 24. Course of ammonia and nitrate in one of the tanks of the alternating nitrification–denitrification process.

Usually the alternating nitrification–denitrification process is timer-controlled. SØRENSEN (1996), however, demonstrated that by online monitoring of nitrate and ammonia and appropriate control measures the effluent quality could be improved.

In Denmark numerous plants are designed for the alternating nitrification–denitrification process. There are also some plants in Germany, a newer one, e.g., in Stuttgart-Feuerbach.

At the Tri-Cycle process three interconnected aeration tanks are consecutively (controlled by a timer) fed with wastewater and return sludge. A cycle begins with a mixing period for denitrification which normally lasts as long as the feeding period ($^1/_3$ of the cycle period). Valves in the connecting pipes and at the inlet and outlet of the tanks guide the mixed liquor through the three tanks. Mixed liquor is discharged from that aeration tank which is aerated for the longest period to the final clarifier. This process is used in at least 6 plants in Germany. Unfortunately there are only brochures of the company (GVA, Wülfrath, Germany) which developed this process.

The Jülich Wastewater Treatment Process (JARV) was developed to overcome problems of bulking sludge in completely mixed aeration tanks. The key of the process is a rather short feeding period compared to the cycle period (ZANDERS et al., 1987). This requires a balancing tank ahead of the aeration tank and an effluent gate to equalize the flow to the final clarifiers. Balancing is avoided if, e.g., four parallel aeration tanks are consecutively fed with wastewater. The effluent from the aeration tanks may contain considerable ammonia peaks. It is, therefore, advisable to build a small post-aeration basin between the aeration tanks and the final clarifiers. Since wastewater is always introduced only during the denitrification period, all readily biodegradable organics are available for denitrification. Only two plants in Germany are operated by this process.

The differences of the three processes compared to the intermittent nitrification–denitrification process are, that almost all readily biodegradable organics are available for denitrification, that there is a higher concentration gradient by which the sludge volume index may be improved, and that during the discharge period the mixed liquor is aerobic (except at the JARV process if without a post-aeration tank). The anoxic tank fraction (V_D/V) may be estimated as for the pre-anoxic zone denitrification process by Eq. 23.

3.3.7 Special Processes for Low COD/TKN Ratio

Wastewater from the food industry may be characterized by high concentrations of organics as well as considerable nitrogen concentrations. Anaerobic pretreatment in such cases is favorable because of the negligible sludge production and energy requirement. If nitrogen removal is required the COD/TKN ratio, however, may be reduced too for far a sufficient denitrification.

Since the denitrification of nitrite requires about 35% less organics as compared to denitrification of nitrate, a process in which *Nitrosomonas* were inhibited was developed by ABELING (1994) and SEYFRIED and ABELING (1992). The pre-anoxic zone process was operated with a control system to keep the pH by dosing of NaOH in the aerobic tank at about pH=8 and to keep ammonia by appropriate aeration control at $S_{NH_4}=10$ mg L^{-1}. According to the findings of ANTHONISEN et al. (1976) and NYHUIS (1985) (see Fig. 10), these are conditions under which *Nitrobacter* are inhibited. Since there is always some ammonia and some nitrite in the effluent, a second stage using a fixed film reactor was implemented.

A different method to remove high nitrogen concentrations without any organic substrate was shown by JETTEN et al. (1997). In a continuous flow stirred reactor as a first stage about 50% of ammonia is converted to nitrite, and in a second-stage fixed film reactor by autotrophic bacteria nitrite is converted to nitrogen gas (N_2) using ammonia as electron donor. But this, however, is not regarded as an activated sludge process.

3.3.8 Post-Denitrification with External Organic Carbon

There are not many plants around the world using external carbon as the sole carbon source for denitrification in single-stage activated sludge plants. One such plant was constructed to treat the whole wastewater of the Salzgitter Steel Works, Germany.

The wastewater originates from blast furnace gas treatment (high ammonia loads), the coke oven gas treatment (phenols, cyanides, ammonia, etc.), and several other discharges. The total flow of about 50,000 m^3 d^{-1} contains about 35 mg L^{-1} ammonia nitrogen. The alkalinity due to the use of soft water for cooling is low. Therefore, at first experiments were carried out using the pre-anoxic zone process in order to gain as much alkalinity as possible. Since the organic carbon content of the wastewater was too low, methanol had to be added to ensure sufficient nitrate removal. The process was very unstable due to the toxic components of the wastewater.

The process, therefore, was changed to post-denitrification. The results of the experiments were satisfying (ZACHARIAS, 1996; ZACHARIAS and KAYSER, 1995). The full-scale plant is shown in Fig. 25. It is possible to operate the first two aeration tanks (#1 and #2) in parallel, in order to prevent a stronger concentration

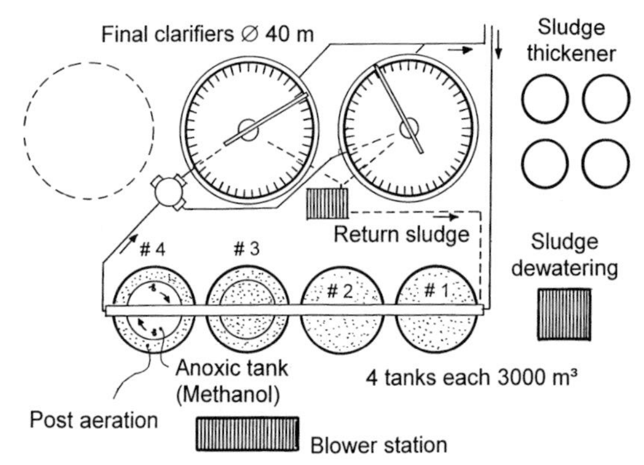

Fig. 25. Flow diagram of the wastewater treatment plant of the Salzgitter Steel Works (post-denitrification) (ZACHARIAS, 1996).

gradient of the toxic substances. This fortunately has not yet been necessary. The center zone of tank #4 (1,500 m^3) serves as denitrification zone, into which methanol is dosed, controlled by the nitrate concentration measured in the tank. In the final aerated zone (1,500 m^3) accidental overdosed methanol will be oxidized. Tank #3 for maintenance purposes is constructed like tank #4. During normal operation both parts of this tank are aerated.

In order to remove 1,250 kg nitrate nitrogen per day 3,500 kg methanol per day are consumed which is in the practical range of 2.5–3.0 kg methanol per kg nitrate nitrogen to be denitrified (ZACHARIAS, 1996).

3.4 Interactions between the Biological Reactors and the Final Clarifiers

The biological reactor and the final clarifier, linked by the return sludge flow, form a unit process. The conditions shown in Fig. 26 for rectangular tanks with scrapers are similar to circular tanks with scrapers. For steady state conditions the return sludge suspended solids concentration (SS_{RS}) is a function of $MLSS$ and the return sludge flow ratio (Q_{RS}/Q):

$$SS_{RS} = MLSS \cdot \frac{Q + Q_{RS}}{Q_{RS}} = MLSS \cdot \left(1 + \frac{Q}{Q_{RS}}\right) \tag{47}$$

Since $MLSS$ will be kept constant by withdrawing of excess sludge, the return sludge suspended solids concentration (SS_{RS}) increases with decreasing return sludge flow (Q_{RS}). The return sludge flow is the sum of the flow of thickened sludge (Q_{sl} with SS_{sl}) and the short circuit flow (Q_{short} with $MLSS$). Since the return sludge flow is set by the return sludge pumping rate and the flow of thickened sludge to the hopper depends on the speed and other factors of the scraper as well as the height of the bottom sludge layer (h_s) the short circuiting flow becomes:

$$Q_{short} = Q_{RS} - Q_{sl} \tag{48}$$

As long as $Q_{RS} > Q_{sl}$ there will be some short circuit flow but if

$$Q_{sl} \cdot SS_{sl} \geq Q_{RS} \cdot SS_{RS} \tag{49}$$

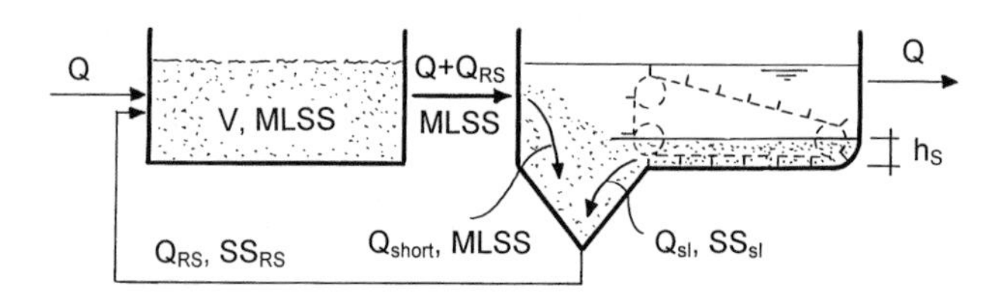

Fig. 26. Interaction between biological reactor and final clarifier.

the sludge in the clarifier may accumulate (increasing h_s) if the wastewater flow and hence the solids load of the final clarifier increases. In practice, therefore, Q_{RS} is recommended to be higher than Q_{sl}.

For return sludge pumping two strategies are common:

- Constant flow of Q_{RS} at least at dry weather periods.
- Constant ratio Q_{RS}/Q.

Due to the diurnal wastewater flow fluctuation at constant Q_{RS} the bottom sludge layer height h_S and SS_{RS} will fluctuate nearly parallel to Q, and $MLSS$ will fluctuate inversely to the wastewater flow variation. If the return sludge ratio Q_{RS}/Q is kept constant the fluctuations are similar but not as pronounced as at constant Q_{RS}. This is due to the fact that the scraper flow Q_{sl} has a constant maximum once the bottom sludge layer height rises beyond a certain h_S. The mass of sludge to be transferred by the scraper to the hopper ($Q_{sl} \cdot SS_{sl}$) at constant Q_{sl} can only increase by increasing SS_{sl} which is the case if the bottom sludge layer height h_s is elevated. If, e.g., at sudden storm water flows the return sludge flow is immediately increased in order to keep Q_{RS}/Q constant, the short circuit flow will increase, and SS_{RS} will decrease accordingly. It, therefore, makes sense to keep the ratio of Q_{RS} to the 1–2 h sliding average of Q constant.

The main design parameter for horizontal flow clarifiers is the sludge volume surface load q_{SV} expressed as L of sludge per m^2 of clarifier surface (A_{clar} in m^2) and hour (Q_h is the hourly peak flow in m^3 h^{-1}):

$$q_{SV} = \frac{Q_h \cdot MLSS \cdot ISV}{A_{clar}} \quad [\mathrm{L\ m^{-2}\ h^{-1}}] \qquad (50)$$

These and all further design calculations for clarifiers have been compiled by EKAMA et al. (1997). The construction of the mixed liquor inlet as well as of the clear water outlet and the sludge collection system are also important for the performance of clarifiers.

If aeration tanks with a depth of more than 6 m precede the final clarifier, mixed liquor may float due to over-saturation with nitrogen. Measures to prevent floating are (ATV, 1996):

- Stripping of nitrogen gas by an overflow weir cascade at the aeration tank outlet or by intensive coarse bubble aeration in a shallower outlet zone of the aeration tank.
- Applying deep final clarifiers at which mixed liquor is introduced near the bottom.

4 Design Procedure

Detailed design information may be taken from handbooks like RANDALL et al. (1992), ATV (1997), etc. In addition for final clarifier design EKAMA et al. (1997) is recommended. The main design steps for single-stage activated sludge plants comprise:

(1) Determination of the design loads (e.g., BOD_5, COD, suspended solids, nitrogen, phosphorus), the average alkalinity and the wastewater flow for dry weather (daily and peak) and the peak storm water flow [m^3 h^{-1}]. Existing data should be checked for annual fluctuations of the wastewater temperature, the loads, and the flows. If nitrification is required, the design load should be selected in combination with the wastewater temperature. It may be helpful to plot the daily values of:

$$Q \cdot C_{COD} \cdot 1.1^{(15-T)}$$

The time period with the highest two week average may be selected to determine the critical sludge age and the sludge production. The highest oxygen demand, however, may have to be calculated for another (e.g., summer) period.

(2) Required effluent quality as percent removal or as concentrations ($C_{BOD,e}$, $C_{COD,e}$, $S_{NH4,e}$, $S_{NO3,e}$, $S_{P,e}$, SS_e). Mode of inspection, e.g., grab sample or 24 h composite sample. Value to be kept, e.g., as annual average, at 80% of all days or at 80% of grab samples.

(3) Selection of the process configuration. Primary sedimentation? Aerobic selector? Anaerobic mixing tank for enhanced phosphate removal? Phosphate precipitation, type of precipitant? Process for nitrogen removal? Dosage of external organic carbon for nitrogen removal, continuously or temporarily? Necessity of dosage of N, P, and trace elements in case of specific industrial wastewater? Sludge disposal, separate stabilization or co-stabilization of sludge?

(4) An aerobic selector may be designed for a volumetric load of 20 kg COD m^{-3} d^{-1}. For enhanced biological phosphate removal an anaerobic contact period of 0.5–1.0 h (based on $Q+Q_{RS}$) may be assumed.

(5) Selection of the aerobic sludge age (SRT_{aer}) considering the degree of treatment, e.g., organic carbon removal only; nitrification at any time, at certain time periods, or at periods beyond a certain reactor temperature (e.g., $T \geq 10\,°C$); co-stabilization of sludge. The following values may be used: Organic carbon removal only: $SRT=4$ d; nitrification at $T \geq 10\,°C$: $SRT_{aer}=$ 8–10 d.

(6) For denitrification only: Estimation of the nitrate to be denitrified (Eq. 22). Assumption of anoxic tank volume fraction V_D/V (Eqs. 23 or 24 or values from Fig. 11). Calculation of the sludge age by Eq. 25.

(7) For co-stabilization of sludge (extended aeration) and nitrogen removal a sludge age of $SRT=25$ d may be assumed.

(8) Calculation of the remaining alkalinity by Eq. 26 and, if aeration tanks deeper than 6 m are planned, checking of the resulting pH according to Sect. 2.5.2.

(9) Calculation of the daily mass of excess sludge $M_{exc,\,c}$ by Eq. 12 for organic carbon removal. For settled municipal wastewater as a rough estimate one may assume 0.4 kg excess sludge solids per kg COD influent. The biomass resulting from the growth of nitrifiers may be neglected. If simultaneous phosphate precipitation is applied one can assume 2.5 or 4.0 kg excess sludge solids per kg Fe or Al added, respectively. Biologically removed phosphate forms 3 kg excess sludge solids per kg P removed.

(10) Determination of the mixed liquor suspended solids $MLSS$ considering the sludge volume index ISV to be expected. It is recommended at first to check the final clarifier design in order to assume the appropriate $MLSS$. Generally, $MLSS=3.0\text{–}5.0$ kg m^{-3} are used for design calculations. The higher value should only be used if a sludge volume index of $ISV=100$ mL L^{-1} or lower can be expected.

(11) Calculation of the reactor volume by Eq. 14.

(12) For nitrogen removal only: Estimation of the required internal recycle flow (Eq. 36) or the wastewater distribution at the step-feed process (Eq. 40) or the cycle period of intermittent nitrification–denitrification processes (Eq. 44).

(13) Design of the aeration installation considering the minimum and maximum oxygen uptake rate (Eq. 13). The mass of oxygen uptake for nitrogen removal may be calculated as follows [kg d^{-1}]: $M_{O,N}=Q_d\,\{4.3\cdot(S_{NO_3,D}+S_{NO_3,e}-S_{NO_3,0})\;-2.9\cdot S_{NO_3,D}\}/1{,}000$.

(14) Design of the final clarifier. As a rough estimate for the surface of horizontal flow clarifiers a volumetric sludge surface loading rate of $q_{SV} \leq 450$ L m^{-2} h^{-1} (see Eq. 50) may be assumed. The side wall water depth should not be lower than 3.00 m and the average water depth should be in the range of 3.50–4.50 m (except, e.g., after deep aeration tanks).

5 References

ABELING, U. (1994), Stickstoffelimination aus Industrieabwässern; Denitrifikation über Nitrit. *Veröffentlichungen des Instituts für Siedlungswasserwirtschaft und Abfalltechnik der Universität Hannover* **86**.

ANTHONISEN, A. C., LOEHR, R. C., PRAKASAM, T. B. S., SRINATH, E. G. (1976), Inhibition of nitrification by ammonia and nitrous acid, *J. WPCF* **24**, 835–852.

ARDERN, E., LOCKETT, T. (1914), Experiments on the oxidation of sewage without the aid of filters, *J. Soc. Chem. Ind.* **33**, 523–529.

ATV (1996), Hinweise zu tiefen Belebungsbecken, *Korrespondenz Abwasser* **43**, 1083–1086.

ATV (1997), *ATV Handbuch: Biologische und weitergehende Abwasserreinigung*, 4th Edn. Berlin: Ernst & Sohn.

BARNARD, J. L. (1973), Biological denitrification, *Water Pollut. Control* **72**, 705–720.

BARNARD, J. L. (1974), Cut P and N without chemicals, *Water Wastes Eng.* **11**, 33–36.

BARNARD, J. L. (1992), Design of prefermentation process, in: *Design and Retrofit of Wastewater Treatment Plants for Nutrient Removal* (RANDAL, C. W., BARNARD, J. L., STENSEL, H. D., Eds.), pp. 85–96. Lancaster, PA: Technomic Publishing Co.

BEGERT, A. (1985), Summen- und Gruppenparameter für organische Stoffe von Wasser und Abwasser, *Wiener Mitt. Wasser, Abwasser, Gewässer* **57**, G1–G29.

BÖHNKE, B. (1978), Möglichkeiten der Abwasserreinigung durch das Adsorptions-Belebungsverfahren, *Gewässerschutz – Wasser – Abwasser RWTH Aachen* **25**, 437–466.

BOES, M. (1991), Stickstoffentfernung mit intermittierender Denitrifikation – Theorie und Betriebsergebnisse, *Korrespondenz Abwasser* **38**, 228–234.

BRAUER, H. (1996), Aerobe und anaerobe biologische Behandlung von Abwässern im Hubstrahl-Bioreaktor, in: *Behandlung von Abwässern, Handbuch des Umweltschutzes und der Umweltschutztechnik*, Vol. 4, (BRAUER, H., Ed.), pp. 414–504. Berlin: Springer-Verlag.

BRINGMANN, G. (1961), Vollständige biologische Stickstoffeliminierung aus Klärwässern im Anschluß an ein Hochleistungsnitrifikationsverfahren, *Ges. Ing.* **82**, 233–235.

COOPER, P. F., COLLINSON, B., GREEN, M. K. (1977), Recent advances in sewage effluent denitrification: Part II, *Water Pollut. Control* **76**, 389–398.

EKAMA, G. A., BARNARD, J. L., GÜNTHERT, F. W., KREBS, P., MCCORQUODALE, J. A. et al. (1997), Secondary Settling Tanks: Theory, Modelling, Design and Operation. *IAWQ Scientific and Technical Report No. 6*. London: IAWQ.

EPA (1975), *Nitrogen Removal, Process Design Manual.* Environmental Protection Agency, Cincinnati, OH.

ERMEL, G. (1983), Stickstoffentfernung in einstufigen Belebungsanlagen – Steuerung der Denitrifikation. *Veröffentlichungen des Instituts für Stadtbauwesen TU Braunschweig* **35**.

GUJER, W., KAYSER, R. (1998), Bemessung von Belebungsanlagen auf der Grundlage einer CSB-Bilanz, *Korrespondenz Abwasser* **45**, 944–948.

HEINZ, A., RÖTHLICH, H., LEESEL, T., KOPMANN, T. (1996), Erfahrungen mit ring-lace Festbettreaktoren in der kommunalen Abwasserreinigung, *Wasser Luft Boden* **40**, No. 4, 18–20.

HENZE, M., GRADY, C. P. L. JR., GUJER, W., MARAIS, G. V. R., MATSUO, T. (1987), Activated Sludge Model No. 1. *IAWPRC Scientific and Technical Reports No. 1.* London: IAWPRC.

HENZE, M., GUJER, W., MINO, T., MATSUO, T., WENTZEL, M. C., MARAIS, G. V. R. (1995), Activated Sludge Model No. 2. *IAWQ Scientific and Technical Report No. 3.* London: IAWQ.

IMHOFF, K. (1955), Das zweistufige Belebungsverfahren für Abwasser, *gwf Wasser Abwasser* **96**, 43–45.

JENKINS, D., DAIGGER, G. T., RICHARD, M. G. (1993), *Manual on the Causes and Control of Activated Sludge Bulking and Foaming*, 2nd Edn. Chelsea, MI: Lewis.

JETTEN, M. S. M., HORN, S. J., VAN LOOSDRECHT, M. C. M. (1997), Towards a more sustainable municipal wastewater treatment system, *Water Sci. Technol.* **35**, No. 6, 171–180.

JÜBERMANN, O., KRAUSE, G. (1968), Die Zentralkläranlage der Erdölchemie GmbH und der Farbenfabriken Bayer AG in Dormagen; Ejektorbelüftung in der biologischen Abwasserreinigung, *Chem. Ing. Techn.* **40**, 288–291.

KAPPELER, J., GUJER, W. (1992), Estimation of kinetic parameters of heterotrophic biomass under aerobic conditions and characterization of wastewater for activated sludge modelling, *Water Sci. Technol.* **25**, No. 6, 125–139.

KAYSER, R. (1990), Process control and expert systems for advanced wastewater treatment plants, in: *Instrumentation and Control of Water and Wastewater Treatment and Transport Systems* (BRIGGS, R., Ed.), pp. 203–210. Oxford: Pergamon Press.

KAYSER, R. (1997), A 131 – Quo vadis, *Wiener Mitt. Wasser, Abwasser, Gewässer* **114**, 149–166.

KAYSER, R., ERMEL, G. (1984), Control of simultaneous denitrification by a nitrate controller, *Water Sci. Technol.* **16**, No. 10–12, 143–150.

KIENZLE, K. H. (1971), Untersuchungen über BSB_5 und Stickstoffelimination in schwachbelasteten

Belebungsanlagen mit Schlammstabilisation. *Stuttgarter Berichte zur Siedlungswasserwirtschaft*, Vol. 47. München: R. Oldenbourg.

KNOWLES, G., DOWNING, A. L., BARRETT, M. J. (1965), Determination of kinetics constants for nitrifying bacteria in mixed culture, *J. Gen. Microbiol.* **38**, 263–278.

KRAUTH, K., STAAB, K. F. (1988), Substitution of the final clarifier by membrane filtration within the activated sludge process; initial findings, *Desalination* **68**, 179–189.

KULPBERGER, R. J., MATSCH, L. C. (1977), Comparison of treatment of problem wastewater with air and high purity oxygen activated sludge systems, *Prog. Water, Technol.* **8**, No. 6, 141–151.

LANG, H. (1981), Nitrifikation in biologischen Klärstufen mit Hilfe des Bio-2-Schlamm-Verfahrens, *Wasserwirtschaft* **71**, 166–169.

LESSEL, T. H. (1991), First practical experiences with submerged rope-type biofilm reactors, *Water, Sci. Technol.* **23**, 825–834.

LUDZACK, F. J., ETTINGER, M. B. (1962), Controlling operation to minimize activated sludge effluent nitrogen, *J. WPCF* **35**, 920–931.

MATSCHÉ, N. F. (1972), The Elimination of nitrogen in the treatment plant of Vienna-Blumental, *Water Res.* **6**, 485–486.

MATSCHÉ, N. F. (1977), Removal of nitrogen by simultaneous nitrification–denitrification in an activated sludge plant with mammoth rotor aeration, *Prog. Water Technol.* **8**, No. 4/5, 625–637.

MIYAJI, Y., IWASAKI, M., SERVIGA, Y. (1980), Biological nitrogen removal by step-feed process, *Prog. Water, Technol.* **12**, No. 6, 193–202.

MOHLMAN, F. W. (1934), The sludge index, *Sewage Works J.* **6**, 119–122.

MORPER, M. R. (1994), Upgrading of activated sludge systems for nitrogen removal by application of the LINPOR® process, *Water Sci. Technol.* **29**, No. 12, 107–112.

NYHUIS, G. (1985), Beitrag zu den Möglichkeiten der Abwasserbehandlung bei Abwässern mit erhöhten Stickstoffkonzentrationen. *Veröffentlichungen des Instituts für Siedlungswasserwirtschaft und Abfalltechnik der Universität Hannover* **61**.

PASVEER, A. (1958), Abwasserreinigung im Oxidationsgraben, *Bauamt und Gemeindebau* **31**, 78–85.

PASVEER, A. (1964), Über den Oxidationsgraben, *Schweizerische Z. Hydrol.* **26**, 466–484.

PASVEER, A., SWEERIS, S. (1962), A New Development in Diffused Air Aeration. *T. N. O. Working Report A 27*. Delft, NL: TNO.

RANDAL, C. W., BARNARD, J. L., STENSEL, H. D. (1992), *Design and Retrofit of Wastewater Treatment Plants for Nutrient Removal.* Lancaster, PA: Technomic Publishing Co.

ROSENWINKEL, K.-H., WAGNER, J. (1998), Stand der Membrantechnik, *Gewässerschutz Wasser Abwasser RWTH Aachen* **165**, 28/1–28/14.

SCHERB, K. (1965), Vergleichende Untersuchungen über das Sauerstoffeintragsvermögen verschiedener Belüftungssysteme auf dem Münchener Abwasserversuchsfeld, *Münchener Beiträge zur Abwasser-, Fischerei- und Flußbiologie* **12**, 330–350.

SCHLEGEL, S. (1983), Nitrifikation und Denitrifikation in einstufigen Belebungsanlagen; Betriebsergebnisse der Kläranlage Lüdinghausen, *gwf Wasser Abwasser* **124**, 428–434.

SCHLEGEL, S. (1986), Der Einsatz von getauchten Festbettkörpern beim Belebungsverfahren, *gwf Wasser Abwasser* **127**, 421–428.

SEYFRIED, C. F., ABELING, U. (1992), Anaerobic–aerobic treatment of high strength ammonium wastewater; nitrogen removal via nitrite, *Prog. Water Technol.* **26**, No. 5/6, 1007–1012.

SEYFRIED, C. F., HARTWIG, P. (1991), Großtechnische Betriebserfahrungen mit der biologischen Phosphatelimination in den Klärwerken Hildesheim und Husum, *Korrespondenz Abwasser* **38**, 184–190.

SØRENSEN, J. (1996), Optimization of a nutrient removing wastewater treatment plant using on-line monitoring, *Water Sci. Technol.* **33**, No. 1, 265–273.

STENSEL, H. D., BARNARD, J. L. (1992), Principles of biological nutrient removal, in: *Design and Retrofit of Wastewater Treatment Plants for Nutrient Removal* (RANDAL, C. W., BARNARD, J. L., STENSEL, H. D., Eds.), pp. 25–84. Lancaster, PA: Technomic Publishing Co.

TAKAHASHI, M., SUZUKI, Y. (1995), Biological enhanced phosphorus removal process with immobilized microorganisms, *Veröffentlichungen des Instituts für Siedlungswasserwirtschaft und Abfallwirtschaft Universität Hannover* **92**, 22/1–22/12.

THOLANDER, B. (1977), An example of design of activated sludge plants with denitrification, *Prog. Water. Technol.* **9**, No. 4/5, 661–672.

THOMAS, E. A. (1962), Verfahren zur Entfernung von Phosphaten aus Abwässern, *Swiss Patent* No. 361 543.

TISCHLER, L. F. (1968), Linear removal of simple organic compounds in the activated sludge process, *Thesis*, The University of Texas, Austin, TX.

ULLRICH, A. J., SMITH, M. W. (1951), The biosorption process of sewage and waste treatment, *Sewage Ind. Wastes* **23**, 1248–1252.

VOGELPOHL, A. (1996), Hochleistungsverfahren und Bioreaktoren für die biologische Behandlung hochbelasteter industrieller Abwässer, in: *Behandlung von Abwässern, Handbuch des Umweltschutzes und der Umweltschutztechnik*, Vol. 4, (BRAUER, H., Ed.), pp. 391–413. Berlin: Springer-Verlag.

VON DER EMDE, W. (1964), Die Geschichte des Belebungsverfahrens, *gwf Wasser Abwasser* **105**, 755–780.

VON DER EMDE, W. (1971), Die Kläranlage Wien Blumental, *Oesterr. Wasserwirtsch.* **23**, 11–18.

WAREHAM, D. G., HALL, K. J., MAVINIC, D. S. (1993), Real-time control of aerobic anoxic sludge digestion using ORP, *ASCE J. Env. Eng.* **119**, 120–136.

WICHT, H. (1995), N_2O-Emissionen durch den Betrieb biologischer Kläranlagen, *Veröffentlichungen des Instituts für Siedlungswasserwirtschaft TU Braunschweig* **58**.

WICHT, H., BEIER, M. (1995), N_2O-Emissionen aus nitrifizierenden und denitrifizierenden Kläranlagen, *Korrespondenz Abwasser* **42**, 404–413.

WUHRMANN, K. (1964), Stickstoff- und Phosphorelimination, Ergebnisse von Versuchen im technischen Maßstab, *Schweiz. Z. Hydrol.* **26**, 520–558.

WUHRMANN, K., VON BEUST, F. (1958), Zur Theorie des Belebtschlammverfahrens II. Über den Mechanismus der Elimination gelöster organischer Stoffe aus Abwasser, *Schweiz. Z. Hydrol.* **20**, 311–330.

ZACHARIAS, B. (1996), Biologische Stickstoffelimination hemmstoffbelasteter Abwässer am Beispiel eines Eisenhüttenwerkes, *Veröffentlichungen des Instituts für Siedlungswasserwirtschaft der TU Braunschweig* **60**.

ZACHARIAS, B., KAYSER, R. (1995), Treatment of steel works effluent with a conventional single sludge system built in cascades, *Proc. 50th Ind. Waste Conf.*, Purdue University (WUKASCH, R. F., Ed.), pp. 705–716. Chelsea, MI: Ann Arbor Press.

ZANDERS, E., GROENEWEG, J., SOEDER, C. J. (1987), Blähschlammverminderung und simultane Nitrifikation/Denitrifikation, *Wasser-Luft-Betrieb* **31**, No. 9, 20–24.

ZEPER, J., DE MAN, A. (1970), New developments in the design of activated sludge tanks with low BOD loadings, in: *Advances in Water Pollution Research*, Vol. 1 (JENKINS, S. H., Ed.), pp. II-8/1–10. Oxford: Pergamon Press.

14 Biological and Chemical Phosphorus Elimination

Georg Schön

Freiburg, Germany

Norbert Jardin

Essen, Germany

List of Abbreviations

Alk	alkalinity [mM]
ASM	activated sludge model
ATP	adenosine triphosphate
β	relative chemical dosage [MMe(MP)$^{-1}$]
COD	chemical oxygen demand [mg L^{-1} COD]
COD_{tot}	total COD concentration [mg L^{-1} COD]
DAPI	4,6-diamidino-2-phenylindole
e	specific phosphorus elimination [–]
EASC	extended anaerobic sludge contact
EBPR	enhanced biological phosphorus removal
ISAH	Institut für Siedlungswirtschaft und Abfalltechnik, Universität Hannover
JHB	Johannesburg process
K_x	potassium content of activated sludge [mg K (g SS)$^{-1}$]
Mg_x	magnesium content of activated sludge [mg Mg (g SS)$^{-1}$]
$\eta_{Hydrolysis}$	yield of hydrolysis [S_{bs} $COD_{tot}{}^{-1}$]
NVS	non-volatile solids [mg L^{-1}]
PE	population equivalent [60 g (PE d)$^{-1}$ BOD_5]
PHB	poly-3-hydroxy butyric acid
PHV	poly-3-hydroxy valeric acid
P_{tot}	total phosphorus concentration [mg L^{-1}]
P_x	phosphorus content of activated sludge [mg P (g SS)$^{-1}$]
S_A	short-chain organic acid [mg L^{-1} COB]
S_{AL}	aluminium [mg L^{-1} Al]
S_{bs}	easily biodegradable substrate [mg L^{-1} COB]
S_{Fe_2}	iron(II) [mg L^{-1} Fe]
S_{Fe_3}	iron(III) [mg L^{-1} Fe]
S_I	inert soluble substrate [mg L^{-1} COD]
SS	suspended solids [mg L^{-1}]
TS	total solids [mg L^{-1}]
UCT	University of Cape Town process
VD	volume for denitrification [m^3]
VN	volume for nitrification [m^3]
VSS	volatile suspended solids [mg L^{-1}]
WAS	waste activated sludge
WWTP	wastewater treatment plant
X_I	inert particulate substrate [mg L^{-1} COB]
X_{PP0}	polyphosphate concentration before P release [mg L^{-1} P]
X_S	soluble particulate substrate [mg L^{-1} COB]

1 Introduction

Municipal sewage plants with biological purification steps were, until some years ago, usually designed and optimized to achieve the most complete possible removal of organic pollutants from the wastewater. The remaining inorganic nutrients which find their way into natural waters are, however, a major worldwide problem. This increased input of nitrogen and phosphorus may lead to a disturbance of the ecological balance in lakes, slow flowing rivers, and water reservoirs as well as in the shallow regions of marine waters along seacoasts. An overly rich provision of inorganic nutrient salts leads to eutrophication of the waters. This can set off an "algal bloom" – a massive development of algae. The well-known consequences are: foam and slime disturbances on bathing beaches and the formation of toxins by cyanobacteria, which can lead to skin problems and breathing difficulties in humans, and death of animals. The O_2 respiration of the algae at night and especially the later biomass decomposition by bacteria primarily cause an oxygen shortage in the waters and a release of H_2S, among other effects. Secondarily, an O_2 shortage causes a reduction of nitrate to toxic nitrite and release of phosphate from the sediment. The processing of drinking water from eutrophied waters additionally demands considerably more effort for purification. In addition to the elimination of the organic pollutants one of the most important goals in modern wastewater technology today is, therefore, to remove a substantial proportion of the nitrogen and phosphorus.

Nitrogen can easily be eliminated microbiologically by means of nitrification and denitrification. Thereby, the major part of the nitrogen is converted into the gaseous state (N_2), thus being effectively removed.

In natural waters, inorganic nitrogen compounds (e.g., nitrate and ammonium) are not obligatory for the growth of cyanobacteria (blue-green algae), because they can assimilate molecular nitrogen into cell material. These microorganisms also possess a photoautotrophic metabolism and utilize carbon dioxide (CO_2) as a carbon source. Thus, in fresh water the concentration of phosphate is of crucial importance for their growth rate and the increase in biomass. Eutrophication can only be prevented at a concentration of phosphate $<10\ \mu g\ L^{-1}$ (DRYDEN and STERN, 1968).

In contrast to the nitrogenous compounds, phosphorus can only be removed from wastewater in its solid state after sedimentation. This can be achieved either through its incorporation into the biomass of the sewage sludge or by chemical precipitation as a phosphate salt. In conventionally operated wastewater treatment plants with a biological purification step and an influent phosphorus concentration of 7–12 mg L^{-1}, normally only 30–40% of the phosphorus is removed with the primary sludge and due to the growth of the biomass (NESBITT, 1969; JARDIN, 1995). But, to avoid a massive development of the algae in natural waters, over 90% would have to be removed (SCHAAK et al., 1985). This means that in the final effluent of the sewage plants only 0.5–1.0 mg P L^{-1} can be permitted. In order to attain such a low phosphorus concentration in the effluent (<1 mg L^{-1}), various chemical precipitation procedures have been developed. However, through special operating procedures, it is also possible to achieve a microorganism-mediated enhanced elimination of phosphorus. To obtain an efficient P uptake by the biomass the activated sludge must flow in a cyclic process through an aerobic and an anaerobic phase. In the aerobic phase specific bacteria in the sludge flocs (called poly-P bacteria) store phosphorus intracellularly in a high concentration as polyphosphate (LEVIN and SHAPIRO, 1965; YALL et al., 1970; MEGANCK and FAUP, 1988; ATV, 1989, 1994; BEVER and TEICHMANN, 1990; TOERIEN et al., 1990; SCHÖN, 1994; VAN LOOSDRECHT et al., 1997). The anaerobic phase is necessary for the enrichment of the poly-P bacteria and their storage of organic reserve material (see Sect. 2.2.3).

2 Enhanced Biological Phosphorus Removal

2.1 Historical Development

In 1955, the first publication appeared on the subject of a phosphorus removal above and beyond the "normal" extent in batch cultures of sewage sludge (GREENBERG et al., 1955). Several authors independently reported that variations in the amount of air introduced were responsible for the strongly varying phosphorus values observed from time to time in the final effluent of wastewater treatment plants (SRINATH et al., 1959; ALARCON, 1961).

The enhanced removal of phosphorus by the activated sludge was confirmed through extensive experiments on aerobic phosphorus uptake and anaerobic release and was clarified as a biochemical process (LEVIN and SHAPIRO, 1965). On the basis of their experimental results LEVIN and SHAPIRO (1965) developed the flow chart of a modified activated sludge process from which the "Phostrip" process arose.

In 1967 high efficiency rates of phosphorus removal in activated sludge sewage plants were reported in San Antonio, Texas (VACKER et al., 1967). Fluctuations in the inflow, low concentrations of oxygen, recycling of phosphorus containing sludge liquor, the occurrence of nitrification, and a partial stabilization of the sludge were all recognized as factors influencing the removal of phosphorus. Similar observations were made at other sewage plants in the USA (MILBURY et al., 1971; GARBER, 1972; YALL et al., 1970).

In the 1970s intensive investigations to explore the basic principles of the enhanced biological phosphorus removal were made, especially in South Africa. In particular, in the studies of BARNARD (1974) it was shown that an enhanced biological P elimination was possible not only in highly loaded but also in lowly loaded processes. He developed a process for the simultaneous removal of nitrogen and phosphorus which he named "bardenpho", a combination of the abbreviations for *BAR*NARD, *den*itrification, and *pho*sphorus elimination.

Basic microbiological investigations led to the isolation and identification of *Acinetobacter* from polyphosphate accumulating sludge (FUHS and CHEN, 1975). Bacteria of this genus have been regarded since then as being primarily responsible for the enhanced biological phosphorus removal. However, these organisms seem not be involved to any substantial extent in P elimination in full scale wastewater treatment plants (see Sect. 2.4). With the help of electron microscopy and other physical-chemical research methods it was possible to demonstrate polyphosphate inclusions in the bacterial cells of the activated sludge (BUCHAN, 1983). Several phosphate compounds could be detected in the active biomass (FLORENTZ et al., 1984). The increase of condensed phosphate under aerobic conditions and the decrease under anaerobic conditions were shown. Other research groups investigated the intracellular phosphate compounds by chemical fractionation (MINO et al., 1984; ARVIN, 1985). With these methods, the release and re-uptake of phosphate by the activated sludge could be followed under a variety of environmental conditions.

Today there is no doubt about the fact that specific bacteria are responsible for the extensive phosphate uptake in activated sludge (Sect. 2.4).

2.2 Microbiological Aspects

2.2.1 Phosphorus Uptake by Microorganisms

All organisms contain phosphorus in numerous cell structures. It is an integral part of the cellular metabolism (e.g., phospholipids, nucleic acids), needed for the energy supply (e.g., ATP), and for biosynthesis (e.g., phosphorylated sugars). The phosphorus content (P) constitutes approximately 3% of the cell dry mass. Thus, under good growth conditions, dense populations of microorganisms can accumulate a considerable amount of phosphorus in a relatively short time. Besides this growth-dependent assimilation of phosphorus in many microorganisms an additional uptake can occur, going far beyond the amount necessary for growth and multiplication (HAROLD, 1966; KULAEV and VAGABOV, 1983). Such organisms can accumulate up to 12% phosphorus on a dry weight basis. The additional phosphorus is stored by the cells in the form of polyphosphate.

Polyphosphates are salts of the polyphosphoric acids, which primarily form long chains of phosphate residues, $(-PO_3H_n)$, with n being approximately 100. Their negative charges are neutralized by cations (K^+, Mg^{2+}, Ca^{2+}) (Kornberg, 1995).

$$ {}^{-}O-\underset{O^-}{\underset{|}{\overset{O}{\overset{\|}{P}}}}-O-\underset{O^-}{\underset{|}{\overset{O}{\overset{\|}{P}}}}-\left[O-\underset{O^-}{\underset{|}{\overset{O}{\overset{\|}{P}}}}\right]_n-O-\underset{O^-}{\underset{|}{\overset{O}{\overset{\|}{P}}}}-O^- \qquad (1) $$

Polyphosphate

The individual monomers are coupled together by energyrich acid–anhydride bonds, in the same way as in adenosine triphosphate (ATP). Hence, in heterotrophic microorganisms the synthesis of polyphosphates can only occur when a suitable organic substrate is available as an energy source for these cells.

The deposition of polyphosphate is dependent on a number of environmental factors, e.g., phosphate and substrate content of the medium, as well as the concentrations of various metal ions. Polyphosphate can be deposited in several different cell compartments, in bacteria predominantly in inclusion bodies (as volutin, i.e., metachromatic granula) in the cytoplasm, but also in the intracytoplasmic space, between the cell membrane and the cell wall (Halvorson et al., 1987; Streichan and Schön, 1991). Beside phosphorus and the metal ions, the polyphosphate granules contain small amounts of additional components, e.g., proteins and lipids. These poly-P granula can be detected by staining with methylene blue (Neisser staining), toluidine blue (as violet inclusions) and 4,6-diamidino-2-phenylindole (DAPI, as bright yellow fluorescent inclusions in the blue fluorescing cell). In electron microscopic thin sections poly-P granula appear as electron dense (black) inclusions in the cells (Fig. 1).

Polyphosphates are often used by the cells as a phosphorus source and as an energy source under energy limiting conditions (Harold, 1966; Kulaev and Vagabov, 1983; Kornberg, 1995). Most probably, they are also active as regulators in metabolism, e.g., to maintain the content of phosphate ions and the concentration of cations in the cell at a constant level.

2.2.2 Phosphorus Uptake by Activated Sludge in Wastewater Treatment Plants

The major phosphorus compounds in wastewater are dissolved (*ortho*-)phosphate (PO_4^{3-}), inorganic polyphosphate, and organically bound phosphorus. Polyphosphate and other condensed phosphorus compounds are normally hydrolyzed quite rapidly by exoenzymes. Consequently, in the aeration tank most of the dissolved phosphorus is found in the form of phosphate. In conventional wastewater treatment plants only a portion of the dissolved phosphate is fixed by the microorganisms for the synthesis of biomass. Theoretically, for aerobic growth a balanced ratio of C:N:P would be approximately 100:14:3. In municipal wastewater after primary clarification the actual ratios are approximately 60:12:3, so that the P content of the wastewater exceeds the amount necessary for normal growth by about 40% (Schönborn, 1986).

By using special operating procedures in which the activated sludge in the treatment plant is subjected to alternating anaerobic and aerobic conditions – temporally or spatially – the biocenosis is altered and polyphosphate accumulating bacteria are enriched in the activated sludge. Thereby, phosphorus becomes concentrated in the biomass (Levin and Shapiro, 1965; Yall et al., 1970; Meganck and Faup, 1988; ATV, 1989, 1994; Toerien et al., 1990; Schön, 1994). Substantially more phosphorus is removed from the wastewater with the excess sludge than in a conventionally operated treatment plant.

In wastewater treatment plants phosphate can also be bound by physical-chemical processes, in addition to its uptake for normal growth and for the storage of polyphosphate. This non-biological fixation is favored by the metabolic activity of the microorganisms and depends on the composition of the wastewater, primarily on the calcium content and the pH value (Arvin and Kristensen, 1985; Streichan and Schön, 1991). In most treatment plants the biologically induced chemical precipitation appears to play only a minor role in the aerated zone. In contrast, under anaerobic or anoxic conditions, the proportion of

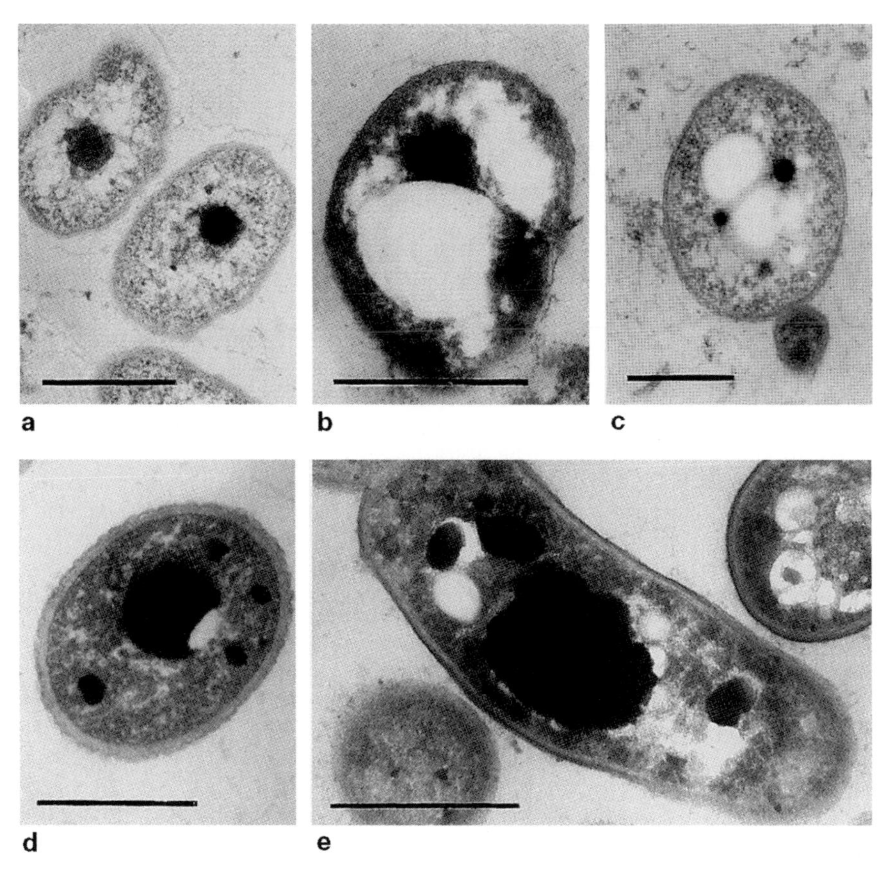

Fig. 1a–e. TEM electron micrographs of polyphosphate accumulating bacteria from different sewage plants, (**a–c**) from plants with nitrification, (**d, e**) from highly loaded plants. Bars represent 0.5 µm; dense (black) bodies: polyphosphate; bright (white) bodies: poly-β-hydroxy butyrate (Photograph by Dr. J. GOLECKI, Freiburg, Germany; from SCHÖN, 1996).

the chemically fixed phosphate may reach higher values due to the increased concentration of phosphate. The activated sludge of conventionally operated treatment plants usually shows a phosphorus content of about 1–2%. For sludge with enhanced biological phosphorus removal the P content usually is above 2% and can reach values of 5%. The effluent phosphorus concentration in bio-P plants is, therefore, reduced to values of less than 1 mg L^{-1}.

2.2.3 Anaerobic P Release and Aerobic P Uptake by Activated Sludge

A key function for the enhanced phosphorus removal in the aeration tank is the incorporation of an anaerobic zone within the conventional plant layout (FUHS and CHEN, 1975; NICHOLLS and OSBORN, 1979; RENSINK et al., 1986).

During the anaerobic phase the phosphorus bound aerobically as polyphosphate is partially released as phosphate into the sludge liquor (Fig. 2).

The P concentration in the final effluent is nonetheless strongly reduced, since the subsequent aerobic phosphate uptake is higher than the former P release. This anaerobic phosphate release, which is a prerequisite for an effective subsequent aerobic P elimination (Sect. 2.2.6) is inhibited by nitrate. Therefore, in nitrifying sewage plants a denitrification must be carried out, so that the return sludge discharged into the anaerobic zone is nearly free of nitrate. If these prerequisites are not completely fulfilled, the P elimination is usually unsatisfactory.

In the anaerobic zone a series of biochemical processes occur. However, their mechanisms are only partially known. The obligatorily aerobic, poly-P bacteria cannot grow anaerobically, but they take up substrate (e.g., acetate) and store it as lipid reserve material. In this biosynthetic process the polyphosphate in the cells is utilized as energy source, and phosphate is released (NICHOLLS and OSBORN, 1979; WENTZEL et al., 1986). The degradation of poly-P under anaerobic conditions and the synthesis of reserve material can be described as follows (HENZE et al., 1997):

$$
\begin{aligned}
2\,C_2H_4O_2 + (HPO_3)_n\,[\text{poly-P}] \\
+ H_2O \rightarrow (C_2H_4O_2)_2 \\
[\text{stored organic material}] \\
+ PO_4^{3-} + 3\,H^+
\end{aligned}
\tag{2}
$$

The increase of soluble phosphate from polyphosphate hydrolysis is, therefore, correlated with the synthesis of osmotic inert poly-β-hydroxy alkanoates [e.g., poly-3-hydroxy butyric acid (PHB) or poly-3-hydroxy valeric acid (PHV)] as storage material (FUKASE et al., 1984; SCHÖN et al., 1993). The higher this anaerobic phosphate release and thus the formation of lipid storage material, the better the

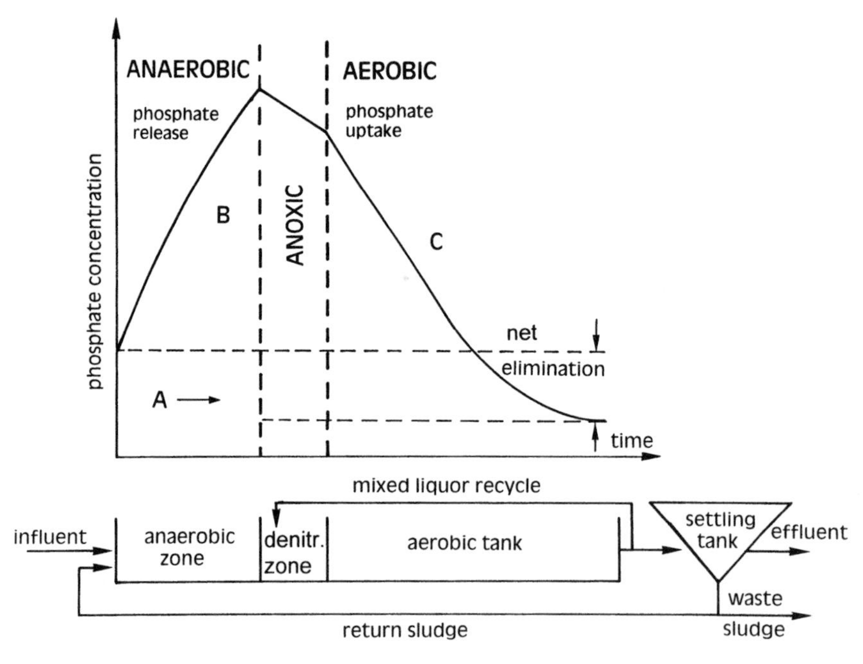

Fig. 2. Anaerobic phosphate release and aerobic or anoxic phosphate uptake by activated sludge from a sewage plant with enhanced biological phosphorus removal. Since more phosphate is taken up aerobically than is released anaerobically, the phosphorus becomes enriched in the biomass and can be removed with the excess sludge. In the anoxic region the rate of phosphate uptake is usually much lower than in the aerated zone. Phosphate concentration in wastewater: (A) inflowing phosphate, (B) released phosphate from the returned sludge, (C) phosphate taken up by the activated sludge.

subsequent phosphorus elimination under aerobic conditions (WENTZEL et al., 1985). The deposition of PHB can easily be recognized by staining the sludge flocs with the fluorescent dye Nile blue. The cells show a bright orange fluorescence.

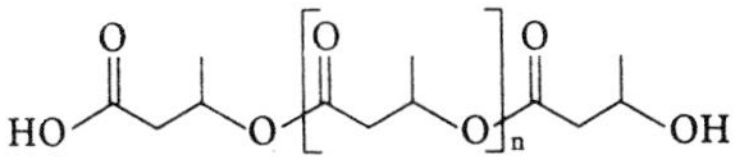

HO–[structure]–OH (3)

Poly(3-hydroxy butyric acid) (PHB)

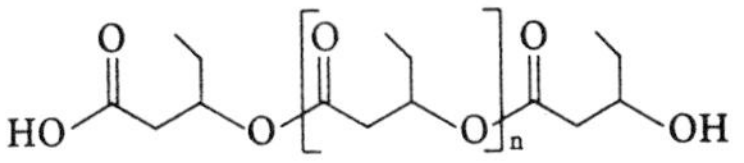

HO–[structure]–OH (4)

Poly(3-hydroxy valeric acid) (PHV)

Substrates for lipid synthesis include short-chain organic acids like acetic and propionic acid. They are mainly produced by fermentation processes in the sewerage system or are excreted by facultative anaerobic bacteria in the anaerobic zone (O_2 and NO_3 free) as fermentation end products from easily fermentable substrates.

In the anaerobic zone organic substrates are also adsorbed on the sludge flocs or the slime layer of the bacteria. These substrates can be rapidly utilized by the floc bacteria under aerobic or anoxic conditions (SCHÖN et al., 1993).

An efficient P elimination depends on the availability of an adequate supply of easily utilizable substrates (MARAIS et al., 1983; GERBER et al., 1986). In simplified form the aerobic accumulation of polyphosphate can be described as follows (HENZE et al., 1997):

$$\begin{aligned} C_2H_4O_2 + 0.16\,NH_4^+ + 1{,}2\,O_2 \\ + 0.2\,PO_4^{3-} \rightarrow 0.16\,C_5H_7NO_2 \\ + 1.2\,CO_2 + 0.2\,(HPO_3)_n\ \text{[poly-P]} \\ + 0.44\,OH^- + 1.44\,H_2O \end{aligned} \quad (5)$$

Under anoxic conditions P removal can also be observed (YALL et al., 1970; SCHÖN and STREICHAN, 1989; KERN-JESPERSEN and HENZE, 1993). The accumulation of poly-P under anoxic conditions can be described (simplified) as follows (HENZE et al., 1997):

$$\begin{aligned} C_2H_4O_2 + 0.16\,NH_4^+ + 0.2\,PO_4^{3-} \\ + 0.96\,NO_3^- \rightarrow 0.16\,C_5H_7NO_2 \\ + 1.2\,CO_2 + 0.2\,(HPO_3)_n\ \text{[poly-P]} \\ + 1.4\,OH^- + 0.48\,N_2 + 0.96\,H_2O \end{aligned} \quad (6)$$

The rate of the anoxic phosphate uptake is generally lower compared to oxic conditions. Recently, however, processes have been proposed in which a high P elimination can be maintained by cycling between anaerobic and denitrifying conditions (KUBA et al., 1997; VAN LOOSDRECHT et al., 1997).

2.2.4 Aerobic Polysaccharide Storage

Besides polyphosphate, under aerobic conditions polysaccharides are also stored in the biomass (ARUN et al., 1988; MATSUO et al., 1992; SATOH et al., 1992; SCHÖN et al., 1993). Thin-layer chromatography and enzymatic analyses have confirmed that the storage material consists primarily of glycogen-like polyglucose.

The importance of the polysaccharides for the poly-P bacteria is still not completely clarified. Probably they are important for the survival of the obligate aerobic poly-P bacteria under anaerobic conditions and provide the reduction equivalents for the synthesis of the lipid storage material (SATOH et al., 1992).

2.2.5 Dependence of Phosphorus Elimination on the Phosphate Concentration

The maximum of phosphate uptake also depends on the phosphate content in the wastewater. In highly loaded activated sludge, bacteria with very large polyphosphate granula can be observed (Fig. 1d, e). The storage capacity in the lowly loaded sludge appears, in contrast, to be less pronounced since here usually only small polyphosphate inclusions can be observed in the bacteria (Fig. 1a–c).

With increasing phosphate concentration (up to a threshold value of ca. 60 mg L^{-1} P), both, the rate of the biological phosphate uptake and the maximum phosphorus elimination increase. Activated sludge from highly loaded plants (without nitrification) shows a substantially higher uptake rate than sludge from lowly loaded plants (with nitrification).

A low concentration of phosphate in the influent wastewater or a simultaneous chemical precipitation will reduce the absolute amount of phosphate which can be fixed biologically.

2.2.6 Inhibition of the Phosphate Release by Nitrate

The phosphate release and hence the synthesis of lipid storage material are inhibited under oxygen free conditions by nitrate (HASCOET and FLORENTZ, 1985). Since the anaerobically stored lipids play a major role in the subsequent aerobic phosphate uptake and polyphosphate storage, the aerobic phosphorus elimination is also reduced. Hence, in plants with a high nitrate load or with nitrate formation through a nitrification the entry of nitrate into the anaerobic zone must be kept at the lowest level possible through an effective denitrification.

It is assumed that under anoxic conditions a competition between "normal" denitrifiers and "poly-P bacteria" for the available substrate is responsible for the nitrate inhibition (MOSTERT et al., 1988). An inhibition of the enzymes involved in phosphate metabolism has also been discussed (LÖTTER and VAN DER MERWE, 1987). It is noteworthy that nitrogen monoxide (NO), an intermediate product of denitrification, strongly inhibits phosphate release in the activated sludge (KORTSTEE et al., 1994). It is also possible that the increase of the redox potential by nitrate influences the P release (IWEMA and MEUNIER, 1985). However, the nitrate effect could also involve regulatory processes which depend on the energy content and the reduction state ($NADH/NAD^+$ ratio) of the cells, as well as on the growth conditions. Most likely, several factors are involved in the interference with the phosphate release, depending on the composition of the influent wastewater and the bacterial flora.

2.3 Biochemical Models of Enhanced Phosphorus Uptake and Release

On the basis of the findings during laboratory scale and pilot scale experiments, observations on full scale plants, and knowledge about interactions in bacterial metabolism, several models for enhanced phosphorus metabolism have been developed (COMEAU et al., 1986; MINO et al., 1987; WENTZEL et al., 1991; MINO et al., 1994) (Fig. 3a–c).

In the aerated stage the bacteria utilize the anaerobically stored lipid reserve material and easily biodegradable exogenous substrates as energy and carbon source (Fig. 3a). Through consumption of the endogenous storage material the cells can also adapt quickly to the aerobic milieu and immediately begin with respiration and growth. The energy (ATP) yield from the respiration of endogenous and exogenous substrates serves, at the same time, for the uptake of phosphate, its storage as polyphosphate, and the synthesis of polyglucose (glycogen).

In the anaerobic stage probably several different metabolic processes occur, which are important for the subsequent enhanced phosphorus elimination in the aerated zone (Fig. 3b, c). The fermentative metabolism of facultative anaerobic bacteria produces predominantly short-chain organic acids, such as acetic and propionic acid, which are taken up by the poly-P bacteria, converted and stored as lipid reserve materials (e.g., poly-β-3-hydroxy butyric acid, PHB). The energy for the lipid synthesis is provided by the polyphosphate previously stored under aerobic conditions. Soluble phosphate is again released into the mixed liquor through the partial degradation of the polyphospate. The polyphosphates also most likely serve as an energy source for the survival of the cells under anaerobic conditions.

In anoxic zones (i.e., NO_3 containing, without O_2) the P dynamics can proceed either as in the aerobic or in the anaerobic stage, depending on the bacterial species involved and the metabolic conditions.

Through the storage of organic reserve material under anaerobic conditions, the obligate aerobic poly-P bacteria appear to have growth

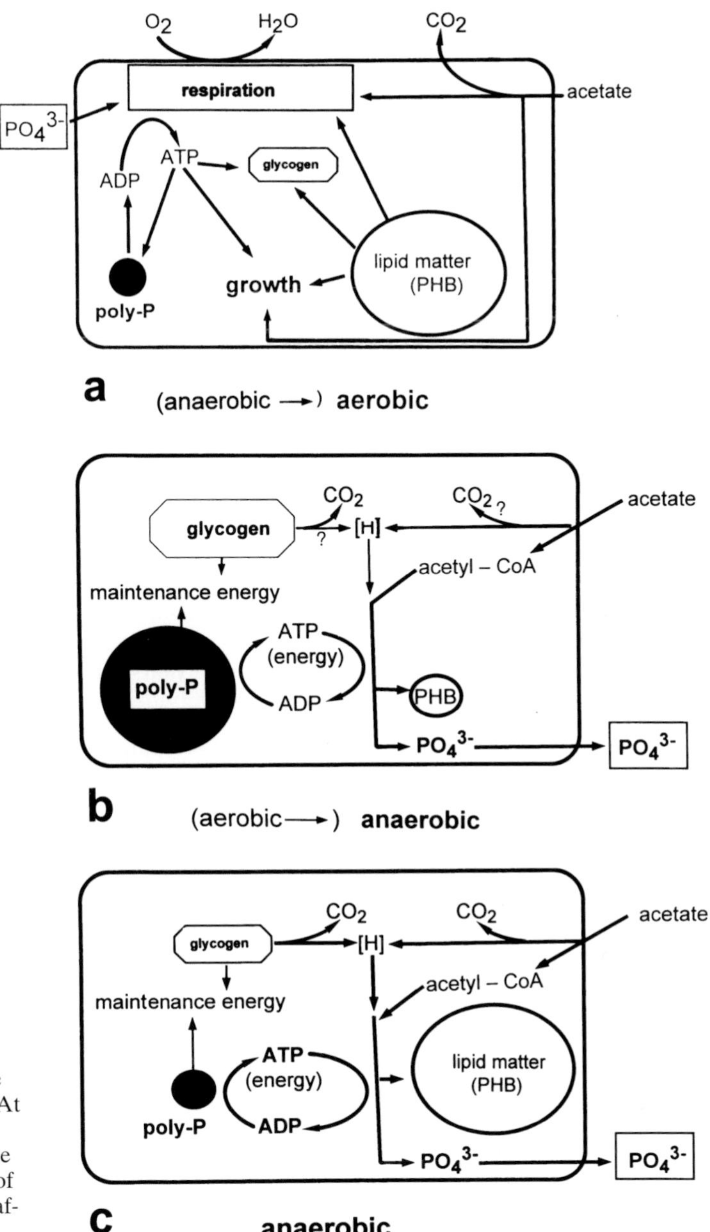

Fig. 3a–c. Bio-P models. Storage material and P metabolism. (**a**) At the beginning of the aerobic phase, (**b**) at the beginning of the anaerobic phase, (**c**) at the end of the anaerobic phase (modified after WENTZEL et al., 1991).

advantages over other obligate aerobes, which depend exclusively on exogenous substrates for their aerobic growth. These metabolic characteristics are probably responsible for the fact that the polyphosphate storing bacteria become enriched in activated sludge under alternating anaerobic-aerobic operating conditions.

2.4 Polyphosphate Accumulating Activated Sludge Bacteria

Until now it has not been possible to find any direct relationship between biological phosphorus elimination and the enrichment of particular species of poly-P bacteria in large scale wastewater plants. Despite numerous investigations it is still unclear which specific poly-P bacteria are predominantly responsible for the enhanced phosphorus elimination.

Based on conventional isolation techniques many authors have concluded that representatives of the genus *Acinetobacter* play a dominant role in the biological enhanced elimination of phosphorus in the activated sludge (e.g., FUHS and CHEN, 1975; DEINEMA et al., 1980, 1985; LÖTTER and MURPHY, 1985; WENTZEL et al., 1986; STEPHENSON, 1987).

Therefore, polyphosphate accumulating strains of these bacteria have been preferred objects for investigations of the phosphate metabolism (KORTSTEE et al., 1994). The models developed by WENTZEL et al. (1986, 1991) on the P dynamics in the activated sludge are mainly based on bacteria of this genus.

However, investigations by CLOETE and STEYN (1988) showed that *Acinetobacter* cannot solely be responsible for the phosphorus elimination in wastewater treatment plants. On the basis of immunological tests on the *Acinetobacter* populations in different activated sludges, they estimated that strains of this genus were responsible for only 5–15% (up to a maximum of 34%) of the phosphate uptake. Furthermore, in electron microscopic investigations of activated sludge from various plants with excess biological P elimination, morphologically very different gram-positive and gram-negative cells with polyphosphate inclusions were observed, including filamentous forms (STREICHAN et al., 1990). Assays of the *Acinetobacter* population via the polyamine pattern of the floc bacteria (AULING et al., 1991) or with specific rRNA targeted oligonucleotide probes (WAGNER et al., 1994; BOND et al., 1995) indicate that this genus plays only a minor role in the overall process of phosphorus elimination in most full scale plants.

A number of other gram-negative, but also gram-positive bacteria isolated from wastewater deposit polyphosphates in high concentrations, e.g., *Pseudomonas, Arthrobacter, Corynebacterium,* and *Microlunatus* species (SURESH et al., 1985; HIRAISHI et al., 1989; KÄMPFER et al., 1990; NAKAMURA et al., 1991; UBUKATA and TAKII, 1994; NAKAMURA et al., 1995).

Electron microscopic and population dynamic investigations indicate that the spectrum of the polyphosphate storing bacteria can be quite variable depending on the operating procedures and the wastewater composition. For full scale plants the importance of the genus *Acinetobacter* in biological phosphorus elimination appears to have been overestimated. Representatives of a variety of other genera of bacteria probably participate to a greater extent in phosphorus elimination. In general, gram-positive bacteria apparently have a greater importance in wastewater treatment plants than had been previously assumed.

Along with the unicellular polyphosphate storing bacteria, filamentous forms have been observed more and more frequently in recent years. These cause floating sludge and scum both in plants with enhanced biological phosphorus removal as well as in those with chemical phosphate precipitation (SCHÖN, 1994). Filamentous forms primarily represent *Microthrix parvicella*, but sometimes also nocardioform bacteria. The reasons for the massive occurrence of these filamentous bacteria are largely unknown. Fatty substances, possibly also tensides, favor the growth of the nocardioforms (LEMMER, 1985) and *Microthrix* (SLIJKHUIS, 1983). It is biologically interesting that these floating sludge formers can also store organic reserve materials along with polyphosphate (SCHÖN et al., 1993). Thus they possess two important metabolic properties which usually characterize the polyphosphate storing bacteria. The extent to which these bacteria are directly involved in enhanced phosphorus elimination is still unclear.

The activated sludge in sewage plants represents a mixed culture of bacteria from various physiological groups, which all influence each other (BRODISCH, 1985). Thus it cannot be excluded that the strains of poly-P bacteria, which in pure culture do not completely correspond to the "hypothetical poly-P bacterium", behave metabolically differently in activated sludge in symbiotic growth together with a variety of other types of floc bacteria than under the conditions of a pure culture. Most likely, the observed changes of the concentrations of organic storage materials and of phosphate in the activated sludge upon changing from aerobic to anaerobic conditions are not caused by one physiological bacterial type, but rather by the mutual influences of various bacterial species in the biocenosis of the activated sludge on each other's metabolism.

2.5 Process Design

As the discussion of the microbiological principles has shown, the integration of an anaerobic zone or anaerobic times during the biological wastewater treatment is a prerequisite to meet the requirements for the enhanced biological phosphorus removal process. Two distinct process designs for the implementation of anaerobic phases can be distinguished:

(1) the so-called mainstream process in which the entire inflow and the activated sludge are exposed to anaerobic conditions and
(2) the sidestream process, which combines the enhanced biological phosphorus removal process with physico-chemical phosphorus elimination.

In contrast to the mainstream process, in the latter plant design for phosphorus removal, only part of the activated sludge is held for some time under anaerobic conditions.

2.5.1 Mainstream Process

The flow scheme and the reactor configuration of a plant with the EBPR process in the mainstream configuration and nitrogen elimination are schematically shown in Fig. 4a.

Besides this basic principle, numerous alternatives have been developed to optimize the side conditions of the enhanced phosphorus elimination, mainly to minimize the nitrate and oxygen inputs into the anaerobic zone and to ensure that nearly all of the easily biodegradable substrate is available for the enhanced biological phosphorus removal. Some different mainstream processes are shown in Fig. 4b–d.

Based on a survey made in 1994, the most commonly used process for enhanced biological phosphorus removal in Germany is the Phoredox process, as diagrammed in Fig. 5.

Different approaches for specifying the dimensions of the anaerobic zone have been developed. Besides the use of dynamic simulation models like the activated sludge model No. 2 (ASM No. 2) (HENZE et al., 1995a), experience in Germany over the last century has shown that, given the varying process stability of enhanced biological phosphorus removal, it is advisable to use more empirical approaches for the process design and the dimensioning of the anaerobic zone (ATV, 1994; Anonymous, 1995).

In these design procedures the anaerobic zone is an integrated part of the overall volume necessary for nitrogen elimination, and its activity will mainly depend on wastewater temperature. The relation between the actual operating temperature and the volume necessary for nitrification and denitrification is shown in Fig. 6 for a wastewater treatment plant with 100,000 population equivalents.

Here, above the design temperature of 10 °C, additional volume is available for enhanced biological phosphorus removal. As the design temperature is only reached in winter (for 10–15% of the year) this additional volume can be used more than 80% of the time, during which the phosphorus will be eliminated mainly by biological processes. To achieve the effluent criteria for phosphorus during winter, when the anaerobic zone cannot be used for EBPR, chemicals for phosphate precipitation are usually added.

In plants with wastewater conditions favorable for the EBPR process, year-round biological phosphorus removal is possible. The major design criterion for the size of the anaerobic zone in these cases is the contact time, which should exceed 0.8 h.

With the mainstream process for enhanced biological phosphorus removal an average effluent concentration of total phosphorus below 1 mg L^{-1} can be achieved under optimal influent and operating conditions.

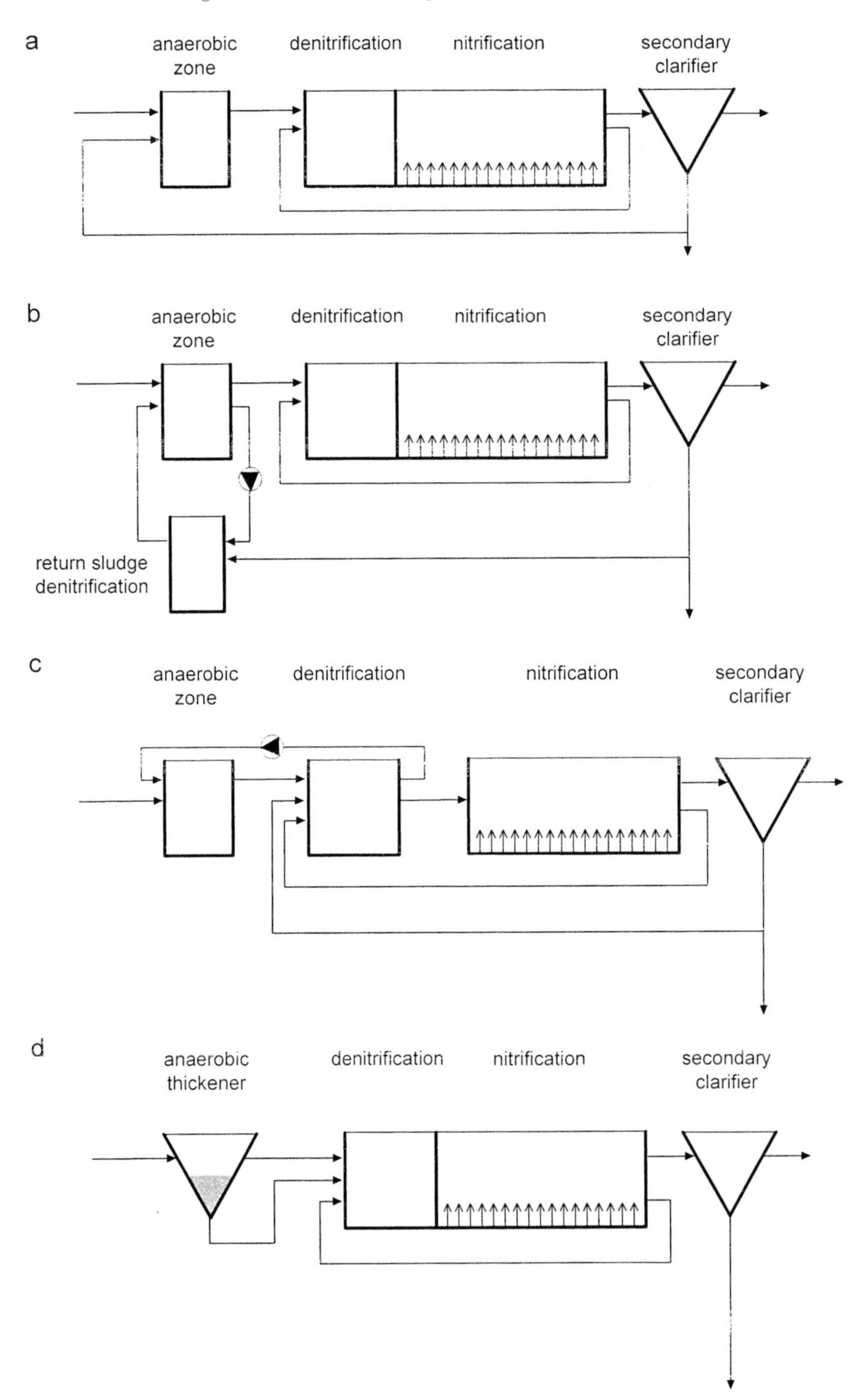

Fig. 4a. Schematic of the mainstream process (Phoredox), (**b–d**) selection of different mainstream process schemes: ISAH process (**b**), modified UCT process (**c**), EASC process (**d**).

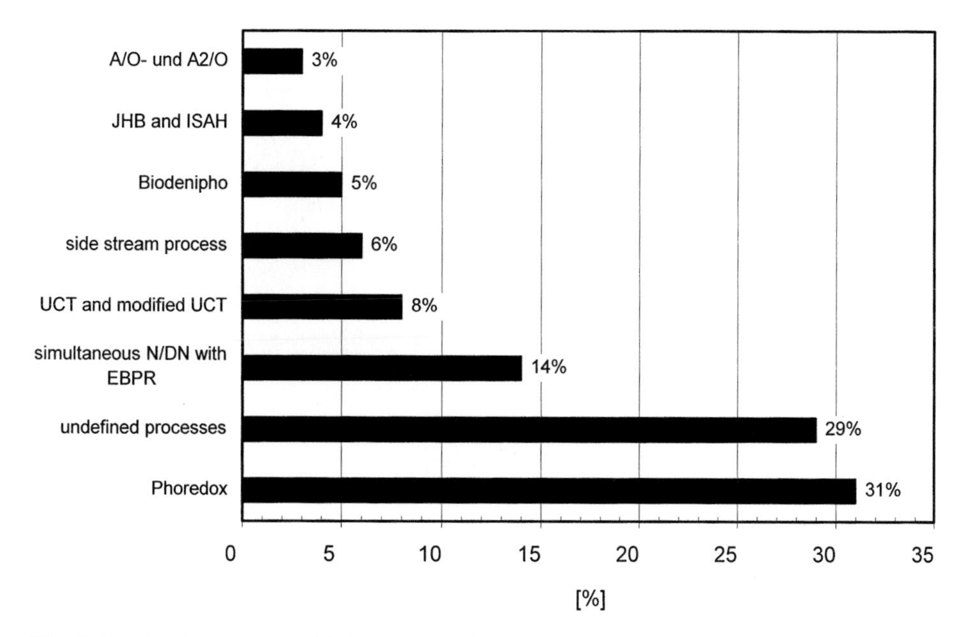

Fig. 5. Relative frequency of EBPR processes in Germany, 1994 (SEYFRIED and SCHEER, 1995).

2.5.2 Sidestream Process

The flow scheme and the reactor configuration of the sidestream process are schematically shown in Fig. 7. Using this process configuration, only a part of the return sludge (approximately 20%) is discharged to an anaerobic reactor for phosphate release (stripper). The hydraulic detention time in the stripper can be up to 24 h. During this time, most of the incorporated phosphate is released into solution. By operating the anaerobic stripper as a sedimentation reactor the released phosphate will be discharged with the overflow to a precipitation stage whereas the phosphorus poor sludge of the underflow will be recycled back to the activated sludge stage. Calcium or aluminum are usually used for the precipitation of the phosphorus rich overflow.

Besides the conventional use of the stripper for the anaerobic release of phosphate, in advanced plant layouts part of the wastewater influent is fed into a so-called pre-stripper to reduce the input of nitrate into the stripper and hence, to increase the phosphate release rate.

Today the dimensioning of a sidestream process is mainly based on operating experience or pilot plant studies, like those used in planning the large wastewater treatment plant at Darmstadt (Germany) with 240,000 PE.

Compared to the mainstream process, enhanced biological phosphorus removal in the sidestream process is usually more stable and lower effluent concentrations are possible. On the other hand, investment costs are substantially higher for sidestream technology.

2.6 Practical Aspects

2.6.1 Influence of Wastewater Quality on P Elimination Efficiency

The maximal phosphate uptake during aerobic conditions mainly depends (besides other operating factors) on the availability of substrate for the phosphate accumulating organisms under anaerobic conditions.

For storing polyalkanoates in bacterial cells the substrate must be present in the form of

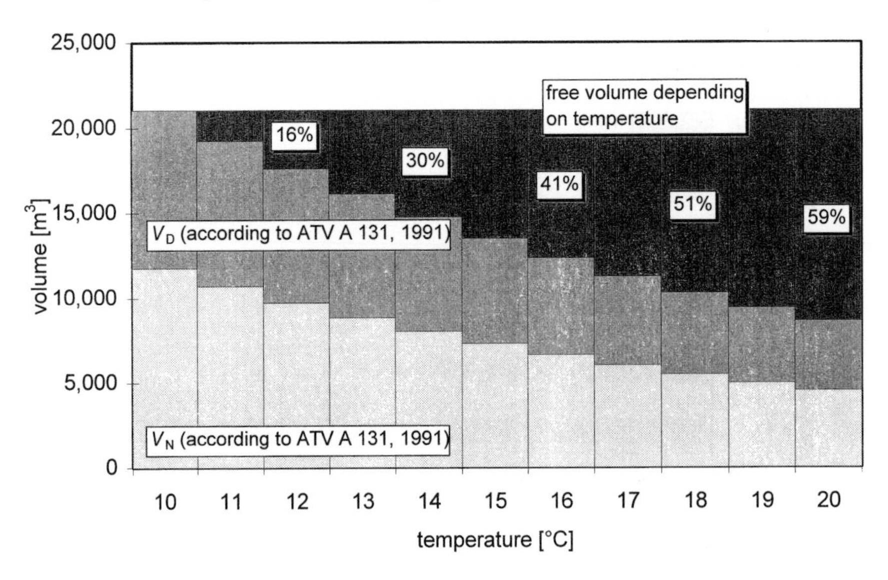

Fig. 6. Dependence of operating temperature on the volume for nitrification and denitrification.

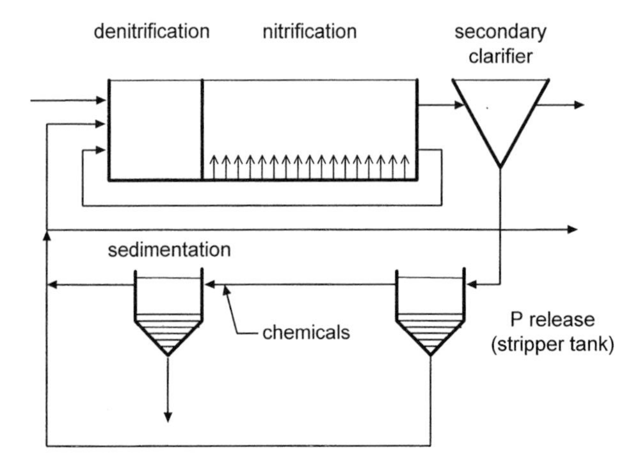

Fig. 7. Schematic of the sidestream process (Phostrip).

short-chain organic acids, either as part of the raw wastewater or be produced under anaerobic conditions by hydrolysis and fermentation.

To approach the EBPR with biokinetic modeling, e.g., with the so-called activated sludge model No. 2 (HENZE et al., 1995a), the COD of the influent must be divided into different fractions:

(1) short-chain organic acids (S_A),
(2) easily biodegradable substrate (S_{bs}),
(3) inert soluble substrate (S_e),
(4) soluble particulate substrate (X_S),
(5) inert particulate substrate (X_e).

In contrast to former simulation models, here the soluble biodegradable substrate has been further divided in two subfractions, the easily fermentable substrate (S_{bs}) and the fraction of short chain organic acids (S_A). The latter substrate is assumed to be directly metabolized by the polyphosphate accumulating microorganisms, while the easily biodegradable substrate must be fermented under anaerobic conditions.

The amount of the easily biodegradable substrate (S_{bs}) can be measured in a respiration test with either oxygen or nitrate as a terminal electron acceptor. From the integrated nitrate or oxygen consumption during the respiration test, the easily biodegradable substrate concentration can be calculated (EKAMA et al., 1986, KRISTENSEN et al., 1992; Anonymous, 1995; KAPPELER and GUJER, 1992; HULSBEEK, 1995).

The concentration of short chain organic acids (S_A) is usually directly determined with conventional titration procedures or by gas chromatography. Several full scale investigations have shown that a stable EBPR process can be expected above about 50 mg L^{-1} of short chain organic acids (Anonymous, 1995).

2.6.2 Measures to Enhance the Amount of Easily Biodegradable Substrate

As mentioned above, the amount of easily biodegradable substrate is one of the key factors determining the degree of the enhanced biological phosphorus elimination. Therefore, several attempts have been made to increase the easily biodegradable substrate concentration in the raw wastewater. Besides the addition of external substrate, e.g., acetate or other short chain organic acids, optimization of wastewater quality has mainly been achieved through the production of internal substrates via primary sludge wastewater hydrolysis and fermentation.

Here, two different approaches for wastewater and/or sludge hydrolysis can be distinguished.

(1) In the EASC (extended anaerobic sludge contact) process the anaerobic zone is constructed as a sedimentation reactor in which part of the raw wastewater is settled and held under anaerobic conditions to stimulate hydrolysis and fermentation reactions (SCHÖN-BERGER, 1990). Since the sedimentation reactor is integrated into the mainstream of the wastewater, the efficiency of sludge hydrolysis is limited.

(2) To enhance sludge hydrolysis and fermentation a complete separation of wastewater flow and sludge hydrolysis must be achieved. Different process designs have been developed for this purpose, mainly based on separating primary sludge sedimentation (in the primary clarifier) from sludge hydrolysis/fermentation in a separate reactor, as shown in Fig. 8.

Based on several full scale studies, the hydrolysis efficiency or the hydrolysis yield ($\eta_{hydrolysis}$) can be estimated at 10–15%. The fraction of easily biodegradable substrate available for denitrification or enhanced biological phosphorus removal is in the range of 70–90%, based on the total amount of soluble substrate (JÖNSSON et al., 1996; URBAIN et al., 1997; SCHÖNBERGER, 1990; JÖRGENSEN, 1990; SCHLEGEL, 1989; PITMAN, 1995; PROHASKA BRINCH et al., 1994; ANDREASEN et al., 1997; Anonymous, 1995).

The substrate consists mainly of short chain organic acids and, hence, is equally good for phosphate release under anaerobic conditions like a synthetic substrate, e.g., acetate. To enhance or stabilize biological phosphorus uptake long-term addition of easily biodegradable substrate, has proven most effective rather than a short-term equalization (hourly or daily) of short-chain organic acids concentration in the influent (TEICHFISCHER, 1995; WITT, 1997).

The main operational parameters determining hydrolysis/fermentation efficiency are the solids retention time and the temperature in the reactor. Solids retention times of more than 1d at temperatures of 20 °C and up to 4 d at lower temperatures (T = 10 °C) are necessary for optimal operation. For these long solids retention times it is necessary to establish an internal sludge recycle to wash the organic acids out of the sludge layer.

The theoretical potential for the internal production of easily biodegradable substrate can be estimated for average conditions, assuming a primary sludge production of 40 g TS $(PE\ d)^{-1}$, a hydrolysis efficiency of 12%, and a S_A fraction of 80%. The corresponding S_A production is then 4 g COD $(PE\ d)^{-1}$, assuming a COD:TS ratio of 1. With a specific wastewater inflow of 250 L $(PE\ d)^{-1}$ an additional S_A concentration of only 16 mg L^{-1} can be expected. Further assuming a specific phosphate uptake ratio of 5 mg P $(100\ mg\ COD)^{-1}$, the surplus phosphorus elimination due to primary sludge hydrolysis will be only 0.8 mg P L^{-1}. This simple calculation clearly shows that the absolute amount of short chain organic acid generation is comparably low, and hence the additional phosphorus elimination is limited.

Besides the production and release of short chain organic acids into the wastewater, part of the organic nitrogen will also be hydrolyzed and recycled back to the wastewater stream, resulting in a nitrogen feedback.

The quantity of the nitrogen feedback mainly depends on the operating conditions, but usually does not exceed 5% of the nitrogen load of the influent.

Another operating problem may arise from the odor due to the solubilization of the primary sludge. It is advisable to construct the hydrolysis reactor as a closed system to avoid negative environmental effects.

Finally, in plants with anaerobic stabilization the decrease in gas production must be considered. Experience in Denmark shows that the gas production can be reduced as much as 25% (ANDREASEN et al., 1997).

2.6.3 Measures for Minimizing Nitrate and Oxygen Input into the Anaerobic Zone

In plants with the EBPR process any input of nitrate and oxygen into the anaerobic zone may result in a deterioration of the phosphate release due to the consumption of easily biodegradable substrate by nitrate or oxygen respiration. In most cases the subsequent phosphate uptake under aerobic conditions usually will not equal that under conditions without any nitrate and/or oxygen input into the anaerobic zone.

It should be noted, however, that anoxic phosphate uptake can contribute to some extent to overall phosphorus elimination under special wastewater and operating conditions (CARLSSON, 1996; KUBA et al., 1996). Nevertheless, a prior release of phosphate under strictly anaerobic conditions is of crucial importance for the enhanced biological phosphorus removal process.

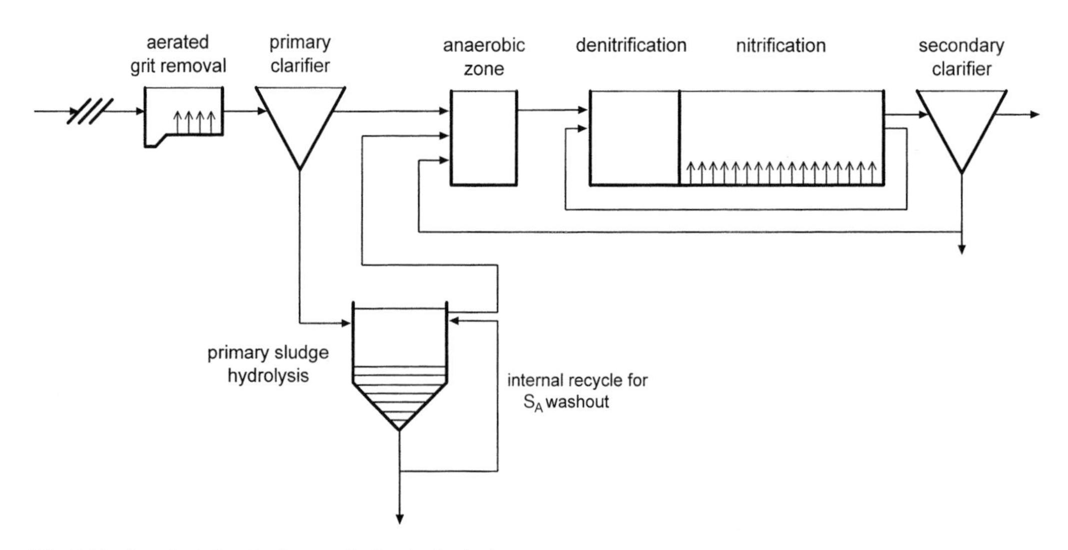

Fig. 8. Basic principle of primary sludge hydrolysis.

Nitrate mainly arises from two sources:

(1) In regions with elevated nitrate concentrations in the groundwater the influent of the wastewater treatment plant usually also shows a significant nitrate content.
(2) Very often, nitrate is discharged into the anaerobic zone via the underflow recycle from the secondary clarifier.

The significance of nitrate input into the anaerobic zone is evident if it is considered that under full scale conditions 4–6 mg easily biodegradable COD are eliminated for every mg of nitrate. Thus, for 10 mg L^{-1} nitrate in the secondary effluent and a return sludge recycle ratio of 1 (typical for full scale plants with nitrogen elimination), 5 mg L^{-1} nitrate will be present in the anaerobic tank. Due to this nitrate discharge, 20–30 mg L^{-1} easily biodegradable COD will be metabolized without phosphate release.

To prevent a nitrate recycle via the return sludge three different approaches have been developed. Using the Johannesburg (JHB) process, the return sludge will be held for some time under anoxic conditions to denitrify the nitrate load in the return sludge. Because only negligible amounts of substrate are available in the return sludge, only the endogenous respiration can be used for denitrification. Depending on the operating conditions, the specific denitrification rate is in the range of 0.4–0.8 mg $NO_3{-}N$ (g VSS h)$^{-1}$.

To enhance the denitrification rate a part of the inflow can be used as a substrate. That results in a reduction of the return sludge denitrification volume. Like the modified UCT process, the ISAH configuration uses the addition of substrate to increase the denitrification rate (Fig. 4b). The maximal denitrification rate achievable with such process schemes mainly depends on the wastewater composition and the fraction of wastewater which is fed into the anoxic return sludge denitrification zone.

Besides a possible deterioration due to nitrate discharge into the anaerobic zone, oxygen input can also reduce the phosphate release under anaerobic conditions because part of the easily biodegradable substrate is metabolized using oxygen as electron acceptor. Stoichiometrically, some 3 g COD will be lost per mg of oxygen in the anaerobic zone. Therefore, an oxygen content of 6 mg L^{-1} in raw influent will result in an oxygen concentration of 3 mg L^{-1} in the anaerobic zone, assuming no oxygen is left in the return sludge and a recycle ratio of 1. Consequently, 10 mg L^{-1} easily biodegradable COD will be metabolized under aerobic conditions without phosphate release.

To optimize the conditions in the anaerobic zone with respect to oxygen discharge, it is necessary to prevent wastewater oxygenation ahead of the anaerobic zone. Excessive turbulence in the anaerobic zone itself or in streams (wastewater, return sludge) entering the anaerobic zone should, therefore, be avoided. Further measures are, e.g., use of centrifugal pumps instead of screw pumps for pumping return sludge, to minimize the hydraulic head loss at weirs ahead of the anaerobic zone, and, finally, to reduce the air supply if aerated grit chambers are used for grit removal. Air supply rates of not more than 0.1–0.2 m_N^3 (m^3 h)$^{-1}$ are advisable if an anaerobic zone follows the aerated grit removal.

2.6.4 Sludge Production

An often claimed advantage of the EBPR process compared to physico-chemical phosphorus elimination is the insignificant additional production of waste activated sludge (e.g., ATV, 1989; WITT and HAHN, 1995). This conclusion is mainly based on the fact that no additional chemicals need to be added to the wastewater for phosphorus elimination. But, irrespective of the nature of the phosphorus elimination mechanism, phosphorus storage is accomplished by its uptake as phosphate (PO_4), so an additional dry solids production can be expected.

Theoretically, the EBPR process can affect WAS production mainly by three mechanisms:

(1) uptake and storage of phosphate as polyphosphate (poly-P) leads to additional dry solids being incorporated into the sludge. From the average composition of the poly P, poly-P synthesis should be accompanied by an uptake of

3 g of total solids per g of phosphorus (JARDIN and PÖPEL, 1994).

(2) The development of poly-P accumulating microorganisms (poly-P bacteria) may change the overall growth kinetics of the activated sludge as shown by the experimental studies of WENTZEL et al. (1990), who observed a much lower decay rate of the poly-P bacteria compared to the non-poly-P microorganisms. The slower decay of the poly-P bacteria will result in an additional organic sludge production, calculated by WENTZEL et al. (1990) to be 10% at a sludge age of 20 d.

(3) Depending on the wastewater characteristics and the operating conditions (e.g., anaerobic hydraulic retention time), a so-called anaerobic stabilization (RANDALL et al., 1987) can occur, which describes substrate utilization by means of fermentative processes. Anaerobic processes are usually associated with a reduced yield of biomass so that the incorporation of an anaerobic zone may lead to a lower overall sludge production.

While the storage of polyphosphate will mainly affect inorganic sludge production the latter two mechanisms will primarily influence organic sludge production.

In an extensive experimental investigation it was found that magnesium and potassium concentrations were significantly correlated with phosphorus uptake. From the linear regression between the cations and the phosphorus content of the excess sludge shown in Figs. 9 and 10 a molar uptake ratio of 0.30 M Mg $(\mathrm{M\ P})^{-1}$ and 0.26 M K $(\mathrm{M\ P})^{-1}$ can be calculated. These compare well to values reported in the literature (e.g., WENTZEL et al., 1992; ARVIN and KRISTENSEN, 1985).

Thus, polyphosphate formation is apparently the main mechanism responsible for the enhanced phosphorus elimination under the experimental conditions of this study.

On the basis of pilot plant studies it was concluded that the influence of the EBPR operation on WAS production is mainly due to an increase of the inorganic sludge production and only to a minor extent to additional organic solids production.

Based on modeling analysis this would indicate that the decay rate for the poly-P bacteria differs only slightly from the decay rate of non-poly-P bacteria, i.e., that the amount of organic mass loss of poly-P bacteria is very similar to that of non-poly-P organisms. Very similar results showing a negligible difference were obtained by MCCLINTOCK et al. (1992), HULSBEEK (1995) and ANTE et al. (1995). They determined the decay rates of the poly-P and non-poly-P bacteria by direct experimental in-

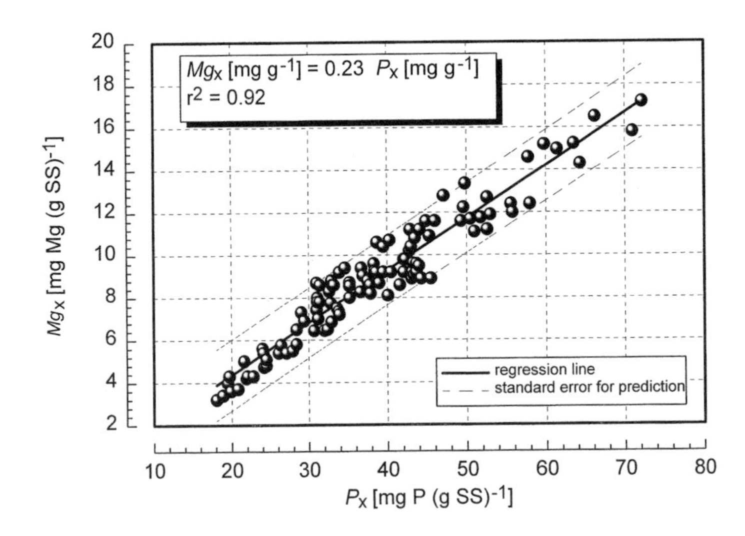

Fig. 9. Relationship between phosphorus and magnesium content in activated sludge of an EBPR pilot plant.

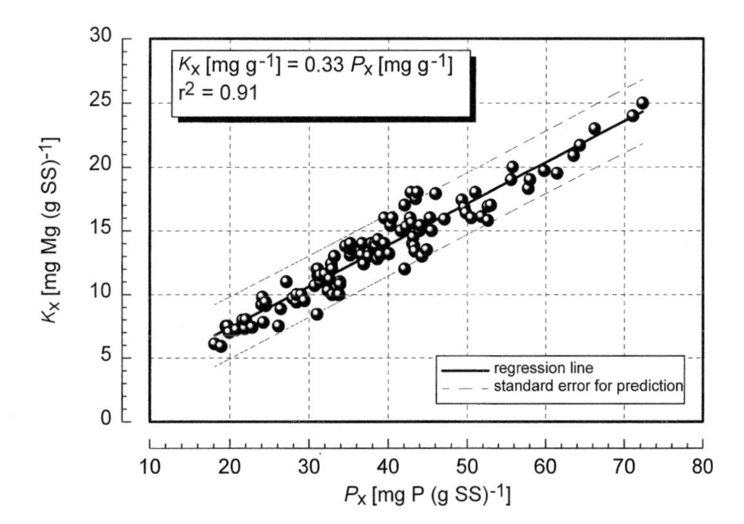

Fig. 10. Relationship between phosphorus and potassium content in activated sludge of an EBPR pilot plant.

vestigation (McClintock et al., 1992) or by data fitting using steady state (Hulsbeek, 1995) or dynamic modeling (Ante et al., 1995). Although McClintock et al. (1992) found a slightly higher value for organic sludge production in an EBPR system compared to a conventional system, the difference of the decay rates between poly-P and non-poly-P bacteria was statistically not significant.

Considering the uptake of inorganic solids during the EBPR process (Fig. 11), an additional sludge production of 3 g of total solids per g of phosphorus due to the enhanced biological phosphorus removal should be calculated for the dimensioning of the overall production.

For assessing this additional sludge production in connection with total sludge yield, a simple estimation for average conditions can be performed. Based on a specific phosphorus load in the raw wastewater of 2.5 g P (PE d)$^{-1}$ and assuming that 0.3 g P (PE d)$^{-1}$ is removed with the primary sludge and an additional 0.5 g P (PE d)$^{-1}$ with the WAS for the growth of the biomass, 1.3 g P (PE d)$^{-1}$ must be eliminated by the enhanced phosphorus removal to ensure an effluent concentration of $<$2 mg L^{-1} P_{tot} [0.4 g P (PE d)$^{-1}$]. The 1.3 g P (PE d)$^{-1}$

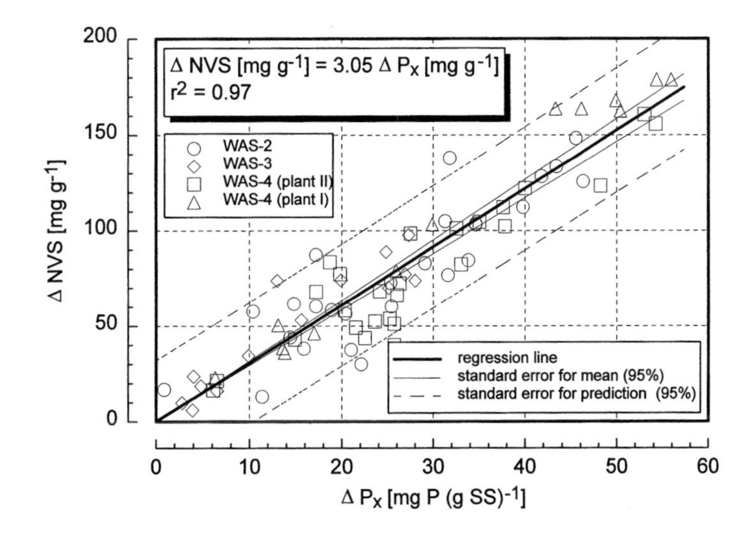

Fig. 11. Relationship of the difference in total and organic sludge production between a plant with and a plant without EBPR (data points are mean values of the different modes of operation).

will result in 3.9 g TS (PE d)$^{-1}$ which is an additional dry solids production of 11% of the WAS production [WAS production was assumed to be 35 g TS (PE d)$^{-1}$].

2.6.5 P Release and P Feedback during Sludge Treatment

To quantify the potential of phosphate release during thickening of excess sludge, pilot plant investigations with different thickening systems have shown that the release and consequently the feedback of phosphorus is very low.

In view of the short sludge retention times in mechanical thickening systems like centrifuges, screening drums or flotation units, P release should be very low. Experimental studies have shown that the 90% percentile of the P feedback from thickening excess sludge of the EBPR plant is estimated as 2.0%, based on the influent P load for flotation, and 2.2% for a centrifuge. Fig. 12 shows the cumulative frequency of the P feedback during the thickening of excess sludge from the EBPR process.

In contrast to mechanical thickening systems, a substantial P feedback can be expected using gravity thickeners which are usually operated at a sludge retention time of more than 0.5 d. The phosphate profiles suggest a similar behavior in all modes of operation: a substantial P release in the sludge layer combined with a very small transfer of soluble phosphate into the supernatant. Fig. 13 shows a typical phosphate profile in a gravity thickener at different sludge retention times.

It can be seen that although about 95% of the stored poly-P was released within 2.8 d, the transfer of released phosphate into the supernatant was very low. Nevertheless, failure of normal operation of the gravity thickener (e.g., if flotation occurs in the gravity thickener), can cause vertical mixing and result in significant phosphate concentrations in the supernatant, leading to a strong increase of P feedback.

Theoretically, most of the phosphorus eliminated in the form of poly-P should be released during the anaerobic treatment of excess sludge (PÖPEL and JARDIN, 1993). In contrast, at most large plants in Germany the soluble phosphorus concentration of digested sludge is often very low (SEYFRIED and HARTWIG, 1991; BAUMANN and KRAUTH, 1991) although from other plants additional phosphorus loads of up to 100% have been reported (PITMAN et al., 1991; SEN and RANDALL, 1988; MURAKAMI et al., 1987). Therefore, conditions can evidently exist where

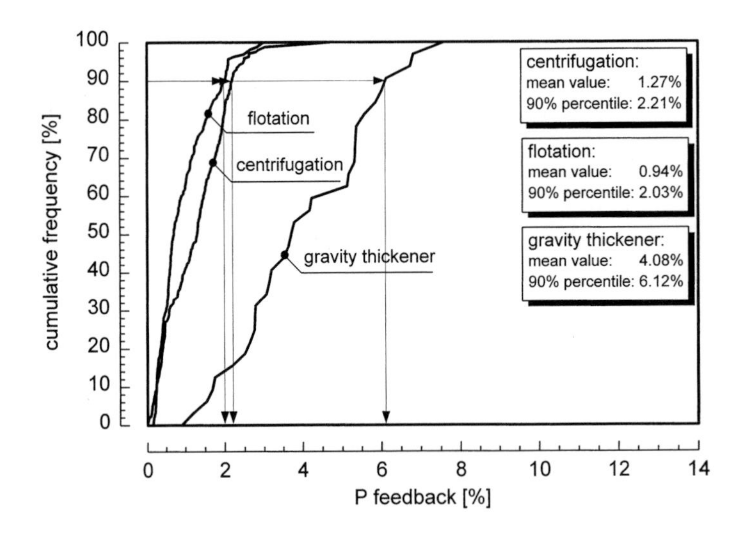

Fig. 12. Cumulative frequency of the P feedback during excess sludge thickening (P feedback is based on the influent load of the thickening system).

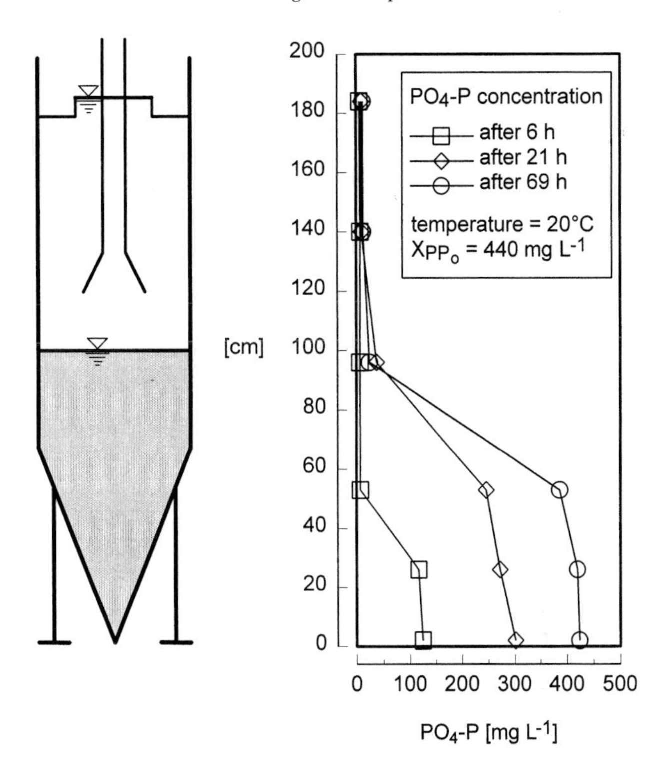

Fig. 13. Phosphate profile in a gravity thickener.

(1) only a part of the stored phosphorus is released to the bulk liquid of the sludge treatment system or
(2) some of the released phosphorus is chemically fixed as metal phosphate precipitates or by other mechanisms.

Pilot plant experiments clearly show that nearly all of the polyphosphate in the excess sludge will be hydrolyzed during the anaerobic or aerobic thermophilic digestion (JARDIN and PÖPEL, 1996; WILD et al., 1996; ASPEGREN, 1995). But only part of the phosphorus remains in solution even though a quantitative release of phosphate could be observed.

The discrepancy between phosphate release during polyphosphate hydrolysis and a comparable low P feedback is mainly due to physico-chemical refixation reactions taking place in the stabilizing reactor. Besides a precipitation of magnesium, ammonium, and phosphate in the form of struvite, detergent zeolites also participate in the refixation reactions (JARDIN, 1995; WILD et al., 1996; ASPEGREN, 1995).

Therefore, the overall P feedback is substantially lower than would be expected assuming a quantitative release of phosphate during polyphosphate hydrolysis. On this basis the P feedback in large wastewater treatment plants with the enhanced biological phosphorus removal process and a corresponding phosphorus content in the range of 2.5–3% in the excess sludge is usually less than 10% of the incoming load. Under these conditions, no additional measures for treating the process water from sludge dewatering are necessary.

On the other hand, some plants with particularly suitable wastewater conditions for the EBPR process show substantially higher phosphorus contents and consequently a higher recycle load of phosphorus. To avoid a deterioration of the EBPR process due to this P feedback a separate treatment of these process waters may be considered, preferably by using chemicals for phosphate precipitation.

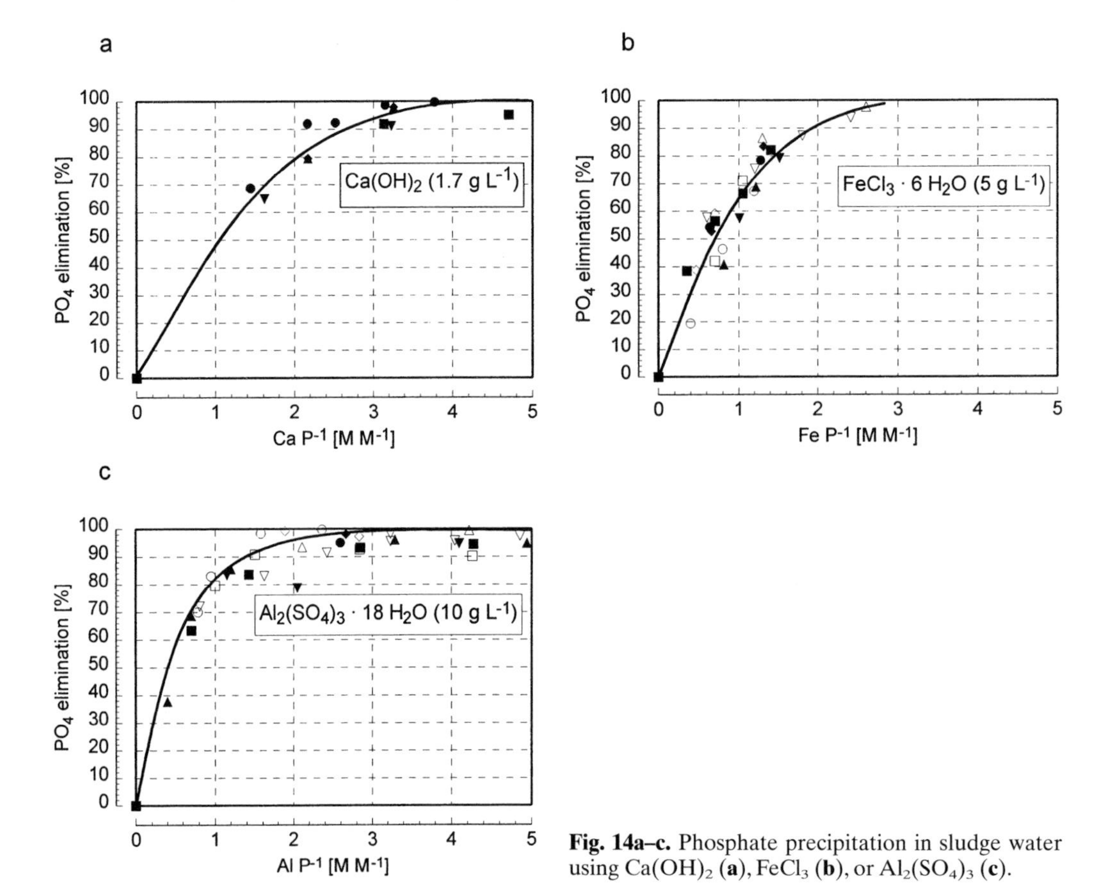

Fig. 14a–c. Phosphate precipitation in sludge water using $Ca(OH)_2$ (**a**), $FeCl_3$ (**b**), or $Al_2(SO_4)_3$ (**c**).

In principle, all common chemicals for phosphate precipitation could be used. However, in view of the relatively high ammonium concentrations and the high alkalinity of the process water, precipitation with calcium may require inordinately large amounts of lime, and in using iron, the reduction of Fe^{3+} to Fe^{2+} must be considered.

Fig. 14a–c shows the results of PO_4 precipitation in sludge water using different chemicals. As can be seen, aluminum proved most effective, with usually over 80% of the soluble phosphate precipitated at a molar dosage of 1 M Al (M P)$^{-1}$. For calcium and iron an 80% elimination was achieved only when molar dosages of 2 M Ca (M P)$^{-1}$ or 1.5 M Fe (M P)$^{-1}$ were reached.

3 Chemical Phosphorus Removal

3.1 Basic Principle

3.1.1 Physico-Chemical Background

Besides the conventional biological uptake of phosphorus for microbial growth and the enhanced biological phosphorus removal process in numerous full scale wastewater treatment plants, phosphorus is eliminated by precipitation reactions.

For this purpose multivalent metal ions like iron, aluminum, or calcium salts are usually added to the wastewater to precipitate phosphate. The precipitation reaction between phosphate and Me(III) is given in principle by:

$$Me^{3+} + PO_4^{3-} = MePO_4 \tag{7}$$

From the solubility equilibrium constants and considering the dissociation of phosphate depending on the pH, it is possible to calculate the total phosphate concentration for different metal phosphates under defined conditions, as shown in Fig. 15. From this graph the optimal pH ranges for precipitation with the different metals can be derived. Whereas iron and aluminum precipitation is favored under slightly acidic conditions (pH 5–6), an effective precipitation of calcium phosphates is only achieved at substantially higher pH values than those usually found in domestic wastewaters. Therefore, the specific dosage of calcium depends primarily on the buffer capacity of the wastewater and only slightly on the phosphate concentration of the influent.

Due to the acid–base equilibrium constant of phosphate, most of the total phosphate is dissolved in the form of $H_2PO_4^-$ and HPO_4^{2-} at neutral pH values. For the reaction of phosphate with aluminium the precipitation equation can be formulated as:

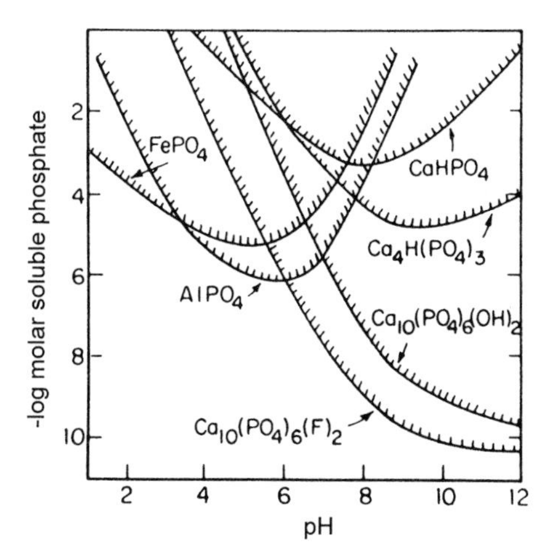

Fig. 15. Solubility of metal phosphates (STUMM and MORGAN, 1981).

$$Al(H_2O)_6^{3+} + H_2PO_4^- = AlPO_4 + 6H_2O + 2H^+ \tag{8}$$

Besides the formation of phosphates, hydroxides are also produced, thereby substantially increasing the overall sludge production:

$$Al(H_2O)_6^{3+} = Al(H_2O)_3(OH)_3 + 3H^+ \tag{9}$$

As is shown in Eqs. (8–9), besides the formation of insoluble phosphates and hydroxides, the alkalinity will decrease due to the production of H^+, giving a lower pH. The extent of the pH decrease mainly depends on the buffer capacity of the wastewater and can reach critical values for the nitrification processes when operating in wastewaters with a limited alkalinity. Alternatively, by using alkaline precipitants a critical decrease of the pH value can be avoided.

The overall formation of metal phosphates is a multi-step process.

(1) After dosing the chemical into the wastewater, rapid mixing is essential for a high efficiency of metal phosphate formation and for minimizing metal hydroxide production. Based on the specific energy input, the intensity of mixing should be in the range of 100–150 W m^{-3}.
(2) After mixing metal phosphates, metal hydroxides and – depending of the chemical used – also carbonates are formed very rapidly.
(3) In a further process stage part the negative load of the colloids is neutralized (destabilization) and as a result aggregation of particles occurs (coagulation).
(4) To facilitate the aggregation of larger particles (macroflocs) a flocculation stage characterized by a much lower specific energy input is mandatory. The energy input in this stage should be around 5 W m^{-3} at a hydraulic retention time of some 20–30 min.
(5) Finally, the aggregated particles must be separated either in a sedimentation stage or by filtration (ATV, 1992).

3.1.2 Chemicals for Phosphorus Precipitation

The most commonly used chemicals for phosphate precipitation are shown in Tab. 1. When using divalent Fe(II), it is necessary to oxidize it to Fe(III) prior to the precipitation process. This can be achieved very easily if Fe(II) is added just ahead of an aerated reactor, e.g., prior to the aerated grit chamber or ahead of the aerated activated sludge tank.

When operating with chemical phosphorus elimination in wastewaters with low alkalinity (Alk <5 mM), care must be taken regarding the nitrification. At the end of the biological treatment process the alkalinity should not be below 1.5 mM to avoid a critical drop of pH. In such cases the use of aluminates may be advisable. Sometimes an increase of the nitrification rate was observed when using alkaline aluminates (FETTIG et al., 1996).

3.2 Process Design

3.2.1 General

The different process designs for physico-chemical phosphate removal are mainly distinguished by the site of chemical dosage and the location where the metal phosphates are removed. Thus, as an integrated part of the overall wastewater treatment, physico-chemical phosphorus elimination can be realized before (pre-precipitation), during (simultaneous precipitation), or after (post-precipitation) biological treatment as schematically shown in Fig. 16a–c.

3.2.2 Pre-Precipitation

When using the pre-precipitation process the chemicals will be added either prior to the grit removal or directly ahead of the primary clarifier. Fig. 16a shows a diagram of the pre-precipitation process.

The main advantage of the pre-precipitation process is the additional removal of organic matter prior to the biological stage and, there-

Tab. 1. Selection of Chemicals for Phosphate Precipitation (ATV, 1992)

Metal	Formula	Form of Delivery	Typical Dosing Rates[a] (mg chemical L^{-1})
Iron(II)	$FeSO_4$	solution	74
	$FeSO_4 \cdot 7\,H_2O$	granulate	135
Iron(III)	$FeClSO_4$	solution (40%)	97
	$FeCl_3$	solution (30–40%)	79
	$FeCl_3 \cdot 6\,H_2O$	or granulate	131
Aluminium	$Al_2(SO_4)_3 \cdot x\,H_2O +$ $Fe_2(SO_4)_3 \cdot x\,H_2O$	granulate	
	$AlCl_3 \cdot x\,H_2O$	solution (30–40%)	65 $AlCl_3$
Polyaluminium chloride	$(Al_2(OH)_nCl_{6-n})_m$	solution (5–10%)	
Sodiumaluminate	$NaAl(OH)_4$	solution	57 $NaAl(OH)_4$
Calcium	CaO	powder	50–150 CaO
	$Ca(OH)_2$	powder	50–150 $Ca(OH)_2$

[a] $\beta=1.5$, $P_i=10$ mg L^{-1} P.

fore, a reduction of the organic load and the oxygen demand under aerobic conditions. Problems with the pre-precipitation may arise in plants with denitrification because part of the organic substrate, needed for denitrification, will be lost due to the co-precipitation of organic compounds. In addition, in some plants the pre-precipitation may result in an increase of the sludge volume index of the activated sludge and possibly lead to problems with flotating sludge in the secondary clarifier.

All chemicals can be used with the pre-precipitation process except Fe(II) salts, which must be oxidized, e.g., in an aerated grit chamber, as mentioned above.

3.2.3 Simultaneous Precipitation

Simultaneous precipitation is the most commonly used process configuration. The chemicals are usually added ahead of or directly in the activated sludge tank. Alternatively, the addition of metal salts is also possible in the return sludge. The simultaneous precipitation process is shown schematically in Fig. 16b.

Except calcium, all other metal salts can be used for the simultaneous phosphorus elimination process. In combination with an efficient secondary clarifier and an effluent concentration of suspended solids <10 mg L^{-1}, corresponding to 0.2–0.3 mg L^{-1} P_{tot} (P content of the activated sludge approximately

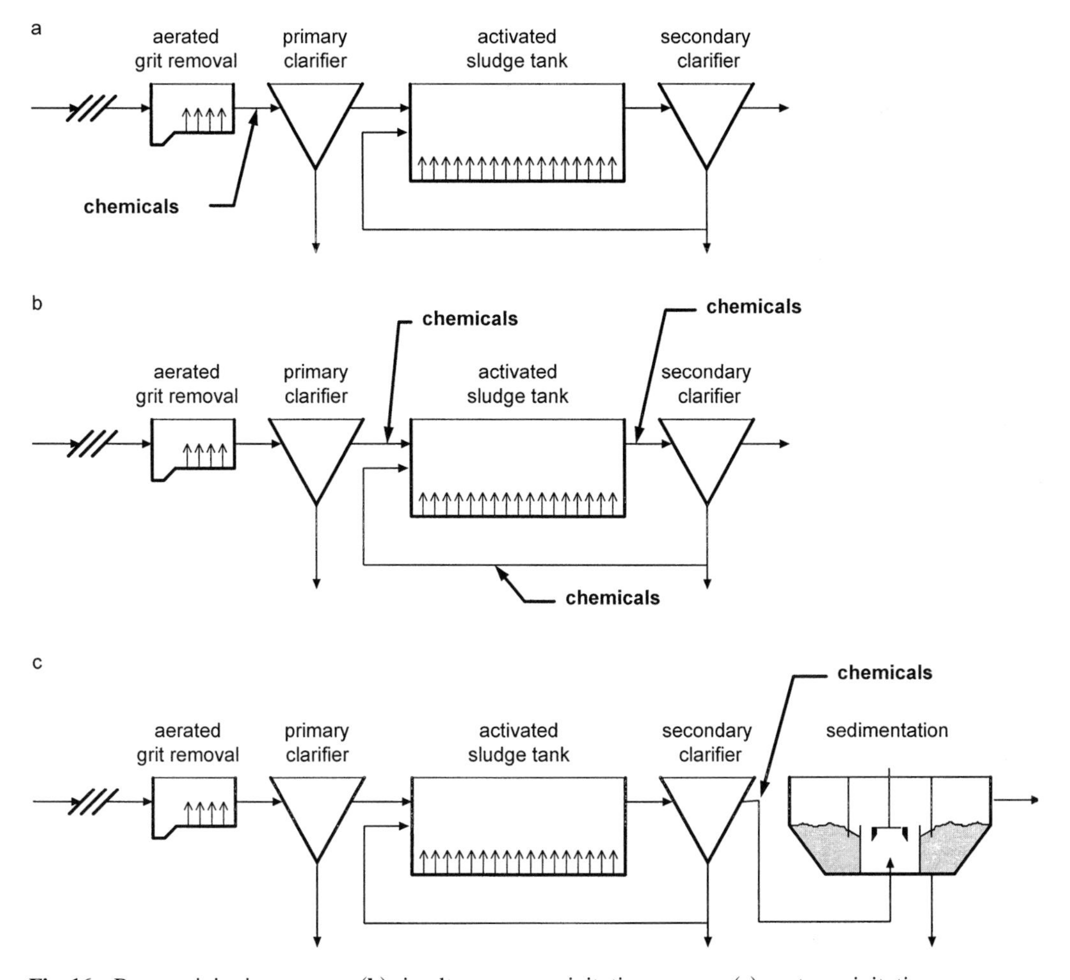

Fig. 16a. Pre-precipitation process, (**b**) simultaneous precipitation process, (**c**) post-precipitation process.

2–3%) an average effluent concentration of total phosphorus below 1 mg L^{-1} P_{tot} can be achieved. Fig. 17 shows operational results of the full scale wastewater treatment plant at Duisburg-Kasslerfeld (450,000 PE) operated by the Ruhrverband (Essen, Germany). As can be seen, the effluent criterion of 1 mg L^{-1} P_{tot}, monitored by the so-called qualified grab sample, is usually achieved.

3.2.4 Post-Precipitation

The post-precipitation process is practiced very rarely. It consists of three stages: dosing, mixing, and separation, often combined in one construction (Fig. 16c).

Most often, calcium is used as the major chemical for phosphate precipitation in the post-precipitation process. Usually no interactions between the biological and the chemical stage of the treatment plant need to be considered, so a separate optimization of the two processes is possible. Therefore, low effluent phosphorus concentrations are possible, especially when flotation is used as the final separation stage.

3.2.5 Filtration

For severe effluent demands involving phosphorus, an additional elimination can be achieved by the combination of flocculation and filtration. The effluent phosphorus concentration is very often below 0.5 mg L^{-1} P_{tot} when using the flocculation–filtration process.

3.3 Practical Aspects

3.3.1 Influence on Biological Wastewater Treatment

The biological wastewater treatment process will be influenced by physico-chemical phosphorus elimination using simultaneous precipitation in several respects:

(1) Due to the precipitation of phosphate and metal hydroxides, substantial amounts of sludge are produced in addition to the biological waste activated sludge generation (for details, see Sect. 3.3.2). As a consequence, the sludge age will decrease and, possibly, the nitrification process will become unstable if the actual sludge age falls below the required mean cell residence time to avoid wash out of nitrifying organisms. For the process design this means that the volume of the activated sludge tank has to be increased compared to the situation without physico-chemical phosphorus elimination assuming the same suspended solids concentration. Fortunately, the dosage of iron or aluminum in the simultaneous precipitation mode usually tends to reduce the

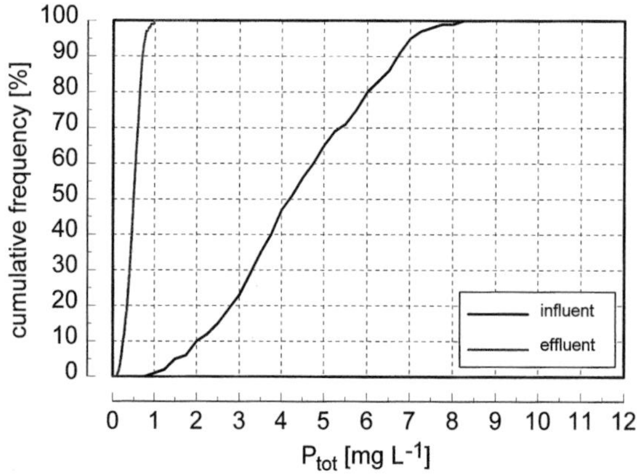

Fig. 17. Frequency distribution of the effluent phosphorus concentration of the WWTP Duisburg-Kaßlerfeld.

sludge volume index. By operating the secondary clarifier with a higher solids flux the suspended solids concentration in the activated sludge tank can be increased and, therefore, no additional tank volume is required under average conditions.

(2) The addition of chemicals during the biological treatment may change the nitrifying activity and, possibly, leads to a decrease of the overall nitrification rate. Particularly, the usage of $Fe(II)SO_4$ often impairs the nitrification process. A decrease in the nitrification rate of up to 35% has been observed (PÖPEL, 1991). On the other hand, the usage of $Fe(III)Cl_3$ can increase the nitrification rate to values of up to 40% (PÖPEL, 1991).

(3) As has been shown in Eqs. (8, 9) the addition of metal salts usually reduces the buffer capacity of the wastewater with a possible drop of the pH. The change in the alkalinity (Δ Alk) can be calculated as follows:

$$\Delta \text{Alk} = 0.11\, S_{\text{Al}} - 0.04\, S_{\text{Fe}_2} - 0.06\, S_{\text{Fe}_3} \tag{10}$$

Practical experience has shown that the alkalinity should not fall below 1.5 mM L^{-1} after the biological treatment to avoid a significant deterioration of the nitrification process. When treating soft wastewater lime or sodium hydroxide dosage may be required to increase the alkalinity.

3.3.2 Slude Production

The following theoretical estimation of the additional sludge production as a result of physico-chemical phosphorus elimination will be restricted to iron and aluminum the most often used precipitants. Assuming that addition of iron or aluminum will result in the formation of only metalphosphates ($MePO_4$) and metalhydroxides [$Me(OH)_3$], the additional sludge production can be determined with some simple calculations (for details, see PÖPEL, 1995).

Defining the specific P elimination as e, with

$$\text{e} = \frac{P_\text{o} - P_\text{e}}{P_\text{o}} \tag{11}$$

and the relative chemical dosage, β, as

$$\beta = \frac{\text{molar dosage of chemical [M L}^{-1}\text{]}}{\text{molar concentration of phosphorus at the dosing point [M L}^{-1}\text{]}} \tag{12}$$

the additional sludge production based on the phosphorus concentration of the influent is given by

$$\frac{\Delta \text{TS}}{P_\text{o}} = \text{e} \cdot \frac{\text{molar weight of MePO}_4}{31} + (\beta - \text{e}) \cdot \frac{\text{molar weight of Me(OH)}_3}{31} \tag{13}$$

The corresponding specific sludge production for iron and aluminum is given by Tab. 2. Here, the additional sludge production was calculated for typical conditions, i.e., P elimination of 90% ($e = 0.9$) and a specific chemical dosage of 1.5 M Me $(\text{M P})^{-1}$.

Besides the formation of metal phosphates and metal hydroxides, a co-precipitation of colloids can also be expected. The extent of this co-precipitation mainly depends on the amount of filterable solids at the dosing point of the chemicals. With simultaneous precipitation, a co-precipitation of colloids of approximately 10% can be assumed. Considering this additional sludge, the overall sludge production will be:

for iron
$\Delta \text{TS} = 7.10\ \text{g TS (g P)}^{-1}$
$\Delta \text{TS} = 2.62\ \text{g TS (g Fe)}^{-1}$

for aluminum
$\Delta \text{TS} = 5.57\ \text{g TS (g P)}^{-1}$
$\Delta \text{TS} = 4.26\ \text{g TS (g Al)}^{-1}$

Assuming that approximately 1.3 g P $(\text{PE d})^{-1}$ must be eliminated by physico-chemical precipitation, an additional WAS production of 9.2 and 7.2 g TS $(\text{PE d})^{-1}$ can be expected for iron and aluminum, respectively. Based on

Tab. 2. Theoretical Estimation of the Additional Sludge Production with the Physico-Chemical Phosphorus Elimination (PÖPEL, 1995)

Chemical		Formula	Additional Sludge Production for Average Conditions ($e=0.9$; $\beta=1.5$) Based on P_o [g TS (g P)$^{-1}$]	Based on Me_{dosed} [g TS (g Me)$^{-1}$]
Iron	based on P_o	Δ TS = 1.42 e + 3.45 β	6.45	
	based on Me_{dosed}	Δ TS = 1.91 + 0.788 e/β		2.39
Aluminium	based on P_o	Δ TS = 1.42 e + 2.52 β	5.06	
	based on Me_{dosed}	Δ TS = 2.89 + 1.630 e/β		3.87

an average "biological" sludge production of 35 g TS (PE d)$^{-1}$ the additional WAS production will be 26% and 21% for iron and aluminum, respectively.

WEDI and NIEDERMEYER (1992) conducted extensive laboratory and full scale investigations to estimate the sludge production when using iron or aluminum as precipitants. They established a similar relationship to the theoretical one derived above for the relative chemical dosage (β) and sludge production, as shown in Fig. 18. The theoretically calculated sludge production is also depicted for a relative phosphorus elimination (e) of 0.9 and assuming a co-precipitation of 10%.

As can be seen, sludge production measured in laboratory tests shows only minor deviations from the sludge production calculated theoretically, especially at higher relative chemical dosages. For iron a specific sludge production of some 2.5–3 g TS (g Fe)$^{-1}$ can be expected for average conditions when using chemicals primarily for phosphate elimination. With aluminum, the specific sludge production is considerably higher, at approximately 4–5 g TS (g Al)$^{-1}$.

Under Scandinavian conditions, where chemicals are often used not only for phosphorus elimination but also as coagulants for the elimination of organic compounds, the specific sludge production is significantly higher. ØDEGAARD and KARLSSON (1994) calculated the specific sludge production, based on the results of extensive investigations of large WWTPs in Norway, to be 3.6 g TS (g Fe)$^{-1}$ and 7 g TS (g Al)$^{-1}$ (mean values). Similar results were found by BALMÉR (1994) in Swedish plants. However, great care must be taken when comparing full scale data with theoretical or laboratory derived approaches because

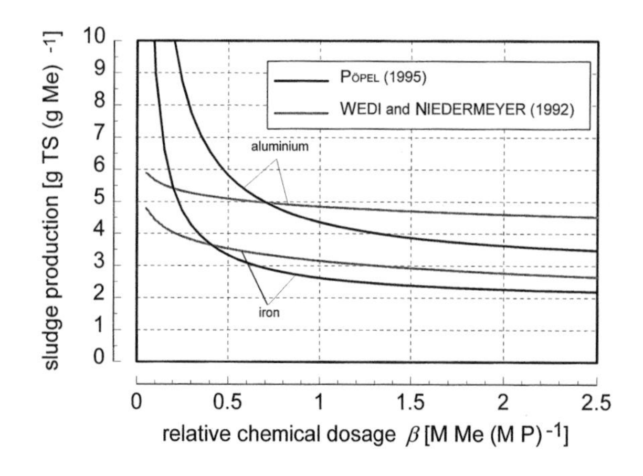

Fig. 18. Comparison of sludge production based on theoretical calculations (PÖPEL, 1995) and laboratory tests (WEDI and NIEDERMEYER, 1992).

several factors must be considered to establish a specific relationship between chemical dosage and sludge production, e.g., the chemical characteristics of the wastewater (pH, alkalinity), the extent of phosphorus elimination, and other operating conditions.

Under average conditions and assuming that chemicals are only used for P elimination, a specific sludge production of 6.5–8 g TS $(g\ P)^{-1}$ for iron and 5–6.6 g TS $(g\ P)^{-1}$ for aluminum can be expected.

4 References

ALARCON, G. O. (1961), Removal of phosphorus from sewage, *Thesis.* The John's Hopkins University, Baltimore, MD, USA.

ANDREASEN, K., PETERSEN, G., THOMSEN, H., STRUBE, R. (1997), Reduction of nutrient emission by sludge hydrolysis, *Water Sci. Technol.* **35**, 79–85.

Anonymous (1995), Vermehrte biologische Phosphorelimination in der Abwasserreinigung – Abschlußbericht eines Erfahrungsaustausches deutschsprachiger Hochschulen. *Mitteilungen der Oswald-Schulze-Stiftung*, Heft 19. Gladbeck: Oswald-Schulze-Stiftung.

ANTE, A., HESSE, H., VOß, H. (1995), Mikrokinetisches dynamisches Modell der Bio-P, *Veröffentlichungen des Institutes für Siedlungswasserwirtschaft und Abfalltechnik der Universität Hannover*, Heft 92. 15/1–15/21.

ARUN, V., MINO, T., MATSUO, T. (1988), Biological mechanism of acetate uptake mediated by carbohydrate consumption in excess phosphorus removal systems, *Water Res.* **22**, 565–570.

ARVIN, E. (1985), Biological removal of phosphorus from wastewater, *CPR Crit. Rev. Environ. Control* **1**, 25–64.

ARVIN, E., KRISTENSEN, G. H. (1985), Exchange of organics, phosphate and cations between sludge and water in biological phosphorus and nitrogen removal process, *Water Sci. Technol.* **17**, 147–162.

ASPEGREN, H. (1995), Evaluation of a high loaded activated sludge process for biological phosphorus removal. Department of Water and Environmental Engineering, Lund Institute of Technology/Lund University, *Report 1004*.

ATV (1989), *Arbeitsbericht der ATV-Arbeitsgruppe 2.6.6:* Biologische Phosphorentfernung. *Korrespondenz Abwasser* **36**, 337–348.

ATV (1992), *ATV-Arbeitsblatt A 202:* Verfahren zur Elimination von Phosphor aus Abwasser. Hennef: Gesellschaft zur Förderung der Abwassertechnik e.V.

ATV (1994), Biologische Phosphorentfernung bei Belebungsanlagen, *Merkblatt M 208*. Hennef: Gesellschaft zur Förderung der Abwassertechnik e.V.

AULING, G., PILZ, F., BUSSE, H.-J., KARRASCH, S., STREICHAN, M., SCHÖN, G. (1991), Analysis of the polyphosphate accumulating microflora in phosphorus-eliminating anaerobic-aerobic activated sludge systems by using diaminopropane as a biomarker for rapid estimation of *Acinetobacter* spp., *Appl. Environ. Microbiol.* **57**, 3585–3592.

BALMÉR, P. (1994), Chemical treatment – Consequences for sludge biosolids handling, in: *Chemical Water and Wastewater Treatment III* (KLUTE, R., HAHN, H. H., Eds.), pp. 319–327. Berlin, Heidelberg: Springer-Verlag.

BARNARD, J. L. (1974), Cut P and N without chemicals, *Water Wastes Eng.* **11**, 33–36.

BAUMANN, P., KRAUTH, K. H. (1991), Untersuchung der biologischen Phosphatelimination bei gleichzeitiger Stickstoffelimination auf der Kläranlage Waiblingen, *Korrespondenz Abwasser* **38**, 191–198.

BEVER, J., TEICHMANN, H. (Eds.) (1990), *Weitergehende Abwasserreinigung*. München, Wien: R. Oldenbourg Verlag.

BOND, P. L., HUGENHOLTZ, P., KELLER, J., BLACKALL, L. L. (1995), Bacterial community structures of phosphate-removing and non-phosphate-removing activated sludges from sequencing batch reactors, *Appl. Environ. Microbiol.* **61**, 1910–1916.

BRODISCH, K. (1985), Zusammenwirken zweier Bakteriengruppen bei der biologischen Phosphateliminierung, *gwf-Wasser/Abwasser* **126**, 237–240.

BUCHAN, L. (1983), Possible biological mechanism of phosphorus removal, *Water Sci. Technol.* **15**, 87–103.

CARLSSON, H. (1996), Biological phosphorus and nitrogen removal in a single sludge system, *Thesis*. Department of Water and Environmental Engineering, Lund University, Sweden.

CLOETE, T. E., STEYN, P. L. (1988), The role of *Acinetobacter* as a phosphorus removing agent in activated sludge, *Water Res.* **22**, 971–976.

COMEAU, Y., OLDHAM, W. K., HALL, K. J. (1986), Biological model for enhanced biological phosphorus removal, *Water Res.* **20**, 1511–1521.

DEINEMA, M. H., HABITS, L. H. A., SCHOLTEN, A., TURKSTRA, E., WEBERS, H. A. A. M. (1980), The accumulation of polyphosphate in *Acinetobacter* spp., *FEMS Microbiol. Lett.* **9**, 275–279.

DEINEMA, M. H., VAN LOOSDRECHT, M., SCHOLTEN, A. (1985), Some physiological characteristics of *Acinetobacter* spp. accumulating large amounts of phosphate, *Water Sci. Technol.* **17**, 119–125.

DRYDEN, F. D., STERN, G. (1968), Renovated waste water creates recreational lake, *Environ. Sci. Technol.* **2**, 268–278.

EKAMA, G. A., DOLD, P. L., MARAIS, G. V. R. (1986), Procedures for determining influent COD fractions and the maximum specific growth rate of heterotrophs in activated sludge systems, *Water Sci. Technol.* **18**, 91–114.

FETTIG, J., MIETHE, M., KASSEBAUM, F. (1996), Coagulation and precipitation by an alkaline aluminium coagulant, *Proc. 7th Gothenburg Symposium*, pp. 107–117. Heidelberg: Springer-Verlag.

FLORENTZ, M., GRANGER, P., HARTEMANN, P. (1984), Use of 31-P nuclear magnetic resonance spectroscopy and electron microscopy to study phosphorus metabolism of microorganisms from wastewater, *Appl. Environ. Microbiol.* **47**, 519–525.

FUHS, G. W., CHEN, M. (1975), Microbiological basis of phosphate removal in the activated sludge process for the treatment of wastewater, *Microb. Ecol.* **2**, 119–138.

FUKASE, T., SHIBATA, M., MIYAJI, Y. (1984), The role of an anaerobic stage on biological phosphorus removal, *Water Sci. Technol.* **17**, 69–80.

GARBER, W. F. (1972), *Phosphorus Removal by Chemical and Biological Mechanisms. Applications of New Concepts of Physical Chemical Wastewater Treatment.* Vanderbilt University Conf. 1972. Oxford: Pergamon Press.

GERBER, A., MOSTERT, E. S., WINTER, C. T., DE VILLIERS, R. H. (1986), The effect of acetate and other short-chain carbon compounds on the kinetics of biological nutrient removal, *Water SA* **12**, 7–12.

GREENBERG, A. E., KLEIN, G., KAUFFMANN, W. J. (1955), Effect of phosphorus removal on the activated sludge process, *Sewage Ind. Wastes* **27**, 277.

HALVORSON, H. O., SURESH, N., ROBERTS, M. F., COCCIA, M., CHIKARMANE, H. M. (1987), Metabolically active surface polyphosphate pool in *Acinetobacter iwoffi*, in: TORRIANI-GORINE A. et al., *Phosphate Metabolism and Cellular Regulation in Microorganisms*, pp. 220–224. Washington D.C.: American Society for Microbiology.

HAROLD, F. M. (1966), Inorganic polyphosphates in biology: Structure, metabolism and function, *Bacteriol. Rev.* **30**, 772–794.

HASCOET, M. C., FLORENTZ, M. (1985), Influence of nitrate on biological phosphorus removal from wastewaters, *Water SA* **11**, 1–8.

HENZE, M., GUJER, W., MINO, T., MATSUO, T., WENTZEL, M. C., MARAIS, G. V. R. (1995a). Activated Sludge Model No. 2, *IAWQ Scientific and Technical Reports*, No. 3. London: IAWQ.

HENZE, M., GUJER, W., MINO, T., MATSUO, T., WENTZEL, M. C., MARAIS, G. V. R. (1995b). Wastewater and biomass characterization for the Activated Sludge Model No. 2: Biological phosphorus removal, *Water Sci. Technol.* **31**, 2, 13–23.

HENZE, M., HARREMOES, P., LA COUR, C., JANSEN, J., ARVIN, E. (1997), *Wastewater Treatment, Biological and Chemical Processes*, p. 95. Berlin, Heidelberg, New York: Springer-Verlag.

HIRAISHI, A., MASAMUNE, K., KITAMURA, H. (1989), Characterization of the bacterial population structure in an anaerobic-aerobic activated sludge system on the basis of respiratory quinone profiles, *Appl. Microbiol.* **55**, 897–901.

HULSBEEK, J. (1995), Bestimmung von Parametern zur Beschreibung der Prozesse bei der biologischen Stickstoff- und Phosphoreliminierung, *Veröffentlichungen des Institutes für Siedlungswasserwirtschaft und Abfalltechnik der Universität Hannover*, Heft 92, 12/1–12/20.

IWEMA, A., MEUNIER, P. (1985), Influence of nitrate on acetic acid induced biological phosphate removal, *Water Sci. Technol.* **17**, 289–294.

JARDIN, N. (1995), Untersuchungen zum Einfluß der erhöhten biologischen Phosphorelimination auf die Phosphordynamik bei der Schlammbehandlung, *Schriftenreihe des Instituts für Wasserversorgung, Abwasserbeseitigung, Abfalltechnik und Umwelt- und Raumplanung der TH Darmstadt*, Vol. 87.

JARDIN, N., PÖPEL, H. J. (1994). Phosphate fixation in sludges from enhanced biological P-removal during stabilization, in: *Chemical Water and Wastewater Treatment III*, (KLUTE, R., HAHN, H. H., Eds.) 353–372. Berlin, Heidelberg: Springer-Verlag.

JARDIN, N., PÖPEL, H. J. (1996), Behavior of waste activated sludge from enhanced biological phosphorus removal during sludge treatment, *Water Environ. Res.* **68**, 965–973.

JÖNSSON, K., JOHANSSON, P., CHRISTENSSON, M., LEE, N., LIE, E.,WELANDER, TH. (1996), Operational factors affecting enhanced biological phosphorus removal at the waste water treatment plant in Helsingborg, Sweden, *Water Sci. Technol.* **34**, 1–2, 67–74.

JÖRGENSEN, P. E. (1990), Biological hydrolysis of sludge from primary precipitation, in: *Chemical Water and Wastewater Treatment* (HAHN, H. H., KLUTE, R., Eds.), pp. 499–510. Berlin, Heidelberg, New York: Springer-Verlag.

KÄMPFER, P., EISENTRÄGER, A., HERGT, V., DOTT, W. (1990), Untersuchungen zur bakteriellen Phosphateliminierung. I. Mitteilung: Bakterienflora und bakterielles Phosphatspeicherungsvermögen in Abwasserreinigungsanlagen, *gwf-Wasser/Abwasser* **131**, 156–164.

KAPPELER, J., GUJER, W. (1992), Estimation of kinetic parameters of heterotrophic biomass under aerobic conditions and characterization of wastewater of activated sludge modeling, *Water Sci. Technol.* **25**, 125–139.

KERRN-JESPERSEN, J. P., HENZE, M. (1993), Biological phosphorus uptake under anoxic and aerobic conditions, *Water Res.* **27**, 617–624.

KORNBERG, A. (1995), Inorganic polyphosphate: toward making a forgotten polymer unforgettable, *J. Bacteriol.* **177**, 491–496.

KORTSTEE, G. J. J., APPELDOORN, K. J., BONTING, C. F. C., VAN NIEL, E. W. J., VAN VEEN, H. W. (1994), Biology of polyphosphate-accumulating bacteria involved in enhanced biological phosphorus removal, *FEMS Microbiol. Rev.* **15**, 137–153.

KRISTENSEN, G. H., JØRGENSEN, P. E., HENZE, M. (1992), Characterization of functional microorganism groups and substrate in for activated sludge and wastewater by AUR, NUR and OUR, *Water Sci. Technol.* **25**, 43–57.

KUBA, T., VAN LOOSDRECHT, M. C. M., HEIJNEN, J. J. (1996), Effect of cyclic oxygen exposure on the activity of denitrifying phosphorus removing bacteria, *Water Sci. Technol.* **34**, 1–2, 33–40.

KUBA, T., VAN LOOSDRECHT, M. C. M., BRANDSE, F., HEIJNEN, J. J. (1997), Occurrence of denitrifying phosphorus removing bacteria in modified UCT-type wastewater treatment plants, *Water Res.* **31**, 777–787.

KULAEV, I. S., VAGABOV, V. M. (1983), Polyphosphate metabolism in microorganisms, *Adv. Microbiol. Physiol.* **24**, 83–171

LEMMER, H. (1985), Wachstumsverhalten von Actinomyceten (*Nocardia*) in Kläranlagen mit Schwimmschlammproblemen, *Korrespondenz Abwasser* **32**, 965–971.

LEVIN, G. V., SHAPIRO, J. (1965), Metabolic uptake of phosphorus by wastewater organisms, *J. Water Pollut. Control Fed.* **37**, 800–821.

LÖTTER, L. H., MURPHY, M. (1985), The identification of heterotrophic bacteria in an activated sludge plant with particular reference to polyphosphate accumulation, *Water SA* **11**, 179–184.

LÖTTER, L. H., VAN DER MERWE, E. H. M. (1987), The activities of some fermentation enzymes in activated sludge and their relationship to enhanced phosphorus removal, *Water Res.* **21**, 1307–1310.

MARAIS, G. V. R., LOEWENTHAL, R. E., SIEBRITZ, I. P. (1983), Observations supporting phosphate removal by biological excess uptake. A review, *Water Sci. Technol.* **15**, 15–41.

MATSUO, T., MINO, T., SATO, H. (1992), Metabolism of organic substances in anaerobic phase of biological phosphate uptake process, *Water Sci. Technol.* **25**, 6, 83–92.

MC CLINTOCK, S. A., PATTARKINE, V. M., RANDALL, C. W. (1992), Comparison of yields and decay rates for a biological nutrient removal process and a conventional activated sludge process, *Water Sci. Technol.* **26**, 2195–2198.

MEGANCK, M. T. J., FAUP, G. M. (1988), Enhanced biological phosphorus removal from waste waters, *Biotreatment Syst.* **3**, 111–204.

MILBURY, W. F., MCCAULY, D., HAWTHORNE, C. H. (1971), Operation of conventional activated sludge for maximum phosphorus removal, *J. Water Pollut. Control Fed.* **43**, 1890–1901.

MINO, T., KAWAKAMI, T., MATSUO, T. (1984), Location of phosphorus in activated sludge and function of intracellular phosphates in biological phosphorus removal process, *Water Sci. Technol.* **17**, 93–106.

MINO, T., ARUN, V., TSUZUKI, Y., MATSUO, T. (1987), Effect of phosphorus accumulation on acetate metabolism in the biological phosphorus removal process, *Advances in Water Pollution Control*, Vol. 4 (RAMADORI, R., Ed.), pp. 27–38. Oxford: Pergamon Press.

MINO, T., SATOH, H., MATUO, T. (1994), Metabolism of different bacterial populations in enhanced biological phosphate removal processes, *Water Sci. Technol.* **29**, 67–70.

MOSTERT, E. S., GERBER, A., VAN RIET, C. J. J. (1988), Fatty acid utilization by sludge from full-scale nutrient removal plants, with special reference to the role of nitrate, *Water SA* **14**, 179–184.

MURAKAMI, T., KOIKE, S., TANIGUCHI, N., ESUMI, H. (1987), Influence of return flow phosphorus load on performance of the biological phosphorus removal process, in: *Biological Phosphate Removal from Wastewaters* (RAMADORI, R., Ed.), pp. 237–247. Oxford: Pergamon Press.

NAKAMURA, K., MASUDA, K., MIKAMI, E. (1991), Isolation of a new type of polyphosphate-accumulating bacterium and its phosphate removal characteristics, *J. Ferment. Bioeng.* **71**, 259–264.

NAKAMURA, K., HIRAISHI, A., YOSHIMI, Y., KAWAHARASAKI, M., MASUDA, K., KAMAGATA, Y. (1995), *Microlunatus phosphorus* gen. nov., sp. nov., a new gram-positive polyphosphate-accumulating bacterium isolated from activated sludge, *Int. J. Syst. Bacteriol.* **45**, 17–22.

NESBITT, J. B. (1969), Phosphorus removal, the state of the art, *J. Walter Pollut. Control Fed.* **41**, 701–713.

NICHOLLS, H. A., OSBORN, D. W. (1979), Bacterial stress, a prerequisite for biological removal of phosphorus, *J. Walter Pollut. Control Fed.* **51**, 557–569.

ØDEGAARD, H., KARLSSON, I. (1994), Chemical wastewater treatment – value for money, in: *Chemical Water and Wastewater Treatment III* (KLUTE, R., HAHN, H. H., Eds.), pp. 191–209. Berlin, Heidelberg: Springer-Verlag.

PITMAN, A. R. (1995), Practical experiences with biological nutrient removal on full-scale wastewater treatment plants in South Africa, *Veröffentlichungen des Institutes für Siedlungswasserwirtschaft und Abfalltechnik der Universität Hannover*, Heft 92, 6/1–6/21.

PITMAN, A. R., DEACON, S. L., ALEXANDER, W. V. (1991), The thickening and treatment of sewage sludges to minimize phosphorus release, *Water Res.* **25**, 1285–1294.

PÖPEL, H. J. (1991), Grundlagen der chemischen Phosphorelimination, pp. 15–50, *Schriftenreihe des Instituts für Wasserversorgung, Abwasserbeseitigung, Abfalltechnik und Umwelt- und Raumplanung der TH Darmstadt*, Vol. 51.

PÖPEL, H. J. (1995), Schlammanfall bei der chemisch-physikalischen Phosphorelimination, pp. 17–30, *Schriftenreihe des Instituts für Wasserversorgung, Abwasserbeseitigung, Abfalltechnik und Umwelt- und Raumplanung der TH Darmstadt*, Vol. 84.

PÖPEL, H. J., JARDIN, N. (1993), Influence of enhanced biological phosphorus removal on sludge treatment, *Water Sci. Technol.* **28**, 1, 263–271.

PROHASKA BRINCH, P., RINDEL, K., KALB, K. (1994), Upgrading to nutrient removal by means of internal carbon from sludge hydrolysis, *Water Sci. Technol.* **29**, 12, 31–40.

RANDALL, C. W., BRANNAN, K. P., BENEFIELD, L. D. (1987), Factors Affecting Anaerobic Stabilization during Biological Phosphorus Removal, in: Biological Phosphate Removal from Wastewaters, *Proc. IAWPRC Specialized Conf.* (RAMADORI, R., Ed.), held in Rome, Italy, 28–30 September 1987, pp. 111–122.

RENSINK, J. H., DONKER, H. J. G. W., SIMONS, T. S. J. (1986), Biologische Phosphorelimination bei niedrigen Schlammbelastungen, *gwf-Wasser/Abwasser* **127**, 449–453.

SATOH, H., MINO, T., MATSUO, T. (1992), Uptake of organic substrates and accumulation of poly-hydroxyalkanoates linked with glycolysis of intracellular carbohydrates under anaerobic conditions in the biological excess phosphate removal processes, *Water Sci. Technol.* **26**, 933–942.

SCHAAK, F., BOSCHET, A. F., CHEVALIER, D., KERLAIN, F., SENELIER, Y. (1985), Efficiency of existing biological treatment plants against phosphorus pollution, *Tech. Sci. Municip.* **80**, 173–181.

SCHLEGEL, S. (1989), Untersuchungen zur Versäuerung des Vorklärschlammes mit dem Ziel einer besseren P-Elimination, pp. 77–88, *Schriftenreihe des Instituts für Siedlungswasserwirtschaft der TU Braunschweig*, Heft 47.

SCHÖN, G. (1994), Biologische Phosphorentfernung bei der Abwasserreinigung im Belebungsverfahren, *BioEngineering* **4**, 23–32.

SCHÖN, G. (1996), Polyphosphatsspeichernde Bakterien und weitergehende biologische Phosphorentfernung in Kläranlagen, in: *Ökologie des Abwassers* (LEMMER et al., Eds.). Berlin: Springer-Verlag.

SCHÖN, G., STREICHAN, M. (1989), Anoxische Phosphataufnahme und Phosphatabgabe durch belebten Schlamm aus Kläranlagen mit biologischer Phosphorentfernung, *gwf-Wasser/Abwasser* **130**, 67–72.

SCHÖN, G., GEYWITZ-HETZ, S., VALTA, A. (1993), Weitergehende biologische Phosphorentfernung und organische Reservestoffe im belebten Schlamm, in: Biologische Phosphoreliminierung aus Abwässern, Kolloquium an der TU Berlin, 27./28. 9. 1993, pp. 181–194, *Schriftenreihe Biologische Abwasserreinigung der Technischen Universität Berlin.*

SCHÖNBERGER, R. (1990), Optimierung der biologischen Phosphorelimination bei der kommunalen Abwasserreinigung, p. 93, *Berichte aus Wassergütewirtschaft und Gesundheitsingenieurwesen*, TU München.

SCHÖNBORN, W. (1986), Historical developments and ecological fundamentals, in: *Biotechnology*, 1st Edn., Vol. 8 (REHM, H.-J., REED, G., Eds.), pp. 3–42. Weinheim: VCH.

SEN, D., RANDALL, C. W. (1988), Factors controlling the recycle of phosphorus from anaerobic digesters sequencing biological phosphorus removal systems, *Hazard. Ind. Waste* **20**, 286–298.

SEYFRIED, C. F., HARTWIG, P. (1991), Großtechnische Betriebserfahrungen mit der biologischen Phosphorelimination in den Klärwerken Hildesheim und Husum, *Korrespondenz Abwasser* **38**, 185–191.

SEYFRIED, C. F., SCHEER, H. (1995), Bio-P in Deutschland, *Veröffentlichungen des Institutes für Siedlungswasserwirtschaft und Abfalltechnik der Universität Hannover*, Heft 92, 9/1–9/26.

SLIJKHUIS, H. (1983), *Microthrix parvicella*, a filamentous bacterium isolated from activated sludge: cultivation in a chemically defined medium, *Appl. Environ. Microbiol.* **46**, 832–839.

SRINATH, E. G., SASTRY, C. A., PILLAI, S. C. (1959), Rapid removal of phosphorus from sewage by activated sludge, *Experienta* **15**, 339–340.

STEPHENSON, T. (1987), *Acinetobacter:* 1st role in biological phosphate removal, in: *Biological Phosphate Removal from Wastewaters, Advances in Water Pollution Control*, Vol. 4 (RAMADORI, R., Ed.), pp. 313–316. Oxford: Pergamon Press.

STREICHAN, M., SCHÖN, G. (1991), Periplasmic and intracytoplasmic polyphosphate and easily washable phosphate in pure cultures of sewage bacteria, *Water Res.* **25**, 9–13.

STREICHAN, M., GOLECKI, J. R., SCHÖN, G. (1990), Polyphosphate-accumulating bacteria from sewage plants with different processes for biological phosphorus removal, *FEMS Microbiol. Ecol.* **73**, 113–124.

STUMM, W., MORGAN, J. J. (1981), *Aquatic Chemistry*, 2nd Edn. New York: John Wiley & Sons.

SURESH, N., WARBURG, R., TIMMERMANN, M., WELLS, J., COCCIA, M. et al. (1985), New strategies for the isolation of microorganisms responsible for phosphate accumulation, *Water Sci. Technol.* **17**, 99–111.

TEICHFISCHER, T. (1995), Möglichkeiten zur Stabilisierung des Bio-P Prozesses, *Veröffentlichungen des Institutes für Siedlungswasserwirtschaft und Abfalltechnik der Universität Hannover*, Heft 92, 20/1–20/20.

TOERIEN, D. F., GERBER, A., LÖTTER, L. H., CLOETE, T. E. (1990), Enhanced biological phosphorus removal in activated sludge systems, *Adv. Microbiol. Ecol.* **11**, 173–230.

UBUKATA, Y., TAKII, S. (1994), Induction ability of excess phosphate accumulation for phosphate removing bacteria, *Water Res.* **28**, 247–249.

URBAIN, V., MANEM, J., FASS, S., BLOCK, J.-C. (1997), Potential of *in situ* volatile fatty acids production as carbon source for denitrification, *Proc. 70th WEFTEC Conf.* Vol. 1, Part II, pp. 333–339, *Water Environment Federation*, Alexandria, VA.

VACKER, D., CONNELL, C. H., WELLS, W. N. (1967), Phosphate removal through municipal wastewater treatment at San Antonio, Texas. *J. Water Pollut. Control Fed.* **39**, 750–771.

VAN LOOSDRECHT, M. C. M., HOOIJMANS, C. M., BRDJANOVIC, D., HEIJNEN, J. J. (1997), Biological phosphate removal processes, *Appl. Microbiol. Biotechnol.* **48**, 289–296.

WAGNER, M., ERHART, R., MANZ, W., AMANN, R., LEMMER, H., WEDI, D., SCHLEIFER, K.-H. (1994), Development of an rRNA-targeted oligonucleotide probe specific for the genus *Acinetobacter* and its application for *in situ* monitoring in activated sludge, *Appl. Environ. Microbiol.* **60**, 792–800.

WEDI, D., NIEDERMEYER, R. (1992), Berechnungsvorschlag zur Phosphorfällung aus kommunalen Abwässern mit sauren Metallsalzen, *gwf-Wasser/Abwasser* **133**, 557–566.

WENTZEL, M. C., DOLD, P. L., EKAMA, G. A., MARAIS, G. V. R. (1985), Kinetics of biological phosphorus release, *Water Sci. Technol.* **17**, 57–71.

WENTZEL, M. C., LÖTTER, L. H., LOEWENTHAL, R. E., MARAIS, G. V. R. (1986), Metabolic behavior of *Acinetobacter* spp. in enhanced biological phosphorus removal – a biochemical model, *Water SA* **12**, 209–224.

WENTZEL, M. C., EKAMA, G. A., DOLD, P. L., MARAIS, G. V. R. (1990), Biological phosphorus removal – steady state process design, *Water SA* **16**, 1, 29–48.

WENTZEL, M. C., LÖTTER, L. H., EKAMA, G. A., LOEWENTHAL, R. E., MARAIS, G. V. R. (1991), Evaluation of biochemical models for biological excess phosphorus removal, *Water Sci. Technol.* **23**, 567–576.

WENTZEL, M. C., EKAMA, G. A., MARAIS, G. V. R. (1992), Processes and modelling of nitrification denitrification biological excess phosphorus removal systems – a Review, *Water Sci. Technol.* **25**, 59–82.

WILD, D., KISLIAKOVA, A., SIEGRIST, S. (1996), P-fixation by Mg, Ca and zeolite a during stabilization of excess sludge from enhanced biological P-removal, *Water Sci. Technol.* **34**, 1–2, 391–398.

WITT, P. CH. (1997), Untersuchungen und Modellierungen der biologischen Phosphatelimination in Kläranlagen, *Schriftenreihe des Instituts für Siedlungswasserwirtschaft der Universität Karlsruhe*, Vol. 81.

WITT, P. CH., HAHN, H. H. (1995), Bio-P und Chem-P: Neue Erkenntnisse und Versuchsergebnisse, *Veröffentlichungen des Institutes für Siedlungswasserwirtschaft und Abfalltechnik der Universität Hannover*, Heft 92, 5/1–5/23.

YALL, I., BOUGHTON, W. H., KNUDSON, C., SINCLAIR, N. A. (1970), Biological uptake of phosphorus by activated sludge, *Appl. Microbiol.* **20**, 145–150.

15 Continuous Flow and Sequential Processes in Municipal Wastewater Treatment

EBERHARD MORGENROTH
Lyngby, Denmark

PETER A. WILDERER
Munich, Germany

List of Abbreviations

A	cross sectional area of the PFR
C_S, $C_{S,in}$, ΔC_S	substrate concentration, influent substrate, substrate removed
C_X	sludge concentration
CFSTR	continuous flow stirred tank reactor
f_{DN}	fraction of total electron acceptor demand associated with anoxic reactions
f_{exr}	$=\Delta V/(V_0+\Delta V)$, volumetric exchange ratio in the SBR
M_X	mass of sludge
$MLSS$	mixed liquor suspended solids
n	number of SBRs
$NH_{4,e}$, $N_{org,e}$, N_{SL}	ammonia and organic nitrogen in the effluent and nitrogen eliminated with the sludge wastage
PFR	plug flow reactor
Q	flow rate
$r_{S,V}$	volumetric conversion rate
SBR	sequencing batch reactor
SBBR	sequencing batch biofilm reactor
SF_V, SF_X	safety factors
SVI	sludge volume index
t_c, t_f, t_r, t_D, t_N, t_s, t_w, t_d, t_i	phase duration with c: cycle, f: fill, r: react, D: denitrification (stirred), N: nitrification (aerated), s: settle, w: wastage, d: draw, i: idle
TKN_0	total Kjeldahl nitrogen
V	volume
V_0, ΔV, V_R	minimal volume, exchanged volume, total volume of the SBR (see Fig. 1)
V_D, V_N	reactor volume in a continuous flow plant dedicated to denitrification and nitrification
x	distance in the PFR
$Y_{H,net}$	net sludge yield including decay processes and accumulation of inert solids
$\alpha \cdot Q$	recycle in a continuous flow system
η	correction factor anoxic respiration
θ_H	hydraulic retention time
θ_X	solids retention time
$\theta_{X,effective}$	$=\theta_X \cdot (t_c - t_s - t_d)/t_c$, effective solids retention time
$\theta_{X,aerobic}$	$=\theta_X \cdot t_N/t_c$, aerobic solids retention time
τ	$=x \cdot A/Q$, residence time in the PFR

1 Introduction

The sequencing batch reactor (SBR) is one of the fill-and-draw versions of the activated sludge process. In contrast to the continuous flow alternative, metabolic reactions and solid–liquid separation are executed in one tank and in a well defined and continuously repeated time sequence. The wastewater to be treated is batchwise introduced into that tank, treated for a distinct period of time, and finally removed from the tank. In SBRs the liquid volume in the tank varies, while it remains constant in continuous flow systems. The first activated sludge plants ever built were in accordance with the SBR concept (Ardern and Lockett, 1914). However, they were soon converted into continuous flow systems for the lack of proper process control technology and also clogging of the aerators. With the improved technical means available today, continuous flow and SBR systems are equally applicable and competitive. Either system alternative provides specific advantages over the other. The decision which of the alternatives should be applied in a specific case must be based on a thorough investigation. This chapter describes the current status of the SBR technology. Similarities and differences between SBRs and different types of continuous flow systems are discussed. Different design approaches for SBR systems are presented.

Historically, the activated sludge technology began with the investigation of fill-and-draw reactors conducted by ARDERN and LOCKETT (1914) at the Manchester laboratories. In these reactors microorganisms were purposely kept in suspension. After the completion of the reaction the particles were allowed to settle before removing the treated wastewater to retain the microorganisms in the system. It was observed that bacteria grew in readily settleable, floc-like structures, which were given the name activated sludge. The first full-scale activated sludge plant was built at Salford (DUCKWORTH, 1914; MELLING, 1914) as an upscale of the Manchester experimental setup. This plant was an SBR plant as we would call it today. The wastewater to be treated was batchwise introduced into the reactor during a distinct period of time. The reactor content was then aerated for a certain time. Then, the sludge flocs were allowed to settle, and the clear supernatant was drained from the reactor to provide room for another batch of wastewater. Good effluent quality was achieved but operation of the SBR turned out to be difficult at that time. No reliable timers were available to control the duration of the different SBR phases. Also problems with clogging of diffusers occurred. The plant was therefore converted into a continuous flow system at a later time. In the 1970s the idea of operating activated sludge systems in a batch rather than in a continuous flow mode was picked up by IRVINE (IRVINE and BUSH, 1979). He and his group of researchers kept investigating the potentials of controlled unsteady state processes (IRVINE and KETCHUM, 1989) for many years, thus contributing largely to the renaissance and to the advanced status of the SBR technology of today (IRVINE et al., 1997).

2 Current Development Status of SBR Technology

In contrast to continuous flow systems the sequence of processes in an SBR is time rather than space oriented. Metabolic reactions and solid–liquid separation take place at different times in the same reactor. The SBR process is characterized by a series of process phases (fill, react, settle, decant) each lasting for a defined period of time. If no wastewater is available to be treated (e.g., at industrial application sites at night) the SBR can rest in an idle phase. The different phases of SBR operation are summarized in Fig. 1.

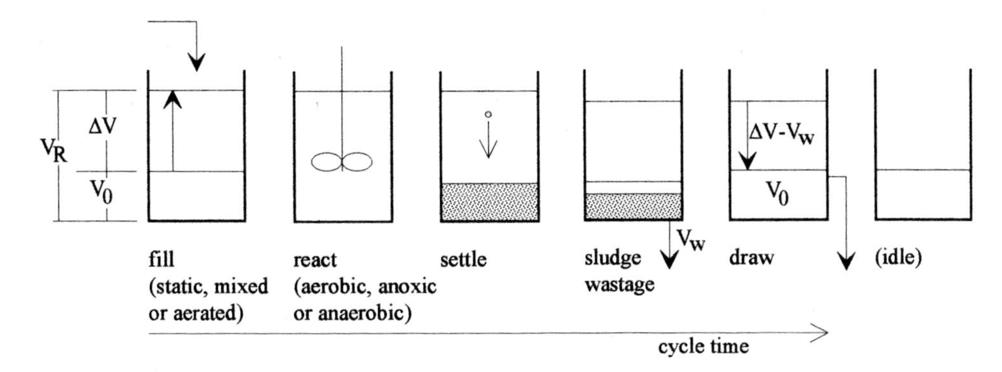

Fig. 1. The different phases of an SBR cycle. Different phases are repeated and order may be altered depending on the process objective.

The sum of the phases makes up a process cycle that is progressively repeated. During each cycle dynamic conditions are prevailing. Control and periodic repetition of the short-term unsteady state allow for, in the long run, enhancement of certain effects such as enzymatic activity, accumulation of metabolic products, selection and enrichment of specific groups of microorganisms (IRVINE et al., 1997). Similar effects are achieved in continuous flow activated sludge systems with a plug flow or cascade flow bioreactor (Fig. 2). There, the activated sludge circulating in the system is also subjected to short-term unsteady state conditions. The sludge is exposed to varying process conditions in subsequent tanks or zones for a distinct hydraulic retention time corresponding to the duration of the respective phase in the SBR.

Since the fill phase is only a fraction of the cycle time it is usually necessary to provide more than one SBR to handle the continuous influx of wastewater. SBR plants consist of a number of SBRs to be operated in parallel. A general scheme of an SBR plant is shown in Fig. 3. The system consists of primary treatment, influent holding tank (optional), several SBRs, and an effluent buffer tank (optional). If an influent holding tank is used the conditions in the SBR are uncoupled from the variations of the influent wastewater as soon as the fill phase is ended. The duration of each process phase can either be fixed or can be adjusted to the actual loading conditions and effluent requirements (variable cycle time). Without a holding tank the number of available SBRs determines the duration of the fill phase and the duration of each cycle (e.g., with two SBRs the fill period makes up 50% of the cycle time). An effluent buffer tank can be used to smooth the variations in the effluent flow rate, which may be necessary for large SBR systems discharging into small receiving water bodies.

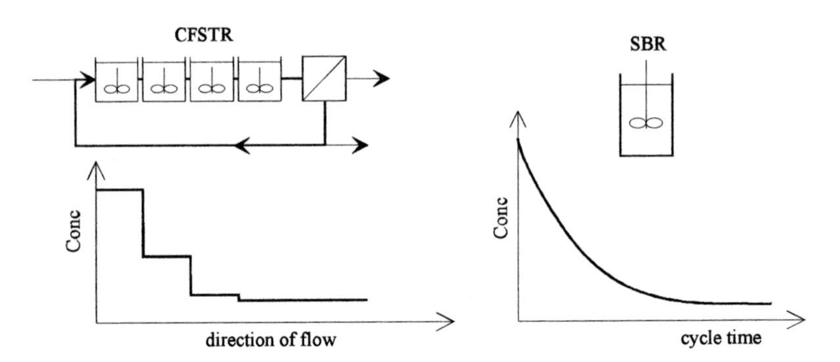

Fig. 2. Concentration profiles of, e.g., organic substrate in a continuous flow stirred tank reactor (CFSTR) or a sequencing batch reactor (SBR). Hydraulic retention time in the cascade of CFSTRs is comparable to cycle time in the SBR.

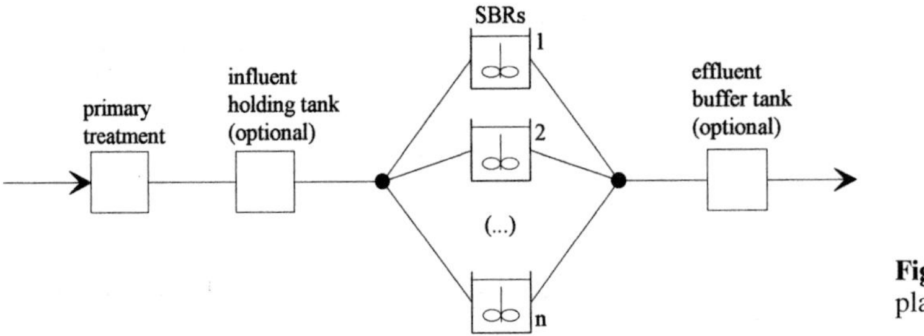

Fig. 3. Scheme of an SBR plant.

2.1 Operation

The operation of the SBR is determined by the very process objective at the site. For instance, control of filamentous bulking requires a pronounced steepness of the concentration gradients. The bacteria are exposed to high substrate concentrations at the beginning of the cycle followed by a period of very low substrate concentrations at the end of the cycle. During the high substrate conditions the bacteria accumulate storage products, which they utilize during the periods of low external substrate availability. This repeated change of high and low substrate concentrations was shown to improve settleability (CHIESA et al., 1985). High initial substrate concentrations can be achieved by using short fill phases and large volumetric exchange ratios ($f_{exr} = \Delta V/(V_0 + \Delta V)$). If, on the other hand, a long fill phase and a small f_{exr} is used then the substrate concentration will not vary significantly over the cycle time, which may be beneficial to avoid substrate inhibition when the incoming wastewater is highly concentrated (Fig. 4). Multiple fill phases are beneficial in nutrient removal systems (Fig. 5c, d). For example, the COD in a second fill phase can be used for denitrification of the nitrogen introduced in the first fill phase (OLES and WILDERER, 1991).

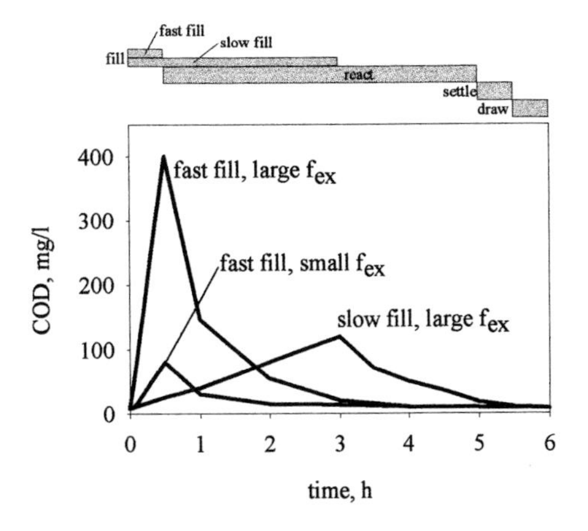

Fig. 4. Concentration gradients for substrate (in this case COD) depend both on the exchange ratio and on the volumetric exchange ratio (f_{exr}).

Fig. 5 contains a number of operation schedules, each designed to meet specific treatment objectives. For COD removal from regular municipal wastewater the SBR cycle may consist of a short fill phase followed by an aerated phase, settle, and draw (a). The same cycle can be used to achieve nitrification. Then, however, the sludge age has to be increased (b). To achieve nitrogen removal an unaerated phase has to be introduced in to the SBR cycle (c). OLES and WILDERER (1991) proposed an SBR cycle with three fill phases, where the second and the third fill is used to supply electron donors required for denitrification. To achieve enhanced biological phosphate removal an anaerobic phase must be included in the operation (d). The first fill phase is used to denitrify any nitrate remaining in the reactor after draw. Then the readily biodegradable COD of the second fill phase is made available for P-release and PHB accumulation. The COD of the third fill phase is used as a carbon source for denitrification.

2.2 Mix

For denitrification and biological phosphorus removal periods without the presence of oxygen are required. However, mass transfer of wastewater components (e.g., organic material, nitrate) to the bacteria needs to be ensured by an unaerated mixing of the SBR. Different mixing devices are shown in Fig. 6. In the jet mixer (a), pumps force the activated sludge suspension through jet nozzles. A rotor floating on the surface (b) can be used for mixing. Mixing can also be provided by using a submerged propeller (c). Mixing devices should be selected to achieve good mixing with limited introduction of oxygen due to any turbulence at the surface.

2.3 Aeration

Oxygen has to be introduced into the system for the oxidation of organic and inorganic wastewater compounds. In addition the transfer of oxygen into the water aeration usually provides sufficient mixing and removal of reaction products (e.g., CO_2) from the water. A

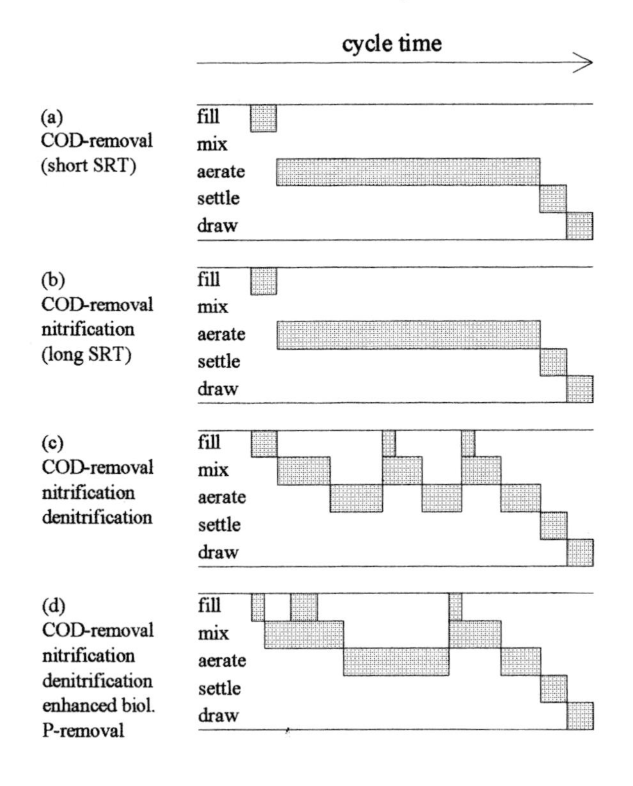

Fig. 5. Examples of SBR cycles for the removal of COD, N, and P. Cycle design depends on the wastewater composition. Note that the given cycles are just examples of the many possible design options.

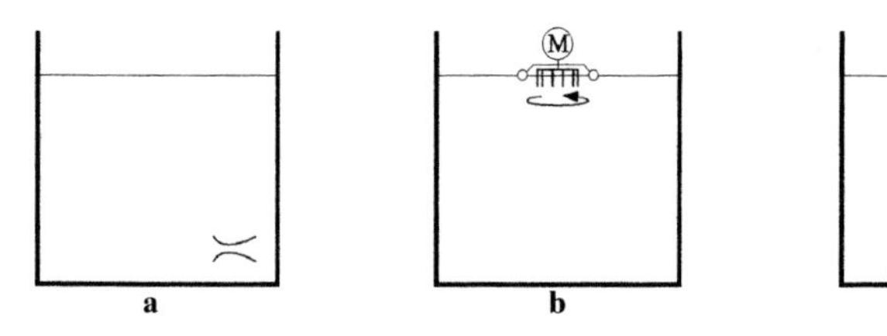

Fig. 6. Unaerated mixing can be provided using **a** submerged jets, **b** floating mixing device (low speed), **c** submerged horizontal (or vertical) propeller.

number of devices are available to aerate the reactor (Fig. 7). Some of the devices described in Fig. 6 can also be used for unaerated mixing (Fig. 7a, b). If air is introduced into the sucction point of the jet (a) it can be used for aeration. If the floating rotor is operated, at a higher speed, then it acts as a surface aerator (b). Air or pure oxygen can be introduced using bubble aeration (c). When selecting an aeration device two aspects are of importance: energy efficiency and reliability. An energy efficient aeration device should be selected as aer-

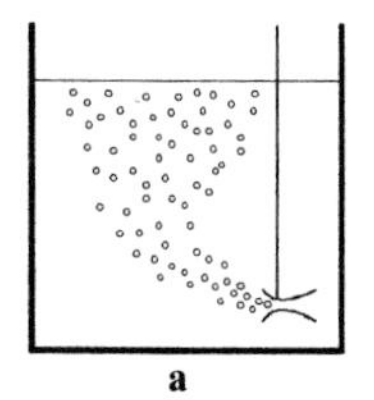

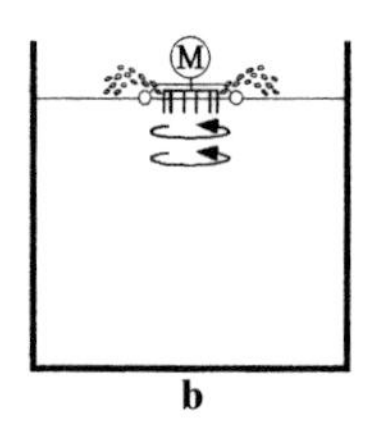

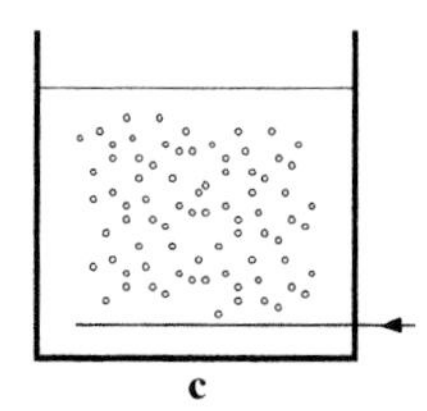

Fig. 7. Aeration of the SBR using **a** jet aeration, **b** floating surface aerator (high speed), **c** compressed air.

ation is among the most energy intensive operations in wastewater treatment systems, consuming 50–90% of the total energy costs (WESNER et al., 1978). Jet aeration systems are generally very reliable but are not very energy efficient. On the other hand, compressed air aeration is generally very energy efficient but the type of diffuser needs to be selected carefully to avoid clogging.

2.4 Draw

In continuous flow activated sludge systems the volume of water in a sedimentation tank remains constant. The rate of clear effluent going over a fixed weir in a continuous system is determined only by the hydraulic loading of the sedimentation tank. In SBRs, however, reaction and solid–liquid separation is performed in one variable volume reactor. After the completion of the reaction phases the sludge is allowed to settle in the SBR. During the following draw phase the water level will change and, thus, either a movable weir or submerged decanting devices must be installed. A great variety of different decanters are being used to remove the treated supernatant from the reactor many of which are proprietary (Fig. 8). Three categories can be distinguished. The decanters shown in (a) and (b) remove the supernatant from the surface of the reactor. In (a) a motor is used to rotate the trough with the overflow weir. Thus, the speed of rotation to lower the weir is used to control the rate of draw. In (b) a decanter is shown that floats on the surface of the water. The draw rate is determined by the hydrostatic pressure at the inlet of the decanter. In (c) a decanter is shown that is installed at a fixed depth in the SBR. Decanters drawing the effluent from the water surface (a, b) have the advantage that the draw phase can begin before the sludge has completely settled. A submerged decanter (c) can draw clear effluent only after the sludge blanket is below the decanter. A motorized unit (a) or decanters controlled by a pump allow the operator to control the effluent flow rate. However, the more mechanical devices are required, the more mechanical problems can occur during the operation. A fixed depth decanter (c) has no moving parts in the SBR and, thus, may permit a more reliable operation. When designing a decanter, precautions to account for foaming sludges must be included. Decanters are often constructed to draw water from below the water surface to prevent foams from getting into the effluent. However, if foams are present in the system they will accumulate on the water surface and must be removed periodically.

2.5 On-Line Control of SBR Systems

The duration of the SBR phases is controlled by a timer that can easily be adjusted to the current situation. Thus, the SBR offers a greater flexibility of changing operating conditions than the continuous flow system. In a continuous flow system the residence times in each reactor are determined by the influent flow rate and the size of the tank, which usually cannot be changed easily. The different control variables in an SBR system are summarized in Tab. 1.

To make use of the various control variables in the SBR reliable information is required on how the process performance can be improved by varying the operating conditions. Compared to continuous flow systems the SBR has the advantage of repeated transient states in the system even when applying steady loads.

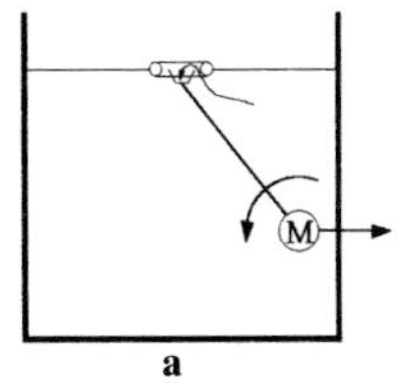

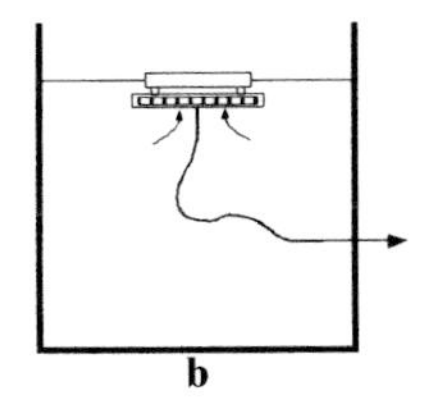

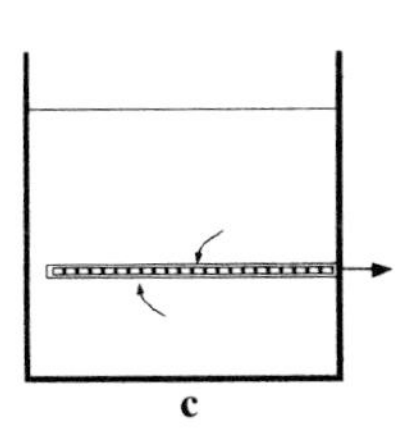

Fig. 8. Typical decanter mechanisms: **a** motorized unit rotates the header pipe, **b** floating header pipe, **c** fixed depth decanter (KETCHUM, 1997).

Tab. 1. Control Variables and Process Information in an SBR

Control Variables	Process Information
Total cycle time Fill time Different combinations of reacting phases Sedimentation time Volumetric exchange ratio (f_{exr}) Number of SBRs in operation	Influent flow load measurements. Every SBR cycle is similar to batch experiments performed to evaluate the kinetic response of activated sludge systems. Thus, by using on-line sensors the inherent dynamics of the SBR can be used to evaluate the current state of the system.

By using on-line measurements the basic process characteristics can be monitored in the SBR during every single cycle. Thus, every cycle in the SBR is comparable to the batch experiments that are performed to evaluate kinetic parameters of the activated sludge. Based on the obtained kinetic information of the system, on-line measurement can directly be used to optimize the SBR operation. If, for example, due to a peak load of nitrogen the ammonia concentration is still above a certain set point at the end of the aeration period, then the cycle duration may be extended to achieve complete nitrification. The cycle time can be increased as long as capacity is available either in one of the other SBRs or in the influent holding tank. On the other hand, if all ammonia has been oxidized before the termination of the aerobic phase, then the time for aeration can be reduced to save energy.

2.6 Other Sequentially Operated Wastewater Treatment Systems

In addition to the classical SBR systems as invented by ARDERN and LOCKETT (1914) a number of other variable volume processes have evolved. In the 1950s PASVEER (1959) proposed fill-and-draw operation of aeration ditches, thus making final clarifiers unnecessary. In contrast to the SBR concept the Pasveer ditches are continuously fed, also during the settle and decant phase. The CAST process (cyclic activated sludge system) may be considered as a further development of the Pasveer system. It is similar to the SBR in that different processes take place at different times in the cycle. However, the CAST reactor is divided into three sections to provide a spatial separation of a separate fill-zone and a draw-zone. Sludge is pumped from the final section of the tank into a small influent chamber to achieve a selector effect (DEMOULIN et al., 1997). The Bio-Denitro or Bio-Denipho processes (EINFELDT, 1992) are operated sequentially in that the flow directions and the aeration are sequentially changed. However, Bio-Denitro or Bio-Denipho processes are constant volume processes and solid–liquid separation is performed in a separate sedimentation tank. The sequential operation of the SBR can also be applied in biofilm systems (WILDERER, 1992). In the sequencing batch biofilm reactor (SBBR) similar phases as described in Fig. 1 are applied in a reactor packed with biofilm support media. Since the biomass is attached to the support no sedimentation phase is required. Thus, the volumetric exchange ratio is not limited by the sludge volume (see Fig. 10) and up to 100% of the bulk liquid in the reactor can be exchanged at the end of each cycle.

3 Comparison of Continuous Flow Plants and SBRs

In the following sections mass balances for continuous flow and sequentially operated systems are developed. Similarities and differences between these systems are then discussed.

3.1 Mass Balances

Depending on the mode of operation SBR systems are comparable to plug flow reactors (PFR) or continuous flow stirred tank reactor (CFSTR) systems (IRVINE and KETCHUM, 1989). In Tab. 2 mass balance equations for SBRs and continuous flow systems are given. It can be seen that the mass balance equation for the SBR with rapid fill is the same as the mass balance equation of the PFR, where the residence time in the PFR (τ) compares to the reaction time in the SBR (t). The mass balance equation for the SBR with slow fill over an extended period of the cycle time and for small volumetric exchange ratios (f_{exr}) resembles the mass balance equation of the CFSTR. In Fig. 9, two SBR cycles are shown with their corresponding continuous flow system. The equivalent *flow rate* of sludge recirculation in the PFR system ($\alpha \cdot Q$) is comparable to the *volume* of water remaining in the SBR after draw (V_0). Thus, the volumetric exchange ratio f_{exr} can be expressed as a function of the comparable recirculation rate ($\alpha \cdot Q$) in the continuous flow system:

$$f_{exr} = \frac{1}{1+\alpha} \tag{1}$$

From Eq. (1) and Fig. 10 it can be seen that normal recycle ratios ranging from 1 to 4 (including return sludge and internal recycles) in the continuous flow system are comparable to f_{exr} in the range from 0.5 to 0.2. Only large f_{exr} lead to mixing characteristics similar to a plug flow reactor. On the other hand, an SBR operated with a very low exchange ratio of $f_{exr}=0.1$ is comparable to a PFR operated with a recycle ratio of $\alpha=9$. A PFR system operated with such a high recycle ratio does not have a significant concentration gradient along the length of the reactor and, thus, has similar characteristics as a CFSTR.

3.2 Application of Continuous Flow and Sequentially Operated Processes

In the discussion above it was pointed out that the microorganisms in both the SBR and the continuous flow activated sludge system are periodically exposed to a series of process conditions. Thus, the activated sludge that develops should possess very similar physical and metabolic properties (WANNER, 1992). Theoretically, both systems should be applicable for

Tab. 2. Mass Balance Equations for SBRs and Continuous Flow Reactors

a SBR with rapid fill (after the end of fill)[a]:

$$\frac{dC_S}{dt} = r_{S,V}$$

b PFR:

$$\frac{\partial C_S}{\partial t} = -\frac{Q}{A} \cdot \frac{\partial C_S}{\partial x} + r_{S,V}$$

which can be simplified to [b, c]

$$\frac{dC_S}{d\tau} = r_{S,V}$$

c SBR with slow fill (during fill)[d]:

$$\frac{dC_S}{dt} = \frac{Q}{V_0 + Q \cdot t}(C_{S,in} - C_S) + r_{S,V}$$

d CFSTR:

$$\frac{dC_S}{dt} = \frac{Q}{V}(C_{S,in} - C_S) + r_{S,V}$$

[a] Initial condition: $C_S(t=0) \approx C_{S,in} \cdot f_{exr} + C_S(t=t_c) \cdot (1-f_{exr})$
[b] Assuming steady state and substituting the residence time $\tau = x \cdot A/Q$
[c] Initial condition: $C_S(\tau=0) = C_{S,in}/(1+\alpha) + C_S(\tau=V/Q) \cdot \alpha/(1+\alpha)$
[d] Initial condition: $C_S(t=0) = C_S(t=t_c)$

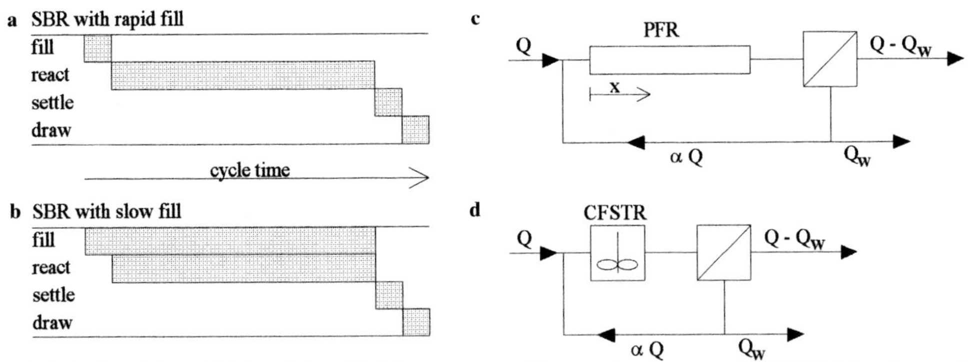

Fig. 9. SBRs with rapid (**a**) and slow fill (**b**) are comparable to a plug flow reactor (PFR) (**c**) and a continuous flow stirred tank reactor (CFSTR) system (**d**), respectively. The mass balance equations for systems **a–d** are given in Tab. 2 **a–d**, respectively.

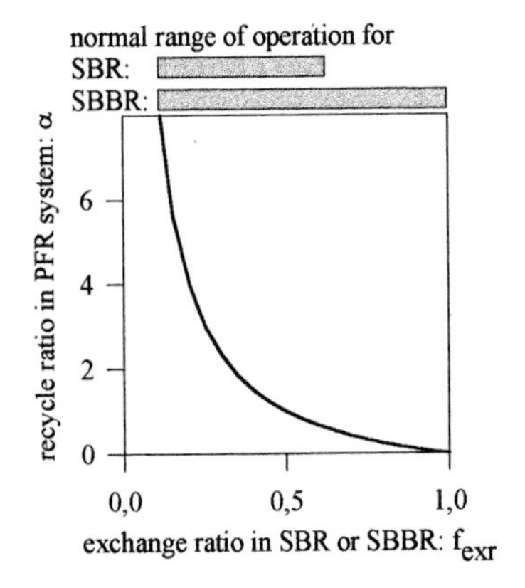

Fig. 10. Volumetric exchange ratios in the SBR and in the SBBR can be compared to the recycle ratio in a continuous flow system. The maximum exchange ratio (f_{exr}) in an SBR is limited by the settled sludge volume in the reactor. In an SBBR the biomass is fixed to the support media and up to 100% of the treated water can be exchanged.

identical purposes. In practice, however, both systems have their specific niches. The SBR has the advantage of a more flexible operation and true plug flow characteristics. Since the treatment process proceeds over time in one reactor, process control can be conveniently executed by means of autoanalyzers (e.g., for dissolved oxygen, pH, nitrate) and computers. Sedimentation takes place under quiescent conditions and is not affected by any currents within the clarifier. Some processes can only be executed if operated in a time oriented mode, such as biological phosphorus removal in biofilms in the SBBR (GONZALEZ-MARTINEZ and WILDERER, 1991). On the other hand, continuous flow systems are more flexible with respect to recycles of sludge and wastewater. Recycle flows can be directed to any reactor in the system (as is required, e.g., in the UCT process), whereas the comparable operation is not possible in the SBR. Operation of the SBR is time oriented and the only possible "recycle" is the volume of sludge and water left in the reactor after draw. The continuous flow system has advantages under highly variable influent conditions (e.g., effluent of combined sewers). It can be operated without the aid of computer timers. The mainstream of wastewater passes through the plant by gravity, whereas pumping is required for filling and/or draining the SBR.

4 Design Approaches for the SBR

For the design of SBR systems similar principles can be applied as for the design of continuous flow activated sludge systems. The following parameters need to be determined during a design calculation:

- **Operating strategy:** The duration and order of the different phases have to be determined according to the process objectives. The influent can be introduced in a short or long fill period, and the SBR cycle may contain one or more fill phases (see Fig. 5). Fixed cycle times can be defined or the cycle times can be adjusted in real time depending on wastewater composition and concentration, using on-line sensors.

- **Sludge age:** For the calculation of the effective sludge age it is assumed that during settle and draw the biomass is dormant. Thus, the effective sludge age is smaller then the solids retention time ($\theta_{X,\text{effective}}$) is smaller than the solids retention time (u_X): $\theta_{X,\text{effective}} = \theta_X \cdot (t_c - t_s - t_d)/t_c$.
- **Volumetric exchange ratio (f_{exr}) and duration of the cycle (t_c):** The amount of wastewater to be treated is determined by the volumetric exchange ratio and duration of the cycle (Tabs. 3 and 4) resulting in a hydraulic retention time (θ_H) of $\theta_H = t_c/f_{exr}$. To achieve a high initial substrate concentration and a large concentration gradient over the cycle time a large volumetric exchange ratio is required.
- **Duration of different phases:** The duration of the different SBR phases (see Fig. 1) is determined according to design equations (Tabs. 3 and 4) and can be optimized using off-line computer simulation or on-line control.
- **Number and volume of SBRs:** The number of SBRs can be chosen according to process needs. Both cost and operational flexibility increase with the number of reactors.

4.1 Application of Experience from Continuous Flow Plants

The design of an SBR can be based on available procedures for continuous flow systems as shown inTab. 3. It is assumed that the mass

Tab. 3. Design Procedure Based on Design of Continuous Flow Plant (ATV, 1997)

Step	Equation
1. INPUT: Design information from continuous flow system, operating strategy for the SBR.	
2. Estimate effective fraction of t_c.	$\text{effective fraction} = \frac{t_c - t_s - t_d}{t_c}$
3. Effective mass of sludge in the SBR equals the mass of sludge in the continuous flow system.	$M_{X,\text{SBR}} = M_{X,\text{Conti.}} \cdot \frac{t_c}{t_c - t_s - t_d}$
4. Select *MLSS* in SBR alter fill and, thus, $n \cdot V_R$.	$n \cdot V_R = \frac{M_{X,\text{SBR}}}{C_X}$
5. Select V_0 depending on the volume of the settled sludge.	$\min.(n \cdot V_0) = M_{X,\text{SBR}} \cdot SVI \cdot SF_V$ *or* $f_{exr,\max} = 1 - C_X \cdot SVI \cdot SF_V$
6. Select exchange ratio *or* cycle time.	
7. Select number of SBRs (n) and calculate missing value (exchange ratio or cycle time).	$f_{exr} = \frac{t_c \cdot Q}{n \cdot V_R} < f_{exr,\max}$
8. For a system with nitrogen removal select unaerated and aerated cycle times according to unaerated and aerated volumes of the continuous flow system.	$\frac{t_D}{t_D + t_N} = \frac{V_D}{V_D + V_N}$
9. Check t_s and t_d based on settling velocity and maximal draw flow rate, respectively.	

and activity of the activated sludge in the SBR will be the same as in the continuous flow system. Higher metabolic activities have sometimes been reported for SBRs compared to continuous flow systems, which may be due to the fact that it is difficult to achieve true plug flow conditions in a continuous flow system. The duration and the order of the SBR phases will be selected proportional to the order and volumes of the reactors or zones in the continuous flow systems (ATV, 1997).

4.2 Design Based on Mass Balances for the SBR

A design procedure directly based on mass balances is presented in Tab. 4. After a selection of the solids retention time (*SRT*) the total mass of sludge in all reactors is calculated based on the influent COD removed. The minimum of the water volume remaining in the reactor after draw (V_0) is determined by the volume of the settled sludge. The exchanged volume (ΔV) depends on the selected cycle time. The total volume of the SBR (V_R) is obtained by adding V_0 and ΔV.

4.3 Dynamic Simulation to Optimize Phase Durations

Design equations should be based on simplifying assumptions to facilitate calculations. These assumptions are appropriate for an initial design purpose. However, today tools for the numerical simulations of complex wastewater treatment systems are available (HENZE et al., 1987). These tools should be used to verify and optimize the preliminary design under constant and under dynamic loading conditions. Under constant influent conditions cycle times and operating strategy can be optimized with respect to, e.g., denitrification, nitrification, and COD removal. Dynamic simulations can be used to evaluate the effect of influent variations and to test possible operational strategies during these transient situations.

5 Discussion

From the analysis above it can be seen that SBRs and continuous flow systems are comparable in many ways. On the other hand, both systems have their specific niches of application. Comparing the required volume of an SBR and a continuous flow system the volume of the SBR will be larger than the corresponding aeration tanks in the continuous flow system. However, if the volume of the SBR is compared to the total volume of the continuous flow plant (aeration tanks plus secondary clarifier), the SBR usually requires less volume. Sedimentation proceeds in the SBR are unaffected by currents as it is the case in continuous flow clarifiers. Sedimentation under quiescent conditions is faster and may permit smaller reactor volumes.

The reduction in size, however, is not the major advantage of the SBR. More important is the inherent flexibility of operation offered by the SBR. The duration of the process phases is an important control variable and can easily be adjusted to actual process conditions without the need of structurally retrofitting existing tanks. Using on-line sensors the inherent process dynamics during an SBR cycle can be used to evaluate kinetic information required for a predictive control strategy. The greater operational flexibility in the SBR constitutes a major advantage over continuous flow systems where most of the operation is already determined by the process configuration. This flexibility, on the other hand, also means that more automation is required to operate an SBR system. Some practitioners today consider the process flexibility and automation as a negative feature of the SBR.

An SBR plant consisting of a number of reactors and holding tanks can be designed in a rather compact manner. Modular design allows adjustment of the plant's capacity by adding additional reactors, or taking reactors temporarily out of operation when the influent conditions change. This aspect is of specific interest to industry and also to some municipalities (e.g., tourist resorts, industrial dischargers with campaign operation, expanding villages).

Tab. 4. Design Procedure Based on Mass Balances for the SBR (for General information on using mass balances for the design of activated sludge systems see, e.g., ORHON and ARTAN, 1994)

Step	Equation
1. INPUT: operating strategy for SBR.	
2. Select effective solids retention time ($\theta_{X,effective}$).	
3. Estimate effective fraction of t_c.	$\text{effective fraction} = \frac{t_c - t_s - t_d}{t_c}$
4. Calculate mass of sludge in all reactors.	$M_{X,SBR} = Y_{H,net} \cdot Q \cdot \Delta C_S \cdot \theta_{X,effective} \cdot \frac{t_c}{t_c - t_s - t_d}$
5. Calculate volume of settled sludge in all reactors.	$n \cdot V_0 = M_{X,SBR} \cdot SVI \cdot SF_V$
6. Select exchange ratio *or* cycle time.	
7. If exchange ratio was selected, then calculate cycle time from:	$t_c = \frac{n \cdot V_0 \cdot f_{exr}}{Q \cdot (1 - f_{exr})}$
8. Calculate exchange volume of all reactors.	$n \cdot \Delta V = Q \cdot t_c$
9. Calculate total volume of the SBRs.	$n \cdot V_R = n \cdot V_0 + n \cdot \Delta V$
10. Select number of SBRs (n).	
11. Calculate min. aerated fraction of reaction time based on min. aerobic solids retention time for, e.g., nitrification.	$\text{min. aerated fraction} = \frac{\text{min. } \theta_{X,aerobic}}{\theta_X} \cdot SF_X$
12. Choose number of fill phases and distribution of the influent on the fill phases depending on the wastewater composition (e.g., TKN/COD ratio).	e.g., see Fig. 5
13. For a system with nitrogen removal check nitrification capacity and denitrification potential depending on cycle design.	$N_{nit} = TKN_0 - NH_{4,e} - N_{org,e} - N_{SL}$ $N_{DP} = (1 - Y_{H,net}) \cdot C_{S,in} \cdot \frac{1}{2.86} \cdot f_{DN} \cdot \eta$
14. Check reacting time for nitrogen removal using denitrification rates.	$t_D = \frac{N_{DP} \cdot n \cdot \Delta V}{r_{DN,X} \cdot M_{X,SBR}}$
15. Check t_s and t_d based on settling velocity and maximum draw flow rate, respectively.	

6 Conclusions

(1) SBRs were the first applications of activated sludge technology. They were converted into continuous flow systems, because of various problems at that time (e.g., lack of reliable timer control, clogging of aerators). With modern technology these problems can be overcome.

(2) Today the SBR offers a great flexibility in the operation of wastewater treatment plants, where different phases (fill, react, settle, draw, idle) can be combined in many different combinations. Using on-line measurement the

cycle time of the SBR can be adjusted continuously based on the wastewater composition.

(3) Various mechanical devices are available for draw, mix, and aerate. Most of these devices are proprietary.

(4) The design of SBR systems can be derived from empirical procedures used for the design of continuous flow activated sludge plants. For better reliability of the calculations, however, SBRs should be designed directly on the basis of mass balances. Cycle times can be optimized using mathematical simulations.

(5) Comparison of the mass balances for SBRs and continuous flow plants reveals that an SBR with a short fill phase and a large volumetric exchange ratio is comparable to a plug flow reactor, whereas an SBR with a slow fill and a small volumetric exchange ratio can be compared to a continuous flow stirred tank reactor.

(6) Even though SBR and continuous flow systems are comparable, both technologies offer their specific advantages. The SBR permits a more flexible operation to accommodate variable influent conditions. The inherent transient conditions in the SBR can be monitored using on-line measurements and, thus, permit predictive control. A continuous flow system may be advantageous for complex recycles of sludge and wastewater within the system, and for handling a highly variable influx from combined sewers.

7 References

ARDERN, E., LOCKETT, W. T. (1914), Experiments on the oxidation of sewage without the aid of filters, *J. Soc. Chem. Ind.* **33**, 523–536.

ATV (1997), Belebungsanlagen im Aufstaubetrieb (Activated sludge variable volume reactors), *Merkblatt ATV-M 210.* Hennef: Abwassertechnische Vereinigung.

CHIESA, S. C., IRVINE, R. L., MANNING, J. F. JR. (1985), Feast/famine growth environments and activated sludge population Selection, *Biotechnol. Bioeng.* **27**, 562–568.

DEMOULIN, G., GORONSZY, M. C., WUTSCHER, K., FORSTHUBER, E. (1997), Co-current nitrification/denitrification and biological P-removal in cyclic activated sludge plants by redox controlled cycle operation, *Water Sci. Technol.* **35**, 215–224.

DUCKWORTH, W. H. (1914), Aeration experiments with activated sludge, *Proc. Annu. Meet. Manchester District Branch of the Assoc. Managers of Sewage Disposal Works* 50.

EINFELDT, J. (1992), The implementation of biological phosphorus and nitrogen removal with the Bio-Denipho process on a 265 000 PE treatment plant, *Water Sci. Technol.* **25**, 161–168.

GONZALEZ-MARTINEZ, S., WILDERER, P. A. (1991), Phosphate removal in a biofilm reactor, *Water Sci. Technol.* **23**, 1405–1415.

HENZE, M., GRADY, C. P. L., GUJER, W., MARAIS, G. V. R., MATSUO, T. (1987), Activated sludge model No. 1, *IAWPRC Scientific and Technical Reports,* No. 1, IAWQ London.

IRVINE, R. L., BUSH, A. W. (1979), Sequencing batch biological reactors – an overview, *J. Water Polut. Control. Fed.* **51**, 235–243.

IRVINE, R. L., KETCHUM, L. H. (1989), Sequencing batch reactors for biological wastewater treatment, *CRC Crit. Rev. Environ. Control* **18**, 255–294.

IRVINE, R. L., WILDERER, P. A., FLEMMING, H. C. (1997), Controlled unsteady state processes and technologies – an overview, *Water Sci. Technol.* **35**, 1–10.

KETCHUM, L. H. J. (1997), Design and physical features of sequencing batch reactors, *Water Sci. Technol.* **35**, 11–18.

MELLING, S. E. (1914), Purification of salford sewage along the line of the Manchester experiments, *J. Soc. Chem. Ind.* **33**, 1124–1127.

OLES, J., WILDERER, P. A. (1991), Computer aided design of sequencing batch reactors based on the IAWPRC activated sludge model, *Water Sci. Technol.* **23**, 1087–1095.

ORHON, D., ARTAN, N. (1994), *Modeling of activated sludge systems*. Lancaster: Technomic.

PASVEER, A. (1959), Contribution to the development in activated sludge treatment, *J. Proc. Inst. Sewage Purif.* **4**, 436–455.

WANNER, J. (1992), Comparison of biocenoses from continuous and sequencing batch reactors, *Water Sci. Technol.* **25**, 239–249.

WESNER, G. M., CULP, G. L., LINECK, T. S., HINDRICHTS, D. J. (1978), Energy conservation in municipal wastewater treatment, *U. S. Environmental Protection Agency Report*, EPA 430/9-77-011.

WILDERER, P. A. (1992), Sequencing batch biofilm reactor technology, in: *Harnessing Biotechnology for the 21th Century* (LADISCH, M. R., BOSE, A., Eds.), pp. 475–479. Washington, DC: American Chemical Society.

16 Design of Nitrification/Denitrification in Fixed Growth Reactors

BERND DORIAS
Stuttgart, Germany

GÜNTER HAUBER
Sydney, Australia

PETER BAUMANN
Stuttgart, Germany

List of Abbreviations

A_{TF}	surface area
B_A	specific surface loading
B_d	daily BOD_5 load
BOD_5	biochemical oxygen demand
B_R	volumetric loading rate
K_m	half time constant for ammonia nitrogen
$NH_4^+ - N_{eff}$	ammonia nitrogen in trickling filter effluent
q_A	specific hydraulic surface load
Q_d	daily flow
Q_{Rec}	recirculated flow
SK	flushing force
V_N	nitrification rate
$V_{N,Max}$	maximum nitrification rate
WWTP	wastewater treatment plants

1 Introduction

In Germany wastewater treatment plants (WWTP) are generally dimensioned according to ATV (Abwassertechnische Vereinigung) standards. For trickling filters and rotating biological contactors (fixed growth reactors) the applicable ATV standard is A-135 *"Principles for the dimensioning of biological filters and biological contactors with a population equivalent greater than 500 p.e."* (ATV, 1989a). This code gives design information for trickling filters and rotating biological contactors for wastewater treatment with and without nitrification. Nitrogen elimination, however, is neither part of this ATV standard nor is there a generally agreed design procedure to date.

Today direct nitrogen elimination in trickling filters and in rotating biological contactors is often applied as state-of-the-art technology. Hence, it is necessary to review ATV standard A-135. The design for denitrification has already been presented in several publications (ATV, 1994a; DORIAS, 1994a, b, 1996a, b). It is based on empirical results meanwhile accepted by experts in the field throughout Germany.

The design for nitrification is complex. The monitoring values for nitrification in Germany have been considerably tightened since the introduction of the original ATV standard (ATV, 1989a). Furthermore, in Germany increased short-term effluent concentrations, e.g., for NH_4^+-N, are not allowed. For processes with a short liquid contact time or low hydraulic buffer capacity like trickling filters and rotating biological contactors (when compared with activated sludge systems) high diurnal variations can cause increased short-term effluent values. Measurements taken in existing trickling filter and rotating biological contactor units show that the design criteria for nitrification as recommended in the ATV standard A-135 (ATV, 1989a) are not adequate to meet these short-term NH_4^+-N requirements.

At present there are neither enough evaluation data available to give clear and satisfactory answers to the open questions regarding nitrification nor to obtain new ideas for improved design recommendations. As soon as such results are available the ATV standard A-135 (ATV, 1989a) will be reviewed regarding its recommendations for nitrification.

This paper presents the current knowledge in Germany in relation to this issue. It refers to actual measurements taken at trickling filter plants. Other alternative information can be drawn from the American EPA manual *"Manual Nitrogen Control"* (EPA, 1993).

2 Trickling Filters

2.1 Process Development

The trickling filter process is considered to be one of the oldest biological wastewater treatments. It is based on experiences made in the middle of the last century with soil filtration. The deciding fact was the observation that microorganisms living in the soil were able to decompose organic matter. CORBETT in Salford (DUNBAR, 1974) using coarse granular filter material made a further development around the turn of the century. With his alteration it was possible to obtain improved filter aeration, which resulted in an improved process efficiency via a continuous distribution of primary treated wastewater over the filter material.

From the early 1950s and onwards trickling filter units in Germany were gradually replaced by the new evolving activated sludge process. This development was further accelerated by the need for direct nitrogen elimination since 1990. This was not considered to be economically feasible with trickling filter units. Meanwhile it has however been proven that trickling filters might economically be viable for direct nitrogen elimination.

2.2 Conditions for High Process Efficiency

The following conditions must be met to obtain the highest possible process efficiency for a trickling filter:

(1) Feed stream consists of primary treated wastewater to avoid clogging by coarse solids.
(2) Filter media must perfectly be conditioned and placed.
(3) The distribution of wastewater has to be evenly sprayed over the surface of the trickling filter. This is particularly important for plastic media with limited internal horizontal cross flow.
(4) Sufficient flushing capacity must exist to wash out surplus sludge, and to maintain a minimum wetting rate particularly for plastic media where otherwise a short circuiting current can occur. If necessary this must be guaranteed by recirculation.
(5) The respective minimal hydraulic loading rate has to be maintained.
(6) Feed and recirculation flows given to maintain minimum surface loading must be controlled.
(7) The trickling filter media have to be adequately supplied with air through the floor (with the exception of denitrification trickling filters, of course).
(8) There must exist sufficient wall insulation to protect against excessive temperature losses and severe winter conditions. Closed designs may be needed and different recirculation pumps may have to be operated during the warm and cold seasons, respectively. In this case artificial ventilation combined with exhaust air treatment might be advisable to avoid nuisance odors.

2.3 Filter Media

The main part of a trickling filter is the inserted packing media. Filter media may be distinguished as mineral packing material, e.g., rocks or lava, or synthetic material. When selecting the filter media it must be made sure that it allows the free passage of wastewater over the trickling filter as well as free air flow (for aerobically operated filters), and the removal of biological surplus sludge with the water flow. Clogging of void spaces reduces the biofilm surface area and thus the purification process and can even lead to the risk of a complete failure.

In conventional trickling filter plants in Germany, rocks or slugs with a grain size of 40–80 mm and 80–150 mm, respectively, were used for packing media and for support material on top of the filter floor. A grain size of 40–80 mm provides a specific surface area of approximately 90 m^2 m^{-3} and a void space ratio of approximately 50%. During normal operation $^2/_3$ of the surface may be considered as utilizable. Presently there is a lack of knowledge with respect to this design information. More detailed recommendations in regard to the utilizable filter surface area can only be made with certain restrictions until new investigation results are available.

There is a large variety of media specifically designed for the first biological stage or the partial treatment of high strength wastewaters in different structural configurations (specific surface of about 200 m^2 m^{-3}, void space ratio of approximately 95%). Plastic sheets or regularly shaped packing systems can help to form channels for higher safety against plugging and to allow free air penetration as well as flushing out of surplus sludge.

In terms of the surface area of the packing material one must distinguish between the theoretical specific surface area and the biologically active surface area (Anonymous, 1989). To describe the process capacity the biologically active surface area should be used. This area is generally less than the theoretical surface area of the packing material because:

(1) Assuming that all applied wastewater comes in contact with the entire surface of the filter media represents the ideal case which in general cannot be realized. This is a function of the hydraulic load and the structure and slope of the packing material.
(2) Particularly in the case of random packed media the usable surface is reduced by the surface touching between the individual media elements.

Depending on the media and packing system configuration and surfaces in use different liquid contact times and contact effects may occur between wastewater and biofilm for equal specific surface areas. Assuming that these effects are accounted for by a media effectiveness factor the surface loading rate would represent the ideal design criteria to determine the required trickling filter media surface area and thus the filter volume.

There is no standard procedure to determine this utilization factor, though. Manufacturer guidelines are more or less based on random determination methods and estimates. For the future further research work must be undertaken to develop this knowledge and to determine the actual and active surface area of plastic filter media. Due to this lack of knowledge design recommendations in regards to the useable surface area of filter media can only be applied with some restrictions until new and further full scale test results are available.

2.4 General Guidelines

The volumetric loading rate B_R in kg BOD_5 per m^3 filter volume and day [kg $(m^3\ d)^{-1}$] is the governing design parameter to determine the trickling filter volume as outlined through the German design norms in ATV A-135 "design of trickling and submerged filters" (ATV, 1989a). The trickling filter volume to be provided can then be determined as the daily BOD_5 load B_d to the filter divided by the volumetric loading rate B_R:

$$V = B_d/B_R \tag{1}$$

For random packed media the filter influent BOD_5 concentration must be smaller than 150 mg L^{-1} to avoid clogging. This may be achieved by recirculation. A recirculation of less than 100% is usually sufficient to achieve this including a safety factor for diurnal variations in concentration. Plastic media typically exhibit a lower tendency for clogging, and higher BOD_5 values may be accepted.

The surface area A_{TF} of a trickling filter and the depth or height (H) of the necessary filter can then be calculated as:

$$H = V(A_{TF})^{-1} \tag{2}$$

$$A_{TF} = (Q_d + Q_{Rec})\ (q_A) - 1 \tag{3}$$

with

Q_d being the daily flow and Q_{Rec} the recirculated flow, respectively.
q_A is the specific hydraulic surface load in $[m^3\ (m^2\ h)^{-1}]$.

The design then has to optimize height vs. surface area to achieve the best result. In general, trickling filter depths between 2.80 m and 4.20 m have shown to give the best results.

To avoid clogging a sufficient flushing force (Spülkraft) is necessary for the transportation of surplus sludge. The flushing force is influenced by the hydraulic surface load and by the design of the influent distributor area. The flushing force SK can be determined as

$$SK = q_A(ab)^{-1}\ 1{,}000 \quad \text{in [mm (sprinkler wing)}^{-1}] \tag{4}$$

with

a: number of rotary sprinkler arms,
b: revolutions per h.

Flushing force values of 4–20 mm per sprinkler have proven to be satisfactory.

2.5 Wastewater Treatment without Nitrification

For the design of random packed trickling filters the following recommendations are given in the ATV standard A-135 (ATV, 1989a): BOD_5 space loading $B_R = 0.4$ kg $(m^3\ d)^{-1}$, hydraulic surface load $q_A = 0.5$–1.0 m h^{-1}. (It is presently under discussion to increase the authorized values for q_A up to 4.0 m h^{-1}).

According to ATV A-135 (ATV, 1989a) for plastic media an increase of the BOD_5 volumetric loading rate due to the greater specific surface area up to $B_R = 0.8$ kg $(m^3\ d)^{-1}$ is permissible. The maximum values should only be used if they are based on clear reliable tests or if they are proven on directly comparable half or full scale units (see Sects. 2.3, 2.6.1).

2.6 Wastewater Treatment with Nitrification

2.6.1 Single-Stage Trickling Filter Units

Nitrification depends on temperature. In Germany the design temperature is generally fixed at 10 °C. For the design of random packed trickling filters the ATV standard A-135 gives the following recommendation (ATV, 1989a): BOD_5 volume load $B_R = 0.2$ kg $(m^3\ d)^{-1}$, hydraulic surface load $q_A = 0.4–0.8$ m h^{-1}. (As for trickling filters without nitrification considerations are under way to increase the permissible value for q_A to as high as 4.0 m h^{-1}).

ATV standard A-135 (ATV, 1989a) allows BOD_5 space loadings for plastic media depending on the specific surface area of the media of up to $B_R = 0.4$ kg $(m^3\ d)^{-1}$. However, there are concerns that in these cases and for load periods with low wastewater temperatures ($T < 12$ °C) nitrification may be less than required. It is thought that this is due to a channel effect caused by irregular bacterial growth. This can occur even shortly after the beginning of the operation whereby a large part of the wastewater trickles through the filter within a very short period of time. In that case particularly ammonia cannot sufficiently be nitrified. At present it is not clear how this irregular radial flow around the media could reasonably be minimized or even prevented.

It is, therefore, recommended to apply higher B_R design values for plastic media only in such cases where these are based on clear and repeatable tests or have been proven on directly comparable semi or full scale units. It is furthermore recommended not to build trickling filters too shallow. The lower the depth of trickling filters the greater the negative consequences of channel formation.

2.6.2 Two-Stage Plants

Up to now no design guidelines exist for two-stage plants. Most of the design information for nitrifying trickling filters is based on the nitrification velocity on the active surface. There are important differences though, as can be seen in Tab. 1: Parameters quoted in the literature.

There is a general consensus that nitrification rates depend to a high degree on temperature and on the NH_4^+-N effluent concentration to be achieved. The temperature factor has frequently been determined to be 1.06 (MÜLLER, 1984; WILD et al., 1971). The dependence of the

Tab. 1. Values Quoted in the Literature for Surface Related Nitrification Rates for Second Stage Filters

Nitrification Velocity [mg $(m^2\ h)^{-2}$]	[g $(m^2\ d)^{-1}$]	Temperature [°C]	NH_4^+-N [mg L^{-1}]	Other Conditions	Reference
	0.86[a]	10	5	after intermediate treatment	WOLF (1989)
15–25	0.36–0.6	≥10–12	<5		RHEINHEIMER et al. (1988)
	10	10	5	without screening	ATV (1989b)
	2.0	10	5	after intermediate treatment	ATV (1989b)
~40	3.0	10	5	after filtration	ATV (1989b)
	~0.96	7–11	≥4	BOD_5	EPA (1975)
	~0.8	10	>5	as 3rd treatment stage	GUJER and BOLLER (1986)

[a] Value from code for household wastewater, lava fill material with $A_R = 60\ m^2\ m^{-3}$.

nitrification rate V_N on the NH_4^+-N effluent concentration is typically accepted as:

$$V_N = V_{N,Max} / NH_4-N_{eff} / (K_m + NH_4-N_{eff}) \quad (5)$$

with: $V_{N,Max}$: maximum nitrification rate [g $(m^2\, d)^{-1}$], K_m half time constant for ammonia nitrogen in mg L^{-1}, $NH_4^+-N_{eff}$ ammonia nitrogen in trickling filter effluent in mg L^{-1}.

Values of 1 mg L^{-1} (GUJER and BOLLER, 1986) to 2 mg L^{-1} (WOLF, 1989) are promoted for the half time constant K_m which may be considered as a safety factor in the design.

2.7 Wastewater Treatment with Nitrification and Denitrification

2.7.1 Introduction

Basically there are three different process methods to integrate denitrification in trickling filters (ATV, 1994a, b):

(1) simultaneous denitrification in the trickling filter with nitrate return to the feed stream,
(2) predenitrification in an "anoxic" zone being either an attached growth reactor, e.g., a trickling filter, or an activated sludge system,
(3) postdenitrification with the addition of an external carbon source in either an attached growth reactor, e.g., a trickling filter, or in an activated sludge system.

2.7.2 Simultaneous Denitrification

Simultaneous denitrification in trickling filter units means that nitrification as well as denitrification processes take place within one trickling filter unit. This is called a single stage process. The exact conditions for this to occur in terms of packing material, surface loads, inflow concentration etc., are not known.

What is known, is the fact that denitrification takes place particularly in the upper trickling filter area, where organic carbon compounds are readily available. Of course, a necessity for this is that oxidized nitrogen (typically largely nitrate NO_3^{2-}) must be available in sufficient quantities. This can be controlled via recirculation.

Present investigation results (DORIAS, 1996a) suggest use of a denitrification capacity of up to 0.05 kg NO_x-N_{den} (kg $BOD_5)^{-1}$.

2.7.3 Predenitrification in Prelocated Trickling Filters

In general, existing trickling filters can be used for direct denitrification, with only minor modifications required and a corresponding operating mode.

By stopping the filter ventilation by covering the trickling filter and by submerging the filter outlet and the lower air orifices, it is possible to create anoxic conditions inside the trickling filter. If at the same time recirculated effluent containing nitrate is fed to the trickling filter together with primary clarified effluent all conditions for a successful denitrification exist. Fig. 1 shows a trickling filter modified for denitrification.

Submerging the trickling filter media completely might cause blockages, and could lead to problems. It should, therefore, not be done.

The clarifier effluent water of a denitrifying trickling filter is fed to an intermediate clarifier, or directly to a postlocated treatment unit. These in general are trickling filters or activated sludge units (Fig. 2).

The following design considerations are given on the basis of investigation results obtained in three wastewater treatment plants:

(1) hydraulic surface load $q_A < 4$ m h^{-1},
(2) the recirculation feedback ratio should not exceed 300% to minimize oxygen transfer to the trickling filter,
(3) the specific denitrification rate is a function of the BOD_5 surface load and may be determined from Tab. 2.

These values are valid for 10 °C and a desired NO_x-N effluent concentration in the denitrifying filter of (≤ 1 mg L^{-1}). Intermediate values can be interpolated (Fig. 3).

It should be noted that the denitrification rates listed in Tab. 2 where exceeded in approx-

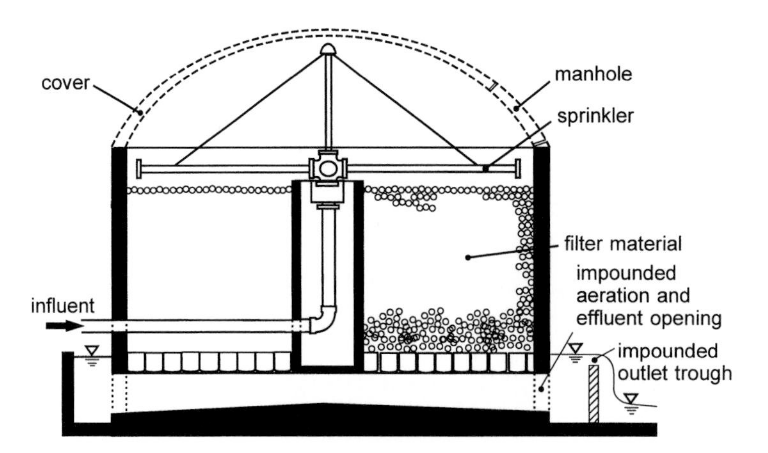

Fig. 1. Trickling filter modified for denitrification.

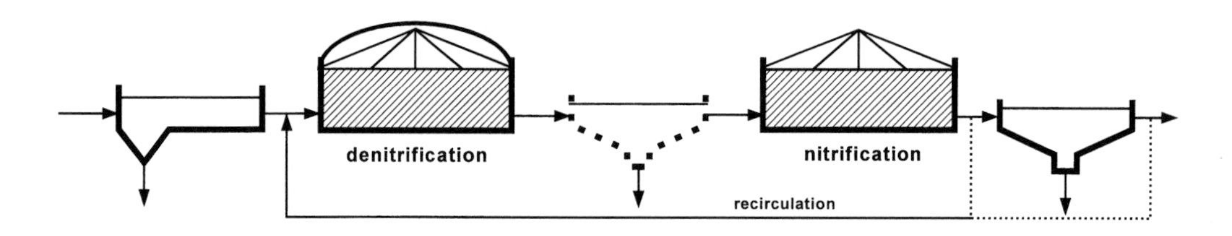

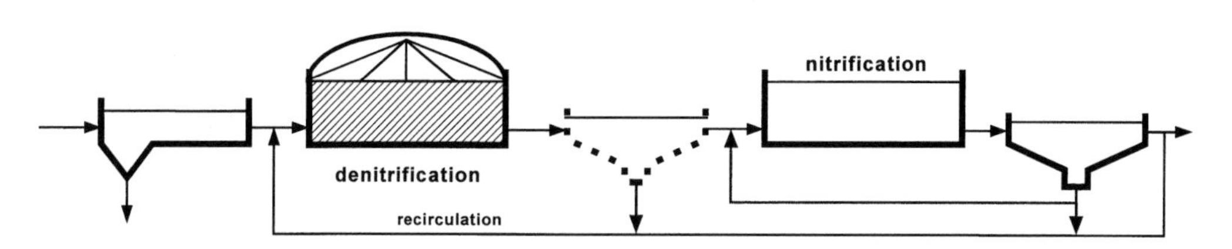

Fig. 2. Flow chart: predenitrification in a modified trickling filter and nitrification in a trickling filter or with activated sludge system.

Tab. 2. Denitrification Rates as a Function of the BOD_5 Surface Load

BOD_5 Surface Load $[g\ (m^2\ d)^{-1}]$	Denitrification Capacity $[kg\ NO_x{-}N_{den}(kg\ BOD_5)^{-1}]$
3	0.14
6	0.11
10	0.09

imately 90% of all measurements taken. The design considerations thus include a sufficient safety factor. If a plant must be designed for a wastewater temperature other than 10 °C the temperature correction factor of 1.06 as determined by DORIAS (1996a) should be applied.

To avoid clogging of the influent distributor a protective screen should be installed before the denitrifying trickling filter. The screen opening must be chosen based on local conditions and experiences, e.g., potential for leaves or small animals falling into the feed stream. In most cases a mesh size of 1–2 mm is adequate. If a fine screen is installed at the inlet works – or if clogging (e.g., by leaves) of the influent distributor arms occurs only during certain seasons – a mechanically cleaned screen might be sufficient, especially for smaller wastewater treatment plants. For new trickling filters or when exchanging the influent distributor special attention should be given to a non-clogging detailed design and construction of this part.

If the second stage nitrifying unit is an activated sludge type process the nitrate recirculating flow should be taken out of the secondary clarifier effluent. This must be taken into consideration when designing the final clarification step. However, when using suitable plastic media a recirculation directly from the activated sludge stage might be possible, but only after corresponding preliminary tests (DORIAS, 1996a).

For postdenitrifying trickling filters a recirculation directly from the effluent of the filter is possible. In addition, it might be advantageous to pass the recirculated flow through the primary clarifier. The separation of surplus sludge in the primary clarifier results in a lower solids load onto the denitrifying trickling filter and thus to a higher safety against clogging. A further advantage is the reduced oxygen transfer into the denitrifying trickling filter.

2.7.4 Denitrification in a Prelocated Activated Sludge Stage

In this method a non-aerated basin is used receiving recirculated nitrate containing water and the return sludge from intermediate sedimentation. Denitrification takes place in these reactors under anoxic conditions. In the trickling filter of the second stage nitrification proceeds (Fig. 4). The full scale applicability of this process for nitrogen elimination has been proven since 1988 in several wastewater treat-

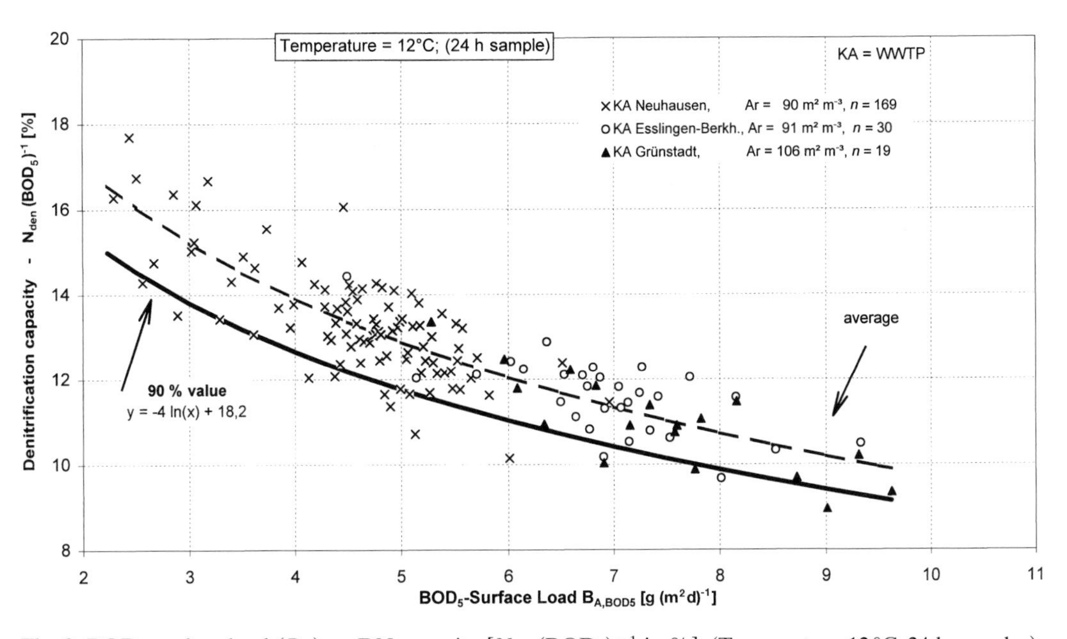

Fig. 3. BOD_5 surface load (B_A) vs. DN-capacity [$N_{DN}(BOD_5)^{-1}$ in %], (Temperature 12 °C, 24 h samples) (DORIAS, 1996a).

ment plants (ATV, 1994a; KRAUTH and ZANDER, 1989; ZANDER, 1988; SCHLEYPEN and NORDMANN, 1994; Anonymous, 1994).

In Tab. 3 guideline values (ATV, 1994a) are given for the design of the necessary anoxic denitrification volume.

The nitrogen elimination rates are higher than in the primary denitrification cells of single-stage activated sludge units. The following reasons may be responsible for the relatively high efficiency of a pure anoxic stage (DORIAS, 1996a):

(1) high proportion of denitrificants in the heterotrophic biomass: in a pure anoxic stage it is assumed that all bacteria are conditioned for denitrification, whereas in a single-stage unit the proportion of denitrificants amounts to only 70%;
(2) more carbon available: in a pure anoxic stage the particulate organic matter may to a large extent be used for denitrification, whereas in the predenitrification compartment in a single-stage activated sludge unit generally only a part of this matter is available;
(3) adaption time of denitrificants: in a pure anoxic stage no adapation time of the denitrificants from aerobic to anoxic conditions is necessary. Thus denitrification can start immediately by all denitrificants.

No primary clarification is necessary in this process configuration other than a well working sand and grit trap. It is advisable to avoid deposits in the anoxic stage and divide the denitrification tank into cascades.

To avoid the extra hydraulic load onto the final clarifier the recirculation flow must be taken directly from the effluent of the nitrifying trickling filter or at least from the sludge hopper of the final clarifier.

Consideration should be given to the eventuality that nitrification is incomplete or that insufficient nitrate may be transported to the denitrification zone. This could happen due to a defective recirculation water pump, disturbance of the nitrification in the trickling filter or problems with the activated sludge stage leading it into an anaerobic condition with strong odor development. Therefore, it should be possible to aerate the last part of the denitrification basin for a short and limited time to overcome such process disturbances.

In order to save electricity and to avoid an unnecessarily high oxygen transfer into the denitrification stage only the quantity of nitrate that may actually be respired is to be recirculated. The regulation of the recirculation flow should, therefore, be effected by on-line measurements of the nitrate concentration in the effluent of the denitrification zone. A controlled nitrate feed further eliminates the risk of an uncontrolled denitrification in the intermediate settling tank and the resulting formation of floating sludge.

Tab. 3. Guideline Values for the Denitrification Capacity in Non-Aerated Basins

Reactor Kept Under Anoxic Conditions [sludge age in days]	Denitrification Capacity [kg NO_X-N_{den} per kg BOD_5]
2	0.11
3	0.13
4	0.15

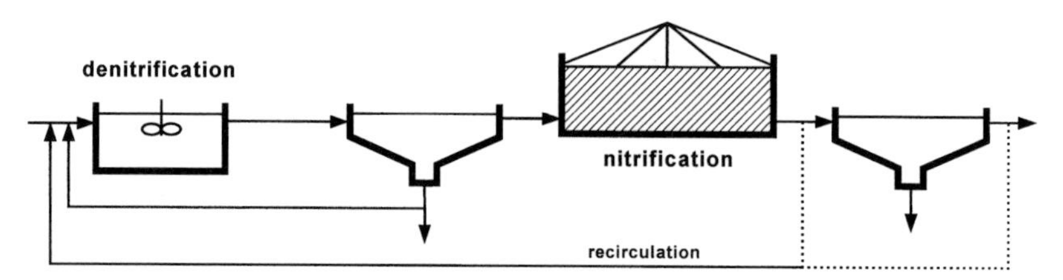

Fig. 4. Flow chart: predenitrification in an anoxic activated sludge system and nitrification with trickling filter.

2.7.5 Postdenitrification

Postdenitrification can be achieved either via the activated sludge process or by means of biofilm reactors. If nitrogen is eliminated exclusively via postdenitrification the following technical disadvantages must be considered:

(1) due to the lack of denitrification in the first stage, a lowering of the alkalinity and, therefore, possibly also a drop of the pH value in weakly buffered wastewaters;
(2) increased hydraulic load of the final clarifier during the denitrification in the biofilter when backwash occurs.

2.7.5.1 Postdenitrification by Means of the Activated Sludge Process

In this process mode an additional anoxic reactor is located between the trickling filters and the final clarifier. Denitrification is effected either by only utilizing endogenous respiration, or by additional dosage of internal or external carbon sources (Fig. 5).

In the case of endogenous respiration denitrification takes place only at a low rate of approximately 0.3–0.5 mg NO_X-N (g MLVSS h)$^{-1}$ which requires large tank volumes. To improve the denitrification rate an internally developed carbon source, such as thickened activated sludge (STÜBNER, 1993) or hydrolyzed or acidified primary sludge (ATV, 1994b) could be used. To further increase the denitrification rate the use of an external carbon supply may also be necessary. In case of external carbon sources like ethanol or acetic acid it is possible to use for design of the denitrification tank a denitrification rate of approximately 2.5 mg NO_X-N (g MLVSS h)$^{-1}$.

Apart from the direct costs of an external carbon source additional costs for sludge treatment and disposal of the resulting surplus sludge must be considered (BAUMANN and KRAUTH, 1993). Also, in the literature, in most cases, a succeeding aerobic stage is also considered necessary (CHRISTENSEN and HARREMOËS, 1977; KIENZLE, 1980) to safely remove surplus carbon and to avoid undesired floating sludge formation due to wild denitrification in the succeeding final clarifier of the activated sludge tank.

2.7.5.2 Postdenitrification with Biofilm Reactors

Biofilm reactors include all processes with fixed biomass including fluidized beds and biofilters. A description of the different processes can be found in STROHMEIER (1994).

With all these processes high degradation rates due to high biomass concentrations are possible. Practical experience, however, shows that in general external carbon sources have to be added (STROHMEIER, 1994).

In case of biofilters, in general, the hydraulic filter load is the decisive design parameter (STROHMEIER, 1994) due to their high degradation capacities of up to 3.5 kg (m^3 d)$^{-1}$ NO_3-N.

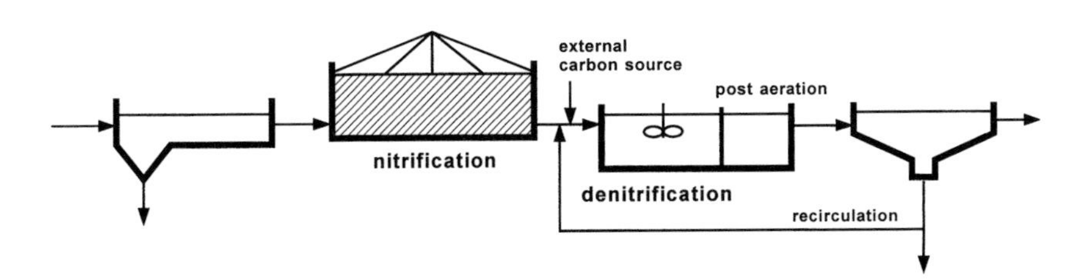

Fig. 5. Flow chart: postdenitrification using external carbon sources.

3 Rotating Biological Contactors

3.1 Process Development

To reduce energy demand of biological wastewater treatment the use of submerged drums was suggested for the first time around the turn of this century. A drum shaped body made of wooden bars was stuffed with brushwood, then placed halfway into the water and put into rotation. Similar rotating drum screens which do not contain a fixed synthetic packing material are still used up to the present day.

Only in the 1950s more interest was shown in rotating biological contactors in Germany, and particularly in rotating disc filters. Since the drums consist of discs which are placed at equal intervals (approximately 20 mm) from each other on a common drive shaft problems as for the brushwood filled drums could be avoided.

Because of their light weight discs made of plastic material such as polystyrol have proven to be most suitable. For this reason it was possible to align the discs on self-supporting shafts up to a length of 7 m without a heavy or bulky overall construction. Today such rotating biological contactors are mainly used in smaller wastewater treatment plants.

3.2 General Design Parameters

According to ATV standard A-135 (ATV, 1989a) the decisive parameter for the design of rotating biological contactors is the BOD_5 surface loading. The total necessary growth surface area A is determined from the daily BOD_5 load and the specific surface loading B_A.

$$A = B_d(B_A)^{-1} \quad \text{in } [m^2] \tag{6}$$

3.3 Wastewater Treatment without Nitrification

According to ATV standard A-135 (ATV, 1989a) the following values are suggested for wastewater treatment without nitrification: at least two drums in series: $B_A = 8$ g $(m^2\ d)^{-1}$ at least three drums in series: $B_A = 10$ g $(m^2\ d)^{-1}$.

With a 24 h flow and load balancing-equalization it is permissible to increase the specific surface load B_A up to 12 g $(m^2\ d)^{-1}$.

3.4 Wastewater Treatment with Nitrification

For wastewater treatment with nitrification the following values are suggested. According to ATV standard A-135 (ATV, 1989a) for rotating biological contactors and a $N-BOD_5$ ratio of ≤ 0.3: at least three drums in series: $B_A = 4$ g $(m^2\ d)^{-1}$ at least four drums in series: $B_A = 5$ g $(m^2\ d)^{-1}$.

Particular attention must be paid to the biologically active surface area when designing these units, and that there is no exact method for its determination. In case of rotating biological contactors other than rotating disc filters with a sufficient interval between the discs (recommended minimum interval 20 mm), the design must either be technically determined or estimated on the basis of existing units. As long as no other information is available, it is presently recommended to reduce the available surface area by 30%.

3.5 Wastewater Treatment with Nitrification and Denitrification

To technically integrate the nitrification in rotating biological contactors, only the prelocated denitrification with an anoxically operated rotating disc filter has sufficiently been tested. This process configuration, just as in the case of covered trickling filters, is based on cutting off ventilation as far as possible by covering and submerging the discs. The nitrate feed that must be eliminated may come from the effluent of a nitrifying biological contactor.

The following design recommendations for denitrification in rotating disc filters are made:

(1) To avoid too much oxygen transfer into the rotating disc filter a recirculation ratio of 300% should not be exceeded.
(2) The relevant surface denitrification rate is a function of the BOD_5 surface loading and may be determined by using the values in Tab. 4.

These values are for 12 °C for an NO_3-N concentration in the outlet of the denitrifying disc component of ≥ 2 mg L^{-1}.

Tab. 4. Denitrification Rates as a Function of the BOD_5 Surface Load

BOD_5 Surface Load $[g\ (m^2\ d)^{-1}]$	Denitrification Capacity $[kg\ NO_X-N_{den}(kg\ BOD_5)^{-1}]$
10	0.12
15	0.09
20	0.07

A denitrifying disc with a BOD_5 surface loading of 16 g $(m^2\ d)^{-1}$ may be expected to yield a denitrification rate of approximately 9% according to Tab. 4. By adding nitrate to the surplus sludge, and a simultaneous denitrification in the discs under aerobic conditions, an additional nitrogen elimination of 3% of the added BOD_5 may be expected.

4 References

Anonymous (1989), *Deutscher Normenausschuß* – Füllstoffe aus Kunststoff für Tropfkörper, Anforderungen, Prüfungen, DIN 19557 – Teil 2, Nov. 1989, Deutsches Institut für Normung e.V. – DIN – Normenausschuß Wasserwesen (Ed.), Berlin/West, No. 37.

Anonymous (1994), *Informationsbroschüre* zur Kläranlage, Abwasserzweckverband "Oberes Striegistal" (Ed.).

ATV (1989a), Grundsätze für die Bemessung von Tropfkörpern und Tauchkörpern mit Anschlußwerten über 500 Einwohnerwerten, *ATV-Arbeitsblatt A* **135** (ATV Standard ATV A 135 E: Principles for the dimensioning of biological filters and biological contactors with connection values over 500 population equivalents).

ATV (1989b), Mehrstufige biologische Kläranlagen, Arbeitsbericht der ATV-Arbeitsgruppe 2.6.5., *Korrespondenz Abwasser* **2**, 181–189.

ATV (1994a), Denitrifikation bei Tropfkörperanlagen, Arbeitsbericht der ATV-Arbeitsgruppe 2.6.3., *Korrespondenz Abwasser* **11**, 2077–2081.

ATV (1994b), Umgestaltung zweistufig biologischer Kläranlagen zur Stickstoffelimination, Arbeitsbericht des ATV-Fachausschusses 2.6., *Korrespondenz Abwasser* **1**, 95–100.

BAUMANN, P., KRAUTH, K. (1993), Dosierung von Kohlenstoffverbindungen bei der weitergehenden Stickstoffelimination, *Korrespondenz Abwasser* **11**, 1792–1805.

CHRISTENSEN, M. H., HARREMOËS, P. (1977), Biological denitrification sewage, a literature review, *Prog. Water Technol.* **8**, 509–555.

DORIAS, B. (1994a), Denitrifikation mit Tropf- und Tauchkörpern, *Berichte der Abwassertechnischen Vereinigung* zur ATV-Bundestagung 1994, Hennef, Gesellschaft zur Förderung der Abwassertechnik e.V., Vol. 44, pp. 621–638.

DORIAS, B. (1994b), Denitrifikation mit Festbettreaktoren, *Stuttgarter Berichte zur Siedlungswasserwirtschaft* Vol. 128, pp. 101–124. München: Oldenbourg.

DORIAS, B. (1996a), Stickstoffelimination mit Tropfkörpern, *Thesis*, Universität Stuttgart, *Stuttgarter Berichte zur Siedlungswasserwirtschaft* Vol. 138. München: Oldenbourg.

DORIAS, B. (1996b), Denitrifikation im Tropfkörper, *Korrespondenz Abwasser* **7**, 1237–1244.

DUNBAR, W. P. (1974), *Leitfaden für die Abwasserreinigung*, 3rd. Edn. München: Oldenbourg.

EPA (U.S. Environmental Protection Agency) (1975), *Process Design Manual for Nitrogen Removal, Technology Transfer.*

EPA (U.S. Environmental Protection Agency) (1993), *Manual Nitrogen Control*, EPA/625/R-93/010.

GUJER, W., BOLLER, M. (1986), Design of a nitrifying tertiary trickling filter based on theoretical concepts, *Water Res.* **20**, 1353–1362.

KIENZLE, K. H. (1980), Vorschläge zur Gestaltung und Bemessung von Denitrifikationsbecken, *Korrespondenz Abwasser* **7**, 441–446.

KRAUTH, K., ZANDER, S. (1989), Vorgeschaltete Denitrifikation durch Kombination einer anoxischen Belebungsstufe mit einer Tropfkörperstufe, *Abschlußbericht* zum BMFT-Forschungsvorhaben 02 WA 8511, erstellt am Institut für Siedlungswasserbau, Wassergüte- und Abfallwirtschaft der Universität Stuttgart.

MÜLLER, W.-R. (1984), Beitrag zur Nitrifikation in Festbetten am Beispiel eines abwärts durchstömten Sandfilters, *Stuttgarter Berichte zur Siedlungswasserwirtschaft* Vol. 82, pp. 67–72.

REINHEIMER, R., HEGEMANN, W., RAFF, J., SEKOULOV, I. (1988), Stickstoffkreislauf im Wasser. München, Wien: Oldenbourg.

SCHLEYPEN, P., NORDMANN, W. (1994), Stickstoff- und Phosphorelimination an einer zweistufigen Belebungs-Tropfkörperanlage, *Korrespondenz Abwasser* **12**, 2242–2249.

STROHMEIER, A. (1994), Verfahren der nachgeschalteten Denitrifikation, *Berichte der Abwassertechnischen Vereinigung e.V.* zur ATV-Bundestagung 1994, Gesellschaft zur Förderung der Abwassertechnik e.V., Nr. 44, pp. 639–658.

STÜBNER, E. (1993), Untersuchungen zur Denitrifikation mit Belebtschlamm als H-Donator, *Thesis*, Technische Universität Hamburg-Harburg, *Hamburger Berichte zur Siedlungswasserwirtschaft* Vol. 10.

WILD, H. E., SAWYER, C. N., MCMAHON, T. C. (1971), Factors affecting nitrification kinetics, *J. Water Pollut. Contr. Fed.* **43**, 1845–1854.

WOLF, P. (1989), Bemessung von Festbettreaktoren zur Stickstoffoxidation, *Berichte aus Wassergütewirtschaft und Gesundheitsingenieurwesen*, TU München, Vol. 91, pp. 89–115.

ZANDER, S. (1988), Ausbau der Tropfkörperanlage Spenge (anoxische Belebungsanlage als 1. Stufe, Tropfkörper als 2. Stufe), *Stuttgarter Berichte zur Siedlungswasserwirtschaft* Vol. 103, 1988, pp. 103–110. München: Oldenbourg.

17 Submerged Fixed-Bed Reactors

JUDITH M. SCHULZ genannt MENNINGMANN
Dinslaken, Germany

1 Process Definition

Biological treatment by means of sessile microorganisms is mainly based on the natural characteristic of many types of bacteria and protozoa to colonize on surfaces. Biofilm operation is also the oldest form of bioprocessing (BISHOP and KINNER, 1986). An overview of this phenomenon is given in Chapter 4, this volume.

Due to the more stringent requirement for wastewater treatment and the increased demand for space saving and low-maintenance processes, approximately 10 years ago, biofilm processes or so-called "fixed-bed reactors" began to be used more and more frequently for wastewater treatment. In wastewater engineering, the term "fixed-bed reactor" in accordance with the German Standard DIN 4045, is defined as being a "reservoir containing substrata colonized by microorganisms which cause biochemical degradation processes". Since these processes take place as a result of biofilm activity they are also called "biofilm reactors" in European Standard EN 1085. Within this group, submerged fixed-bed reactors can be more closely defined as "permanently submerged surfaces with a regular structure and a theoretical percentage of available void space of over 85%".

Fig. 1 shows the different categories of biofilm processes, including the submerged fixed-bed reactor referred to by SEYFRIED (1997).

This chapter deals exclusively with fixed-bed reactors of the "submerged" type in accordance with the above categorization.

Fig. 2 illustrates the submerged fixed-bed principle in wastewater engineering. The reactors comprise aerators (usually membrane-type tubular aerators), which are installed under the substratum to introduce the oxygen and circulate the water, and the submerged fixed bed itself, which is fixed in the reactor.

If denitrification is required, the necessary forced flow can be produced either by means of agitators or, alternatively, by nozzles. As a matter of principle, there is no sludge recircu-

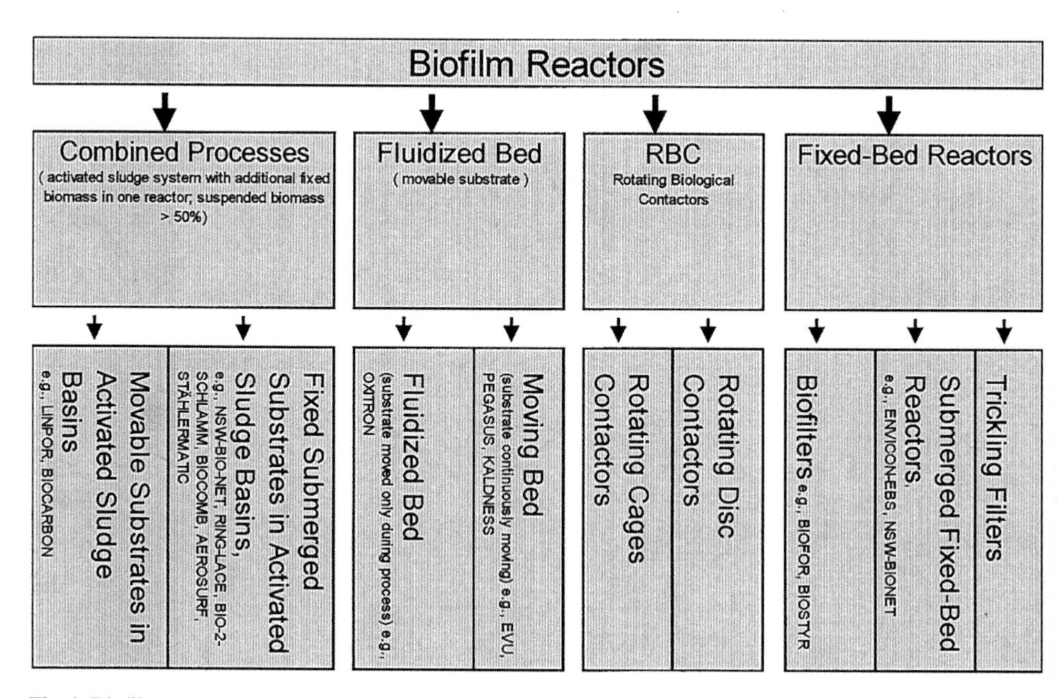

Fig. 1. Biofilm reactors.

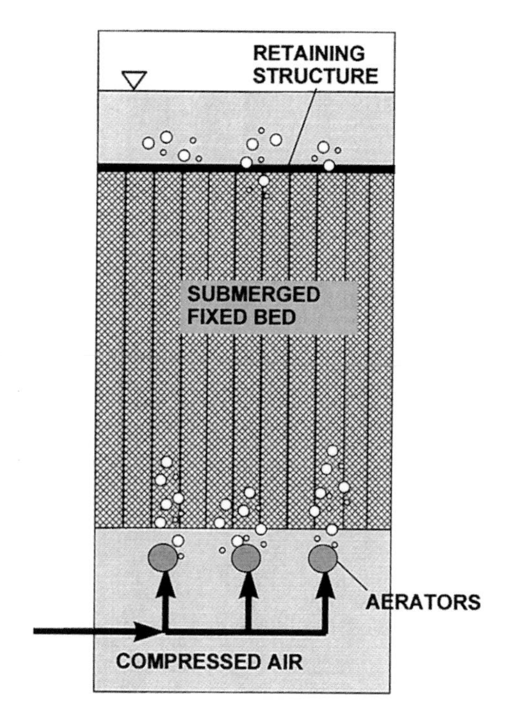

Fig. 2. Submerged fixed-bed reactor.

lation. Surplus sludge is sedimented in a secondary sedimentation tank and subsequently disposed of.

Considering the question of the ideal field of application there are at least three significant advantages offered by biofilm processes in general – and submerged fixed-bed reactors in particular – in contrast to the activated sludge process, namely:

- low space requirement,
- low maintenance costs in terms of personnel and time,
- high flexibility *vis à vis* load surges.

Operational experience has shown that either compactness or low maintenance costs can be achieved. Biofilm processes either have large specific surfaces and high volumetric degradation efficiency, but require a higher degree of maintenance ("biofilters"), or they are very rugged and require little maintenance, but more space. As far as its requirements in terms of volume and retention times are concerned, the submerged fixed-bed reactor of the type described here is more closely related to the activated sludge process than to biofilters. The higher costs involved in installing the fixed-bed reactor are compensated for by a distinctly lower space requirement and the absence of sludge recirculation to control the biomass. Therefore, it is, generally speaking, ideal for

- the complete, decentralized biological treatment of domestic wastewater from 4 to approximately 8,000 PE (population equivalents), as well as of specific commercial and industrial effluents;
- the pretreatment of highly polluted industrial effluents (prior to, e.g., an activated sludge process);
- final treatment, e.g., in the form of a nitrification stage.

Several examples of ideal cases of application are given at the end of this chapter.

2 Application of Submerged Fixed-Bed Reactors

As can be seen from Fig. 2, the process comprises three variable components:

- substratum (carrier, fixed-bed),
- aerators,
- reactor volume/design.

All three have a decisive influence on the degradation efficiency and must be aligned with each other in a manner appropriate to the effluent to be treated. Since sludge cannot be recirculated with this process, once the plant has been constructed, little can be done to influence degradation efficiency, since the biomass cannot be set to a specific value. Only aeration intensity can be adjusted and/or the effluent can be recirculated. For this reason, all factors must be taken into account in the planning stage and be carefully considered in relation to each other.

2.1 The Substratum

The suitable substratum for the specific case of application is of particular importance for this process. The submerged fixed-bed reactor operates with block-like structures in the form of plastic grids or laminated plastic sheets, the specific surfaces of which range between 100 and 400 $m^2\ m^{-3}$.

Carrier materials of this type are formed from, e.g., extruded HDPE grids (ENVICON, NSW) like the BIOPAC® shown in Fig. 3, or laminated PVC sheets (MUNTERS, RICHTER). The substrata are usually installed in uniformly distributed blocks above the aerators and are held firmly between the supporting and the retaining structure. It is particularly important that, when densely colonized, the substratum still allows a high rate of flow. This not only ensures that flow resistance is kept to a minimum, but also prevention of blockages. Intensive mixing of the gas and water phases, as well as optimum contact with the biofilm, is achieved and air pockets are prevented while, simultaneously, oxygen is utilized to an optimum degree (ATV, 1996).

Substrata may have a smooth or rough surface structure. Smooth surfaces may retain complex biofilms less efficiently than fissured, rough surfaces, because they offer fewer opportunities for the biofilm to adhere and the smoother the substratum the higher the danger that portions of the biofilm will become detached as a result of the turbulence and cause blockages at other points. Therefore, in the case of high flow rates, particularly when these are characterized by surges, special attention should be paid to the micro-roughness of the substratum (VAN LOOSDRECHT et al., 1995).

It must also be borne in mind that material- and production-related release agents or pigments in the substratum material may inhibit the growth of the biofilm. A longer conditioning time is needed before growth develops on materials of this type; they are often avoided for a prolonged period by particularly sensitive bacteria such as nitrificants. The physical and biological interactions occurring between bacteria and surfaces have been treated thoroughly by BERKELEY et al. (1980). Technical aspects, such as break resistance and resistance to physical forces, e.g., pressure, must also be taken into account when choosing substrata for installation in larger-scale reactors.

Inert materials, such as polyethylene, have neutral properties and can, in principle, be considered suitable for biological processes, as is recommended by the German ATV (ATV, 1996).

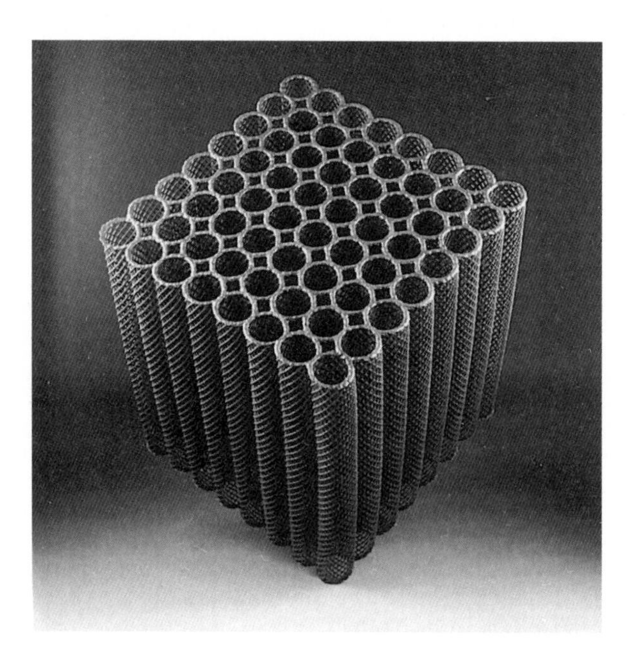

Fig. 3. BIOPAC® (ENVICON, Germany): extruded grid substratum.

Fig. 4 shows a substratum optimally covered with a mushroom-type biofilm as described in Fig. 6.

The fixed-bed substratum must be held down by means of a retaining structure to prevent it from floating. The substratum carrying the biofilm rests on a supporting structure. The weight of the biofilm can be calculated according to Tab. 2. Low-loaded biofilms weigh between 70 and 150 kg m^{-3}, whereas high-loaded biofilms and such with filamentous components may weigh up to 350 kg m^{-3}. The weight of the substratum ranges from 40 to 80 kg m^{-3}, depending on type, material, and specific surface. Experience gained to date has shown that the stack height should not exceed 4 m. The ideal stack height depends on the stability of the substratum and on the degree to which water can be circulated in the reactor.

The majority of biofilm reactors (biofilters, trickling filters) are designed on the basis of the space loading (kg BOD_5 m^{-3} fixed-bed volume d^{-1}). The conventional design of submerged fixed-bed reactors, which is based solely on the potentially available substratum surface (growth surface), is derived from the rotating disc contactor. However, in the case of this system, the increase in size of the surface originally available is solely biofilm-related so that, mathematically, at the most, the organic load on the surface is reduced, and the design is without risk. The design of grid-type substrata blocks of the submerged fixed-bed reactor is also based on the product-specific potential growth surface. At the same time, however, it must be borne in mind that the active surface initially increases due to the growth of the furry biofilm. In case of an inadequate forced flow, overloading, or the selection of a too high

Fig. 4. Mushroom-type biofilm on HDPE grid (BIOPAC®).

specific surface, it can decrease again due to overgrowth on the substratum structure. In such cases only a reduced surface is available, so that the surface load is far higher than originally calculated. If this is not taken into consideration, breakdown can occur as a result of overloading. The microorganisms then typically react by increasing the number of suspended cells. This results in distinct turbidity combined with high effluent values. Simultaneously, the oxygen supply to the biofilm is reduced due to the smaller surface area, resulting in anaerobic processes and the release of organic acids.

It is, therefore, extremely important to base the choice of the submerged fixed bed and, thus, the specific surface, on the load to be expected and, in addition, to take the fixed bed-related space loading (kg BOD_5 m^{-3} fixed-bed volumel d^{-1}) into consideration as a design aid. Tab. 1 should be of assistance in this regard.

In cascaded plants, it is quite possible to install substrata with increasing specific surfaces along the flow path in different basins.

Fig. 5 shows a possible relationship between space loading and degradation efficiency, determined in a laboratory-scale plant (V=60 L) with domestic wastewater.

2.2 Aeration

Aeration is normally effected by means of uniformly distributed fine-bubble membrane-type tubular aerators. It is important that the secondary biological sludge which is detached from the fixed bed as a result of the forced flow is constantly kept floating and is thus conveyed to the secondary sedimentation stage. An inadequate or wrongly directed forced flow, e.g., due to the installation of inappropriate aerators (disc aerators) or an unfavorable reactor design, leads to uncontrollable sludge deposits in the fixed-bed reactor (BISHOP and KINNER, 1986).

The aeration of submerged fixed-bed reactors according to this method fulfills two tasks of equal importance. On one hand, the aera-

Tab. 1. Relationship between BOD Load and Specific Surface of the Substratum

BOD Load	Specific Surface of the Substratum (m^2 m^{-3})
Low (nitrification)	≥200
Medium	~150
High	~100

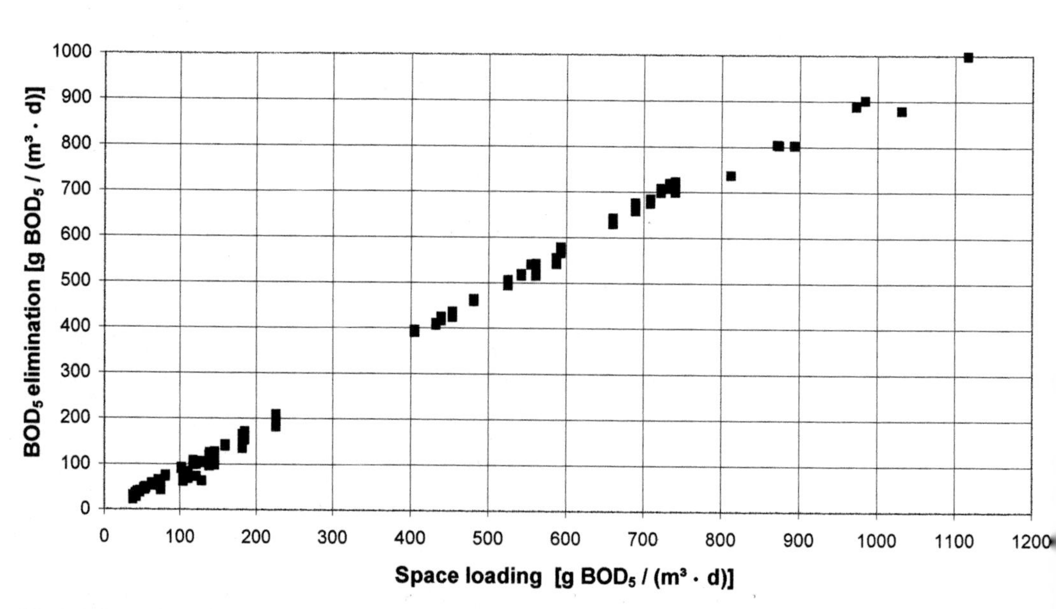

Fig. 5. BOD elimination as a function of space loading.

tors introduce the atmospheric oxygen needed for biological metabolism and, on the other hand, they generate the flow needed to ensure that the fixed-bed reactor does not become blocked. At the same time, the fixed-bed reactor acts as a flow barrier. The coarser the grid, the greater the similarity of flow to that of a completely mixed basin. Depending on the specific surface of the substratum, between 5 $Nm^3 h^{-1} m^{-2}$ and 15 $Nm^3 h^{-1} m^{-2}$ air must be introduced. According to SCHLEGEL (1988), oxygen input is intensified by the fixed bed due to the fact that large bubbles burst and the retention time of the bubbles is longer. Therefore, it is in principle very easy to retrofit existing activated sludge plants with an increase of the density of aerators and without a changing of the reactor geometry.

It has proved practical for reasons of cost to minimize aeration in everyday operation after a short start-up phase. In this context, the energy input for plants can be considerably reduced if, instead of continuous aeration, intermittent aeration is chosen. Preference should be given to short intervals. In addition, in low-load periods, e.g., during the night, oxygen input can be further reduced provided that on the basis of measurements the plant will not become anaerobic and the minimum requirements will be met. Particularly in the case of high-loaded plants, supplementary flushing by means of a higher oxygen input at regular intervals can increase the stability of the process and have a positive effect on the clarity of the effluent.

2.3 Reactor Volume/Design

It is known that the hydraulic regime can strongly influence the system (BISHOP and KINNER, 1986). The overall hydraulic mode may be complete mix, plug flow, or some hybrid of the two as in cascade reactors. When designing submerged fixed-bed reactors, not only the substratum, but also the reactor geometry must be aligned to the organic fixed-bed load for the specific application. Since the thickness and structure of the biofilm is influenced to a high degree by the concentration of biodegradable substances, the degree of dilution, which is a function of the volume of the basin, is a determining factor. Under no circumstances should an attempt be made to balance out high concentrations in small reactors by choosing a substratum with a particularly high specific surface. Another aspect to be considered is the technical flow control of the biofilm. This, together with stability factors, limits the maximum height of the fixed bed to approximately 4 m.

It is in principle possible to install fixed beds in all types of conventional basins. In this respect, the most important question to be clarified is whether a completely mixed basin, a plug flow basin, or a cascaded reactor offers the ideal solution. This depends on the boundary conditions.

In the case of municipal wastewater, the usual treatment objective is optimum carbon removal with simultaneous nitrification. To achieve this, fixed-bed reactors are usually loaded with 4–6 g BOD_5 $m^{-2} d^{-1}$.

In order to effect nitrogen removal as reliably as possible, it must be taken into account that load surges can have a negative influence on nitrification and that the water flowing in, particularly in the case of the mixed system, has widely fluctuating properties. In such cases, it is recommended to divide the reactors into several cascades. This results in a higher fixed-bed load in the first cascades and a lower load in the last cascades. This effect allows a higher mean loading of the plant with simultaneous maximum stability of the biological degradation.

As in the trickling filter process, where the different stages of biological degradation take place along a vertical axis (WANNER and GUIJER, 1984), the division of the basin into 3–6 cascades has a positive effect on the biological efficiency, leading to a horizontal segmentation of different biocenotic areas. This results in greater operational reliability and an increased degradation rate. A reactor of this type responds very flexibly to hydraulic and organic load surges. This enables the surge loads typical for combined water systems to be balanced out to a high degree.

Higher fixed-bed loads are primarily used for pretreatment, e.g., that of industrial effluents. In this context, it must be borne in mind that even easily biodegradable substances can have an inhibiting effect on microbes if a certain concentration limit is exceeded. A good example is sugar, which is certainly very easily biodegradable, but in high concentrations is a conservation agent (candied fruits, jam). If an industrial effluent contains a component which, due to its concentration, has an inhibiting effect, it must be ensured that this critical value is not exceeded in the reactor. The reactor must be designed with a correspondingly large volume.

3 Development and Structure of Biofilms in Submerged Fixed-Bed Reactors

The colonization processes which take place on a substratum have already been described in Chapter 4, this volume.

Numerous experts, such as those referred to by IAWQ (1993), TOETTRUP et al. (1993), BOLLER et al. (1993), BISHOP and KINNER (1986) and MOSER (1985), prefer thin biofilms (<100 μm) for aerobic applications. Although high degradation rates have been achieved with this technique even with considerably thicker biofilms, i.e., up to 1,500 μm. However, viewed under a microscope, the biofilms are so fluffy and fissured and (LEWANDOWSKI et al., 1994; HAMILTON and CHARACKLIS, 1989) a far larger surface is created than in the case of thin biofilms, and nutrients rarely have to penetrate the substratum deeper than 200 μm. The biofilm – liquid film interface, however, is not smooth or uniform (BISHOP and KINNER, 1986). It seems very logical that, with this type of process, a minimum quantity of biomass is needed, i.e., in the order of that required in the case of activated sludge systems. There is a close relationship between the organic load and the biomass, as is shown in Tab. 2. The organic proportion of the sludge is approximately 65–85%. Porosity increases enormously with increasing biofilm thickness, due to the formation of typical filaments, as can be seen in Fig. 6.

This behavior ensures that there is an adequate nutrient gradient even in thick biofilms so that sensitive processes, such as nitrification, are not inhibited. In parallel to this, a substantial increase in the biofilm surface is achieved, which is reflected in both the adsorption potency and the degradation efficiency. The experience gained with the fixed-bed reactor technique described here has shown that, in practice, thicker biofilms can also be used successfully without a risk of damage to the plant. Fig. 7 is a microscopic image of a *Zoogloea*-type biofilm.

The biofilms can be described as "natural populations" because natural selection occurs and determines which organisms can survive (MOSER, 1985).

There are generally three types of organisms composing the biofilm as known from other systems: bacteria, protozoa, and metazoa. The role of each group is described by MOSER (1985).

The biofilms on permanently submerged surfaces differ in some points from those in the case of trickling filters and rotary disc contactors. There have been observed no Psychodidae or other insects, because these have rare

Tab. 2. Relationship of the Biofilm Thickness as a Function of the Load

B_F	B_L	W	Application
<3	1.5	0.015	nitrification and complete carbon elimination
3–6	1–4	0.017	nitrification and simultaneous carbon elimination
6–12	4–10	0.02	carbon elimination with partial nitrification
12–30	10–20	0.03	carbon elimination

B_F: BOD_5 load (g BOD_5 m^{-2} d^{-1})
B_L: thickness of biofilm or length of biofilm filaments (mm)
W: weight factor (kN m^{-2}) according to Deutscher Normenausschuß Wasserwesen, 1989

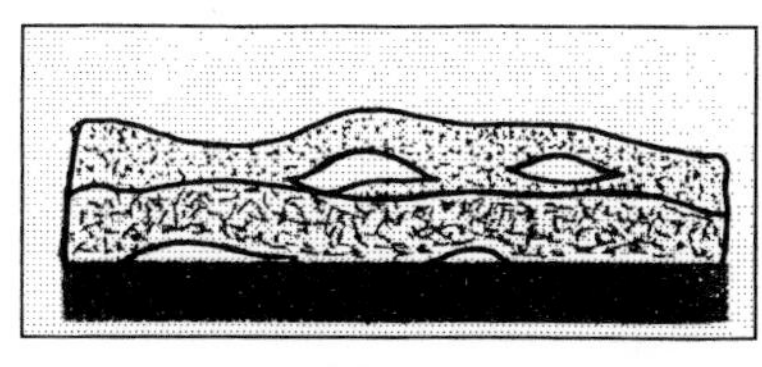

LAYER-TYPE
Layers of microorganisms of one and the same kind are often found in high-turbulence reactors and water with a constant composition
ZOOGLOEA-TYPE
Antler-shaped bacterial agglomerates, very common in domestic wastewater and reactors with medium turbulence and high organic load

MUSHROOM-TYPE
Mushroom-shaped and round bacterial agglomerates; a typical form in low-turbulence reactors

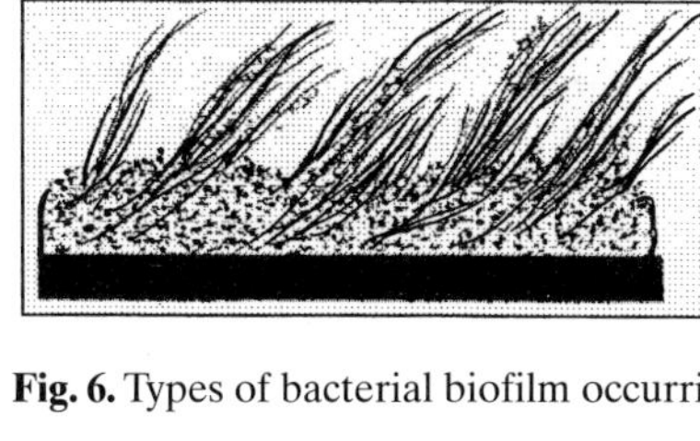

FUR-TYPE
Mixed biocenosis of sessile filamentous and non-filamentous bacteria or fungi. The length of the filaments increases with reducing turbulence. This type is very common in all kinds of wastewater

Fig. 6. Types of bacterial biofilm occurring on submerged surfaces.

opportunity of depositing their eggs. Very large colonies of Peritricha are typical in submerged fixed-bed domestic wastewater treatment plants due to the moderate turbulence and the constant supply of oxygen and bacteria. *Carchesium*, *Zoothamnium*, *Epistylis*, *Opercularia*, and *Vorticella*, in particular, often stabilize the biofilm with the skeleton of their stalks. Some species of filamentous bacteria or fungi may also play the same role. These filaments are surrounded by a bacterial community connected by extracellular polymeric substances (EPS). The surface of the biofilm is the habitat of other protozoa, chiefly the grazing species, as well as of rotifers (*Rotaria*, *Cephalobdella*) and worms (*Nais*, *Chaetogaster*, Tubificidae, Nematoda sp.), which feed on the bacterial matrix, thus, loosening it. Similar biocenotic communities have been described for the rotating biological contactor (RBC) by BISHOP and KINNER (1986) and for activated sludge systems by CURDS and COCKBURN (1970).

Although these organisms remove a portion of the sludge, only in rare cases do they have a damaging effect on the biofilm and are rather to be seen as positive due to the reduction of the sludge and the loosening effect. An increase in load and number of non-aeration intervals has been found helpful when these organisms occur in masses. The bacteria grow in the submerged biofilm in a variety of forms, the most significant of which are shown in Fig. 6. Mixed forms frequently occur. Among the factors which determine the growth form and biofilm structure are the species of bacteria, the degree of turbulence, and the compositon of the wastewater (Fig. 6). Tab. 2 provides an overview of the biofilm thickness to be ex-

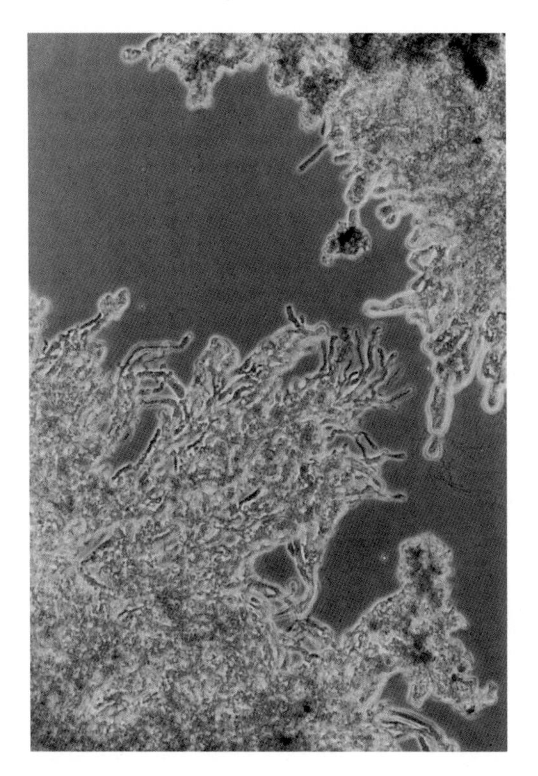

Fig. 7. Microscopic photograph of a *Zoogloea*-type biofilm.

pected as a function of the load, and applies to the turbulences usual in the case of aeration without back-washing.

The surplus sludge production of the submerged biofilms is influenced to a high degree by the preset fixed-bed load. As a rule, sludge production is considerably lower than in the case of an activated sludge process with the same load. In municipal plants with 4–6 BOD_5 m^{-2} d^{-1} and cascades, it is usually between 0.01 and 0.1 kg MLVSS per kg BOD_5 eliminated. In industrial high-load plants with 20–30 g BOD_5 sludge production in the order of 0.6 kg MLVSS per kg BOD_5 m^{-2} d^{-1} eliminated has been measured (SCHLEGEL, 1998). At the same time, it could be shown that the sludge quantity can be distinctly reduced as a result of pretreatment in a fixed-bed reactor. Considering the microbiological and ecological aspects, this can probably primarily be attributed to

- sludge removal by macro-invertebrates (Rotaria, worms),
- the lower separation rate of sessile bacteria (AUDIC et al., 1984),
- the reduced selection rate of fast-growing bacteria to the benefit of slower-growing species (HAMILTON and CHARACKLIS, 1989),
- further mineralization due to the high oxygen content (ABBASSI et al., 1996).

4 Examples of Practical Applications

Practical application ranges from three-compartment septic tanks to large-scale plants with several thousands of cubic meters of fixed beds in correspondingly dimensioned concrete basins.

Apart from the domestic and industrial applications described in the following examples, fixed-bed reactors can also be used for any type of biodegradable sewage. In this connection, all types of fixed-bed reactors can be employed, not only for pretreatment, in the form of high-load units, and for complete biological treatment, but also in the form of a downstream unit to achieve more advanced treatment objectives.

4.1 Domestic Wastewater

An important aspect of the *in situ* treatment of wastewater in both centralized and decentralized plants is that fixed-bed reactors can be adapted to all types of wastewater profile, which makes them an ideal solution when looking towards the future. There are special problems connected with small, decentralized wastewater treatment plants, such as inhomogeneous influent patterns, extreme sensitivity to temperature, and a lack of maintenance personnel. For the decentralized treatment of the wastewater from 4 to 1,000 PE, in particular, there is a need for systems such as fixed-bed reactors, which require little maintenance and are very easily controlled.

With fixed-bed reactors designed for approximately 4–6 g BOD_5 m^{-2} d^{-1}, carbon and nitrogen can be simultaneously eliminated without difficulty. A reduction in the COD of 90% and in the BOD_5 of 95% is possible. Ammonium-N is normally nitrified to over 95% ($t>12°C$), (BOD load $<4 \text{ g m}^{-2} \text{ d}^{-1}$).

Reactors with a sessile biomass have been found to be ideal for nitrification purposes. As early as 1988, SCHLEGEL published the operating results of a fixed-bed nitrification plant which showed that approximately 1.5 g $NH_4{-}N \text{ m}^{-2} \text{ d}^{-1}$ can be reliably nitrified. Results of studies published in Munich in 1991 (Technische Universität München, 1991) and observations of fluidized beds by KUGEL and HELLFEIER in 1995 verify that systems with sessile nitrificants respond to fluctuating loads and low temperatures in an exceptionally stable manner. High denitrification rates can be achieved with intermittent aeration, the recirculation of water containing nitrates, or the addition of carbon from an external source. When compared with other types of plants, the surface-related degradation efficiency must be taken into account. With continuous aeration and thick biofilms, simultaneous denitrification is about 40% due to the partially anoxic conditions within the biofilm layers. It is possible to achieve 70% nitrogen elimination with a recirculation of only 100%, and to adhere to the limit of 18 mg L^{-1} total N recommended for domestic wastewater treated in much larger plants (SCHULZ et al., 1998).

Simultaneous phosphorus removal of 20–50% is also frequently achieved, as has been confirmed not only by internal measurements but also by a study carried out by the Institute of Technology in Munich in 1991 (Technische Universität München, 1991). Purely biological phosphorus elimination is probably not possible due to the absence of sludge recirculation and the necessary alteration between aerobic and anaerobic conditions. However, simultaneous P eliminaton has been observed in a large number of plants. Further research is required to describe the reasons for this. Possibly crystallization of phosphate compounds plays a role, because crystalline structures frequently occur in biofilms.

Fig. 8 illustrates the ENVICON 3K PLUS as an example of very small fixed-bed reactors (4 to 200 PE) in three-compartment septic tanks. Fig. 9 shows a mobile container-type wastewater treatment plant based on the submerged fixed-bed reactor principle.

4.2 Industrial Applications

The submerged fixed-bed reactor technique is applied successfully in the treatment of many different types of industrial effluents, e.g.: toluene-polluted groundwater, effluents from latex and tenside processing, from fish, juice, and beer processing, from the beverage industry and from textile and carpet production, as well as from the production of amides and aromatics. The submerged fixed-bed process is suitable for the treatment of all types of biodegradable wastewaters, even if the COD:BOD ratio is poor, or there are fluctuations in the influent. The microbiological reason for this suitability is the agglomeration of specific slow-growing sessile specialized microorganisms. A few possible applications are described in the following sections.

4.2.1 A Submerged Fixed-Bed Reactor as a Pretreatment Stage – Effluent from a Vegetable and Ice-Cream Processing Plant

A wastewater treatment plant, designed for 25,000 PE, needed a more efficient method to pretreat the large volumes of effluent produced by an industrial-scale vegetable and ice-cream processing plant. To achieve this and to improve nitrification, the wastewater treatment plant would have had to expand its capacity to approximately 30,000 PE. A cost analysis showed that in this case the installation of a fixed-bed reactor would cost 50% less than a conventional expansion of the existing activated sludge plant. Fig. 10 shows the degree of COD elimination as a result of the fixed-bed reactor referred to by SCHLEGEL (1998).

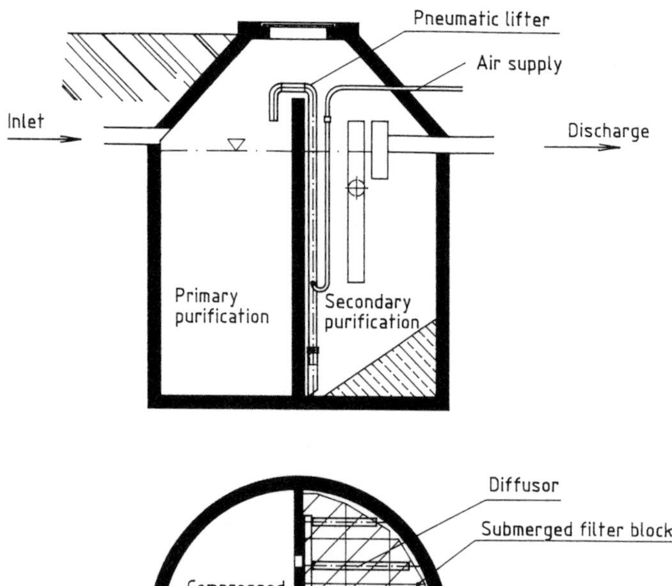

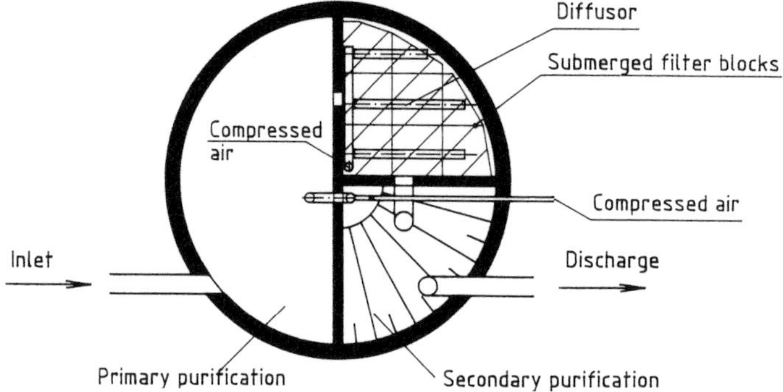

Fig. 8. ENVICON 3K PLUS in a three-compartment septic tank.

Fig. 9. A container-type wastewater treatment plant.

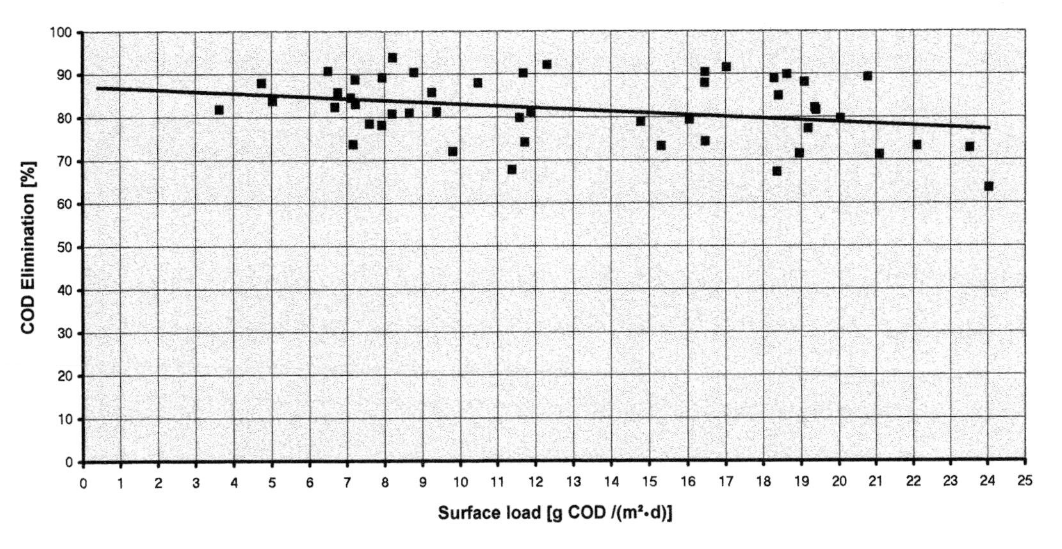

Fig. 10. COD elimination in the effluent from a vegetable and ice-cream processing plant as pretreatment stage.

4.2.2 A Fixed-Bed Reactor as the Sole Biological Treatment Stage Presented by the Example of a Detergent and Cosmetics Production Plant

A plant required a completely biological solution for the treatment of tenside-polluted effluents from the production of detergents, cleansing agents, and cosmetics. It was decided to install a fixed-bed reactor with the following key data:

- primary sedimentation area 3.5 m^3
- fixed-bed area 6.5 m^3
- secondary sedimentation area 3.5 m^3
- fixed-bed surface 975 m^2

The COD in the influent of the reactor was about 5,000 mg L^{-1}, and tensides approximately 2,000 mg L^{-1} (methylene blue active substances, MBAS). Antispumin was continuously metered in to reduce foam formation.

Fig. 11 shows the COD and tenside (MBAS) elimination efficiency of the fixed-bed reactor. From this, it is clear that using fixed-bed technology complete biological treatment of industrial effluents of these types is possible with low maintenance. In this case, sludge production was only a fraction of that in an activated sludge plant with the same load.

4.2.3 A Fixed-Bed Reactor as the Final Treatment Stage Presented by the Example of a Slaughterhouse

The slaughterhouse had a physicochemical wastewater treatment plant with grease trap, flotation tank, and proteolysis, as well as a decanter, in order to pretreat the effluent produced when slaughtering pigs and cattle, prior to discharge into the sewerage system. The highly concentrated effluent conveyed to the physico-chemical plant was found to have a COD between 4,600 and 6,700 mg L^{-1}. The effluent flowing out of the decanter and conveyed without sedimentation to the biological treatment unit still had a residual COD content between 1,200 and 3,000 mg L^{-1} and had to be reduced to COD values <800 mg L^{-1}. The fixed-bed reactor installed corresponded to that described in Sect. 4.2.2. The BOD_5 load exceeded 12 g m^{-2} d^{-1}. Phosphorus was metered in to compensate for nutrients. On the average more than 80% of the COD in the influent of the fixed-bed reactor was eliminated. Fig. 12 shows that the BOD_5 value aimed at was achieved over the entire period.

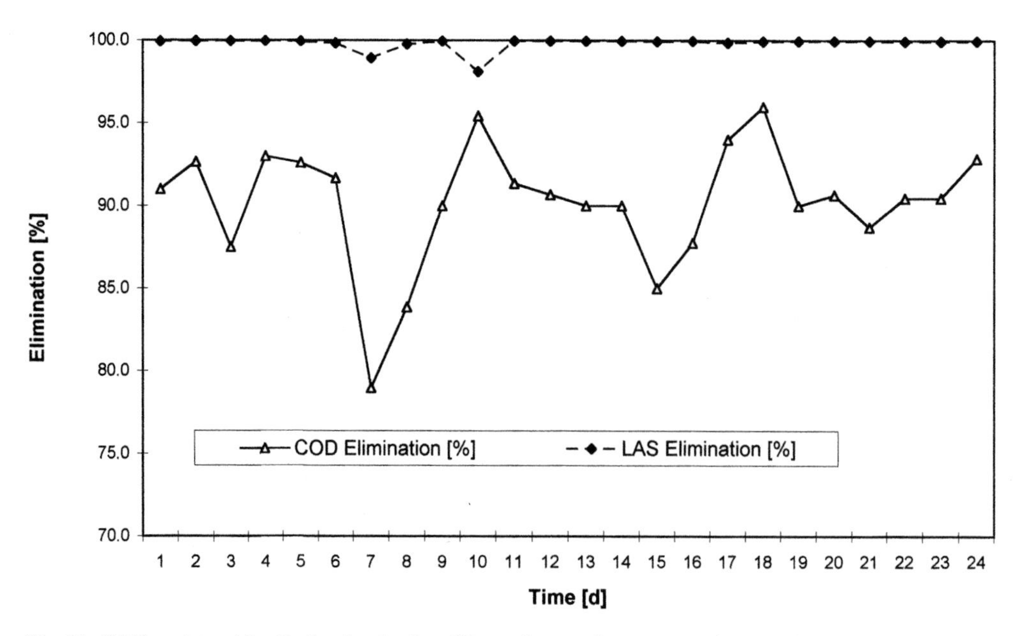

Fig. 11. COD and tenside elimination in the effluent from a detergent and cosmetics production plant.

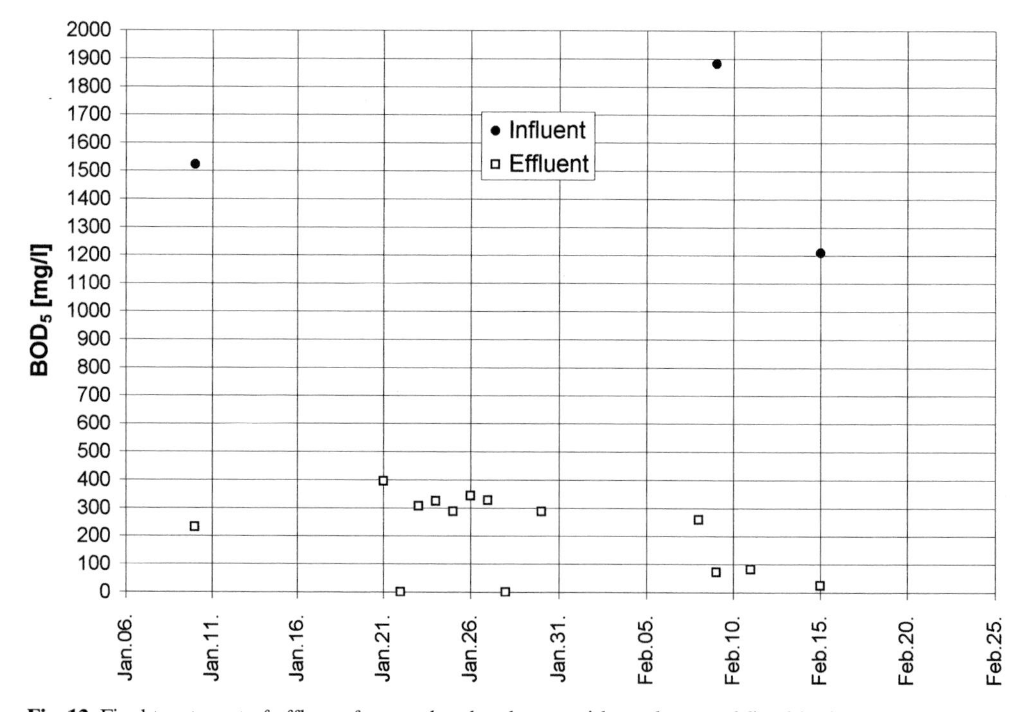

Fig. 12. Final treatment of effluent from a slaughterhouse with a submerged fixed-bed reactor.

5 References

ABBASSI, B., GIACOMAN-VALLEJOS, G., RÄBIGER, N. (1996), Prozeßführung der biologischen Abwasserreinigung mit minimierter Überschußschlammproduktion, *3. GVC-Kongreß*: Verfahrenstechnik der Abwasser- und Schlammbehandlung, 14–16 Oct, 1996, Würzburg, Germany.

ATV (1996), Arbeitsbericht der ATV Arbeitsgruppe 2.6.3. "Tauch- und Tropfkörper" im Fachausschuß 2.6 "Aerobe biologische Abwasserreinigungsverfahren", "Anlagen mit getauchten Festbetten", *Korrespondenz Abwasser* **43**, 2013–2023.

AUDIC, J. M., FAUP, G. M., NAVARRO, J. M. (1984), Specific activity of *Nitrobacter* through attachment granular media, *Water Res.* **18**, 745–750.

BERKELEY, R. C., LYNCH, J. M., MELLING, J., RUTTER, P. R., VINCENT, B. (1980), *Microbial Adhesion to Surfaces*. Chichester: Ellis Horwood.

BISHOP, P. L., KINNER, N. E. (1986), Aerobic fixed-film processes, in: *Biotechnology*, Vol. 8 (REHM, H.-J., REED, G., Eds.), pp. 113–176. Weinheim: VCH.

BOLLER, M., GUJER, W., TSCHUI, M. (1993), Parameters affecting nitrifying biofilm reactors, *2nd Int. Spec. Conf. Biofilm Reactors*, Paris.

CURDS, C. R., COCKBURN, A. (1970), Protozoa in biological sewage treatment process I and II, *Water Res.* **4**, 225–249.

HAMILTON, W. A., CHARACKLIS, W. G. (1989), Relative activities of cells in suspension and in biofilms, in: *Structure and Function of Biofilms* (CHARACKLIS, W. G., WILDERER, P. A., Eds.), pp. 199–220. Chichester: John Wiley & Sons.

IAWQ (1993), *First Int. Spec. Conf. Microorganisms in Activated Sludge and Biofilm Processes* (27/28 September), Paris.

KUGEL, G., HELLFEIER, G. (1995), Versuche mit Wirbelbettreaktoren zur Stickstoffelimination, *Korrespondenz Abwasser* **5**, 770–780.

LEWANDOWSKI, Z., STOODLEY, P., ALTOBELLI, S., FUKUSHIMA, E. (1994), Hydrodynamics and kinetics in biofilm systems – recent advances and new problems, *Water Sci. Technol.* **29**, 223–229.

MOSER, A. (1985), Special cultivation techniques, in: *Biotechnology*, Vol. 2 (REHM, H.-J., REED, G., Eds.), pp. 311–347. Weinheim: VCH.

Normenausschuß Wasserwesen (NAW) (1989), DIN 19557 "Kläranlagen" Füllstoffe aus Kunststoff.

SCHLEGEL, S. (1988), Der Einsatz von getauchten Festbettkörpern (fixed-bed filters) zur Nitrifikation, *Korrespondenz Abwasser* **35**, 120–126.

SCHLEGEL, S. (1998), Kostengünstige Vorbehandlung industriellen Abwassers vor der Indirekteinleitung. Paper presented at ATV Seminar: "Leistungsverbesserung bestehender Kläranlagen" (2/3 March), Essen.

SCHULZ genannt MENNINGMANN, J., RICHARDT, M., CHROMIK, R. (1998), Denitrifikation mit getauchten Festbettreaktoren in der dezentralen Abwasserreinigung, *Entsorgungs Praxis* **1/2**, 37–41.

SEYFRIED, C. F. (1997), Verfahrenstechnischer Überblick zur Biofilmtechnik. ATV Seminar 28/97 "Einsatz von Biofilmreaktoren", Oldenburg.

Technische Universität München (1991), Lehrstuhl und Prüfamt für Wassergütewirtschaft und Gesundheitsingenieurwesen der Technischen Universität München: Abschlußbericht über Untersuchungen an Vario Compact Kläranlagen (VCK) (January 1991).

TOETTRUP, H., ROGALLA, F., VIDAL, A., HARREMOËS, P. (1993), The treatment trilogy for floating filters, *2nd Int. Spec. Conf. Biofilm Reactors*, Paris.

VAN LOOSDRECHT, M. C. M., EIKELBOOM, D., GJALTEMA, A., MULDER, A., TIJHUIS, L., HEIJNEN, J. J. (1995), Biofilm structures, in: *Int. IAWQ Conf. Workshop Biofilm Structure, Growth and Dynamics* (30 August–1 September), Noordwijkerhout.

WANNER, O., GUIJER, W. (1984), Competition in biofilms, *Water Sci. Technol.* **26**, 27–44.

18 Experience with Biofilters in Wastewater Treatment

CARIN SIEKER
Berlin, Germany

MATTHIAS BARJENBRUCH
Rostock, Germany

1 Introduction

Biofilters are compact wastewater treatment units, which combine biological degradation of dissolved wastewater components and elimination of suspended solids. They have proved their capability of treating carbon and nitrogenous pollution, achieving high degradation rates. However, the observed performance of biofilters varies considerably in practice. The object of this paper is to discuss the experience gained with biofilters in treating municipal wastewater and to outline the limits of their multifunctional use.

In past years, higher requirements in wastewater treatment, especially in the treatment of nitrogen pollution, have caused an advanced development in the field of biofilm technology. In the beginning of the 1980s in France, the filtration process and the biofilm technology were at first combined by using aerated sand filters as the main biological stage for the degradation of COD. In 1986, a dry bed filter was put into operation as a second stage for advanced COD elimination. Later on, biofilters were built for advanced treatment (residue nitrification and denitrification). At the recent state of development, biofilters are also used as the main biological stage in municipal wastewater treatment with complete nitrification and denitrification of wastewater. The first plant of this kind was built in Cergy, France, in 1992 (ROGALLA et al., 1994).

The advantages of biofilters are, in particular, the high removal rates and the absence of secondary clarifiers, both causing a significant reduction of volume and space. This is favorable, especially when only little space is available, when the ground is expensive or difficult for building, or when treatment units have to be integrated into existing plants.

2 Fundamentals

Biofilters can be classified as biofilm systems. The performance of biofilm systems depends on the diffusional resistance to the penetration of substrates into the biofilm (HARREMOES, 1978). According to this classical theory, the removal rate is limited by the substrate that least penetrates the biofilm. The ratio of the different compounds and the thickness of the biofilm will determine the removal rate which can be obtained.

For biofilters this "biofilm theory" can only partly explain the influence of mass transport. The flow through porous media is often accompanied by hydraulic channeling, which can lead to stagnating areas, e.g., in dead end pores. In these sheltered areas, the mass transfer rate is reduced and consequently the nutrient supply of the biofilm is limited. The maximum removal rate which can be achieved with biofilters, therefore, strongly depends on the extent of these stagnating areas.

The short-term influence of different air and water velocities on the nitrification rate is documented by TSCHUI et al. (1994). The studies showed that the nitrification rate increases with increasing water velocities. The increase is especially pronounced in the range between 4 and 6 m h^{-1}. Air velocity tests in the laboratory-scale units revealed a gradual increase of the biological activity with air rates up to 30 m h^{-1} and higher. The greater activity at higher water and air velocities can basically be explained by stronger turbulence in the filter. Increasing turbulence reduces stagnating areas and subsequently improves the oxygen supply of the biomass. This has an effect on the mostly O_2-limited nitrification rate.

3 Mode of Operation and Possible Applications of Biofilters

Biofilters can be applied for either biodegradation of organics or denitrification or nitrification. Biological phosphorus removal has been studied only on a pilot scale up to now (GONCALVES et al., 1994). Therefore, phosphorus is now chemically eliminated.

Particulate wastewater components, suspended solids which are generated by precipitation/flocculation and biological surplus sludge are retained mainly by physical processes within the filter bed by using granular material with diameters between 2–8 mm. However, an effluent concentration below 5 mg L^{-1} total suspended solids (TSS), which can be achieved with conventional gravity filters, cannot be guaranteed with biofilters.

Biofilters show a wide range of application. They are very flexible in their utility and can easily be integrated into the wastewater treatment process. They can be used for (Fig. 1):

- post-nitrification combined with the elimination of suspended solids and eventually precipitation/flocculation,
- post-denitrification combined with the elimination of suspended solids and eventually precipitation/flocculation,
- biological main stage (COD removal and/or nitrification and/or denitrification) combined with the elimination of suspended solids.

4 Process and Operational Conditions of Biofilters

Biofilters are usually divided into single reactor units, each of them comprised of inflow and outflow pipelines including fittings, filter bottoms, filter material, and related equipment for filter washing. The process techniques of biofilters can be designed in different ways. Tab. 1 gives global examples of how to distinguish between the different process techniques. Fig. 2 shows some of the common biofilter systems.

The biofilters require pretreated wastewater with a reduction of suspended solids. High influent concentrations of suspended solids lead to clogging, which causes an uneven distribution of substrates and increases the frequency of washing. Suspended solids in the inflow of biofilters should be kept below 50–70 m L^{-1}. When filters are applied as a full biological

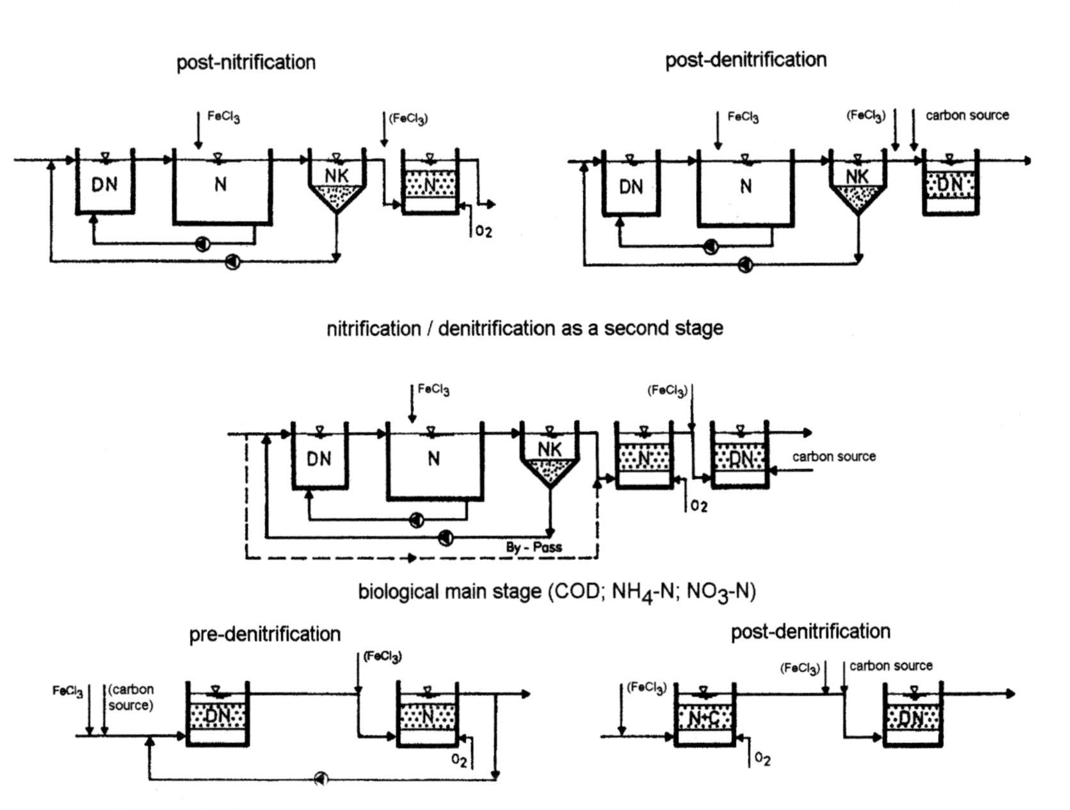

Fig. 1. Possible applications of biofilters.

Tab. 1. Process Modes of Biofilter Systems

Flow Direction	Direction of Process Air	Type of Flow through Pore Volume	Density of Material	Filter Washing
Downward flow	co-current	flooded	floating $\rho < 1.0\ g\ cm^{-3}$	continuous
Upward flow	counter--current	dry bed filter	not floating $\rho > 1.0\ g\ cm^{-3}$	intermittent

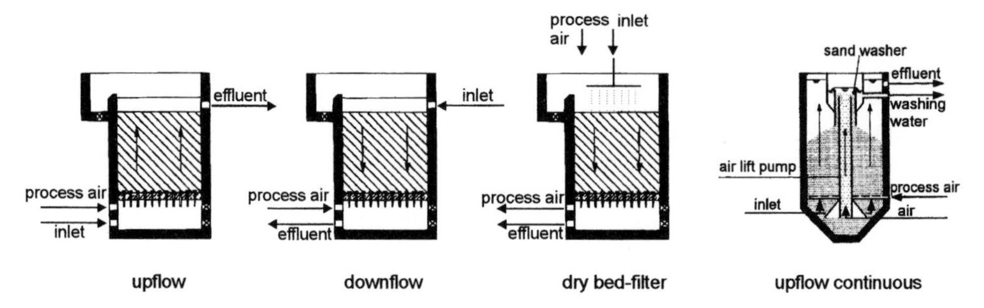

Fig. 2. Common biofilter systems.

stage, an intensive primary treatment with precipitation/flocculation is required and the primary settlement tank is often equipped with a lamella separator.

Biofilters which are operated intermittently have to be regularly washed to avoid clogging. The sludge water is usually recirculated to the primary stage. It normally amounts to 5–15% of the dry weather inflow when the biofilter is used as post-treatment stage. However, it can increase to 30–40% of the dry weather inflow when the biofilter is operated as the full treatment stage. A large amount of sludge water is produced during a short period of time since the washing in most cases occurs discontinuously. These sludge water peaks lead to high hydraulic loadings in the receiving treatment stage, e.g., the primary settling tank. It can, therefore, be necessary to include a sludge water storage tank to achieve a more or less steady sludge water recirculation.

5 Design and Operation

The information found in the literature varies considerably with regard to the volumetric rates that can be achieved with biofilters. Tab. 2 shows the range of observed volumetric rates given in the literature and the common design loading for biofilters.

For nitrifying plants, the obtained differences depend on various parameters, e.g., the COD load and the aerating conditions.

Basically, the parameters that affect the degradation rate in biofilters can be summarized as follows:

- Constructive factors, e.g., used filter material and the distribution system of air and water,
- operating conditions, e.g., the kind of washing, applied air and filter velocities, and loading of the biofilter,
- influent conditions, e.g., high loads of suspended solids, high carbon loads, nutrient supply, COD : N, inflow fluctuation, and temperature,
- application of the biofilter, e.g., full biological treatment, post-treatment.

Tab. 2. Observed Volumetric Rates and Common Design Loading for Biofilters

Degradation of		Max. Volumetric Loading [kg m^{-3} d^{-1}]	Common Design Loading [kg m^{-3} d^{-1}]
Carbon	BOD_5	4–7	3–5
	COD	7–10	5–8
Ammonia	NH_4-N	0.1–1.5 (max. 2.0)	0.4–0.8
Nitrate	NO_x-N	0.8–4.0 (max. 5.0)	0.7–1.2

6 Nitrification Rates

Most biofilters that are designed as the main stage to treat carbon compounds also have nitrification capacity. The nitrification rate achieved in biofilters with high carbon loading is normally much lower in comparison with tertiary treatment processes, where the COD concentration in the influent is low. CANLER and PERRET (1993) obtained removal rates of 0.36 kg m^{-3} d^{-1} NH_4-N in plants having a C:N ratio of about 10. In tertiary Biostyr systems, VEDRY et al. (1994) found removal rates of 1.7–1.8 kg m^{-3} d^{-1} NH_4-N at a temperature of 19 °C. Similar values were found by TSCHUI et al. (1994) in pilot-scale plants. They distinguished between Biocarbon and Biostyr systems with maximum volumetric rates of 0.7 ± 0.1 and 1.5 ± 0.2 kg m^{-3} d^{-1} NH_4-N, respectively, at a temperature of 10 °C.

The predictions as to how tertiary nitrification stages respond to influent peaks after a long term of ammonia deficiency vary significantly. In tertiary nitrification steps, the influent NH_4-N-concentration can be very low over a long period of time, because the ammonia in the wastewater is largely nitrified in the preceding biological stage. It seems, therefore, a logical consequence that due to decay, grazing effects, and washout processes the concentration of active nitrifiers in the systems decreases and, consequently, the removal capacity drops. Under these conditions, even a slight increase of the NH_4-N-loading can immediately cause higher effluent concentrations. Fig. 3 shows the operating results of a biofilter (A = 3,504 m^2) which is used for post-nitrification.

Over several days the filter received very low loadings below 0.03 kg m^{-3} d^{-1} NH_4-N. Then the loading abruptly increased to 0.7 kg m^{-3} d^{-1} NH_4-N. Although normally a removal capacity of 0.7 kg m^{-3} d^{-1} NH_4-N can easily be achieved with nitrifying biofilters, in this case, the maximum rate achieved reached only 0.2 kg m^{-3} d^{-1} NH_4-N. It took about 4 days until the nitrification capacity recovered and values of 0.4 kg m^{-3} d^{-1} NH_4-N were observed.

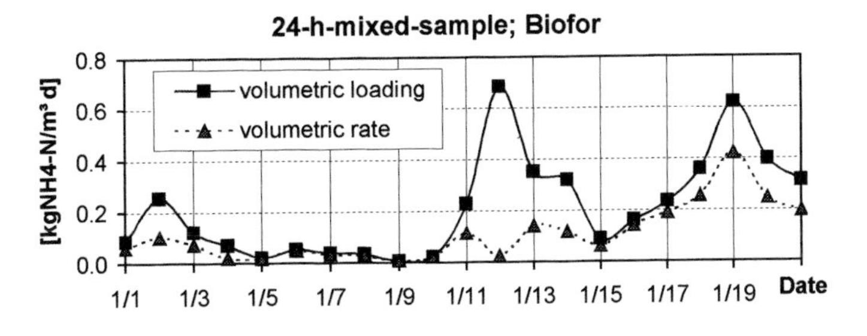

Fig. 3. Effect of a long-term ammonia deficiency on nitrification (A = 3,504 m^2) (BARJENBRUCH and SEYFRIED, 1995).

7 Denitrification Rates

In biofilters, denitrification can be used in either the pre- or post-situated mode. The achievable denitrification rate is influenced by the available carbon source, the C:N-ratio, and the oxygen load in the inflow (e.g., high oxygen concentrations in the recirculation flow). Tab. 3 gives an overview of the observed denitrification capacities of biofilters.

8 Hydraulic Residence Time, Flow Patterns, and Dynamic Behavior of Biofilters

Due to the high volumetric rates which can be achieved with biofilters, the filters are dimensioned a lot smaller than activated sludge systems. For a given degree of treatment, biofilters require one third the aeration volume of activated sludge units (SMITH and HARDY, 1992). This causes short residence times and implies small volumes for the compensation of influent peaks. The filter media usually have a porosity of about 40%. Due to stagnation areas and hydraulic channeling, the effective porosity amounts to only 25–30%. Besides, the flow pattern of biofilters shows a predominant plug flow with relatively small dispersion in the horizontal direction. This additionally decreases the buffer capacity of the biofilter. Compared to activated sludge systems, the buffer capacity in biofilters is very small. Therefore, influent peaks can only be eliminated by increasing the biological activity of the filter. To find out how biofilters react under unsteady influent conditions, investigations were carried out at the full-scale Biostyr plant in Nyborg. Fig. 4 shows influent/effluent measurements.

The high influent peaks during weekdays are caused by the sludge water recirculation. The experimental results show that if the loading of the Biostyr filter exceeds 0.7 kg m^{-3} d^{-1} TKN it immediately affects the effluent quality. The shape of the effluent peaks is very similar to the shape of the influent peaks. This indicates low buffering in the Biostyr filter. A fictitious activated sludge unit was dimensioned and the effluent concentrations were simulated with the ASM 1 model to compare the dynamic behavior of the Biostyr filter with an activated sludge plant. The activated sludge system achieves much flatter and wider effluent peaks and the mean residence time amounts to 20 h.

Tab. 3. Denitrification Performance of Biofilters

Flow	v_F[a]	Substrate	COD:N	Concentration in	out	r_{NO_3-N}[b]	Reference
–	[m h^{-1}]	–	–	[mg L^{-1} NO_3–N]		[kg m^{-3} d^{-1}]	
Down	4.5	methanol	<3.8	6–25	2–10	–	WILDERER et al. (1994)
Down	7.5	methanol	4.1	20	0.2–0.4	1.96	KRAFT (1994)[c]
Up	–	methanol	–	–	–	2.5	STROHMEIER (1994)
Up	3–5	raw sewage	5.2	–	12	1.0	ROGALLA et al. (1992)
Up	6.6	methanol	>5.0	10.5	1	1.0	KOOPMANN et al. (1990)[c]
Up	4.6	acetol 100	6.1	21.9	7.9	0.8	BARJENBRUCH (1997)

[a] Filter velocity
[b] Denitrification rate
[c] Pilot scale

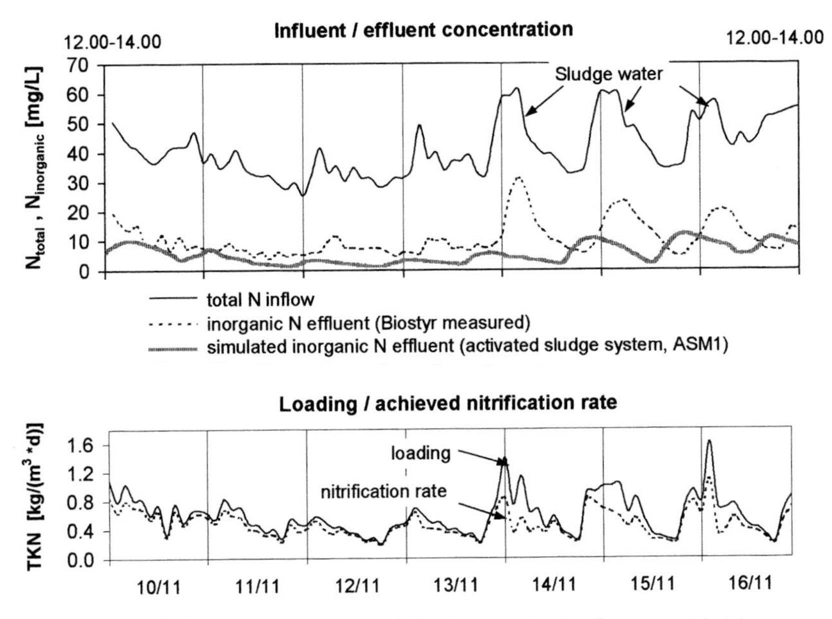

Fig. 4. Influent/effluent measurement of the Biostyr plant in Nyborg with 2 h composite samples (Temp. 12 °C) compared to the simulated effluent conditions of a fictitious activated sludge plant.

9 Conclusion

Biofilters have proved their capability of treating carbon and nitrogenous pollution and for achieving high degradation rates. At the same time, they show a reliable elimination of suspended solids, but the multifunctional use is limited. The achievable degradation capacity can vary extremely depending on numerous factors including operational conditions such as the applied air and water velocity. High air and water velocities have a positive effect on the degradation activity of the filter, but they can have a negative influence on the filtration capacity.

In countries in which the effluent standards are based on grab samples and not on daily average samples, the dynamic behavior of the treatment unit becomes important. As a result of the compact dimensioning, the absence of final effluent tanks, and the predominant plug flow, biofilters are provided with a very small volume for balancing influent peaks and, therefore, have little buffer capacity. Influent peaks can only be eliminated by increasing the activity of the filter. As soon as the loading exceeds the maximum degradation rate, the peak passes right through the filter.

This implies that the dimensioning and the operation of the biofilter becomes a very important factor in ensuring proper treatment performance.

10 References

Barjenbruch, M. (1997), Leistungsfähigkeit und Kosten von Filtern in der kommunalen Abwasserreinigung. *Veröffentlichung des Instituts für Siedlungswasserwirtschaft und Abfalltechnik der Universität Hannover* **97**.

Barjenbruch, M., Seyfried, C.-F. (1995), Experiences gained with biologically intensified filtration in Germany, *Proc. FILTECH Conf.*, pp. 191–202. Karlsruhe.

Canler, J. P., Perret, J. M. (1993), Biological aerated filters assessment of the process based on 12 sewage treatment plants. *EWPCA – Proc. Spec. Conf. Biofilm Reactors*, Paris.

GONCALVES, R. F., LE GRAND, L., ROGALLA, F. (1994), Biological phosphorus uptake in submerged biofilters with nitrogen removal, *Water Sci. Technol.* **29**, 135–144.

HARREMOES, P. (1978), Biofilm kinetics, *Water Pollut. Microbiol.* **2**, 82–109.

KOOPMANN, B., STEVENS, C. M., SONDERLICK, C. A. (1990), Denitrification in a moving bed upflow sandfilter, *Res. J. WPCF* **62**, 239–245.

KRAFT, A. (1994), Simultane Denitrification im Abwasserfilter, *awt-Abwassertechnik* **1**, 35–39.

ROGALLA, F., BADARD, M., HANSEN, F., DANSHOLM, P. (1992), Upscaling a compact nitrogen removal process, *Water Sci. Technol.* **26**, 1067–1076.

ROGALLA, F., SALZER, C., LAMOUCHE, A., SPECHT, W. (1994), Biofilter zur Stickstoffelimination: Erfahrungen aus der Praxis, *ATV-Proceedings* „Konzepte und Verfahren zur Stickstoffelimination", Nürnberg (6.–7. April).

SMITH, A. T., HARDY, J. P. (1992), High rate sewage treatment using biological aerated filters, *JIWEM* **6**, 179–183.

STROHMEIER, A. (1994), Einsatzmöglichkeiten und großtechnische Erfahrungen mit der Biofiltration zur N- und P-Entfernung, *ÖWAV-Proceedings „Abwasserreinigungskonzepte – Internationaler Erfahrungsaustausch über neue Entwicklungen"*, Vienna.

TSCHUI, M., BOLLER, M., GUJER, W., EUGSTER, J., MÄDER, C., STENGEL, C. (1994), Tertiary nitrification in aerated pilot biofilters, *Water Sci. Technol.* **29**, 53–60.

VEDRY, B., PAFFONI, C., GOUSAILLES, M., BERNARD, C. (1994), First months operation of two biofilter prototypes in the waste water plant at Acheres, *Water Sci. Technol.* 29, 39–46.

WILDERER, P., BÖHM, B., EICHINGER, J. (1994), Denitrifikation in einer nachgeschalteten Sandfilteranlage als kostengünstige Ausbaualternative für bestehende Klärwerke, 24. Abwassertechnisches Seminar, *Berichte aus Wassergüte- und Abfallwirtschaft der Technischen Universität München* **117**.

19 Special Aerobic Wastewater and Sludge Treatment Processes

UDO WIESMANN
JUDY LIBRA
Berlin, Germany

List of Abbreviations

INS	naphthalene-1-sulfonic acid
2,6NDS	naphthalene-2-6-disulfonic acid
2NS	naphthalene-2-sulfonic acid
5AS	5-aminosalicylate
6A2NS	6-amino-naphthalene-2-sulfonic acid
ANS	amino-naphthalene sulfonic acids
ASR	airlift suspension reactor
COD	chemical oxygen demand
CP	chlorophenol
CSTR	completely stirred tank reactor
DCA	1,2-dichloroethane
DCM	dichloromethane
DMSO	dimethylsulfonoxide
DNT	dinitrotoluene
DOC	dissolved organic carbon
EPA	environmental protection agency
GAC	granular activated carbon
HCB	hexachlorobenzene
HNS	hydroxy-naphthalene sulfonic acids
HPLC	high-pressure liquid chromatography
HRT	hydraulic retention time
LDS	lignin degrading systems
MCB	monochlorobenzene
NP	nitrophenol
NS	naphthalene sulfonic acids
PAH	polycyclic aromatic hydrocarbons
PCP	pentachlorophenol
PW	photoprocessing waste
SBR	sequencing batch reactors
SCP	single cell protein
SOB	sulfur oxidizing bacteria
SRB	sulfur and/or sulfate reducing bacteria
SS	suspended solids
TCB	trichlorobenzene
TeCB	tetrachlorobenzene
TNT	2,4,6-trinitrotoluene
VSS	volatile suspended solids

1 Introduction

The first aerobic industrial wastewater treatment processes were developed in the 1960s, and modeled on municipal wastewater treatment plants: Effluent from the production processes was piped to a central treatment plant, where it was treated, often with pH correction and without presedimentation, in open activated sludge basins before being discharged into a receiving water. With time, the disadvantages of this technology became obvious

(1) the chemical oxygen demand (COD) of the effluent was considerably higher than that allowed by law, resulting in high costs due to high discharge fees or the necessity of adding activated carbon to the process,
(2) a part of the organic volatile compounds were stripped, leading to air pollution problems and the necessity to cover the basins,
(3) certain effluents, e.g., from production units of the chemical industry with a high concentration of non-biodegradable compounds, were only diluted and
(4) effluents with toxic compounds reduced the removal rate in activated sludge plants.

Many more reasons can be cited for the need to change this end-of-the-pipe technology. New technologies are necessary, especially when new production units are to be constructed or old units have to be modernized. Furthermore, the local circumstances have to be taken into consideration. In industrialized countries a large percentage of industrial effluent is discharged into municipal wastewater treatment systems. However, numerous large cities all over the world do not have a municipal wastewater treatment system, so that industrial wastewater treatment is absolutely necessary to guarantee a minimum quality of the raw water, in order to produce clean water for drinking and sanitary purposes. Since clean water is also an essential raw material for industry and agriculture, many national economies will be influenced by the need for water in the near future.

Therefore, the water consumption by industry must be reduced. From a fundamental point of view there are three main possibilities to conserve water:

(1) We have to develop new production processes with a lower water consumption.
(2) The water needed for cooling and processing has to be recycled.
(3) An economic solution has to be developed for the treatment of several effluents with different types and extent of pollution.

The treatment of wastewater and process water should be carried out in most cases in the immediate proximity of the production unit. Both production and waste management have to be designed and optimized by a single team, and both have to be operated by one team. Wastewater treatment must be integrated into the process!

Some questions to be considered are:

(1) Is it best to pretreat a highly concentrated effluent before mixing it with other lower loaded wastewater?
(2) Is it best to treat only one of three effluents and to recycle this water?
(3) Can we save raw materials or recover product from the effluent directly following the production process?

Different kinds of physical, chemical and biological processes must be tested to find the best solution. Biological and chemical oxidation and reduction processes are of special interest if almost complete transformation to compounds such as CO_2, H_2O, N_2 and CH_4 is possible. The pollution in the effluents from such optimized processes will no longer be characterized by thousands of different chemicals, but rather only three or four single compounds may often dominate, and the knowledge of stoichiometry as well as kinetic coefficients may be helpful for a better understanding and optimization of the biodegradation processes.

Therefore, in Sects. 2, 3 and 4 biokinetics are discussed in some detail.

2 Biological Degradation of Special Compounds

2.1 Chlorinated *n*-Alkanes, Particularly Dichloromethane and 1,2-Dichloroethane

2.1.1 Properties, Use, Environmental Problems, and Kinetics

Chlorine, a by-product from the electrolysis of chloroalkali in the beginning of this century, was used for the production of chlororganic compounds as a result of its high reactivity with organics. From the 114 organic priority pollutants listed by the U.S. Environmental Protection Agency in 1977, 22 are chlorinated alkanes (PATTERSON, 1985). Some of them are cited in Tab. 1. Because of their tendency to bioaccumulate in animals and their toxicity, they must be almost completely removed from all liquid and gaseous effluents.

Inspection of the compound properties listed in Tab. 1 shows that as the chlorine content increases, so does the Henry coefficient, whereas solubility and biodegradability show a reversed trend. Therefore, tri- and tetrachloroethene are normally air pollutants and removed from water and air by activated carbon adsorption, so that the discussion here will be concentrated on dichloromethane (DCM) and 1,2-dichloroethane (DCA).

For DCM and DCA the maximum concentrations in drinking water are 10 $\mu g\ L^{-1}$ (WHO, 1987). Toxicity experiments with rats led to a LD_{50} of 3 g kg^{-1} live weight (WHO, 1987). Mutagenic effects in rats could not be

Tab. 1. Some Chlorinated *n*-Alkanes, Properties and Use (PATTERSON, 1985)

Compound	c^{*a} [mg L^{-1}]	H^b —	$LD_{50}{}^c$ [mg kg^{-1} bw]^d	Use	EPA-Nr.^{e}
Dichloromethane CH_2Cl_2	16,700	0.13	–167 2,400	• solvent for intermediate products of chemical industry • cleaning agent for metal surfaces • propellent for polyurethane production	
1,2-Dichloroethane $C_2H_4Cl_2$	8,700	0.046	730	• intermediate product in the production of polyurethane • solvent • fuel additive	10
Trichloromethane $CHCl_3$	7,800	0.14	>450 –800	• solvent in pharmaceutical industry	
1,1,1-Trichloroethane $C_2H_3Cl_3$	4,800	0.21	>10,000 –14,300	• solvent for dyes and ink • cleaning agent for cars and textiles	11
Trichloroethene C_2HCl_3	1,000	0.49	>4,200 –7,200	• solvent for dyes and ink • cleaning agent for cars and textiles	87
Tetrachloroethene C_2Cl_4	150	1.2	>4,000 –5,000	• solvent for dyes and ink • cleaning agent for cars and textiles	85

[a] Solubility, 20°C.
[b] Henry coefficient, 20°C.
[c] Rat, oral.
[d] bw: body weight.
[e] EPA list of organic pollutants.

proven (JONGEN et al., 1978). LOEW et al. (1984), and RANNUG (1980) indicate that DCA may cause cancer. Data on worldwide production of DCA from 1960–1981 showed an increasing production from $93–825 \cdot 10^3$ t a^{-1}, however, production has decreased in Germany from $137–67 \cdot 10^3$ t a^{-1} in the 3 year period of 1990–1993 (HERBST, 1995). In contrast to DCM, the production of DCA has not been remarkably reduced in Germany (1986: $1.65 \cdot 10^6$ t a^{-1}; 1992: $1.51 \cdot 10^6$ t a^{-1}) (HERBST, 1995). The study of the mechanism and the kinetics of DCM degradation started in the late 1970s. The catabolic reactions were first published by LEISINGER (1988).

For catabolism and anabolism the following stoichiometry can be written

$$CH_2Cl_2 + Y_{O2/S}\,O_2 \rightarrow Y_{CO2/S}\,CO_2 + 2\,HCl + Y_{H2O/S}\,H_2O + Y_{B/S}\,B \quad (1)$$

with B as a symbol for biomass.

The influence of DCM concentration on the specific growth rate can be described by Haldane kinetics and the influence of oxygen by Monod kinetics.

$$\mu = \mu_{max} \frac{c_S}{K_S + c_S + \frac{c_S^2}{K_i}} \cdot \frac{c'}{K' + c'} \quad (2)$$

The kinetic and stoichiometric coefficients are listed in Tab. 2. The variability of the values can be explained by the use of different methods for kinetic measurements and different cultures. Some authors used pure cultures, some mixed. We can conclude from a mean K_S value of 10 mg L^{-1} DCM that very low DCM concentrations below 1 mg L^{-1} can only be obtained by a physical-chemical second step, e.g., adsorption on activated carbon or ozonation.

A hypothetic pathway for the catabolism of 1,2-dichloroethane oxidation was published by JANSSEN et al. (1985).

Two authors (RUDEK, 1992; HERBST, 1995) observed double Haldane kinetics with regard to DCA and oxygen concentration (Tab. 2), resulting in

$$\mu = \mu_{max} \frac{c_S}{K_S + c_S + \frac{c_S^2}{K_i}} \cdot \frac{c'}{K' + c' + \frac{c'^2}{K_i'}} \quad (3)$$

The seldom observed oxygen inhibition may have a practical importance if pure oxygen is used in order to avoid DCA losses by stripping effects.

2.1.2 Treatment of Wastewater Containing DCM or DCA

DCM balances in activated sludge systems showed that DCM is almost totally stripped with air (GERBER et al., 1979). Trickling filters are also unsuitable. Only 11% of the DCM added were biodegraded, 75% were desorbed into the air and 14% remained in the effluent (WINKELBAUER and KOHLER, 1989).

In further lab and pilot scale studies, fixed and fluidized bed reactors with solid particles as support material for bacteria were used. In fixed bed reactors a high fluid recycle rate is necessary for aeration in external absorption tanks (KÄSTNER, 1989) or for the addition of H_2O_2 by mixing (STUCKI et al., 1992). Similarly, in two-phase fluidized bed reactors a high recycle rate must be used for oxygen addition and fluidization (GÄLLI, 1986; STUCKI et al., 1992; BURGDORF et al., 1991; NIEMANN, 1993; HERBST, 1995). In order to avoid substrate desorption totally, oxygen can be added in the absorption tank using non-porous membranes (HERBST, 1995). A special biomembrane reactor to mineralize DCA was tested with success (FREITAS DOS SANTOS and LIVINGSTON, 1995). The DCA was separated from a biofilm by a membrane. Concentration is decreased inside the membrane and the biofilm to such a low value that almost none was stripped by air bubbles.

Because of the high bacteria concentration growing on the surface of small sand particles (GÄLLI, 1985; STUCKI et al., 1992; NIEMANN, 1993), in porous glass particles (BURGDORF et al., 1991) or in dense flocs weighted by $CaCO_3$ (HERBST, 1995; HERBST and WIESMANN, 1996) high mean reaction rates of up to 1400 mg DCM L^{-1} h^{-1} and 400 mg DCA L^{-1} h^{-1} could be realized. Because of the formation of HCl, a relatively high amount of NaOH or $Ca(OH)_2$ must be dosed.

Tab. 2. Kinetic and Stoichiometric Coefficients for Aerobic and Anoxic Degradation of DCM and DCE (HERBST, 1995, expanded)

	Author	Year	Bacteria	$Y_{O2/S}$ [mol mol^{-1}]	$Y_{B/S}$ [gSS (gDCM)$^{-1}$]	μ_{max} [h^{-1}]	K_S [mg DCM L^{-1}]	K_i [mg DCM L^{-1}]	K' [mg O_2 L^{-1}]
DCM	BRUNNER	1982	*Pseudomonas*		0.158	0.11	17–42.5	995	
	DIKS and OTTENGRAF	1991	*Hyphomicrobium*		0.17	0.11		300	0.055
	HAUSCHILD et al.	1992	*Pseudomonas*			0.22	4	1,470	
	NIEMANN	1993	mixed culture	0.62			3–10	200–2,500	0.5–0.9
	HERBST	1995	mixed culture	0.56	0.105	0.037	11.6	405	
	FREEDMAN et al.	1997	*Acinetobacter* sp.		0.118[c]	0.046			
DCA	FREEDMAN et al.	1997	(anoxic)	0.58[d]	0.087	0.037			
	JANSSEN et al.	1985	*Xanthobacter*				109		
	SALLIS et al.	1990	*Rhodococcus*			0.104	26		
	RUDEK	1992	*Xanthobacter*	1.12	0.12	0.14	[a]	[b]	0.055
	FREITAS DOS SANTOS and LIVINGSTON	1995	*Xanthobacter autotrophicus*	1.65	0.166	0.19	7.6		0.1
	HERBST	1995	mixed culture	1.7	0.17	0.19	57	125	0.5

[a] Substrate inhibition is detected, but not quantified.
[b] Oxygen inhibition is detected, but not quantified.
[c] [g VSS (g DCM)$^{-1}$].
[d] $Y_{NO_3/S}$ [mol mol^{-1}].

2.2 Chlorobenzene

2.2.1 Properties, Use, Environmental Problems, and Kinetics

Of the 12 chlorobenzenes, 6 are of technical importance (MCB, 1,2-DCB, 1,4-DCB, 1,2,4-TCB, 1,2,4,5-TeCB and HCB). The compounds with less chlorine are characterized by a high water solubility and a high vapor pressure or Henry coefficient (Tab. 3).

The chlorobenzenes become more and more toxic as the number of chlorine atoms increases. The annual worldwide production for 1985–1990 can be found in Tab. 4 (DOTT, 1992). German production decreased in the 1980s and 1990s from year to year as a result of stronger regulations. Emissions in North-Rhine-Westphalia (Bayer Leverkusen/Germany) for 1992 were estimated at 2886 $t\ a^{-1}$ MCB and 1055 $t\ a^{-1}$ DCB (DÖPPER and STOCK, 1995). Three years earlier (1989) these emissions for Hessen (Hoechst AG/Germany) were higher by a factor of nearly 10 (DÖPPER and STOCK, 1995). These reductions result from a decreased production and higher efficiencies in wastewater treatment.

Tab. 3. Some Chlorobenzenes, Properties, and Use for T = 25°C

	c^* [$mg\ L^{-1}$]	H —	LD_{50} [$mg\ kg^{-1}$]	Use	EPA-Nr.
Monochlorobenzene (MCB)	460[a] T = 20–30°C	0.16[a]	rat: 2,900[a]	• intermediate in the production of dyes, agrochemical, synthetics, aids for textile industry[a]	7
1,2-Dichlorobenzene (1,2-DCB)	130[a] T = 20–30°C	0.10[a]	rat: 1,500	• intermediate in the production of 3,4-dichloroaniline • decolorizor, degreaser[a]	
1,4-Dichlorobenzene (1,4-DCB)	73[a] T = 20–30°C	0.079[a]	rat: 500	• deodorizer, moth repellent • intermediate in production of dyes[a]	
1,2,4-Trichlorobenzene (1,2,4-TCB)	30[b]	0.059		• intermediate in the production of trichloronitrobenzene etc. • solvent	8
1,2,4,5-Tetrachlorobenzene (1,2,4,5-TeCB)	0.5–2.4[c] T = 25°C	0.12[c] $kPam^3\ mol^{-1}$		• intermediate in the production of tetrachloronitrobenzene and trichlorophenol[c]	
Hexachlorobenzene (HCB)	0.006[b]	0.071		• intermediate in the production of pentachlorothiophenol and PCP production (in Germany until 1985)[c]	9

[a] RIPPEN (1991).
[b] PATTERSON (1985).
[c] NOWAK (1994).
For symbols used, see Tab. 1.

Tab. 4. Amounts of Annual Production of Chlorobenzenes

	Worldwide (1985–1990) 1000 t a^{-1} [a]	Production in BRD 1000 t a^{-1} [b]	Production in BRD 1000 t a^{-1} [c]
Monochlorobenzene (MCB)	500	75 (1990)	60–70 (1989)
1,2-Dichlorobenzene (1,2-DCB)	80	12 (1991)	12 (1989)
1,4-Dichlorobenzene (1,4-DCB)	80	15–20 (1984)	15–20 (1989)
1,2,4-Trichlorobenzene (1,2,4-TCB)	30	5	>5 (1983)
1,2,4,5-Tetrachlorobenzene (1,2,4,5-TeDB)	5		3 (<1986)
Hexachlorobenzene (HCB)		1.5	1.5 (<1993)

[a] Dott (1992).
[b] Schäfer (1994).
[c] Nowak (1994).

2.2.2 Principles of Biological Degradation

For anaerobic dechlorination a suitable auxiliary substrate is necessary (acetate, acetone and others), beginning from HCB different anaerobic cultures are needed to obtain MCB. MCB can not be dechlorinated anaerobically (Nowak, 1994).

In some papers the production of the stable intermediate 1,2,3-TCB was observed (Bosma et al., 1988; Ramanand et al., 1993). An example for the formation of intermediates during the anaerobic degradation of a mixture of 1,2,3-TCB, 1,3,4-TCB and 1,3,5-TCB in batch experiments (20 mg L^{-1} of each compound, 1 L flask tests) is presented in Fig. 1. The mixed culture was obtained from methanogenic sediments of the river Saale near Jena (Germany). Acetone was used as an auxiliary substrate. It is remarkable that all three DCBs were analyzed in different amounts, presumably as a result of different degradation rates. Special cultures of aerobic bacteria seem capable of mineralizing all chlorinated benzenes from 1,2,4,5-TeCB (Sander et al., 1991) and 1,2,3,4-TeCB (Feidicker, 1993) to MCB. Prieger-Kraft (1995) proved that both compounds were mineralized as the only source of carbon and energy. For anoxic conditions no dechlorination could be observed. Presently there is no

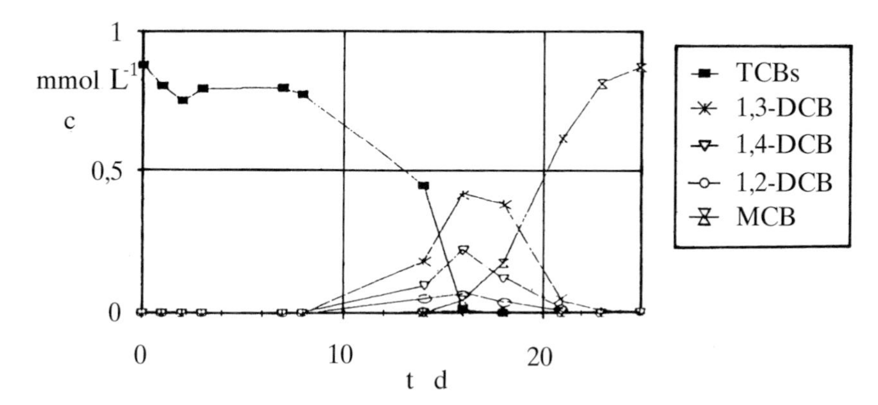

Fig. 1. Formation of isomeric dichlorobenzenes and monochlorbenzene by anaerobic degradation of a mixture from all 3 TCB (20 mg L^{-1} of each) pH 7, 28°C, auxiliary substrate: acetone (Kirsch, 1995).

confirmation of the reduction of PCB and HCB aerobically. In contrast to the anaerobic dechlorination, the ring of MCB is opened and chloro-*cis*, *cis*-muconic acid is formed. Only then the chlorine atom can be separated.

Aerobic pure cultures use chlorinated benzenes as the only source of carbon and energy with $\mu_{max}=0.55\ h^{-1}$ using MCB (REINIKE and KNACKMUSS, 1984), with $\mu_{max}=0.18\ h^{-1}$ using 1,2-DCB (HAIGLER et al., 1988) and with $\mu_{max}=0.07\ h^{-1}$ using 1,3-DCB (DE BONT et al., 1986).

2.2.3 Treatment of Wastewater with Chlorobenzene

In all known continuous biological processes

(1) only lab scale experiments were carried out, and
(2) the pure or mixed bacterial cultures used were immobilized on solid particles.

Obviously dechlorination of HCB and PCB is only possible by using anaerobic conditions and auxiliary substrates. FATHEPURE et al. (1988) used a biofilm reactor. HCB could be reduced to 1,2,3,4-TeCB with methanol and to 1,2,3-TCB and 1,2-DCB with acetate as an auxiliary substrate. NOWAK (1994) was successful in reducing 1,2,3,4-TeCB, all three TCB and all three DCB to MCB in a fluidized bed reactor with a mixture of methanol, acetate, acetone, ethanol, and propanol and a mixed culture from a sediment of the river Saale (Germany).

An aerobic mineralization of MCB, the three DCBs and 1,2,4-TCB was achieved in fixed bed reactors by BOUWER and MCCARTHY (1982) with small glass spheres, by VAN DER MEER et al. (1987) with sand particles from the river Rhine, and by SCHÄFER (1994) with polyurethane foam particles with and without activated carbon as carriers for mixed cultures, or *Pseudomonas* sp. (VAN DER MEER et al., 1987). In the experiments of SCHÄFER (1994) no auxiliary substrates as energy or carbon source were needed.

2.3 Chlorophenols

Of the 10 chlorophenols, 5 are produced as intermediates or end products in the chemical industry (Tab. 5). Compared with chlorobenzenes, chlorophenols have a higher solubility in water and a lower Henry coefficient of nearly two orders of magnitude. Because of its high toxicity and its low biodegradability, pentachlorophenol (PCP) was used during the 1970s as a herbicide, fungicide, and disinfectant worldwide in amounts of several 10,000 t a^{-1}. As a result of stronger regulations, production has decreased during the 1980s (1983: BRD 1,800 t a^{-1}, 1989: EU 1,000 t a^{-1}, RIPPEN, 1991).

KHANDKER (1992) studied the aerobic degradation of 4-MCP as the only source of carbon and energy by a culture with three bacteria [$Y_{B/S}=0.63$ gSS $(gMCP)^{-1}$, $\mu_{max}=0.15\ h^{-1}$, $K_S=21.1$ mgMCP$(L)^{-1}$]. The relatively high K_S points to a large region of substrate limitation.

The cometabolic degradation of phenol as the real source of carbon and energy and nearly all chlorinated phenols as cometabolites was studied by BANERJEE et al. (1984) using a pure bacteria culture. With increasing number of chlorine atoms, the reaction rate decreases and the partition coefficient K_{OW} (octanol–water) increases dramatically. These experiments were carried out in the region of zero order reaction with respect to each single concentration, starting from phenol to 2,3,4,5-tetrachlorophenol. The authors used an interesting penetration model based on the lipophilic and hydrophilic properties of an outer lipid and an inner porous layer of the cell wall. Because of the increasing K_{OW} with increasing number of chlorine atoms, diffusion through the lipid layer becomes more and more rate limiting, resulting in an extremely low degradation rate for 2,3,4,5-tetrachlorophenol, which has an extremely high octanol–water partition coefficient.

The oxidation of 2-CP, 3-CP, 4-CP as well as 2,4-DCP by H_2O_2 was catalyzed by an immobilized peroxidase (SIDDIQUE et al., 1993). High oxidation rates were obtained, but high molecular polymerization products arose, which precipitated as small solid particles.

Tab. 5. Some Chlorophenols, Properties and Use

	c* [mg L^{-1}] 20°C	H 20°C	LD$_{50}$ [mg kg^{-1}]	Use	EPA-Nr.
2-Monochlorophenol (2-CP)	29,000	0.49	rat: 670	• intermediate in the production of phenols, phenol resins, dyes • bactericide, fungicide	24
4-Monochlorophenol (4-CP)	27,000	$0.025 \cdot 10^{-3}$	rat: 250–670	• intermediate in the production of 2,4-DCP, TCP and TeCP • solvent for mineral oil industry	
2,4-Dichlorophenol (2,4-DCP)	4,400	$0.131 \cdot 10^{-3}$	rat: 580	• intermediate in the production of 2,4-dichlorophenoxy-acetate • herbicide, toxin for moths	31
2,4,5-Trichlorophenol (2,4,5-TCP)	1,200	$0.28 \cdot 10^{-3}$	rat: 820	• intermediate in the production of 2,4,5-trichlorophenoxy-acetate • herbicide, fungicide, preservation aid	8
Pentachlorophenol (PCP)	19	$0.029 \cdot 10^{-3}$	rat: 50	• herbicide, fungicide • wood preserver • disinfectant	

For symbols used, see Tab. 1.

2.4 Nitroaromatics, Particularly 4-Nitrophenol and 2,4-Dinitrotoluene

2.4.1 Properties, Use, Environmental Problems, and Kinetics

The most important compound is 2,4,6-trinitrotoluene (TNT). It has been produced in large amounts since 1900 and particularly used as a warfare agent. The soil of nearly all production sites is polluted by TNT and the cleaning of such soil will be a responsibility for long into the future. Because of its relatively low solubility in water of 130 mg L^{-1} TNT causes mainly soil pollution problems.

In contrast, several other nitroaromatics are characterized by a higher solubility in water (Tab. 6). 2,4-Dinitrotoluene (2,4-DNT) and 4-nitrophenol 4-NP, e.g., are intermediates in the production of TNT (2,4-DNT) as well as pesticides, azo and sulfur dyes, and chemicals for the photoindustry (4-NP). Therefore, they are more likely to be found in wastewaters and our knowledge about their biodegradability will be discussed below.

In 1983 20,500 t 4-NP were produced worldwide, 16,500 t of that in Germany and Denmark (BUA, 1992). In 1988 the production of the pesticides parathion and methyl-parathion was shut down in Germany, resulting in a decreasing production of only 2,000 t annually. In

Tab. 6. Some Nitroaromatic Compounds, Properties and Use

	c^* [mg L^{-1}]	H —	Use	EPA-Nr.[b]
Nitrobenzol $C_6H_5NO_2$	2,000 20°C	0.0029	• intermediate in the production of aniline • intermediate in the production of dyes • rubber, pharmaceutical products, photochemicals	
p-Nitrophenol $C_6H_4OHNO_2$	13,700 20°C	$0.021 \cdot 10^{-6}$ 25°C	• intermediate in the production of pesticides, azo and sulfur dyes, as well as chemicals for photo industry	58
2,4-Dinitrophenol $C_6H_3OHN_2O_4$	200 12.5°C	$0.65 \cdot 10^{-6}$	• pesticide, fungicide • intermediate in the production of dyes, explosives, and chemicals for photo industry	
2,4-Dinitrotoluene $C_7H_6N_2O_4$	250 20–25°C	$(2.4\text{–}200) \cdot 10^{-6}$	• intermediate in the production of polyurethane and of 2,4,6-dinitrotoluene	
2,4,6-Trinitrotoluene	130 20°C	$0.30 \cdot 10^{-8}$	• explosive • intermediate in the production of dye and pharmaceutics	

[b] EPA list (Environmental Protection Agency of USA) of organic priority pollutants.
For symbols used, see Tab. 1.

contrast, RIPPEN (1991) reported an annual worldwide production of 60,000 t a^{-1} and a German production of 10,000 t a^{-1}. Usually 4-NP is produced in batch reactors to fill purchase orders. After each production run, the reactor, pipes and stirring vessels are washed with water and the remaining raw materials, products and byproducts enter the wash water, which is eventually discharged to the large wastewater treatment plant. Because of the multiple dilution processes, the concentration of 4-NP in water is normally relatively low and special cultures of bacteria cannot grow in activated sludge plants. Therefore, 4-NP must be removed from the concentrated industrial effluents by biological or physical-chemical processes before discharge.

The human toxicity is based on damage to the function of liver, kidney and the central nervous system (THIEM and BOOTH, 1979). The respiration rate of activated sludge is inhibited by 50% for a 4-NP concentration of 110 mg L^{-1} (PAGGA et al., 1982).

With 660,000 t a^{-1} worldwide and 97,000 t a^{-1} in Germany the annually produced amount of 2,4-DNT is much higher than that of 4-NP (RIPPEN, 1991). 2,4-DNT is mainly used as an intermediate in the production of TNT and polyurethane. RIPPEN (1991) reported concentrations of 9.7 mg L^{-1} in TNT production effluents and about 14 mg L^{-1} in wastewaters of special production processes for organic chemicals.

Two different pathways are known for aerobic biodegradation of 4-NP.

(1) The elimination of NO_2^- as a first step and the formation of hydrochinone by using O_2 and 2 NAD(P)H (SPAIN et al., 1979) and
(2) the formation of 4-nitrocatechol by using O_2 and 2 NAD(P)H as a first step and the elimination of NO_2^- as a second step (GUILLAUME et al., 1963; RAYMOND and ALEXANDER, 1971; EHRING et al., 1992).

Biodegradation experiments with anaerobic bacteria have been unsuccessful (BATTERSBY and WILSON, 1989; HU and SHIEH, 1987).

Tab. 7 shows the published stoichiometric coefficient for the description of the aerobic catabolism and anabolism. It follows from these stoichiometric coefficients that 67.5% of the carbon are used for catabolism (formation of CO_2) and 24% of the nitrogen for anabolism [$Y_{NO2/S}=0.76$ mol N/(mol S–N)$^{-1}$], resulting in a formation of 76% as NO_2.

The low value of $\mu_{max}=0.003$ h^{-1} (HEINZE, 1997) results in a high generation time of $t_G=9.6$ d, but this value obtained with a mixed culture differs considerably from that, which was obtained by SCHMIDT et al. (1987) with *Pseudomonas* sp.

The formation of 4-methyl-5-nitrocatechol is postulated as the first step in the degradation of 2,4-dinitrotoluene (2,4-DNT) (SPANGGORD et al., 1991), which can then be mineralized by several further steps.

In contrast, the fungi *Phanerochaete chrysosporium* form the intermediate 2-amino-4-nitrophenol, and they are able to mineralize it (VALLI et al., 1992). BAUSUM et al. (1992) carried out measurements with 2,4-DNT labeled by ^{14}C. The measurement of $^{14}CO_2$ made it possible to determine $Y_{CO2/S}=0.64$ mol CO_2 (mol C)$^{-1}$ (Tab. 7). The same authors used first order kinetics to describe the influence of 2,4-DNT concentration, but they obtained a relatively high scattering of the reaction rate. Further kinetic results were published by HEINZE et al. (1995) and HEINZE (1997) (Tab. 7). It is remarkable that the maximum growth rate $\mu_{max}=0.1$ h^{-1} is higher by a factor of 33 than that for the mineralization of 4-NP. Additionally, the higher $Y_{B/S}$ and the lower $Y_{O2/S}$ show that a higher amount of carbon can be used for anabolism, although two NO_2^- ions have to be eliminated from one 2,4-DNT molecule!

2.4.2 Treatment of Wastewater Containing 4-NP or 2,4-DNT

Lab scale experiments with an activated sludge reactor and an influent concentration of 400 mg L^{-1} 4-NP were successfully carried out by JAKOBEZYK et al. (1984). If very low ef-

Tab. 7. Stoichiometric and Kinetic Coefficients for Aerobic Degradation of 4-NP (HEINZE, 1996, 1997)

	Author	Year	Bacteria	$Y_{O2/S}$ [mol O_2(molC)$^{-1}$]	$Y_{B/S}$ [g SS(g DOC)$^{-1}$]	$Y_{NO2/S}$ [mol N(mol S–N)$^{-1}$]	$Y_{CO2/S}$ [mol CO_2(molC)$^{-1}$]	μ_{max} [h^{-1}]	K_S [mg L^{-1} DOC]
4-NP	JENSEN and LAUTRUP-LARSEN	1967	*Pseudomonas* sp.			0.52			
	JAKOBEZYK et al.	1984	mixed culture			0.76			
	SCHMIDT et al.	1987	*Pseudomonas* sp.					0.31 ± 0.002	1.1 ± 0.2
	OU and SHARMA	1989	*Pseudomonas* sp.				0.675		
	HEINZE	1997	mixed culture	0.82	0.41	0.76		0.003	<1
2,4-DNP	BAUSUM et al.	1992	mixed culture				0.64		
	HEINZE	1997	mixed culture	0.71	0.62	0.80		0.1	2.6–4.9

fluent concentrations in the μg L^{-1} range are required, bioreactors (e.g., fixed bed reactors) with activated carbon as a support material for bacteria must be used (SPEITEL et al., 1989). For the treatment of highly loaded synthetic wastewater (630–2,500 mg L^{-1} 4-NP) HEITKAMP et al. (1990) used a fixed bed reactor with an immobilized culture of *Pseudomonas* sp. For a mean detention time of only 2.3 h, a mineralization efficiency of 93% was obtained. These experiments show that high reaction rates can be obtained if slowly growing bacteria are immobilized. An effluent from a 2,4-DNT production plant of BASF Schwarzheide GmbH has been treated since 1995 by a two-step anoxic/aerobic activated sludge plant. Possibly, in the first step 2,4-DNT is reduced to amino-nitrotoluene and 2,4-diaminotoluene, which are mineralized in the second step (SOCHER, lecture held 1997 at Schwarzheide). A two-step anaerobic/aerobic process was proposed by BERCHTOLD et al. (1995). In the first step ethanol was used as an energy and carbon source. The resulting 2,4-diaminotoluene was mineralized in batch experiments.

2.5 Polycyclic Aromatic Hydrocarbons and Mineral Oils

2.5.1 Properties, Use, and Environmental Problems

Polycyclic aromatic hydrocarbons (PAH) are mainly contained in the tar of hard coals and in all kinds of mineral oils. During the various preparation processes for intermediate products in the chemical industry and for numerous carriers of energy, PAH enter the wastewater, groundwater, and solid wastes. Because of their toxicity 13 PAH were included into the EPA list. Four of them and two further PAH are presented in Tab. 8. Two characteristics are of notable importance, their low solubility in water, but high solubility in mineral oil, and their toxicity (LD_{50}): with increasing number of rings and molecular weight

(1) the solubility in water decreases and
(2) the toxicity increases dramatically.

Tab. 8. Some Polycyclic Aromatic Compounds, Properties, and Use (RIPPEN, 1990)

	c* [mg L^{-1}]	LD_{50} [mg kg^{-1}]	Use	EPA-Nr.[b]
Naphthalene	25 (20°C)	rat: 1,780	• raw material for the production of dyes	
Anthracene	0.048 (20°C)	*Daphnia magna*: 3 mg L^{-1}	• intermediate in the production of dyes and anthrachinone	78
Phenanthrene	0.95 (20°C)	rat: 700	• raw material for the production of explosives, pharmaceutical products, drugs, herbicides, tanneries	81
Benzo(e)pyrene	0.0038 (25°C)[a]			
Benzo(k)fluorenthene	0.0006 (25°C)[a]			75
Acenaphthene	(25°C)	rat: 10,000	• raw material for the production of textile pigment dyes and synthetics, insecticides and fungicides	77

[a] SIMS and OVERCASH (1983).
[b] EPA-list (USA environmental protection agency).
For symbols used, see Tab. 1.

Therefore, the toxic PAH are mostly dissolved in other organics such as mineral oils. First, mineral oils and their biodegradability will be discussed before continuing with the biodegradability of PAH.

2.5.2 Mineral Oils

2.5.2.1 Composition of Mineral Oils and Standard Emulsion of *n*-Dodecane

Mineral oils are composed of the following compounds (BERWICK, 1984):

(1) saturated hydrocarbons,
- (•) *n*-alkanes,
- (•) branched alkanes,
- (•) cyclic alkanes;

(2) unsaturated hydrocarbons;
(3) heterocyclic alkanes;
(4) asphalts.

The saturated hydrocarbons can be mineralized by bacteria and yeasts and have been used as an energy and carbon source for the production of **s**ingle **c**ell **p**rotein (SCP). The *n*-alkanes are very important components, with their content in distillation products increasing as boiling points decrease. The solubility in water is only about 0.002 mg L^{-1} for compounds with the number of C-atoms ≥12 (MACKAY and SHIU, 1981). For solutions with concentrations above the solubility, the oil phase can be dispersed as small droplets using an energy input, for instance a rotor–stator system. A stable oil–water emulsion can be produced with a high energy input and the addition of an emulsifier.

An emulsion

(1) produced from *n*-dodecane (750 mg L^{-1}),
(2) with an alkylethoxylate (Eumulgin ET5, Henkel) (150 mg L^{-1}) and
(3) with a rotational speed of 10,000 min^{-1} for 1 min was called a standard emulsion (CUNO, 1996). The distribution of oil droplets of this standard emulsion is presented by Fig. 2. It could be demonstrated by biodegradation experiments of emulsions with mean droplet diameters between 1 and 15 μm, that bacterial growth rate was totally independent of the droplet size (Fig. 3).

These results can only be understood, if the following mechanism of the oil degradation is assumed: After collisions the bacteria were

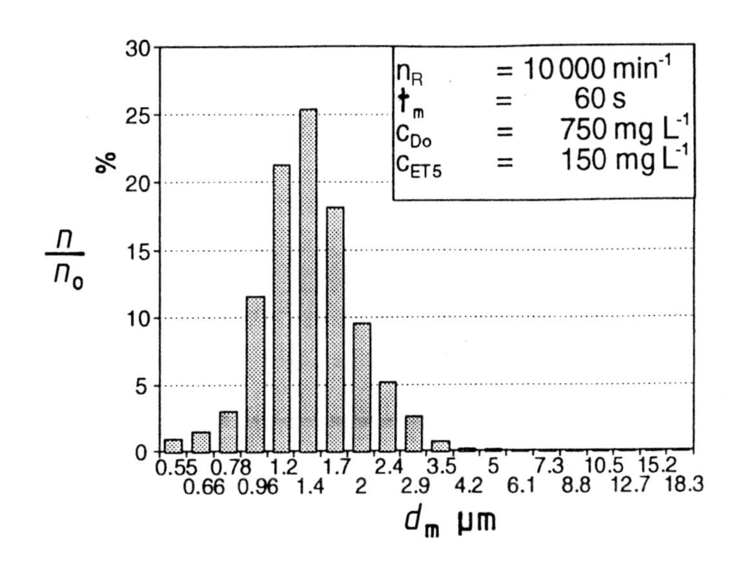

Fig. 2. Oil droplet distribution of the standard emulsion (CUNO, 1996), n_R number of revolutions, c_{DO} concentration of *n*-dodecane, *n* number of droplets of each size range, n_0 total number of droplets, t_m mixing time, c_{ET5} concentration of the emulsifier, d_m mean diameter of droplets.

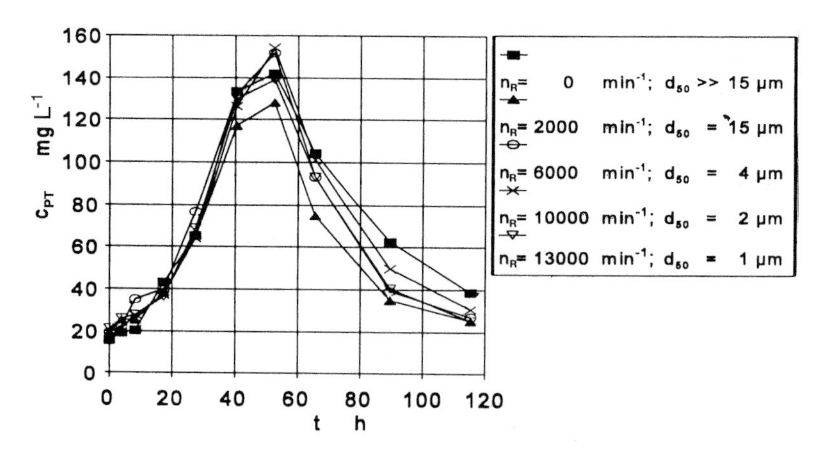

Fig. 3. Concentration of protein for *n*-dodecane emulsions with different mean diameter of the droplets (CUNO, 1996), *n* number of revolutions, d_{50} diameter of 50% of the droplet distribution, *t* time, c_{Pr} concentration of protein.

covered by oil layers, then oil diffuses into the cell. Probably the microkinetics of oil degradation rate is limiting.

The following results of a kinetic biodegradation study were obtained with the standard emulsion mentioned above.

2.5.2.2 Kinetics of Aerobic Oil Mineralization

Some results of stoichiometric and kinetic coefficients are shown in Tab. 9. The survey is not complete. Further data for other *n*-alkanes are presented by CUNO (1996). Unfortunately no further results were published for *n*-dodecane. Therefore, a direct comparison is only possible with some limitation.

2.5.3 Biodegradation of PAH

2.5.3.1 PAH Dissolved in Water

Because of the low solubility in water and the frequently rate limiting dissolution from crystals, PAH biodegradation can be better studied by using solution agents such as dime-

Tab. 9. Stoichiometric and Kinetic Coefficients for Aerobic Degradation of *n*-Alkanes

Author	Year	Bacteria	Substrate	Temperature [°C]	$Y_{B/S}$ [gSS(gDOC)$^{-1}$]	μ_{max} [h^{-1}]	K_S [mgDOC L^{-1}]	k_d
BOYLES	1984	*A. calcoaceticus*	*n*-hexadecane	30°C		1.01	—	—
BURY and MILLER	1993	*Pseudomonas aeruginosa*	*n*-decane[a]	30°C	1.3[b]	0.26	118[c]	—
		Orchrobactrum anthropi	*n*-decane[a]		1.4[b]	0.16	176[c]	—
CUNO	1996	mixed culture	*n*-dodecane[a]	20°C	1.34	0.17	22.9	0.04

[a] With emulsifier.
[b] Calculated from g VSS (g S)$^{-1}$.
[c] Calculated from mg S L^{-1}.

thylsulfonoxide (DMSO). With the help of DMSO the solubility of acenaphthene can be increased remarkably. Fig. 4 shows the results of a batch experiment with a mixed culture of PAH degrading bacteria. The concentration of the culture using acenaphthene as an energy and carbon source could not be measured. Therefore, in the solution of the model

$$\frac{dc_B}{dt} = \mu_{max} \frac{c_S}{K_S + c_S} c_B - k_d c_B \tag{4}$$

$$\frac{dc_S}{dt} = -\frac{\mu_{max}}{Y_{B/S}} \frac{c_S}{K_S + c_S} c_B \tag{5}$$

assumptions must be made for $Y_{B/S}$ and k_d. The line in Fig. 4 shows the solution, which agrees with the experimental data very well. Tab. 10 shows some further kinetic coefficients for the aerobic biodegradation of acenaphthene and other PAH. The maximal growth rate decreases from naphthalene over phenanthrene, acenaphthene to anthracene. Further results for PAH with 4 rings were published by CUNO (1996).

The cometabolism of 2 PAH by a pure culture was studied by SPRINGFELLOW and AITKEN (1995). *Pseudomonas stutzeii* is able to use both naphthalene and phenanthrene as the only source for carbon and energie. The influence of both substrates on the oxygen consumption rate could be described by using a model for a competitive metabolism

$$\frac{r_{O_2}}{r_{O_2,max}} = \frac{c_S}{K_S\left[1+\frac{c_i}{K_i}\right]+c_S} \tag{6}$$

with c_S concentration of phenanthrene and c_i concentration of naphtalene.

The measured K_S and K_i are given in Tab. 10.

2.5.3.2 PAH Dissolved in *n*-Dodecane Standard Emulsion

An interesting question regarding the biodegradation of *n*-dodecane droplets (standard emulsion) and the biodegradable PAH dissolved inside the droplets is:

Will both be oxidized with the same reaction rate?

Indeed, the results shown in Fig. 5 confirm this assumption. Although *n*-dodecane, acenaphthene and anthracene are biodegraded by bacteria with different growth rates (Tabs. 9 and 10) the results in Fig. 5 can be described approximately by Eqs. (4, 5) using the same kinetic coefficients for all three substrates.

A possible model to gain insight into these results is the transfer of very small oil droplets with dissolved PAH into the bacterial cell. PAH, e.g., pyrene, dissolved in non-biodegradable, and in water insoluble substances such as heptamethylnonane can be used by bacteria such as *Rhodococcus* sp. at a considerably higher rate than those dissolved in the aqueous medium (BOUCHEZ et al., 1997). PAH need an agent for dissolution, which may be biodegradable (*n*-alkanes) or not (heptamethylnonane). Naphthalene and phenanthrene as

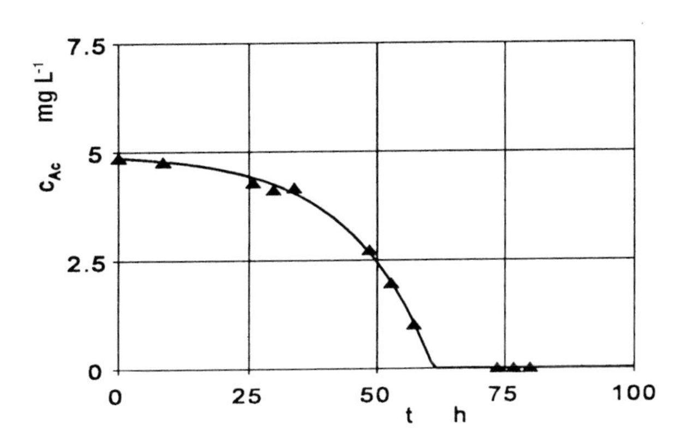

Fig. 4. Biodegradation of acenaphthene dissolved in water by using a solution agent DMSO and modeling by using Eqs. (4,5) (CUNO, 1996), 5% (vol.) dimethylsulfoxide DMSO.

Tab. 10. Stoichiometric and Kinetic Coefficients for Aerobic Degradation of Some PAH

Author	Year	Bacteria	Substance	$Y_{B/S}$ [gVSS(gS)$^{-1}$]	μ_{max} [h^{-1}]	K_S [mgSL^{-1}]	k_d [h^{-1}]	K_i [mgSL^{-1}]
WODZINSKI and JOHNSON	1968	*Pseudomonas* sp.	naphthalene	0.5	0.30			
GUHA and JAFFÉ	1996	mixed culture	phenanthrene	0.39	0.0011	0.09	0.0016[b]	
SPRINGFELLOW and AITKEN	1995	*Pseudomonas stutzeri*	phenanthrene			0.24		1.28[c]
KOMATSU et al.	1993	*Pseudomonas* sp.	acenaphthene		0.12			
CUNO	1996	mixed culture	acenaphthene	0.7[a]	0.062	0.065	0.0002	
BREURE et al.	1990	*Pseudomonas* sp.	anthracene		0.003		0.002	
CUNO	1996	mixed culture	anthracene		0.023	0.18		

[a] $\frac{\text{g Pr}}{\text{g S}}$.

[b] m (maintenance coefficient).

[c] K_i coefficient of competitive metabolism by naphthalene.

well as the dissolution agent hexadecane can also be mineralized by sulfate reducing bacteria, which could be proven with ^{14}C labeled substances (COATES et al., 1997; ZHANG and YOUNG, 1997). With ^{13}C-bicarbonate the latter authors could show that carboxylation is an initial key reaction for anaerobic sulfate reduction.

2.6 Naphthalene Sulfonic Acids

2.6.1 Properties, Use, and Environmental Problems

Naphthalene sulfonic acids (NS) and their substituted analogs can be commonly found in

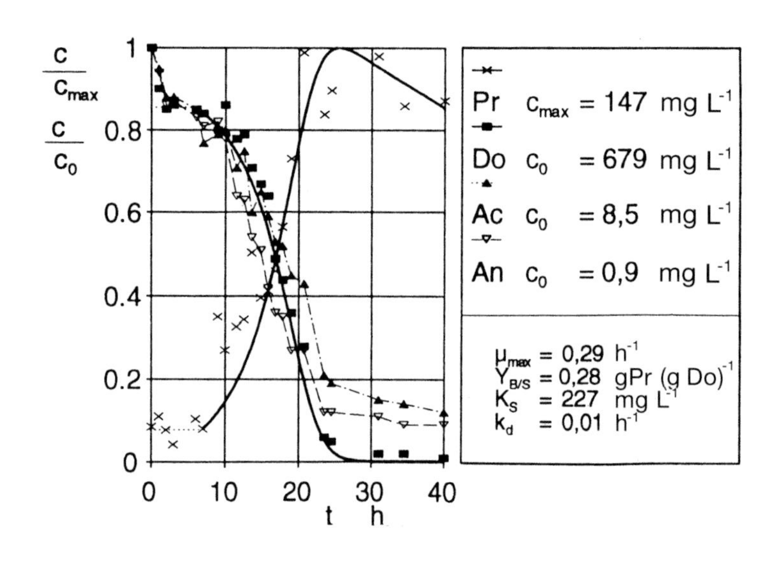

Fig. 5. Biodegradation of *n*-dodecane standard emulsion with dissolved PAH, Pr protein, Do *n*-dodecane, Ac acenaphthene, An anthracene, $\frac{c}{c_{max}}$ calculated by using Eqs. (4) and (5).

wastewaters from the production of naphthols, azo dyes, wetting agents, dispersants, and anionic surfactants (NÖRTEMANN and HEMPEL, 1991). Wastewaters from these production areas can have organic loads up to 4300 mg L^{-1} COD (KRULL and HEMPEL, 1994). The decolorization of azo dyes, most often achieved through the chemical or biological reduction of the azo bond, results in the production of corresponding sulfonated aromatic amines. Most azo dyes usually contain an amino-naphthalene sulfonic acid (ANS) (HAUG et al., 1991), with the water soluble dyes containing 1–3 sulfonic acids groups. Naphthalene sulfonates and aromatic amines are accumulated in waters and sediments because of their poor biodegradability when discharged into natural water bodies (VIDAL et al., 1993; NELSON and HITES, 1980).

2.6.2 Biodegradation

The degradation pathway for naphthalene sulfonic acids begins with deoxygenation in the 1,2-position by unspecific naphthalene dioxygenases and the spontaneous elimination of the sulfonic acid group. The resulting 1,2-dihydroxynaphthalene from naphthalene-1-sulfonic acid (1NS) and naphthalene-2-sulfonic acid (2NS) can be further degraded by *Pseudomonas testosteroni* A3 using the "classical" naphthalene pathway (BRILON et al., 1981). However, amino or hydroxy substituted NS (ANS, HNS), as often found in the metabolites from the anaerobic reduction of azo dyes, were readily cooxidized by 2NS or 1NS degrading microorganisms but produced no growth. Instead, labile and toxic metabolites were produced (NÖRTEMANN and HEMPEL, 1991). NÖRTEMANN et al. (1986) showed that some ANS and HNS can totally be degraded via interspecies transfer within a mixed culture isolated from the Elbe (Germany). VIDAL et al. (1993) obtained similar results with a four-membered bacterial community able to degrade 2NS isolated from Rio Reconquista (Argentina). Three of the isolated bacteria formed a brown-dark pigment when grown alone on 2NS, resulting in an inhibition of growth. The presence of a *Pseudomonas aeruginosa* strain avoided the production of pigment and resulted in a complete degradation of 2NS.

2.6.3 Kinetics of Biodegradation

DIEKMANN et al. (1988) investigated the degradation of 6-amino-naphthalene-2-sulfonic acid (6A2NS) by mixed cultures via 5-aminosalicylate (5AS) interspecies transfer in a chemostat. Experimental data showed the conversion of 6A2NS into 5AS to be the rate determining step. Both the simple Monod model and an extended model, taking the interspecies transfer of 5AS into account, were able to describe the experimental results. The kinetic coefficients for 6A2NS as well as for naphthalene-2-6-disulfonic acid (2,6NDS) are listed in Tab. 11.

Tab. 11. Stoichiometric and Kinetic Coefficients for Aerobic Biodegradation of Naphthalene Sulfonic Acids

Author	Year	Organism	Substrate	Y_{BS} [gSS(gDOC)$^{-1}$]	μ_{max} [h^{-1}]	K_S [mgDOCL^{-1}]	m_S [gDOC(hgSS)$^{-1}$]
KRULL et al.	1994	*P. testosteroni* A3	1NS/ 2NS	0.84	0.503	3.0	0.034
KRULL and HEMPEL	1994	defined mixed culture	1NS/1,6 2,6NDS	0.37	0.037	41	0.004
DIEKMANN et al.	1988	defined mixed culture	6A2NS	1.47	0.20	2	0.009
DIEKMANN and HEMPEL	1989	defined mixed culture	6A2NS	1.28	0.20	2	0.007

2.6.4 Treatment of Wastewater Containing NS

In batch and continuous flow experiments with four naphthalene mono- and disulfonic acids using specially adapted bacteria, DA CANALIS et al. (1992) showed that 2NS inhibits the degradation of the other 3 sulfonic acids. Building on these results, KRULL and HEMPEL (1993) used a two-stage airlift–loop reactor system to treat a wastewater containing a complex mixture of naphthalene mono- and disulfonic acids with high amounts of inorganic salts. The two different specially adapted microorganisms were immobilized on sand and the two stages were separated by a settling tank. In the first stage naphthalene-2- and -1-sulfonic acid were metabolized by *Pseudomonas testosteroni* A3 at residence times as low as 1.5 h. The total degradation of the remaining naphthalene-1-sulfonic acid and an overall degradation of recalcitrant naphthalene disulfonic acids of 71% were obtained by a defined mixed culture in the second reactor. Overall degradation increased up to 84% in investigations with salt free wastewater, showing that inorganic salts considerably inhibit biological degradation of the recalcitrant substances in the second reactor.

2.7 Dyes

2.7.1 Properties, Use, and Environmental Problems

Dyes are used in many industrial products, e.g., food, textiles, cosmetics. In the textile area alone over 4,000 compounds are commercially available. Various dye classification schemes are possible: based on the chemical structure of the chromophore, e.g., azo, nitro, phthalocyanine, triarylmethine etc., or the application class, e.g., direct, disperse, reactive, etc. (SCHÖNBERGER and KAPS, 1994). More than half of the dyes used commercially are azo dyes. Many dyes are water insoluble or dispersed as colloids. Most of the reactive dyes are very water soluble.

Over 700,000 t a^{-1} of dyes are produced worldwide (ZOLLINGER, 1987). SCHÖNBERGER (1996) estimated that a total of 13,200 t were used in the German textile industry in 1993. Comparing trends in the German textile industry, KAPS et al. (1990) found that the use of reactive dyes, the majority being azo dyes, is still increasing. In 1988, 37% of all dyes being used were already reactive dyes. On the average 10–15% of the dyes used in the textile industry find their way into the wastewater (VAIDYA and DATYE, 1982). Most textile processing wastewater in Germany is discharged with little pretreatment into municipal wastewater treatment plants (SCHÖNBERGER, 1996). There, abiotic mechanisms such as adsorption onto the sludge can remove insoluble dyes. The water soluble reactive dyes are generally not removed.

Dyes are not only aesthetically unappealing in streams, they either tend to accumulate in the waters and sediments of the receiving ecosystem (NELSON and HITES, 1980; MICHAELS and LEWIS, 1985, 1986) or, in the case of azo dyes, can be reduced anaerobically, producing possibly toxicologically relevant aromatic amines (BROWN and LABOURER, 1983a, b). The dyes themselves have a relatively low oral toxicity in rats. CLARK and ANKLIKER (1984) found that 82% of 4461 dyes tested had an $LD_{50} > 5$ g kg^{-1}. In fish toxicity tests, SEWEKOW (1988) found approximately 60% of 3000 dyes tested had an $LD_{50} > 100$ mg L^{-1}; 13% had an $LD_{50} < 10$ mg L^{-1}. However, some metabolites from the anaerobic reduction of azo dyes, aromatic amines, have been found to be carcinogenic and mutagenic (DELCLOS et al., 1984). Dyes containing lipophilic aromatic amines that have been found to be mutagenic or carcinogenic have been banned in Germany.

2.7.2 Biodegradation

The aerobic degradation of azo dyes by adapted bacteria has been reported (MEYER et al., 1979; KULLA et al., 1984; BLÜMEL et al., 1997), but the relevance of such specialized bacteria to textile wastewater treatment is probably limited. The aerobic decolorization of a number of texile dyes, mainly azo dyes, by different species of white-rot fungi has successfully been demonstrated (SPADARO et al., 1992; OLLIKKA et al., 1993; HEINFLING et al., 1997). Their lignin degrading systems (LDS) are relatively non-specific and so are able to oxidize and decolorize a variety of compounds of high molecular weight and even partly mineralize dyes which are resistant to bacterial degradation.

The most common approach to biological treatment of azo dyes is the combination of anaerobic reduction with aerobic degradation of the aromatic amines. This will be discussed in more detail.

2.7.2.1 Anaerobic Reduction of Azo Dyes

It has long been known that azo dyes can be reduced by anaerobic bacteria in the human intestine to aromatic amines (DIECKHUES, 1960). The existing hypothesis for the decolorization of azo dyes under anaerobic conditions is the reduction of the dyes through azo reductases, which have mainly flavoproteins $FMNH_2$/$FADH_2$ as cofactors (RAFII et al., 1990). The reduction equivalents gained by the oxidation of an auxiliary substrate through NAD(P)H reduce the azo bond to form aromatic amines, and thus decolorize the solution (Fig. 6).

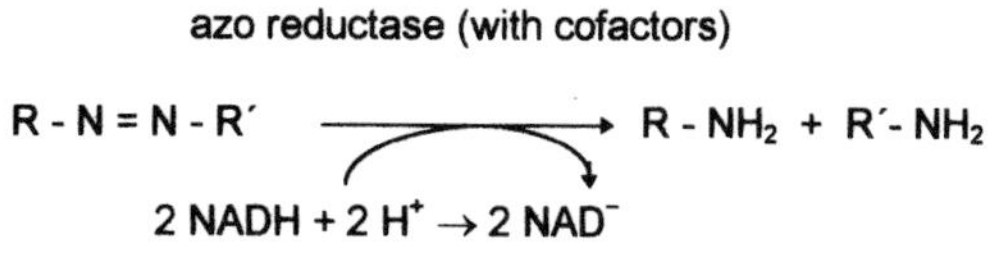

azo reductase (with cofactors)

$$R-N=N-R' \longrightarrow R-NH_2 + R'-NH_2$$

$$2\,NADH + 2\,H^+ \rightarrow 2\,NAD^-$$

Fig. 6. Scheme for the anaerobic decolorization of azo dyes (ZIMMERMANN et al., 1982).

Putting it into the perspective of the classical three-step anaerobic degradation of complex carbon substrates, the anaerobic reduction of the dye is thought to take place in the first of the three steps by the acidogenic bacteria.

The hypothesis for the biological reduction of azo dyes through the redox equivalents varies with respect to the enzyme (azo reductase) participation in the decolorization. The enzymatic reduction theory was proposed by ZIMMERMAN et al. (1982), RAFII et al. (1990), HAUG et al. (1991), and CHUNG and STEVENS (1993), with FMN, FAD, and riboflavin as cofactors. Although most of the microorganisms reported to produce azo reductase are facultative anaerobic bacteria (CHUNG and STEVENS, 1993), some obligate anaerobic bacteria producing azo reductase have been isolated from human intestinal microflora (RAFII et al., 1990). RAFII et al. (1990) found that the enzyme was extracellular and did not require induction by an azo dye for production.

LIBRA et al. (1997) presented evidence that the sulfur reducing bacteria (SRB) can play an important role in the reduction of azo dyes. The azo bond is most likely chemically reduced through the sulfide produced by the SRB from the sulfate often found in dye wastewaters. The biomediated chemical reduction rate was much faster than the rate found for fermenting bacteria.

Of course various other compounds besides hydrogen sulfide can chemically reduce azo dyes. BLÜMEL et al. (1997) found that elemental iron can be used as a reducing agent. Organic compounds produced by bacteria such as ascorbic acid and cysteine are also possible reducing agents (LIBRA et al., 1997).

2.7.2.2 Aerobic Degradation of Metabolites

Various authors have investigated the pathways of aromatic amine degradation in pure cultures (MEYER et al., 1979; NÖRTEMANN et al., 1986; HAUG et al., 1991). Determination of the degree of degradation in mixed cultures, however, is difficult. An auxiliary substrate is usually added to facilitate anaerobic decolorization, often in concentrations (measured as DOC) very much higher than that of the metabolites. Reduction then in the effluent DOC is usually not a reliable indication of metabolite degradation. It is possible to follow the metabolites analytically, e.g., with HPLC, however, some of the metabolites are instabile in oxygen, and through autooxidation are not measurable for a longer time although no change in the DOC concentration occurs (SOSATH and LIBRA, 1997).

Using a commercial preparation of a common stable metabolite of reactive dyes, 2-[(4-aminophenyl)-sulfonyl]ethanol, SOEWONDO (1997) showed that the metabolite not only was no longer detectable with HPLC, but that it was degraded by an adapted mixed culture.

2.7.3 Treatment of Wastewater Containing Textile Dyes

Treatment systems based on anaerobic reduction and aerobic degradation of azo dyes can be carried out sequentially in separate continuous flow reactors, simultaneously in one reactor with anaerobic and aerobic zones, or sequentially in one batch reactor. GLÄSSER (1992) used two reactors in series with bacteria immobilized on sand to degrade the azo dye Mordant Yellow 3. Sequential two-stage continuous flow treatment of highly concentrated synthetic azo dye wastewater (C.I. Reactive Violet 5) and wastewater from a textile finishing company containing mainly C.I. Reactive Black 5, decolorized the wastewater (98%, 60% respectively), but the DOC of the effluent did not decrease below the DOC of the dye (LIBRA and SOSATH, 1997). An ozonation stage followed by an aerobic polishing stage was added, resulting in further decolorization and degradation of the dyes. A similar treatment scheme with four steps was applied treating a highly concentrated textile finishing wastewater, however, in a sequencing batch system (LIEBELT, 1997; HEMMI et al., 1997). Biological removal accounted for most of the DOC removal (94%) in contrast to the results of LIBRA and SOSATH (1997) with only 50–60% of the DOC removal due to the 3 biological stages.

KUDLICH et al. (1996) carried out anaerobic reduction and aerobic degradation simultaneously with a culture of *Shingomonas* sp. BN6 immobilized in alginate, with the anaerobic reduction taking place towards the center of the immobilized biomass spheres and the aerobic degradation more at the outer edge. HARMER and BISHOP (1992) found the decolorization in an aerobic biofilm reactor could be influenced by changing the bulk dissolved oxygen concentration.

3 Aerobic Oxidation of Inorganic Substances

3.1 Nitrification

3.1.1 Properties, Use, and Environmental Problems

Nitrogen is dissolved in various industrial effluents mostly as ammonium. Compared with municipal wastewater with concentrations of 40–60 mg L^{-1} NH_4–N, these concentrations are much higher (Tab. 12). The following problems arise if wastewater with NH_4^+ is discharged into rivers, lakes and the sea:

(1) Ammonium is oxidized by chemolithoautotrophic bacteria to nitrite and nitrate leading to a decrease in dissolved oxygen concentration and finally fish death.
(2) Ammonium and ammonia are in chemical equilibrium. With increasing temperature and pH more and more ammonia is produced which is toxic to fishes.
(3) Nitrate as the oxidation product of ammonium stimulates the growth of algae and thereby eutrophication.
(4) High concentrations of nitrate and nitrite in drinking water cause methemoglobinemia in babies and favor formation of carcinogenic nitrosamines.

For these reasons ammonium, nitrite, and nitrate must be removed from wastewaters.

3.1.2 Stoichiometry and Yield Coefficients

Chemolithoautotrophic bacteria oxidize ammonia to nitrite (e.g., *Nitrosomonas euro-*

Tab. 12. Origin of Some Highly Loaded Wastewaters and Ammonia Concentrations (WIESMANN, 1994)

Author	Year	Industry/Products	Concentration [$mg\ L^{-1}\ NH_4-N$]
KOZIOROWSKI and KUCHARSKI	1972	cokery	800–1000
PASCIK	1982	oil refinery	450– 630
HUTTON and LA ROCCA	1975	fertilizer	200– 940
BRAUN	1982	livestock cattle	500–2300
BASU	1974	distillery	114– 380
ADAMS and ECKENFELDER	1977	cellulose, paper	260
BROWN	1975	pharmaceuticals	475

paea) and nitrite to nitrate (e.g., *Nitrobacter winogradskyi*). *Nitrosomonas* is a $1\times1.5\ \mu m$ short bacillus (LOVELESS and PAINTER, 1968). All nitrifiers using NH_4^+ show the following catabolism

$$NH_4^+ + \frac{3}{2}O_2 \rightarrow NO_2^- + 2H^+ + H_2O, \quad \Delta G^{0\prime} = -240\text{–}350\ kJ\ (mol)^{-1} \tag{7}$$

and catabolism as well as anabolism

$$55\,NH_4^+ + 76\,O_2 + 109\,HCO_3^- \rightarrow C_5H_7NO_2 + 54\,NO_2^- + 57\,H_2O + 104\,H_2CO_3 \tag{8}$$

The "molecular mass of *Nitrosomonas*" $C_5H_7NO_2$ means the average elementary composition of 113 g mol^{-1}. From Eq. (8) 4 yield coefficients can be deduced.

$$Y^0_{B/NH_4-N} = \frac{113}{14\cdot55} = 0.15\ g\ oTS\ (g\ NH_4-N)^{-1} \tag{9}$$

$$Y^0_{B/O_2} = \frac{113}{32\cdot76} = 0.047\ g\ oTS\ (g\ O_2)^{-1} \tag{10}$$

$$Y^0_{O_2/NH_4-N} = \frac{76\cdot32}{55\cdot14} = 3.16\ g\ O_2\ (g NH_4-N)^{-1} \tag{11}$$

$$Y^0_{HCO_3^- - NH_4-N} = \frac{109\cdot61}{55\cdot14} = 8.6\ g\ HCO_3^-\ (g NH_4-N)^{-1} \tag{12}$$

If maintenance metabolism is considered using the model of HERBERT (1958), there are obtained for the real growth rate

$$r_B = \mu c_B - k_d c_B \tag{13}$$

and for the real oxygen consumption rate

$$r_{O_2} = \frac{1}{Y_{B/O_2}}(\mu c_B + k_e c_B) \tag{14}$$

From these equations it follows for the apparent yield coefficients

$$Y_{B/NH_4-N} = Y^0_{B/NH_4-N}\left(1 - \frac{k_d}{\mu}\right) \tag{15}$$

and $$Y_{O_2/NH_4-N} = Y^0_{O_2/NH_4-N}\left(1 + \frac{k_e}{\mu}\right) \tag{16}$$

with μ specific growth rate, k_d decay coefficient, and k_e coefficient for endogeneous respiration.

Nitrobacter sp. depends on the product of ammonia oxidation and shows the following catabolism

$$NO_2^- + \frac{1}{2}O_2 \rightarrow NO_3^-, \Delta G^{0\prime} = -65\text{–}90\ kJ \tag{17}$$

and catabolism as well as the anabolism

$$200\,NO_2^- + NH_4^+ + 4\,H_2CO_3 + HCO_3^- + 96\,O_2 \rightarrow C_5H_7NO_2 + 3\,H_2O + 200\,NO_3^- \tag{18}$$

From Eq. (18) the following yield coefficients can be calculated

$$Y^0_{B/NO_2-N} = \frac{113}{200\cdot14} = 0.04\ g\ oTS\ (g\ NO_2-N)^{-1} \tag{19}$$

$$Y^0_{B/O_2} = \frac{113}{96\cdot32} = 0.036\ g\ oTS\ (g\ O_2)^{-1} \tag{20}$$

$$Y^0_{O_2/NO_2-N} = \frac{96\cdot32}{200\cdot14} = 1.1\ g\ oTS\ (g\ O_2)^{-1} \tag{21}$$

For the apparent yield coefficients equations according to Eqs. (15, 16) can be obtained.

Comparing Eqs. (7) and (17) the growth rate of *Nitrosomonas* sp. must be obviously larger by a factor of 4 as that of *Nitrobacter* sp. resulting from the higher energy yield which can be gained from the oxidation of NH_4^+ compared with NO_2^-. This is also recognizable if the following yield coefficients are compared and added

$$Y^0_{O_2/NH_4-N+NO_2-N} = 3.16 + 1.1 = 4.26 \; \frac{g\,O_2}{g\,N}. \quad (22)$$

3.1.3 Kinetics

The real substrates are NH_3 and HNO_2. Their concentrations can be calculated from dissociation equilibria given by

$$c(HN_3-N) = \frac{c(NH_4-N)\cdot 10^{pH}}{K_a} \quad (23)$$

$$\text{and} \quad c(HNO_3-N) = \frac{c(NO_2-N)}{K_b\,10^{pH}} \quad (24)$$

$$\text{with} \quad K_a = \exp\left(\frac{6344}{273+T}\right) \quad (25)$$

$$\text{and} \quad K_b = \exp\left(\frac{2300}{273+T}\right) \quad (26)$$

as equilibrium constants (WIESMANN, 1994). For higher concentrations of NH_3 or HNO_2, substrate inhibition can be observed and described by Haldane kinetics (HALDANE, 1965). For the specific growth rates of *Nitrosomonas* sp. and *Nitrobacter* sp. it follows

$$\mu = \mu_{max} \frac{c(NH_3-N)}{K_{SH} + c(NH_3-N) + \dfrac{c(NH_3-N)^2}{K_{iH}}} \cdot \frac{c'}{K'+c'} \quad (27)$$

$$\text{and} \quad \mu = \mu_{max} \frac{c(HNO_2-N)}{K_{SH} + c(HNO_2-N) + \dfrac{c(HNO_2-N)^2}{K_{iH}}} \cdot \frac{c'}{K'+c'}. \quad (28)$$

The rates for growth of *Nitrosomonas* sp., ammonium oxidation and oxygen consumption can be obtained by using Eqs. (9, 10, 27, 28)

$$r_{BG} = \mu\, c_{BNS} \quad (29)$$

$$r_{NH_4-N} = \frac{\mu}{Y^0_{B/NH_4-N}}\, c_{BNS} \quad (30)$$

$$r_{O_2G} = \frac{\mu}{Y^0_{B/O_2}}\, c_{BNS} \quad (31)$$

Corresponding relations follow for *Nitrobacter* sp. Some important properties of nitrification can be discussed (Tabs. 13, 14) with the help of Eqs. (23–31):

(1) With increasing pH, $c(NH_3)$ increases and $c(HNO_2)$ decreases. Finally, the growth rate of *Nitrosomonas* sp. is reduced by substrate inhibition and that of *Nitrobacter* sp. by substrate limitation.
(2) With increasing temperature, $c(NH_3)$ decreases and $c(HNO_2)$ increases.
(3) As a result of a higher K' (s. Tab. 13) nitrite enrichment can be observed for decreasing oxygen concentrations below 2–3 mg L^{-1}.

3.1.4 Nitrification in Large Scale Plants

The development of large scale nitrification technology can be demonstrated by discussing two examples.

Mixed effluents of a chemical plant (Du Pont due Nemours & Co. Inc.; Charleston, SC)

Tab. 13. Kinetic Coefficients of NH_4 Oxidation by *Nitrosomonas* sp. in Activated Sludge

Author	Year	T [°C]	μ_{max} [h^{-1}]	K_{SH} [mg L^{-1}]	K_{iH} [mg L^{-1}]	K′ [mg L^{-1}]
KNOWLES et al.	1965	30	0.051	0.084		
BERGERON	1978	25	0.0064	0.2	35	1.8
NYHUIS	1985	15–17	0.04	0.056	33	0.5
DOMBROWSKI	1991	25	0.014	0.71	540	0.29
BATEN	1993	30		0.34	46	1.9

with concentrations of c_{SO}=2,180 mg L^{-1} COD and c_{NO}=200 mg L^{-1} NH_4—N were treated in open totally mixed activated sludge tanks with two bioreactors, each of 2,842 m^3 (SHADE, 1977).

For pH stabilization at 7.2 and nutrient supply NaOH and P salt were added. For a mean sludge age t_{vB}=13–16 d and a mean hydraulic residence time t_v=2.2 d ammonia concentration could be reduced to c_{Na}<1 mg L^{-1} NH_4—N resulting in a nitrification rate of r_N=0.07 kg NH_4—N (m^3 d$)^{-1}$. These high t_v and low r_N are caused by the following reasons:

(1) Because of the high COD content, the activated sludge contains about 95% heterotrophs and only 5% nitrifiers. In order to stabilize the mean sludge concentration at 3.5 g TS L^{-1} a relatively high amount of surplus sludge must be removed, which contains about 0.175 g TS L^{-1} nitrifiers. The reaction rate is limited by a relatively small concentration of nitrifiers!

(2) Total bacteria concentration is relatively low. Its increase is limited by the low thickening ratio in sedimentation tanks of activated sludge plants. In order to increase bacteria concentration, bacteria must be immobilized on solid support material.

These two disadvantages can be avoided by using cascades of fluidized bed reactors (HEIJNEN, 1988). A waste stream from yeast production was treated anaerobically in the first two stages. This effluent was loaded particularly by 120 mg L^{-1} NH_4—N and 50–300 mg L^{-1} COD and treated in an airlift reactor. 200 g L^{-1} of sand particles of 0.1–0.3 mm diameter were used. In the lab scale reactor biofilms of a thickness of 50–100 μm forming a total biomass of 15–25 g L^{-1} VSS could be established, resulting in a nitrification rate of r_N=1.8 kg N (m^3 d), which is higher than that of the activated sludge plant mentioned above by a factor of 26. Hydraulic mean residence time could be reduced to only 2.0 h.

Tab. 14. Kinetic Coefficients of NO_2 Oxidation by *Nitrobacter* sp. in Activated Sludge

Author	Year	T [°C]	μ_{max} [h^{-1}]	K_{SH} [mg L^{-1}]	K_{iH} [mg L^{-1}]	K′ [mg L^{-1}]
KNOWLES et al.	1965	30	0.058	$1.9\cdot10^{-4}$		
BERGERON	1974	25	0.005	$2.5\cdot10^{-4}$	35	1.4
NYHUIS	1985	15–17	0.016	$1.7\cdot10^{-4}$	0.15	0.75
DOMBROWSKI	1991	25	0.019	$0.39\cdot10^{-4}$	0.25	1.10

3.2 Sulfide Oxidation

3.2.1 Properties, Use, and Environmental Problems

Hydrogen sulfide H_2S is highly soluble in water (6,720 mg L^{-1}) but already a small amount, approximately 1 mg L^{-1}, is toxic for fish and algae. Its very high Henry coefficient and human toxicity (maximum workplace concentration: 10 ppm) makes monitoring and good ventilation essential for areas with potential H_2S emissions (ROTH, 1990).

Sulfide is produced under anaerobic conditions by sulfate reducing bacteria (SRB). This can take place in sewer systems, anaerobic digestors or in anaerobic treatment of wastewaters. H_2S is very corrosive to gas piping. Since the products of combustion can damage the engine and severely corrode exhaust gas heat recovery equipment, it must be removed from biogas. H_2S not only damages equipment in the biogas system, at concentrations exceeding 200 mg L^{-1}, it can also inhibit methanogenic activity. Wastewaters with high sulfate concentrations, e.g., textile dyes production wastewaters, textile dye baths, and acid mine drainage, can have concentrations ranging as high as 16 g L^{-1} $SO_4{-}S$. Moreover, tannery wastewaters contain both sulfide and sulfate in concentrations ranging from 200–1,500 mg L^{-1} S and 300–1,000 mg L^{-1} $SO_4{-}S$, respectively (GENSCHOW and HEGEMANN, 1995). Anaerobic treatment of these wastewaters can suffer severe inhibition of carbon removal and biogas production.

Similar to the oxidation of ammonia to nitrate, where the oxidation product itself can be a problem that requires further action, the product of sulfide oxidation, sulfate, can also cause problems. Sulfur oxidizing bacteria can grow at very low pH values <1.0, and when this takes place within the sewer system, e.g., the sulfuric acid produced can rapidly destroy concrete sewer pipes and pumping installations.

3.2.2 Biological Oxidation

Two types of sulfur oxidizing bacteria (SOB) can be differentiated according to their energy source: the phototrophic green and purple sulfur bacteria and the chemotrophic colorless sulfur bacteria (GAUDY and GAUDY, 1988). However, since most all of the green and purple sulfur bacteria are obligate anaerobes, we will confine the discussion to the aerobic colorless bacteria. Of the 11 genera that make up the colorless sulfur bacteria family, *Thiobacillus* sp. is the most thoroughly studied and most often used in technical applications. All *Thiobacillus* sp. use oxygen as terminal electron acceptor, but one species, *T. denitrificans* is capable of using nitrate as terminal electron acceptor by reducing it to nitrogen gas. Most of the species are chemolithotrophic autotrophs. The conditions for growth for the various species range from acidic to neutral, with most of them able to oxidize elemental sulfur, hydrogen sulfide, and thiosulfate according to the following catabolic reactions:

$$S + H_2O + 1.5\,O_2 \rightarrow SO_4^{2-} + 2\,H^+ \qquad (32)$$

$$H_2S + 2\,O_2 \rightarrow SO_4^{2-} + 2\,H^+ \qquad (33)$$

$$S_2O_3^{2-} + 2\,O_2 + H_2O \rightarrow 2\,SO_4^{2-} + 2\,H^+ \qquad (34)$$

Elemental sulfur can be formed from the oxidation of sulfide and thiosulfate and stored outside the cell, before being further oxidized to sulfate.

3.2.3 Treatment of Wastewater Containing Sulfides

JANSSEN et al. (1995) found that the formation of sulfur and sulfate from the oxidation of sulfide in a fed-batch bioreactor could be controlled instantaneously and reversibly by the amount of oxygen supplied. At sulfide loading rates of up to 2.33 mmol $(L\ h)^{-1}$, both products can be formed at oxygen concentrations below 0.1 mg L^{-1}. Because the microorganisms tend to form sulfate rather than sulfur, the oxygen concentration is not appropriate to optimize the sulfur production.

Sulfur oxidizing bacteria can be used to oxidize the thiosulfate found in photoprocessing waste (PW), when combined with granular activated carbon (GAC) to adsorb the toxic/refractory compounds that inhibit thiosulfate oxidization (LIN et al., 1997). Continuous flow treatment of a 4- to 8-times dilution of PW using a system of sulfur oxidizing bacteria combined with granular activated carbon, which simulates a typical activated sludge wastewater treatment system, was able to oxidize thiosulfate loads of 0.8–3.7 kg $(L\ d)^{-1}$ $S_2O_3^{2-}$ at hydraulic retention times (HRT) of 7.7–1.9 d, respectively. As expected, continuous treatment led to breakthrough and the GAC had to be renewed. FOX and VENKATASUBBIAH (1996) coupled anaerobic-aerobic treatment of high sulfate wastewater with sulfate reduction and biological sulfide oxidation for a pharmaceutical wastewater with a COD concentration of 40000 mg L^{-1} and a sulfate concentration of 5000 mg L^{-1}. Recycling anaerobic effluent through a sulfide oxidizing biological system reduced inhibition in the anaerobic reactor by both reducing inhibitory sulfide concentrations within the reactor and by diluting the influent. The major product of the biological oxidation of sulfide by a *Thiobacillus* sp. appeared to be elemental sulfur. Diluting the influent to 40% and using a HRT of 1 d, COD removal efficiencies were greater than 50% and the conversion of influent sulfate was greater than 95% with effluent sulfide concentrations of less than 20 mg L^{-1}.

Some interesting processes based on sulfide oxidation from related fields are:

(1) the biological removal of hydrogen sulfide from gas streams with
   - (•) bioscrubbers in which the bacteria are suspended in the liquid phase, here the scrubbing solution,
   - (•) biofilters in which bacteria are fixed on solid support material in the gas phase;

(2) the bioleaching of heavy metals and stabilization of municipal sludges.

For example, CHUNG et al. (1996) isolated a heterotrophic *Pseudomonas putida* CH11 from livestock farming wastewater and applied it for the treatment of H_2S containing gas.

BENMOUSSA et al. (1994) studied the bioleaching of heavy metals and stabilization of municipal sludges. Acidification of the medium to very acidic levels (pH 1.5) through the introduction of elemental sulfur and inoculum, resulted in the solubilization of metals initially present in the sludge as well as an appreciable reduction in phosphorus (52%) and in the volatile suspended solids concentration (40–50%). In addition, the sludge which initially had a highly repulsive odor was rendered odorless. This microbial leaching procedure with *Thiobacillus* strains could eventually replace conventional sludge stabilization processes and hence considerably reduce the time and cost of treatment.

4 Aerobic Wastewater Treatment Processes

4.1 Distinctions to Sewage Treatment

Comparing the aerobic treatment of industrial effluents with municipal wastewater, four main differences have to be considered:

(1) The COD is often considerably higher than that of municipal effluents. Therefore, an effective high oxygen transfer and high mean residence times are needed. For COD concentrations $>3,000$ mg L^{-1}, an anaerobic treatment may be more economical.

(2) A higher degree of COD may be nonbiodegradable. Therefore, a combination with chemical or physico-chemical processes is occasionally necessary.

(3) Some of the COD may be desorbed by aeration, so that a waste air cleanup process must be installed. In order to reduce gas flow rates, the use of pure oxygen may also be useful.

(4) The pH of the influent is often too low or too high, resulting in measures for pH correction and control.

4.2 Different Reaction Systems

4.2.1 Activated Sludge Processes

After World War II the activated sludge process, used mainly for the treatment of municipal wastewater before, found more and more application for industrial effluents. By now, the large and often highly specialized demand has resulted in the development of a variety of types of bioreactors and processes: Activated sludge reactors may be constructed

(1) as a CSTR, a plug flow reactor or a cascade (Fig. 7a, b),
(2) as a system with low sludge age and high degradation rate and a high amount of surplus sludge, or with high sludge age, a low degradation rate and a low amount of surplus sludge,
(3) with a low (3–5 m) or a high (15–22 m) height of the reactor,
(4) without or with suspended activated carbon,
(5) with surface aerators or diffused aeration,
(6) with oxygenation by air or by pure oxygen (Fig. 7c),
(7) as a continuous flow process bioreactor or sequencing batch reactor,
(8) as a one-step or a two-step process (Fig. 7d),
(9) with or without nitrification.

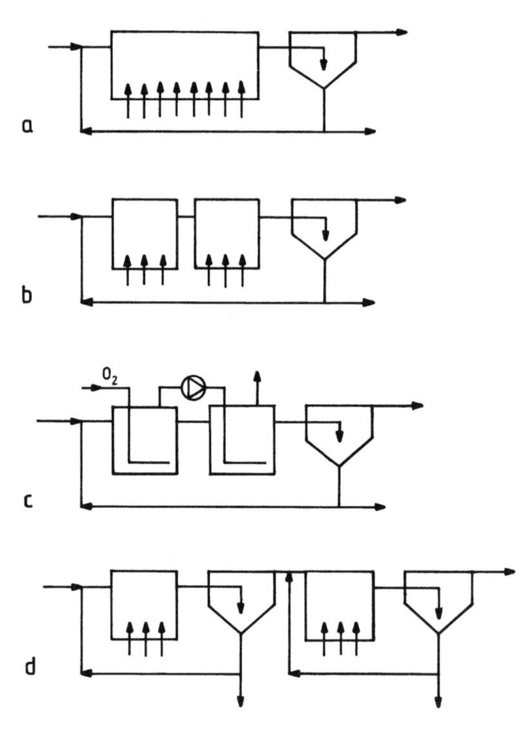

Fig. 7a–d. Activated sludge systems (ASS), **a** conventional process, **b** 2-step cascade for aeration, **c** 2-step cascade for oxygenation, **d** 2-step process with separate sludge systems.

All activated sludge processes are characterized by at least one disadvantage: For an increasing flow rate, either the overflow rate of the settler increases, causing a sludge loss with the overflow, or the underflow rate is increased by the operators, which results in a lower sludge recycling. Therefore, the increasing flow rate reduces the efficiency of the bioreactor in two ways, by the reduction of the residence time *and* the bacteria concentration. In order to minimize this problem, a large settler with a low overflow velocity is frequently used.

4.2.2 Fixed Bed Reactors

Activated sludge processes are characterized by two important disadvantages, the necessity of a sedimentation tank with sludge return and a limited bacteria concentration of only 3–6 g VSS L^{-1} inside the reactor. Both can be avoided by using fixed bed reactors with solid particles such as sand, coke, activated carbon or synthetic particles as support material for bacteria. For lower COD concentrations, a direct aeration or oxygenation is possible with the wastewater flowing upwards or downwards (Fig. 8a). The air must be distributed very carefully and the solid support must be packed equally in order to avoid channeling of the gas and to make a well-balanced oxygen uptake possible. Besides the three-phase flow system with the solid phase as a fixed bed, a more even flow can be achieved if the oxygen is added in a recycle flow in such a way that nearly all the oxygen is absorbed, resulting in an almost totally mixed fixed bed reactor as a two-phase system (Fig. 8b). With the exception

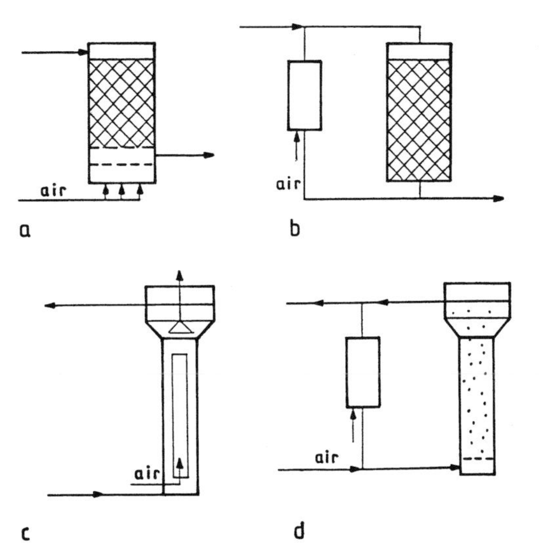

Fig. 8a–d. Fixed bed and fluidized bed reactors, **a** direct aeration in fixed bed reactor, **b** aeration in external absorber in fixed bed reactor, **c** direct aeration in airlift fluidized bed, **d** aeration in external absorber in fluidized bed reactor.

of systems with a very low COD concentration, all fixed bed reactors must be backwashed from time to time to avoid blockages by growing bacteria or other solid particles, which were not previously removed. Although the bacterial concentration in fixed bed reactors can be increased by a factor of two compared to activated sludge reactors, the biofilm often is relatively thick, resulting in oxygen limitation.

4.2.3 Fluidized Bed Reactors

All these disadvantages can be avoided in fluidized bed reactors. Frequently, sand particles with diameters of only 1–4 mm are used in order to reduce the energy for fluidization and to increase the outer surface per mass. The specific surface is relatively high, resulting in a high bacterial concentration of up to 25 g VSS L^{-1}. Nevertheless, the biofilm is thin and the concentration drop inside these films is limited. Additionally, the outer mass transfer resistance can be limited because of the relatively effective flow conditions. Because of the high wastewater up-flow velocity needed for fluidization, a large amount of water must be recycled. In the two-phase fluidized bed reactor, oxygen is transferred to the recirculation stream in an external absorber (Fig. 8c), resulting in relatively low frictional forces at the surfaces of the biofilms. The three-phase fluidized bed reactor is directly aerated (Fig. 8d), and flotation effects with higher frictional forces may occur. Porous synthetic particles may be used with success. In this case a part of the immobilized bacteria cannot be removed by frictional forces.

At the top of a three-phase fluidized bed reactor, the gas has to be separated first, otherwise no effective separation of solid particles is possible. Compared with activated sludge or fixed bed reactors a remarkably higher specific reaction rate can be obtained, allowing a smaller reaction volume to be used. But the process is more susceptible to changing flow rates and loads and, therefore, it needs more servicing.

4.3 A Simple Model for the Activated Sludge Process

The model described here is only valid for completely stirred tank reactors (CSTR) and limited to the reaction system. The properties of the sedimentation tank are not considered. The simplest model describes only carbon removal and bacterial growth in non-dynamic systems without oxygen limitation. It consists of the balances of the organic substrate (LAWRENCE and MCCARTY, 1970; BENEFIELD and RANDALL, 1980)

$$0=\dot{V}_{\mathrm{M}}(c_{\mathrm{SM}}-c_{\mathrm{Sa}})-\frac{\mu_{\max}c_{\mathrm{Sa}}c_{\mathrm{Ba}}}{Y^{0}_{\mathrm{B/S}}(K_{\mathrm{S}}+c_{\mathrm{Sa}})}V_{\mathrm{R}} \tag{35}$$

and the bacteria (Fig. 9)

$$0=\dot{V}_{\mathrm{M}}(c_{\mathrm{BM}}-c_{\mathrm{Ba}})+\left(\frac{\mu_{\max}c_{\mathrm{Sa}}c_{\mathrm{Ba}}}{K_{\mathrm{S}}+c_{\mathrm{Sa}}}-k_{\mathrm{d}}c_{\mathrm{Ba}}\right)V_{\mathrm{R}} \tag{36}$$

for Monod kinetics and bacteria decay where c_{S} is the concentration of the organic substrate

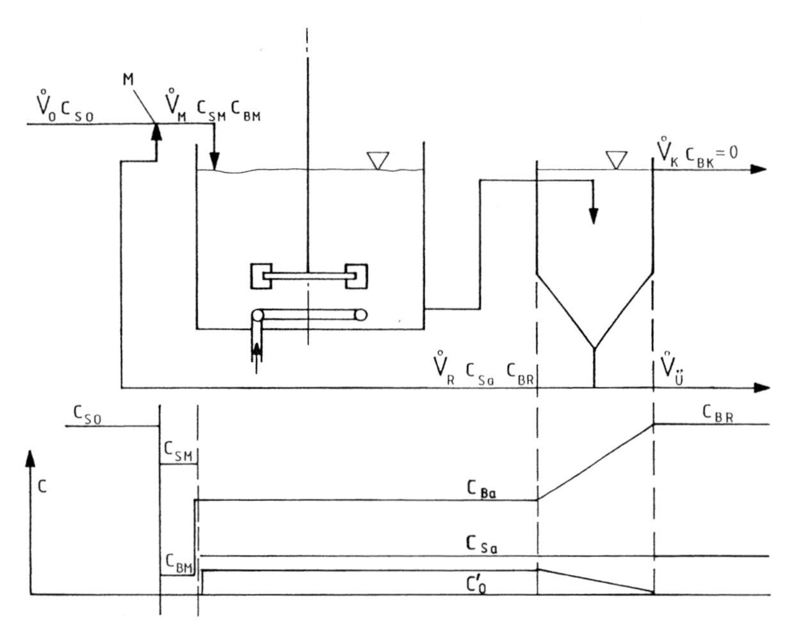

Fig. 9. Carbon removal in a CSTR with sedimentation and sludge return.

in mg L^{-1} COD or DOC and c_B is the concentration of heterotrophic aerobic bacteria in g L^{-1} VSS (volatile suspended solids).

From balances on the mixing point M we obtain

$$\dot{V}_M = \dot{V}_0 + \dot{V}_R \tag{37}$$

$$c_{SM} = \frac{c_{So} + n_R c_{Sa}}{1 + n_R} \tag{38}$$

and

$$c_{BM} = \frac{n_R n_E c_{Ba}}{1 + n_R} \tag{39}$$

where the recycle number n_R is

$$n_R = \dot{V}_R / \dot{V}_0 \tag{40}$$

and the thickening number n_E is

$$n_E = c_{BR} / c_{Ba} . \tag{41}$$

From Eqs. (35–41) the substrate concentration of the effluent follows

$$c_{Sa} = \frac{K_S k_d t_{vB} + K_S}{t_{vB}(\mu_{max} - k_d) - 1} \tag{42}$$

where the sludge age or the mean residence time of bacteria can be described as

$$t_{vB} = \frac{t_v}{1 + n_R - n_E n_R} . \tag{43}$$

From the introduction of the sludge age it follows that c_{Sa} now depends only on one operating parameter t_{vB}! For $c_{Sa} = c_{So}$ the critical sludge age t_{vBK} is

$$t_{vBK} = \frac{K_S + c_{So}}{c_{So}(\mu_{max} - k_d) - K_S k_d} . \tag{44}$$

The bacteria are only able to grow for

$$t_{vB} > t_{vBK} .$$

For $t_{vB} < t_{vBK}$

all bacteria are washed out!

Example
An effluent of 2,4-DNP with 100 mg L^{-1} DOC has to be treated in an activated sludge plant to an effluent concentration of 1 mg L^{-1} DOC 2,4-DNP. From Sect. 2.3, Tab. 2, $\mu_{max}=0.1\ h^{-1}$, $K_S=2.6–4.9\approx 3.8$ mg L^{-1} DOC, and $k_d=0.001\ h^{-1}$ are assumed.

From Eq. (48) t_{vB} is obtained

$$t_{vB}=\frac{K_S}{c_{Sa}(\mu_{max}-k_d)-K_S k_d}\approx 38\ h. \qquad (45)$$

With typical values of $n_R=0.4$ and $n_E=3$ from Eq. (43) $t_v=7.6$ h can be calculated.

The critical sludge age calculated from Eq. (44) is $t_{vBK}=10.4$ h and the critical mean residence time $t_{vk}=2.08$ h.

This simple model can be enlarged stepwise to include a variety of deviations from the former assumptions:

(1) non-steady state,
(2) different kind of substrates,
(3) different kind of bacteria,
(4) oxygen limitation,
(5) inhibition by different substances,
(6) nitrification, denitrification, etc.

This model expands to the activated sludge model No. 1 (GUJER, 1985; HENZE et al., 1986) with 11 balance equations and to the activated sludge model No. 2 (HENZE et al., 1995) with 19 balance equations. The kinetics used are based on the Monod kinetic. Nevertheless, they are very complicated. Eq. (4) of the activated sludge model No. 2, the balance of heterotrophic organisms growing on fermentable substrates, contains a specific growth rate with 6 limitations! It is understandable that a high number of coefficients have to be adjusted which makes it easier to describe experimental or practical results but increases the numerical calculation procedure!

5 Thermophilic Aerobic Treatment of High-Strength Wastewater and Sludge

5.1 Thermophilic Bacteria

All bacteria can be grouped according to their maximum specific growth rate as a function of temperature in psychrophilic, mesophilic and thermophilic strains with maximal growth between 15–20, 20–37 and 50–65°C (MÜLLER, 1980). The use of thermophilic bacteria in wastewater and sludge treatment processes is characterized by three important advantages

(1) higher substrate removal rates,
(2) lower amounts of surplus sludge, and
(3) partial disinfection of the remaining organic solids.

Thermophilic bacteria were described in detail by CROSS (1968) and DEGRYSE (1976). They are mainly gram-positive, aerobic, mostly rod-like and spore forming bacteria.

The most important genus is *Bacillus* with *B. stearothermophilus* (65°C) and *B. coagulans* (60°C) (WOLF and BARKER, 1968). At least 7 other genera are able to grow at a temperature of 55°C (WALKER and WOLF, 1971). Information about a division of *B. stearothermophilus* and *B. coagulans* into lower groups is given by STELDERN et al. (1974).

5.2 Aerobic Thermophilic Wastewater Treatment

In thermally insulated bioreactors with volumes >1 m^3 the temperature of the aerated high-strength wastewater can be increased by self-heating up to 65°C. In this temperature region only thermophilic bacteria can grow. But the aeration rate must be optimized, otherwise the temperature decreases as a result of the heat loss by evaporation and by convective

Tab. 15. Studied Thermophilic Wastewater Treatment Systems with Sludge Recycle in Lab and Pilot Scale (selection from BLOCK, 1992)

Author	Year	Reactor	Heating	Operation	Substrate	t_v [d] per step	Temperature [°C] per step	pH per step	Basis	c_{SO}[a] [g L^{-1}]	B_R[a] [mg $(L\ h)^{-1}$]	α % (summarized)	r_S [mg $(L\ h)^{-1}$ per step]
LOLL	1976	aeration by circulation 1 m^3 per step	auto-thermic	continuous 3 steps	sugar beat 20°C, pH = 4.1	1.67	43/51/51	5.9/8.6/9.3	BOD_5 COD	18.3 26.2	 654/417/200	50/80/96 37/70/82	230/138/70 240/215/83
					pig growth 17°C, pH = 7.0	1.67	50/42/38	9.0/9.3/9.4	COD COD	17.2 22.5	 563/292/233	78/95/97 48/58/61	338/70/10 270/58/18
JACKSON	1982	submers + recycle 750 L	heated	continuous 1 step	paper industry pH = 7.0	0.5–10 (sludge age)	46–48	7.2	BOD_5	2.5		80–90	82
SCHEFFLER et al.	1982	ASR + recycle 20 L	heated	continuous 1 step	yeast production pH = 4.4–5.6	5–15 (sludge age)	50–56		BOD_5 COD	16–20 25–30	 208–417	 75–85	 333 (B_R = 417)
KIESE et al.	1986	ASR + recycle 28 L per step	heated	continuous 2 steps	yeast production pH = 4.4–5.6		50–55		BOD_5 COD	13–22 19–28	 542–1125	 70–85	 758 (B_R = 1125)
GARIÉPY et al.	1989	CSTR 1.5 L	heated	semi-continuous	slaughter-house	0.25–1.25	58		COD	2.3–2.8	75–217	69–94	276
CUMMINGS and JEWELL	1977	semi-continuous 29.7 m^3	auto-thermic		dairy	1.31 2.93 3.43 5.47 9.48	31 49		COD	27.0	20.1 482 9.0 216 8.2 197 5.0 120 2.6 67	0.33 0.63 0.64 0.62 0.74	

[a] COD.

cooling. Wastewater produced by filtration and/or centrifugation of thermally conditioned sludges from municipal wastewater treatment plants has a COD of 5,000–12,000 mg L^{-1}, with about 50% of the COD present as lower fatty acids. It can be treated at T=60°C by aerobic thermophilic bacteria (BLOCK, 1992), which can only use the fatty acids as carbon and energy source. Strains with lower growth rates, capable of degrading the remaining organic compounds, have to be immobilized or recycled effectively. Otherwise, only compounds such as fatty acids can be mineralized.

Tab. 16 shows some results for kinetic coefficients and $Y_{B/S}$. Remarkably high values were measured for μ_{max}, when the easily metabolized substrates hexadecane (MATELES et al., 1965) or glucose (MATSCHÉ and ANDREWS, 1973) were used. At temperatures close to 60°C, μ_{max} reaches a maximal value of nearly 1 h^{-1} or a generation time of 40 min. The decay coefficient k_d increases with temperature at a faster rate compared to μ_{max}. For 64.5°C k_d is only 30% lower than μ_{max}.

Although the substrate is consumed by the growing bacteria at a high rate, the effective growth rate is low as a result of bacterial decay and new substrate is formed from lysis products. The apparent yield coefficient is

$$Y_{B/S}=Y^0_{B/S}\left(1-\frac{k_d}{\mu_{max}}\right) \tag{46}$$

which gives

$$Y_{B/S}\approx 0.1 \text{ g SS (g COD)}^{-1}$$

for $Y^0_{B/S}=0.45$, $\mu_{max}=0.66$ h^{-1} and $k_d=$ 0.441 h^{-1} (Tab. 16, MATSCHÉ and ANDREWS, 1973; T=64.5°C), which means that only ~20% of the bacterial mass or surplus sludge is formed compared to that at mesophilic temperatures.

5.3 Thermophilic Autothermic Sludge Stabilization and Hygienization

Large amounts of organic sludge are produced in the food and drinks industry. Furthermore, surplus sludge is produced in large

Tab. 16. Kinetic and Stoichiometric Coefficients for Aerobic Thermophilic Wastewater Treatment (selection from BLOCK, 1992)

Author	Year	Reactor	Culture	Substrate	Temperature [°C]	μ_{max} [h^{-1}]	K_S [mg L^{-1}]	k_d [h^{-1}]	k_e [h^{-1}]	$Y_{B/S}$ [g SS (g S)$^{-1}$]
MATELES et al.	1967		*B. stearothermophilus*	hexadecane	37	0.25				
					45	0.49				
					55	1.16				
					60	1.14				
					65	1.06				
					70	0.85				
MATSCHÉ and ANDREWS	1973	CSTR	*Bacillus* sp.	glucose	45.5	0.41		0.027	0.01	0.46 COD
					52	0.53		0.047	0.02	0.44
					56	0.69		0.071	0.03	0.45
					60	0.90		0.118	0.04	0.45
					62	0.96				
					64.5	0.66		0.441	0.06	
GARIÉPY et al.	1989	CSTR	mixed culture	slaughterhouse	45	0.24	46	0.022		0.35 COD
				effluents	52	0.25	30	0.013		0.30
					58	0.42	990	0.033		0.32

quantities in biological wastewater treatment plants. In order to relieve landfills, the sludge has to be stabilized, disinfected, and recycled as fertilizer for agriculture if it cannot be used as food for swine [FRG (1994), Gesetz zur Vermeidung, Verwertung und Beseitigung von Abfällen vom 27.9.1994 with legal force for the FRG, and Verordnung (EWG) Nr. 259/93 the EU decree from 1.2.1993]. Most published results are for the stabilization of primary and surplus sludge from wastewater treatment plants. Tab. 17 shows typical results (YI, 1985). Frequently, the thermally insulated reactors are two-step semicontinuous systems. For initial concentrations of 30 g L^{-1} VSS about 30–50% are removed at total mean residence times between 3 and 10 d. Especially in the larger tanks with volumes >20 m^3, high temperatures of 61–65°C can be obtained (BREITENBÜCHER, 1982). Nevertheless, the large systems are frequently not totally mixed and oxygen limitation occurs, resulting in a lower space loading and substrate removal. Using a highly efficient injector aeration system, no significant increase in the substrate removal rate was obtained above a specific power input of 6–8 kW m^{-3} (PONTI et al., 1995). A mean residence time of less than 1 d allowed organic matter removal up to 40% with a specific power consumption of 10 kWh kg^{-1} COD oxidized!

Obviously, an optimization problem has to be solved in large scale autothermic stabilization tanks, and the local costs for electric power may be important. If the sludge loading rate decreases a lower energy input may be sufficient. Disinfection during thermophilic stabilization processes was studied by several authors. At 48.5°C the viable eggs of *Ascaris suum* could be reduced in sewage sludge by 84.6–91.6%. At 52.5–62.5°C, a nearly 100% lethal effect on eggs of the model nematode *Toxocara canis* could be observed (PLACHY et al., 1993). The description of the kinetics is based on the rate limiting hydrolysis of solid particles by bacterial exoenzymes. Normally, this is described by first order kinetics with regard to the concentration of organic matter.

$$r = k c \tag{47}$$

Fig. 10 shows a plot of $\ln k$ vs. T^{-1} (YI, 1985) for mesophilic and thermophilic aerobic sludge stabilization. Although the results are derived from altogether 15 authors and refer to surplus sludge (10 authors) as well as primary and surplus sludge (5 authors), two different linear reaction ranges can be distinguished for mesophilic and thermophilic organisms! Kinetic coefficients were published by KAMBHU (1971), MATSCHÉ and ANDREWS (1973), and GARIEPY et al. (1989).

6 Process Combinations

6.1 Anaerobic Wastewater Treatment Followed by Aerobic Treatment

The effluent of an anaerobic treatment plant is normally characterized by a COD, which is

Tab. 17. Process Parameter for Thermophilic Autothermic Sludge Stabilization (YI, 1985)

Author	Year	Reactor	V_R [m^3]	Substrate	c_0 [g L^{-1} VSS]	B_R [kg m^{-3} d^{-1} VSS]	t_v [d]	α —	T [°C]
SMITH et al.	1975	CSTR	0.2	primary and surplus sludge	29.2	7.2	4.0	0.30	50.2
MATSCH and DRNEVICH	1977	semicontinuous 2 steps	—	primary and surplus sludge	23.7–30.7	5.1–8.5	3.0–5.0	0.29–0.45	45–57
BREITENBÜCHER	1982	semicontinuous 2 steps	2×24	surplus sludge	28.5	3.6	8.0	0.34	61–65

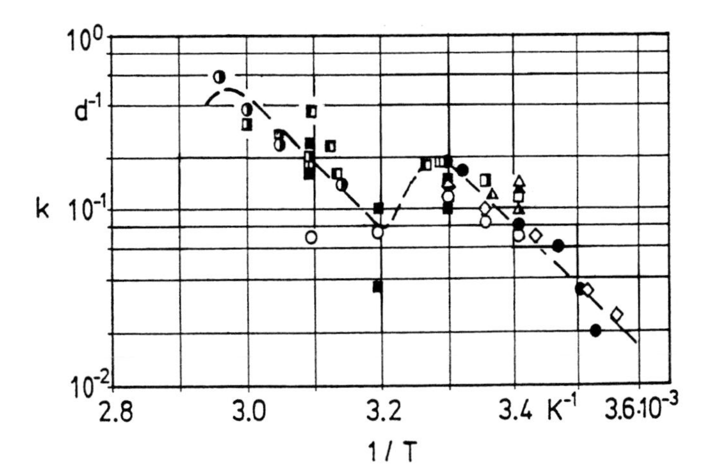

Fig 10. Reaction rate coefficient k of first order aerobic sludge stabilization as a function of temperature for the mesophilic and thermophilic range (YI, 1985).

too high to be directly discharged into a receiving water. Therefore, it is frequently discharged into a municipal sewage system and posttreated after mixing in an activated sludge plant (Fig. 11a) or in a trickling filter. In cases without such a possibility, the COD of the anaerobically treated effluent has to be reduced in a second aerobic stage in order to meet local laws. Frequently, the anaerobically treated effluent is mixed with other wastewaters with lower COD, resulting in a higher flow rate. Therefore, the loading of the aerobic stage in such cases is in the range of the anaerobically treated wastewater. An example was given by Sulzer (Winterthur) of a treatment plant for a paper and chemical pulp factory, which discharges into the Danube (Germany). The first anaerobic bioreactor has to be heated to 32–35°C, otherwise the reaction rate will be too low for a high COD reduction and biogas formation. The anoxic and aerobic stage operates at 15–25°C.

After acidification in a closed vessel and methanization in 3 fixed bed reactors, the wastewater is aerated in 4 closed activated sludge reactors after mixing with several other effluents. Each activated sludge reactor is followed by a sedimentation tank with a sludge recycle system and surplus sludge removal.

The flow rate and influent COD for the anaerobic stage are 500 $m^3 h^{-1}$ and 3,300 mg L^{-1} compared to 1,083 $m^3 h^{-1}$ and 1,200 mg L^{-1} for the aerobic stage.

The anaerobic pretreatment results in three advantages compared to an aerobic pretreatment:

(1) only a small amount of surplus sludge is produced,
(2) the biogas can be used as an energy carrier,
(3) a large amount of energy is saved that would be necessary for aeration.

6.2 Anaerobic Wastewater Treatment Followed by Denitrification and Nitrification

Effluents polluted by high amounts of organics and ammonia result from the industrial production of fertilizers and the refining of coal and mineral oil. In some effluents from the agriculture industry, ammonia is formed from the nitrogen contained in organic compounds during anaerobic or aerobic biodegradation (swine and cattle breeding, rendering plants). In some cases an anaerobic pretreatment combined with a nitrification stage may be a suitable process. If denitrification is necessary, a part of the aerobically treated water (e.g., 75%) must be recycled to a denitrification stage between the anaerobic and aerobic reactors in order to supply the heterotrophic denitrifying bacteria with organics (Fig. 11b). For a 75% denitrification, the recycle flow must be increased by a factor of 3. For cases with a high ammonia concentration, e.g., of 200 mg L^{-1} NH_4-N, a nitritate concentration of 40 mg L^{-1} NO_3-N would remain by using such a process configuration. In order to meet European laws, total nitrogen concentration (NO_3-N, NH_4-N, NO_2-N, organic N) must be reduced to 18 mg L^{-1} N before discharging treated effluents into receiving waters. In such cases a second denitrification stage may be used after the nitrification reactor, and a N-free carbon and energy source has to be added, e.g., methanol, in order to avoid further nitrogen addition (Fig. 11b).

For lower concentrations of carbon and ammonia an oxidation ditch may show advantages: The oxygen input by brush rotators can be adjusted such that an oxic period is followed by an anoxic one before the circulated wastewater arrives at the second rotator. An oxic and anoxic process following each other can also be realized in sequencing batch reactors (SBR). Typical for this process is the change of the filled reaction volume: The cycle begins with the addition of wastewater for carbon and ammonia oxidation. The volume is then increased after nitrification by addition of wastewater to begin denitrification. After sedimentation the volume will be reduced to a level, such that the bacteria remain in the reactor to start the next cycle with a relatively high bacteria concentration.

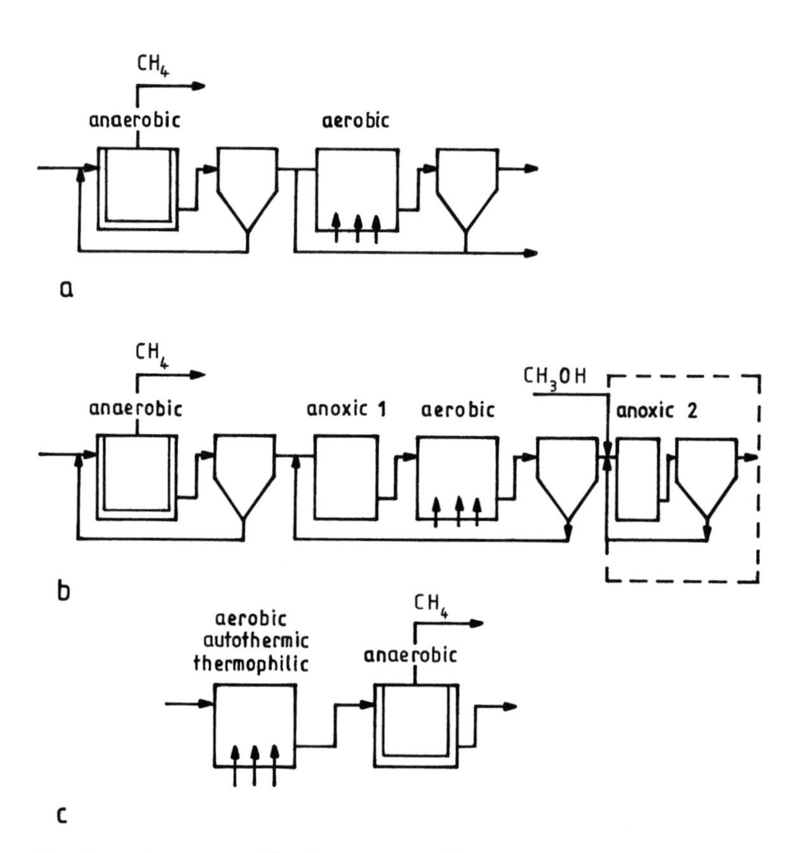

Fig. 11a–c. Process combinations, **a** anaerobic and aerobic wastewater treatment, **b** anaerobic wastewater treatment, denitrification and nitrification with and without a second denitrification step, **c** aerobic autothermic thermophilic and anaerobic mesophilic sludge stabilization – batch and semibatch processes.

6.3 Thermophilic Aerobic Autothermic Treatment Followed by Mesophilic Anaerobic Sludge Stabilization and Hygienization

Surplus sludge treatment is normally associated with municipal wastewater treatment, however, the treatment of industrial wastewaters with biological aerobic processes can also produce a large amount of sludge. Wastewaters containing suspended solids and colloids that require similar treatment can often be found in the food industry (beverage, fruits, vegetable, potato etc.). These organic residues (surplus sludges and organic wastes) must be treated and dewatered before landfilling, use in agriculture or burning. In some cases a hygienization is necessary.

The thermophilic autothermic stabilization and hygienization is one type of aerobic sludge treatment. The primary or secondary dewatered sludge with solid concentration of 2–5% is mixed and carefully aerated in thermally insulated reactors with a high enough volume (low relation of surface to volume). After the addition of seed sludge containing thermophilic aerobic bacteria, the temperature increases to values of 55–65°C as a result of catabolic energy production (see Sect. 5). After several hours the sludge may be hygienized and partially stabilized. The second stabilization step follows in a mesophilic anaerobic bioreactor with a higher volume, because of the longer time required for a nearly complete stabilization (Fig. 11c). The biogas obtained is used for heating this second stage to nearly 35°C. Other combinations are also used (mesophilic anaerobic, thermophilic aerobic; thermophilic anaerobic, mesophilic anaerobic; only thermophilic autothermic aerobic, only mesophilic anaerobic in one or two steps). Especially the thermophilic processes must be optimized. Process engineering aspects such as energy balances by using on-line measuring methods and modeling may be important.

7 Conclusions

In all probability industrial wastewater technology will go the way from the treatment of a mixture of all effluents to the treatment of single or mixtures of a few effluents. This development will be catalyzed by the necessity to conform to new laws and regulations, and by the intention to save money. The best laws are ineffective if the authorities do not have enough well-educated personnel with adequate pay. And the best laws and the best authority will be relatively ineffective if the economic situation of the companies does not really allow the needed investments. It is in this area of conflict that environmental protection has to be realized. It is obvious that a high amount of creativity and energy is necessary in order to go this way with success.

8 References

ADAMS, C. E., ECKENFELDER, W. W. (1977), Nitrification design approach for high strength ammonia wastewaters, *J. WPCF* **49**, 413–421.

BANERJEE, S., HOWARD, P. H., ROSENBERG, A. M., DOMBROWSKI, A. E., SIKKA, H., TULLIS, D. L. (1984), Development of a general kinetic model for biodegradation and its application to chlorophenols and related compounds, *Environ. Sci. Technol.* **18**, 416–422.

BATEN, R. (1993), Die biologische Reinigung ammoniumreichen Abwassers – Regelung eines mehrstufigen Prozesses. *Thesis*, Institute of Biotechnology, RWTH Aachen, Germany, Berichte des Forschungszentrums Jülich Nr. 3063.

BASU, A. (1974), Characterization of distillery wastewater, *J. WPCF*, 2184–2190.

BATTERSBY, N. S., WILSON, V. (1989), Survey of the anaerobic biodegradation potential of organic chemicals in digesting sludge, *Appl. Environ. Microbiol.* **55**, 433–439.

BAUSUM, H. T., MITCHELL, W. R., MAJOR, M. A. (1992), Biodegradation of 2,4- and 2,6-dinitrotoluene by freshwater microorganisms, *J. Environ. Sci. Health* **A 27**, 663–695.

BENEFIELD, L. D., RANDALL, C. W. (1980), *Biological Process Design Wastewater Treatment.* Englewood Cliffs, NJ: Prentice-Hall.

BENMOUSSA, H., TYAGI, R. D., CAMPBELL, P. G. C. (1994), Heavy metal bioleaching and stabilization of municipal sludges: Effect of the form of elemental sulfur used as substrate, *Rev. Sci. de l'Eau* **7**, 235–250.

BERCHTOLD, S. R., VANDERLOOP, S. L., SUIDAN, M. T., MALONEY, S.W. (1995), Treatment of 2,4-dinitrotoluene using a two-stage system: Fluidized-bed anaerobic granular activated carbon reactors and aerobic activated sludge reactors, *Water Environ. Res.* **67**, 1081–1091.

BERGERON, P. (1978), Untersuchungen zur Kinetik der Nitrifikation, *Karlsruher Berichte zur Ingenieurbiologie* **12**.

BERWICK, P. G. (1984), Physical and chemical conditions for microbial oil degradation, *Biotechnol. Bioeng.* **26**, 1294–1305.

BLOCK, J. (1992), Untersuchungen zur Kinetik der Abwasserreinigung durch aerob thermophile Bakterien, *VDI-Forschungsberichte*, Reihe 15: Umwelttechnik, Nr 98.

BLÜMEL, S., KUDLICH, M., STOLZ, A. (1997), Mögliche Strategien für den mikrobiellen Abbau sulfonierter Azofarbstoffe, *7. Kolloquium des Sonderforschungsbereiches 193: Treatment of Wastewaters from Textile Processing*, 17–18 November 1997, Technical University, Berlin, Germany.

BOSMA, T. N. P., VAN DER MEER, J. R., SCHRAA, G., TROS, M. E. (1988), Reductive dechlorination of all trichloro- and dichlorobenzene isomers, *FEMS Microbiol. Ecol.* **53**, 223–229.

BOUCHEZ, M., BLANCHET, D., VANDECASTEELE, J. R. (1997), An interfacial uptake mechanism for the degradation of pyrene by a *Rhodococcus* strain, *Microbiology* **143**, 1087–1092.

BOUWER, E. J., MCCARTY, P. L. (1982), Removal of trace chlorinated organic compounds by activated carbon and fixed film bacteria, *Environ Sci. Technol.* **16**, 836–843.

BOYLES, D. T. (1984), Biodegradation of topped Kuwait crude, *Biotechnol. Lett.* **6**, 31–36.

BRAUN, R. (1982), *Biogas-Methangärung organischer Abfallstoffe – Grundlagen und Anwendungsbeispiele*. Wien: Springer-Verlag.

BREITENBÜCHER, K. (1982), Aerob-thermophile Stabilisierung von Abwasserschlämmen, *Korrespondenz Abwasser* **29**, 203–207.

BREURE, A. M., VOLKERING, F., VAN ANDEL, J. G. (1990), Biodegradation of polycyclic aromatic hydrocarbons, *DECHEMA Biotechnology Conferences 4*. Weinheim: VCH.

BRILON, C., BECKMANN, W., KNACKMUSS, H.-J. (1981), Catabolism of naphthalenesulfonic acids by *Pseudomonas* sp. A3 and *Pseudomonas* sp. C22, *Appl. Environ. Microbiol.* **42**, 44–55.

BROWN, G. E. (1975), Land application of high nitrogenous industrial wastewater, *Proc. Nat. Conf. Management and Disposal of Residues from the Treatment of Industrial Wastewaters*, Washington, DC.

BROWN, D., LABOUREUR, P. (1983a), The degradation of dyestuffs: Part I. Primary biodegradation under anaerobic conditions, *Chemosphere* **12**, 397–404.

BROWN, D., LABOUREUR, P. (1983b), The aerobic biodegradability of primary amines, *Chemosphere* **12**, 405–414.

BRUNNER, W. (1982), Bakterieller Abbau von Dichlormethan und 1,2 Dichlorethan, *Thesis*, ETH Zürich.

BUA (Beratergremium für umweltrelevante Altstoffe) (Ed.) (1992), 2-Nitrophenol, 4-Nitrophenol, *BUA-Stoffbericht 75*. Weinheim: VCH.

BURGDORF, R., DONNER, C., WEBER, A., WIESMANN, U. (1991), Bakterien knacken CKW – Biologischer Abbau im Wirbelschichtreaktor, *Chem. Ind.* **11**, 13–18.

BURY, S. J., MILLER, C. A. (1993), Effect of micellar solubilization on biodegradation rates of hydrocarbons, *Environ. Sci. Technol.* **27**, 104–110.

CHUNG, K.-T., STEVENS, S. E., Jr. (1993), Degradation of azo dyes by environmental microorganisms and helminths, *Environ. Toxicol. Chem.* **12**, 2121–2132.

CHUNG, Y.-C., HUANG, C., TSENG, C.-P. (1996), Biodegradation of hydrogen sulfide by a laboratory-scale immobilized *Pseudomonas putida* CH11 biofilter, *Biotechnol. Prog.* **12**, 773–778.

COATES, J. D., WOODWARD, J., ALLEN, J., PHILIP, P., LOVLEY, D. R. (1997), Anaerobic degradation of polycyclic aromatic hydrocarbons and alkanes in petroleum-contaminated marine harbor sediments, *Appl. Environ. Microbiol.* **63**, 3589–3593.

CROSS, T. (1968), Thermophilic actinomycetes, *J. Appl. Bact.* **31**, 36–53.

CUMMINGS, R. J., JEWELL, W. J. (1977), Thermophilic aerobic digestion of dairy waste, in: *Food, Fertilizer and Agricultural Residues* (LOEHR, R. C., Ed.), pp. 637–657. Ann Arbor, MI: Ann Arbor Science Press.

CUNO, M. (1996), Untersuchungen zum biologischen Abbau von Mineralölen und polyzyklischen aromatischen Kohlenwasserstoffen, *Thesis*, Technical University, Berlin.

DA CANALIS, C., KRULL, R., HEMPEL, D. C. (1992), BakteriellerAbbau komplexer Naphthalinsulfonsäuregemische im Airlift-Schlaufenreaktor, *gwf Wasser/Abwasser* **133**, 226–230.

DE BONT, J. A. M., VORAGE, M. J. A. W., HARTMANS, S., VAN DEN TWEEL, W. J. J. (1986), Microbial degradation of 1,3-dichlorobenzene, *Appl. Environ. Microbiol.* **52**, 677–680.

DEGRYSE, E. (1976), Bacterial diversity at high temperature, *Experimentia* (Suppl.) **26**, 401–410.

DELCLOS, K. B., TARPLEY, W. G., MILLER, E. C., MILLER, J. A. (1984), 4-Aminobenzene and N,N-dimethyl-4-aminobenzene as equipotent hepatic carcinogens in male 57BL/GXC3H/He F1 mice and characterization of N-(deoxyguanosine-8-yl)-4-aminobenzene as the major persistent hepatic

DNA-bound dye in these mice, *Cancer Res.* **44**, 2540–2550.

DIECKHUES, B. (1960), Untersuchungen zur reduktiven Spaltung der Azofarbstoffe durch Bakterien, *Zbl. Bakteriol. Parasitenkd. Infektionskr. Hyg. (Abt. 1)* Orig. **1**, 244–249.

DIEKMANN, R., HEMPEL, D. C. (1989), Production rates of CO_2 by suspended and immobilized bacteria degrading 6-amino-2-naphthalenesulphonic acid, *Appl. Microbiol. Biotechnol.* **32**, 113–117.

DIEKMANN, R., NÖRTEMANN, B., HEMPEL, D. C., KNACKMUSS, H.-J. (1988), Degradation of 6-aminonaphthalene-2-sulphonic acid by mixed cultures: Kinetic analysis, *Appl. Microbiol. Biotechnol.* **29**, 85–88.

DIKS, R., OTTENGRAF, P. (1991), Verification of a simplified model for the removal of dichloromethane from waste gases using a biological trickling filter, in: *Bioprocess Engineering* Vol. 6, No. 3 (BRAUER, H., Ed.), pp. 93–99. Berlin, Heidelberg: Springer-Verlag.

DÖPPER, M., STOCK, W. (1995), Herkunft, Eintrag und Verbleib von Chloraromaten in der Umwelt, in: Mikrobielle Eliminierung chlororganischer Verbindungen Vol. 6, *Schriftenreihe Biologische Abwasserreinigung Sonderforschungsbereich 193*, Berlin.

DOMBROWSKI, T. (1991), Kinetik der Nitrifikation und Reaktionstechnik der Stickstoffeliminierung aus hochbelasteten Abwässern, *VDI-Fortschrittberichte* Reihe 15, Nr. 87.

DOTT, W. (1992), Chlorierte Kohlenwasserstoffe – Produktion, Anwendung und Toxizität, in: Biologischer Abbau von Chlorkohlenwasserstoffen (Kolloquium an der TU Berlin 21–22 November 1991) Vol. 1, *Schriftenreihe Biologische Abwasserreinigung Sonderforschungsbereich 193*, Berlin.

EHRING, B., PRASSLER, J., WALLNÖFER, P. R., ZIEGLER, W. (1992), Mikrobielle Transformation von *o*-Nitroanilin und *p*-Nitrophenol, *UWSF-Z. Umweltchem. Ökotox.* **4**, 81–85.

EPA (U.S. Environmental Protection Agency) (1975), Development document for effluent limitation guidelines, in: *New Source Performance Standards for the Pressed and Blow Glass Segment of the Glass Manufacturing Point Source Category*, U.S. EPA 440/1-75/034-a.

FATHEPURE, B. Z., TIEDJE, J. M., BOYD, S. A. (1988), Reductive dechlorination of hexachlorobenzene to tri- and dichlorobenzenes in anaerobic sewage sludge, *Appl. Environ. Microbiol.* **54**, 327–330.

FEIDIEKER, D. (1993), Untersuchungen zur mikrobiologischen Sanierbarkeit eines mit Chlorkohlenwasserstoffen kontaminierten Industriestandortes, *Thesis*, Technical University, Berlin.

FOX, P., VENKATASUBBIAH, V. (1996), Coupled anaerobic–aerobic treatment of high-sulfate wastewater with sulfate reduction and biological sulfide oxidation, *Water Sci. Technol.* **34**, 359–366.

FREEDMAN, D. L., SMITH, C. R., NOGUERA, D. R. (1997), Dichloromethane biodegradation under nitrate-reducing conditions, *Water Environ. Res.* **69**, 115–122.

FREITAS DOS SONTOS, L. M., LIVINGSTON, A. G. (1995), Novel membrane bioreactor for detoxification of VOC wastewaters: Biodegradation of 1,2-dichloroethane, *Water Res.* **29**, 179–194.

FRG (1994), Gesetz zur Vermeidung, Verwertung und Beseitigung von Abfällen vom 27.09.1994, *Bundesgesetzblatt* Teil I, Z 5702 A, Bonn 06. 10. 94.

GÄLLI, R. (1986), Optimierung des mikrobiellen Abbaus von Dichlormethan, in: *BUA-Stoffbericht 6.* Weinheim: VCH.

GARIÉPY, S., TYAGI, R. D., COULLIARD, D., TRAN, F. (1989), Thermophilic process for protein recovery as an alternative to slaughterhouse wastewater treatment, *Biol. Wastes* **29**, 93–105.

GAUDY, A. F., GAUDY, E. T. (1988), *Elements of Bioenvironmental Engineering*, pp. 347–355. San Jose, CA: Engineering Press.

GENSHOW, E., HEGEMANN, W. (1995), Beeinflussung des anaeroben Abbaus von Gerbereiabwasser durch Chlorid und Sulfat, *5. Kolloquium des Sonderforschungsbereiches 193: Abwässer aus der Zellstoffindustrie und der Lederherstellung*, 14–15 November 1994, Technical University Berlin.

GERBER, V., GOROBETS, L., LOAKIMIS, E. (1979), Effects of urea and dichloromethane on biochemical treatment of refinery wastewater, in: *Chemistry and Technology of Fuels and Oils* Vol. 1,5 No. 3–4, New York: Consultens Bureau Enterprise. pp. 269–272.

GLÄSSER, A. (1992), Entwicklung eines zweistufigen Verfahrens zum Totalabbau von Azofarbstoffen, *Thesis*, Universität Gesamthochschule Paderborn, pp. 55–59.

GUHA, S., JAFFÉ, P. (1996), Determination of Monod kinetic coefficients for volatile hydrophobic organic compounds, *Biotechnol. Bioeng.* **50**, 693–699.

GUILLAUME, J., TACQUET, A. M., KUPERWASER, B. (1963), Oxydation du paranitrophénol par certaines mycobactéries, *Comptes Rendus Hebdomadaires des Séances de l'Academie des Sciences* **256**, 1634–1637.

GUJER, W. (1985), Ein dynamisches Modell für die Simulation von komplexen Belebtschlammverfahren, *Thesis*, ETH Zürich.

HAIGLER, B. E., NISHINO, S. F., SPAIN, J. C. (1988), Degradation of 1,2-dichlorobenzene by a *Pseudomonas* sp., *Appl. Environ. Microbiol.* **54**, 294–301.

HALDANE, J. B. S. (1965), *Enzymes.* Cambridge, MA: MIT Press.

HARMER, C., BISHOP, P. (1992), Transformation of azo dye OA-7 by wastewater biofilms, *Water Sci. Technol.* **26**, 155–187.

HAUG, W., SCHMIDT, A., NÖRTEMANN, B., HEMPEL, D. C., STOLZ, A., KNACKMUSS, H.-J. (1991), Mineralization of the sulfonated azo dye Mordant Yellow 3 by a 6-aminonaphthalene-2-sulfonate-degrading bacterial consortium, *Appl. Environ. Microbiol.* **57**, 3144–3149.

HAUSCHILD, I., HORN, T., RUDEK, M., SAFAK, T., STARNICK, J. (1992), The effect of substrate and product inhibition on the microbial degradation of chlorinated hydrocarbons, in: *DECHEMA Biotechnology Conferences 5*, pp. 1061–1067. Weinheim: VCH.

HEIJNEN, J. J. (1988), Large-scale anaerobic–aerobic treatment of complex industrial waste water using immobilized biomass in fluidized bed and air lift suspension reactors, *GVC-Tagung Verfahrenstechnik Abwasserreinigung*, 17–19 October 1988, Baden-Baden, Germany, p. 203.

HEINFLING, A., BERGBAUER, M., SZEWZYK, U. (1997), Biodegradation of azo and phthalocyanine dyes by *Trametes versicolor* and *Bjerkandera adusta*, *Appl. Microbiol. Biotechnol.* **48**, 261–266.

HEINZE, L. (1997), Mikrobiologischer Abbau von 4-Nitrophenol, 2,4-Dinitrophenol und 2,4-Dinitrotoluol in synthetischen Abwässern, *Thesis*, Technical University, Berlin.

HEINZE, L., BROSIUS, M., WIESMANN, U. (1995), Biologischer Abbau von 2,4-Dinitrotoluol in einer kontinuierlich betriebenen Versuchsanlage und kinetische Untersuchungen, *Acta Hydrochim. Hydrobiol.* **23**, 254–263.

HEITKAMP, M. A., CAMEL, V., REUTER, T. J., ADAMS, W. J. (1990), Biodegradation of *p*-nitrophenol in an aqueous waste stream by immobilized bacteria, *Appl. Environ. Microbiol.* **56**, 2967–2973.

HEMMI, M., KRULL, R., HEMPEL, D. C. (1997), Developing a sequenced batch process for the biological and chemical purification of residual dyehouse liquors, *7. Kolloquium des Sonderforschungsbereiches 193: Treatment of Wastewaters from Textile Processing*, 17–18 November 1997, Technical University Berlin, Germany.

HENZE, M., GRADY, C. P. L., Jr., GUJER, W., MARAIS, G. v. R., MATSUO, T. (1986), Activated Sludge Model No. 1, *IAWPRC: Scientific and Technical Report No. 1*, 1 Queen Anne's Gate, London.

HENZE, M., GUJER, W., MINO, T., MATSUO, T., WENZEL, M. C., MARAIS, G. v. R. (1995), Activated sludge model No. 2, *IAWPRC: Scientific and Technical Report No. 3*, 1 Queen Anne's Gate, London.

HERBERT, D. (1958), Some principles of continuous culture, in: *Recent Progress in Microbiology* (TUNEVALL, Ed.), pp. 381–396. Stockholm: Almquist & Wiksell.

HERBST, B. (1995), Biologischer Abbau von Dichlormethan und Dichlorethan durch Mischkulturen im Wirbelschichtreaktor, *VDI Fortschrittsberichte* Reihe 17: Biotechnik, Nr. 126.

HERBST, B. (1995), Biologischer Abbau von Dichlormethan und Dichlorethan durch Mischkulturen im Wirbelschichtreaktor, *Thesis*, Technical University, Berlin.

HERBST, B., WIESMANN, U. (1996), Kinetics and reaction engineering aspects of the biodegradation of dichloromethane and dichloroethane, *Water Res.* **30**, 1069–1076.

HU, L. Z., SHIEH, W. K. (1987), Anoxic biofilm degradation of monocyclic aromatic compounds, *Biotechnol. Bioeng.* **30**, 1077–1083.

HUTTON, W. C., LA ROCCA, S. A. (1975), Biological treatment of concentrated ammonia wastewaters, *J. WPCF* **47**, 989–997.

JACKSON, M. L. (1982), Thermophilic treatment of a high-biochemical oxygen demand wastewater: Laboratory, Pilot Plant and Design, *37. Ind. Waste Conf.*, Purdue University. Ann Arbor, MI: Ann Arbor Science.

JAKOBEZYK, J., KWIATKOWSKI, M., OSTROWSKA, J. (1984), Untersuchung der biologischen Zersetzung aromatischer Nitroverbindungen in Abwässern, *Chem. Tech.* **36**, 112–116.

JANSSEN, A. J. H., SLEYSTER, R., VAN DER KAA, C., JOCHEMSEN, A., BONTSEMA, J., LETTINGA, G. (1995), Biological sulfide oxidation in a fed-batch reactor, *Biotechnol. Bioeng.* **47**, 327–333.

JANSSEN, D. B., SCHEPER, A., DIJKHUIZEN, L., WITHOLT, B. (1985), Degradation of haloorganic compounds by *Xanthobacter autotrophicus* GJ10, *Appl. Environ. Microbiol.* **49**, 673–677.

JENSEN, H. L., LAUTRUP-LARSEN, G. (1967), Microorganisms that decompose nitro-aromatic compounds, with special reference to dinitro-*ortho*-cresol, *Act. Agric. Scand.* **17**, 115–126.

JONGEN, W., ALINK, G., KOEMAN, J. (1978), Mutagenic effect of dichloromethane on *Salmonella typhimurium*, *Mutation Res.* **56**, 245–248.

KAMBHU, K. (1971), Thermophilic aerobic digestion for the biological treatment of wastes, *Thesis* (unpublished), Clemson University/USA.

KAPS, U., KOPP, M., RICHTER, K. (1990), Quantitative Untersuchungen zur Erfassung der Umweltexposition im Bereich der in der Textilveredlung eingesetzten Chemikalien, *UBA Forschungsbericht* 10602061, Umweltbundesamt, Berlin.

KÄSTNER, M. (1989), Biodegradation of volatile chlorinated hydrocarbons, *DECHEMA Biotechnology Conferences (7. Jahrestreffen der Biotechnologen)*, Vol. 3, Part B, May 1989, Frankfurt/M., pp. 909–912.

KHANDKER, S. A. K. (1992), Mikrobieller Abbau von aromatischen und chloraromatischen Verbindungen unter aeroben und denitrifizierenden Bedingungen, *Thesis*, Technical University, Berlin.

KIESE, S., SCHEFFLER, U., PILEPP, E. (1986), Hochlastverfahren zur innerbetrieblichen biologischen Abwasserreinigung, *Chem. Tech.* **15**, 24–30.

KNOWLES, G., DOWNING, A. L., BARRET, M. J. (1965), Determination of kinetic constants for nitrifying bacteria in mixed culture, with the aid of an electronic computer, *J. Gen. Microbiol.* **38**, 263–278.

KOMATSU, T., OMORI, T., KODAMA, T. (1993), Microbial degradation of the polycyclic aromatic hydrocarbons acenaphthene and acenaphthylene by a pure bacterial culture, *Biosci. Biotech. Biochem.* **54**, 864–865.

KOZIOROWSKI, B., KUCHARSKI, J. (1972), *Industrial Waste Disposal.* Oxford: Pergamon Press.

KRULL, R., HEMPEL, D. C. (1993), Biodegradation of naphthalene-sulphonic acid-containing sewages in a two-stage treatment plant, *Bioproc. Eng.* **10**, 229–234.

KRULL, R., HEMPEL, D. C. (1994), Biodegradation of naphthalenesulphonic acid containing sewages in a two-stage treatment plant, *Bioproc. Eng.* **10**, 229–234.

KUDLICH, M., BISHOP, P., KNACKMUSS, H.-J., STOLZ, A. (1996), Synchronous anaerobic and aerobic degradation of the sulfonated azo dye Mordant Yellow 3 by immobilized cells from a naphthalenesulfonate degrading mixed culture, *Appl. Microbiol. Biotechnol.* **46**, 597–603.

KULLA, H. G., KRIEG, R., ZIMMERMAN, T., LEISINGER, T. (1984), Experimental evolution of azo dye-degrading bacteria, in: *Current Perspectives in Microbial Ecology* (KLUG, M. J., REDDY, C. A., Eds.), pp. 663–667. Washington, DC: American Society of Microbiology.

LAWRENCE, A. W., MCCARTY, P. L. (1970), Unified basis for biological treatment design and operation, *J. San. Eng. Div. ASCE* **96**, 757 ff.

LEISINGER, T. (1988), Mikrobieller Abbau problematischer Komponenten in Abluft, in: *Biosensors and Environmental Biotechnology*, Serie: Biotech Vol. 2 (HOLLENBERG, C., SAHM, H., Eds.), pp. 125–133. Stuttgart: Gustav Fischer.

LIBRA, J. A., SOSATH, F. (1997), Biologisch-chemische Verfahrenskombinationen für die Behandlung von Reaktivfarbstoffen der Textilfärberei: Möglichkeiten und Grenzen, *Proc. GVC/IUV Colloquium Produktionsintegrierter Umweltschutz*, 15–17 Sept. 1997, Bremen, Germany, pp. B41–52.

LIBRA, J. A., YOO, E. S., BORCHERT, M., WIESMANN, U. (1997), Mechanisms of biological decolorization: their engineering application, *7. Kolloquium des Sonderforschungsbereiches 193: Treatment of Wastewaters from Textile Processing*. 17–18 November 1997, Technical University, Berlin, Germany.

LIEBELT, U. (1997), Anaerobe Teilstrombehandlung von Restflotten der Reaktivfärberei, *Thesis*, Technical University Berlin, pp. 53–55.

LIN, B., FUTONO, K., YOKOI, A., HOSOMI, M., MURAKAMI, A. (1997), Effects of activated carbon on treatment of photo-processing waste by sulfur-oxidizing bacteria, *Water Sci. Technol.* **35**, 187–195.

LOEW, G., REGBAGATTI, M., POULSEN, M. (1984), Metabolism and relative carcinogenic potency of chloroethanes: A quantum chemical structure–activity study, in: *Cancer Biochemistry and Biophysics* Vol. 7, pp. 109–132. New York: Gorden and Breach.

LOLL, U. (1976), Purification of concentrated organic wastewaters from the foodstuffs industry by means of aerobic–thermophilic degradation process, Sect. Int. Cong. Ind. Wastewater and Wastes, in: *Progress in Water Technology* Vol. 8 (JENKINS, S. H., Ed.), pp. 373–379.

LOVELESS, J. E., PAINTER, H. A. (1968), The influence of metal ion concentrations and pH value on the growth of a *Nitrosomonas* strain isolated from activated sludge, *J. Gen. Microbiol.* **52**, 1–14.

MACKAY, D., SHIU, W. J. (1981), A critical review of Henry's law constants for chemical environmental interest, *J. Phys. Chem. Ref. Data* **10**, 1175–1199.

MATELES, R. I., BARUAH, J. N., TANNENBAUM, S. R. (1967), Growth of a thermophilic bacterium on hydrocarbons: a new source of single-cell protein, *Science* **157**, 1322–1323.

MATSCH, L. C., DRNEVICH, R. F. (1977), Autothermal aerobic digestion, *J. WPCF* **49**, 296–310.

MATSCHÉ, N. F., ANDREWS, J. F. (1973), A mathematical model for the continuous cultivation of thermophilic microorganisms, *Biotechnol. Bioeng. Symp.* **4**, 77–90.

MEYER, U., OVERNEY, G., WATTENWYL, A. (1979), Über die biologische Abbaubarkeit von Azofarbstoffen, *Textilveredelung* **14**, 15–20.

MICHAELS, G. B., LEWIS, D. L. (1985), Sorption and toxicity of azo and triphenylmethane dyes to aquatic microbial populations, *Environ. Toxicol. Chem.* **4**, 45–50.

MICHAELS, G. B., LEWIS, D. L. (1986), Microbial transformation rates of azo and triphenylmethane dyes, *Environ. Toxicol. Chem.* **5**, 161–166.

MÜLLER, G. (1980), *Wörterbuch der Biologie – Mikrobiologie*, UTB Vol. 1024. Stuttgart, New York: Gustav Fischer.

NELSON, C. R., HITES, R. A. (1980), Aromatic amines in and near Buffalo river. *Environ. Sci. Technol.* **14**, 1147–1149.

NIEMANN, D. (1993), Biologische Abluftreinigung mit Biofilm-Wirbelschicht-Reaktoren, *Thesis*, ETH Zürich.

NÖRTEMANN, B., HEMPEL, D. C. (1991), Application of adapted bacterial cultures for the degradation of xenobiotic compounds in industrial wastewaters, in: *Biological Degradation of Wastes* (MARTIN, A. M., Ed.), pp. 261–279. London: Elsevier Applied Science.

NÖRTEMANN, B., BAUMGARTEN, J., RAST, H. G., KNACKMUSS, H.-J. (1986), Bacterial communities degrading amino- and hydroxynaphthalenesulfonates. *Appl. Environ. Microbiol.* **52**, 1195–1202.

NOWAK, J. (1994), Umsatz von Chlorbenzolen durch methanogene Mischkulturen aus Saalesediment in Batch- und Reaktorversuchen, *Thesis*, Technical University Berlin (Berichte zur Siedlungswasserwirtschaft Nr. 4).

NYHUIS, G. (1985), Beitrag zu den Möglichkeiten der Abwasserbehandlung bei Abwässern mit erhöhten Stickstoffkonzentrationen, *Veröffentlichungen des Instituts für Siedlungswasserwirtschaft und Abfalltechnik der TU Hannover*, Heft 61.

OLLIKKA, P., ALHONMÄKI, K., LEPPÄNEN, V., GLUMOFF, T., RAIJOLA, T., SUOMINEN, I. (1993), Decolorization of azo, triphenyl methane, heterocyclic, and polymeric dyes by lignin peroxidase isoenzymes from *Phanerochaete chrysosporium*, *Appl. Environ. Microbiol.* **59**, 4010–4016.

OU, L. T., SHARMA, A. (1989), Degradation of methyl parathion by a mixed bacterial culture and a *Bacillus* sp. isolated from different soils, *J. Agric. Food Chem.* **37**, 1514–1518.

PAGGA, U., HALTRICH, W. G., GÜNTHNER, W. (1982), Untersuchungen über die Wirkung von 4-Nitrophenol auf Belebtschlamm, *Vom Wasser* **59**, 51–65.

PASCIK, I. (1982), Biochemische Stickstoffelimination aus ammoniumreichen Raffinerie-Abwässern in der Bayer-Turmbiologie, *Chem. Ing. Tech.* **54**, 1190–1192.

PATTERSON, J. W. (1985), *Industrial Wastewater Treatment Technology*, pp. 305–306. Boston, MA: Butterworth Publishers.

PLACHY, P., JURIS, P., TOMASOVICOVA, O. (1993), Destruction of *Toxocara caninis* and *Ascaris suum* eggs in wastewater sludge by aerobic stabilization under laboratory conditions, *Helminthologia* (Bratislava) **30**, 139–142.

PONTI, C., SONNLEITNER, B., FIECHTER, A. (1995), Aerobic thermophilic treatment of sewage sludge at pilot plant, *J. Biotechnol.* **38**, 183–192.

PRIEGER-KRAFT, A. (1995), Mikrobieller Abbau von 1,2,3,4- und 1,2,3,5-Tetrachlorbenzol, *Thesis*, Technical University, Berlin.

RAFII, F., FRANKLIN, W., CERNIGLIA, C. E. (1990), Azoreductase activity of anaerobic bacteria isolated from human intestinal microflora, *Appl. Environ. Microbiol.* **56**, 2146–2151.

RAMANAND, K., BALBA, M. T., DUFFY, J. (1993), Reductive dehalogenation of chlorinated benzenes and toluenes under methanogenic conditions, *Appl. Environ. Microbiol.* **59**, 3266–3272.

RANNUG, U. (1980), Genotoxic effects of 1,2-dibromethane and 1,2-dichloroethane, *Mutation Res.* **76**, 269–295.

RAYMOND, D. G. M., ALEXANDER, M. (1971), Microbial metabolism and cometabolism of nitrophenol, in: *Pesticide Biochemistry and Physiology I*, pp. 123–130.

REINEKE, W., KNACKMUSS, H.-J. (1984), Microbial metabolism of haloaromatics: Isolation and properties of a chlorobenzene–degrading bacterium, *Appl. Environ. Microbiol.* **47**, 395–402.

RIPPEN, G. (1991), *Handbuch Umweltchemikalien: Stoffdaten, Prüfverfahren, Vorschriften.* Landsberg/Lech: ecomed.

ROTH (1990), *Wassergefährdende Stoffe*. Landsberg/Lech: ecomed.

RUDEK, M. (1992), Mikrobieller Abbau von 1,2-Dichlorethan, *Thesis*, Technical University, Berlin.

SALLIS, P., ARMFIELD, S., BULL, A., HARDMAN, D. (1990), Isolation and characterization of a haloalkane halidohydrolase from *Rhodococcus erythropolis* Y2, *J. Gen. Microbiol.* **136**, 115–120.

SANDER, P., WITTICH, R. M., FORTNAGEL, P., WILKEY, H., FRANKE, W. (1991), Degradation of 1,2,4-trichloro- and 1,2,4,5-tetrachlorobenzene by *Pseudomonas* sp. strains, *Appl. Environ. Microbiol.* **57**, 1420–1440.

SCHÄFER, M. (1994), Untersuchungen zum Abbau chlorierter Benzole durch Mischkulturen in kontinuierlich betriebenen Reaktoren, *Thesis*, Technical University, Berlin (Berichte zur Siedlungswasserwirtschaft, Nr. 3).

SCHEFFLER, U., KIESE, S., PILEPP, E. (1982), Aerob-thermophile Reinigung hochbelasteter Abwässer, *wlb – wasser, luft und betrieb* **26**, 32–36.

SCHMIDT, S. K., SCOW, K. M., ALEXANDER, M. (1987), Kinetics of *p*-nitrophenol mineralization by a *Pseudomonas* sp.: Effects of second substrates, *Appl. Environ. Microbiol.* **53**, 2617–2623.

SCHÖNBERGER, H. (1996), Zur Abwässerfrage der Textilveredlungsindustrie, *Thesis*, Technical University, Berlin.

SCHÖNBERGER, H., KAPS, U. (1994), Reduktion der Abwasserbelastung in der Textilindustrie, *UBA Text* 3/94, Umweltbundesamt, Berlin.

SEWEKO, U. (1988), Naturfarbstoffe – eine Alternative zu synthetischen Farbstoffen. *Melliand Textilberichte* **4**, 271–276, Leverkusen, Germany.

SHADE, H. (1977), Wastewater nitrification, *Chem. Eng. Prog.* **73**, 45–49.

SIDDIQUE, M. H., ST. PIERRE, C. C., BISWAS, N., BEWTRA, J. K., TAYLOR, K. E. (1993), Immobilized enzyme-catalyzed removal of 4-chlorophenol from aqueous solution, *Water Res.* **27**, 883–890.

SIMS, R. C., OVERCASH, M. R. (1983), Fate of polynuclear compounds (PNAS) in soil–plant systems, *Residue Rev.* **88**, 1–68.

SMITH, J. E., Jr., YOUNG, K. W., DEAN, R. B. (1975), Biological oxidation and disinfection of sludge, *Water Res.* **9**, 17–24.

SOEWONDO, P. (1997), Zweistufige anaerobe und aerobe biologische Behandlung von synthetischem Abwasser mit dem Azofarbstoff C.I. Reactive Orange 96, *Thesis*, Technical University, Berlin.

SOSATH, F., LIBRA, J. A. (1997), Biologische Behandlung von synthetischen Abwässern mit Azofarbstoffen, *Acta Hydrochim. Hydrobiol.* **25**, 259–264.

SPADARO, J. T., GOLD, M. H., RENGANATHAN, V. (1992), Degradation of azo dyes by the lignin-degrading fungus *Phanerochaete chrysosporium*, *Appl. Environ. Microbiol.* **58**, 2397–2401.

SPAIN, J. C. (Ed.) (1991), *Biodegradation of Nitroaromatic Compounds*. New York, London: Plenum Press.

SPAIN, J. C., WYSS, O., GIBSON, D. T. (1979), Enzymatic oxidation of *p*-nitrophenol, *Biochem. Biophys. Res. Commun.* **88**, 634–641.

SPANGGORD, R. J., SPAIN, J., NISHINO, S. F., MORTELSMANS, K. E. (1991), Biodegradation of 2,4-dinitrotoluene by a *Pseudomonas* sp., *Appl. Environ. Microbiol.* **57**, 3200–3205.

SPEITEL, G. E., JR., LU, C.-J., TURAKHIA, M., ZHU, X.-J. (1989), Biodegradation of trace concentrations of substituted phenols in granulat activated carbon columns, *Environ. Sci. Technol.* **23**, 68–74.

SPRINGFELLOW, W.T., AITKEN, M. D. (1995), Competitive metabolism of naphthalene, methylnaphthalene, and fluorene by phenanthrene-degrading *Pseudomonas*, *Appl. Environ. Microbiol.* **61**, 357–362.

STELDERN, D., OTTOW, J. C. G., LOLL, U. (1974), Thermophile aerobe Sporenbildner bei der biologischen Reinigung hochkonzentrierter Abwässer, *Z. allg. Mikrobiol.* **14**, 229–236.

STUCKI, G., THÜER, M., BENTZ, R. (1992), Biological Degradation of 1,2-Dichloroethane under Groundwater Conditions, *Water Res.* **26**, 273–278.

THIEM, K.-W., BOOTH, G. (1991), Nitro compounds, aromatic, in: Ullmann's Encyclopedia of Industrial Chemistry Vol. A 17, 5th Edn. Weinheim: VCH.

VAIDYA, A. A., DATYE, K. V. (1982), Environmental pollution during chemical processing of synthetic fibers, *Colourage* **14**, 3–10.

VALLI, K., BROCK, B. J., JOSHI, D. K., GOLD, M. H. (1992), Degradation of 2,4-dinitrotoluene by the lignin-degrading fungus *Phanerochaete chrysosporium*, *Appl. Environ. Microbiol.* **58**, 221–228.

VAN DER MEER, J. R., ROELOFSEN, W., SCHRAA, G., ZEHNDER, A. J. B. (1987), Degradation of low concentrations of dichlorobenzenes and 1,2,4-trichlorobenzene by *Pseudomonas* sp. strain P51 in nonsterile soil columns, *FEMS Microbiol.* **27**, 1538–1542.

Verordnung (EWG) Nr. 259/93 des Rates vom 01.02.1993 zur Überwachung und Kontrolle der Verbringung von Abfällen in der, in die und aus der Europäischen Gemeinschaft (259/93/EWG).

VIDAL, C. M., VITALE, A. A. (1993), 2-naphthalenesulphonic acid-degrading microorganisms, *Revista Argentina de Microbiologia* **25**, 221–226.

WALKER, P. D., WOLF, J. (1971), The taxonomy of *Bacillus stearothermophilus*, *Spore Research* **1**, 247–262.

WHO (World Health Organisation) (1987), *Methylene Chloride Health and Safety Guide, Environmental Health Criteria, No. 32: Methylene Chloride IPCS International Program on Chemical Safety*, p. 42. Stuttgart: Wissenschaftliche Verlagsgesellschaft.

WIESMANN, U. (1994), Biological nitrogen removal from wastewater, in: *Advances in Biochemical Engineering Biotechnology* Vol. 51 (FIECHTER, A., Ed.), pp. 114–154. Berlin, Heidelberg: Springer-Verlag.

WINKELBAUER, W., KOHLER, H. (1989), Biologischer Abbau von leichtflüchtigen Chlorkohlenwasserstoffen im offenen System am Beispiel von Dichlormethan in einem Tauchtropfkörper, *gwf Wasser Abwasser* **130**, 13–20.

WODZINSKI, R., JOHNSON, M. (1968), Yields of bacterial cells from hydrocarbons, *Appl. Microbiol.* **16**, 1886–1891.

WOLF, J., BARKER, A. N. (1968), The genus *Bacillus*: Aids to the identification of its species, in: *Identification Methods for Microbiologists* (GIBBS, M. B., SHAPTON, D. A., Eds.), pp. 93–108. London: Academic Press.

YI, M. E. YUN-SEOK (1985), Reaktionstechnische Modellierung der autothermen, aerob-thermophilen Schlammstabilisierung und Abwasserreinigung, *Thesis*, Technical University, Berlin.

ZHANG, X., YOUNG, L. Y. (1997), Carboxylation as an initial reaction in the anaerobic metabolism of naphtaline and phenanthrene by sulfidogenic consortia, *Appl. Environ. Microbiol.* **63**, 4759–4764.

ZIMMERMAN, T., KULLA, H.G., LEISINGER, T. (1982), Properties of purified orange II azoreductase, the enzyme initiation azo dye degradation by *Pseudomonas* KF46, *Eur. J. Biochem.* **129**, 197–203.

ZOLLINGER, H. (1987), *Color Chemistry – Syntheses, Properties and Applications of Organic Dyes and Pigments*. Weinheim, New York: VCH.

20 Modeling of Aerobic Wastewater Treatment Processes

MOGENS HENZE
Lyngby, Denmark

The trend of improving the effluent quality from wastewater treatment plants leads to an increasing complexity in the design and operation of the plants. In order to design the plants and optimize and control the operation, it is necessary to use dynamic models. Most models used for wastewater treatment plants today are deterministic, the exception being some models for control, which can be of the black-box type. The deterministic models aim to give a realistic description of the main processes of the plant. However, models are never true in the process sense, and will always simplify the complicated processes occurring in a biological treatment plant. The models we have at hand today, include the phenomena for which we think we have a reasonable explanation. A major gap in modeling is lack of knowledge about the development of the microbiology in treatment plants. Thus, phenomena like bulking and foaming are still waiting for more basic understanding before we can develop reliable models for this. Below various elements of modeling are discussed. For further information on modeling in general and on modeling of biofilters the reader is referred to Chapter 13, Volume 4 of the *Biotechnology* series (REHM et al., 1991).

1 Purpose of Modeling

Models are used for several purposes in relation to wastewater treatment plants:

- design,
- control,
- operational optimization,
- teaching,
- organizing tool.

The intended use of a model has implications for its structure. Models used for design and for control need not be identical.

Models are not better than the assumptions with which they have been built. Several factors influence their behavior. The way they handle detailed wastewater composition is important. Wastewater characterization has a strong impact on the real plant behavior as well as on its modeling. Errors in the characterization of the wastewater or changes in the composition of the wastewater can result in erroneous modeling results.

2 Elements of Activated Sludge Models

An activated sludge model consists of different elements:

- transport processes,
- components,
- biological and chemical processes,
- hydraulics.

To this can be added a framework of component conversion and data presentation tools. The needed amount of detail depends on the intended use of the model and on the amount of information on the wastewater and the treatment plant, which is available.

Transport Processes and Treatment Plant Layout

Transport processes deal with movement of water and include no chemical or biological processes. The flow scheme of the treatment plant, which describes the movement of water and sludge, must be known and a model of the flow scheme must be part of the overall model. Often it is necessary to simplify the flow scheme, due to high complexity of the plant or due to limitations to the number of tanks that can be applied in the model. Simplification of a flow scheme can be very beneficial in order to improve the overview of the situation in the plant. Parallel tanks can be modeled as one tank, and tanks in series might also be modeled as one. However, one must be aware of the fact that the more simplified the flow scheme model is, the bigger the deviation from the real world one might get with the model calculations.

The flow pattern of wastewater and sludge must be known. Not all treatment plant operators can give this information directly, at least not the correct one.

Aeration

Aeration in each of the tanks must be described, either with a fixed oxygen concentration, or as an aeration coefficient, K_La, operating uncontrolled or controlled together with a control strategy. Aeration of tanks without mechanical aeration should also be taken into account. The oxygen penetration through the surface of, e.g., denitrification tanks can be significant.

The use of a fixed oxygen concentration is a simplification that will not allow for simultaneous nitrification–denitrification processes to be modeled, nor the impact of oxygen in recycle flows. Thus the K_La model is recommended for professional modeling.

Components

These are soluble and suspended substances, which one wants to model. They are substances present in the raw wastewater as well as substances found within the treatment plant. The components to model depend on the purpose of the model. If the purpose is to model a nitrifying plant, one phosphorus component will suffice, but if modeling of biological phosphorus removal is the objective, then at least two phosphorus components are needed: polyphosphate in the biomass (X_{PP}) and a soluble component, orthophosphate (S_{PO_4}).

Processes

These include modeling of chemical and biological processes. The model to be used will always be a strong simplification of reality. *In many cases these simplified models will give reasonable modeling results. In other cases not.* For example, a simple phosphorus precipitation model, like the one applied in ASM2 (HENZE et al., 1995) can model the part of the phosphorus removal that is not related to assimilation and biological excess uptake. On the other hand, a model without accounting for the denitrification that occurs in nitrification tanks, the so-called simultaneous denitrification, will give too high effluent nitrate concentrations. Thus the degree of simplification must be considered. If one wants to model denitrification, there must as a minimum be one process for this (anoxic growth of heterotrophic biomass). But it is also possible to apply many processes for the description of denitrification. This will be the case for a model that is aimed at predicting intermediates like nitrite, dinitrogen oxide, nitrous oxide, etc., in the denitrification process.

Hydraulic Patterns

The hydraulics of the various tanks has to be modeled. Any tanks can be modeled as ideal mix, while some might need to be modeled as a series of ideally mixed tanks. More complex models of the hydraulics are seldom needed.

3 Presentation of Models

The complex interrelationship between the various components and processes are best presented in a matrix. A very simple model for an activated sludge process for removal of organic matter is shown in the matrix in Tab. 1. With the table it is possible to get a quick overview over the conversions in the described process.

Tab. 1. Simple Model Matrix for Activated Sludge Organic Removal[a]

Components	S_{O_2}	S_S	X_H	Process Rate
Processes:				
Aerobic growth	$-\frac{1}{Y_H}+1$	$-\frac{1}{Y_H}$	$+1$	$\mu_{m,H}\frac{S_S}{K_S+S_S}X_H$
Decay		$+1$	-1	$b_H X_H$

[a] S_{O_2} Soluble oxygen, g O_2 m^{-3}; S_S soluble organic substrate, g COD m^{-3}; X_H biomass, g COD m^{-3}; Y_H yield coefficient, g COD g^{-1} COD; $\mu_{m,H}$ maximum growth rate, d^{-1}; K_S half saturation constant, g COD m^{-3}; b_H decay constant, d^{-1}.

Mass Balances

The rows in the matrix give the mass balance of the process. The second row in Tab. 1 shows, e.g., that decay removes biomass, X_H (the stoichiometric coefficient is negative, -1) and produces soluble substrate, S_S (the stoichiometric coefficient is $+1$).

Rates

The rate equation of the processes are is found in the right-hand column. To find the rate for the change of biomass in relation to decay, r_{XH}, the rate equation must be multiplied by the stoichiometric coefficient in the matrix. This gives

$$r_{XH} = -1 \cdot b_H X_H \quad (1)$$

Component Participation

The columns in the matrix give an overview in which processes the various components participate (according to the model). The second column shows that the substrate, S_S, is removed by aerobic growth and is produced by decay.

4 The Activated Sludge Models No. 1, 2 and 3 (ASM1, ASM2, ASM3)

Most models for activated sludge processes are based on the IAWQ Activated Sludge Model No. 1, called ASM1 (HENZE et al., 1987). This model is used as a platform for further model development and is today the international standard for advanced activated sludge modeling. As the ASM only contains a very simplified model, most modelers expand the ASM1, depending on the degree of complexity needed to model the actual problem. In some cases more complex kinetic equations are needed, in other cases more processes or more components have to be included. For most cases nutrient limitation of the growth processes is included.

4.1 Activated Sludge Model No. 1 (ASM1)

This model describes activated sludge processes with nitrification and denitrification. It is the experience through more than 10 years of use, that the model gives a good description of the processes, as long as the wastewater has been characterized in detail and is of domestic or municipal origin. The model has not been developed to fit industrial wastewater treatment. When used for industrial wastewater, great care should be taken in the calibration and the interpretation of the results.

The ASM1 can calculate several details in the plant:

- oxygen consumption in the tanks,
- concentration of ammonia and nitrate in the tanks and in the effluent,
- concentration of COD in the tanks and in the effluent,
- MLSS in the tanks,
- solids retention time,
- sludge production.

As with all models, ASM1 will also give erroneous results, if it is fed with erroneous or too simplified information. Tab. 2 shows the process matrix for ASM1. All processes, reaction kinetics, mass balances, and stoichiometry in the model are included in the matrix. If understood, the matrix notation is a useful tool to get an easy overview of the model.

In the ASM1 the processes for organic matter removal and nitrification are slightly coupled as seen in Fig. 1.

4.2 Activated Sludge Model No. 2 (ASM2)

This model incorporates biological phosphorus uptake (HENZE et al., 1995). The combination of this with denitrification results in a model much more complex than the ASM1. To describe the life of PAOs (phosphate accumulating organisms) internal storage compounds of phosphorus and organic matter are needed. The modeling of the growth of the hetero-

Tab. 2. The Process Matrix for the Activated Sludge Model No. 1 (ASM1) Based on HENZE et al. (1997)[a]

	1 S_I	2 S_S	3 X_I	4 X_S	5 $X_{B,H}$	6 $X_{B,A}$	7 X_P	8 S_{O_2}	9 S_{NO_3}	10 S_{NH_4}	11 S_{ND}	12 X_{ND}	13 S_{ALK}	Process Rate[b]
1. Aerobic growth of heterotrophs		$-1/Y_H$			1			$-(1-Y_H)/Y_H$		$-i_{XB}$			$-i_{XB}/14$	$\mu_{max,H} \cdot M_2 \cdot M_8 \cdot X_{B,H}$
2. Anoxic growth of heterotrophs		$-1/Y_H$			1				$-(1-Y_H)/2.86\ Y_H$	$-i_{XB}$			$((1-Y_H)/14-2.86\ Y_H)-i_{XB}/14$	$\mu_{max,H} \cdot M_2 \cdot I_8 \cdot M_9 \cdot \eta_g \cdot X_{B,H}$
3. Aerobic growth of autotrophs						1		$-(4.57-Y_A)/Y_A$	$1/Y_A$	$i_{XB}-1/Y_A$			$-i_{XB}/14-1/(7\ Y_A)$	$\mu_{max,A} \cdot M_{10} \cdot M_8 \cdot X_{B,A}$
4. Decay of heterotrophs				$1-f_P$	-1		f_P					$i_{XB}-f_P \cdot i_{XP}$		$b_H \cdot X_{B,H}$
5. Decay of autotrophs				$1-f_P$		-1	f_P					$i_{XB}-f_P \cdot i_{XP}$		$b_A \cdot X_{B,A}$
6. Ammonification of soluble organic nitrogen										1	-1		1/14	$k_s \cdot S_{ND} \cdot X_{B,H}$
7. Hydrolysis of entrapped organics		1		-1										$k_h \cdot sat(M_8 + \eta_h \cdot I_8 \cdot M_9)$
8. Hydrolysis of entrapped organic nitrogen											1	-1		rate 7 (X_{ND}/X_S)
Unit	COD	COD	COD	COD	COD	COD	COD	negative COD	N	N	N	N	Mole	

[a] For detailed explanation, see HENZE et al. (1987)

[b] M_x = Monod kinetics for component x $(x/(K+x))$; I_x = inhibition kinetics for component x $(K/(K+x))$; *sat* = saturation kinetics $k_h \dfrac{X_S/X_{B,H}}{K_{-}+(X_S/X_{B,H})}$

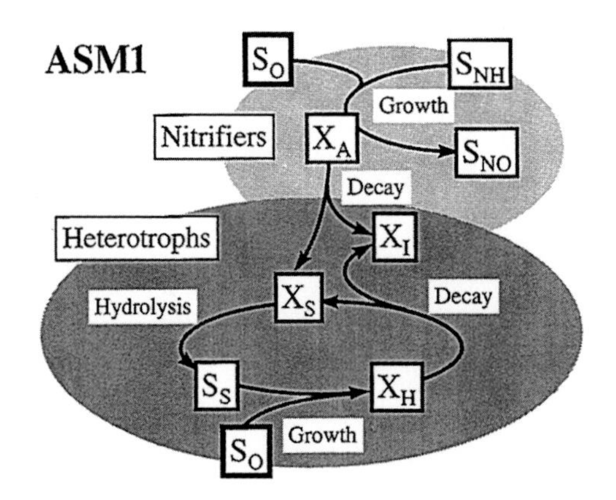

Fig. 1. Processes for heterotrophic and nitrifying bacteria in the Activated Sludge Model No. 1 (ASM1) (GUJER et al., 1999).

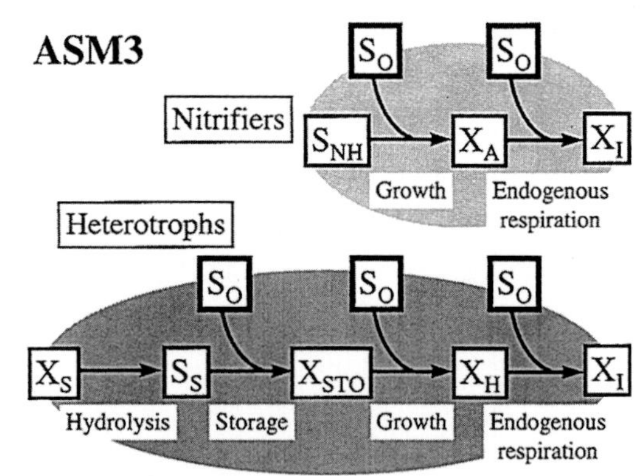

Fig. 2. Processes for heterotrophic organisms in the Activated Sludge Model No. 3 (ASM3) (GUJER et al., 1999).

trophic organisms needs at least three different organic substrates:

- acetic acid-like compounds, S_A,
- fermentable substrate, S_F,
- slowly degradable substrate, X_S.

The ASM2 assumes that growth only occurs on the substrate S_A, and that the other organic substrates are hydrolyzed and fermented in order to be converted to S_A.

4.3 Activated Sludge Model No. 3 (ASM3)

This model represents an alternative process description for heterotrophic bacteria. Under transient loading conditions heterotrophic bacteria can store organic matter as polymeric compounds like polyhydroxyalkanoate (PHA) or glycogen (GLY). This storage can influence the overall process of activated sludge by supplying organic material under starvation conditions. The Activated Sludge Model No. 3 (ASM3) includes this storage in its processes and applies a simplified maintenance/decay process. Fig. 2 shows the processes used to describe heterotrophic bacteria. It can be seen that the substrate flows through the organisms under oxygen consumption for storage, growth, and maintenance.

5 Wastewater Characterization

Wastewater characterization is an important element in dealing with models. The detailed wastewater composition has significant influence on the performance of the treatment processes (HENZE, 1992). The degree of detail needed in the wastewater characterization depends on the objective of modeling. Fig. 3 shows a detailed fractionation of municipal wastewater in relation to biological phosphorus removal. As can be seen from the figure, what is considered soluble in the model (substances with a symbol S) is not equivalent to analytically determined soluble material. This is caused by the fact that part of the particulate organic material is easily degradable. Thus, from a reaction point of view, it cannot be distinguished from the analytically soluble material.

Tab. 3 shows a detailed fractionation of wastewater, which allows for modeling denitrification as well as biological phosphorus removal processes.

The characterization of wastewater can be simplified by combining existing data on BOD, COD, soluble fractions, etc. An example of this is from the Dutch river boards that use the procedure shown below (ROELEVELD and KRUIT, 1998).

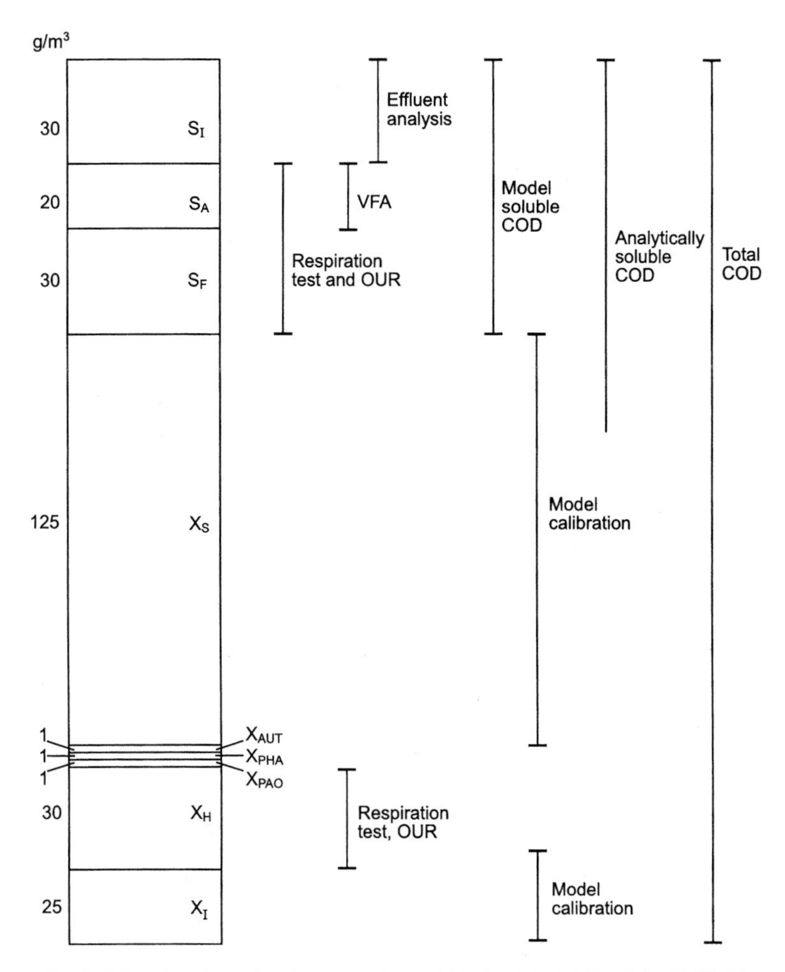

Fig. 3. COD fractionation in the Activated Sludge Model No. 2 (ASM2). The figure shows a typical composition of primary settled municipal wastewater. Various analytical methods which can be used for the determination of single components are shown. Organic inert suspended material, X_I, and organic, slowly degradable suspended material, X_S, can only be determined by modeling and calculation (HENZE et al., 1995).

At present there are no methods to directly determine these two fractions independently. For inert suspended organic matter, X_I, the reason is that its concentration in the sludge has its origin from two sources, the one being a component of the raw wastewater and the second the production during the decay of microorganisms. These two contributions cannot be separated analytically at present. Slowly degradable suspended organic matter, X_S, cannot be determined analytically except from calculation based on determination of the other fractions of organic matter.

Influent:

$COD_{total} = S_S + S_I + X_S + X_I$
$COD_{soluble} = S_S + S_I$ (0.1 µm filter)
$COD_{suspended} = X_S + X_I$

S_I = soluble COD (0.1 µm filter) in the effluent from a low loaded biological treatment plant.
$S_S + X_S$ is determined from a BOD_5 measurement or from a BOD_{20} measurement (or calculation).

Tab. 3. Wastewater Components in ASM2. Observe that Model Soluble Organic Material is Bigger than Analytically Determined

Symbol	Component	Typical Level	Unit
"Soluble" compounds:			
S_F	fermentable organic matter	20–250	g COD m^{-3}
S_A	acetate and other fermentation products	10– 60	g COD m^{-3}
S_{NH_4}	ammonium	10–100	g N m^{-3}
S_{NO_3}	nitrate + nitrite	0– 1	g N m^{-3}
S_{PO_4}	orthophosphate	2– 20	g P m^{-3}
S_I	inert soluble organic matter	20–100	g COD m^{-3}
"Suspended" compounds:			
X_I	inert suspended organic matter	30–150	g COD m^{-3}
X_S	slowly degradable organic matter	80–600	g COD m^{-3}
X_H	heterotrophic biomass	20–120	g COD m^{-3}
X_{PAO}	phosphorus accumulating biomass (PAO)	0– 1	g COD m^{-3}
X_{PP}	stored polyphosphate in PAO	0– 0.5	g P m^{-3}
X_{PHA}	stored organic polymer in PAO	0– 1	g COD m^{-3}
X_{AUT}	nitrifying biomass	0– 1	g COD m^{-3}

$$S_S + X_S = 1.5\ BOD_5$$
$$S_S + X_S = BOD_{20}/0.85$$

A simplified evaluation of the nitrogen components is done based on the content of nitrogen in the organic components:

$$TN = f_{NX} \cdot X_S + f_{NS} \cdot S_S + S_{NH_4}$$

where f_{NX} and f_{NS} are the fractions of nitrogen (g N g^{-1} COD). These will typically be around 0.03 g N g^{-1} COD (HENZE et al., 1997).

6 Model Calibration

Calibration of models is difficult. It is mandatory for a successful calibration that the content and the functioning of the model is understood. When calibrating by hand, it is important to follow a strategy and be aware of which parameters are to be used for which part of the calibration. It is equally important to know the realistic intervals within which the various parameters should vary, in order to still have a realistic model. Tab. 4 shows the steps to follow for calibration of the Activated Sludge Model No. 1. Constants not shown should under normal circumstances not be touched.

Tab. 4. Calibration Steps for the Activated Sludge Model No. 1 (ASM1)

Parameter	Calibration Based on ...	Comments
Heterotrophic growth rate, $\mu_{H,max}$	OUR	often kept unchanged
Half saturation constant, K_S	COD in effluent	used to model the hydraulics of the tanks
Growth rate for nitrifying biomass, $\mu_{A,max}$	NH_4 in effluent	if NH_4 for all conditions is low, μ should not be changed
Half saturation constant, K_{NH_4}	NH_4 in effluent	only if NH_4 is low
Half saturation constant, $K_{O,A}$	NH_4 in effluent	only if S_{O_2} varies
Denitrification factor, η_{NO_3}	NO_3 in effluent	
Half saturation constant, K_{NO_3}	NO_3 in effluent	only if NO_3 is low

For processes without changes in performance or loading calibration is easy, but not very reliable. Often many sets of calibration constants will give a nice model fit. The more dynamic the data for calibration, the better the calibration will be, and the bigger the chance will be for the model to respond correctly to situations outside the range of those used in the calibration. Many treatment plants have only small dynamic changes. For such plants single dynamic events (overloading, mechanic breakdown, snow melt, etc.) will be excellent for calibration purposes.

7 Computer Programs

There are several computer programs with different model content. Many of the models will include one or more of the models in the ASM family. Tab. 5 gives a short overview.

8 Use of Models

Models can be used for many purposes. An important field is control of treatment plants. It is often possible to use simplified models for control. Complex processes, like nitrification–denitrification combined with biological phosphorus removal and external carbon source addition can make it necessary to use even more complex models than those in the ASM family (MEINHOLT et al., 1998).

Optimizing the operation of a treatment plant is another possible use for models. Fig. 4 shows a model calculation for a treatment plant, which receives varying loads of ammonia from the sludge treatment. This often creates a problem with overload of the nitrification process, resulting in breakthrough of ammonia peaks into the effluent from the plant. A model calculation can help find the optimal operational strategy for handling ammonia-rich sidestreams.

Fig. 5 shows a model calculation with ASM1 in relation to start-up of a nitrifying activated sludge treatment process. The calculations are made in order to evaluate a possible effect of inoculation with nitrifying biomass. The results show that the inoculation has no significant impact on the start-up.

Fig. 6 shows a model calculation of a process for denitrification with methanol. The process has an aerobic and an anoxic tank. By using two heterotrophic populations with different growth characteristics during modeling, a result is obtained that explains the differences in methanol removal rate. If the process is run solely aerobic for a period, the group of organisms that under aerobic conditions metabolizes methanol quickly will increase. This group has a slow methanol conversion rate under anoxic conditions. In the second part of the experiment, from day 40 onwards, the process is operated with 30% anoxic reaction time and 70% aerobic reaction time. Under these operational conditions, the other group of organisms will dominate, resulting in higher methanol conversion rates under anoxic conditions.

Tab. 5. Selected Simulation Programs for Activated Sludge Plants

Program	Model Content	
ASIM	ASM1, ASM2, ASM3	• simple program, which allows for quick modeling
EFOR	ASM1, ASM2	• program which has a detailed wastewater calibration model, where available data (including BOD) is used for calculation of both raw and treated wastewater composition • choice between different settler models
GPX	ASM1, ASM2, etc.	• program where different models can be selected for the simulation of both biology and settling
SIMBA	ASM1, ASM2	• program widely used in the Netherlands and Germany based on matlab-simulink

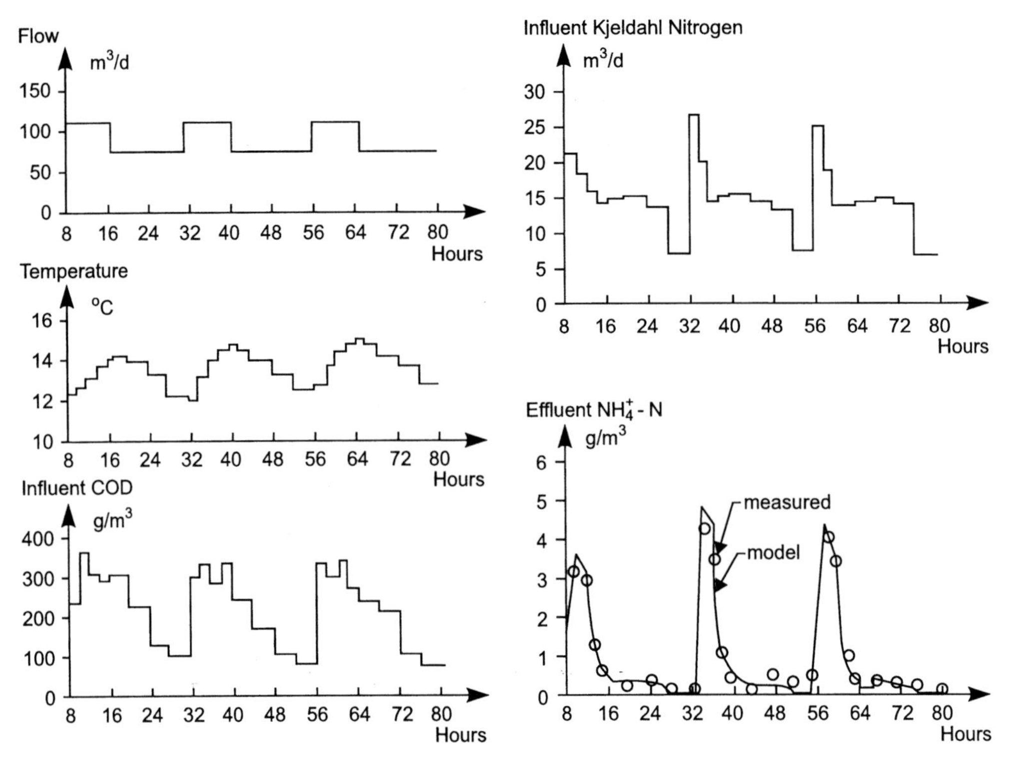

Fig. 4. The use of the ASM1 for modeling of effluent variations for ammonia dynamically loaded with ammonia-rich supernatant (GUJER, 1985).

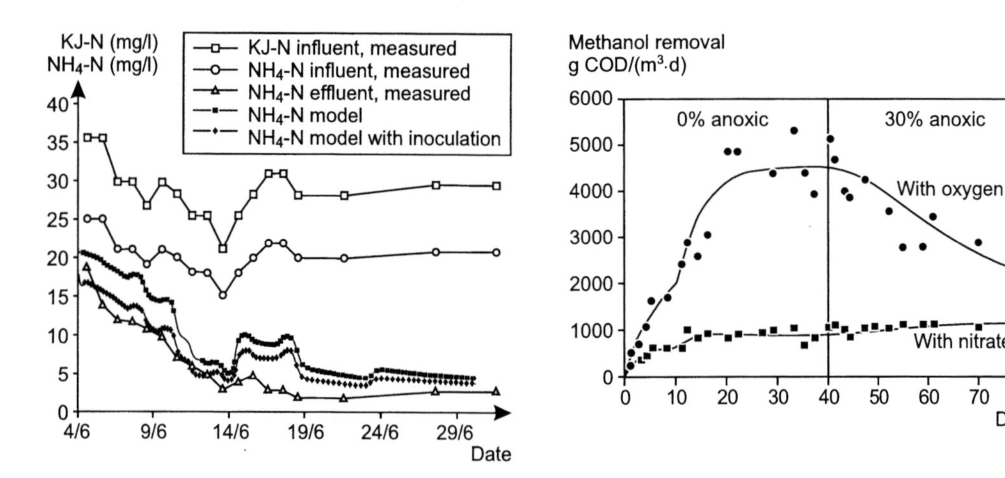

Fig. 5. Model calculation of start-up of an activated sludge process with and without inoculation (FINNSON, 1994).

Fig. 6. Modeling denitrification rate with methanol under varying operational conditions in an activated sludge process (PURTSCHERT and GUJER, 1999).

9 References

FINNSON, A. (1994), Computer simulations of full-scale activated sludge processes. *Thesis*, Kungl. Tekniska Högskolen, Department of Water Ressources Engineering.

GUJER, W. (1985), Ein dynamisches Modell für die Simulation von komplexen Belebtschlammverfahren. ERWAG, Dübendorf, Switzerland.

GUJER, W., HENZE, M. (1991), Activated sludge modelling and simulation, *Water Sci. Technol.* **23**, 1011–1023.

GUJER, W., HENZE, M., LOOSDRECHT, M., MINO, T. (1999), Activated sludge model No. 3, *Water Sci. Technol.* **39**, No. 1, 183–193.

HENZE, M. (1992), Characterization of wastewater for modelling of activated sludge processes, *Water Sci. Technol.* **25**, No. 6, 1–15.

HENZE, M., GRADY, C. P. L., GUJER, W., MARAIS, G. v. R., MATSUO, T. (1987), Activated-Sludge Model No. 1, *IAWPRC Scientific and Technical Reports, No. 1*. London: IAWPRC.

HENZE, M., GUJER, W., MINO, T., MATSUO, T., WENTZEL, M. C., MARAIS, G. v. R. (1995), Activated Sludge Model No. 2, *IAWQ Scientific and Technical Reports, No. 3*. London: IAWQ.

HENZE, M., HARREMOËS, P., LA COUR JANSEN, J., ARVIN, E. (1997), Wastewater Treatment – Biological and Chemical Processes, 2nd Edn. Berlin: Springer-Verlag.

MEINHOLT, J., ARNOLD, E., ISAACS, S., PEDERSEN, H. R., HENZE, M. (1988), Effect of continuous addition of an organic substrate to the anox phase on biological phosphorus removal, *Water Sci. Technol.* **38**, No. 1, 97–105.

PURTSCHERT, I., GUJER, W. (1999), Population dynamics by methanol addition in denitrifying waste water treatment plants, *Water Sci. Technol.* **39**, No. 1, 43–50.

REHM, H.-J., REED, G., PÜHLER, A., STADLER, P. (1991), *Biotechnology* 2nd Edn., Vol. 4, pp. 407–439. Weinheim: VCH.

ROELEVELD, P. J., KRUIT, J. (1998), Richtlinien für die Charakterisierung von Abwasser in den Niederlanden, *Korrespondenz Abwasser* **45**, 465–468.

Metal Ion Removal

21 Metal Removal by Biomass: Physico-Chemical Elimination Methods

GERALD BUNKE
PETER GÖTZ
RAINER BUCHHOLZ

Berlin, Germany

List of Abbreviations

Notation

a	$[\mathrm{L\,L^{-1}}]$	parameter of Langmuir isotherm
a_S	$[\mathrm{cm^2\,cm^{-3}}]$	sorbent particle surface per sorbent particle volume
b	$[\mathrm{L\,g^{-1}}]$	parameter of Langmuir isotherm
c	$[\mathrm{g\,L^{-1}}]$, $[\mathrm{mol\,L^{-1}}]$	metal concentration in solution
D_{ax}	$[\mathrm{cm^2\,s^{-1}}]$	axial dispersion coefficient
D_S	$[\mathrm{cm^2\,s^{-1}}]$	intraparticle surface diffusion coefficient
k_{des}	$[\mathrm{s^{-1}}]$	desorption rate constant
k_L	$[\mathrm{cm\,s^{-1}}]$	film diffusion coefficient
k_{sorp}	$[\mathrm{L\,g^{-1}\,s^{-1}}]$	sorption rate constant
K	[–]	sorption equilibrium constant
K_F	$[\mathrm{g^{1-n}\,L^n/L}]$	sorption equilibrium constant of Freundlich isotherm
q	$[\mathrm{g\,L^{-1}}]$, $[\mathrm{mol\,L^{-1}}]$	solid phase concentration
r	[cm]	radial distance
R_p	[cm]	particle radius
t	[s]	time
v	$[\mathrm{cm\,s^{-1}}]$	linear rate
z	[cm]	axial distance

Greek

α	[–]	separation factor
ε	[–]	porosity

Subscripts

i	component
j	component
0	initial condition

Superscripts

n	exponent of Freundlich isotherm
–	average value (bar over symbol)
sat	saturation

Abbreviations

BTM	dried biomass
M, Me	metal

1 Introduction

The contamination of wastewaters, river sediments and soil with toxic metals, in particular with heavy metals, is a complex problem. The removal of these contaminations has received much attention in recent years. From an environmental protection point of view, heavy metal ions should be removed at the source in order to avoid pollution of natural waters and subsequent metal accumulation in the food chain.

Conventional methods for removing contaminants are chemical precipitation, chemical oxidation, chemical reduction, ion exchange, adsorption (e.g., activated carbon, zeolite, clay minerals), filtration, electrochemical treatment, and evaporation. All these procedures have significant disadvantages, e.g., incomplete removal, high energy requirements, and production of toxic sludge or other waste products which also require disposal. Often these metals are very expensive and their reclamation would be desireable.

Alternative methods for metal separation from wastewaters have been developed in the last decade. The removal of heavy metals has received special attention. One of these methods is the biosorption of heavy metal ions on biomass (VOLESKY, 1994). Biosorptive methods have the following advantages:

- use of renewable biomaterials, which can reduce production costs,
- fast adsorption kinetics (KUYUCAK and VOLESKY, 1988),
- high selectivity of biosorbents (possibility to recover valuable metals, separation of metal mixtures),
- cleansing of aqueous solution with low metal concentration,
- high capacity by small equilibrium concentration,
- low investment and operation costs,
- easy desorption of metals by pH swing,
- low affinity with competing cations (calcium and magnesium).

Dead biomass, in particular, has high capacities for different heavy metals (VOLESKY, 1990a,b).

These biomaterials are used in the same way as synthetic adsorbents or ion exchangers.

Repeated regeneration is possible (WINTER et al., 1995). Biomass has also a high efficiency with low heavy metal concentrations (BERENDS and HARTMEIER, 1992; WINTER et al., 1994).

The high potential of biological systems for heavy metal adsorption is given in a survey by VOLESKY and HOLAN (1995). Until then the accessible maximum of adsorbed mass of metal per dried biomass was between 0.2–0.4 $g\ g^{-1}$. The marine brown alga *Ascophyllum nodosum* has an excellent capacity for cadmium with a maximum at 0.21 $g\ g^{-1}$ (HOLAN et al., 1993). Other biological materials with even higher capacity – up to 0.9 g metal (mercury) per g dried biomass – are in the test phase (FIESELER, 1998).

Some authors describe technical procedures using algae as biosorbents (DARNALL, 1991; DECARVALHO et al., 1995; LEUSCH et al., 1995). VOLESKY noticed that there are highly effective bioadsorbents for the separation of heavy metals from wastewater. Algae and fungi have the main significance for application in technical processes (KUYUCAK and VOLESKY, 1990; VOLESKY and HOLAN, 1995).

Clarification of the technological aspects of heavy metal separation by biomass is insufficient, although some publications exist (KUYUCAK and VOLESKY, 1990; DARNALL, 1991; HOLAN et al., 1993). A structural study of application for a technical process has not been carried out so far.

2 Place of Origin, Specific Amount and Composition of Wastewaters from Communities, Industries, and Agriculture

Place of Origin

Increased knowledge about ecotoxicological effects and further legislative demands for reductions in industrial emissions require research and development in the area of wastewater treatment. In this context metals – in particular heavy metals – represent an important class of contaminants. The heavy metal contamination of wastewaters is an intricate problem. Wastewaters from varying commercial and industrial sources are contaminated by different heavy metals with occasionally high levels of concentration. Heavy metal contaminated process wastewaters result from several industrial activities, mainly the metallurgical industry (electroplating plants, mine wastewaters, etc.), non-ferrous metallurgy, breweries, production of viscose fibers, and coal power stations.

Other sources of wastewaters which are contaminated with heavy metals are rain water sewage and wastewater from the glass industry.

Among other things, these sources are responsible for the feed of metals into the environment in dissolved form and as dust. In addition to these diffuse sources a large proportion of heavy metal feeds are caused by industrial wastewaters which come in insufficiently cleansed form into the receiving streams.

Practically each metallurgy concern, but also many other industrial producers are point sources of emission. They could be eradicated by useful removal procedures.

Immobilized heavy metals in river sediments, if released, can be a problem for drinking water production because the recovery of drinking water is achieved by riverbank filtration. The loading of surface waters by heavy metals is one of the main environmental problems in developed industrial countries. As a result of earlier, partially uncritical evaluation of industrial wastewater discharges all rivers contain large deposits of heavy metals bound in river sediments. Without any doubt this sediment of harmful pollutants cannot be disturbed indefinetely. There is always a risk that immobilized heavy metals could be released into the water and develop their biocidal action if the environmental conditions of the surface water are changed. For this reason every additional feed of heavy metals has to be avoided.

Specific Amount

New data about heavy metals in solutions are provided by FISCHWASSER (1998). In the Federal Republic of Germany special wastes of 1.8 mio t per year are still being produced to date in process solutions in the metal treatment and metal processing industry. These special wastes are among others caused by mordant etching, demetalization, chrome mordant processes and pollute the environment considerably.

About 80 inside-wall purification plants are currently in use for tank tracks and tank waggons in the Federal Republic of Germany. They have a wastewater production of 324 m^3 per week and purification plant (BOCK et al., 1998).

Other data of specific amounts of heavy metals in sewage sludge can be found in KICK (1986).

Composition

Organic compounds with different concentrations can exist besides heavy metal inorganic components. The concentration in which organic compounds occur ranges from extremely high, e.g., in brewery wastewaters (slop of beet molasses), to extremely low in wastewaters of adaptive metallurgy.

Slop of beet molasses has a very complex composition. It is a substrate containing to a high degree heavily degradable substances, e.g., betaine or Maillard products (FIEBIG, 1988). The mean composition of an undiluted slop of beet molasses without heavy metals has been reported in RAMMHOLDT (1994). Only materials, with concentrations higher than 1 g L^{-1} will be dealt with here. This applies to overall nitrogen content, phosphate content, sulfate concentration, magnesium content, total sulfur concentration, chloride content, nitrate content, potassium concentration, and sodium content.

Nickel electroplating solutions for use in metal surface protection are polluted with oxidation products of different organic additives. These concentrations are very low compared to the concentrations in the slop of beet molasses. Other wastewaters from precious metal powder production are polluted with hydrazine (N_2H_4) up to 1.3 g L^{-1} (FINKELDEI and PETERSEN, 1998).

The contaminations to be found in the wastewaters in addition to metals are composed very differently. The composition depends on the relevant process.

2.1 Analysis

In nature there exist 67 metals. They are subdivided into light metals (density <4.5 g cm^{-3}) and heavy metals (density >4.5 g cm^{-3}). Some heavy metals are, as trace elements, indispensable for humans, animals, and plants. Other heavy metals add to the pollution of the environment as radioactive nuclides or toxic substances. They are noxious, if specific thresholds are exceeded in the air, water, and soil. In order to analyze the metals a qualified random sample needs to be carried out. In Germany, metal concentrations in the watery phase are determined according to the instructions of the *"Allgemeine Rahmen-Abwasser-Verwaltungsordnung"* (ARV-89, 1989).

Atomic absorption spectroscopy (GF-AAS) is used for the estimation of metal concentration. Normally, a three-time determination is carried out for the formation of measured values. Another method is X-ray fluorescence spectrometry. Plasma emission spectroscopy (ICP/OES) is an additional analytical method. Photometry is used as an analytical method alongside the different spectroscopy methods. Quantitative metal determination by means of titration is also possible. Electrochemical methods, as voltammetry or polarography, can be used. Chromatographic analytical methods for the determination of metals are given in SCHWEDT (1980).

2.2 Maximum Permissible Concentrations Acceptable for Direct and Indirect Discharge to a Receiving Aquatic System, Limiting Values for Drinking Water

The permitted discharge values for discharges in surface waters (direct discharge) and discharges into the sewage system (indirect discharge) are laid down in national, European, and international guidelines. In Germany, the ARV-89 prescribes the maximum permissible concentrations for the discharge of process wastewaters into the receiving streams.

Tab. 1 summarizes the permitted discharge values for contaminated waters from ground water redevelopments by disturbances and accidents (LÜHR, 1992), the limiting values of drinking water of the "*Trinkwasserverordnung*" (TrinkwV, 1990) valid in the Federal Republic of Germany, and the maximum permissible concentrations in drinking water corresponding to the guidelines of the European Community (E.G.-TW, 1980) and the World Health Organization of the United Nations (WHO Guidelines, 1984) (see HEIN and SCHWEDT, 1991).

The WHO nominates guidelines for organic and inorganic water contaminations. The prescribed values correspond to reference values and lead values for the provision of limiting values for national legislation. These values should be the guide values for cleansing processes.

3 Metal Removal by Biomass – Leaching, Biosorption

Leaching

Valuable knowledge of the use of microorganisms for the extraction of pure metals has been gained in the field of mining. Metal extraction from weakly concentrated ores is predominantly carried out by microorganisms (ATLAS and BARTH, 1987). These organisms are able to survive in extreme acid environments and can remobilize metals from metal sulfides. In the literature this process is called "leaching". It is also applied more and more for the recovery of values from slag and sludge with sufficiently high heavy metal concentrations (BURGSTALLER and SCHINNER, 1993; VASQUEZ and ESPEJO, 1997). The same process was successfully used for the recovery of heavy metals from mining wastes (HSU and HARRISON, 1995). Attempts have also been made on a laboratory scale to apply this process to clarification sludge (ZANELLA et al., 1993). On the

Tab. 1. Permissible Concentrations for Direct Discharges in Receiving Streams; Limiting Values of Drinking Water

Metal		Contaminated Water	Drinking Water	Drinking Water	Drinking Water
		ARV-89 Discharge Guide Value [mg L^{-1}]	TVO-BRD Limiting Value [mg L^{-1}]	E-G.-TW Maximum Concentration [mg L^{-1}]	WHO-TW Guide Value [mg L^{-1}]
Al	aluminum	2.0 total	0.2	0.2	0.05
Sb	antimony	0.3 total	0.01	0.01	
Ba	barium	2.0 total	1.0		
Pb	lead	0.1 total	0.04	0.05	0.05
Cd	cadmium	0.005 total	0.005	0.005	0.005
Ca	calcium	0.005–5.0	400.0		
Cr	chromium	0.5 total	0.05	0.05	0.05
Fe	iron	0.5 total	0.2	0.2	0.3
K	potassium	0.005–5.0	12.0	12.0	
Cu	copper	0.3 total	3.0		1.0
Mn	manganese	0.5 manganese (II)-sulfate	0.05	0.05	0.1
Mg	magnesium	0.005–0.05	50.0	50.0	
Na	sodium	0.005–5.0	150.0	175.0	200.0
Ni	nickel	0.5 total	0.05	0.05	
Hg	mercury	0.005 total	0.001	0.001	0.001
Ag	silver	0.1 total	0.01	0.01	
Zn	zinc	0.5 total	0.1–5.0	0.1–5.0	5.0

one hand, leaching of the clarification sludge takes place throughout direct enzymatic sulfide oxidation by the microorganisms (*Thiobacillus* strains). On the other hand the metals are leached indirectly. In this case leaching takes place through trivalent iron after oxidation of the water-insoluble metal sulfides to water-soluble metal sulfate. Iron introduced with softening chemicals into the wastewater is used for the growth of *Thiobacillus ferrooxidans*. *T. ferrooxidans* is a chemolithoautotrophic, gram-negative, polar-flagellated bacterium. It is able to oxidize sulfur compounds and Fe^{2+} ions (KONISHI et al., 1992). The heavy metals are removed from the purified sludge. Enrichment with iron occurs at the same time. Utilization of the sludge without further separation steps is, therefore, not possible.

Biosorption

Conventional processes for heavy metal removal often are neither effective nor economical. Biosorption is a promising alternative. However, by the use of microorganisms as an adsorbent for heavy metals it is possible to achieve high capacity and selectivity. For this reason extensive considerations have been made regarding the utilization of this phenomenon for technical purposes. A general overview of the achieveable ratio of the adsorbent mass, its adsorbent capacity (enrichment factor in g metal per kg tried biomass per g metal per L solution: g Me kg^{-1} BTM per g Me^{-1} L) and the separable metal mass is given by RÖHRICHT et al. (1990) (Fig. 1).

The authors pointed out that it is possible to remove solved metals in the concentration range of 2–10 mg L^{-1} at reasonable cost by known physico-chemical procedures. Below this concentration range biosorption can be used effectively. It is then particularly effec-

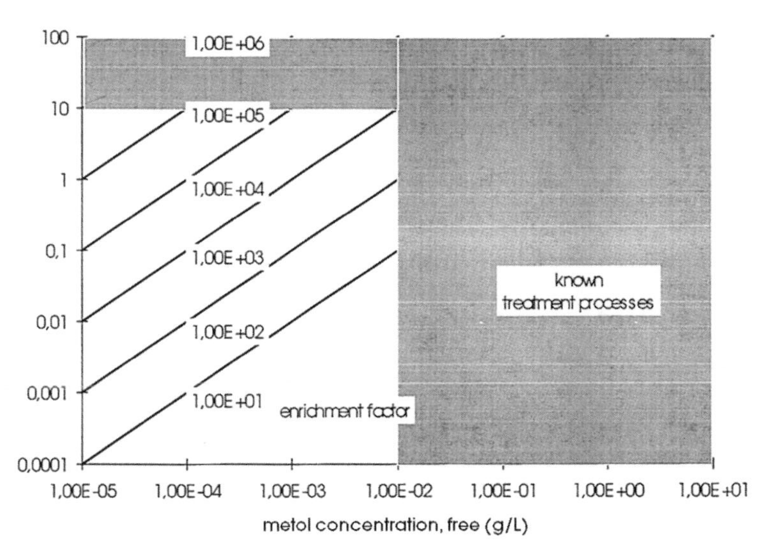

Fig. 1. Metal capacity of bioadsorbents as a function of dissolved metal concentration in the thermodynamical equilibrium by different metal enrichment factors (RÖHRICHT et al., 1990).

tive, if high enrichment factors are achieveable. Biomass with enrichment factors $>10^4$ are particularly interesting for technical process developments.

3.1 Bacteria, Algae, Fungi, other Biological Materials

The most severe environmental contamination by heavy metals is caused by industrial wastewaters. An effective and affordable method to cleanse these wastewaters is needed for removal of these contaminations. It is known that many microorganisms can be utilized in the enrichment of heavy metals. For many decades this capability of bacteria has been used, for e.g., for the deferrization and demanganization of groundwater in drinking water treatment on a technical scale. The microorganisms which are principally useful for biosorption of metals belong to bacteria, fungi, and algae. Other biological materials can also be applied for the removal of heavy metals from the water phase.

In Fig. 2 sorption isotherms are shown for four heavy metals using free or suspended bacteria as biosorbent at pH 7 according to RÖHRICHT et al. (1990). The authors got straight lines on the double-logarithmic scale for the investigated concentration range. This behavior can be described with the Freundlich model, which will be presented below. The differences in sorption capacities for these metals at low metal concentratios in the liquid phase can be clearly recognized.

3.1.1 Bacteria

The different microorganisms strongly differ in their affinity to metals (HORIKOSHI et al., 1981). RÖHRICHT et al. (1990) examined nine strains of bacteria and three yeasts for heavy metal removal. The results confirm the statement regarding affinity to different metals mentioned above. The authors reported that a clear classification of high and low activities of a certain group cannot be recognized yet. A clear classification was also not possible using literature data. Various experiments were carried out with either metabolically active bacteria or with resting bacteria. Many of these were characterized by slime production and/or floc formation (VOLESKY, 1990a). Bacterial cultures of *Zoogloea ramigera*, *Sphaerotilus natans*, *Citrobacter* sp., *Azotobacter* sp., and

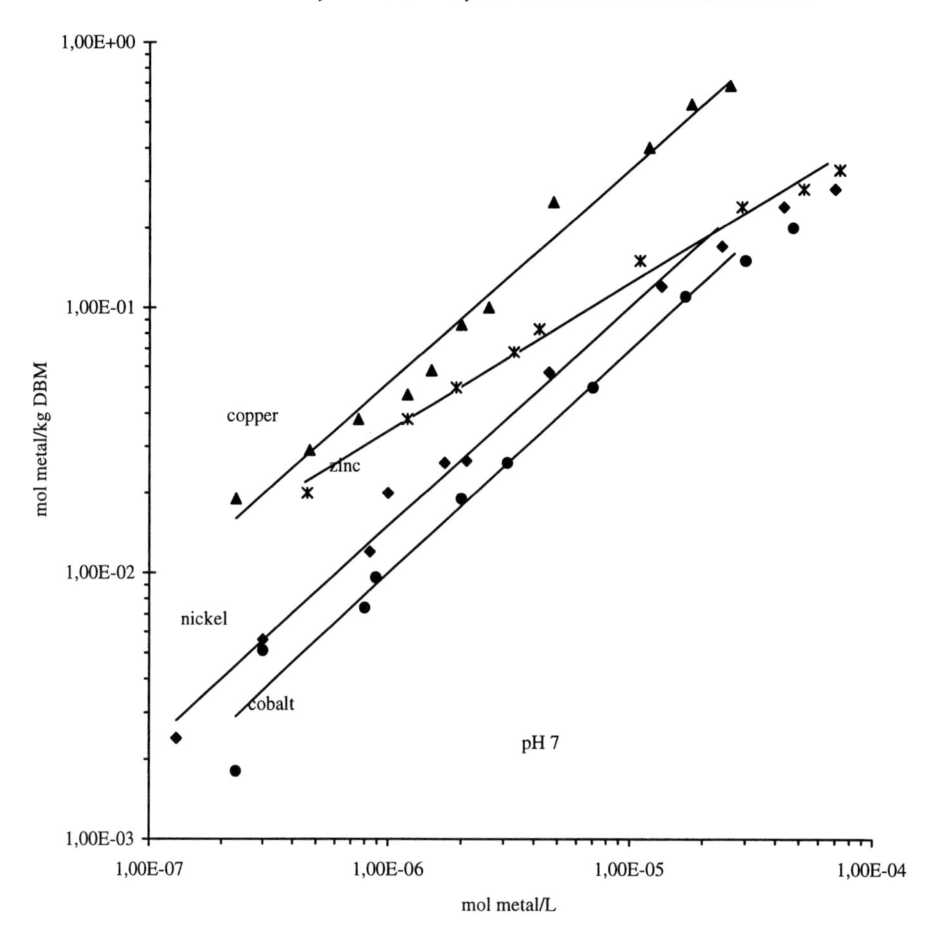

Fig. 2. Sorption isotherms of free suspended *Bacillus subtilis* (DSM 2019) for different metals at pH 7 (RÖHRICHT et al., 1990).

Klebsiella aerogenes were examined for their uptake of different metals. The results proved the importance of cell capsule expolymers for metal loading.

In addition to dead biomass, which functions as ion-exchange material, another kind of metal deposition using microorganisms has been known for many decades. These microorganisms, typically from oligotrophic biotopes, are able to efficiently precipitate iron, manganese and other metals as metallic oxides. This capability of bacteria is used for deferrization and demanganization of groundwater on a technical scale. Bacteria involved in these processes mostly belong to the species *Hyphomicrobium*, *Pedomicrobium*, *Gallionella*, *Leptothrix*, and *Siderocapsa*. It is noticeable that in these species often EPS structures (exocellular polymer substances, e.g., slime sheaths and capsules) are involved in the precipitation of metals (CORZO et al., 1994; GEDDIE and SUTHERLAND, 1993; GEESEY and JANG, 1989; MITTELMANN and GEESEY, 1985).

The bacteria mentioned above which have a capsule of slime layers or liberate polymers into the surrounding solution are very effective biosorbents (VOLESKY, 1990a).

In contrast to the biosorption mentioned above, precipitation of these bacteria takes place during the growth of cells. In examples of the species *Siderocapsa* and *Leptothrix* it is even found that the cells leave their slime sheaths after a large amount of metallic oxide was deposited on the slime sheaths.

Most representatives of these species are typically oligotrophic bacteria and they live under extremely poor nutrient conditions. Meanwhile, representatives of species *Leptothrix* and *Hyphomicrobium* were also found in organic highly loaded wastewater and activated sludge. *Sphaerotilus natans* is also a typical organism of wastewaters capable of metal deposition.

The majority of bacteria mentioned are sessile microorganisms and, therefore, are typically found in the biofilms of corresponding biotopes. First experiments for their application in biofilm reactors and for the exploration of interactions of these organisms with different metals have been made by HSIEH et al. (1985) and FERRIS et al. (1989).

Additional physiological groups of bacteria, e.g., sulfate reducers, were also tested for their capability of metal ion removal (FORTIN et al., 1994). Furthermore, first studies of the composition of the populations of microorganisms in mineral wastewaters were made by SCHIPPERS et al. (1995). The intensive interactions of bacteria with metals taking place in such waters were examined.

3.1.2 Algae

Microalgae are used for heavy metal removal. Algae are well known for their capacity to bind heavy metals to their cell walls (VOLESKY, 1990a; BERENDS and HARTMEIER, 1992). Microalgae can be used as adsorbents in heavy metal removal in vital as well as in devitalized form. Some of these biological adsorbents are able to remove heavy metals selectively from the water phase. Heavy metals can be recycled in a highly concentrated form by suitable desorption processes.

Cultivation of microalgae. Microalgae cultivated in the laboratory for reproducibility were used for heavy metal removal from model wastewater (WILKE et al., 1998). Microalgae from different origins were cultivated. They were obtained from central strain collections of the German Federal States (Göttingen, SAG) and from research institutes.

Method. Biomass production takes place in a batch reactor (glass cylinder) continuously provided with compressed air. Nutrient solutions necessary to grow the special algae are provided. The growth of biomass in the batch reactor is followed by measuring of pH value and biomass concentration. Biomass is separated from nutrient solution by centrifugation.

Biomass is washed with deionized water to achieve the same initial conditions for all successive experiments. Washing is needed to remove salt which comes from the nutrient solution. High concentrations of salt ions in the solution influence the adsorption capacity, especially in the case of marine algae cultivation. The washing process is very important because competitive ions from the cultivation media can significantly influence the sorption capacity.

Next the biomass is freeze-dried and used for the subsequent heavy metal adsorption experiments. The following strains were cultivated by this method:

- *Lyngbya taylorii*,
- *Chlorella kessleri*,
- *Microcystis* sp.,
- *Chlorella vulgaris*,
- *Spirulina maximum*,
- *Anabaena cylindrica*,
- *Spirulina platensis*,
- *Microcystis aeruginosa*,
- *Phaeodactylum tric.*,
- *Porphyridium cruentum*,
- *Tetraselmis* sp.,
- *Chlorella* sp.,
- *Spirulina laxissima*,
- *Dunaliella salina*,
- *Dunaliella bioculata*.

The influence of cultivation conditions on the biosorption capacity of heavy metals by microalgae is investigated using different cultivation solutions, variations of carbon dioxide and temperature.

Immobilization of biomass. Immobilization is needed for the effective use of biomass in packed bed columns as an adsorbent or ion exchanger (GREENE and BEDELL, 1990; KRATOCHVIL and VOLESKY, 1998). The cell diameter of the used microalgae ranges between 1–100 μm. The immobilization of microalgae into a hollow sphere with an external sodium cellulose sulfate membrane is one method for the development of an industrial process. The immobilized microalgae hollow spheres have diameters of about 2–3 mm. These diameters normally cause pressure drops in fixed bed columns. An immobilization method was used by WILKE et al. (1998). The equipment is illustrated in Fig. 3.

Method. For the immobilization procedure a sodium cellulose sulfate solution was mixed with freeze-dried microalgae for 2 h. The resulting mixture was dropped into a Polymin P solution, pH 7, containing NaCl. The positively charged sodium cellulose sulfate/microalgae solution immediately precipitates in the precipitation bath.

The resulting hollow spheres are surrounded by a thin membrane. After stirring for 1 h the immobilization process is complete.

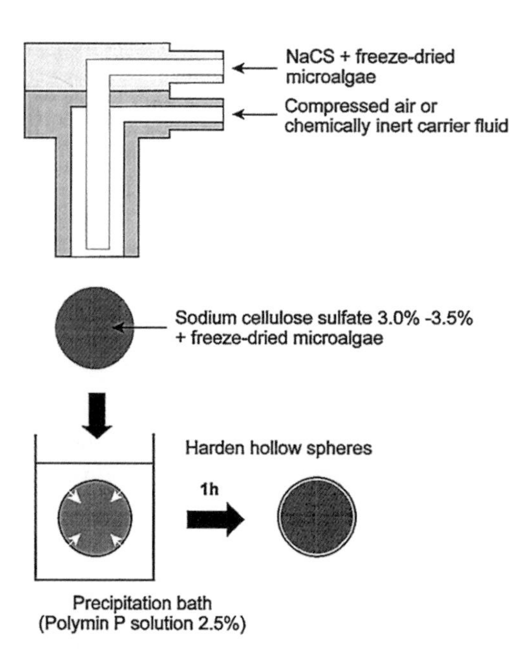

Fig. 3. Preparation of immobilized biosorbent (WILKE et al., 1998).

3.1.3 Fungi

Fungi also have excellent potential for metal biosorption. The genera *Rhizopus* and *Penicillium* and the family of the *Mucorales* particularly demonstrate this potential (VOLESKY, 1990 a).

Fungi are also used for ore dressing. Metals can be separated very efficiently from ores by *Sulfolobus* (bioleaching). Iron can also be released from high-iron ores with the help of heterotrophic bacteria and fungi. Fungi of the genera *Aspergillus* and *Penicillium* are able to achieve this.

Microbial leaching has been attributed to the formation of oxalic acid and citric acid by the microorganism. If high microbial leaching rates are desired a carbon source has to be added (e.g., molasses).

The possibility to extract the aluminum content from the aluminum silicate with microorganisms appears interesting. The fungi *Aspergillus niger*, *Scopulariopsis brevicaulis* and *Penicillium expansum* are especially suitable. Fungi form acids or lyes and are able to pulp rock chemically.

SINGER et al. (1982) reported that the aluminium included in the fly-ash from power stations could be extracted by microbial leaching. *Aspergillus niger*, which *in situ* produces citric acid, can be used for this purpose.

Microbial leaching of sulfide ore can also be achieved with fungi. The efficiency factor, however, is considerably lower than with bacteria (WAINWRIGHT and GRAYSTON, 1989).

Mycelia of *Aspergillus ochraceus* and *Penicillum funicolosum* can sorb uranium. The removal of metal ions from wastewaters can be accomplished with fungal biomass as an adsorbent (GADD, 1990).

One aim is wastewater purification and the other the recycling of valuable metals. A selectivity for thorium exists in acid environments so that the purification of appropriate industrial wastewaters is possible.

Radium and thorium were recycled by means of *Rhizopus arrhizus* mycelia. The metal uptake is higher in this case than for activated carbon. Anion exchange resins have only half the capacity for thorium in comparison with fungal biomass. *Rhizopus arrhizus* mycelia are highly selective for these metals. The loading is not disturbed in the presence of other metal ions or harmful substances (WAINWRIGHT, 1995).

Desorption can be done by means of hydrogen carbonate solutions, EDTA solutions, and ammonium carbonate solutions.

Industrial processes using fungal biomass as adsorbent, which can compete with the physico-chemical processes, are in spite of the existing advantages not available.

3.1.4 Other Biological Materials

Yeasts are able to leach zinc, copper, and iron from filter dust biologically (WENZEL et al., 1990). LEE et al. (1997) investigated the removal of copper, lead, and cadmium by using apple residues. Cation exchange capacities were estimated. The influence of different parameters (pH, ionic strength, ligands and co-ions, feed concentration, and chemical treatment) was studied in batch and column experiments. The authors reported on agricultural wastes such as tree bark, peanut skin, etc. which were used to remove heavy metals from water and which were described in the literature.

3.2 Mechanism of Heavy Metal Removal: Adsorption, Accumulation

It is well known that some strains of microorganisms are able to remove heavy metals from aqueous solutions. The metal uptake can be divided in two different mechanisms:

- process of bioaccumulation,
- process of biosorption.

Bioaccumulation. Bioaccumulation is defined as the active uptake of metals by vital organisms into living cells. For this purpose the microorganisms consume energy. The process of bioaccumulation is very slow compared to the biosorption process.

Biosorption. The expression of "biosorption" is a generic term for the passive attachment (sorption) of heavy metals to biological molecules based on complexation, chelate formation, or ion exchange. Biosorption can be characterized as the reversible and fast reaction of metal ions with functional groups of cell wall polymers. The cell walls of algae consist of a multitude of polysaccharides (polynannuronic acid, polyguloronic acid, fucoidan) which have negatively charged ligands like functional groups (carboxyl, sulfate groups).

Adsorption phenomena in extracellular biopolymers are often also classified as biosorption (PONS and FUSTE, 1993). The expression of biosorption is used differently in the literature. As a rule the term biosorption is understood as the whole of the accumulation proceedures in respective living or dead biological cell material (MURALEEDHAVAN and VENKOBACHAR, 1990).

3.3 Distribution Equilibrium

Basic Concepts of Distribution Equilibrium

A short review of the constants, coefficients, and equations used to describe the distribution equilibrium will be helpful in discussing results on metal sorbent systems. The preference of the sorbent for one of two metals in the solution can be expressed by the following separation factor (Eq. 1):

$$\alpha_{ji} = \frac{q_j c_i}{q_i c_j} \tag{1}$$

Assuming a constant temperature and a very low concentration of metal in the solution, a linear equation between both metal concentrations in the solution and in the sorbent phase (linear sorption isotherm) exists in equilibrium (Eq. 2):

$$q_i = K_i c_i \tag{2}$$

Here is K_i defined as the equilibrium constant of the sorption process. A linear function for the sorption isotherm only exists for a small metal concentration range. The metal concentration in the solution is usually so high that the isotherm is nonlinear. The solution usually contains more than one component. The complex interaction between the dissolved components and the surface of the sorbent result in different isotherms. A classification of the different forms was proposed by GILES et al. (1974). One of the most frequently used equations for the description of nonlinear form of the isotherm is the Freundlich model (Eq. 3):

$$q_i = K_{F,i} c_i^n \quad (3)$$

Another possibility to describe the adsorption process is given by the Langmuir equation. At low metal concentrations in the solution the Langmuir equation passes into the linear equation, and in high concentration, the result is a constant maximum load. For two competitive components the following Langmuir equation arises (Eq. 4):

$$q_i = \frac{a_i c_i}{1 + b_1 c_1 + b_2 c_2} \quad i = 1, 2 \quad (4)$$

This equation can be extended to more than two components. A multitude of isotherm models for the description of the sorption equilibrium exists in the literature. None of these models has the significance of the Freundlich or Langmuir models.

The constants, coefficients, and equations for ion exchange, complexation, or adsorption are used depending upon which mechanisms are responsible for the interaction between the metal component in the solution and the surface of the sorbent.

Screening of Heavy Metal Loading

The ability of microalgae to adsorb heavy metals with high capacity and/or selectivity is tested in the screening phase.

Adsorption Experiments

Method. A selected algae biomass is put in a metal salt solution with constant concentration. Biomass suspension is centrifuged after a shaking time of 30 min. Metal concentrations of culture supernatant are analyzed by AAS. The heavy metal loading of algae is estimated by decomposition of biomass.

KLIMMEK and STAN (1998) reported results regarding the loading possibility for four selected heavy metals by 16 different algae (15 microalgae, one macroalga). Free microalgae were used for sorption in batch experiments. The initial concentrations c_0 were 400 mg L^{-1} for lead and for cadmium, nickel, and zinc 100 m L^{-1}. The results of this screening are shown in Fig. 4.

A wide range of sorption capacity of the different strains of microalgae and of the four selected metals can be observed. The sorption of lead shows the highest capacity for most strains, even if the adsorption is calculated in mole heavy metal per dried biomass. The lead uptake by the marine cyanobacterium *Lyngbya taylorii* amounts to 175 mg lead per g dried biomass. This value still does not represent the full saturation loading of the sorption isotherm. This blue-green alga shows the highest uptake of nickel and cadmium and, in the case of zinc, almost the highest uptake.

Several metals occur simultaneously in real wastewaters. For this reason the investigation of metal ion competition during sorption is of interest.

WOLF et al. (1977) investigated the iron exchange in peat. In a ternary system including H^+, Ca^{2+} and a heavy metal (Pb^{2+}, Cu^{2+}, Cd^{2+}, or Zn^{2+}) relative selectivity of the peat for the different ions was observed.

The ion exchange selectivity or competitive exclusion was demonstrated. The selectivity was higher for H^+ compared to Ca^{2+}. The replacement of stronger bound H^+ ions with less tightly bound Ca^{2+} ions facilitates the sorption of heavy metal ions. Heavy metals can displace Ca^{2+} more easily than H^+.

The results showed that the more Ca^{2+} added, the fewer exchange sizes were there for heavy metal loading. This means that peat has a higher exchange capacity for heavy metals than for Ca^{2+}. Similar results were observed by BUNZL et al. (1976).

IRVING and WILLIAMS (1948) observed that stability complexes of bivalent ions follow the order

$$Pb^{2+} > Cu^{2+} > Ni^{2+} > Co^{2+} > Zn^{2+} >$$
$$Cd^{2+} > Fe^{2+} > Mn^{2+} > Mg^{2+} .$$

HELFFERICH (1962) stated that the ion exchanger prefers the counter ion of higher valence. KHAN (1969) reported that the pH drop upon addition of metal cations to humic acid followed the order from higher to lower stability:

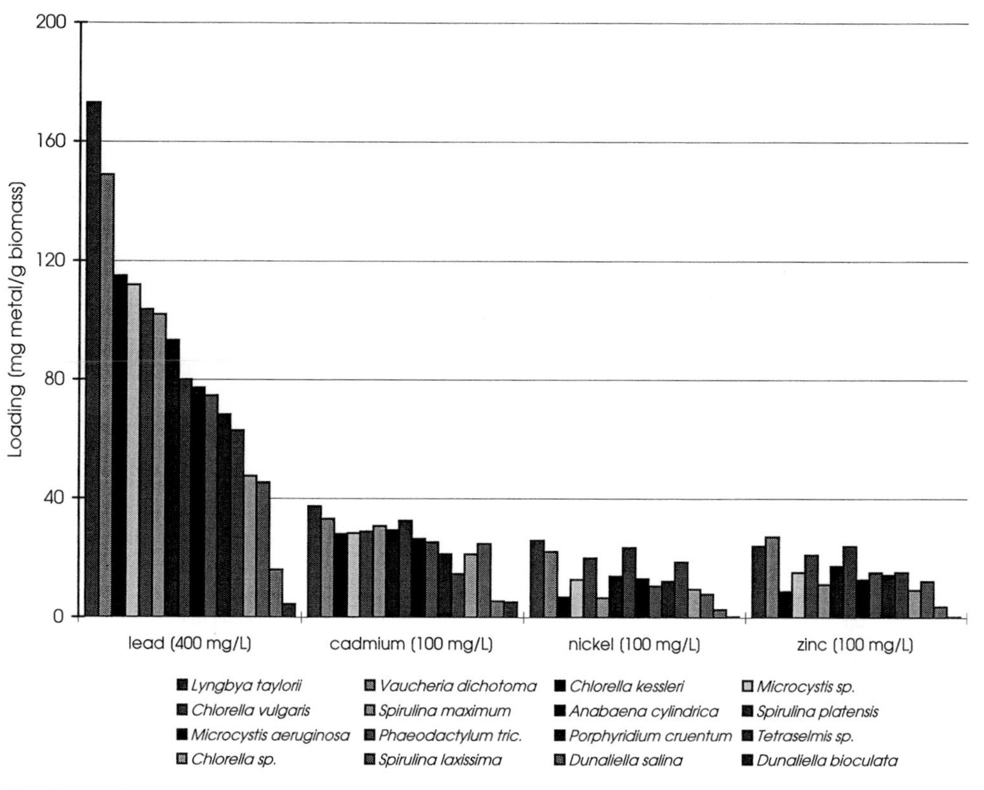

Fig. 4. Metal uptake of 16 algae in the screening phase (initial concentration of batch experiments in brackets, 20 mg algae in 9 mL metal salt solution, pH 5–6) (KLIMMEK and STAN, 1998).

$$Fe^{3+} > Al^{3+} > Cu^{2+} > Zn^{2+} > Ni^{2+} > Co^{2+} > Mn^{2+}.$$

TAKAMATSU and YOSHIDA (1978) observed that the order of stability constants for three metal–humic complexes was:

$$Cu^{2+} > Pb^{2+} \gg Cd^{2+}.$$

BUNZL et al. (1976) found that the distribution coefficients for ion exchange in peat followed the order:

$$Pb^{2+} > Cu^{2+} > Cd^{2+} > Zn^{2+} > Ca^{2+}.$$

GAMBLE et al. (1983) used humid acid and found that bivalent metal ions follow an Irring–Williams type series.

This information can be found in KADEC and KEOLEIAN (1986). All these authors tried to find a rule for the selectivity of metal cation interactions with organic material. This overview shows that there is no absolute rule for the selectivity of metal cation uptake.

New research describes another method (KRATOCHVIL and VOLESKY, 1998). Biosorbents can be prepared in different ionic forms similar to synthetic ion exchange resins. Biosorbents can be protonated (H form). They can be saturated with Ca^{2+}, Mg^{2+}, Na^{+}, etc. by washing with mineral acids, salt solutions and/or bases. As a result it is possible to selectively separate the heavy metals from the water phase. The desorption chemicals are selected so that an optimal process is possible.

Adsorption Equilibrium of Immobilized Microalgae

WILKE et al. (1998) ascertained that high loading of biomass can also be achieved by the use of immobilized microalgae. A matrix material to be selected for immobilization should include a high loading capacity for heavy metals.

Investigations of the influence of calcium as a competing ion showed a decrease of the adsorption capacity with an increase of the calcium concentration.

3.4 Kinetics

Kinetics of Heavy Metal Loading by Use of Non-Immobilized Microalgae

The expression "sorption kinetic" is to be understood as the velocity of adjustment of distribution equilibrium between metal concentration of biomass surface and metal concentration in the surrounding fluid phase.

Kinetic models contain equations for the description of the finite quick kinetic of mass transfer processes between fluid phase and fixed sorbent phase as well as in both phases. These equilibrium systems result from the formation of balance equations. These equations usually form partial differential equations, which are solved with analytical or numerical methods by using necessary initial and boundary conditions.

Assuming mass transfer from the bulk phase to the external surface of the sorbent particle (film diffusion) to be the rate-limiting step, the following differential equation for the estimate of average sorption changing of sorbent particle in time is obtained (Eq. 5):

$$\frac{\delta \bar{q}_i}{\delta t} = k_{L,i} a_S (c_i|_{r=R_P} - c_i(\bar{q})) \tag{5}$$

Mass transfer is caused as a result of the difference between actual particle random concentration $c_i|_{r=R_P}$ and equilibrium concentration. In addition to the sorption equilibrium, some sorbent parameters and the film diffusion coefficient k_L have to be known in order to describe the passage of time of sorption according to the application of the film diffusion model. The film diffusion coefficient can be determined experimentally.

Many empirical equations also exist in the literature for the calculation of k_L (SONTHEIMER et al., 1985).

With increasing sorbent particle size diffusion processes in the pore system play a role. Quantitative analysis of intraparticular diffusion requires knowledge of pore systems and effective diffusion coefficients of sorbed substances.

Mass transfer in fluid filled pores can take place by diffusion of solved components in the pore liquid inside the particles. This transport mechanism is known as pore diffusion.

Transport in pores can also take place on inside surfaces in a sorbed state. This transport mechanism is known as surface diffusion.

All of these models are based on the basic principle of the 2nd Fick's equation. The equation for spherical adsorbent particles under the use of loading difference as driving force is shown as an example. Surface diffusion coefficients are assumed to be constant (Eq. 6):

$$\frac{\delta \bar{q}_i}{\delta t} = D_{S,i} \left(\frac{\delta^2 q_i}{\delta r^2} + \frac{2}{r} \frac{\delta q_i}{\delta r} \right) \tag{6}$$

The surface diffusion coefficient $D_{s,i}$ can be estimated by the adaption of the calculated concentration–time curves to experimental values.

A further value influencing the kinetics in the case of finite sorption rate is the introduction of the sorption equilibrium into active centers. This influence can be taken into account by the use of two rate constants, one for loading and another for desorption, and by assuming a saturation loading value (Eq. 7):

$$\frac{\delta \bar{q}}{\delta t} = k_{sorp} c (q_{sat} - \bar{q}) - k_{des} \bar{q} \tag{7}$$

A combination of the described kinetic resistances is in many cases necessary to achieve a satisfactory model appropriation. Such questions were dealt with by ROSEN (1952), BABCOCK et al. (1966) and SLONINA and SEIDEL (1991).

If only ion exchange processes take place, then the corresponding appropriate transport models are used as described in KADLEC and KEOLEIAN (1986).

Method. An equal amount of metal salt solution with a constant concentration c_0 is put in a number of Erlenmeyer flasks. The same biomass is put into all sample containers. The experiments are stopped by centrifugation after a shaking time of, e.g., 3, 5, 15, 30 and 60 min, respectively, and the metal concentration of the culture supernatant is analyzed by AAS. Heavy metal loading of biomass is estimated by decomposition. By this a number of measuring points are achieved for the kinetic curve.

Less then 5 min is sufficient for adjustment of the distribution equilibrium in a shaking flask by using cadmium, nickel, lead, and zinc. These results are only valid for non-immobilized microalgae (KLIMMEK and STAN, 1996).

Kinetics of Heavy Metal Adsorption by Use of Immobilized Microalgae

Additional transport resistances are developed due to immobilization, and the results are that, compared to non-immobilized microalgae, the heavy metal loading slows down during the adsorption phase and the heavy metal elution also becomes slower during the desorption phase.

Kinetic measurements for the estimation of time needed to achieve distribution equilibrium by the use of microalgae showed that the adsorption equilibrium is only reached after more than 90 min.

3.5 Desorption

Regeneration of Microalgae

A possibility for the regeneration of loaded microalgae is the desorption of metal ions in a shaking flask experiment.

Method. The supernatant is decanted after centrifugation of microalgae. In this tube the biomass is resuspended in a 0.1 N HCl solution. After a shaking time of 10 min the biomass suspension is centrifuged and the desorbed heavy metal concentration of culture supernatant is analyzed. The concentration in the liquid points out the mass of desorbed heavy metal ions. Now the regenerated biomass can be used for the next adsorption process or the residual loading can be estimated by the decomposition of algal biomass. Both, HCl solutions (pH swing procedure) and EDTA as complexing agent are to be used for the desorption of metal ions.

Desorption of Immobilized Microalgae

The results of regeneration experiments demonstrate that loaded immobilized microalgae biomass is able to regenerate by pH swing. Other parameters for the research of regeneration possibilities are concentrations of salt and complexing agent and temperature.

3.6 Metal Ion Removal in Fixed Bed Columns

Fixed Bed Columns

The construction of fixed bed columns for metal ion removal requires distribution equilibrium measurements to determine adsorption isotherms and kinetic studies to estimate the time which is needed to achieve this distribution equilibrium.

The above mentioned transport resistance appears, if immobilized biosorbents are used in batch experiments (film diffusion, intraparticular diffusion in biosorbent particles, finite sorption rate to reach the sorption equilibrium). Two more transport mechanisms appear (mass transfer via convection or via axial dispersion), if immobilized biosorbents are used in column experiments. The convective mass transfer through the column is put in a diffusion transport, which causes a change of concentration profile by back mixing in axial direction z. Extensive investigations of axial dispersion are published by WESTERTERP et al. (1984).

Eq. (8) describes the mass transfer in the fluid phase in case of additional transfer resistance:

$$\frac{\delta c}{\delta t} + \frac{1-\varepsilon}{\varepsilon}\,\frac{\delta \bar{q}}{\delta t} + v\,\frac{\delta c}{\delta z} - D_{\mathrm{ax}}\,\frac{\delta^2 c}{\delta z^2} = 0 \qquad (8)$$

In this equation the accumulation in the biosorbent phase appears decoupled from the accumulation in the fluid phase. The balance equation for the sorbent phase is to be solved as an additional equation to describe the change of average loading $\bar{q}$ in time. The effects described above can be the rate-limiting steps.

RUTHVEN (1984) presented various models for the design of fixed bed sorption processes and also discussed solutions for these processes. Similar models were used by CHRIST et al. (1994) for ion exchange in fixed bed columns. The major advantages of these models are that they can predict the performance of a fixed bed column process under various conditions.

3.7 Cyclic Adsorption and Desorption

Cyclic investigations in fixed bed columns are carried out to estimate the working lifetime of the immobilized biomass. The concentration–time function is measured at the end of the column under dynamic conditions. The regeneration of the loaded biosorbent can be realized by pH swing. It is necessary to determine whether, after a number of cycles, the adsorption capacity remains nearly unchanged.

Biosorbents should be recycled for economic reasons. In order to execute this process the sorbed metals have to be desorbed. The more adsorption–desorption cycles are possible without destruction of biomass, the more economical the process.

3.8 Applications

The equipment which could possibly be integrated in a production process is of main significance in the future. Such processes could be used for valuable product recycling and, e.g., for the retention of dangerous materials at the place of origin of wastewater emissions. Only there could the recycling of process water for repeated use in the factory be possible. For this intelligent and custom-made environmental process engineering solutions are necessary. A flow diagram of such a process is shown in Fig. 5.

For example, wastewater from an adaptive metallurgy concern is cleaned in the sorption column. Provided there is additional organic or inorganic waste, the decontaminated water is either recycled or removed from this process. One part of the decontaminated water can be used for the preparation of desorption solu-

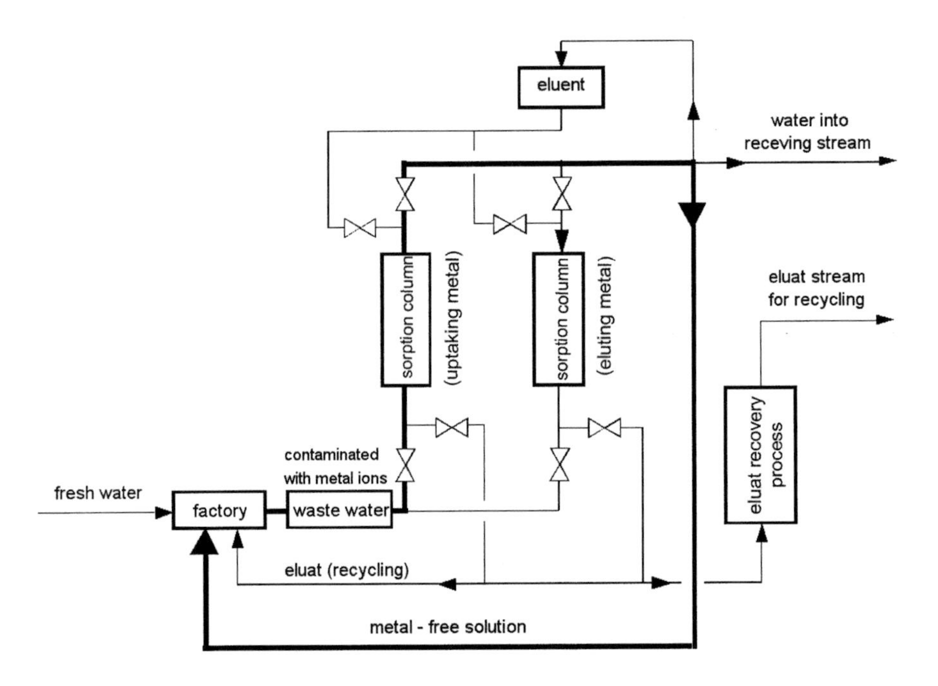

Fig. 5. Schematic diagram of a split treatment.

tion. Highly concentrated eluates are produced during the regeneration of biosorbent. These eluates are led back into the process (e.g., into an electroplating plant). This is the preferred process to aim for. If there are eluates with many components and a recycling process is not possible recovery is realized in a special step. The eluates are removed in this case.

Deficits in the area of biosorption of heavy metals exist in transferring laboratory results into applicable processes. Research is, therefore, aimed at solving this problem. In the light of recent developments in research, some technical solutions have been developed for heavy metal removal from wastewater. By new methods of isolation, enrichment and cultivation of relevant organisms the biological potential available for heavy metal removal will be increased and optimized for technical applications.

Some patents exist for the use of dead biomass for metal removal from the water phase. A granulated *Bacillus* preparation is used for the AMT–BIOCLAIM process (BRIERLEY et al., 1986; BRIERLEY, 1990). A biosorbent exists in the USA, produced from algae (Algasorb™). This sorbent consists of immobilized algal cells in silicon (GREEN et al., 1987).

WAINWRIGHT (1995) reported on a semi-industrial process of metal biosorption by means of fungal biomass (Biosorbent M). The process is used for the separation of radioactive nuclides from wastewaters.

3.8.1 Batch Reactor, Fluidized Bed Reactor

Algal and fungal biomass for laboratory investigations are usually cultivated in batch processes. The estimation of distribution equilibrium between metal and biosorbent occurs mostly in batch processes.

3.8.2 Fixed Bed Reactor

Fluidized bed reactors and sponge reactors are also used (WHITE and GADD, 1990). WHITE and GADD reported that 90–95% of the thorium included in the solution could be removed with an air lift bioreactor. *Rhizopus arrhizus* and *Aspergillus niger* were the most effective fungi tested.

Packed adsorption columns, or fixed bed reactors, are primarily applicable, if biosorbent systems are used. Wastewater which has to be purified passes through the packed bed either from bottom to top or in the reverse direction. The desorption solution passes through the packed bed in either direct current or in reverse current. The reverse current has its advantages in an incomplete loading of biosorbent. Multicolumn units are used for the realization of the continuous removal of metals.

Two interconnected columns in series are applied for the removal of metals in the process of fungal biomass, as presented by WAINWRIGHT (1995).

Immobilized biosorbent particles can be packed in sorption columns. The use of particles in such a process has many advantages and is perhaps the most effective possibility for the continuous removal of metals from wastewaters (KRATOCHVIL and VOLESKY, 1998).

4 Physico-Chemical Elimination Methods

Conventional methods of heavy metal removal from wastewaters are based on methods of chemical precipitation and sludge separation, chemical oxidation or reduction, ion exchange, adsorption, reverse osmosis, membrane technology, filtration, electrochemical treatment, and evaporation. Here continuous work on increasing the efficiency of these processes takes also place.

4.1 Physico-Chemical Precipitation Methods

These procedures have advantages in highly concentrated wastewaters (>100 mg L^{-1}). In lowly concentrated wastewaters (<100 mg L^{-1}) these procedure often are too ineffective or too expensive. For this reason, in Germany the threshold values of heavy metal discharges in waters for many industry branches have to be increased or removed in certain cases.

Very low wastewater threshold values can only be maintained by using these processes at high cost (RÖHRICHT et al., 1990).

Metal cations may be removed from the wastewater after conversion into suitable oxidation states as insoluble products.

Neutralizing precipitation and carbonate precipitation are established and often utilized in precipitation processes. Their effectiveness, however, is strongly dependent on the attendant substances. In the presence of complexing agents precipitation can, e.g., take place only incompletely. Sulfide precipitation, however, may lead to satisfactory results in the presence of complexing substances. Only sulfide precipitation on a technical scale is of insignificant importance. This is the result of high toxicity and low threshold of odor concentration of H_2S. Expensive equipment results from this. Additionally, many precipitation agents have been developed recently so that this process is not essential.

4.2 Ion Exchange, Adsorption, Membrane Plants

Ion exchangers are used for the cleansing of weakly concentrated metal solutions. In the presence of complexing agents the separation of cations is mostly incomplete. Additionally, an irreversible damage of ion exchangers can arise from organically highly loaded wastewaters.

Different highly porous substances can be used as adsorbent materials. Activated carbon is often applied. The advantage of activated carbon is that heavy metals are found in concentrated form in ash after burning.

Also zeolites show high efficiency for the removal of heavy metals (ZAMZOW et al., 1990). A comparison of the results using different adsorbent materials (e.g., biomass, activated carbon, zeolite, clay) and their applications for special technical processes is lacking.

Reverse osmosis, electrodialysis, and fluid membrane processes are part of membrane separation processes. All these processes are extremely sensitive in the presence of contaminations. Therefore, they are often unsuitable for the cleansing of wastewaters from the food industry.

4.3 Other Processes, Electrolysis, Spray Evaporation

As a result of the high energy cost usually electrolysis is only applied in the production of precious metals. In addition, the concentrations should be in the upper milligram to gram range. Electrolysis gains importance as a result of increasing cost for the disposal of highly loaded wastewaters.

Spray evaporation is an innovative process for the treatment of watery fluids resulting from metal working. Spray evaporation separates old emulsions without auxiliary contrievances for the breaking of emulsions. Water is separated from the emulsified or solved components by evaporation in a closed cyclic process. Water is condensed. The sum of metals in condensate may be held below 0.05 mg L^{-1}. Thus, the effluent threshold values can be retained.

5 Summary

Industrial use of metals is widespread due to their physical and chemical properties. As a side effect they are introduced into the global ecosystem. A major part of the resulting impact on the environment stems from introducing wastewater containing metals into rivers, lakes, or groundwater. Therefore, the amount of toxic metals immobilized in river sediments is still high. In order to fight damage to the environment by metals, there are national and international limits and regulations for the discharge of contaminated water. These limit concentrations are regularly reduced due to the health hazards from metal emissions. The use of biomass for reducing the amount of metals in wastewater and for metal recovery is a research topic worldwide. Bacteria, algae, and fungi are used in vital and non-vital forms as biosorbents. Other biological materials (yeast, plant residues, etc.) showed promising results in metal removal applications. Biosorbents can be used for the removal of metal contaminations from wastewaters, surface waters, and groundwaters. Water purification is possible up to the demanding thresholds allowed for drinking water. Biological materials have to be modified by immobilization so that they will be insoluble in water and remain in the apparatus. Particle form, particle diameter, and particle consistency may be adjusted by immobilization, so that technical flow problems can be restricted (HARTMEIER, 1986).

Although the use of biomass in these processes yields superior results compared to physical and chemical methods, there is still no large-scale industrial application.

Leaching of metals from low-grade ores can be accomplished by microorganisms. It is also increasingly applied for the recovery of values from slag and sludge, if a sufficiently high heavy metal concentration exists.

6 Outlook

Conventional methods of heavy metal ion removal are comparable with procedures using biosorbens.

State-of-the-art heavy metal removal from wastewaters with biological adsorbent procedures are shown, which are useable on a technical scale. It is also clearly pointed out which problems in the secure construction of plants exist at present. Detailed knowledge about the interactions between biomass and adsorption performance does not exist yet. Results in the literature and of present research conducted allow clear objectives.

Further necessary investigations of the internal and external mass transfer for different matrices (influence of biomass on mass transfer) are to be carried out. A prediction of the performance of biosorption by division into specific classes of material systems (combination of adsorbate, adsorbent, and characteristics of solution) is to be undertaken.

Experimental investigations in connection with subsequent mathematical modeling of biosorption processes leads to the expectation of progress in understanding this kind of substance separation problem.

Complex mathematical models are used for the description of the complicated process connections for real separation processes. Analytical solutions are useful in special cases. These solutions are clear and easy to handle. Fast computers exist for the mathematical description of technically interesting processes which can also be used for complicated mathematical connections. These complicated equation systems are to be described with non-linear partial differential equation systems. Solutions of such partial differential equation systems require numerical solution procedures.

Costs for wastewater cleansing should not increase production costs considerably. The development of processes for effective and inexpensive removal of heavy metals from wastewater to very low concentrations is, therefore, of great importance.

7 References

ARV-89 (1989), *Allgemeine Rahmen-Verwaltungsvorschrift über die Mindestanforderungen an das Einleiten von Abwasser in Gewässer* – Rahmen-Abwasser-VwV vom 8.9.1989 (GMBI. p. 518), geändert durch allgemeine Verwaltungsvorschrift vom 19.12.1989 (GMBI. p. 798).

ATLAS, R. M., BARTH, R. (1987), Microorganisms in mineral and energy recovery and fuel and biomass production, in: *Microbial Ecology* (ATLAS, R. M., BARTH, R. Eds.), pp. 442–463. Menlo Park, CA: Benjamin/Comings Publishers.

BABCOCK, R. E., GREEN, D. W., PERRY, R. H. (1966), Longitudinal dispersion mechanisms in packed beds, *AIChE J.* **12**, 922–927.

BERENDS, A., HARTMEIER, W. (1992), Biosorption von Schwermetallen im Trinkwasserbereich, *Wasser & Boden* **44**, 508–510.

BOCK, D., MENZEL, U., ROTT, U. (1998), Abwasseranfall und -zusammensetzung sowie Brauchwassernutzung am Beispiel der Tankwagenreinigung, in: *Preprints, Colloquium Produktionsintegrierter Umweltschutz, Abwässer der metallverarbeitenden Industrie und des Transportgewerbes*, Bremen, Sept. 1–2 (IUV Universität Bremen, GVC-VCI, Eds.). pp. D-45–D-50, Bremen, Germany.

BRIERLEY, J. L. (1990), Production and application of a *Bacillus*-based product for use in metal biosorption, in: *Biosorption of Heavy Metals* (VOLESKY, B., Ed.), pp. 305–311. Boca Raton, FL: CRC Press.

BRIERLEY, J. L., BRIERLEY, C. L., GOYAK, G. M. (1986), AMT–BIOCLAIM: A new wastewater treatment and metal recovery technology, in: *Fundamental and Applied Biohydrometallurgy* (LAWRENCE, R. W., BRANION, R. M. R., EBNER, H. G., Eds.), pp. 291–304. Amsterdam: Elsevier.

BUNZL, K, SCHMIDT, W., SANSONI, W. (1976), Kinetics of ion exchange in soil organic matter. IV. Adsorption and desorption of Pb^{2+}, Cu^{2+}, Cd^{2+}, Zn^{2+} and Ca^{2+} by peat, *J. Soil Sci.* **27**, 32–41.

BURGSTALLER, W., SCHINNER, F. (1993), Leaching of metals with fungi, *J. Biotechnol.* **27**, 81–116.

CHRIST, R. H, MARTIN, J. R., CARR, D., WATSON, J. R., CLARKE, H. J., CHRIST, D. R. (1994), Interactions of metals and protons with algae. 4. Ion exchange vs. adsorption models and reassessment of Scatchard plots; ion-exchange rates and equilibria compared with calcium alginate, *Environ. Sci. Technol.* **28**, 1859–1866.

CORZO, J., LEON-BARRIOS, M., HERNANDO-RICO, V., GUTIERREZ-NAVARRO, A. M. (1994), Precipitation of metallic cations by the acidic exopolysaccharides from *Bradyrhizobium japonicum* and *Bradyrhizobium (Chamaecytisus)* strain BGA-1, *Appl. Environ. Microbiol.* **60**, 4531–4536.

DARNALL, D. W. (1991), Removal and recovery of heavy metals from wastewaters using a new biosorbent AlgaSORB, Innovative Hazard, *Waste Treat. Technol. Ser.* **3**, 65–72.

DECARVALHO, R. P., CHONG, K. H., VOLESKY, B. (1995), Evaluation of the Cd, Cu and Zn biosorption in two-metal systems using algal biosorbent, *Biotechnol. Prog.* **11**, 39–44.

E.G.-TW (1980), Richtlinien des Rates vom 15.07. 1980 über die Qualität von Wasser für den menschlichen Gebrauch, in: *Amtsblatt der Europäischen Gemeinschaften Nr. L 229 vom 30.8. 1980*, pp. 11–29.

FERRIS, F. G., SCHULTZE, S., WITTEN, T. C., FYTE, W. S., BEVERIDGE, T. J. (1989), Metal interactions with microbial biofilms in acidic and neutral pH environments, *Appl. Environ. Microbiol.* **55**, 1249–1257.

FIEBIG, R. (1988), Versuche zur ein- und zweistufigen anaeroben Behandlung von Rübenmelasseschlempe, *Thesis*, Technical University, Berlin.

FIESELER, C. (1998), Biosorbens A 0105, EISU Innovative Gesellschaft für Technik und Umweltschutz mbH, in: *Entsorga '98*, May 12–16, Köln, Germany.

FINKELDEI, C.-H., PETERSEN, K. (1998), Verfahren zur Behandlung hochkonzentrierter hydrazinhaltiger Abwässer aus der Edelmetallpulverfertigung, in: *Preprints, Colloquium Produktionsintegrierter Umweltschutz, Abwässer der metallverarbeitenden Industrie und des Transportgewerbes*, Bremen, Sept. 1–2 (IUV Universität Bremen, GVC-VDI, Eds.), pp. D-37–D-38. Bremen, Germany.

FISCHWASSER, K. (1998), Möglichkeiten und Grenzen interner Stoffkreisläufe in der Metallbe- und -verarbeitung, in: *Preprints, Colloquium Produktionsintegrierter Umweltschutz, Abwässer der metallverarbeitenden Industrie und des Transportgewerbes*, Bremen, Sept. 1–2 (IUV Universität Bremen, GVC-VDI, Eds.), pp. A-3–A-18. Bremen, Germany.

FORTIN, D., SOUTHAM, G., BEVERIDGE, T. J. (1994), Nickel sulfide, iron–nickel sulfide and iron sulfide precipitation by a newly isolated *Desulfotomaculum* species and its relation to nickel resistance, *FEMS Microbiol. Ecol.* **14**, 121–132.

GAMBLE, D. S., SCHNITZER, M., KERNDORF, H. (1983), Multiple metal ion exchange equilibria with humic acid, *Geochim. Cosmochim. Acta* **47**, 1311–1323.

GEDDIE, J. L., SUTHERLAND, I. W. (1993), Uptake of metals by bacterial polysaccharides, *J. Appl. Bacteriol.* **74**, 467–472.

GEESEY, G. G., JANG, L. (1989), Interactions between metal ions and capsular polymers, in: *Metal Ions and Bacteria* (BEVERIDGE, T. J., DOYLE, R. J., Eds.), pp. 325–357. New York: John Wiley & Sons.

GILES, C. H., SMITH, D., HUITSON, A. (1974), A gereral treatment and classification of the solute adsorption isotherm, *J. Coll. Interface Sci.* **47**, 755–765.

GREEN, B., MCPHERSON, R., DARNALL, D. (1987), Algal sorbents for selective metal ion recovery, in: *Metals Speciation, Separation and Recovery* (PATTERSON, J. W., PASSINO, R., Eds.), pp. 315–338. Chelsea: Lewis Publisher.

GREENE, B., BEDELL, G. W. (1990), in: *An Introduction to Applied Phycology* (AKATSUKA, I., Ed.), pp. 137–149. Amsterdam: SPB Academic Publishers.

HARTMEIER, W. (1986), *Immobilisierte Biokatalysatoren.* Berlin, Heidelberg, New York: Springer-Verlag.

HEIN, H., SCHWEDT, G. (1991), Richtlinien und Grenzwerte Luft-Wasser-Boden-Abfall, Ein Arbeitsmittel von *Umweltmagazin*, pp. 2-2–2-4. Würzburg: Vogel.

HELFFERICH, F. (1962), *Ion Exchange.* New York: McGraw-Hill.

HOLAN, Z. R., VOLESKY, B., PRASETYO, I. (1993), Biosorption of cadmium by biomass of marine algae, *Biotechnol. Bioeng.* **41**, 819–829.

HORIKOSHI, A., NAKAJIMA, A., SAKAGUCHI, T. (1981), Studies on the accumulation of heavy metal elements in biological systems, *Eur. J. Appl. Microbiol. Biotechnol.* **12**, 90–96.

HSIEH, K. M., LION, L. W., SHULER, M. L. (1985), Bioreactor for the study of defined interactions of toxic metals and biofilms, *Appl. Environ. Microbiol.* **50**, 1155–1161.

HSU, C.-H., HARRISON, R. G. (1995), Bacterial leaching of zinc and copper from mining wastes, *Hydrometallurgy* **37**, 169–179.

IRVING, H., WILLIAMS, R. J. P. (1948), Order of stability of metal complexes, *Nature* **162**, 746–747.

KADLEC, R. H., KEOLEIAN, G. A. (1986), Metal ion exchange on peat, in: *Peat and Water, Aspects of Water Retention and Dewatering in Peat* (FUCHSMAN, C. H., Ed.), pp. 61–93. London, New York: Elsevier Applied Science Publishers.

KHAN, S. U. (1969), Interaction between the humic acid fraction of solids and certain metallic cations, *Soil Sci. Soc. Am. Proc.* **33**, 851–854.

KICK, H. (1986), Agriculture applications of sludges and wastewaters, in: *Biotechnology* 1st Edn., Vol. 8 (REHM, H.-J., REED, G., Eds.), pp. 307–304. Weinheim: VCH.

KLIMMEK, S., STAN, H.-J. (1998), Charakterisierung der Biosorption von Schwermetallen an Algen, in: *Preprints, Colloquium Produktionsintegrierter Umweltschutz, Abwässer der metallverarbeitenden Industrie und des Transportgewerbes*, Bremen, Sept. 1–2 (IUV Universität Bremen, GVC-VDI, Eds.), pp. D-45–D-50. Bremen, Germany.

KONISHI, Y., KUBO, H., ASAI, S. (1992), Bioleaching of zinc sulfide concentrate by *Thiobacillus ferrooxidans*, *Biotechnol. Bioeng.* **39**, 66–74.

KRATOCHVIL, D., VOLESKY, B. (1998), Advances in the biosorption of heavy metals, *TIBTECH* **16**, 291–300.

KUYUCAK, N., VOLESKY, B. (1988), Biosorbens for recovery of metals from industrial solutions, *Biotechnol. Lett.* **10**, 137–142

KUYUCAK, N., VOLESKY, B. (1990), Biosorption by algal biomass, in: *Biosorption of Heavy Metals* (VOLESKY, B., Ed.), pp. 173–198. Boca Raton, FL: CRC Press.

LEE, S. H., JUNG,. C. H., CHUNG, H., LEE, M. Y., YANG, J.-W. (1997), Removal of heavy metals from aqueous solution by apple residues, *Process Biochem.* **33**, 205–211.

LEUSCH, A., HOLAN, Z. R., VOLESKY, B. (1995), Biosorption of heavy metals (Cd, Cu, Ni, Pb, Zn) by chemical reinforced biomass of marine algae, *J. Chem. Technol. Biotechnol.* **62**, 2279–2288.

LÜHR, H.-P. (1992), Einleiterwerte für kontaminierte Wässer, in: *IWS-Schriftenreihe* **14**, pp. 35–55. Berlin: Gustav Schmidt-Verlag.

MITTELMANN, M. W., GEESEY, G. G. (1985), Copper-binding characteristics of exopolymers from a freshwater sediment bacterium, *Appl. Environ. Microbiol.* **49**, 846–851.

MURALEEDHARAN, T. R., VENKOBACHAR, C. (1990), Mechanism of biosoption of copper(II) by *Ganoderma lucidum*, *Biotechnol. Bioeng.* **35**, 320–325.

PONS, M. P., FUSTE, M. C. (1993), Uranium uptake by immobilized cells of *Pseudomonas* strain EPS 5028, *Appl. Microbiol. Biotechnol.* **39**, 661–665.

RAMMHOLDT, T. N. (1994), Schwermetallproblematik in Brennereiprozessen, Entwicklung eines biologischen Verfahrens zur Abtrennung von Schwermetallen aus Melasseschlempe, *Thesis*, Technical University, Berlin.

RÖHRICHT, M., DAGINNUS, K., FAHR, C., PRALLE, K., WEPPEN, P., DECKWER, W.-D. (1990), Biosorber zur Entfernung von giftigen Schwermetallen, *BioTec* **7**, 59–63.

ROSEN, J. B. (1952), Kinetics of a fixed bed system for solid diffusion, *J. Chem. Phys.* **20**, 387–394.

RUTHVEN, D. M. (1984), *Principles of Adsorption and Adsorption Processes*. New York: John Wiley & Sons.

SCHIPPERS, A., HALLMANN, R., WENTZIEN, S., SAND, W. (1995), Microbial diversity in uranium mine waste heaps, *Appl. Environ. Microbiol.* **61**, 2930–2935.

SCHWEDT, G. (1980), Chromatographische Methoden in der anorganischen Analytik, in: *Chromatographische Methoden* (KAISER, R. E., Ed.). Heidelberg, Basel, New York: Hüthig.

SINGER, A., NAVROT, J., SHAPIRA, R. (1982), Extraction of aluminium from fly ash by commercial and microbiologically produced citric acid, *Eur. J. Appl. Microbiol. Biotechnol.* **16**, 288–330.

SLONINA, P., SEIDEL, A. (1991), Breakthrough curves of insulin, [D-Phe[6]]-gonadotropin-releasing hormone and phenylalanine methyl ester on copolymers of alkylacrylate and divinylbenzene, *J. Chromatogr.* **537**, 167–180.

SONTHEIMER, H., FRICK, B. R., FETTIG, J., HÖRNER, G., HUBELE, C., ZIMMER, G. (1985), *Adsorptionsverfahren zur Wasserreinigung*, DVGW-Forschungsstelle am Engler-Bunte-Institut der Universität Karlsruhe, Germany.

TAKAMATSU, T., YOSHIDA, T. (1978), Determination of stability constants of metal–humic acid complexes by potentiometric titration and ion-selective electrodes, *Soil Sci.* **125**, 377–386.

TrinkwV (Trinkwasserverordnung) (1990), Verordnung über Trinkwasser und über Wasser für Lebensmittelbetriebe vom 5. Dezember 1990, in: *Bundesgesetzblatt*, Jahrgang 1990, Nr. 66, Teil I, pp. 2612–2629.

VASQUEZ, M., ESPEJO, R. T. (1997), Chemolithotrophic bacteria in copper ores leached at high sulfuric acid composition, *Appl. Environ. Microbiol.* **63**, 332–334.

VOLESKY, B. (1990a), Removal and recovery of heavy metals by biosorption, in: *Biosorption of Heavy Metals* (VOLESKY, B., Ed.), pp. 7–44. Boca Raton, FL: CRC Press.

VOLESKY, B. (1990b), Biosorption by fungal biomass, in: *Biosorption of Heavy Metals* (VOLESKY, B., Ed.), pp. 139–172. Boca Raton, FL: CRC Press.

VOLESKY, B. (1994), Advances in biosorption of metals: Selection of biomass types, *FEMS Microbiol. Rev.* **14**, 291–302.

VOLESKY, B., HOLAN, Z. R. (1995), Biosorption of heavy metals, *Biotechnol. Progr.* **11**, 235–250.

WAINWRIGHT, W. (1995), Neue industrielle Verfahren mit Pilzen, Pilze und Umweltbiotechnologie, in: *Biotechnologie mit Pilzen,* pp. 77–112. Berlin, Heidelberg: Springer-Verlag.

WAINWRIGHT, M., GRAYSTON, S. J. (1989), Accumulation and oxidation of metal sulfides by fungi, in: *Metal–Microbe Interactions* (POOLE, R. K., GADD, G. M., Eds.), pp. 119–130. Oxford: IRL Press.

WENZEL, R., BURGSTALLER, W., SCHINNER, F. (1990), Extraction of zinc, copper and lead from filter dust by yeast, *Biorecovery* **2**, 1–33.

WESTERTERP, K. R., VAN SWAAIJ, W. P. M., BEENACKERS, A. A. C. M. (1984), *Chemical Reactor Design and Operation.* New York: John Wiley & Sons.

WHITE, C., GADD, G. M. (1990), Biosorption of radionuclides by fungal biomass, *J. Chem. Technol. Biotechnol.* **49**, 331–343.

WHO Guidelines (1984), *Guidelines for Drinking Water Quality* (World Health Organization, Ed.), Geneva.

WILKE, A., BUNKE, G., GÖTZ, P., BUCHHOLZ, R. (1998), Removal of heavy metal ions from wastewater by adsorption on microalgae, in: *Proceedings, MinChem'98,* The Sixth Symposium on Mining Chemistry, Siofok, Hungary, September 27–30 (LAKATOS, I., Ed.), pp. 331–336. Budapest: Miskolc.

WINTER, C., WINTER, M., POHL, P. (1994), Cadmium adsorption by non-living biomass of the semimacroscopic brown alga, *Ectocarpus siliculosus*, grown in axenic mass culture and localization of the adsorbed Cd by transmission electron microscopy, *J. Appl. Phycol.* **6,** 479–487.

WINTER, C., WINTER, M., POHL, P. (1995), Cadmium adsorption by non-living biomass of the semimacroscopic brown alga, *Ectocarpus siliculosus*, grown in axenic mass culture and localisation of the adsorbed Cd by transmission electron microscopy, *J. Appl. Phycol.* **7**, 227–237.

WOLF, A., BUNZL, K., DIERL, F., SCHMIDT, W. F. (1977), The effect of Ca^{2+} ions on the adsorption of Pb^{2+}, Cu^{2+} and Zn^{2+} by humic substances, *Chemosphere* **5**, 207–213.

ZAMZOW, M. J., EICHBAUM, B. R., SANDGREN, K. R., SHANKS, D. E. (1990), Removal of heavy metals and other cations from waste water using zeolites, *Separation Sci. Technol.* **25**, 1555–1569.

ZANELLA, A., BRUNNER, F., SCHINNER, F. (1993), Leaching of sewage sludge with *Thiobacillus ferrooxidans:* Determination of physiological activity via reduction of a tetrazolium salt, in: *Biohydrometallurgical Technologies – Bioleaching Processing* (TORMA, A. E., Ed.), pp. 589–594. Jackson Hole: TMS.

Anaerobic Processes

22 Anaerobic Metabolism and its Regulation

Michael J. McInerney
Norman, OK, USA

1 Introduction

Methanogenesis is an important terminal electron accepting process in many anaerobic environments where the supply of oxygen, nitrate, oxidized forms of sulfur, iron, and manganese are limited (FERRY, 1993). Examples of such environments include freshwater and some marine sediments, flooded soils, wet wood of trees, tundra, landfills, and sewage digestors. In these environments, a complex microbial community consisting of many interacting microbial species completely degrades natural polymers such as polysaccharides, proteins, nucleic acids, and lipids to methane and carbon dioxide. Because of the large amounts of organic matter that are degraded in the natural environments, methanogenesis is an important process in cycling of carbon and other elements in nature and may be responsible for up to 60% of atmospheric methane (CONRAD, 1996). The amount of energy released during methanogenesis is relatively low compared to other terminal electron accepting processes. For example, the conversion of hexoses to methane and carbon dioxide releases only 15% of the energy that would be released by the aerobic mineralization of hexose (SCHINK, 1997). Thus, the amount of biomass produced per unit of substrate degraded is much less than that of other terminal electron accepting processes. For this reason, methanogenesis has been used as the treatment of choice for sewage and other complex wastes since sludge yields are low and most of the energy in the original substrates is retained in the energy-rich fuel, methane. Anaerobic digestion is often a net energy producer, resulting in significantly lower operating costs compared to aerobic treatment (LETTINGA, 1995).

Although the low cell yields associated with anaerobic digestion make it attractive for wastewater treatment, it is also one of its main disadvantages because large reactor volumes and long retention times are needed to achieve the required treatment efficiency (MCCARTY, 1971). Thus, only relatively high strength wastes such as sewage are treated by anaerobic digestion. In addition, anaerobic digestion is often perceived as being an unstable process (ANDERSON et al., 1982; MCCARTY, 1971; SPEECE, 1983). However, great advances have been made in the past 20 years in our understanding of the biochemistry and energetics of anaerobic metabolism. This has allowed the delineation of the most sensitive steps in the process and the development of strategies to enhance the operational stability of anaerobic reactors. The result has been the development of novel reactors where the slow growing organisms are retained in the reactor even at high volume loadings. With these advances, it is now believed that almost any type of waste can be treated (LETTINGA, 1995).

This chapter will provide a brief overview of the microbiology of anaerobic digestion and the factors that regulate this process. The constraints that thermodynamics places on key reactions will be discussed to illustrate its role in the control of the flow of carbon and energy during anaerobic digestion. In addition, the properties of microorganisms involved will be discussed to understand how population dynamics influences the process. There is a number of excellent reviews and books where these issues are treated in greater detail (DOLFING, 1988; FERRY, 1993; HOBSON et al., 1974; LETTINGA, 1995; LOWE et al., 1993; MALINA and POHLAND, 1992; MCCARTY and SMITH, 1986; MCINERNEY and BRYANT, 1981; SCHINK, 1988, 1997; SPEECE, 1983).

2 Trophic Structure of Anaerobic Ecosystems

In anaerobic digestion, organic matter is completely degraded to methane and carbon dioxide in discrete steps by the concerted action of several different metabolic groups of microorganisms (Fig. 1) (MCINERNEY and BRYANT, 1981). In the first step, fermentative bacteria hydrolyze the polymeric substrates such as polysaccharides, proteins, and lipids and ferment the hydrolysis products to acetate and longer-chain fatty acids, CO_2, formate, H_2, NH_4^+ and HS^-. Acetogeneic bacteria O-demethylate pectins and low molecular weight ligneous materials and ferment hydroxylated

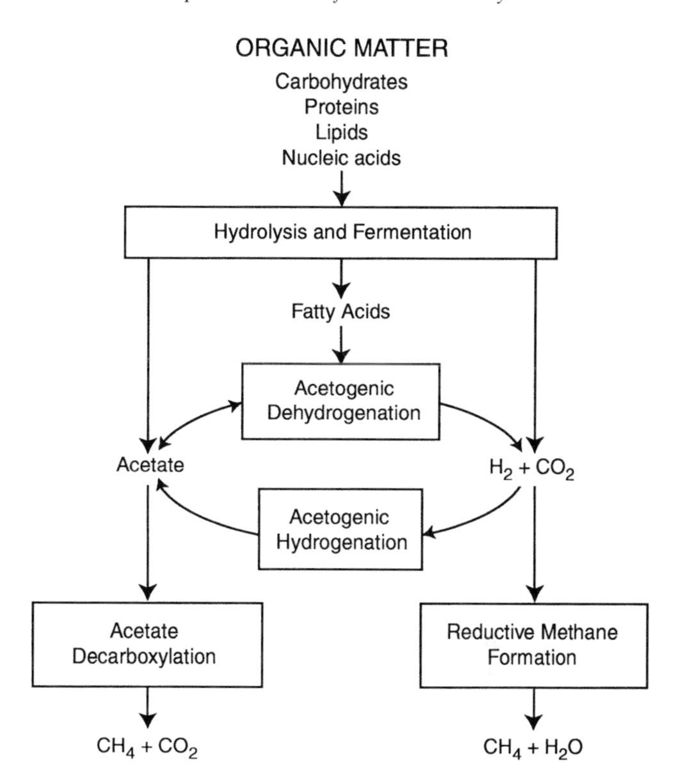

Fig. 1. Schematic representation of the different metabolic steps involved in the complete degradation of organic matter to methane and carbon dioxide (modified from McINERNEY and BRYANT, 1981).

and methoxylated aromatic compounds with the production of acetate (DOLFING, 1988). In the next step, a group of organisms, usually called syntrophic or proton reducing acetogenic bacteria, degrades propionate and longer-chain fatty acids, alcohols, and some amino acids and aromatic compounds to the methanogenic substrates, H_2, formate, and acetate. The degradation of these compounds with H_2 production is thermodynamically unfavorable unless the concentration of H_2 (or formate) is kept low by H_2 using bacteria such as methanogens (McINERNEY and BRYANT, 1981). Because of the diverse number of organisms involved in these reactions and their ability to perform other types of metabolisms such as fermentation or sulfate reduction (SCHINK, 1997; McINERNEY, 1992), the organisms that participate at this second step will be called syntrophic metabolizers. The final step involves two different groups of methanogens, the hydrogenotrophic methanogens which use the H_2 and formate produced by other bacteria to reduce CO_2 to CH_4 and the acetotrophic methanogens which metabolize acetate to CO_2 and CH_4.

In the gastrointestinal tract of animals, organic matter is only partially converted to CO_2 and CH_4 (WOLIN, 1982). Acetate and longer-chain fatty acids accumulate and are absorbed and used by the animal as energy sources. Only fermentative bacteria and H_2 or formate using methanogens seem to be involved in this partial methane fermentation. In termite guts (BREZNAK, 1994), the cecum of rodents and the large intestine of most humans (MILLER and WOLIN, 1994), the reduction of carbon dioxide to acetate by acetogenic bacteria seems to be the dominant electron accepting reaction. In these latter ecosystems, the community structure appears to be similar to that of the rumen, except that H_2 using acetogenic bacteria replace the H_2 using methanogens as the terminal electron accepting group.

In ecosystems with high levels of sulfate, such as marine and estuarine sediments and petroleum reservoirs, sulfate reduction rather than methanogenesis is the terminal electron accepting reaction. Organic matter is completely oxidized to CO_2 with the reduction of sulfate to sulfide. Again, the degradation process involves the concerted efforts of several metabolic groups of bacteria with the sulfate reducing bacteria apparently performing the functions of the syntrophic metabolizers and the hydrogenotrophic and acetotrophic methanogens (McINERNEY, 1986; OUDE ELFERINK et al., 1994; STAMS, 1994) since the degradation of propionate and longer-chain fatty acids to carbon dioxide in marine sediments does not require interspecies H_2 transfer (BANAT and NEDWELL, 1983). However, it is likely that the use of H_2 by sulfate reducers influences product formation of fermentative bacteria in a manner analogous to that found in methanogenic environments.

This brief survey of the trophic structure of anaerobic ecosystems reveals that, in general, anaerobic metabolism proceeds in a stepwise manner where several metabolic groups of bacteria interact in the mineralization process. This is in contrast to metabolism under aerobic and denitrifying conditions where a single species is usually able to mineralize completely a compound when the electron acceptor (e.g., oxygen or nitrate) is in excess. The degree of mutual interdependence between the different trophic groups in anaerobic communities varies considerably depending on the genetic capabilities of the organisms and the constraints that kinetics and thermodynamics place on key reactions. For some interactions, energy limitations are such that neither partner can operate without the activity of the other organism. Perturbations may result in the accumulation of intermediates such as H_2 and acetate that exceed the narrow limits that are needed for the degradation of key intermediates such as fatty acids to be thermodynamically favorable. Other perturbations may stimulate fermentative metabolism or enhance the growth of certain fermentative populations, resulting in the production of acetate or other organic acids at rates faster than the other trophic groups can degrade these acids. Thus, the efficiency of anaerobic digestion will depend on the dynamics and kinetics of key populations within the reactor as well as on thermodynamic limitations. The challenge of anaerobic digestion is to maintain the appropriate balance between the different groups of bacteria so that rates of production of key intermediates match consumption rates and the pool sizes are and within the narrow limits that thermodynamics places on these reactions.

3 Methanogens

3.1 Physiology of Methanogens

Methanogens are a taxonomically and phylogenetically diverse group of microorganisms (BOONE et al., 1993; JONES et al., 1987; ZINDER, 1993) that all gain energy for growth from the reactions that lead to the production of methane. As a group, methanogens use a small number of compounds, H_2 or one-carbon atom compounds (BOONE et al., 1993; ZINDER, 1993). This specialization makes methanogens dependent on other organisms for the supply of substrates in most anaerobic environments. Without methanogens, effective degradation of organic matter would cease due to the accumulation of nongaseous products of fermentation which have almost the same energy content as the original substrate.

The ability of methanogens to use H_2 plays a key regulatory role that controls the types of products made by fermentative bacteria and sets the thermodynamic conditions required for the degradation of fatty and aromatic acids. The favorable thermodynamics of H_2 use by methanogens (Tab. 1) allows them to metabolize H_2 to very low partial pressures. Methanogens are able to metabolize H_2 down to H_2 partial pressures ranging from 3–7 Pa (LOVLEY, 1985; CORD-RUWISCH et al., 1988). Methanogens also have a high affinity for H_2 use, with apparent K_M values in the range of 5–13 μM (670–1,700 Pa) (Tab. 2) (ROBINSON and TIEDJE, 1984; WARD and WINFREY, 1985; ZINDER, 1993). With H_2 partial pressures ranging from 2–1,200 Pa in digestors (CORD-RUWISCH et al., 1997; GUWY et al., 1997; MOSEY and FERNANDEZ, 1989; STRONG and CORD-RUWISCH, 1995), this means that the methanogens are undersaturated with respect to H_2 use. Thus, in-

Tab. 1. Reactions Involved in Syntrophic Metabolism[a]

Reaction			$\Delta G^{0\prime}$ [kJ per reaction]
Methanogenic reactions			
$4H_2 + HCO_3^- + H^+$	→	$CH_4 + 3H_2O$	−135.6
$Acetate^- + H_2O$	→	$CH_4 + HCO_3^-$	− 31.0
Syntrophic reactions without H_2 use by methanogens			
$Lactate^- + 2H_2O$	→	$Acetate^- + HCO_3^- + H^+ + 2H_2$	− 4.2
$Ethanol + H_2O$	→	$Acetate^- + H^+ + 2H_2$	+ 9.6
$Butyrate^- + 2H_2O$	→	$2\,Acetate^- + H^+ + 2H_2$	+ 48.3
$Propionate^- + 3H_2O$	→	$Acetate^- + HCO_3^- + H^+ + 3H_2$	+ 76.1
$Benzoate^- + 7H_2O$	→	$3\,Acetate^- + HCO_3^- + 3H^+ + 3H_2$	+ 70.6[b]
$Acetate^- + 4H_2O$	→	$2HCO_3^- + H^+ + 4H_2$	+104.6
Syntrophic reactions with H_2 use by methanogens			
$2\,Lactate^- + H_2O$	→	$2\,Acetate^- + HCO_3^- + CH_4 + H^+$	−143.6
$2\,Ethanol + HCO_3^-$	→	$2\,Acetate^- + H_2O + H^+ + CH_4$	−116.4
$2\,Butyrate^- + HCO_3^- + H_2O$	→	$4\,Acetate^- + H^+ + CH_4$	− 39.4
$4\,Propionate^- + 3H_2O$	→	$4\,Acetate^- + HCO_3^- + H^+ + 3CH_4$	−102.4
$4\,Benzoate^- + 19H_2O$	→	$12\,Acetate^- + HCO_3^- + 9H^+ + 3CH_4$	−124.4

[a] Calculated from the data in THAUER et al. (1977).
[b] Calculated using the free energy of formation for benzoate given in KAISER and HANSELMANN (1982).

Tab. 2. Selected Kinetic Data for the Use of Hydrogen, Formate and Acetate by Methanogenic Bacteria

Substrate	Organisms	Apparent K_m	Reference	Threshold	Reference
Hydrogen	most methanogens	3–13 μM	KRISTIJANSSON et al. (1982), ROBINSON and TIEDJE (1984)	3–12 Pa	CORD-RUWISCH et al. (1988), LOVLEY (1985), LOVLEY et al. (1984)
Formate	many methanogens	5–580 μM	LOVLEY et al. (1984), SCHAUER et al. (1982)	15–26 μM	SCHAUER et al. (1982)
Acetate	*Methanosarcina* sp.	3.0–4.5 mM	SCHÖNHEIT et al. (1982), WESTERMANN et al. (1989)	0.6–1.2 mM	WESTERMANN et al. (1989), FUKUZAKI et al. (1990a)
	Methanosaeta sp	~0.1–1.2 mM	HUSER et al. (1982), MIN and ZINDER (1989), OHTSUBO et al. (1992), AHRING and WESTERMANN (1987a)	5–75 μM	WESTERMANN et al. (1989), MIN and ZINDER (1989), AHRING and WESTERMANN (1987a), JETTEN et al. (1990)

creases in H_2 production may not result in higher H_2 partial pressures because the H_2 consumption rate by methanogens would also increase (KASPAR and WUHRMANN, 1978a; ROBINSON and TIEDJE, 1982; SHEA et al., 1968; STRAYER and TIEDJE, 1978). Because of the large capacity for H_2 use by methanogens, H_2 concentrations are normally very low in well-operated anaerobic digestors, even though large amounts of H_2 are produced. As will be discussed in Sects. 4 and 5, the ability of methanogens to maintain low levels of H_2 affects the types of products formed by fermentative bacteria and is essential for the degradation of fatty and aromatic acids by syntrophic associations.

Formate is a common fermentation product, especially by bacteria that use pyruvate–formate lyase in their metabolism, and may be an essential intermediate for syntrophic metabolism (THIELE and ZEIKUS, 1988; DONG et al., 1994). Many methanogens are able to use formate and it serves as a source of electrons for methane formation equivalent to H_2. The affinity for formate use varies for different methanogens (Tab. 2). The apparent K_M values for formate use was 5–26 μM for two rumen methanogens, 580 μM for *Methanobacterium formicicum*, and 0.22 mM for *Methanospirillum hungatei* (LOVLEY et al., 1984; SCHAUER et al., 1982). SCHAUER et al. (1982) reported that the lowest concentration of formate metabolized was 26 μM for *M. formicicum* and 15 μM for *M. hungatei.*

Acetate is a major product of fermentative metabolism and is quantitatively the most important substrate for methane production. About 60–90% of the methane produced in anaerobic digestors are derived from the methyl group of acetate (BOONE, 1982; MACKIE and BRYANT, 1981; MOUNTFORT and ASHER, 1978; SMITH and MAH, 1966). At thermophilic temperatures or at high ammonium levels, the oxidation of the methyl group of acetate to H_2 may be the predominant route for acetate metabolism (see Sect. 5). Acetate using methanogens include members of the genera *Methanosarcina* and *Methanosaeta* (BOONE et al., 1993). *Methanosarcina* sp. have faster growth rates, higher apparent K_M values for acetate use, and higher threshold acetate values than *Methanosaeta* sp. (Tab. 2) (ZINDER, 1993). The differences in the apparent K_M values for acetate use have been attributed to differences in respective enzymes used to activate acetate (JETTEN et al., 1990). Since *Methanosarcina* sp. and *Methanosaeta* sp. have different threshold values for acetate use but use the same reaction for acetate metabolism (Tab. 1), the threshold values cannot represent a thermodynamic limitation. Acetate threshold values may result when a critical or inhibitory concentration of undissociated acetic acid is reached which, for *Methanosarcinca* sp., is between 4 and 7 μM (FUKUZAKI et al., 1990a).

Consistent with the known growth characteristics of the acetotrophic methanogens, a drop in acetate concentrations below 1 mM was correlated with a displacement of *Methanosarcina* sp. by *Methanosaeta* sp. in a thermophilic digestor (ZINDER et al., 1984). Generally, high-rate anaerobic sludge blanket reactors usually have granules composed of *Methanosaeta* sp. rather than *Methanosarcina* sp. (GROTENHUIS et al., 1991; WU et al., 1993). Interestingly, RASKIN et al. (1996) found that *Methanosaeta* sp. and *Methanosarcina* sp. were present at approximately equal levels in glucose degrading anaerobic biofilm reactors even though acetate levels were below the threshold values for *Methanosarcina* sp. The maintenance of *Methanosarcina* in these biofilms could not be attributed to the utilization of other substrates such as methanol, suggesting that there may be *Methanosarcina* sp. that have a lower threshold value than has been reported in the literature. STRAYER and TIEDJE (1978) found that acetate was converted to methane at or near the maximal rate in eutrophic lake sediments since the addition of acetate to these sediments did not result in a corresponding increase in the rate of acetate utilization or methane production. This suggests that there is little additional capacity to metabolize acetate if the acetate production rates increase. A similar conclusion was reached by HICKEY and SWITZENBAUM (1991) for anaerobic sewage digestors. This is probably the reason why acetate concentrations in digestors increase when organic or volume loading rates increase.

Methanol, methylamine and trimethylamine also serve as substrates for *Methanosarcina* sp. and other methylotrophic methanogens (BOONE et al., 1993). These compounds may be important substrates for methanogenesis in marine systems (OREMLAND and POLCIN, 1982). Some hydrogenotrophic methanogens can oxidize secondary alcohols to ketones and primary alcohols to carboxylic acids (WIDDEL, 1986). Additional methanogenic substrates are discussed by ZINDER (1993) and BOONE et al. (1993).

3.2 Effect of Sulfate on Methanogenesis

Anaerobic digestion has been successfully applied to a variety of waste streams with high sulfate levels such as molasses-based fermentation industries (e.g., distilleries and yeast, glutamate, and citric acid production plants), pulp and paper processing facilities, and edible oil refineries (COLLERAN et al., 1995). Under most conditions, sulfate reducers appear to outcompete methanogens for H_2 but methanogenesis appears to be the dominant process for acetate metabolism (COLLERAN et al., 1995; VISSER et al., 1993; YODA et al., 1987; ISA et al., 1986). The thresholds for H_2 use for sulfate reducers (1–10 nM) are much lower than for methanogens (about 23–75 nM) (CORD-RUWISCH et al., 1988), suggesting that sulfate reducers can lower the H_2 concentration to levels too low for methanogens to use. The greater affinities for H_2 use by sulfate reducers compared to methanogens (ROBINSON and TIEDJE, 1984) are also consistent with the observation that sulfate reducers outcompete methanogens for H_2.

However, it is not clear why methanogenesis from acetate persists in reactors with high sulfate levels. Kinetic studies suggest that sulfate reducers should outcompete methanogens for acetate (SCHÖNHEIT et al., 1982). However, this conclusion was reached using only one methanogen (*Methanosarcina barkeri*) and one sulfate reducer and a different conclusion may be reached if more species are tested. One explanation for the persistence of methanogenesis may be that they are less sensitive to sulfide toxicity than sulfate reducers. VISSER et al. (1996) found that inhibition by sulfide should allow acetotrophic methanogens to outcompete acetate using sulfate reducers if the pH was less than 6.9 and to coexist at pH values between 6.9 and 7.7. Another possibility is that a long period of time (more than 1 year) would be needed for acetate using sulfate reducers to establish a population level comparable to acetotrophic methanogens (VISSER et al., 1993). However, this would not explain the persistence of acetotrophic methanogenesis in reactors that have received sulfate for long periods of time. IZA et al. (1986) proposed that acetotrophic methanogens may have more favorable colonization or adhesion properties which allow them to be maintained in high-rate reactors under conditions where sulfate reducers would be washed out. Finally, it may be that the acetate using sulfate reducers in granules and biofilms are sulfate limited which allows acetotrophic methanogens to outcompete them for acetate (NIELSEN, 1987). The above discussion indicates that the coexistence of two species is probably due to a multitude of factors, only one of which is the competition for common substrates.

4 Regulation of Fermentative Metabolism

The first step in anaerobic digestion is the hydrolysis and fermentation of polymeric substrates. For many wastes, particularly those with high cellulose content, the hydrolysis step itself may be rate limiting (PFEFFER and LIEBMAN, 1976; EASTMAN and FERGUSON, 1981). Polysaccharides such as cellulose, hemicellulose, pectin, and starch are hydrolyzed to lower molecular weight materials such as sugars and oligosaccharides (uronic acids and methanol are also formed from pectin) which are then fermented to the products shown in Fig. 2 (BEGUIN and AUBERT, 1994; LJUNGDAHL and ERIKSSON, 1985). Proteins are hydrolyzed to peptides and amino acids which are also fermented to the products shown in Fig. 2 as well as to other organic compounds (MCINERNEY, 1988). Isobutyrate, isovalerate and D-2-methylbutyrate are produced from the branched-chain amino acids, valine, leucine, and isoleucine, respectively. Phenylacetate, phenylpropionate, indole, and indolacetate are produced from aromatic amino acids. Glycerides, phospholipids, and other fats are hydrolyzed to long-chain fatty acids, glycerol, and sugars; only the latter two are substrates for fermentative bacteria (MCINERNEY, 1988; MACKIE et al., 1991).

The concentration of H_2 (or formate) in the ecosystem will play an important role in regulating the kinds and amounts of products made by fermentative bacteria. Many fermentative

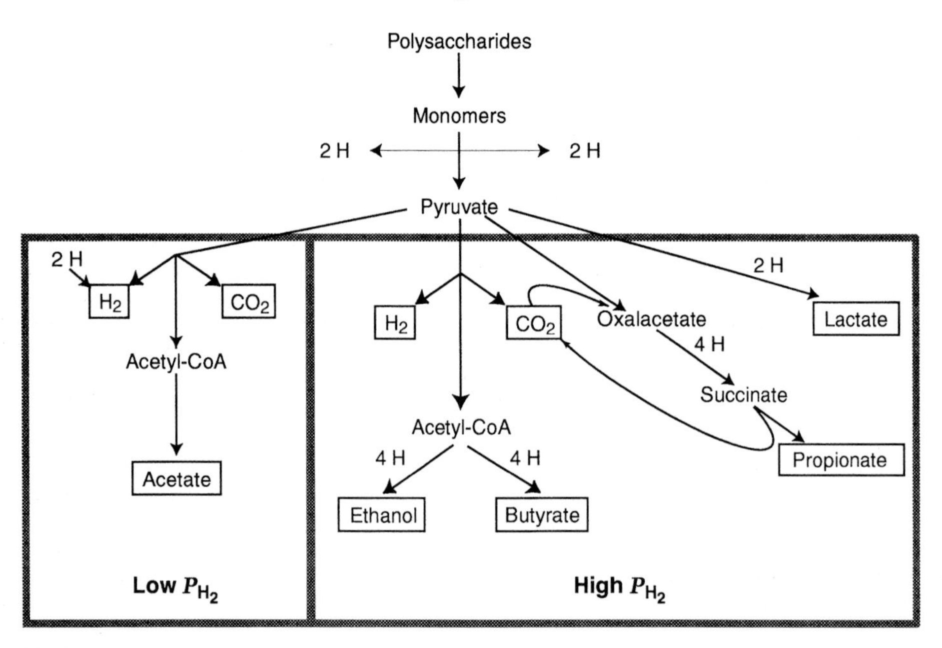

Fig. 2. Flow of reducing equivalents (H) during the fermentation of polysaccharides under low or high partial pressures of hydrogen (P_{H_2}). Products of fermentative metabolism are enclosed in boxes (modified from McINERNEY and BRYANT, 1981).

bacteria have the ability to reoxidize their reduced electron carriers such as $NADH_2$ (nicotinamide adenine dinucleotide) by reducing protons to H_2 (WOLIN, 1982). This reaction is favorable only when the partial pressure of H_2 is low, about 100 Pa (WOLIN, 1982). In well operated digestors with a low H_2 partial pressure, fermentative bacteria will reoxidize $NADH_2$ mainly by H_2 production. This allows them to metabolize pyruvate almost exclusively to acetate, CO_2, and H_2 (Fig. 2). Consistent with this concept, ZINDER (1986) found that radioactive glucose was almost exclusively metabolized to acetate and little or no radioactive carbon was detected in other intermediates such as propionate, butyrate, or lactate. While the oxidation of glucose to acetate, CO_2 and H_2 is favorable under standard conditions, no mesophilic organism is known that can catalyze this reaction in pure culture. KRUMHOLZ and BRYANT (1986) isolated the mesophilic bacterium, *Syntrophococcus sucromutans*, which metabolizes glucose only in the presence of a H_2 using bacterium or an external electron acceptor since this organism lacks the ability to reoxidize its reduced electron carriers through the formation of reduced products. Several moderately thermophilic bacteria have been described that are able to convert sugars to acetate, CO_2 and H_2 (DIETRICH et al, 1988; SOUTSCHEK et al., 1984). As discussed below, H_2 formation is more favorable at higher temperatures which may explain why these organisms can grow by this reaction.

If the H_2 partial pressure is high, as often occurs with shock loads or when methanogenesis is inhibited (STRONG and CORD-RUWISCH, 1995; CORD-RUWISCH et al., 1997; GUWY et al., 1997; MOSEY and FERNANDEZ, 1989), the production of H_2 from $NADH_2$ becomes unfavorable (WOLIN, 1982) and the flow of electrons shifts from H_2 production to the formation of reduced products such as ethanol, lactate, propionate, butyrate, etc. (Fig. 2). Numerous studies show that little or no ethanol or lactate, much less propionate and butyrate, and more acetate and CO_2 are made when fermentative organisms are grown in association with H_2 using bacteria (BAUCHOP and MOUNTFORT, 1981; McINERNEY and BRYANT, 1981; STAMS,

1994; WOLIN, 1982). The shift in electron flow is probably a consequence of the inhibition of $NADH_2$–ferredoxin oxidoreductase activity which is involved in H_2 production from $NADH_2$ (TEWES and THAUER, 1980) and/or the regulation of the synthesis of enzymes involved in the formation of reduced fermentation products (STAMS, 1994).

A mathematical model based on the thermodynamics of NADH oxidation was developed that accurately predicts the changes in volatile fatty acids in response to changes in operational parameters such as decreases in the hydraulic retention time or increases in the organic loading rate (MOSEY, 1983). The simulation results show that under high loading rates or shortened retention times, the concentrations of propionate and butyrate increase in response to increases in the H_2 partial pressure. This mimics the situation found in real digestors. The regulatory role of H_2 also explains the accumulation of volatile fatty acids when the acidification phase of anaerobic digestion is separated from the methanogenic phase as occurs in phase-separated reactors (COHEN et al., 1979; HARPER and POHLAND, 1986; FOX and POHLAND, 1994; MASSEY and POHLAND, 1978). The high H_2 partial pressures found in acidification reactors would make the production of H_2 from $NADH_2$ energetically unfavorable and force fermentative bacteria to reoxidize their reduced electron carriers by producing propionic, butyric and lactic acids rather than acetic acid.

Some fermentative anaerobes have the ability to produce formate by coupling the reduction of CO_2 to the oxidation of $NADH_2$ or reduced ferredoxin (MILLER and WOLIN, 1973). This reaction could be used in place of H_2 production from pyridine nucleotide linked or ferredoxin-linked hydrogenase reactions. Here, the reducing equivalents generated in fermentative metabolism are transferred to methanogens in the form of formate rather than H_2. This would provide an ecologically relevant explanation for the widespread ability of methanogens to use formate and would allow fermentative bacteria that lack hydrogenases to enter into interspecies interactions (MCINERNEY and BRYANT, 1981).

An alternative explanation for the increased production of propionate and butyrate when reactors are subjected to higher loading rates or shortened retention times is based on the changes in the populations of individual species within the reactor rather than as a consequence of the thermodynamics of their reactions (MCCARTY and MOSEY, 1991; MOSEY and FERNANDES, 1989). This model proposes that clostridial-like bacteria predominate in reactors operated at modest loading rates where the low H_2 partial pressures (200–1,000 Pa) allow these organisms to produce acetate, CO_2, and H_2 and little or no butyrate from carbohydrates. The response to inhibition or overload in a reactor is often the rapid accumulation of acetic and propionic acids. MCCARTY and MOSEY (1991) propose that this is the result of the rapid growth of enteric or propionic acid bacteria in response to high substrate concentrations. These bacteria are not affected by the H_2 partial pressure and produce acetate, propionate and formate in fixed ratios. The production of these acids leads to a drop in pH which selects for the growth of more acid-tolerant, butyric acid producing bacteria, resulting in the production of butyric acid and H_2. This scenario matches reasonably well with the changes in volatile acids detected in glucose-fed reactors in response to changes in retention time and pH (ZOETEMEYER et al., 1982). However, direct experimental evidence for the proposed population changes is not available.

5 Regulation of Syntrophic Metabolism

The degradation of alcohols, fatty acids, amino acids, aromatic compounds, and certain organic acids involves the syntrophic association of a syntrophic metabolizer and a H_2 using bacterium such as a methanogen or a sulfate reducer (SCHINK, 1992; STAMS, 1994). It is at this stage of the process where thermodynamics plays an important role in regulating the efficiency of anaerobic digestion. As shown in Tab. 1, the degradation of ethanol, propionate, and butyrate to acetate, CO_2, H_2 is thermodynamically unfavorable under standard conditions. Hydrogen production from these compounds can only occur if a H_2 using bacterium is present to maintain the concentrations of H_2 at a very low level. The effect of the partial pressure of H_2 on the energetics of ethanol, butyrate and propionate degradation at concentrations of reactants and products reflective of those found in digestors is shown in Fig. 3. Ethanol degradation is thermodynamically favorable at a partial pressure of less than 20 kPa. However, much lower H_2 partial pressures are needed for the degradation of butyrate (about 80 Pa) and propionate (about 4 Pa) to be thermodynamically favorable. Consistent with these thermodynamic predictions, small changes in H_2 partial pressure were shown to inhibit the degradation of butyrate and benzoate by syntrophic cocultures (AHRING and WESTERMANN, 1988; DWYER et al., 1988; SCHINK, 1997) and propionate degradation in methanogenic mixed cultures (SMITH and MCCARTY, 1989).

Changes in acetate concentration also affect the thermodynamics of syntrophic metabolism (KASPAR and WUHRMANN, 1978b; WARIKOO et al., 1996; BEATY and MCINERNEY, 1989). Fig. 4 shows the boundary lines that delineate acetate and H_2 concentrations where the degradation of ethanol, propionate, and butyrate are thermodynamically favorable. When H_2 partial pressure is low (about 3 Pa), which is close to the threshold value for H_2 use by methanogens, ethanol and propionate degradation are favorable at any of the acetate concentrations shown (up to 100 mM) and butyrate degradation is favorable up to about 60 mM acetate. However, if conditions in the digestor result in a higher partial pressure of H_2 (e.g., about 30 Pa), then acetate concentrations will have to be much lower for the degradation of propionate and butyrate to be thermodynamically favorable, about 0.1 and 5 mM, respectively. One can see that achieving the low acetate concentrations needed for propionate degradation to be thermodynamically favorable may be a problem in digestors where *Methanosarcina* sp. are the predominant acetate user since their threshold for acetate degradation ranges from 0.6–1.2 mM (Tab. 2). Thus, an increase in the concentration of either H_2 or acetate will make syntrophic metabolism more sensitive to an increase in the concentration of the other product.

Consistent with thermodynamic predictions, we observed a much higher benzoate threshold concentration in the presence of 65 mM acetate in methanogenic cocultures where the H_2 partial pressure was about 2 Pa compared to sulfate reducing cocultures where the H_2 partial pressure was about 0.2 Pa (JACKSON and MCINERNEY, 1997; WARIKOO et al., 1996). The accumulation of volatile fatty acids in digestors or methanogenic mixed cultures was in response to changes in acetate and H_2 concentrations and was thermodynamically controlled (HICKEY and SWITZENBAUM, 1991; SMITH and MCCARTY, 1989). The extent of butyrate and benzoate metabolism was greater

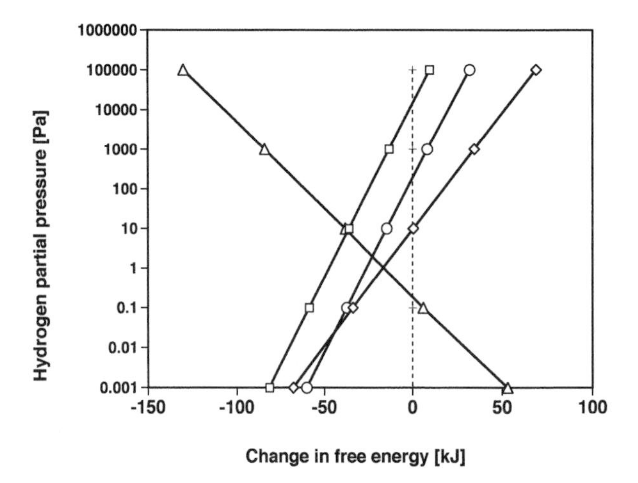

Fig. 3. The effect of hydrogen partial pressure on the free energy change for hydrogenotrophic methanogenesis (triangles) and the degradation of ethanol (squares), propionate (diamonds) and butyrate (circles) by syntrophic associations. Stoichiometry of the reactions is given in Tab. 1. The free energy changes were calculated assuming that the concentrations of ethanol, propionate, butyrate, and acetate were 0.1 mM, that the concentration of bicarbonate was 50 mM, and that the partial pressure of CH_4 was 50 kPa.

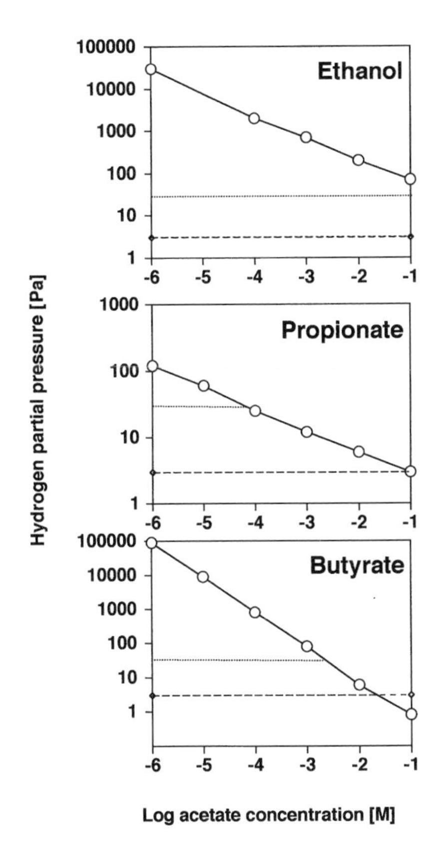

Fig. 4. Boundary lines delineating the acetate concentrations and hydrogen partial pressures at which the degradation of ethanol, propionate and butyrate are thermodynamically favorable. In each figure, the line with the circles represents the acetate and hydrogen levels where the free energy change for the degradation of the substrate is at equilibrium ($\Delta G' = 0$), assuming a substrate concentration of 0.1 mM. The area above this line is where the free energy change is positive; the area under this line is where the free energy change is negative. The horizontal dashed and dotted lines represent conditions where the hydrogen partial pressure is 3 and 30 Pa, respectively.

when an acetate using organism was present to prevent the accumulation of acetate (BEATY and MCINERNEY, 1989; WARIKOO et al., 1996). Also, faster rates of metabolism and higher growth yields per mole of substrate degraded were observed when an acetate using bacterium was present (AHRING and WESTERMANN, 1987b; BEATY and MCINERNEY, 1989; STAMS et al., 1992). In addition to the effect that acetate has on the extent of syntrophic metabolism, acetate also affects the kinetics of syntrophic metabolism (BEATY and MCINERNEY, 1989; DOLFING and TIEDJE, 1988; FUKUZAKI et al., 1990b; WARIKOO et al., 1996). Thus, in order to adequately describe the kinetics of syntrophic metabolism, one must include terms to account for acetate acting as an uncompetitive or noncompetitive inhibitor (WARIKOO et al., 1996; FUKUZAKI et al., 1990b) and as a thermodynamic regulator (HOH and CORD-RUWISCH, 1996; WARIKOO et al., 1996).

While much of the above discussion has focused on the role of H_2 in syntrophic metabolism, formate may also be an important intermediate involved in the transfer of reducing equivalents between the syntrophic partners. THIELE and ZEIKUS (1988) proposed that formate rather than H_2 was the interspecies electron carrier involved in ethanol degradation based on radioisotopic studies and thermodynamic considerations. BOONE et al. (1989) used a diffusion model to analyze the flux of H_2 and formate in a butyrate degrading syntrophic coculture. The model calculated that formate could be expected to transfer electrons 98-fold more rapidly than H_2 and that the rate of H_2 diffusion was too slow to account for the rate of butyrate degradation. However, other studies (GOODWIN et al., 1991; SMITH, 1992) found that the diffusion of H_2 in microbial flocs was rapid enough to account for the observed rate of syntrophic metabolism. Interspecies formate transfer has been demonstrated in the syntrophic degradation of amino acids since *Desulfoarculus baarsii* present in these cultures is unable to use H_2 (ZINDEL et al., 1988). More recently, the need for formate transfer for the syntrophic degradation of propionate was implicated because propionate degradation did not occur when a methanogen that uses only H_2 was present (DONG et al, 1994). Thus, it seems that both H_2 and formate participate in interspecies electron transfer reactions and that individual species differences may determine which of the two compounds is used in a given syntrophic interaction.

Examples of the reactions catalyzed by syntrophic associations are given in Tab. 1. The first known syntrophic association was the degradation of ethanol by *S organism* (BRYANT et al., 1967). Since then, other organisms that can syntrophically oxidize ethanol have been described including several *Desulfovibrio* sp., a *Bacteroides* sp., *Pelobacter carbinolicus*, *Pelobacter acetylenicus*, *Pelobacter venetianius*, and *Thermoanaerobium brockii* (BRYANT et al., 1977; BEN-BASSAT et al., 1981; DWYER and TIEDJE, 1986; SCHINK, 1992).

Members of the genera *Syntrophomonas* and *Syntrophospora* and *Thermosyntropha lipolytica* in addition to several unnamed species syntrophically β-oxidize butyrate and longer-chain, saturated fatty acids and form a major line of descent in the gram-positive phylum (LOROWITZ et al., 1989; MCINERNEY, 1992; MCINERNERY et al., 1979; ROY et al., 1986; STIEB and SCHINK, 1986b; SVETLITSHNYI et al., 1996; ZHAO et al., 1993). Some of these organisms are also able to metabolize branched-chain fatty acids such as isobutyrate and 2-methylbutyrate (STIEB and SCHINK, 1986a; THOLOZAN et al., 1988a; WU et al., 1994). Isobutyrate is isomerized to butyrate prior to C—C bond cleavage. The syntrophic metabolism of propionate is found in members of the genus *Syntrophobacter* as well as several unnamed strains (BOONE and BRYANT, 1980; MUCHA et al., 1988; STAMS et al., 1992; WALLRABENSTEIN et al., 1995b). Propionate can also be reductively carboxylated to butyrate in methanogenic digestors (THOLOZAN et al., 1988b). Several species of the genus *Syntrophus* can syntrophically metabolize benzoate and other aromatic compounds (AUBURGER and WINTER, 1995; MOUNTFORT and BRYANT, 1982; WALLRABENSTEIN et al., 1995a; WARIKOO et al., 1996).

Two organisms have been obtained in co-culture that syntrophically oxidize acetate (SCHNÜRER et al., 1996; ZINDER and KOCH, 1984). This route for acetate metabolism has been observed at thermophilic temperatures where the acetate concentration is below 1 mM (PETERSON and AHRING, 1991) and at mesophilic temperatures where high ammonia levels inhibit acetotrophic methanogenesis (SCHNÜRER et al., 1994).

Interspecies electron transfer is important for the degradation of certain amino acids in anaerobic environments (MCINERNEY, 1988; ÖRLYGSSON et al., 1994; STAMS, 1994). Several bacteria have been described that can degrade amino acids when grown in association with H_2 using bacteria (ZINDEL et al., 1988; WILDENAUER and WINTER, 1986).

Even when the concentrations of acetate and H_2 are low, the amount of energy available to each organism to support ATP formation and growth is very low (SCHINK, 1997). This is probably the reason for the slow growth rates and low cell yields observed for syntrophic associations. Doubling times for methanogenic, propionate degrading syntrophic associations range from 4–34 d (BOONE and BRYANT, 1980; STAMS et al., 1992; WALLRABENSTEIN et al., 1995b) while those for methanogenic, butyrate degrading syntrophic associations range from 1.1–3.4 d (MCINERNEY et al., 1979; STIEB and SCHINK, 1985b; ROY et al., 1986). Thermodynamic calculations based on the amount of en-

Tab. 3. Kinetic Coefficients for the Anaerobic Conversion of Various Substrates Calculated from Bioenergetic Principles[a]

Substrate	Maximum Specific Substrate Utilization Rate [gCOD · gVSS^{-1} d^{-1}]	Yield [gVSS · gCOD^{-1}]	Predicted Minimum Retention Time [d^{-1}]
Carbohydrates	15.9	0.35	0.2
Protein	11.3	0.20	0.4
Long-chain fatty acids	8.5	0.04	3.2
Propionate	8.4	0.04	3.3
Butyrate	8.7	0.06	2.0
Acetate	8.4	0.03	3.9
Hydrogen gas	29.2	0.03	1.1

[a] Data from PAVLOSTATHIS and GIRALDO-GOMEZ (1991) and MCCARTY (1971).

ergy available per electron equivalent predict that carbohydrate and protein fermentors and hydrogenotrophic methanogens can be maintained in completely mixed reactors operated at retention times less than 1–2 d (Tab. 3) (McCARTY, 1971; PAVLOSTATHIS and GIRALDO-GOMEZ, 1991). However, the minimum retention times for acetotrophic methanogens and fatty acid degrading syntrophic associations are higher, about 2–4 d (Tab. 3), which is in agreement with observed rates of growth of syntrophic associations reported above. This analysis shows that syntrophic metabolism of fatty acids is the rate limiting step for the degradation of many types of wastes. New technologies circumvent the problem of the slow growth of acetotrophic methanogens and syntrophic associations by using immobilization to maintain these populations in the reactor.

At the loading rates frequently reported for anaerobic waste treatment systems, the minimal distance between syntrophic partners for effective degradation has been estimated to be about 11 μm (McCARTY and SMITH, 1986). By reducing this distance, one should be able to minimize diffusion limitations and maximize the rate of metabolite transfer. The cell densities of methanogenic granular sludge are extremely high which means that intercellular distances are very low (GROTENHUIS et al., 1991; WU et al., 1991, 1993). This allows for the rapid transfer of metabolites between syntrophic partners and explains the high rates of volatile fatty acid degradation that have been reported for granular sludge.

6 Environmental Factors

Fermentation and growth of anaerobic microorganisms are dependent on the optimal supply of nutrients. The reader is referred to excellent reviews on the requirements for nitrogen, phosphorous and trace elements for anaerobic digestion (MALINA and POHLAND, 1992; SPEECE, 1983; LETTINGA, 1995) and the effects of inhibitors on anaerobic digestion have also been reviewed (COLLERAN et al., 1995; KUGELMAN and CHIN, 1971; McCARTNEY and OLESZKIEWICZ, 1991; PARKIN et al., 1983; SPEECE, 1983). Sects. 6.1–6.2 will provide a brief overview of the importance of temperature and pH on anaerobic growth and metabolism.

6.1 Temperature

Temperature is an important factor that affects the rate of growth and metabolism of microorganisms and consequently the population dynamics in a reactor. Methanogens have been found in environments over a wide range of temperatures from tundra soils at 6 °C to geothermal areas with temperatures above 100 °C (BOONE et al., 1993; KOTSYURBENKO et al., 1996; ZINDER, 1993). The upper temperature limit for the methanogenesis from H_2 and CO_2 should not be a limit for anaerobic digestion since *Methanopyrus kandleri* has a temperature optimum near 100 °C and can grow at 110 °C (KURR et al., 1991). There are several other H_2 using methanogens with temperature optima from 83–85 °C (BOONE et al., 1993).

Acetotrophic methanogenesis is one of the most sensitive reactions to temperature increases. None of the described species of acetotrophic methanogens can grow above 70 °C (BOONE et al., 1993; VAN LIER et al., 1994; LETTINGA, 1995). Temperature optima for thermophilic species of *Methanosaeta* are between 60 and 65 °C while thermophilic *Methanosarcina* sp. have temperature optima between 50 and 58 °C and maximum growth temperatures between 60 and 65 °C (BOONE et al., 1993; ZINDER, 1993). ZINDER et al. (1984) and ANGELIDAKI and AHRING (1994) found a sharp decrease in CH_4 production and a large increase in acetate production when the temperature was increased. AHRING (1995) proposes that syntrophic acetate oxidation becomes increasingly more important relative to acetotrophic methanogenesis at thermophilic temperatures.

The degradation of propionate and butyrate may also be sensitive to temperatures above 70 °C. Maximum growth temperatures of 60 °C were observed for syntrophic propionate degradation (STAMS et al., 1992) and about 70 °C for syntrophic butyrate degradation (HENSON and SMITH, 1985; SVETLITSHNYI et al., 1996). Propionate degradation was believed to be the most sensitive step to increasing temperatures, exhibiting a maximum temperature between 70 and 80 °C (VAN LIER et al., 1993; WIEGANT et al., 1985; LEPISTO and RINTALA, 1996). The temperature optima for propionate degradation in granular sludge (between 55 and 60 °C) correspond well with the known temperature optimum of the thermophilic syntrophic propionate degrader (VAN LIER et al., 1993; STAMS et al., 1992). Also, thermophilic temperatures may make the anaerobic digestion more susceptible to inhibition by the un-ionized form of ammonia (ANGELIDAKI and AHRING, 1994).

Of the many methanogenic bacteria that have been described, only a few have optimal growth temperatures below 30 °C (BOONE et al., 1993) and only one psychrophilic methanogen, *Methanogenium frigidum*, has been described (FRANZMANN et al., 1997). Temperatures below 15 °C have been shown to greatly reduce the rate of methanogenesis in lake sediments, rice paddies, tundra soils and anaerobic reactors (CONRAD et al., 1987; REBAC et al., 1997; KOTSYURBENKO et al., 1996; NACHAIYASIT and STUCKEY, 1997; VAN LIER et al., 1997). The optimal temperature for methanogenesis in these systems is often in the mesophilic range (28–35 °C). CONRAD et al. (1989) concluded that nearly all of the H_2-turnover in Lake Constance sediments incubated at 4 °C was due to acetogenesis rather than methanogenesis. This may explain in part the reduction in efficiency that is observed when reactors are operated at temperatures below 20 °C.

Finally, temperature can have a significant effect on the partial pressure of H_2 in reactors and, thus, on the energetics and kinetics of syntrophic metabolism (ZINDER, 1993; SCHINK, 1997; STAMS, 1994). Thermodynamic calculations indicate that reactions which are endergonic under standard conditions (e.g., the oxidation of propionate to acetate, CO_2 and H_2) would become energetically more favorable at higher temperature, and exergonic reactions (e.g., hydrogenotrophic methanogenesis) become less favorable at higher temperatures. Consistent with these thermodynamic predictions, CONRAD and WETTER (1990) found that the minimum thresholds for methanogenesis and acetogenesis were higher with increasing temperature, and WESTERMANN (1994) found an 8- to 18-fold increase in H_2 partial pressure as the incubation temperature of two swamp slurries increased from 2–37 °C. However, WESTERMANN (1994) found that the free energy associated with syntrophic propionate and butyrate degradation was exergonic and fairly constant over the temperature range tested even though the H_2 partial pressure varied considerably. This is consistent with a concept by DOLFING (1992) called a "zero-sum society" where the increase in energy available to the consumer of an intermediate following an increase in concentration of that intermediate is offset by the decrease in energy available to the producer of this intermediate. The effect of increasing temperature on syntrophic metabolism is probably a kinetic one in that diffusion rates of intermediates will be higher at higher temperatures and the elevated H_2 partial pressures in thermophilic systems may generate steeper H_2 gradients between the H_2 producer and the H_2 consumer, both of which would act to relieve the diffusion limitation of syntrophic reactions (STAMS, 1994).

6.2 Effect of pH and Volatile Acid Toxicity

Most methanogens have pH optima near neutrality although there are examples of methanogens that can exist in more extreme environments (BOONE et al., 1993; JONES et al, 1987). GOODWIN and ZEIKUS (1987) showed that acetogenesis and methanogenesis could occur in bog sediments at pH values as low as 4.0 although the rates of fermentation, methanogenesis and acetogenesis were severely inhibited (GOODWIN et al., 1988). In general, the growth of mesophilic fatty acid oxidizing bacteria is also restricted to pH values near neutrality (6.3–8.1) (BEATY and MCINERNEY, 1989; ROY et al., 1986; STIEB and SCHINK, 1986b) with one report of a thermophilic fatty acid degrader being able to grow at a slightly more alkaline pH (SVETLITSHNYI et al., 1996). Similarly, degradation of acetate and propionate by sludge enrichments was restricted to pH values near neutrality (BOONE and XUN, 1987; FUKUZAKI et al., 1990b). While the above discussion shows that the metabolism of volatile fatty acids and methanogenesis is limited by extremes in pH, the fermentative metabolism of carbohydrates and proteins occurs over a wide pH range (ELEFSINIOTIS and OLDHAM, 1994; ZOETEMEYER et al., 1982).

The production of volatile fatty acids can be toxic to microorganisms and may inhibit acetotrophic methanogenesis and syntrophic metabolism. High concentrations of acetate and propionate inhibit their own degradation by sludge enrichments (FUKUZAKI et al., 1990a, b; PAVLOSTATHIS and GIRALDO-GOMEZ, 1991). Acetate also noncompetitively inhibits propionate degradation with a K_i of 8.3 mM (FUKUZAKI et al., 1990b) and uncompetitively inhibits benzoate degradation with a K_i of 10 mM (WARIKOO et al., 1996). Volatile fatty acids can enhance the inhibitory effect of pH on methane production and volatile fatty acid degradation in anaerobic digestors (MCCARTY and MCKINNEY, 1961). As the pH decreases, the concentration of the undissociated from of the acid (HA) increases relative to the ionized form (A^-). Undissociated short-chain organic acids can readily diffuse across biological membranes and uncouple or dissipate the proton motive force (BARONOFSKY et al., 1984). Organisms such as acetotrophic methanogens and syntrophic metabolizers would be most sensitive to the uncoupling effects of the undissociated organic acids since so little energy is available from their energy yielding reactions (SCHINK, 1997). FUKUZAKI et al. (1990a, b) found that the concentration of the undissociated form of acetate controlled the extent of acetate degradation by acetotrophic methanogens and severely inhibited the rate of propionate degradation by sludge enrichments. As discussed in Sect. 5, the accumulation of acetate also places thermodynamic constraints on the extent of syntrophic metabolism (KASPAR and WUHRMANN, 1978b). Acetate was shown to inhibit the extent of benzoate degradation by syntrophic cocultures (SCHÖCKE and SCHINK, 1997; WARIKOO et al., 1996) and butyrate degradation by fluidized bed reactors (LABIB et al., 1992) even though the H_2 partial pressure was very low.

Volatile acid accumulation has been used as an indicator of unbalanced conditions within the reactor. ANDERSON et al. (1982) suggested that digestor failure occurs when the concentrations of the undissociated volatile fatty acids (expressed as acetic acid) reach a level of 500 μM (30 mg^{-1}) regardless of the pH. HILL et al. (1987) suggest that concentrations of acetate greater than 13 mM or a propionate to acetate ratio greater than 1.4 indicate unbalanced conditions. Other studies suggest that propionate is a better indicator of stress (KASPAR and WUHRMANN, 1978b). AHRING et al. (1995) found that changes in the concentrations of volatile fatty acids, in particular, butyrate and isobutyrate concentrations, were good indicators of impending reactor instability.

7 Types of Reactors

The last 20 years have resulted in the development of many new technologies for the anaerobic treatment of waste streams. Most of the new developments are based on the retention of slow growing bacteria through immobilization. These new reactors are able to operate at very high volume loading rates and still maintain high treatment efficiencies. Thus, these new reactor designs should remove the stigma associated with anaerobic digestion and make it possible to treat almost any type of waste (LETTINGA, 1995).

7.1 Conventional Stirred Anaerobic Reactor

The conventional process involves a large holding tank into which wastes are fed intermittently or continuously. In older designs, the waste was not mixed or heated and retention times of 30–60 d were needed. Newer designs involve the use of completely mixed reactors operated at 35 °C which reduces the retention time to about 15 d (SWITZENBAUM, 1983; SPEECE, 1983). These processes are best suited for high strength wastes [>2,000 mg L^{-1} biological oxygen demand (BOD)]. The long retention times dictate the need for large reactor volumes and generally preclude the processing of rapid waste flow.

7.2 Phase Separated Systems

As discussed in Sects. 4 and 5, it is important to maintain a balance between the fast growing fermentative bacteria and the more slow growing syntrophic associations and acetotrophic methanogens so that volatile fatty acids do not accumulate. In phase separated systems, the fermentative process is separated from the methanogenic phase in order to optimize the reaction rates of both processes (COHEN et al., 1979; FOX and POHLAND, 1994; HARPER and POHLAND, 1986; MASSEY and POHLAND, 1978). Since the methanogenic reactor would always be receiving high concentrations of volatile fatty acids from the first reactor, the populations of syntrophic metabolizers in the methanogenic reactor would always be high. This would make the system less sensitive to shock loadings compared to conventional completely mixed systems where the population levels of syntrophic metabolizers may be too low to handle a rapid accumulation of propionate or longer-chain fatty acids (HARPER and POHLAND, 1986). Phase separation can be achieved kinetically by maintaining short hydraulic retention times in the first stage and longer retention times in the second stage. It can also be achieved through the use of membranes or dialysis systems to retain the biomass in the first stage or by ion exchange resins to separate the volatile fatty acids. The use of inert support materials in the second reactors allows high volume loading (AOKI and KAWASE, 1991).

7.3 Fixed Film or Anaerobic Filter Reactors

These reactors use inert support materials to provide a surface for the growth of anaerobic bacteria and to reduce turbulence to allow unattached populations to be retained in the system (SPEECE, 1983, SWITZENBAUM, 1983). Generally, these reactors operate in an upward flow mode at hydraulic retention times on the order of 3 to several 100 h and at loading rates of 0.4–27 kg m^{-3} d^{-1} (SWITZENBAUM, 1983). Full scale reactors generally run at hydraulic retention times of 1–10 d and loading rates of 4–16 kg m^{-3} d^{-1}. These reactors can be operated in a recycle mode to dilute high concentrations of substrates or toxic materials or to control pH. These types of reactors are often used to treat wastes with low levels of organic contaminants such as chlorinated solvents. KAWASE et al. (1989) achieved loading rates of up to 65 kg m^{-3} d^{-1} with a fixed bed reactor with a ceramic porous matrix operated at 54 °C.

7.4 Fluidized/Expanded Bed Reactors

A fluidized bed reactor differs from a fixed bed reactor in that a majority of the biomass is immobilized to the support matrix as opposed to the fixed bed reactor where biomass is entrapped as well as attached to the surface (SPEECE, 1983; SWITZENBAUM, 1983). The particle size of the material is also smaller, usually around the size of sand grains. Their small size allows them to become suspended with fluid flow. In expanded bed reactors, the volume of the inert supporting materials expands to twice its volume with flow while in fluidized bed reactors the expansion of the support material is much less. The advantages of these systems are that very high biomass concentrations (10–40 kg m^{-3}) can be maintained even at very fast hydraulic retention times (IZA, 1991; SWITZENBAUM, 1983). Loading rates of 0.65–60 kg m^{-3} d^{-1} have been reported (IZA, 1991; SWITZENBAUM, 1983).

7.5 Anaerobic Rotating Biological Contactor

In these reactors, the waste stream is passed through rotating disks or drums with attached bacteria. The reactor operates in a plugged flow mode where acidogenic fermentation occurs in the first portion of the reactor followed by methanogenic reactions in latter portions of the reactor (YEH et al., 1997). These reactors can be operated at short hydraulic retention times and have high specific surface areas and biomass concentrations.

7.6 Anaerobic Baffle Reactor

In anaerobic baffle reactors, wastewater is forced to flow under and over a series of baffles as it passes through the reactor (SPEECE, 1983). Each baffle acts as a solids separator to maintain the bacteria within each compartment of the reactor. Thus, bacteria can move vertically within each compartment but have little horizontal movement through the system (BACHMANN et al., 1985; GROBICKI and STUCKEY, 1992). This provides for a relatively simple and inexpensive mechanism for biomass retention. Organic loading rates as high as 36 kg m^{-3} d^{-1} have been achieved with chemical oxygen demand (COD) removal rates of 24 kg m^{-3} d^{-1} (BACHMANN et al., 1985).

7.7 Upflow Anaerobic Sludge Blanket Reactor (UASB)

In the UASB system, liquid waste moves upward through a thick blanket of anaerobic granular sludge, about 0.5–2.5 mm in diameter (LETTINGA et al., 1980; LETTINGA and HULSHOFF POL, 1991; LETTINGA, 1995). The granular sludge contains high concentrations of naturally immobilized bacteria and has excellent settling properties. Thus, high concentrations of biomass can be achieved without support materials which reduces the cost of construction of these types of reactors. Mixing is achieved by the generation of methane within the blanket as well as by hydraulic flow. A settler–gas separator is used to separate gas from the effluent and also creates a quiescent zone where particles flocculate and settle. The granulation process is very important for the start up of these reactors and the factors controlling this process have been reviewed (LETTINGA, 1995). A large number of full scale plants are in operation and have been shown to treat soluble wastes at very short hydraulic retention times (in hours) and at organic loading rates up to 40 kg m^{-3} d^{-1} (LETTINGA et al., 1980).

More recent design innovations include expanded granular bed systems where the 3-phase separator (gas, liquid, biomass) has been improved to handle hydraulic velocities up to 15 m h^{-1} compared to about 1 m h^{-1} for separators in the UASB system (ZOUTBERG and DE BEEN, 1997). This will allow the UASB concept to be used at flow rates that are usually only achieved with fixed bed or fluidized bed reactors. The expanded granular bed system should be advantageous when treating very low strength wastes (COD concentration much less than 1 g L^{-1}) and at low operating temperatures (10 °C) (REBAC et al., 1997; NACHAIYASIT and STUCKEY, 1997; VAN LIER et al., 1997).

Another innovation is the development of a staged upflow anaerobic reactor (LETTINGA, 1995; VAN LIER, 1996; VAN LIER et al., 1994) where various stages of the degradation process are separated into a series of compartments (usually 5) by baffles much like the anaerobic baffle reactors discussed in Sect. 7.6. The use of a stage reactor design allows better biomass retention compared to normal UASB design and provides for a more stable performance since the environmental conditions in each compartment select for populations most adapted to those conditions. These reactors can be operated at temperature and loading rates that are not possible with normal UASB systems (VAN LIER et al., 1994, 1996, 1997).

8 Concluding Remarks

The use of H_2 and acetate by methanogens plays a critical role in regulating the efficiency of anaerobic digestion. Low partial pressures of H_2 allow fermentative bacteria to produce more acetate, CO_2 and H_2 and less ethanol, propionate and longer-chain fatty acids. Low hydrogen partial pressures are also absolutely required for the degradation of key intermediates such as fatty and aromatic acids. More recently, it has been shown that acetate use controls the extent and rate of fatty and aromatic acid degradation by syntrophic associations and prevents the accumulation of the undissociated acetic acid to toxic levels. Perturbations which result in the accumulation of acetate or H_2, such as the inhibition of methanogenesis or an increase in the loading rate, can initiate a cascade of events which may lead to digestor failure. Volatile fatty acids accumulate due to their enhanced production by fermentative bacteria and the inhibition of syntrophic metabolism. This will result in a more acidic pH where the concentration of the undissociated form of the fatty acid may be toxic. New reactor technologies use a variety of approaches to maintain high biomass concentrations of acetotrophic methanogens and organisms capable of syntrophic metabolism in the reactor. By having high biomass concentrations of these organisms, the reactor always has a high potential to degrade acetate and longer-chain fatty acids regardless of the operating conditions and consequently is less susceptible to perturbation.

9 References

AHRING, B. K. (1995), Methanogenesis in thermophilic biogas reactors, *Antonie Van Leeuwenhoek* **67**, 91–102.

AHRING, B. K., WESTERMANN, P. (1987a), Kinetics of butyrate, acetate, and hydrogen metabolism in a thermophilic, anaerobic, butyrate-degrading triculture, *Appl. Environ. Microbiol.* **53**, 434–439.

AHRING, B. K., WESTERMANN, P. (1987b), Thermophilic anaerobic degradation of butyrate by a butyrate-utilizing bacterium in coculture and triculture with methanogenic bacteria, *Appl. Environ. Microbiol.* **53**, 429–433.

AHRING, B. K., WESTERMANN, P. (1988), Product inhibition of butyrate metabolism by acetate and hydrogen in a thermophilic coculture, *Appl. Environ. Microbiol.* **54**, 2393–2397.

AHRING, B. K., SANDBERG, M., ANGELIDAKI, I. (1995), Volatile fatty acids as indicators of process imbalance in anaerobic digestors, *Appl. Microbiol. Biotechnol.* **43**, 559–565.

ANDERSON, G. K., DONNELLY, T., MCKEOWN, K. J. (1982), Identification and control of inhibition in the anaerobic treatment of industrial wastewaters, *Proc. Biochem.* **17**, 28–32.

ANGELIDAKI, I., AHRING, B. K. (1994), Anaerobic thermophilic digestion of manure at different ammonia loads: effect of temperature, *Water Res.* **28**, 727–731.

AOKI, N., KAWASE, M. (1991), Development of high-performance thermophilic two-phase digestion process, *Water Sci. Technol.* **23**, 1147–1156.

AUBURGER, G., WINTER, J. (1995), Isolation and physiological characterization of *Syntrophus buswellii* strain GA from a syntrophic benzoate-degrading, strictly anaerobic coculture, *Appl. Microbiol. Biotechnol.* **44**, 241–248.

BACHMANN, A., BEARD, V. L., MCCARTY, P. L. (1985), Performance characteristics of the anaerobic baffled reactor, *Water Res.* **19**, 99–106.

BANAT, I. M., NEDWELL, D. B. (1983), Mechanism of turnover of C_2–C_4 fatty acids in high-sulfate and low-sulfate anaerobic systems, *FEMS Microbiol. Lett.* **17**, 107–110.

BARONOFSKY, J., SCHREURS, W. J. A., KASHKET, E. R. (1984), Uncoupling by acetic acid limits growth of and acetogenesis by *Clostridium thermoaceticum*, *Appl. Environ. Microbiol.* **48**, 1134–1139.

BAUCHOP, T., MOUNTFORT, D. O. (1981), Cellulose fermentation by a rumen anaerobic fungus in both the absence and the presence of rumen methanogens, *Appl. Environ. Microbiol.* **42**, 1103–1110.

BEATY, P. S., MCINERNEY, M. J. (1989), Effects of organic acid anions on the growth and metabolism of *Syntrophomonas wolfei* in pure culture and in defined consortia, *Appl. Environ. Microbiol.* **55**, 977–983.

BEGUIN, P., AUBERT, J. P. (1994), The biological degradation of cellulose, *FEMS Microbiol. Rev.* **13**, 25–58.

BEN-BASSAT, A., LAMED, R., ZEIKUS, J. G. (1981), Ethanol production by thermophilic bacteria: metabolic control of end product formation in *Thermoanaerobium brockii*, *J. Bacteriol.* **146**, 192–199.

BOONE, D. R. (1982), Terminal reactions in the anaerobic digestion of animal waste, *Appl. Environ. Microbiol.* **43**, 57–64.

BOONE, D. R., BRYANT, M. P. (1980), Propionate-degrading bacterium *Syntrophobacter wolinii* gen. nov., sp. nov., from methanogenic ecosystems, *Appl. Environ. Microbiol.* **40**, 626–632.

BOONE, D. R., XUN, L. (1987), Effects of pH, temperature, and nutrients on propionate degradation by a methanogenic enrichment culture, *Appl. Environ. Microbiol.* **53**, 1589–1592.

BOONE, D. R., JOHNSON, R. L., LIU, Y. (1989), Diffusion of interspecies electron carriers H_2 and formate in methanogenic ecosystems and its implications in the measurement of K_m for H_2 or formate uptake, *Appl. Environ. Microbiol.* **55**, 1735–1741.

BOONE, D. R., WHITMAN, W. B., ROUVIERE, P. (1993), Diversity and taxonomy of methanogens, in: *Methanogenesis: Ecology, Physiology, Biochemistry, and Genetics* (FERRY, J. G., Ed.), pp. 35–80. London: Chapman & Hall.

BREZNAK, J. A. (1994), Acetogenesis from carbon, dioxide in termite guts, in: *Acetogenesis* (DRAKE, H. L., Ed.), pp 303–330. London: Chapman & Hall.

BRYANT, M. P., WOLIN, E. A., WOLIN, M. J., WOLFE, R. S. (1967), *Methanobacillus omelianskii*, a symbiotic association of two species of bacteria, *Arch. Microbiol.* **59**, 20–31.

BRYANT, M. P., CAMBELL, L. L., REDDY, C. A., CRABILL, M. R. (1977), Growth of *Desulfovibrio* in lactate or ethanol media low in sulfate in association with H_2-utilizing methanogenic bacteria, *Appl. Environ. Microbiol.* **33**, 1162–1169.

COHEN, A., ZOETEMEYER, R. J., VAN DEURSEN, A., VAN ANDEL, J. G. (1979), Anaerobic digestion of glucose with separated acid production and methane formation, *Water Res.* **13**, 571–580.

COLLERAN, E., FINNEGAN, S., LENS, P. (1995), Anaerobic treatment of sulphate-containing waste streams, *Antonie Van Leeuwenhoek* **67**, 29–46.

CONRAD, R. (1996), Soil microoraganisms as controllers of atmospheric trace gases (H_2, CO, CH_4, OCS, N_2O, and NO), *Microbiol. Rev.* **60**, 609–640.

CONRAD, R., WETTER, B. (1990), Influence of temperature on energetics of hydrogen metabolism in homoacetogenic, methanogenic, and other anaerobic bacteria, *Arch. Microbiol.* **155**, 94–98.

CONRAD, R., GOODWIN, S., ZEIKUS, J. G. (1987), Hydrogen metabolism in a mildly acidic lake sediment (Knaack Lake), *FEMS Microbiol. Ecol.* **45**, 243–249.

CONRAD, R., BAK, F., SEITZ, H. J., THEBRATH, B., MAYER, H. P., SCHÜTZ, H. (1989), Hydrogen turnover by psychrophilic homoacetogenic and mesophilic methanogenic bacteria in anoxic paddy soil and lake sediment, *FEMS Microbiol. Ecol.* **62**, 285–294.

CORD-RUWISCH, R., SEITZ, H.-J., CONRAD, R. (1988), The capacity of hydrogenotrophic anaerobic bacteria to compete for traces of hydrogen depends on the redox potential of the terminal electron acceptor, *Arch. Microbiol.* **149**, 350–357.

CORD-RUWISCH, R., MERCZ, T. I, HOH, C.-Y., STRONG, G. E. (1997), Dissolved hydrogen concentration as an on-line control parameter for the automated operation and optimization of anaerobic digesters, *Biotechnol. Bioeng.* **56**, 626–634.

DIETRICH, G., WEISS, N., WINTER, J. (1988), *Acetothermus paucivorans*, gen. nov., sp. nov., a strictly anaerobic, thermophilic bacterium from sewage sludge, fermenting hexoses to acetate, CO_2 and H_2, *Syst. Appl. Microbiol.* **10**, 174–179.

DOLFING, J. (1988), Acetogenesis, in: *Biology of Anaerobic Bacteria* (ZEHNDER, A. J. B., Ed.), pp. 417–468. New York: Wiley-Liss.

DOLFING, J. (1992), The energetic consequences of hydrogen gradients in methanogenic ecosystems, *FEMS Microbial. Ecol.* **101**, 183–187.

DOLFING, J., TIEDJE, J. M. (1988), Acetate inhibition of methanogenic, syntrophic benzoate degradation, *Appl. Environ. Microbiol.* **54**, 1871–1873.

DONG, X., PLUGGE, C. M., STAMS, A. J. M. (1994), Anaerobic degradation of propionate by a mesophilic acetogenic bacterium in coculture and triculture with different methanogens, *Appl. Environ. Microbiol.* **60**, 2834–2838.

DWYER, D. F., TIEDJE, J. M. (1986), Metabolism of polythylene glycol by two anaerobic bacteria, *Desulfovibrio desulfuricans* and a *Bacteroides* sp., *Appl. Environ. Microbiol.* **52**, 852–856.

DWYER, D. F., WEEG-AERSSENS, E., SHELTON, D. R., TIEDJE, J. M. (1988), Bioenergetic conditions of butyrate metabolism by a syntrophic, anaerobic bacterium in coculture with hydrogen-oxidizing methanogenic and sulfidogenic bacteria, *Appl. Environ. Microbiol.* **54**, 1354–1359.

EASTMAN, J. A., FERGUSON, J. F. (1981), Solubilization of particulate organic carbon during the acid phase of anaerobic digestion, *J. Water Pollut. Control Fed.* **53**, 352–366.

ELEFSINIOTIS, P., OLDHAM, W. K. (1994), Influence of pH on the acid-phase anaerobic digestion of primary sludge, *J. Chem. Technol. Biotechnol.* **60**, 89–96.

FERRY, J. G. (Ed.) (1993), *Methanogenesis: Ecology, Physiology, Biochemistry, Genetics.* New York: Chapman & Hall.

FOX, P., POHLAND, F. G. (1994), Anaerobic treatment applications and fundamentals: substrate specificity during phase separation, *Water Environ. Res.* **66**, 716–724.

FRANZMANN, P. D., LIU, Y., BALKWILL, D. L., ALDRICH, H. C., MACARIO, E. C., DE BOONE, D. R. (1997), *Methanogenium frigidum* sp. nov., a psychrophilic, H_2-using methanogen from Ace Lake, Antarctica, *Int. J. Syst. Bacteriol.* **47**, 1068–1072.

FUKUZAKI, S., NISHIO, N., NAGAI, S. (1990a), Kinetics of the methanogenic fermentation of acetate, *Appl. Environ. Microbiol.* **56**, 3158–3163.

FUKUZAKI, S., NISHIO, N., SHOBAYASHI, M., NAGAI, S. (1990b), Inhibition of the fermentation of propionate to methane by hydrogen, acetate, and propionate, *Appl. Environ. Microbiol.* **56**, 719–723.

GOODWIN, S., ZEIKUS, J. G. (1987), Ecophysiological adaptations of anaerobic bacteria to low pH: analysis of anaerobic digestion in acidic bog sediments, *Appl. Environ. Microbiol.* **53**, 57–64.

GOODWIN, S., CONRAD, R., ZEIKUS, J. G. (1988), Influence of pH on microbial hydrogen metabolism in diverse sedimentary ecosystems, *Appl. Environ. Microbiol.* **54**, 590–593.

GOODWIN, S., GIREALDO-GOMEZ, E., MOBARRY, M., SWITZENBAUM, M. S. (1991), Comparison of diffusion and reaction rates in microbial aggregates, *Microbial Ecol.* **22**, 161–174.

GROBICKI, A., STUCKEY, D. C. (1992), Hydrodynamic characteristics of the anaerobic baffled reactor, *Water Res.* **26**, 371–378.

GROTENHUIS, J. T. C., SMIT, M., PLUGGE, C. M., YUANSBENG, X., VAN LAMMEREN, A. A. M., et al. (1991), Bacteriological composition and structure of granular sludge adapted to different substrates. *Appl. Environ. Microbiol.* **57**, 1942–1949.

GUWY, A. J., HAWKES, F. R, HAWKES, D. L., ROZZI, A. G. (1997), Hydrogen production in a high rate fluidized bed anaerobic digester, *Water Res.* **31**, 1291–1298.

HARPER, S. R., POHLAND, F. G. (1986), Recent developments in hydrogen management during anaerobic biological wastewater treatment, *Biotechol. Bioeng.* **28**, 585–602.

HENSON, J. M., SMITH, P. H. (1985), Isolation of a butyrate-utilizing bacterium in coculture with *Methanobacterium thermoautotrophicum* from a thermophilic digester, *Appl Environ. Microbiol.* **49**, 1461–1466.

HICKEY, R. F., SWITZENBAUM, M. S. (1991), Thermodynamics of volatile fatty acids accumulation in anaerobic digesters subject to increases in hydraulic and organic loading, *J. Water Pollut. Control Fed.* **63**, 141–144.

HILL, D. T., COBB, S. A., BOLTE, J. P. (1987), Using volatile acid relationships to predict anaerobic digestor failure, *Trans. ASAE* **30**, 496–501.

HOBSON, P. N., BOUSFIELD, S., SUMMERS, R. (1974), Anaerobic digestion of organic matter, *CRC Crit. Rev. Environ. Control* **4**, 131–191.

HOH, C. Y., CORD-RUWISCH, R. (1996), A practical kinetic model that considers endproduct inhibition in anaerobic digestion processes by including the equilibrium constant, *Biotechnol. Bioeng.* **51**, 597–604.

HUSER, B. A., WUHRMANN, K., ZEHNDER, A. J. B. (1982), *Methanothrix soehngenii* gen. nov. sp. nov., a new acetotrophic non-hydrogen-oxidizing methane bacterium, *Arch. Microbiol.* **132**, 1–9.

ISA, Z., GRUSENMEYER, S., VERSTRAETE, W. (1986), Sulfate reduction relative to methane production in high-rate anaerobic digestion: microbiological aspects, *Appl. Environ. Microbiol.* **51**, 580–587.

IZA, J. (1991), Fluidized bed reactors for anaerobic wastewater treatment, *Water Sci. Technol*, **24**, 109–132.

JACKSON, B. E., MCINERNEY, M. J. (1997), Comparison of benzoate degradation in methanogenic and sulfate-reducing syntrophic cocultures, *BioFactors* **6**, 53.

JETTEN, M. S. M., STAMS, A. J. M., ZEHNDER, A. J. B. (1990), Acetate threshold values and acetate activating enzymes in methanogenic bacteria, *FEMS Microbiol. Ecol.* **73**, 339–344.

JONES, W. J., NAGLE, JR., D. P., WHITMAN, W. B. (1987), Methanogens and diversity of archaebacteria, *Microbiol. Rev.* **51**, 135–177.

KAISER, J.-L., HANSELMANN, K. W. (1982), Fermentative metabolism of substituted monoaromatic compounds by a bacterial community from anaerobic sediments, *Arch. Microbiol.* **133**, 185–194.

KASPAR, H. F., WUHRMANN, K. (1978a), Kinetic parameters and relative turnover of some important catabolic reactions in digestor sludge, *Appl. Environ. Microbiol.* **36**, 1–7.

KASPAR, H. F., WUHRMANN, K. (1978b), Product inhibition in sludge digestion, *Microbial. Ecol.* **4**, 241–248.

KAWASE, M., NOMURA, T., MAJIMA, T. (1989), An anaerobic fixed bed reactor with a porous ceramic carrier, *Water Sci. Technol.* **21**, 77–86.

KOTSYURBENKO, O. R., NOZHEVNIKOVA, A. N., SOLOVIOVA, T. I., ZAVARZIN, G. A. (1996), Methanogenesis at low temperatures by microflora of tundra wetland soil, *Antonie Van Leeuwenhoek* **69**, 75–86.

KRISTJANSSON, J. K., SCHÖNHEIT, P., THAUER, R. K. (1982), Different K_s values for hydrogen of methanogenic bacteria and sulfate reducing bacteria: an explanation for the apparent inhibition of methanogenesis by sulfate, *Arch. Microbiol.* **131**, 278–282.

KRUMHOLZ, L. R., BRYANT, M. P. (1986), *Syntrophococcus sucromutans*, sp. nov., gen. nov., uses carbohydrates as electron donors and formate, methoxymonobenzoids or *Methanobrevibacter* as electron acceptor systems, *Arch. Microbiol.* **143**, 313– 318.

KUGELMAN, I. J., CHIN, K K. (1971), Toxicity, synergism, and antagonism in anaerobic waste treatment processes, *Adv. Chem. Ser.* **105**, 55–89.

KURR, M., HUBER, R., KÖNIG, H., JANNASCH, H. W., FRICKE, H., TRINCONE, A., KRISTJANSSON, J. K., STETTER, K. O. (1991), *Methanopyrus kandleri*, gen. nov., sp. nov., represents a novel group of hyperthermophilic methanogens, growing at 110 °C, *Arch. Microbiol.* **156**, 239–247.

LABIB, F., FERGUSON, J. F., BENJAMIN, M. M., MERIGH, M., RICKER, N. L. (1992), Anaerobic butyrate degradation in a fluidized-bed reactor: effects of increased concentrations of H_2 and acetate, *Environ. Sci. Technol.* **26**, 369–376.

LEPISTO, R., RINTALA, J. (1996), Conversion of volatile fatty acids in an extreme thermophilic (76–80 °C) upflow anaerobic sludge-blanket reactor, *Biores. Technol.* **56**, 221–227.

LETTINGA, G. (1995), Anaerobic digestion and wastewater treatment systems, *Antonie Van Leeuwenhoek* **67**, 3–28.

LETTINGA, G., HULSHOFF POL, L. W. (1991), UASB-process design for various types of wastewaters, *Water Sci. Techol.* **24**, 87–107.

LETTINGA, G., VAN VELSEN, A. F. M., HOMBA, S. W., ZEEUW, W., DE KLAPWIJK, A. (1980), Use of the upflow sludge blanket (USB) reactor concept for biological wastewater treatment, especially for anaerobic treatment, *Biotechnol. Bioeng.* **22**, 699–734.

LJUNGDAHL, L. G., ERIKSSON, K.-E. (1985), Ecology of microbial cellulose degradation, *Adv. Microbiol. Ecol.* **8**, 237–299.

LOROWITZ, W. H., ZHAO, H., BRYANT, M. P. (1989), *Syntrophomonas wolfei*, subsp., *saponivida* subsp., nov., a long-chain fatty-acid-degrading, anaerobic, syntrophic bacterium; *Syntrophomonas wolfei* subsp. *wolfei* subsp. nov.; and emended descriptions of the genus and species, *Int. J. Syst. Bacteriol.* **39**, 122–126.

LOVLEY, D. R. (1985), Minimum threshold for hydrogen metabolism in methanogenic bacteria, *Appl. Environ. Microbiol.* **49**, 1530–1531.

LOVLEY, D. R., GREENING, R. C., FERRY, J. G. (1984), Rapidly growing methanogenic organism that synthesizes coencyme M and has a high affinity for formate, *Appl. Environ. Microbiol.* **48**, 81–87.

LOWE, S. E., JAIN, M. K., ZEIKUS, J. G. (1993), Biology, ecology, and biotechnological applications of anaerobic bacteria adaptcd to environmental stresses in temperature, pH, salinity, or substrates, *Microbiol. Rev.* **57**, 451–509.

MACKIE, R. I., BRYANT, M. P. (1981), Metabolic activity of fatty acid-oxidizing bacteria and the contribution of acetate, propionate, butyrate and CO_2 to methanogenesis in cattle waste at 40 and 60 °C, *Appl. Environ. Microbiol.* **41**, 1363–1373.

MACKIE, R. I., WHITE, B. A., BRYANT, M. P. (1991), Lipid metabolism in anaerobic ecosystems, *CRC Crit. Rev. Microbiol.* **17**, 449–479.

MALINA, JR., J. F., POHLAND, F. G. (Eds.) (1992), *Design of Anaerobic Processes for the Treatment of Industrial and Municipal Wastes*. Lancaster, PA: Technomic Publishing Company.

MASSEY, M. L., POHLAND, F. G. (1978), Phase separation of anaerobic stabilization by kinetic controls, *J. Water Pollut. Control Fed.* **50**, 2204–2222.

MCCARTNEY, D. M., OLESZKIEWICZ, J. A. (1991), Sulfide inhibition of anaerobic degradation of lactate and acetate, *Water Res.* **25**, 203–209.

MCCARTY, P. L. (1971), Energetics and kinetics of anaerobic treatment, *Adv. Chem. Ser.* **105**, 91–107.

MCCARTY, P. L., MCKINNEY, R. E (1961) Volatile acid toxicity in anaerobic digestion, *J. Water Pollut. Control Fed.* **33**, 223–232.

MCCARTY, P. L., MOSEY, F. E. (1991), Modelling of anaerobic digestion processes (a discussion of concepts), *Water Sci. Technol.* **24**, 17–33.

MCCARTY, P. L., SMITH, D. P. (1986), Anaerobic wastewater treatment, *Environ. Sci. Technol.* **20**, 1200–1206.

MCINERNEY, M. J. (1986), Transient and persistent associations among prokaryotes, in: *Bacteria in Nature*, Vol. 2 (POINDEXTER, J. S., LEADBETTER, E. R., Eds.), pp. 293–338. New York: Plenum Press.

MCINERNEY, M. J. (1988), Anaerobic hydrolysis and fermentation of fats and proteins, in: *Biology of Anaerobic Microorganisms* (ZEHNDER, A. J. B., Ed.), pp. 373–416. New York: Wiley-Liss.

MCINERNEY, M. J. (1992), The genus *Syntrophomonas* and other syntrophic bacteria, in: *The Prokaryotes, A Handbook on the Biology of Bacteria: Ecophysiology, Isolation, Identification, Applications*, 2nd Eds. (BALOWS, A., TRÜPER, H. G., DWORKIN, M., HARDER, W., SCHLEIFER, K.-H., Eds.), pp. 2048–2057. Berlin Heidelberg New York: Springer-Verlag.

MCINERNEY, M. J., BRYANT, M. P. (1981), Review of methane fermentation fundamentals, in: *Fuel Gas Production from Biomass* (WISE, D. L., Ed.), pp. 19–46. Boca Raton, FL: CRC Press.

MCINERNEY, M. J., BRYANT, M. P., PFENNIG, N. (1979), Anaerobic bacterium that degrades fatty acids in syntrophic association with methanogens, *Arch. Microbiol.* **122**, 129–135.

MILLER, T. L., WOLIN, M. J. (1973), Formation of hydrogen and formate by *Ruminococcus albus*, *J. Bacteriol.* **116**, 836–846.

MILLER, T. L., WOLIN, M. J. (1994), Acetogenesis from CO_2 in the human colonic ecosystem, in: *Acetogenesis* (DRAKE, H. L., Ed.), pp. 365–385. New York: Chapman & Hall.

MIN, H., ZINDER, S. H. (1989), Kinetics of acetate ulilization by two thermophilic acetotrophic methanogens: *Methanosarcina* sp. strain CALS-1 and *Methanothrix* sp. strain CALS-1, *Appl. Environ. Microbiol.* **55**, 488–491.

MOSEY, F. E. (1983), Mathematical modeling of the anaerobic digestion process: regulatory mechanisms for the formation of short-chain volatile acids from glucose, *Water Sci. Technol.* **15**, 209–232.

MOSEY, F. E., FERNANDES, X. A. (1989), Patterns of hydrogen in biogas from the anaerobic digestion of milk-sugars, *Water Sci. Technol.* **21**, 187–196.

MOUNTFORT, D. O, ASHER, R. A. (1978), Changes in proportions of acetate and CO_2 used as methane precursors during anaerobic digestion of bovine waste, *Appl. Environ. Microbiol.* **35**, 648–654.

MOUNTFORT, D. O., BRYANT, M. P. (1982), Isolation and characterization of an anaerobic benzoate-degrading bacterium from sewage sludge, *Arch. Microbiol.* **133**, 249–256.

MUCHA, H., LINGENS, F., TRÖSCH, W. (1988), Conversion of propionate to acetate and methane by syntrophic consortia, *Appl. Microbiol. Biotechnol.* **27**, 581–586.

NACHAIYASIT, S., STUCKEY, D. C. (1997), Effect of low temperatures on the performance of an anaerobic baffled reactor (ABR), *J. Chem. Technol. Biotechnol.* **69**, 276–284.

NIELSEN, P. H. (1987), Biofilm dynamics and kinetics during high-rate sulfate reduction under anaerobic conditions, *Appl. Environ. Microbiol.* **53**, 27–32.

OREMLAND, R. S., POLCIN, S. (1982), Methanogenesis and sulfate reduction: competitive and noncompetitive substrates in estuarine sediments, *Appl. Environ. Microbiol.* **44**, 1270–1276.

ÖRLYGSSON, K., HOUWEN, F. P., SVENSSON, B. H. (1994), Influence of hydrogenotrophic methane formation on the thermophilic anaerobic degradation of protein and amino acids, *FEMS Microbiol. Ecol.* **13**, 327–334.

OTHSUBO, S., DEMIZU, K., KOHNO, S., MIURA, I., OGAWA, T., FUKUDA, H. (1992), Comparison of acetate utilization among strains of an acetoclastic methanogen, *Methanothrix soehngenii*, *Appl. Environ. Microbiol.* **58**, 703–705.

OUDE ELFERINK, S. J. W. H., VISSER, J., HULSHOFF POL, L. W., STAMS, A. J. M. (1994), Sulfate reduction in methanogenic bioreactors, *FEMS Microbiol. Rev.* **15**, 119–136.

PARKIN, G. F., SPEECE, R. E., YANG, C. H. J., KOCHER, W. M. (1983), Response of methane fermentation systems to industrial toxicants, *J. Water Pollut. Control Fed.* **55**, 44–53.

PAVLOSTATHIS, S. G., GIRALDO-GOMEZ, E. (1991), Kinetics of anaerobic treatment: a critical review, *Crit. Rev. Environ. Control* **21**, 441–490.

PETERSON, S. P., AHRING, B. K. (1991), Acetate oxidation in a thermophilic anaerobic sewage sludge digestor: the importance of non-acetoclastic methanogenesis from acetate, *FEMS Microbiol. Ecol.* **86**, 149–158.

PFEFFER, J. T., LIEBMAN, J. C. (1976), Energy from refuse by bioconversion, fermentation and residue disposal processes, *Res. Recov. Conserv.* **1**, 295–313.

RASKIN, L., RITTMANN, B. E., STAHL, D. A. (1996), Competition and coexistence of sulfate-reducing and methanogenic populations in anaerobic biofilms, *Appl. Environ. Microbiol.* **62**, 3847–3857.

REBAC, S., VAN LIER, J. B., JANSSEN, M. G. J., DEKKERS, F., SWINKELS, K. T. M., LETTINGA, G. (1997), High-rate anaerobic treatment of malting waste water in a pilot-scale EGSB system under psychrophilic conditions, *J. Chem. Technol. Biotechnol.* **68**, 135–146.

ROBINSON, J. A., TIEDJE, J. M. (1982), Kinetics of hydrogen consumption by rumen fluid, anaerobic digestor sludge, and sediment, *Appl. Environ. Microbiol.* **44**, 1374–1384.

ROBINSON, J. A., TIEDJE, J. M. (1984), Competition between sulfate-reducing and methanogenic bacteria for H_2 under resting and growing conditions, *Arch. Microbiol.* **137**, 26–32.

ROY, F., SAMAIN, E., DUBOURGUIER, H. C., ALBAGNAC, G. (1986), *Syntrophomonas sapovorans* sp. nov., a new obligately proton reducing anaerobe oxidizing saturated and unsaturated long chain fatty acids, *Arch. Microbiol.* **145**, 142–147.

SCHAUER, N. L., BROWN, D. P., FERRY, J. G. (1982), Kinetics of formate metabolism in *Methanobacterium formicicum* and *Methanospirillum hungatei*, *Appl. Environ. Microbiol.* **44**, 549–554.

SCHINK, B. (1988), Principles and limits of anaerobic methane degradation: environmental and technological aspects, in: *Biology of Anaerobic Bacteria* (ZEHNDER, A. J. B., Eds.), pp. 771–846. New York: Wiley-Liss.

SCHINK, B. (1992) The genus *Pelobacter*, in: *The Prokaryotes, A Handbook on the Biology of Bacteria: Ecophysiology, Isolation, Identification, Applications* 2nd Edn. (BALOWS, A., TRÜPER, H. G., DWORKIN, M., HARDER, W., SCHLEIFER, K.-H., Eds.), pp. 3393–3399. New York: Springer-Verlag.

SCHINK, B. (1997), Energetics of syntrophic cooperation in methanogenic degradation, *Microbiol. Mol. Biol. Rev.* **61**, 262–280.

SCHNÜRER, A., HOUWEN, F. P., SVENSSON B. H. (1994), Mesophilic syntrophic acetate oxidation during methane formation by a triculture at high ammonium concentration, *Arch. Microbiol.* **162**, 70–74.

SCHNÜRER, A., SCHINK, B., SVENSSON, B. H. (1996) *Clostridium ultunense* sp. nov., a mesophilic bacterium oxidizing acetate in syntrophic association with a hydrogenotrophic methanogenic bacterium, *Int. J. Syst. Bacteriol.* **46**, 1145–1152.

SCHÖCKE, L., SCHINK, B. (1997), Energetics of methanogenic benzoate degradation by *Syntrophus gentianae* in syntrophic coculture, *Microbiology* **143**, 2345–2351.

SCHÖNHEIT, P., KRISTJANSSON, J. K., THAUER, R. K. (1982), Kinetic mechanism for the ability of sulfate reducers to out-compete methanogens for acetate, *Arch. Microbiol.* **132**, 285–288.

SHEA, T. G., PRINORIUS, W. A., COLE, R. D., PEARSON, E. A. (1968), Kinetics of hydrogen assimilation in the methane fermentation, *Water Res.* **2**, 833–848.

SMITH, D. P. (1992), Spherical diffusion model of interspecies hydrogen transfer in steady-state methanogenic reactors, *Water Sci. Technol.* **26**, 2389–2392.

SMITH, D. P., MCCARTY, P. L. (1989), Energetic and rate effects on methanogenesis of ethanol and propionate in perturbed CSTRs, *Biotechnol. Bioeng.* **34**, 39–54.

SMITH, P. H., MAH, R. A. (1966), Kinetics of accetate metabolism during sludge digestion, *Appl. Environ. Microbiol.* **14**, 368–371.

SOUTSCHEK, E., WINTER, J., SCHINDLER, F., KANDLER, O. (1984), *Acetomicrobium flavidum*, gen. nov., sp. nov., a thermophilic, anaerobic bacterium from sewage sludge, forming acetate, CO_2 and H_2 from glucose, *Syst. Appl. Microbiol.* **5**, 377–390.

SPEECE, R. E. (1983), Anaerobic biotechnology for industrial wastewater treatment, *Environ. Sci. Technol.* **17**, 416–427.

STAMS, A. J. M. (1994), Metabolic interactions between anaerobic bacteria in methanogenic environments, *Antonie Van Leeuwenhoek*, **66**, 271–294.

STAMS, A. J. M., GROLLE, K. C. F., FRIJTERS, C. T. M. J., VAN LIER, J. B. (1992), Enrichment of thermophilic propionate-oxidizing bacteria in syntrophy with *Methanobacterium thermoautotrophicum* or *Methanobacterium thermoformicicum*, *Appl. Environ. Microbiol.* **58**, 346–352.

STIEB, M., SCHINK, B. (1986a), Anaerobic degradation of isovalerate by a defined methanogenic coculture, *Arch. Microbiol.* **144**, 291–295.

STIEB, M., SCHINK, B. (1986b), Anaerobic oxidation of fatty acids by *Clostridium bryantii* sp. nov., a spore forming, obligately syntrophic bacterium, *Arch. Microbiol.* **144**, 291–295.

STRAYER, R. F., TIEDJE, J. M. (1978), Kinetic parameters of the conversion of methane precursors to methane in a hypereutrophic lake sediment, *Appl. Environ. Microbiol.* **36**, 330–340.

STRONG, G. E., CORD-RUWISCH, R. (1995), An *in situ* dissolved-hydrogen probe for monitoring anaerobic digesters under overload conditions, *Biotechnol. Bioeng.* **45**, 63–68.

SVETLITSHNYI, V., RAINEY, F., WEIGEL, J. (1996), *Thermosyntropha lipolytica* gen. nov., sp. nov., a lipolylic, anaerobic, alkalitolerant, thermophilic bacterium utilizing short- and long-chain fatty acids in syntrophic coculture with a methanogenic archaeum, *Int. J. Syst. Bacteriol.* **46**, 1131–1137.

SWITZENBAUM, M. S. (1983), Anaerobic fixed film wastewater treatment, *Enzym. Microb. Technol.* **5**, 242–250.

TEWES, F. J., THAUER, R. K. (1980), Regulation of ATP-synthesis in glucose fermenting bacteria involved in interspecies hydrogen transfer, in: Anaerobes and Anaerobic Infections, *Symposia* held at the XII. Int. Congr. Microbiol., Munich, Sept. 3–8 (GOTTSCHALK, G., PFENNIG, N., WENER, H., Eds.), pp. 97–104. Stuttgart, New York: Gustav Fischer.

THAUER, R. K., JUNGERMANN, K., DECKER, K. (1977), Energy conservation in chemotrophic anaerobic bacteria. *Bacteriol. Rev.* **41**, 100–180.

THIELE, J. H., ZEIKUS, J. G. (1988), Control of interspecies electron flow during anaerobic digestion: significance of formate transfer versus hydrogen transfer during syntrophic methanogenesis in flocs, *Appl. Environ. Microbiol.* **54**, 20–29.

THOLOZAN, J. L., SAMAIN, E., GRIVET, J. P. (1988a), Isomerization between *n*-butyrate and isobutyrate in enrichment cultures, *FEMS Microbiol. Ecol.* **53**, 187–191.

THOLOZAN, J.-L., SAMAIN, E., GRIVET, J.-P., MOLETTA, R., DUBOURGUIER, H. C., ALBAGNAC, G. (1988b), Reductive carboxylation of propionate to butyrate in methanogenic ecosystems, *Appl. Environ. Microbiol.* **54**, 441–445.

VAN LIER, J. B. (1996), Limitations of thermophilic anaerobic wastewater treatment and the consequences for process design, *Antonie Van Leeuwenhoek*, **69**, 1–14.

VAN LIER, J. B., HULSBEEK, J., STAMS, A. J. M., LETTINGA, G. (1993), Temperature susceptibility of thermophilic methanogenic sludge: implications for reactor start-up and operation, *Biores. Technol.* **43**, 227–235.

VAN LIER, J. B., BOERSMA, F., DEBETS, M. M. W. H., LETTINGA, G. (1994), High rate thermophilic anaerobic wastewater treatment in compartmentalized upflow reactors, *Water Sci. Technol.* **30**, 251–261.

VAN LIER, J. B., GROENEVELD, N., LETTINGA, G. (1996), Development of thermophilic methanogenic sludge in compartmentalized upflow reactors, *Biotechnol. Bioeng.* **50**, 115–124.

VAN LIER, J. B., REBAC, S., LETTINGA, G. (1997), High-rate anaerobic wastewater treatment under psychrophilic and thermophilic conditions, *Water Sci. Technol.* **35**, 199–206.

VISSER, A., BEEKSMA, I., VAN DER ZEE, F., STAMS, A. J. M., LETTINGA, G. (1993), Anaerobic degradation of volatile fatty acids at different sulphate concentrations, *Appl. Microbiol. Biotechnol.* **40**, 549–556

VISSER, A., HULSHOFF POL, L. W., LETTINGA, G. (1996), Competition of methanogenic and sulfidogenic bacteria, *Water Sci. Technol.* **33**, 99–110.

WALLRABENSTEIN, C., GORNY, N., SPRINGER, N., LUDWIG, W., SCHINK, B. (1995a), Pure culture of *Syntrophus buswellii*, definition of its phylogenetic status, and description of *Syntrophus gentianae* sp. nov., *Syst. Appl. Microbiol.* **18**, 62–66.

WALLRABENSTEIN, C., HAUSCHILD, E., SCHINK, B. (1995b), *Syntrophobacter pfennigii* sp. nov., new syntrophically propionate-oxidizing anaerobe growing in pure culture with propionate and sulfate, *Arch. Microbiol.* **164**, 346–352.

WARD, D. M., WINFREY, M. R. (1985), Interactions between methanogenic and sulfate-reducing bacteria in sediments, *Adv. Aquat. Microbiol.* **3**, 141–179.

Warikoo, V., McInerney, M. J., Robinson, J. A., Suflita, J. M. (1996), Interspecies acetate transfer influences the extent of anaerobic benzoate degradation by syntrophic consortia, *Appl. Environ. Microbiol.* **62**, 26–32.

Westermann, P. (1994), The effect of incubation temperature on steady-state concentrations of hydrogen and volatile fatty acids during anaerobic degradation in slurries from wetland sediments, *FEMS Microbiol. Ecol.* **13**, 295–302.

Westermann, P., Ahring, B. K., Mah, R. A. (1989), Threshold acetate concentration for acetate catabolism by acetoclastic methanogenic bacteria, *Appl. Environ. Microbiol.* **55**, 514–515.

Widdel, F. (1986), Growth of methanogenic bacteria in pure culture with 2-propanol and other alcohols as hydrogen donors, *Appl. Environ. Microbiol.* **51**, 1056–1062.

Wiegant, W. M., Claassen, J. A., Lettinga, G. (1985), Thermophilic anaerobic digestion of high-strength wastewaters, *Biotechnol. Bioeng.* **27**, 1374–1381.

Wildenauer, F. X., Winter, J. (1986), Fermentation of isoleucine and arginine by pure and syntrophic cultures of *Clostridium sporogenes*, *FEMS Microbiol. Ecol.* **38**, 373–379.

Wolin, M. J. (1982), Hydrogen transfer in microbial communities, in: *Microbial Interactions and Communities* (Bull, A. T., Slater, J. H., Eds.), pp. 323–356. New York, London: Academic Press.

Wu, W.-M., Hickey, R. F., Zeikus, J. G. (1991), Characterization of metabolic performance of methanogenic granules treating brewery wastewater: role of sulfate-reducing bacteria, *Appl. Environ. Microbiol.* **57**, 3438–3449.

Wu, W.-M., Thiele, J. H., Jain, M. K., Pankratz, H. S., Hickey, R. F., Zeikus, J. G. (1993), Comparison of rod-versus filament-type methanogenic granules: microbial population and reactor performance, *Appl. Microbiol. Biotechnol.* **39**, 795–803.

Wu, W.-M., Jain, M. K., Zeikus, J. G. (1994), Anaerobic degradation of normal- and branched-chain fatty acids with four or more carbons to methane by a syntrophic methanogenic triculture, *Appl. Environ. Microbiol.* **60**, 2220–2226.

Yeh, A. C., Lu, C., Lin, M. R. (1997), Performance of an anaerobic rotating biological contactor: effects of flow rate and influent organic strength, *Water Res.* **31**, 1251–1260.

Yoda, M., Kitagawa, M., Miyaji, Y. (1987), Long term competition between sulfate-reducing and methane-producing bacteria for acetate in anaerobic biofilm, *Water Res.* **21**, 1547–1556.

Zhao, H., Yang, D., Woese, C. R., Bryant, M. P. (1993), Assignment of fatty acid-β-oxidizing syntrophic bacteria to *Syntrophomonadaceae* fam. nov. on the basis of 16S rRNA sequence analysis, *Int. J. Syst. Bacteriol.* **43**, 278–286.

Zindel, U., Freudenberg, W., Reith, M., Andreesen, J. R., Schnell, J., Widdel, F. (1988), *Eubacterium acidaminophilum* sp. nov., a versatile amino acid degrading anaerobe producing or utilizing H_2 or formate. Description and enzymatic studies, *Arch. Microbiol.* 150, 254–266.

Zinder, S. H. (1986), Patterns of carbon flow from glucose to methane in a thermophilic anaerobic bioreactor, *FEMS Microbiol. Ecol.* **38**, 243–250.

Zinder, S. H. (1993), Physiological ecology of methanogens, in: *Methanogenesis: Ecology, Physiology, Biochemistry, Genetics* (Ferry, J. G., Ed.), pp. 128–206. New York: Chapman & Hall.

Zinder, S. H., Koch, M. (1984) Non-acetoclastic methanogenesis from acetate: acetate oxidation by a thermophilic syntrophic coculture, *Arch. Microbiol.* **138**, 263–272.

Zinder, S. H., Cardwell, S. C., Anguish, T., Lee, M., Koch, M. (1984), Methanogenesis in a thermophilic (58 °C) anaerobic digestor: *Methanothrix* sp. as an important aceticlastic methanogen, *Appl. Environ. Microbiol.* **47**, 796-807.

Zoetemeyer, R. J., van den Heuvel, J. C., Cohen, A. (1982), pH influence on acidogenic dissimilation of glucose in an anaerobic digestor, *Water Res.* **16**, 303–311.

Zoutberg, G. R., de Been, P. (1997), The Biobed EGSB (expanded granular sludge bed) system covers shortcomings of the upflow anaerobic sludge blanket reactor in the chemical industry, *Water Technol.* **35**, 183–188.

23 CSTR Reactors and Contact Processes in Industrial Wastewater Treatment

HELMUT KROISS
KARL SVARDAL
Wien, Austria

List of Abbreviations

BOD	biochemical oxygen demand
COD	chemical oxygen demand
CSTR	completely stirred tank
ESP	excess sludge production
HRT	hydraulic retention time
MCRT	mean cell residence time
MLSS	mixed liquor suspended solids
MLVSS	mixed liquor volatile suspended solids
SRT	sludge retention time
TKN	total Kjeldahl nitrogen
TOC	total organic carbon
UASB	upflow anaerobic sludge blanket
VSS	volatile suspended solids

1 Definition of Anaerobic Reactor Configurations

The main characteristic of CSTR (completely stirred tank) reactors and contact processes is the fact that the biological reactor is assumed to be completely mixed. While such a mixing situation can easily be obtained in the liquid phase, this is not the case for the solid phase. Depending on wastewater composition it cannot be predicted how much energy will be needed to also get complete mixing of the solid fraction (anaerobic sludge) in the reactor. Therefore, in practice there is a grey zone between the CSTR and e.g., the upflow systems where the liquid phase can be nearly completely mixed while the solid phase is not (ATV, 1990).

Completely mixed tanks can also be supplemented with support material (e.g., plastic media) as a basis for fixed growth in order to increase active biomass concentration in the reactor (PASCIK and HENZLER, 1988). Hence, there is also a grey zone between CSTR and expanded or fluidized bed reactors.

This chapter is restricted to reactor configurations, where the liquid phase is assumed to be completely mixed and no fixed or fluidized growth support is added.

2 Process Configurations

2.1 Suspended Growth Reactors

The simplest anaerobic process configuration is a completely mixed suspended growth reactor (Fig. 1). It consists of an empty tank with a mixing device. The main process characteristic is that the hydraulic retention time (HRT) is equal to the mean cell residence time (MCRT) also called sludge age or sludge retention time (SRT). The decisive parameter to describe the biological process behavior is the MCRT.

There is an important relationship between mixing efficiency and suspended solids with

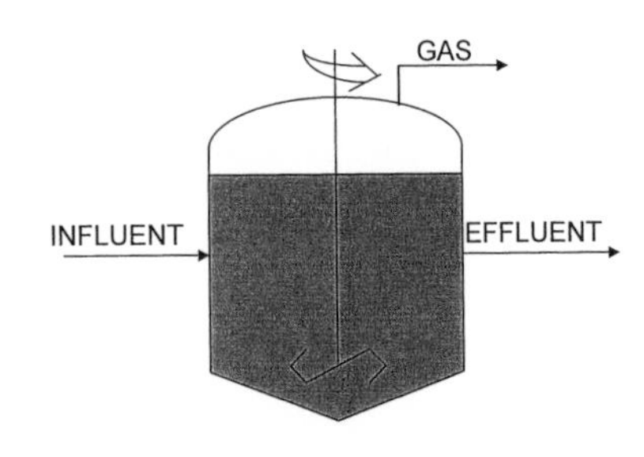

Fig. 1. Completely mixed suspended growth reactor.

their settling behavior. The following types of suspended solids can be distinguished:

(1) organic and inorganic solids contained in the influent, which are not affected by the biological process,
(2) organic particles which are only partly degraded,
(3) inorganic precipitation products which develop in the process (e.g., calcium carbonate, magnesium ammonium phosphate),
(4) the bacterial mass which develops from the degradation of the organic compounds.

As long as mixing can provide equal distribution of the suspended solids in the total reactor volume the concentrations of all compounds are the same in the reactor and the effluent. If this is not the case, an accumulation of (heavy) solids in the reactor can occur, which can finally result in sediments reducing the active volume and hence the HRT (MCRT). The consequence can be a reduced treatment efficiency.

A separation of the solids from the effluent does not influence the anaerobic process concept but can be important for the design and operation of following treatment steps (e.g., aerobic biological posttreatment). Hence one important advantage of the CSTR reactor concept is that treatment efficiency does not depend on solids separation characteristics. This process scheme has been applied for many decades in mesophilic municipal sludge digesters, where it is extremely successful and reliable with HRT (MCRT) between 15 and 30 d.

While for the anaerobic degradation of organic solids the digestion process normally is limited by the rate of hydrolysis, the methane production from easily biodegradable dissolved organic compounds is normally limited by the maximum growth rate of the acetoclastic methanogenic bacteria, sometimes also by the syntrophic acetogenic bacteria.

2.2 Anaerobic Contact Process (Fig. 2)

The basic process idea is the same as for the activated sludge process, i.e., the uncoupling of the HRT from the MCRT in order to reduce the volume requirement for the reactor.

The process consists of a completely mixed reactor, a solid–liquid separation unit (sedimentation, thickening, membrane separation) and a return sludge cycle. In order to control the solids concentration in the system and hence the MCRT an excess sludge removal device is necessary. The control of solids concentration via the effluent suspended solids in the settled effluent is dangerous, as it leads to an accumulation of heavy sediments in the system, slowly replacing the active biomass by uncoupling MCRT and SRT. Therefore, it is recommended to remove the excess sludge directly from the reactor (bottom) or from the return sludge.

The more diluted the wastewater the larger is the difference in volume requirements for the reactor between CSTR and contact process. At the same time the effort for the liquid–solids separation unit will increase.

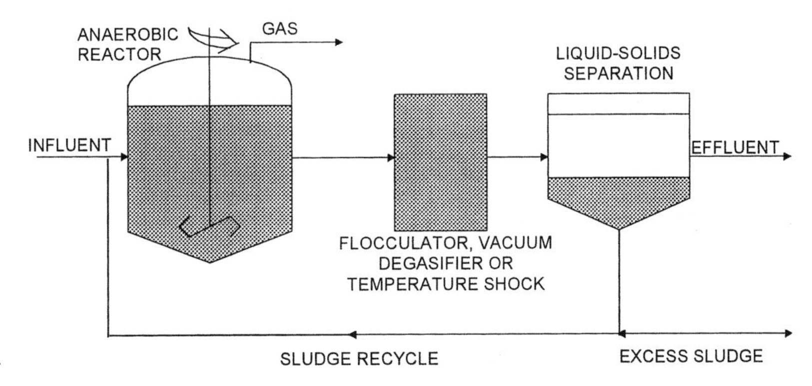

Fig. 2. Anaerobic contact process scheme.

If liquid–solids separation is performed by settling and thickening of the solids, which is the most common solution, the process performance depends on the settling and thickening characteristics of the anaerobic MLSS. These characteristics also depend on the gas production potential in the settler/thickener as small gas bubbles adhere to the flocs of the anaerobic sludge. The best way to control this gas production is to maintain a constant low concentration of biodegradable substrate in the effluent. This process, therefore, has a basic instability in case of overloading or inhibition of anaerobic bacteria resulting in higher effluent concentrations.

Whenever the water surface of the settler/thickener unit is connected to the air atmosphere, gas bubbles will be formed due to the change in the gas–liquid equilibrium. The problems with gas bubble formation can be reduced by vacuum degasifiers and short temperature shocks. The addition of flocculants in order to improve the settling and thickening process can increase the problem, as flotation can occur due to the fact that more gas can be entrapped in larger flocs.

The use of membrane separation units would probably not be affected by gas bubbles but scaling and fouling can create new problems.

Another basic problem of this process configuration with settling/thickening is the fact that it favors a selection of solids based on the settling properties. Inorganic solids like sand, calcium carbonate or large organic particles will accumulate while the active biomass having poor settling properties can be washed out with the effluent. The consequence is that MCRT and SRT do not correspond any more which can result in a failure of the process especially with low concentrated wastewaters.

In order to solve some of the problems mentioned above, especially in the case of increased solids production by precipitation or coarse (inorganic) compounds (sand etc.) in the influent, the reactor concept shown in Fig. 3 has been proved in practice (SVARDAL et al., 1993). It combines the contact process with some ideas from the UASB process. Both processes derive from the Clarigester (ROSS, 1984) type of reactor.

The basic differences of the concept mentioned above to the conventional concept are:

(1) wastewater distribution at the bottom of the reactor by a rotating system which is equipped with a scraper to move the heavy sediments during full operation to a hopper from where they can be pumped off;
(2) circular sedimentation–thickening unit in the upper part of the reactor (avoiding changes in gas–liquid equilibrium) with a scraper connected to the rotating system returning the thickened sludge to the reactor volume.

This system can also be equipped with a second external settling unit and return sludge pumping as realized for the traditional contact process.

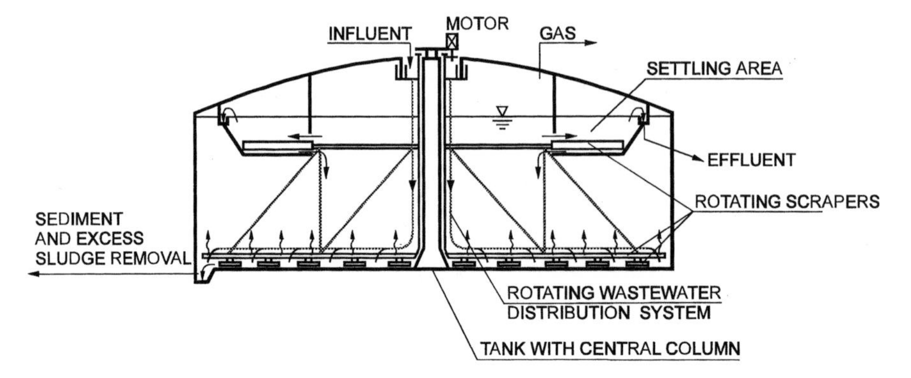

Fig. 3. Contact process reactor type with integrated settling tank and heavy sediment removal.

2.3 Covered Anaerobic Lagoon

Especially where land is available at low cost and under favorable climatic conditions also covered anaerobic lagoons can successfully be applied for industrial wastewater treatment (Fig. 4). This low rate anaerobic process has been further developed by using mixing devices in the reaction zone and sludge recycle systems (LANDINE and COCCI, 1989; cited in MALINA and POHLAND, 1992). Depending on design and equipment the process kinetics of covered anaerobic lagoons are a mixture between CSTR and contact process.

If the floating cover membrane is insulated, the heat losses from the (large) surface area can be minimized even in cold climates. Due to the long HRT a cotreatment of organic solids and dissolved organic pollution will be possible without major problems. For more details see MALINA and POHLAND (1992).

3 Design Considerations for Reactors

3.1 General Remarks

The design principles for the processes described above comprise many different aspects which have to be considered:

(1) Kinetics of the biodegradation processes. The basic model concepts are very similar to the aerobic processes (continuous fermentation, activated sludge).
(2) Mass balance concept based on COD, N, Ca, P, S.
(3) Inhibition by influent wastewater compounds and anaerobic degradation byproducts (organic acids, ammonia, hydrogen sulfide).
(4) Alkalinity, pH.
(5) Process configuration (one-stage, two-stage).
(6) Liquid–solids separation (mass balance, settling, thickening, membrane separation).

Due to the complexity of the design problem it is not advisable to base the design of a full scale plant on theoretical considerations only. A sound design will also have to be based on practical experience. This can be experience obtained from full scale applications or where this is not available from labscale and pilot investigations (Sect. 4).

To apply full scale experience to a new real wastewater problem or to set up a lab scale and pilot plant investigation program for design purposes it is very important to know the relationship of the different decisive parameters and processes according to our understanding, which has considerably improved during the last decades.

The composition and flow of industrial wastewater normally is strongly related to the production processes applied in the mill or factory (ATV, 1993). This close relationship has to be considered for the design in every case of application and also plays a key role in process development from lab scale to full scale.

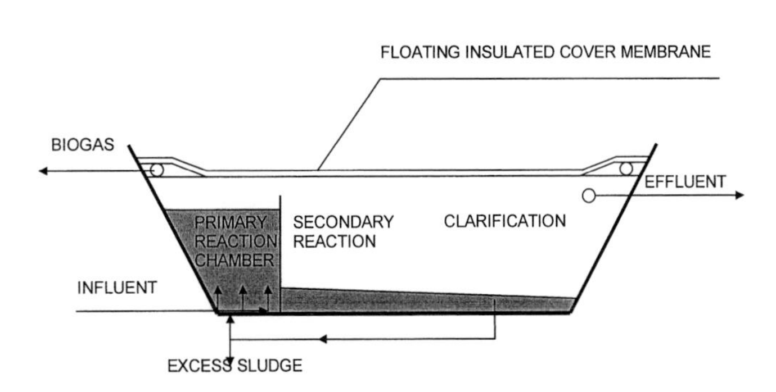

Fig. 4. Covered anaerobic lagoon (with partial mixing in primary and secondary reaction chamber).

3.2 Design Considerations for Methane Reactors

3.2.1 CSTR Process

For the design of anaerobic CSTR processes it is assumed that the effluent concentrations of the degraded dissolved substances are equal to those in the reactor. It is further assumed that mixing provides an intensive contact between bacteria and the dissolved substrates, so that transport and diffusion limitation do not play a decisive role for the degradation process. Under such assumptions the equilibrium between the growth rate of the bacteria (μ) and the substrate concentration can be described by the Monod equation (Eq. 1), with K_s being the half saturation coefficient.

$$\mu = \mu_{\max} \cdot \frac{S}{K_s + S} \tag{1}$$

The first consequence of this model is that stable high treatment efficiency can only be obtained (for steady state design load conditions) if:

$$\mathrm{HRT} = \mathrm{MCRT} \gg \frac{1}{\mu_{\max} - b} \tag{2}$$

with $\mu_{\max}$ the maximum growth rate of the rate limiting bacteria (most acetoclastic methanogenic bacteria) and b their decay rate (ATV, 1994).

For anaerobic breakdown of large molecules (especially organic solids) hydrolysis is often the rate limiting factor (e.g., anaerobic stabilization of organic particulate matter).

The values for the kinetic parameters (Tabs. 1 and 2) are reported in the literature (MALINA and POHLAND, 1992).

The literature review data (Tab. 1) for the kinetic parameters vary considerably by about one order of magnitude, which on one side reflects the still restricted knowledge and on the other side explains the wide range of "possible" design recommendations and results from full scale applications of anaerobic processes. In any case it is not possible to calculate the required volume from these data without having thoroughly investigated the specific local wastewater situation or having enough experience from very similar cases. The data in Tab. 2 can be used for a first estimation of reactor volumes but cannot replace data from full scale plants and labscale investigations.

3.2.2 Anaerobic Contact Process

While for the CSTR process solids management has no influence on the reactor design, it is of great importance for the contact process.

One of the main advantages of anaerobic processes is the low value for the biomass yield Y,

Tab. 1. Literature Data on the Kinetic Parameters of Anaerobic Processes

Substrate	Process Step	k_s [mg COD L^{-1}]	$\mu_{\max}$ [d^{-1}]	Y [g VSS (g COD)$^{-1}$]	b [d^{-1}]
Carbohydrates	acidogenesis	22.5–630	7.2–30	0.14–0.17	6.1
Acetate	acetoclastic methanogenesis	11–421	0.08–0.7	0.01–0.054	0.004–0.037
Hydrogen, CO_2	methanogenesis	$4.8 \cdot 10^{-5}$–0.60	0.05–4.07	0.017–0.045	0.088

Tab. 2. Representative Kinetic Data for Anaerobic Digestion at 35 °C (MALINA and POHLAND, 1992)

Process	k_s [mg COD L^{-1}]	$\mu_{\max}$ [d^{-1}]	Y [g VSS (g COD)$^{-1}$]
Acidogenesis	200	2.0	0.15
Acetoclastic methanogenesis	50	0.4	0.03

reducing the costs for sludge handling and disposal. For the treatment of low concentrated mainly dissolved wastewaters (<2,000 mg L^{-1} degradable COD) by the anaerobic contact process the loss of biomass in the overflow of the solids–liquid separation unit can easily become the limiting factor for the MCRT which controls the process performance (Eq. 3).

$$\text{MCRT} = \frac{V \cdot X_{a;R}}{X_{a;E} \cdot Q_{ES} + X_{a;e} \cdot Q} \gg \frac{1}{\mu_{max} - b} \quad (3)$$

V [m^3] is the volume of the anaerobic reactor, $X_{a;R,E,e}$ [kg m^{-3}] the active acetoclastic methane bacteria concentration in the reactor (R), the excess sludge removed (E) and in the effluent (e); Q_{ES} and Q [m^3 d^{-1}] are the flow of excess sludge removal and influent (effluent).

At degradable COD concentrations in the influent higher than about 2,000 mg L^{-1} the influence of $X_{a;E} \cdot Q$ decreases considerably. In most of the cases X_a can be replaced by X_v, the concentration of VSS or the COD of the total solids. The last two parameters can easily be determined while $X_{a;R;E;e}$ requires complicated activity measurements.

From Eq. (3) the volume of the reactor can be derived from the following parameters:

(1) Necessary MCRT to reach the required treatment efficiency and stability. It depends on the wastewater composition, temperature, load variations, inhibition problems, influent concentration and posttreatment design. The most reliable way to determine this value is by pilot investigations on site with the real wastewater.

(2) Maximum MLSS (X_a) concentration in the reactor. It depends on the solid–liquid separation process, the total solids production and its characteristics which have to be assumed on the basis of the cost minimization goal. Reliable data about the settling and thickening characteristics can be derived from pilot investigations or full scale experience.

(3) Excess sludge production (ESP). For design purposes the biomass production can be assumed to be about 7% of the mean removed COD load over a period of SRT. Non-biodegradable solids in the influent will accumulate in the system over the SRT, if they are not washed out to the effluent. Precipitation product accumulation can be estimated from wastewater composition (SVARDAL, 1991a) and the loads of Ca, Mg, N, P. For hydrolyzable organic solids a hydrolysis rate of about 0.1 d^{-1} can be assumed for a first estimate.

3.3 Mass Balance Concept

One of the most valuable tools for design and operation purposes is the mass balance concept (KROISS, 1985). According to the conservation law for mass and energy the influent loads to a process must be equal to the sum of the end products. There are three ports where masses can leave the process: effluent, excess sludge and biogas. For non-steady state conditions the changes in the masses stored in the process volumes have to be taken into account, too. In anaerobic treatment the important parameters for mass balances are: COD, N, Ca, P, and S. Carbon balances could also be used (while BOD and TOC cannot), but in practice there are analytical and sampling problems mainly due to the role of CO_2 and its dissociation products.

As these balances apply also for operational results from chemical supervision, this tool enables the operator to check the correctness of the data.

For CSTR and contact processes especially the Ca balance is of great importance to check how much of $CaCO_3$ is retained in the reactor or the system due to sedimentation. Ca^{2+} effluent concentrations depend on the pH in the reactors and normally are in the range below about 300 mg L^{-1} (SVARDAL, 1991b). The P requirements for the biological process can be estimated from a COD and P balance. It can be assumed that about 2% of the active biomass (1.45 g COD g^{-1} MLVSS) consist of phosphorus. For design purposes it can be assumed, that all nitrogen containing compounds (TKN load) will be transformed to ammonia nitrogen and all sulfur compounds to hydrogen sulfide during the anaerobic process.

A first evaluation of the methane production can be based on the assumption that about 80% of the COD can be converted to methane (1 kg COD converted corresponds to 0.35 Nm^3 methane). For economic evaluations methane productions have to be based on mean COD loads expected in the foreseeable future and not on design figures.

3.4 Mixing Design

For the mixing of the reactors energy requirements depend on the technical solution, the size of the reactor, the goal of mixing to be achieved and the composition of the suspended solids (MLSS).

The most common mixing technologies are mechanical mixers in the reactor with horizontal or vertical shafts and gas mixing making use of the airlift effect. Mixing by external pumps is neither very effective nor economical. In any case it is important that repair and maintenance of the mixers can be performed under operating conditions.

For the design of mixing devices it has to be considered that for continuous mixing the specific energy input (W m^{-3}) can be much lower than for intermittent mixing, where settled solids have to be resuspended. The specific energy input for complete mixing of the liquid phase is in the range of 1.5–4 W m^{-3} depending on the size of the reactor. If (heavy) sediments play an important role more specific energy input will be necessary to avoid settling. For the resuspension of, e.g., sewage sludge a specific energy input of about 20 W m^{-3} should be provided to achieve complete mixing also for the solid fraction. For the anaerobic biocenosis gentle mixing with reduced shear forces is advisable as syntrophic and methanogenic bacteria need to closely cooperate. Full scale experience at an anaerobic post-treatment plant for paper mill wastewater where calcium carbonate precipitation occurs shows, that even continuous mixing with 80 W m^{-3} could not prevent the buildup of sediments.

The basic goal of mixing must be to have complete mixing of the liquid phase in order to maintain the most important advantage of these reactors, i.e., to avoid any step changes of important parameters as: pH, temperature, concentrations of substrate, and inhibiting compounds. At the same time substrate transport and diffusion resistance are kept at a low level and the active biomass is kept in close contact with the wastewater. In order to avoid the accumulation of (heavy) sediments in the reactor other technical solutions than mixing will be more reliable and economic (e.g., Fig. 3). Another solution tried in practice is the installation of a cyclone (centrifuge) in the return sludge cycle in order to selectively remove inorganic heavy sediments before they accumulate in the reactors (NÄHLE, lecture held at Fa. Philip Müller, Ludwigshafen, Germany).

Gas production by the bacteria also provides a certain mixing energy which can be sufficient to achieve a nearly complete mixing of the liquid phase. The energy depends on the depth of gas production. The mixing effect will increase with the depth of the reactors and the volumetric COD removal efficiency (kg COD m^{-3} d^{-1}) or the area specific gas production (Nm^3 m^{-2} h^{-1}). The specific net energy input is about 2.8 Wh m^{-3} gas per m depth of gas production below the surface.

As a basic rule for anaerobic reactor design and equipment a period of about 10 years of continuous operation should be possible without emptying such a reactor. Opening and emptying of a methane reactor with removal of sediments normally takes several weeks and is a rather expensive procedure.

3.5 One-Stage vs. Two-Stage Anaerobic Systems

In many cases it is economic to put an equalization tank prior to the methane reactor in order to achieve stable efficiency and reliable operation. It will protect the methane reactor from sudden changes in loading and wastewater composition and can prevent inhibitory concentrations of toxic compounds or be part of an early warning system. Equalization tanks can also be designed and operated as CSTR or as contact process for acidification. In such a case the one-stage process becomes a two-stage one.

Whether the one- or the two-stage anaerobic process should be chosen depends mainly on the wastewater composition and biodegradable COD concentration. The basic biological reason for applying the two-stage process is the relationship between the acidifying bacterial community and the pH. At one-stage processes the acidification process has to be performed in the methane reactor, where a pH >7 is favorable for the methane bacteria. Under these conditions acidifying bacteria producing propionic acid will play an important role. The breakdown of propionic acid by the syntrophic bacteria to acetic acid is one of the most susceptible processes for methane production. It needs a very low partial pressure of H_2, which normally can only be achieved having stable low effluent substrate concentrations. Propionic acid inhibits the acetoclastic methane bacteria already at rather low concentrations (>5 mg L^{-1} and undissociated C_2H_5COOH) (KROISS and PLAHL-WABNEGG, 1983). The consequence is that for highly concentrated wastewaters (>5,000 mg COD L^{-1}) the problem of process failure caused by propionic acid inhibition increases using the one-stage process. For low concentrated wastewaters and pH values >7.5 such inhibition problems cannot occur.

At a pH of 5.5–6.5 acidifying bacteria have optimal environmental conditions and will preferably produce acetic and butyric acid (at least for carbohydrate breakdown). Butyric acid conversion to acetic acid is a rather stable process. By the two-step process the instability caused by propionic acid accumulation can be avoided to a large extent.

At pH <4.5 the variety of acidifying bacteria decreases and can result in reduced efficiency. This means that part of the organic pollution has to be acidified in the methane reactor. Using the excess sludge from the methane reactor as (continuous) seed material for the acidifying reactor can considerably improve the performance.

For organic biodegradable compounds containing nitrogen it has to be considered that acidification results in ammonia release. This is of great influence on the pH in the acidifying and the methane reactor. Despite the acid production the pH in an acidification step can be higher than in the influent.

Ammonia concentrations between about 300 and 1,000 mg $NH_4-N\ L^{-1}$ in the methane reactor maintain the pH at >7. Beyond a concentration of 1,500 mg L^{-1} inhibition of acetoclastic methane bacteria by ammonia has to be considered (KOSTER and LETTINGA, 1983; LAY et al., 1998).

Due to the high growth rate of acidifying bacteria mixed equalization tanks also act as acidifying reactors, even if they are not designed for it. Unstable acidification creates risks for the methane reactor.

The gas from acidifying reactors mainly consists of CO_2, H_2 and some CH_4 (H_2S).

Whenever hydrolysis and acidification are the rate limiting steps of the anaerobic breakdown of organic pollution a two-step process is also advisable in order to optimize the process parameters of this first step independent of the methanogenic bacteria. In such a case the first step can also be equipped with a solid–liquid separation unit making it an acidifying contact process in order to reduce the overall volume. The excess sludge handling for this step has to take care of the odor potential of this sludge due to the high concentrations of fatty acids (as well as ammonia and hydrogen sulfide).

In some cases it can be more economical to remove the solids from the equalization–acidification step only to reduce the sludge production in the methane reactor. At a given required MCRT this results in a smaller methane reactor volume as long as the sludge settling and thickening characteristics are not impaired at the same time.

The final decision whether a one- or a two-step process should be selected has to be based on a cost comparison for equally reliable and efficient configurations of the complete wastewater and sludge treatment and disposal scheme and not on the anaerobic treatment step alone.

3.6 Liquid–Solids Separation for the Contact Process

The mass flow scheme for the separation unit (Fig. 5) clearly shows that the main problem is to nearly equal the mass of suspended solids entering the unit from the reactor with the mass flow in the return sludge. Mass flows in the effluent and excess sludge are comparatively small. Normally the decisive factor for the design of a settler/thickener is the required thickening effect. As an increase of the return sludge flow also increases the inflow to the settler/thickener the effect on performance is restricted. Most of the separation units are, therefore, designed according to thickener principles. The design parameter is the surface solids loading, which can vary considerably depending on the sludge composition (50–120 kg m^{-2} d^{-1}). In most of the cases the hydraulic surface loading, responsible for the separation process and hence for the suspended solids concentration in the effluent, is also low enough (0.2–1 m^3 m^{-2} h^{-1}). Suspended solid concentrations in the effluent up to about 200 mg L^{-1} can be considered normal.

It has to be considered that the design of a settling/thickening unit results in an adaption stress to the bacteria during operation. Only those bacteria will have good survival opportunities in the system which are retained in the liquid–solid separation system. If the bacteria able to degrade the pollution of wastewater cannot adapt to it, the process can fail due to washout of the active biomass. In such a case the contact process changes towards the CSTR process. Another example for the close relationship between the separation step and the sludge settling and thickening characteristics of the MLSS is the granule formation at the UASB process, where the floc forming and dispersed growing bacteria have to be washed out. The consequence is that for contact processes it is very difficult to "predict" the settling and thickening properties of the MLSS or to evaluate full scale experience from different plants.

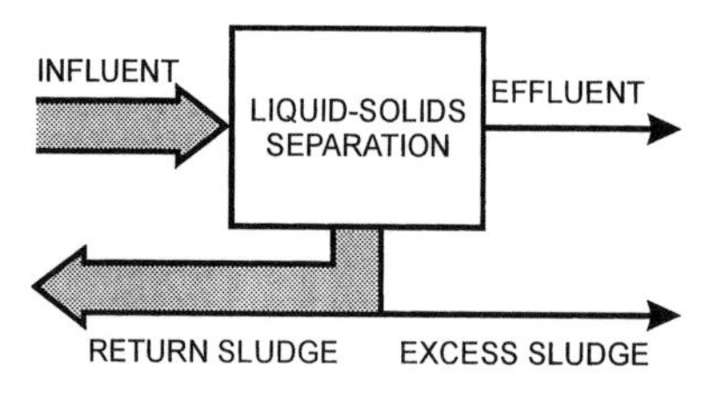

Fig. 5. Solids mass balance for the liquid–solids separation unit.

As the maximum MLSS concentration which can reliably be maintained in the contact reactor directly influences the required volume, it represents a basic parameter and, therefore, needs thorough investigation and consideration with regard to process stability. Bulking of anaerobic sludges is not a common problem, but the differences in settling and thickening properties can vary at least by a factor of 3–5. The volumes of the methane reactor and the settler/thickener can be exchanged to some extent by the choice of the MLSS concentration which leads to an optimization problem.

4 Investigations for Design and Operation

4.1 Design and Normal Operation

There is a consent based on theoretical investigations and on experience in practice that the highest potential in risk and cost minimization for an investment exists in the period before the decision on a detailed design. This has often been neglected as thorough investigations for the process design of anaerobic treatment plants are costly and time consuming. Nevertheless, it has to be stressed, that there are only very rare cases where a successful full scale concept can easily be transferred from one location to another without detailed investigations of the specific local situation. A second important point is that normally there is no operational experience on anaerobic wastewater treatment at an industry, and often the designer is not an expert in the production processes which influence wastewater flow and composition. In order to overcome all these problems the following tools can be applied:

(1) Materials accounting for the whole production process (factory, mill) with special emphasis on water and wastewater (which material flows can enter the wastewater system?)
(2) Wastewater characterization by physical-chemical analysis (flow and concen-

tration variations over time, probability distribution of loads, etc.)

(3) Lab scale investigations, several plants in parallel under defined conditions (biodegradability, inhibition tests, kinetic parameter determination, gas composition, alkalinity requirements, pH control, etc.)

(4) Pilot plant investigations (testing and optimizing of a design and operation concept under full scale conditions, i.e., with the real wastewater on site comprising all possible normal and exceptional conditions; development of operational experience, start-up, analytical supervision, trouble shouting, etc.)

The basic difference between lab scale and pilot scale investigations should not be the size of the reactors but the questions to be answered. Lab scale investigations are characterized by the search for relationships between parameters under controlled conditions. Therefore, the influences of the different parameters have to be separated in order to better understand the processes involved. With this understanding it is possible to develop different process configurations for the full scale wastewater treatment plant. The main goal of pilot scale investigations should be to prove design concepts under most realistic conditions. All possible influences (variability) have to be included into the program otherwise it will not be possible to derive clear cause effect relationships. There always will be an important trial and error component.

One of the main advantages of completely stirred reactors is that lab scale and pilot scale conditions in terms of concentrations can be transferred to full scale without major problems. CSTR process parameters, as e.g., hydraulic detention time, can be directly transferred from pilot to full scale. Often the depth of lab scale or pilot scale reactors will be much lower than in full scale. This has an influence on the gas–liquid equilibrium [e.g., the concentration of CO_2 (H_2S) in the liquid phase of the reactor, pH, precipitation of $CaCO_3$, MAP, etc.]. With decreasing size of lab scale reactors the fixed bacterial growth on the walls will have an increasing influence on some of the results (improved settling and thickening characteristics of the MLSS, increased inhibition resistance, etc.), while in full scale plants the influence of fixed growth normally can completely be neglected.

At the contact process an easy scale-up of settling/thickening units is not possible as hydraulic models cannot comprise all important influencing parameters. This means that hydraulic and solids surface loadings can not be transferred to full scale. From pilot plants the settling/thickening characteristics can be determined and introduced into conventional reactor design. The degassing problems, too, cannot be transferred as the depth of the full scale reactors normally cannot be simulated in pilot or lab scale. For these problems full scale experience will be welcome.

4.2 Start-Up Phase

Many reports on full scale start-up deal with problems caused by a lack of information or wrong operational decisions. The relatively large volumes and the complete mixing result in the advantage that all changes of concentration over time will be slow. But if the process fails, e.g., by an accumulation of fatty acids due to overloading or inhibition, the result is a large volume of a heavily polluted odors liquor which cannot be discharged without causing severe new problems. Remedial actions by changing the chemical composition, pH or temperature in the reactors, e.g., by the addition of chemicals or thermal energy will be expensive and sometimes even time consuming. Such situations normally lead to prolonged start-up periods with all their consequences.

Between the final decision on the process design and the start-up of a full scale plant there is always enough time to investigate the start-up procedure for the full scale plant. The most reliable method to achieve a trouble free start-up is to run pilot plant investigations on site with the real wastewater. One important goal of these investigations is to determine the maximum growth rate of the volumetric COD removal capacity under full scale conditions using the same seed material as planned for full scale. From this information it is possible to make a detailed start-up schedule. After having decided on the amount and quality (source) of the seed material (e.g., digested municipal sludge, excess sludge from another similar treatment plant) the daily increase in COD loading (% d^{-1}) can be fixed in order to reliably avoid process instability. The duration of the start-up period can be calculated in advance assuming an exponential growth of the bacteria, in order to avoid too optimistic forecasts. The increase in COD loading during the start-up phase can vary between about 1 and 6% d^{-1} depending on the specific wastewater characteristics.

5 Aerobic Biological Posttreatment

Anaerobically treated industrial wastewaters always contain easily biodegradable compounds. Their concentrations can be very low but they are never as low as from lowly loaded aerobic treatment plants. In many cases, but always during malfunction the effluent is odorous. If the wastewater contains sulfur compounds the effluent will strongly smell of hydrogen sulfide. In many cases the suspended solids concentration in the effluent will also be much higher than in the effluent of aerobic treatment plants. Most of the nitrogen contained in the influent will leave the anaerobic process as ammonia, a possible source of fish toxicity in a receiving water. The consequence of all these characteristics of anaerobic effluents is, that in most cases aerobic posttreatment will be required before discharge into a receiving water. If the effluent can be discharged into a municipal sewer system, the main problems with anaerobic effluents are odor formation and safety risks due to increased gas production in the sewer system.

As long as the anaerobic step achieves stable high efficiency, the aerobic posttreatment will not cause special problems to achieve low BOD effluent concentrations. If there is a requirement for nitrification and nitrogen removal the situation becomes more difficult. It is known that hydrogen sulfide inhibits nitrite oxidation, which can result in difficulties to maintain very low nitrite effluent concentrations. Because most of the biodegradable carbon compounds are removed anaerobically, nitrogen removal by denitrification needs special investigations (bypass of raw wastewater, external carbon source, etc.). For more details of nitrification and denitrification in an aerobic posttreatment see Chapters 3 and 13, this volume. In most of the cases lab scale or better pilot scale investigations have to be recommended before the design.

Equalization, anaerobic, and aerobic treatment have to be handled as a unit which has to fulfil the requirements set by a permit. This permit consists of effluent concentrations, maybe production specific effluent loads and should always contain the information on the relevant sampling procedures, the probability of compliance and the maximum exceeding of the standards. For design engineers the most difficult question is to find the overall least cost solution in order to meet the requirements. Reliability of a process cannot easily be quantified.

One basic conclusion can be drawn with regard to the relationship between reliability and treatment efficiency. In a multi-stage biological system the stability of a pretreatment step should increase with increasing efficiency. This can easily be demonstrated by the following example:

If an anaerobic pretreatment has a mean treatment efficiency for COD of 60% the aerobic posttreatment will be able to cope with short-term variations between at least 30 and 90% COD removal efficiency. If the mean treatment efficiency will be increased to, e.g., 95% of the biodegradable COD, a drop of the efficiency to only 80% could easily cause severe problems (4 times the mean load for the aerobic plant). It could also be shown that the cost savings by anaerobic pretreatment are strongly reduced if the aerobic step must be designed to cope with unstable anaerobic treatment efficiency. A basic feature of anaerobic processes is that their biological stability increases with increasing treatment efficiency. This means that only in rare cases does the maximizing of volumetric conversion rates (kg COD m^{-3} d^{-1}) lead to an optimal cost efficiency for the overall wastewater treatment system.

The design of aerobic posttreatment is influenced by suspended solids concentration in the effluent of the anaerobic stage. In case of an activated sludge process for posttreatment the basic design parameter is the sludge age (MCRT) which depends on the effluent quality requirements and the load variation characteristics of the anaerobic effluent over time. The volume of the aeration tank depends on excess sludge production and MLSS concentration. Both parameters strongly depend on the quantity and quality (settling and thickening properties) of the suspended solids in the influent. They will also be influenced by the composition of the anaerobic effluent.

6 Final Remarks

Despite the fact that the excess sludge production will be much lower than with aerobic treatment alone the excess sludge problem must also be solved. The sludge treatment and disposal system should always be ready before the start-up of a biological treatment system. The possibilities of sludge disposal at industrial plants are manifold and have to be adapted to the specific local situation.

The main task of industrial enterprises is always to produce sellable products and not to produce wastewater according to the actual capability of a maybe very sophisticated treatment system. Control strategies for the biological processes which mainly influence the influent flow cannot fulfill the requirements of a reliable treatment. Control and automation can help to make optimal use of the volumes and bacteria present in the treatment plant but several functions of the volume cannot be replaced by better control. The relatively large volumes of CSTR and contact processes in this respect are quite favorable for many applications, especially with higher concentrated wastewaters.

7 References

ATV (1990), *Arbeitsbericht des ATV-Fachausschusses 7.5:* Anaerobe Verfahren zur Behandlung von Industrieabwässern, *Korrespondenz Abwasser* **37** (10), 1247–1250.

ATV (1993), *Arbeitsbericht des ATV-Fachausschusses 7.5:* Technologische Beurteilungskriterien zur anaeroben Abwasserbehandlung, *Korrespondenz Abwasser* **40** (2), 217–223.

ATV (1994), *Arbeitsbericht des ATV-Fachausschusses 7.5:* Geschwindigkeitsbestimmende Schritte beim anaeroben Abbau von organischen Verbindungen in Abwässern, *Korrespondenz Abwasser* **41**, 101–107.

KOSTER, I. W., LETTINGA, G. (1983), Ammonium toxicity in anaerobic digestion, *Proc. Eur. Symp. Anaerobic Waste Water Treatment (AWWT)*, Noordwijkerhout, NL (VAN DER BRINK, W. J., Ed.), pp. 58–72. The Hauge, NL: TNO Corporate Communication Department.

KROISS, H. (1985), Anaerobe Abwasserreinigung, in: *Wiener Mitteilungen* (Institute for Waterquality and Waste Management, Ed.), Vol. 62. Vienna, Austria: University of Technology.

KROISS, H., PLAHL-WABNEGG, F. (1983), Sulfide toxicity with anaerobic waste water treatment, *Proc. Eur. Symp. Anaerobic Waste Water Treatment (AWWT)*, Noordwijkerhout, NL (VAN DER BRINK, W. J., Ed.), pp. 72–85. The Hauge, NL: TNO Corporate Communication Department.

LANDINE, R. C., COCCI, A. A. (1989), *The low rate ADI-BVF System: Its advantages, case histories and biogas collection and utilization.* New Brunswick, Canada: ADI Limited, Fredericton.

LAY, J.-J., LI, Y.-Y., NOIKE, T. (1998), The influence of pH and ammonia concentration the methane production in high-solids digestion processes, *Water Environ. Res.* **70**, 1075–1082.

MALINA, J. F., POHLAND, F. G. (1992), Design of anaerobic processes for the treatment of industrial waste water, in: *Water Quality Management Library*, Vol. **7**. Lancaster, USA: Technomic Publishing Company.

PASCIK, I., HENZLER, H.-J. (1988), Anaerobic treatment of the waste water from pulp bleaching plants with immobilized organisms, in: *Advances in Water Pollution Control* (IAWPRC, Ed.), pp. 491–498, *Proc. 5th Int. Symp. Anaerobic Digestion* 1988. Bologna, Italy, Oxford, UK: Pergamon Press.

ROSS, W. R. (1984), The phenomenon of sludge pelletization in the anaerobic treatment of a maize processing waste, *Water SA* **10**, 197.

SVARDAL, K. (1991a), Anaerobe Abwasserreinigung – Ein Modell zur Berechnung und Darstellung der maßgebenden chemischen Parameter, *Wiener Mitteilungen*, Vol. **95** (Thesis).

SVARDAL, K. (1991b), Calcium carbonate prescription in anaerobic waste water treatment, *Water Sci. Technol.* **23**, 1239–1248.

SVARDAL, K., GÖTZENDORFER, K., NOWAK, O., KROISS, H. (1993), Treatment of citric acid wastewater for high quality effluent on the anaerobic-aerobic route, *Water Sci. Technol.* **28**, 177–186.

24 Fixed Film Stationary Bed and Fluidized Bed Reactors

HANS-JOACHIM JÖRDENING
KLAUS BUCHHOLZ
Braunschweig, Germany

Symbols and Abbreviations

Symbol	Name	unit
a	geometrical factor	—
c_i	concentration of the substrate i	mol m^{-3}
c_{ib}	substrate concentration at the biofilm–liquid interface	mol m^{-3}
c_{ip}	substrate concentration at the biofilm–particle interface	mol m^{-3}
c_{is}	substrate concentration in bulk liquid	mol m^{-3}
n	expansion index	—
D	diffusion coefficient	m^2 s^{-1}
D_{eff}	effective diffusion coefficient	m^2 s^{-1}
d_R	reactor diameter	m
d_P	particle diameter	m
$d_{V/S}$	Sauter diameter	m
j	molar substrate flux	mol m^{-2} s^{-1}
k_0	kinetic constant (zero order)	mol m^{-3} s^{-1}
k_1	kinetic constant (1st order)	s^{-1}
k_e	mass transfer coefficient	m s^{-1}
r	reaction rate	mol m^{-3} s^{-1}
Re	Reynolds number	—
Re_P	particle Reynolds number	—
Re_T	particle terminal Reynolds number	—
q_r	amount of particles	—
Sc	Schmidt number	—
Sh	Sherwood number	—
t	time	s
g	gravitational acceleration	m s^{-2}
C_D	drag coefficient	
u	superficial upflow velocity	m s^{-1}
u_T	terminal settling velocity	m s^{-1}
u_T^*	terminal settling velocity in a particle swarm	m s^{-1}
X	biomass concentration	kg m^{-3}
x_b	bioparticle radius	m
x_c	substrate concentration depth	m
x_p	particle radius	m
x_s	radial distance of $c_S > 0$	m
Y	yield coefficient	—
$B_{V,COD}$	volumetric loading rate	kg
$B_{V,inert}$	volumetric loading rate of non-biodegradable solids	kg
ε	voidage	—
Φ	sphericity	—
τ_X	biomass retention time	d
ϕ	Thiele modulus	—
ϕ_{1m}	1st (reaction order) modified Thiele modulus	—
ρ	density	kg m^{-3}
ρ_P	particle density	kg m^{-3}
ρ_L	liquid density	kg m^{-3}
η	effectiveness factor	—
η	dynamic viscosity	kg m^{-1} s^{-1}

1 Introduction

Fixed and fluidized bed reactors offer the advantage of high-load systems, requiring much less volume and space, and hence less investment as compared to conventional systems. Furthermore, these systems tend to operate more stable under transient conditions, like fluctuations of substrates and pH. Such advantages are of interest to those industries which produce large amounts and/or highly concentrated wastewaters, notably the food, paper, and pulp industries. Some fixed film systems reactor configurations are shown schematically in Fig. 1. Although most fixed film systems can be associated with this scheme, the number of variations is high concerning the flow mode (up- and downflow), the fluid distribution system at the reactor inlet, the support material and expansion (for fluidized beds), and the configuration of the reactor outlet (gas–liquid support separation). Recycle is in general provided for dilution of substrate and fluidization.

Several problems, however, have slowed down the application of these systems. These are longer start-up times, if no specific inoculum is available, a more sophisticated process control required, and the cost of the support material (WEILAND and ROZZI, 1991). Proper solutions are available in principle to overcome these problems.

Nevertheless, there has been considerable progress in application due to research and pilot plant investigations as well as data published on industrial systems, many of which are in operation now. A large amount of literature has accumulated, mainly over the last two decades, so that only part of it can be mentioned in this chapter. Earlier overviews are available summarizing a good deal of original papers notably up to the mid 1980s (HENZE, 1983; HENZE and HARREMOES, 1983; NYNS, 1986; HALL and HOBSON, 1988; AUSTERMANN-HAUN et al., 1993). This chapter concentrates on basic principles of anaerobic fixed and fluidized bed systems, and on recent experience accumulated on the lab, pilot, and industrial scale. It should, however, be kept in mind that industrial applications must be based on pilot plant experience at the factory site in every case of new substrate or source of wastewater or any specific problem.

A very broad and most successful application of high-performance anaerobic treatment in fixed and fluidized bed reactors relates to wastewater from industries based on agricultural and forestry products with typically high concentrations of organic substrates readily degraded by anaerobic bacteria (WEILAND and WULFERT, 1986; AUSTERMANN-HAUN et al., 1993). They may result from washing procedures of raw materials, blanching, extraction, fermentation, or enzymatic processing. Original substrates are in most cases carbo-

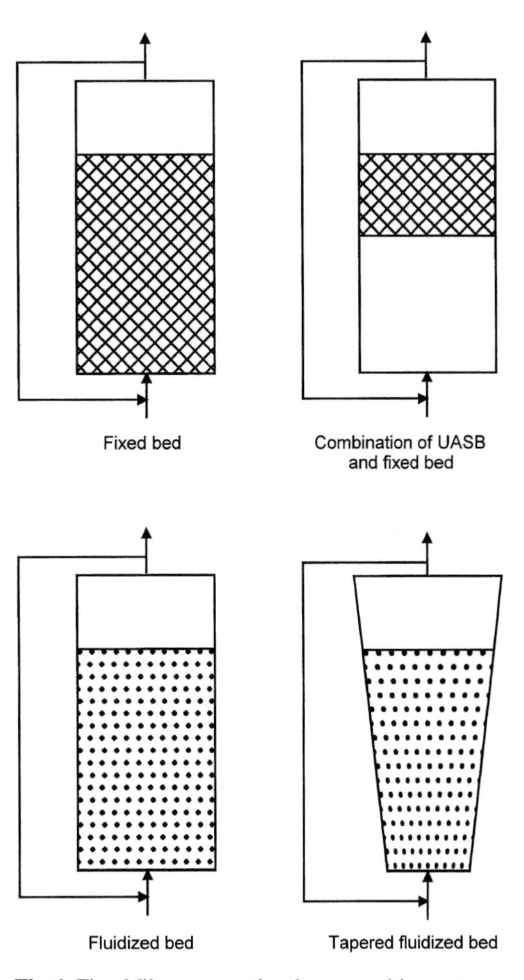

Fig. 1. Fixed film systems for the anaerobic wastewater treatment.

hydrates such as sugar, starch, cellulose, and hemicellulose, protein and fat, which readily undergo bacterial degradation to fatty acids, mainly acetic, propionic, butyric, and lactic acid. The majority of installations are in the potato, starch, and sugar industries, in fruit, vegetable and meat processing, in cheese, yeast, alcohol, citric acid, pectin manufacturing, and in the paper and pulp industries. The concentrations of substrates are typically in the range of 5–50 kg(COD) m^{-3}, which are diluted by recirculation (loop reactor) down to less than 2 kg(COD) m^{-3}.

In some countries like China systems for the treatment of domestic sewage also play a major role (YI-ZHANG and LI-BIN, 1988). Furthermore, considerable efforts were undertaken in order also to treat inhibitory or toxic substances. Thus cyanide, formaldehyde, ammonium, nickel, and sulfide containing wastewater were investigated, and it has been shown that methanogens can accommodate even to rather high concentrations of such toxicants, depending on the retention time (PARKIN and SPEECE, 1983). Furthermore, organochlorine compounds in kraft bleaching effluents and in pesticide containing water could be treated to the stage of mineralization by adapted biofilms, including chloroform, chlorophenols, chlorocatechols, and similar compounds as well as chlorinated resin acids (SALKINOJA-SALONEN et al., 1983). These positive results were obtained on the laboratory scale. The degradation of furfural in sulfite evaporator condensate can proceed to a conversion of about 90% (NEY et al., 1989).

2 Basic Principles

Reactors are tubes with fixed bed internals or fluidized suspended particles, which serve as a support for biomass immobilization. The dimensions range from 10–500 m^3 with a ratio of height to diameter from 1–5. In general an external loop recycles a part of the effluent to the inlet, where mixing with the wastewater provides for its dilution to non-inhibitory substrate concentrations and pH. In a few cases tapered beds have been used; most of the fluidized beds are provided with a settling zone with a larger diameter at the top of the reactor.

2.1 Biofilm Formation

The basis for the use of packed bed and fluidized bed systems is the immobilization of bacteria on solid surfaces. Many species of bacteria (and also other microorganisms) have the ability for adhering to supporting matrices.

While in the aerobic wastewater treatment immobilized bacteria have been used since the beginning of this century, the application of those systems to the anaerobic wastewater treatment is relatively new.

The fundamentals of bacterial adhesion and growth on solid surfaces are discussed in Chapter 4, this volume, thus only some aspects concerning anaerobic fixed films will be considered here. The preconditioning of solid surfaces is influenced by both environmental conditions (e.g., pH, T) and by the surface itself (e.g., hydrophobicity, surface charge) (GERSON and ZAJIC, 1979). With the addition of cationic polymers (STRONACH et al., 1987; DIAZ-BAEZ, 1988) or slime-producing bacteria (DIAZ-BAEZ, 1988) the initial anaerobic biofilm attachment could be improved, but the biofilm development was worse than in systems without these components. JÖRDENING (1987) reported a positive effect with the supplementation of calcium.

The primary adhesion of cells to the surface is due to hydrogen bonds, van der Waals forces and/or electrostatic interactions (DANIELS, 1980). This reversible form of adhesion can become irreversible by the production of exopolymeric substances (EPS), which act as a glue (FLEMMING, 1991). Experiments on the first steps of the formation of anaerobic biofilms gave different results: while LAUWERS et al. (1990) and SANCHEZ et al. (1996) found facultative anaerobic bacteria as primary colonizers, SREEKRISHNAN et al. (1991) observed that the biofilm formation was initiated by methanogenic bacteria.

After a lag phase, which seems to be necessary for an adaptation of the microorganisms to the new environment, exponential growth of bacteria begins. The growth rate is mainly determined by substrate transport and temperature (HEIJNEN et al., 1986; CHARACKLIS, 1990).

2.2 Biofilm Characteristics

In view of the wide range of possible biofilm compositions it is obvious that biofilm thickness does not correspond to the activity of the biocatalyst. HOEHN (1970) reported that the highest biofilm density can be found, if the biofilm thickness corresponds to the active biofilm thickness, i.e., the substrate-penetrated part of the biofilm.

2.3 Kinetics and Mass Transfer

The reaction kinetics for any process change with immobilization of the catalyst. In biofilm reactors the following mass transfer processes have to be considered:

(1) transport of substrate from the fluid to the surface of the support through the boundary layer (external mass transfer),
(2) transport of substrate from the surface into the pores of the biocatalyst,
(3) reaction,
(4) (5) transport of products in the opposite direction of steps 1 and 2.

The transport of substrates and products through the reactor is related to the hydrodynamic characteristics of the system and in general is much faster than steps 2–4. Mass transfer is mostly reduced by diffusion limitation.

Depending on the mode of limitation one can distinguish between film diffusion and pore diffusion and combined limited systems. Fig. 2 illustrates schematically the resulting substrate profiles.

For the description of the activity changes by immobilization an effectiveness factor is usually used, defined as

$$\eta = \frac{\text{observed reaction rate}}{\text{reaction rate at bulk liquid conditions}} = f(\text{S}, \text{Sh}, \phi) \quad (1)$$

2.3.1 External Mass Transfer

Passing a solid surface, the fluid characteristics change from turbulent to laminar flow and produce a boundary layer around the surface. The flux j_1 through the boundary layer is equal to a mass transfer coefficient and the concen-

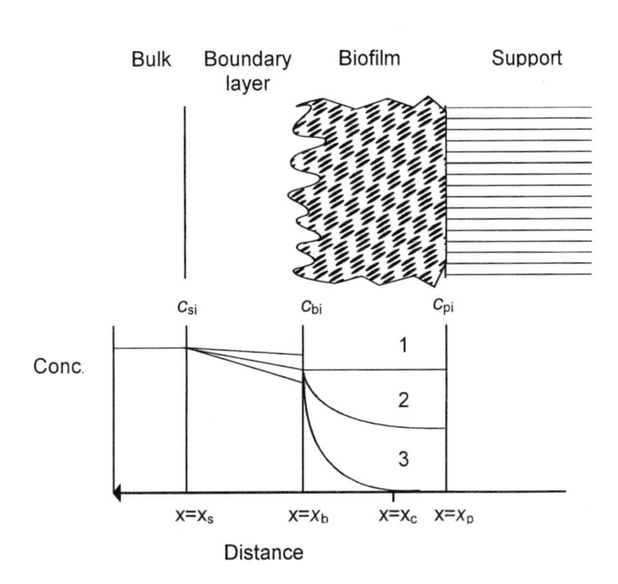

Fig. 2. Substrate concentration profiles at an immobilized biocatalyst surface, (1) reaction rate-controlled system; (2) combination of reaction rate and diffusion controlled system; (3) diffusion-controlled system.

tration gradient from the outer shell to the particle surface:

$$j_1 = k_e (c_{is} - c_{ib}) \tag{2}$$

An analytical solution for k_e can only be given for the ideal case of a single particle at infinite dilution. The mass transfer coefficient is given by:

$$k_e = 2\,\frac{D}{d_p} \tag{3}$$

For more complex systems, such as packed or fluidized bed reactors, the external mass transfer is usually described by dimensionless analysis which leads to the correlation characterized by the Sherwood number:

$$Sh = f(\mathrm{Re}, \mathrm{Sc}) = \frac{k_e\, d_p}{D} \tag{4}$$

$$Re = \frac{d_p\, u\, \rho}{\eta} \tag{5}$$

$$Sc = \frac{\eta}{D\, \rho} \tag{6}$$

MULCAHY and LA MOTTA (1978) calculated the external mass transfer for a fluidized biofilm with a correlation given by SNOWDEN and TURNER (1967):

$$Sh = \frac{0.81}{\varepsilon}\, Re^{0.5}\, Sc^{0.33} \tag{7}$$

2.3.2 Internal Mass Transfer

The diffusional transport of substrate in the biofilm may reduce the reaction rate. For steady state conditions the net diffusion rate is equal to the reaction rate: Writing in a general form the resulting mass balance gives a second order differential equation

$$D_{eff}\left(\frac{d^2 c_i}{dx^2} + \frac{a}{x}\,\frac{dc_i}{dx}\right) - r = 0 \tag{8}$$

where D_{eff} is the effective pore diffusion coeffficient, c_i the concentration of the substrate i, x the position in the porous particle, r the reaction rate, and a a geometrical factor: for a plate $a = 0$, for a cylinder $a = 1$, and for a sphere $a = 2$.

The diffusion coeffficient in the biofilm (D_{eff}) may be smaller than in water. KITSOS et al. (1992) found by comparison of acetate diffusion with lithium diffusion a value of 7%, related to the diffusivity in water ($6.6 \cdot 10^{-10}\ m^2\ s^{-1}$). OZTURK et al. (1989) calculated an effective diffusion coeffficient of $1.7 \cdot 10^{-10}\ m^2\ s^{-1}$ from measurements with inactive anaerobic biofilms. These values are low in comparison to those determined for glucose or oxygen in aerobic biofilms, which are nearly the same as for water (ONUMA et al., 1985; HORN and HEMPEL, 1996). KITSOS et al. (1992) explain this disagreement with principal differences between aerobic and anaerobic biofilms with regard to the symbiosis between the bacterial groups in anaerobic biofilms.

Two boundary conditions can be defined at the support interface and at the interface between the biofilm and the boundary layer.

$$\frac{dc_i}{dx} = 0 \quad x = x_p \tag{9}$$

$$c_i = c_{is} \quad x = x_b \tag{10}$$

If a zero order reaction is assumed, the integration of Eq. (5) with the boundary conditions yields an expression given by SHIEH and KEENAN (1986), i.e.

$$\left(\frac{x_c}{x_b}\right)^3 - 1.5\left(\frac{x_c}{x_b}\right)^2 + \left(\frac{1}{2} - \frac{3}{\Phi^2}\right) = 0 \tag{11}$$

where the Thiele modulus is defined as

$$\Phi = x_b \sqrt{\frac{\rho \cdot k_0}{D_{eff} \cdot c_i}} \tag{12}$$

Hence, the effectiveness factor can be calculated for $c_S > 0$.

For first order reactions SHIEH et al. (1982) show a good agreement between η and the Thiele modulus given by

$$\eta = \frac{\coth(3\,\Phi_{1m})}{\Phi_{1m}} - \frac{1}{3\,\Phi_{1m}^2} \tag{13}$$

with a modified 1st order Thiele modulus Φ_{1m}

$$\Phi_{1m} = (\rho k_1 c_i)^{0.5} \cdot \frac{x_b^3 - x_p^3}{x_b^2} \tag{14}$$

Since anaerobic reactors are generally treated at substrate concentrations which are high in relation to the substrate affinity constant, zero order kinetics are the useful way for calculating diffusion limitation.

2.4 Support Characteristics

Many different support materials were tested for the application in fixed film stationary and fluidized bed reactors. The major factors for bacterial attachment and growth related to both systems are roughness and porosity. The process configurations will be discussed separately, because of differences in the relative importance of these factors and because additional factors may apply to only one of the systems.

2.4.1 Stationary Fixed Film Reactors

Supports in fixed bed reactors must meet specific requirements for scale-up: Biomass tends to accumulate to a major part as suspended particles in the voids of the fixed bed. Biogas must separate from the fluid phase, and its transport must be possible through the full-scale reactor up to several meters and gas hold-up of up to 3%, without major pressure drop. Supports with small dimensions (range of a few mm or below) are, therefore, not suited for this reactor type. Most supports tested with success up to the pilot or industrial scale offer suffficient hydrodynamic radii (in general more than 20 mm) and a very high void volume (over 70%, mostly more than 90%). As a consequence the support surface per volume is rather low, the concentration of biomass in the surface fixed film is low as well, and the suspended biomass in the void volume contributes considerably to the activity (HENZE and HARREMOES, 1983; WEILAND and WULFERT, 1986). The maximal volumetric load is distinctly lower in general as compared to UASB, expanded, or fluidized bed reactors.

Supports utilized in most applications are typically internals, such as Raschig or Pall rings, Berl or Intalox saddles, plastic cylinders, clay blocks or potter clay of dimensions in the range of typically 20–60 mm (HENZE and HARREMOES, 1983; YOUNG and DAHAB, 1983; WEILAND et al., 1988). Trends in application favor supports with a void volume of over 90% and a surface area in the range of 100–300 m^2 m^{-3} (BISCHOFSBERGER, 1993; AUSTERMANN-HAUN et al., 1993).

2.4.2 Fluidized Bed Reactors

The choice of support material determines the process engineering much more than for packed bed reactors. This is due to the fluidization characteristics which depend on the density and the diameter of the support.

For the calculation of the fluidization behavior, considering only ideally spherical particles without a biolayer, one starts with the terminal settling velocity for a single particle at infinite dilution:

$$u_T = \sqrt{\frac{4 g d_p \rho_p - \rho_L}{3 C_D \rho_L}} \tag{15}$$

where g is the gravitational acceleration. The terminal settling velocity u_T depends on the particle diameter d_p, the drag coefficient C_D, and the difference in the densities of the particle ρ_p and the liquid ρ_L. C_D correlates with the particles Reynolds number Re_p

$$Re_p = \frac{u_T d_p}{v_L} \tag{16}$$

Equations describing the relation between C_D and Re_p can be found in the literature. BIRD et al. (1960) gave a generally accepted formula for the intermediate region of Reynolds numbers ($1 < Re_P < 50$) with

$$C_D = \frac{18.5}{Re_P^{0.6}} \tag{17}$$

From these equations the single particle settling velocity can be calculated by iteration.

However, for the calculation of the fluidized bed expansion additional effects of the reactor wall, the characteristics of flow and of adjacent particles have to be considered. Usually the fluidization is then described with empirical equations. The correlation used most frequently for fluidized beds is given by RICHARDSON and ZAKI (1954):

$$u = u_T^* \varepsilon^n \tag{18}$$

where u is the superficial liquid velocity, u_T^* the settling velocity of the particle swarm and equal to $u_T \cdot 10^{-d_P/d_R}$, the bed voidage ε and an expansion index n. n is given as follows, provided that the particle diameter is much smaller than that of the bed:

$n = 4.65$	$Re_T \leq 0.2$
$n = 4.35 \cdot Re_T^{-0.03}$	$0.2 \leq Re_T \leq 1$
$n = 4.45 \cdot Re_T^{-0.1}$	$1 \leq Re_T \leq 500$
$n = 2.39$	$500 \leq Re_T \leq 7{,}000$

For anaerobic fluidized beds n can normally be calculated for Re_T in the range of 1–500.

Most of the materials (e.g., granular activated carbon, pumice, sepiolite) used as supports for anaerobic fluidized beds are not ideal spheres and show a particle diameter distribution. For that the Sauter diameter $d_{V/S}$ should be used for the above calculations, defined as

$$d_{V/S} = \frac{\int x\, q_r(\mathrm{x})\, \mathrm{dx}}{\int q_r(\mathrm{x})\, \mathrm{dx}} \tag{19}$$

where q_r is the amount of particles (of volume or surface fraction) with the diameter x. A sphericity factor Φ can be determined by microscopical comparison of particle shape with model geometrical figures given by RITTENHOUSE (1943). From these values the voidage can be described with a correlation of WEN and XU (1966).

$$\frac{1-\varepsilon}{\varepsilon^3} = 11\,\Phi \tag{20}$$

Even the volume contraction has to be considered for the calculation of ε, if a mixture of different particle sizes is used. A detailed calculation procedure is given by OUCHIJAMA and TANAKA (1981).

The easiest way for determining the fluidization behavior is to make experiments in lab scale reactors. There one has to consider that for obtaining representative data it is necessary to adjust a ratio of 100 as a minimum for the reactor and the particle diameter. However, the growing biofilm and possible precipitations of inert solids may cause significant changes in the fluidization behavior during work. Therefore, it is very important to use real wastewater and to control the fluidization until a dynamic steady state is reached.

3 Reactor Design Parameters

The development of any anaerobic system needs the evaluation of some optimal conditions concerning several topics. For fixed film systems this includes especially the choice of the support, the reactor geometry, the start-up procedure and the handling of excess sludge or inert support.

3.1 Scale-Up

Concepts for scale-up have been summarized by KOSSEN and OOSTERHUIS (1985), where dimensional analysis and rules of thumb may be mentioned, since they provide guidance and recourse to practical experience. Fluid flow and fluidization can be treated with the aid of Reynolds, Peclet, and Froude numbers in order to estimate regimes appropriate for technical-scale operation (MÖSCHE, 1998). However, a rational design seems very difficult due to the high complexity of the systems. Therefore, empirical rules are mostly used in practice for the design of technical reactors (HENZE and HARREMOES, 1983).

The parameter considered most important is the load with biodegradable organics in terms of COD. This must be correlated to the active biomass in the reactor. So a load of 1–1.5 kg (COD) kg^{-1} (VSS) d^{-1} is considered as the upper limit for stable operation, where the following correlation can be used for guidance (HENZE and HARREMOES, 1983):

$$B_{\mathrm{V,COD}} = \frac{X/\tau_{\mathrm{X}} - B_{\mathrm{V,inert}}}{Y} \tag{21}$$

A range of further parameters of high significance should be taken into account: geometrical dimensions and $H\,D^{-1}$; recirculation rate, determining the substrate dilution and pH (and their gradient); residence time and distribution; mixing behavior; flow rate, pressure drop, and energy requirements; fluidization and bed expansion.

The recirculation rate, inlet substrate concentration, and pH and its gradient are correlated to each other, and they are highly important aspects since the stability of the stationary operation greatly depends on it, as subsequently discussed (BURKHARDT and JÖRDENING, 1994; MÖSCHE, 1998).

An example for modeling of an industrial fluidized bed reactor as a guide to scale-up and optimization of its operation has been presented by SCHWARZ et al. (1996, in press). The model comprises those aspects which were most sensitive for the results: material balance equations for substrates and products in the gas and liquid phase, kinetics of the biological degradation, mass transfer between gas and liquid phase, chemical equilibria as well as convection and dispersion (Fig. 3).

Maximum individual reaction rates for the different substrate components turned out to be most sensitive for substrate and product concentrations as well as for pH. With scale-up the ratio of feed and recirculation increases, since the flow rate is limited by the carrier settling velocity. Axial gradients of pH values were calculated and turned out to be most sensitive to the load (Fig. 3) and the buffer capacity. Thus at high loading rates dynamic changes in the feed composition could lead to a pH drop in the lower part of the reactor and to a breakdown of the system. Backmixing turned out to be of minor importance in this system (with 500 m^3, 18 m height of the fluidized bed). A distribution of substrate feeding at two positions, the bottom and the medium height, turned out to overcome these problems. The corresponding increase of the superficial upflow velocity restricts this possibility only for highly concentrated wastewaters. Otherwise, problems caused by a higher segregation of the bed would occur. For pumice as support material an increase of the superficial upflow velocity of 10% in the upper half of the fluidized bed did not cause any problems (MÖSCHE, 1998).

Furthermore, the buffer capacity was significantly influenced by CO_2 and its equilibrium

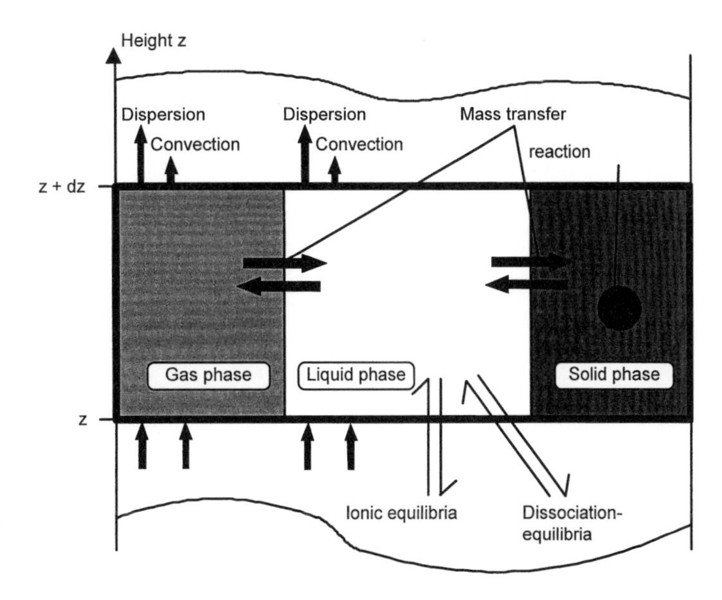

Fig. 3. Finite volume in a fluidized bed reactor and considered fluxes (SCHWARZ et al., 1996).

concentration as a function of pressure and of ions such as calcium present in the wastewater.

3.2 Support

A great number of support materials have been investigated in laboratory scale reactors for the use in packed or fluidized bed systems (HENZE and HARREMOES, 1983). Despite that, the number of supports which are used even for technical-scale systems is rather low. Certainly, one big problem for the realization of new processes is due to the difficulties of cooperation with manufacturers. But often, the problems result from shortcomings in the studies with respect to requirements of large-scale application. For a successful application of any material as support in fixed film stationary or fluidized bed reactors on the technical scale the following general requirements should be fulfilled:

(1) availability of the material in big quantities (>100 m^3),
(2) low cost of the material (related to the achieveable performance; it should be in general lower than 150 \$ m^{-3}),
(3) inert behavior (mechanically and microbially stable) without toxic effects and easy disposal,
(4) low pressure drop (low energy demand for mixing or fluidization).

3.2.1 Stationary Bed Reactors

A selection of supports, which seem to have been applied successfully on the pilot and industrial scale is summarized in Tab. 1; examples are shown in Fig. 4.

The biofilm thickness is of limited significance in fixed bed reactors since it makes up the minor part of biomass. Data published are mostly in the range of 1–4 mm, the upper limit relating to the inflow zone (top of a reactor operated downflow) (SWITZENBAUM, 1983; VAN DEN BERG and KENNEDY, 1983; ANDREWS, 1988, p.790/794).

More important is the biomass concentration in the reactor which obviously is distributed into two fractions, one immobilized in the biofilm, the other being suspended in the void volume of the reactor (HALL, 1982; WEILAND and WULFERT, 1986). The organic dry biomass in pilot reactors is mostly in the range of 5–15 kg m^{-3} (HENZE and HARREMOES, 1983). Gradients are observed, with the maximal concentration near the reactor inlet [e.g., about 15 kg m^{-3} in the lower and 4 kg m^{-3} in the upper part for upstream operation (WEILAND and WULFERT, 1986)]. Gradients were also found by HALL (1982), depending on the flow direction: 6 fixed and 9 suspended for upstream, and 9 fixed and 4 suspended for downstream operation (all in kg m^{-3} organic dry matter). For an industrial plant in starch processing a biomass concentration of 20 kg m^{-3}

Tab. 1. Typical Supports for Fixed Beds

Support	Diameter Range [mm]	Surface [m^2 m^{-3}]	Bed Porosity	Equivalent Pore Diameter	Reference
Raschig rings	10.16	45–49	0.76–0.78		CARRONDO et al. (1983)
Pall rings	90.90	102	0.95	20	YOUNG and DAHAB (1983)
	corrugated modular blocks	98	0.95	46	
Pall rings	25	215			SCHULTES (1998)
Hiflow 90®	90	65	0.965	>30 mm	WEILAND et al. (1988)
Plasdek C. 10®	modular blocks	148	0.96	>30 mm	WEILAND and WULFERT (1986)
Flocor R®	corrugated rings	320	97	>30 mm	
Ceramic Raschig rings	25	190	0.74		ANDREWS (1988)

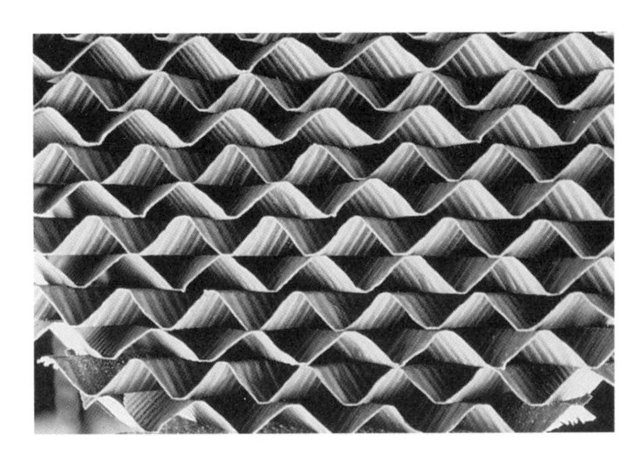

Fig. 4a–c. Typical support materials for anaerobic stationary fixed film reactors: **a** Hiflow 90® (uncovered and biofilm-covered); **b** Flocor R®; **c** Plasdek C®. 10 (WEILAND and WULFERT, 1986).

was reported (SCHRAEWER, 1988). A high biomass concentration was reported for a special support, sinter glass fillings (of high price) up to 65 kg m^{-3} in lab-scale experiments, of which 31 kg m^{-3} were found to be immobilized within the inner pores of the support (NEY et al., 1989).

3.2.2 Fluidized Bed Reactors

The energy demand for the fluidization of the support is often discussed to be very high in contrast to other anaerobic techniques. In general, much higher volumetric flow rates have to be raised in comparison to a CSTR. Nevertheless, the energy demand is relatively low because only the additional pressure drop of the support has to be overcome. Hence, with respect to the higher loading rates, the overall energy demand is in the same range or lower than in CSTRs, depending on the support density. Tab. 2 shows some materials which are tested to be used as support for anaerobic fluidized bed systems. All materials have particle diameters significantly lower than 1 mm. By this, the surface for colonization can be increased, the superficial upflow velocity can be reduced, and diffusion limitation will play no role, even for porous materials.

Porous materials show the advantage of lower superficial upflow velocities than non-porous materials. Furthermore, biomass gradients mainly occur in reactors filled with non-porous materials (ANDERSON et al., 1990; JÖRDENING, 1987). FRANKLIN et al. (1992) reported that a sand fluidized bed reactor contained in the bottom part only uncovered bare sand. He explains this with the extreme shear forces, which have a much bigger influence on bacteria on the surface. In contrast bacteria in pores are protected against shear forces and, therefore, such pronounced gradients are not known for the use of porous materials. In Fig. 5 some porous and non-porous supports are shown.

3.3 Wastewater

Wastewater should be acidified to a high degree (>80%, related to COD). Otherwise, acidifying bacteria could lower the pH, overgrow the methanogenic biofilm and, hence, reduce the methanogenic activity. Thus 2-stage systems are considered superior, since the performance in terms of stability and space-time-yield will be superior to 1-stage systems. The load of reactors with volatile fatty acids (in a system with a separate acidification reactor) can be higher by factors of 4–5, compared to feeding with complex substrates (HENZE and HARREMOES, 1983).

Furthermore, inhibitory substances like sulfur compounds may play a major role, as in yeast processing (FRIEDMANN and MÄRKL, 1994). It is also essential that results concerning load refer to an average of a stable, continuous process rather than to a singular maximum. Even results obtained in the laboratory reactors differ in general from pilot plant and full-scale reactors operating at the factory site

Tab. 2. Support Materials for Anaerobic Fluidized Bed Systems

Support	Diameter $[10^{-3}\,m]$	Density $[kg\,m^{-3}]$	Surface $[m^2\,m^{-3}]$	Porosity —	Upflow Velocity $[m\,h^{-1}]$	Biomass $[kg\,m^{-3}]$	Reference
Sand	0.5	2,540	7,100[a]	0.41	30	4–20	ANDERSON et al. (1990)
Sepiolite	0.53	1,980	20,300[b]			32	BALAGUER et al. (1992)
GAC	0.6					34	CHEN et al. (1995)
Biomass granules					2–6.5		FRANKLIN et al. (1992)
Sand	0.1–0.3	2,600			16	40	HEIJNEN (1985)
Biolite	0.3–0.5	2,000			5–10	30–90	EHLINGER (1994), HOLST et al. (1997)
Pumice	0.25–0.5	1,950	$2.2\cdot10^6$	0.85	10		JÖRDENING (1996), JÖRDENING and KÜSTER (1997)

[a] Calculated from the given data with the assumption of total sphericity.
[b] Calculated with data given in SANCHEZ et al. (1994).

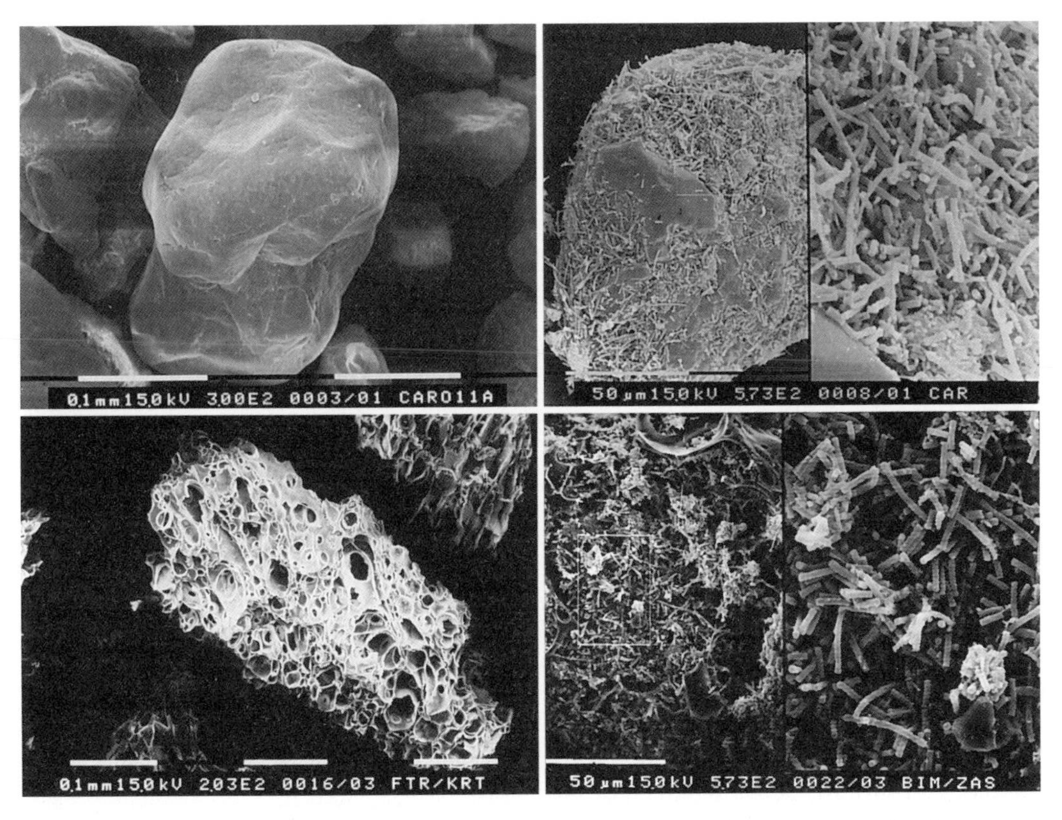

Fig. 5a–d. Support materials for anaerobic fluidized bed reactors (**a** and **b** uncovered and biofilm covered sand, **c** and **d** uncovered and biofilm covered pumice particles).

with variations in substrate quality and concentrations and further fluctuations.

3.3.1 Solids–Stationary Fixed Film Reactors

Suspended solids and even suspended biomass may cause clogging in the reactor; this can be reinforced by extracellular polysaccharides secreted by acidogenic bacteria (EHLINGER et al., 1987). Therefore, backwash and excess sludge removal must be provided for in the reactor design. Gas-phase desorption and transport through the reactor must be possible. Fixed bed systems are not feasable for wastewater with high solids content or components which tend to precipitate, such as calcium ions.

3.3.2 Solids–Fluidized Bed Reactors

Solids from the wastewater may cause clogging, especially at the entrance region of the reactor. HOLST et al. (1997) recommend solid concentrations lower than 0.5 kg m^{-3} with respect to problems of clogging in the distribution system, while MÖSCHE (1998) reports that even solid concentrations up to 1.7 kg m^{-3} did not cause any problems. This difference can be explained by differences in the composition of the solids as well as by the construction of the reactor inlet.

Some inorganic compounds, such as calcium carbonate or ammonium magnesium phosphate, will be precipitated in the reactor mainly onto the support, when the actual concentrations are beyond the equilibrium. At higher concentrations of precipitated solids on the support diffusion limitation occurs. In such cases it is necessary to provide the possibility for removing this material, and substituting it by uncovered new support. Sand and other materials with a high settling velocity in relation to the support have either to be removed before entering the reactor or must be removed by a device at the bottom of the reactor (as described in Sect. 3.4.2.1).

3.4 Reactor Geometry and Technological Aspects

General aspects of reactor design are dealt with in Chapters 17 and 18, this volume.

3.4.1 Fixed Bed Reactors

Upflow reactors tend to be favored, because they allow clumps of biomass to be retained in the filter by gravity, and the start-up may be shorter (e.g., 3–4 months compared to 4–6 months for downflow) (ANDREWS, 1988; WEILAND et al., 1988). The height is limited by the gradients of the biomass and reaction rate. Thus the fixed bed height was chosen not to exceed 7 m, with an overall reactor height of 12 m, a diameter of 14.5 m for a 1,800 m^3 reactor, working with distillery effluents at a load of 4 kg m^{-3} d^{-1} and 90% conversion. The substrate inlet was introduced by 6 inlet devices in order to obtain a distribution at the reactor inlet with a recycle ratio of 5–10. The purified water was collected by 12 tubes at the top of the reactor (WEILAND et. al., 1988). This system exhibited good process stability even for changes in substrate composition due to diverse raw materials used.

Fixed bed reactors do not require major specific design considerations. The ratio of height and diameter is usually in the range of 1–2. The inlet must provide equal distribution of the wastewater by means of distribution devices. This is in general a system of tubes with nozzles, about one for each 5–10 m^2 (LETTINGA et al., 1983). The fluid flow should in general be about 1 m h^{-1}, up to a maximum of 2 m h^{-1} (AUSTERMANN-HAUN et al., 1993).

3.4.2 Fluidized Bed Reactors

Fluidized bed reactors are tall in general and high in relation to agitated tanks or stationary bed reactors. While the height–diameter ratio for laboratory and pilot scale reactors has a range from 5–25, the range for technical plants only varies from 2–5. Fig. 6 shows a technical plant with 500 m^3 volume. The ratio of height and diameter should not be too high with respect to axial concentration gradients which increase with the height of the reactor. But difficulties concerning a uniform fluidization of the support increase with increasing reactor diameter (COUDERC, 1985). Therefore, a compromise for both has to be found. The reactor volume can be calculated as the ratio of the COD load (in terms of kg d^{-1}) and the volumetric loading rate (kg m^{-3} d^{-1}) can be estimated from lab and pilot-scale experiments.

Fig. 6. Technical anaerobic fluidized bed reactor (sugar factory Clauen, Germany).

3.4.2.1 Fluidization of the Support

One of the key factors for the development of a fluidized bed system is the fluidization zone. This zone has to provide a homogenous distribution of support and substrate by the incoming feed to prevent any dead zone formation and to avoid high shear forces. Most of the lab-scale reactors work with inlet tubes in downward direction or sieves; sometimes with glass beads for providing a good upflow distribution. Those are not applicable for full-scale reactors. The use of a multitube system could be a solution for technical reactors, but such a multitube system is expensive and may cause problems with solids (blocking) or precipitation (lime or other inorganic compounds) (IZA, 1991).

Problems of a good distribution system are sometimes discussed in the literature (FRANKLIN et al., 1992; OLIVA et al., 1990), but new developments or improvement of existing distribution systems are rarely described in detail.

One fluidized bed bottom used in a 500 m^3 BMA reactor with sugar factory wastewater is shown in Fig. 7 (JÖRDENING et al., 1996). It has a conical shape with 12 concentric pipes for the incoming water. By means of an inner double cone the superficial liquid velocity is twice that of the reaction zone. So only material with a high settling velocity can settle down in this part. By a flap at the lower end of the cone those particles can be removed.

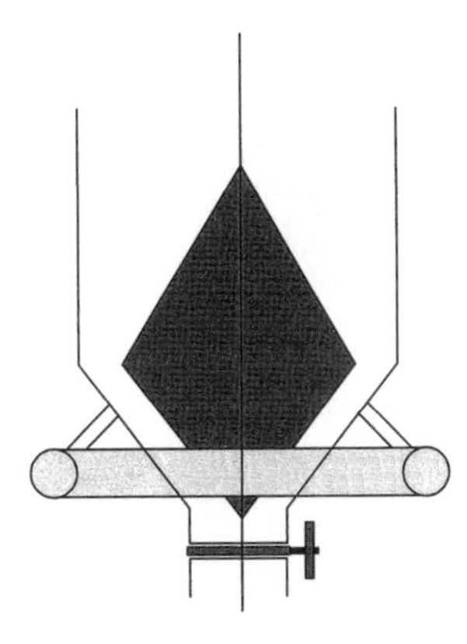

Fig. 7. Flow distribution system for fluidized bed reactors (JÖRDENING et al., 1996).

3.4.2.2 Bed Height and Loss of Support

The fluidized bed height is determined by the flow rate depending on the support. It varies over time with the growth of the biofilm, the gas production rate, and possible precipitates on the support. Changes occur only slightly with time.

Strategies for avoiding support loss are different for several systems at the technical scale:

(1) for the Anitron process (sand as support) the bed height can be controlled by a stationary support–biomass separating device at the maximum bed height which contains the removal of support from the reactor by a centrifugal pump which disrupts the biomass from the support. While the support is led back to the reactor excess biomass is rejected. The maximum bed height is at a minimum of 1.5 m below the overflow section;

(2) for the Gist-Brocades (sand as support) and the Anaflux process (Biolite® as support) the bed height is controlled only with the fluidization velocity. The safety for avoiding biocatalyst loss is provided by 3-phase separation constructions. While the Gist-Brocades system does not contain a unit for removing support, the Anaflux system as well as the Anitron system are provided with a so-called centrifugal pump for removal of excess biomass which is taken from the separator. Therefore, automatic control is not necessary, but daily control via a sample port is useful to avoid support loss.

4 Reactor Operation

4.1.1 Start-Up Procedure

The start-up of fixed bed reactors is governed by several parameters (WEILAND et al., 1988; BURKHARDT and JÖRDENING, 1994):

(1) size and quality of the inoculum, notably the activity of slow growing methanogens;
(2) degree of adaption, mainly the content of bacteria adapted to adhesion;
(3) degree of biomass retention.

The limiting activity, thus slowly growing propionic acid or lactic acid converting bacteria are of crucial importance, and the load must be controlled in order to maintain the substrates below growth-limiting concentrations. The initial load should be low, in the range of 0.1 kg(COD) kg^{-1}(VSS) d^{-1}; the flow rate reduced to 0.4 m h^{-1} at maximum (HENZE and HARREMOES, 1983).The start-up is faciliated by Ca ions in a concentration range of 100–200 g m^{-3}. The load may be increased at a rate of 5–10% d^{-1}. The overall time required for start-up to full load may be 1–3 months. It can be much less with adapted inoculum and a sufficient amount of biomass. In industries working in campaign periods, re-start after storage of the biomass (in the reactor at ambient temperature) is rather straightforward within several days to full load.

The time required for the development of a well-attached film on the support in fluidized beds is sometimes reported to be very long in contrast to suspended systems (SWITZENBAUM and JEWELL, 1980; HEIJNEN et al., 1986).

Because continuous inoculation seems to be impractical for technical systems, batch inoculation is mostly used. This means, that digestor sludge is added and the continuous flow of wastewater to the reactor is started after some days or weeks of adaption. If suspended bacteria are used as an inoculum, the start of the continuous work will cause a significant loss of activity. Hence it is advantageous to use – whenever possible – immobilized inoculum from comparable plants to reduce the time for adaption and loss of inoculum. JÖRDENING and PELLEGRINI (1992) reported, that the use of immobilized inoculum could reduce the start-up time by about 30%. HEIJNEN et al. (1986) showed by a comparison of data from the literature that the loading rate profile is a further key factor concerning a rapid biofilm development: The use of a so-called "maximum load profile" reduces the start-up time up to a half, compared with the so-called "maximum efficiency profile". The "maximum load profile" means the increase of load, if the reactor concentration level of volatile fatty acids is still high ($>$2,000 ppm), the "maximum effficiency profile" means the increase of the load only if the reactor concentration level of volatile fatty acids is decreased to a minimum ($<$100 ppm) and, therefore, the possible conversion is achieved.

4.1.2 Operation Results – Stationary Bed

Many reviews have been published by HENZE and HARREMOES (1983), SWITZENBAUM (1983) and AUSTERMANN-HAUN et al. (1993). Tab. 3 gives selected data for typical substrates and carriers on the lab and pilot scale, which also seem appropriate for scale-up to the industrial level. One example is included representing very high load, however, with an expensive carrier (Siran®). Most typical loading rates (kg m^{-3} d^{-1} COD) are in the range of 3–10 for common wastewater from agricultural and food processes with high conversion (over 80% of degradable COD), but there are also examples for loading rates in the range of 20 and even 40 kg m^{-3} d^{-1} COD. However, these rather exceptional results may not be suitable for scale-up.

A considerable number of technical-scale systems in the range from several 10 m^3 up to several 1,000 m^3 are in operation. The sites are small and medium agricultural installations, typically treating wastewater from poultry, pig, cattle farms and distilleries, medium and large-scale reactors in the food industries such as sugar, potato, starch, and dairy processing, but also the paper and pulp and the chemical industry. Some typical data are summarized in Tab. 4. In many cases acidification takes place in the wastewater system prior to feeding to the methane reactor even if this is not mentioned in detail; notably in the case of substrates easily undergoing microbial fermentation. For example, this is the case for sugar industry wastewater carrying sucrose as the main substrate stored in lagoons before being fed to the methane reactor. More examples and data have been published by AUSTERMANN-HAUN et al. (1993).

Tab. 3. Data of Laboratory and Pilot-Scale Anaerobic Stationary Bed Systems

Substrate	COD_{in} [$kg\ m^{-3}$]	Carrier	Load [$kg\ m^{-3}\ d^{-1}$]	Conversion [%]	Reference
Sucrose 30%, ethanol 65%	3–6	modular blocks pall rings	2–4	83–85	YOUNG and DAHAB (1983)
Distillery effluents[a]	<10	Plasdek® Flocor® Hiflow 90®	8–10	90–95	WEILAND and WULFERT (1986)
Volatile acids	6		1.7–3.4	87–98	YOUNG and MCCARTY (1969)[b]
Protein–carbo-hydrate waste			3.1–3.3	90–50	MUELLER and MANCINI (1975)[c]
Sludge heat treatment liquor	10		6.5	55–65	DONOVAN (1981)[c]
Chemical plant waste			16	65	RAGAN (1981)[c]
Sulfite evaporator condensate[d]	37	Siran® Raschig rings	45	84	NEY et al. (1989)

[a] Acidification prior to methanogenesis.
[b] Test in different reactors.
[c] Taken from SWITZENBAUM (1983).
[d] Sulfite evaporator condensate: acetic acid 425, methanol 75, furfural 28 m (?).

Tab. 4. Data of Industrial-Scale Anaerobic Stationary Bed Systems

Wastewater	Reactor Size [m^3]	Support Material	Load [$kg\ m^{-3}\ d^{-1}$]	Removal [%]	Reference
Meat processing	22[a]	porous glass (Siran®)	10–50	up to 80	BREITENBÜCHER (1994)
Dairy	260	plastic rings (Biofar®)	10		WEILAND et al. (1988)
Dairy	362	plastics (Flocor® and cloisonyle)	10–12	70–80	AUSTERMANN-HAUN et al. (1993)
Sugar	[b]	plastics	[b]	90	CAMILLERI (1988), HENRY and VARALDO (1988)
Sugar	1,400	plastic rings (Flocor®)	13		WEILAND et al. (1988)
Potato processing	660	pallrings (100 mm)	3–6		WEILAND et al. (1988)
Starch	4·300[c]	lava	25	>70	SCHRAEWER (1988)
Soft drinks production	85	modular plastic blocks (Plasdek®)	4–6	90	AUSTERMANN-HAUN et al. (1993)
Distillery	13,000	modular plastic blocks	8–12		WEILAND et al. (1988)
Distillery	92	modular plastic blocks (BIO-NET®)	13	78	AUSTERMANN-HAUN et al. (1993)
Chemistry	1,900	plastics	16–20	90	HENRY and VARALDO (1988)

[a] 2-stage system with acidification (40 m^3) as a first stage.
[b] Up to 2,000 $m^3\ d^{-1}$ wastewater with 16 $t\ d^{-1}$ COD.
[c] 2-stage system with acidification reactor of 1,000 m^3 as the first stage.

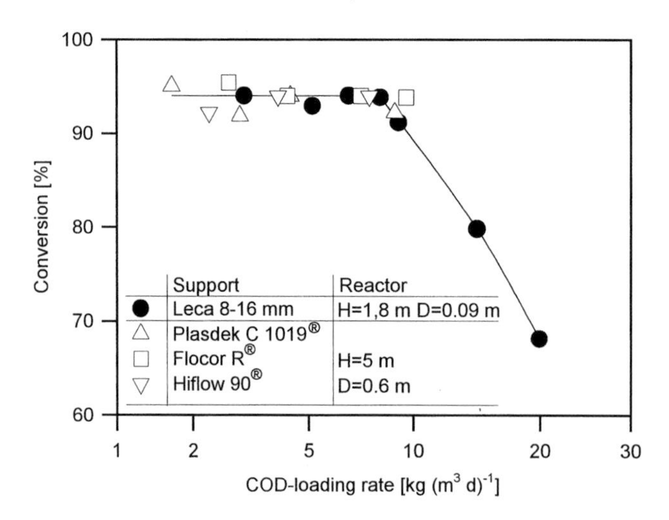

Fig. 8. Influence of the volumetric loading rate on the conversion (WEILAND and WULFERT, 1986).

Tab. 5. Data of Laboratory and Pilot-Scale Anaerobic Fluidized Bed Systems

Waste	Support	Reactor		Loading	Concen-tration	HRT	Removal	Reference
		Volume $[m^3]$	Ratio[a] [m]	$[kg\ m^{-3}\ d^{-1}]$	$[kg\ m^{-3}]$	[h]	[%]	
Spoilt beer	sand	$64 \cdot 10^{-3}$	18	1–14.8	1–12	0.2–7	75–87	ANDERSON et al. (1990)
Synthetic	GAC[b]	$1.2 \cdot 10^{-3}$	5.6	5.8–108	0.5–9	0.5–8	75–98	CHEN et al. (1995)
Synthetic	sand	$70 \cdot 10^{-3}$	20	3–20	13	1.5	98	JÖRDENING (1987)
Fruits and vegetables	biolite		25	38			>80	MENON et al. (1997)
Brewery	sand	$63 \cdot 10^{-3}$	18.7	14.6	1–3.5	5–34	70[b]	OZTURK et al. (1989)

[a] Height/diameter.
[b] Granular activated carbon.

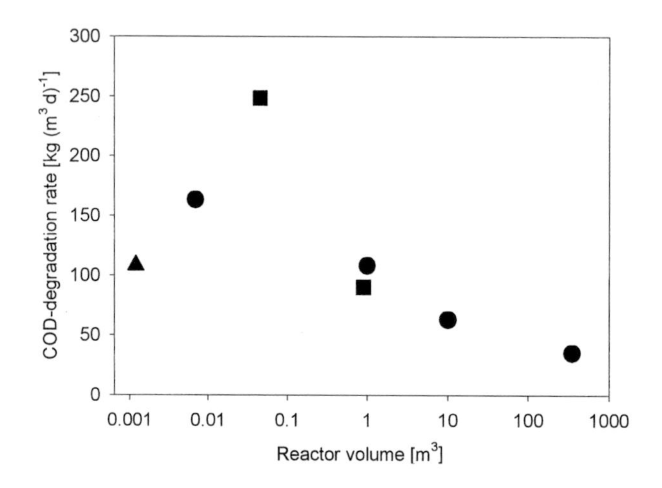

Fig. 9. Decrease of reactor performance with scale up, ■ KEIM et al. (1989), ● BURKHARDT and JÖRDENING (1994), ▲ CHEN et al. (1995).

Overloading may be possible without inactivation of the biomass (if an appropriate pH, at least 6, is held), but the conversion will decrease according to the maximal activity of biomass (in general 1–2 kg COD m^{-3} kg^{-1} dry biomass). A corresponding correlation is shown in Fig. 8 (WEILAND and WULFERT, 1986).

4.1.3 Operational Results – Fluidized Bed Reactors

The performance of fluidized bed reactors on the laboratory and pilot scale is sometimes tremendous: KEIM et al. (1989) used Siran® spheres in a fluidized bed reactor for the treatment of evaporation condensate and reported a loading rate up to 315 kg m^{-3} d^{-1} with 79% removal. JÖRDENING et al. (1991) used another sinter glass (Poraver®) for the treatment of a sugar wastewater with loading rates up to 183 kg m^{-3} d^{-1} with 89% removal. Data of some laboratory and pilot-scale plants are given in Tab. 5.

Despite these results on the laboratory scale, the data reported for the performance on the technical scale, as shown in Tab. 6, are in a significantly lower range of 15–50 kg m^{-3} d^{-1} (see, e.g., EHLINGER, 1994; FRANKLIN et al., 1992; JÖRDENING, 1996; JÖRDENING and KÜSTER, 1997; OLIVA et al., 1990). Some data are summarized in Fig. 9 showing the decrease of performance with increasing reactor volume.

5 Conclusions

As a general conclusion it may be stated that both fixed and fluidized bed reactors are well established on the industrial scale, notably in the food and related industries, like breweries and distilleries, and in the paper and pulp industry where the substrates are of natural origin, but also in others, e.g., the chemical industry.

Tab. 6. Data of Technical-Scale Anaerobic Fluidized Bed Systems

Waste	Company	Support	Reactor		Loading	Concentration	HRT	Removal	Reference
			Volume [m^3]	Ratio[a] [m]	[kg m^{-3} d^{-1}]	[kg m^{-3}]	[h]	[%]	
Soy bean	Dorr–Olivier	sand	304	2.0	14–21	0.8–10		75–80	SUTTON et al. (1982)
Several	Degrement	biolite	210–480		16–21	3.8–5	0.25	50–60	EHLINGER (1994)
Yeast and pharmaceuticals	Gist-Brocades	sand	400	4.4	8–30	1.9–4	0.14–0.41	95–98	FRANKLIN et al. (1992)
Evaporation conditions and osmosis permeability	Gist-Brocades	granular	125		21–35	5–22.5			FRANKLIN et al. (1992)
Sugar beet	BMA	pumice	500	5.0	20–60	1–5	1.5	65–78	JÖRDENING (1996)
Brewery	Degrement	biolite	165	3.2					OLIVA et al. (1990)

[a] Height/diameter.

Specific limitations, however, must be taken into consideration:

(1) Acidification inside the methane reactor requires a higher residence time as compared to 2-stage systems with acidification as a first stage; this also provides a safer and more stable processing, e.g., for the case of a shock load – a quite common situation in practice.
(2) Wastewater free of suspended solids is required for fixed bed reactors. No precipitation of solids should occur inside the reactor.
(3) Advantages for both types of systems are higher density of biomass and, therefore, higher reaction rates and a considerably smaller reactor volume required as compared to conventional systems like stirred tanks. This is especially true for fluidized bed reactors, since they may offer a much higher support surface and, hence, a high biomass concentration per volume. Much less space is required due to a low reactor volume and with a high ratio of height to diameter of the reactors.
(4) For these reasons investment is lower as compared to systems without biomass immobilization.

Acknowledgement
The valuable comments by Prof. Dr. WEILAND are gratefully acknowledged.

6 References

ANDERSON, G. K., OZTURK, I., SAW, C. B. (1990), Pilot-scale experiences on anaerobic fluidized bed treatment of brewery wastes, *Water Sci. Technol.* **22**, 157–166.

ANDREWS, G. F. (1988), Design of fixed film and fluidized bed bioreactors, in: *Handbook on Anaerobic Fermentations* (ERICKSON, L. E., YEE-CHUK FUNG, D., Eds.), pp. 765–802. New York, Basel: Marcel Dekker.

AUSTERMANN-HAUN, U., KUNST, S., SAAKE, M., SEYFRIED, C. F. (1993), Behandlung von Abwässern, in: *Anaerobtechnik* (BÖHNKE, B., BISCHOFSBERGER, W., SEYFRIED, C. F., Eds.), pp. 467–696. Berlin, Heidelberg, New York: Springer-Verlag.

BALAGUER, M. D., VICENT, M. T., PARIS, J. M. (1992), Anaerobic fluidized bed with sepiolite as support for anaerobic treatment of vinasse, *Biotechnol. Lett.* **5**, 333–338.

BIRD, R. B., STEWART, W. E., LIGHTFOOD, E. N. (1960), *Transport Phenomena*. New York: John Wiley & Sons.

BISCHOFSBERGER, W. (1993), Übersicht über anaerobe Verfahrenstechniken, in: *Anaerobtechnik* (BÖHNKE, B., BISCHOFSBERGER, W., SEYFRIED, C. F., Eds.), pp. 96–134. Berlin, Heidelberg, New York: Springer-Verlag.

BREITENBÜCHER, K. (1994), Hochleistung durch mehr Biomasse, *UTA Umwelttechnik Aktuell* **5**, 372–374.

BURKHARDT, C., JÖRDENING, H.-J. (1994), Maßstabsvergrößerung und Betriebsdaten von anaeroben Hochleistungs-Fließbettreaktoren, in: *Anaerobe Behandlung von festen und flüssigen Rückständen* (MÄRKL, H., STEGMANN, R., Eds.), pp. 145–162. Weinheim: VCH.

CAMILLERI, C. (1988), Start-up of fixed film stationary bed anaerobic reactors, in: *Anaerobic Digestion 1988* (HALL, E. R., HOBSON, P. N., Eds.), pp. 407–412. Oxford: Pergamon Press.

CARRONDO, M. J. T., SILVA, J. M. C., FIGUERA, M. I. I., GHANO, R. M. B., OLIVEIRA, J. F. S. (1983), Anaerobic filter treatment of molasses fermentation wastewater, *Water Sci. Technol.* **15**, 117–126.

CHARACKLIS, W. G. (1990), Microbial fouling, in: *Biofilms* (CHARACKLIS, W. G., MARSHALL, K. C., Eds.), pp. 523–584. New York: John Wiley & Sons.

CHEN, S. J., LI, C. T., SHIEH, W. K. (1995), Performance evaluation of the anaerobic fluidized bed system: I. substrate utilization and gas production, *J. Chem. Technol. Biotechnol.* **35B**, 101–109.

COUDERC, J.-P. (1985), Incipient fluidization and particulate systems, in: *Fluidization*, 2nd Edn. (DAVIDSON, J. F., CLIFT, R., Eds.), pp. 1–46. London: Academic Press.

DÁNIELS, S. L. (1980), Mechanisms involved in sorption of microorganisms to solid surfaces, in: *Adsorption of Microorganisms to Surfaces* (BITTON, G., MARSHALL, K. C., Eds.), pp. 7–58. New York: John Wiley & Sons.

DIAZ-BAEZ, M. C. (1988), A study of factors affecting attachment in the start up and operation of anaerobic fluidized bed reactors, in: *Poster-Papers 5th Int. Conf. Anaerobic Digestions* (TILCHE, A., ROZZI, A., Eds.), pp. 105–108. Bologna: Monduzzi Editore.

DONOVAN, E. J. (1981), Treatment of high strength wastes by an anaerobic filter, *Proc. Semin.: Anaerobic Filters: An Energy Plus for Wastewater Treatment*, pp. 179–187, Howey-In-The-Hills, FL. ANL/CSNV-TM-50, 9.–10. 1. 1980.

EHLINGER, F. (1994), Anaerobic biological fluidized beds: operating experiences in France, pp. 315–

323, *Proc. 7th Int. Symp. Anaerobic Digestion*, Cape Town, South Africa, 23.–27. 01. 1994.

EHLINGER, F., AUDIC, J. M., VERRIER, D., FAUP, G. M. (1987), The influence of the carbon source on microbial clogging in an anaerobic filter, *Water Sci. Technol.* **19**, 261–273.

FLEMMING, H.-C. (1991), Biofilme und Wassertechnologie, Teil 1: Entstehung, Aufbau und Zusammensetzung, *gwf Wasser Abwasser* **132**, 197–207.

FRANKLIN, R. J., KOEVOETS, W. A. A., VAN GILS, W. M. A., VAN DER PAS, A. (1992), Application of the biobed upflow fluidized bed process for anaerobic waste water treatment, *Water Sci. Technol.* **25**, 373–382.

FRIEDMANN, H., MÄRKL, H. (1994), Der Einfluß der Produktgase auf die mikrobiologische Methanbildung, *gwf Wasser Abwasser* **6**, 302–311.

GERSON, D. F., ZAJIC, J. E. (1979), The biophysics of cellular adhesion, *American Chemical Society Series* **106**, pp. 29–57.

HALL, E. R. (1982), Biomass retention and mixing characteristics in fixed-film and suspended growth anaerobic reactors, pp. 371–396, *Proc. JAWPR Specialized Semin. Anaerobic Treatment Waste Water in Fixed Film Reactors*, 16.–18. 6. 1982, Copenhagen, Denmark.

HALL, E. R., HOBSON, P. N. (1988), *Anaerobic Digestion 1988*. Oxford: Pergamon Press.

HEIJNEN, J. J. (1985), *US Patent* 4560479.

HEIJNEN, J. J., MULDER, A., ENGER, W., HOEKS, F. (1986), Review on the application of anaerobic fluidized bed reactors in waste water treatment, *Proc. EWPCA Conf. Anaerobic Wastewater Treat.* pp. 159–173.

HENRY, M., VARALDO, C. (1988), Anaerobic digestion treatment of chemical industry wastewaters at the Cuise-Lamotte (Oise) plant of Société Française Hoechst 479, in: *Anaerobic Digestion 1988* (HALL, E. R., HOBSON, P. N., Eds.), pp. 479–486. Oxford: Pergamon Press.

HENZE, M. (Ed.) (1983), *Anaerobic Treatment of Wastewater in Fixed Film Reactors*. Oxford: Pergamon Press.

HENZE, M., HARREMOES, P. (1983), Anaerobic treatment of wastewater in fixed film reactors – a literature review, *Water Sci. Technol.* **15**, 1–101.

HOEHN, D. C. (1970), The Effects of Thickness on the Structure and Metabolism of Bacterial Films, *Thesis*, University of Missouri.

HOLST, T. C., TRUC, A., PUJOL, R. (1997), Anaerobic fluidized beds: ten years of industrial experience, *Water Sci. Technol.* **36**, 415–422.

HORN, H., HEMPEL, D. C. (1996), Modellierung von Substratumsatz und Stofftransport in Biofilmsystemen, *gwf Wasser/Abwasser* **137**, 293–299.

IZA, J. (1991), Fluidized bed reactors for anaerobic wastewater treatment, *Water Sci. Technol.* **24**, 109–132.

JÖRDENING, H.-J. (1987), Untersuchungen an Hochleistungsreaktoren zum anaeroben Abbau von calciumhaltigen Abwässern, *Thesis*, Technical University of Braunschweig, Germany.

JÖRDENING, H.-J. (1996), Scaling-up and operation of anaerobic fluidized bed reactors, *Zuckerindustrie* **121**, 847–854.

JÖRDENING, H.-J., KÜSTER, W. (1997), Betriebserfahrungen mit einem anaeroben Fließbettreaktor zur Behandlung von Zuckerfabriksabwasser, *Zuckerindustrie* **122**, 934–936.

JÖRDENING, H.-J., PELLEGRINI, A. (1992), Wirbelschichtreaktoren in der Abwasserreinigung, *Chem. Ing. Tech.* **64**, 877–878.

JÖRDENING, H.-J., JANSEN, W., BREY, S., PELLEGRINI, A. (1991), Optimierung des Fließbettsystems zur anaeroben Abwasserreinigung, *Zuckerindustrie* **116**, 1047–1052.

JÖRDENING, H.-J., MANSKY, H., PELLEGRINI, A. (1996), *German Patent* 19502615 C2.

KEIM, P., LUERWEG, M., AIVASIDIS, A., WANDREY, C. (1989), Entwicklung der Wirbelschichttechnik mit dreidimensional kolonisierbaren Trägermaterialien aus makroporösem Gas am Beispiel der anaeroben Abwasserreinigung, *Korrespondenz Abwasser* **36**, 675–687.

KITSOS, H. M., ROBERTS, R. S., JONES, W. J., TORNABENE, T. G. (1992), An experimental study of mass diffusion and reaction rate in an anaerobic biofilm, *Biotechnol. Bioeng.* **39**, 1141–1146.

KOSSEN, N. W. F., OOSTERHUIS, N. M. G. (1985), Modelling and scaling-up of bioreactors, in: *Biotechnology*, 1st Edn., Vol. 2 (REHM, H.-J., REED, G., Eds.), pp. 571–605. Weinheim: VCH.

LAUWERS, A. M., HEINEN, W., GORRIS, L. G. M., VAN DER DRIFT, C. (1990), Early stages in biofilm development in methanogenic fluidized-bed reactors, *Appl. Microbiol. Biotechnol.* **33**, 352–358.

LETTINGA, G., HOBMA, L. W., HULSHOFF-POL, L. W., DE ZEEUW, W., DE JONG, P. et al. (1983), Design operation and economy of anaerobic treatment, *Water Sci. Technol.* **15**, 177–196.

MENON, R., DELPORTE, C., JOHNSTONE, D. S. (1997), Pilot treatability testing of food processing wastewaters using the ANAFLUX & trade; anaerobic fluidized-bed reactor, *52nd Ind. Waste Conf.*, Purdue, West Lafayette, IN, USA.

MÖSCHE, M. (1998), Anaerobe Reinigung von Zukkerfabriksabwasser in Fließbettreaktoren, *Thesis*, Technical University of Braunschweig, Germany.

MUELLER, J. A., MANCINI, J. L. (1975), Anaerobic filter kinetics and application, *Proc. 30th Ind. Waste Conf.*, pp. 423–447, Purdue, West Lafayette, IN, USA.

MULCAHY, L. T., LA MOTTA, E. J. (1978), Mathematical model of the fluidized bed biofilm reactor, *Report No. 59-78-2*, Dept. Civil Eng., University of Massachusetts/Amherst, USA.

NEY, U., SCHOBERT, S. M., SAHM, H. (1989), Anaerobic degradation of sulfite evaporator condensate by defined bacterial mixed cultures, *DECHEMA Biotechnol. Conf.*, Vol. 3, Part B, pp. 889–892. Weinheim: VCH.

NYNS, E.-J. (1986), Biomethanization process, in: *Biotechnology* 1st Edn., Vol. 8 (REHM, H.-J., REED, G., Eds.), pp. 207–268. Weinheim: VCH.

OLIVA, E., JACQUART, J. C., PRIVOT, C. (1990), Treatment of waste water at the El Aguila brewery (Madrid, Spain). Methanization in fluidized bed reactors, *Water Sci. Technol.* **22**, 483–490.

ONUMA, M., OMURA, T., UMITA, T., AIZAWA, J. (1985), Diffusion coefficients and its dependency on some biochemical factors, *Biotechnol. Bioeng.* **27**, 1533–1539.

OUCHIJAMA, N., TANAKA, T. (1981), Porosity of a mass of solid particles having a range of sizes, *Ind. Eng. Chem. Fundam.* **20**, 66–71.

OZTURK, I., ANDERSON, G. K., SAW, C. B. (1989), Anaerobic fluidized bed treatment of brewery wastes and bioenergy recovery, *Water Sci. Technol.* **21**, 1681–1684.

PARKIN, G. F., SPEECE, R. E. (1983), Attached versus suspended growth anaerobic reactors: Response to toxic substances, *Water Sci. Technol.* **15**, 261–289.

RAGAN, J. L. (1981), Celanese experience with anaerobic filters, *Proc. Semin.: Anaerobic Filters: An Energy Plus for Wastewater Treatment*, pp. 129–135, 9.–10. 01. 1980, Howey-In-The-Hills, FL, ANL/CSNV-TM-50.

RICHARDSON, J. F., ZAKI, W. N. (1954), Sedimentation and fluidization: Part 1, *Trans. Inst. Chem. Eng.* **32**, 35–53.

RITTENHOUSE, G. (1943), A visual method of estimating two-dimensional sphericity, *J. Sediment. Petrol.* **13**, 79–81.

SALKINOJA-SALONEN, M. S., NYNS, E.-J., SUTTON, P. M., VAN DEN BERG, L., WHEATLEY, A. D. (1983), Starting up of an anaerobic fixed film system, *Water Sci. Technol.* **15**, 305–308.

SANCHEZ, J. M., ARIJO, S., MUNOZ, M. A., MORINIGO, M. A., BORREGO, J. J. (1994), Microbial colonization of different support materials used to enhance the methanogenic process, *Appl. Microbiol. Biotechnol.* **41**, 480–486.

SANCHEZ, J. M., RODRIGUEZ, F., VALLE, L., MUNOZ, M. A., MORINIGO, M. A., BORREGO, J. J. (1996), Development of methanogenic consortia in fluidized bed batches using sepiolite of different particle size, *Microbiologia* **12**, 423–434.

SCHRAEWER, R. (1988), Das Anfahrverhalten von anaeroben Bioreaktoren zur Reinigung hochbelasteter Stärkeabwässer, *Starch/Stärke* **40**, 347–352.

SCHULTES, M. (1998), Füllkörper oder Packungen? *Chem. Ing. Tech.* **70**, 254–261.

SCHWARZ, A., YAHYAVI, B. M., MÖSCHE, M., BURKHARDT, C., JÖRDENING, H.-J. et al. (1996), Mathematical modeling for supporting scale-up of an anaerobic wastewater treatment in a fluidized bed reactor, *IAWQ Conf.*, Singapore.

SCHWARZ, A., REUSS, M., MÖSCHE, M., JÖRDENING, H.-J., BUCHHOLZ, K. (in press), Modellgestützte Betrachtungen zur Reaktionstechnik und Fluiddynamik im industriellen Anaerob-Fließbettreaktor, in: *Technik anaerober Prozesse* (MÄRKL, H., STEGMANN, R., Eds.). Weinheim: Wiley-VCH.

SHIEH, W. K., KEENAN, J. D. (1986), Fluidized bed biofilm reactor for wastewater treatment, *Adv. Biochem. Eng. Biotechnol.* **33**, 132–169.

SHIEH, W. K., MULCAHY, L. T., LAMOTTA, E. J. (1982), Mathematical model for the fluidized bed biofilm reactor, *Enzyme Microb. Technol.* **4**, 269–275.

SNOWDON, C. B., TURNER, J. C. R. (1967), Mass transfer in liquid-fluidized beds of ion-exchange resin beads, in: *Proc. Int. Symp. Fluidization* (DRINKENBURG, A. A. H., Ed.), pp. 599–608. Eindhoven.

SREEKRISHNAN, T. R., RAMACHANDRAN, K. B., GHOSH, P. (1991), Effect of operating variables on biofilm formation and performance of an anaerobic fluidized-bed bioreactor, *Biotechnol. Bioeng.* **37**, 557–566.

STRONACH, S. M., DIAZ-BAEZ, M. C., RUDD, T., LESTER, J. H. (1987), Factors affecting biomass attachment during start-up and operation of anaerobic fluidized beds, *Biotechnol. Bioeng.* **30**, 611–620.

SUTTON, P. M., LI, A., EVANS, R. R., KORCHIN, S. (1982), Dorr–Oliver's fixed film and suspended growth anaerobic systems for industrial wastewater treatment and energy recovery, *37th Ind. Conf.*, Purdue, West Lafayette, IN, USA.

SWITZENBAUM, M. S. (1983), A comparison of the anaerobic filter and the anaerobic expanded/fluidized bed process, *Water Sci. Technol.* **15**, 345–358.

SWITZENBAUM, M. S., JEWELL, W. J. (1980), Anaerobic attached film expanded-bed reactor treatment, *J. WPCF* **52**, 1953–1965.

VAN DEN BERG, L., KENNEDY, K. J. (1983), Dairy waste treatment with anaerobic stationary fixed film reactor, *Water Sci. Technol.* **15**, 359–368.

WEILAND, P., ROZZI, A. (1991), The start-up, operation and monitoring of high-rate anaerobic treatment systems: discussers report, *Water Sci. Technol.* **24**, 257–277.

WEILAND, P., WULFERT, K. (1986), Festbettreaktoren zur anaeroben Reinigung hochbelasteter Abwässer – Entwicklung und Anwendung, *BTF* **3**, 152–158.

WEILAND, P., THOMSEN, H., WULFERT, K. (1988), Entwicklung eines Verfahrens zur anaeroben Vorreinigung von Brennereischlempen unter Einsatz eines Festbettreaktors, in: *Verfahrenstechnik der mechanischen, thermischen, chemischen und biologischen Abwasserreinigung, Part 2: Biologische*

Verfahren (GVC-VDI, Ed.), pp. 169–186. Düsseldorf: VDI-Gesellschaft für Verfahrenstechnik und Chemieingenieurswesen.

WEN, C. Y., YU, Y. H. (1966), A generalized method for predicting the minimum fluidization velocity, *AIChE J.* **12**, 610–612.

YI-ZHANG, Z., LI-BIN, W. (1988), Anaerobic treatment of domestic sewage in China, in: *Anaerobic Digestion 1988* (HALL, E. R., HOBSON, P. N., Eds), pp. 173–184. Oxford: Pergamon Press.

YOUNG, J. C., DAHAB, M. F. (1983), Effect of media design on the performance of fixed-bed anaerobic reactors, *Water Sci. Technol.* **15**, 369–384.

YOUNG, J. C., MCCARTY, P. L. (1969), The anaerobic filter for waste treatment, *J. Water Pollut. Control Fed.* **41**, R161.

25 Possibilities and Potential of Anaerobic Wastewater Treatment Using Anaerobic Sludge Bed (ASB) Reactors

GATZE LETTINGA
LOOK W. HULSHOFF POL
JULES B. VAN LIER
GRIETJE ZEEMAN
Wageningen, The Netherlands

List of Abbreviations

AAFFEB	anaerobic attachment fixed film expanded bed
AnFB	anaerobic fluidized bed
AnWT	anaerobic wastewater treatment
ASB	anaerobic sludge bed reactor
EGSB	expanded granular sludge bed
FB	fluidized bed
GSS	gas–solid separator
IC	internal circulation
IC-UASB	internal circulation UASB
OLR	organic loading rate
SSAR	staged sludge anaerobic reactor system
UASB	upflow anaerobic sludge bed
UFB	upflow fluidized bed
VFA	volatile fatty acids

1 Introduction

Anaerobic wastewater treatment (AnWT), although a relatively young technology, is becoming increasingly popular worldwide. This can mainly be attributed to the important fundamental benefits of AnWT over conventional aerobic wastewater treatment (AeWT), especially in the light of sustainability and robustness (see below). When accepting that AnWT basically is a pretreatment method, today little if any serious drawbacks can be brought up anymore against AnWT. Although highly susceptible to a variety of xenobiotics, in fact like most other organisms increasingly evidence is obtained that many of these compounds are anaerobically biodegradable, even including many very toxic compounds like nitro-aromatics. Moreover, the drawback of its slow rate of "first" start-up vanished almost completely, because presently sufficient quantities of high quality anaerobic excess sludges from existing full scale installations are becoming available for rapid start-up of new installations.

The benefits of AnWT over conventional aerobic treatment systems are:

- Treatment costs are low, because the AnWT systems are technically simple and relatively inexpensive, while generally a supply of high grade energy is not needed, which means a high extent of *robustness*.
- A useful energy carrier is produced, which means a high extent of *sustainability*.
- AnWT can be applied at practically any place and at any scale, which implies an extraordinary flexibility.
- Modern AnWT reactors frequently can accommodate (very) high space loading rates, leading to low space requirements.
- The volume of excess sludge produced in AnWT generally is very small, because of the very low growth yields of anaerobic organisms, and the excellent thickening characteristics of the anaerobic sludge.
- The excess sludge generally is very well stabilized.
- Anaerobic organisms can be preserved unfed for long periods of time, i.e., exceeding 1 year.
- AnWT is the core method for *integrated environmental protection systems*, viz. by combining it with the proper posttreatment methods useful products like ammonia or sulfur can be recovered.

The available modern high rate AnWT reactor systems offer tremendous potentials for treating a large variety of wastewaters, including high, medium and (very) low strength, hot and cold, complex and simple wastewater. Particularly the well known "upflow anaerobic sludge bed" (UASB) and its recent modification, viz. the "expanded granular sludge bed" (EGSB) reactor are popular systems.

2 AnWT Reactor Technology

For achieving high loading rates in AnWT merely the following conditions should be met, viz.:

(1) A high retention of viable, i.e., highly active anaerobic sludge under operational conditions
(2) Accomplishment of a sufficient contact between the retained sludge and the wastewater in order to prevent intergranular and external transport limitation problems
(3) Maintenance of proper environmental conditions, such as temperature, pH, presence and availability of nutrients and trace elements, absence of too high concentrations of toxic and/or inhibitory compounds

Although condition (3) looks self-evident, it should be taken into account that in practice the specific activity of anaerobic bacterial consortia can become quite detrimentally affected by metabolic intermediates and end products. The requirement of "favorable environmental conditions" applies to all (frequently quite) different organisms participating in the degradation process, especially of more complex compounds. The micro niches within the immobilized bacterial aggregates contain the various consortia, and generally very specific environmental conditions prevail in these micro environments. This feature points to the extraordinary importance of a *proper immobilization of the required balanced consortia or microecosystems.*

In the following the discussion will be focused on the *ASB technology* (LETTINGA et al., 1980, 1983a, b, 1993; LETTINGA and HULFSHOFF POL, 1986; LETTINGA, 1995). The ASB reactor concept is based on the insight that anaerobic sludge inherently has or will attain excellent settling characteristics, provided mechanical agitation is kept at a minimum. As a matter of fact mechanical mixing can be omitted in ASB reactors when the imposed OLR can be maintained beyond a certain minimum value which provides sufficient mixing in the system (viz. to meet criterium 2) as a result of the biogas production. If condition (2) cannot be met, e.g., when the OLR imposed on the systems necessarily has to be maintained at a very low level, a gentle mode of mechanical mixing certainly is beneficial, preferentially in an intermit mode. The need for that depends on the evenness where the feed is distributed over the bottom of the reactor and the liquid upflow velocities prevail in the system. Sludge aggregates, which become dispersed under the influence of the biogas produced and the imposed superficial liquid velocity, should be retained within the reactor. For this purpose the biogas should be separated from the system, which is accomplished in ASB reactors by a gas collector system situated in the upper part of the reactor. This gas collector in fact creates a built-in settler device in the top section of the reactor. In this settler section smaller and bigger sludge aggregates (flocs and particles) can coalesce and/or settle out, and subsequently slide back into the digester compartment beneath the *gas–solid separator* (GSS) device. The biogas only can be collected in and released satisfactorily from the gas dome by maintaining a sufficient liquid–gas interface here and by preventing the occurrence of heavy foaming and scum layer formation. The GSS device represents an essential accessory of ASB reactors. Since EGSB reactors merely use granular sludge aggregates and are operated at much higher superficial velocities, quite specific designs for the GSS device may be required. This particularly is true when the settleability of the sludge aggregates for some reason is (or becomes) rather poor, or when little if any sludge washout can be admitted. The latter situation prevails when treating very low strength wastewaters or wastewaters which require the growth-in of extraordinarily small amounts of specific bacteria or bacterial consortia. The smaller the amount of in-growing viable sludge per m^3 wastewater, the higher the requirements put on the GSS device. The relatively plain traditional concepts used in UASB reactors or the first generations EGSB reactors then do not satisfy. A promising system, comprehensively investigated since the last decade at the authors' department and since recently at bigger pilot plant scale in cooperation with potential users of the concept, comprises a *screen drum device*. The design of the sieve drum and its position in the reactor (Fig. 1) depends on a variety of factors, such as nature, strength and composition of the wastewater, characteristics of the granular sludge, operational conditions, etc. Comprehensive investigations have been made with low strength cold solutions of VFA, with or without carbohydrates, and low strength cold wastewaters containing a fraction of dispersed matter, viz. malting wastewater (REBAC et al., 1995, 1997), and with lipid containing wastewaters as well (PETRUY and LETTINGA, 1997; HWU and LETTINGA, 1997; HWU et al., 1997).

The main differences between EGSB and UASB systems in fact comprise the exclusive use of granular sludge in EGSB reactors and the substantially higher superficial liquid velocities (v_f) applied in systems, viz. 5–10 m h^{-1}, sometimes even up to 20 m h^{-1}. In UASB reactor values for v_f are up to 2 m h^{-1}. As a result of the high values for v_f in EGSB reactors, under operational conditions the sludge bed is slightly expanded, which leads to an excellent contact between sludge and wastewater. EGSB reactors are significantly taller than UASB reactors, sometimes up to 20 m. This obviously leads to substantial savings in space. In the meantime the EGSB system has found practical application in the

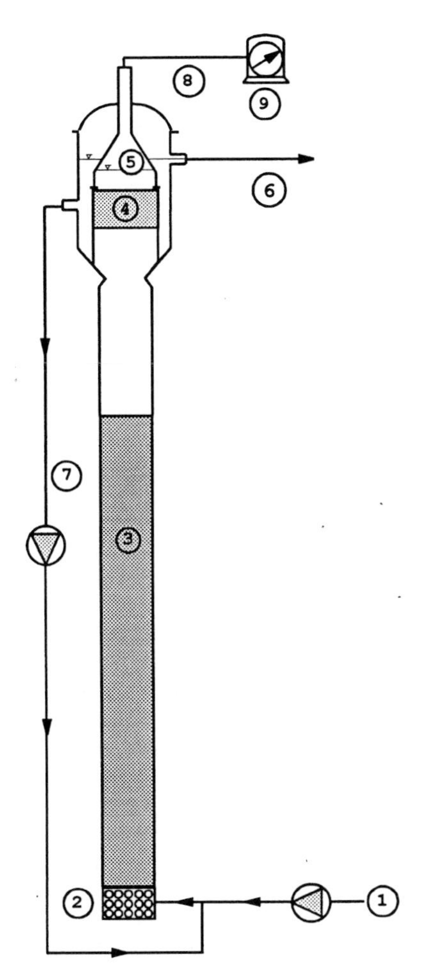

Fig. 1. An EGSB reactor equipped with a sieve drum GSS device in the upper part of the reactor. A settler compartment like in UASB reactors in fact is not needed anymore.
1 feed, 2 feed distribution, 3 expanded sludge bed, 4 sieve drum, 5 gas–liquid separation, 6 effluent, 7 effluent recirculation, 8 biogas, 9 wet test gasometer.

so-called upflow fluidized bed (UFB) system (FRANKIN ET AL., 1992; VERSPRILLE et al., 1994; ZOUTBERG et al., 1997), and in the IC-UASB reactor (internal recirculation UASB). In the latter system (VELLINGA et al., 1986; YSPEERT et al., 1993) halfway the reactor gas is separated using a set of gas collectors, and then the gas–liquid–solids mixture is led upwards via a pipe to an expansion vessel where the gas is separated from the mixed liquor, which then drops back to the bottom of the reactor via another pipe. The extent of liquid–sludge recirculation depends on the gas production. Since 1990, the commercial EGSB systems (IC-EGSB and UFB-EGSB) found increasing application, particularly for brewery and potato processing wastewater, and also for wastewaters of chemical industries and paper mills. Over 60 full scale installations have been put in operation so far. However, these commercial EGSB systems have not yet been applied situations, which require an exceptionally good retention of viable biomass. Very likely the reactor then should be equipped with an improved sieve drum GSS device, as described above.

Contrary to UASB reactors in EGS systems virtually all the retained sludge is exposed to the wastewater, consequently employed, while also a substantially greater fraction of the immobilized organisms (consequently present inside the granules) participate in the degradation process. Clear evidence has been obtained that the prevailing substrate affinity of the granular sludge in these systems is extraordinaryly high.

Contrary to the real *anaerobic fluidized bed* (AnFB) *systems*, which use a mineral support material for bacterial attachment, such as fine sand (0.1–0.3 mm), basalt, biolite, or plastic carrier material (LI and SUTTON, 1981; HEIJNEN et al., 1992; EHLINGER, 1994) a complete fluidization of the sludge aggregates is not pursued in EGSB systems. It simply is not needed. Real FB systems are practically not feasible, because it is impossible to control the process of film buildup, viz. film attachment and de-attachment. In the dynamic process of film formation and de-attachment, aggregates of very different size and density will always be present, and consequently irrevocably a serious segregation of the various particles will occur. As a result the heavier (e.g., bare) carrier particles will accumulate in the lower part of the reactor as a stationary bed, while light (fluffy) aggregates such as the de-attached biofilms will dominate in the upper part, if not be washed out. The AnFB systems installed so far, like for instance the anaflux process (HOLST et al., 1997), in fact resemble more an expanded bed reactor than a real AnFB system. So, e.g., the anaflux system, although employing porous carrier particles of 0.5 mm diameter (density close 2.0) for bacterial attachment, is operated at superficial velocities of 5–10 m h^{-1}, consequently far lower than needed for fluidization of the carrier particles. However, contrary to commercially applied EGSB reactors it is considered necessary in the anaflux system to apply a rather elaborate procedure in order to retain as much as possible of the segregating immobilized viable biomass. In this procedure, the light aggregates (viz. the particles with too thick biofilms) are extracted from the internal sludge–solids separator (situated at top of the reactor) and then subjected to sufficient turbulent forces in some external device to shear off part of the biofilm. Next, both the media and the "free" biomass are returned to the reactor. So far over 25 anaflux systems have been installed for treating various types of wastewaters.

As a matter of fact the idea of using inert carrier materials (e.g., glass, plastic, ion exchanger beads, or diatomaceous earth) in expanded bed systems, was elaborated by JEWELL (1983) in the (AAFFEB) reactor, and by IZA et al. (1988) and BINOT et al. (1983) in the so-called *anaerobic attached fixed film expanded bed.* The AAFEB concept has not been applied at full scale, despite the satisfactory results obtained at pilot plant scale. The AAFEB concept in fact was overhauled by the UASB and EGSB system.

3 The Proper Application of High Rate ASB Reactors

The proper design and application of anaerobic treatment systems require adequate knowledge and understanding of the relevant process technological aspects of anaerobic reactor systems, and the (bio-)chemistry, the microbiology and ecology of the anaerobic digestion process, i.e.,

- anaerobic sludge immobilization, viz. granulation and granular sludge deterioration,
- kinetics of anaerobic conversion reactions in immobilized biomass,
- proper process conditions to be imposed on the system,
- proper environmental conditions for anaerobic degradation.

It is beyond the scope of this contribution to discuss all these aspects. However, considering its importance, something should be elucidated about the phenomenon of granular sludge formation and its maintenance; consequently the first and secondary start-up of ASB installations.

3.1 First and Secondary Start-Up of ASB-Reactors

The required high retention of highly active sludge in ASB systems can only be accomplished by the *immobilization of well balanced bacterial ecosystems.* As various syntrophic conversion reactions generally proceed in anaerobic treatment we need immobilization of the proper consortia. Only in this way will it be possible to minimize the detrimental effect of higher concentrations of specific intermediates and to optimize the environmental conditions, e.g., pH and redox potential.

The phenomenon of *granulation* is of big importance for the first start-up of an UASB reactor, i.e., the start-up using a flocculent anaerobic sludge of relatively poor specific methanogenic activity. This first start-up may require several months (e.g., DE ZEEUW, 1987; HULSHOFF POL and LETTINGA, 1986; HULSHOFF POL et al., 1987). Granulation in UASB reactors, but also in systems using some kind of mobile inert carrier, originates mainly from the fact that bacterial growth is delegated to a limited number of growth nuclei, consisting of inert organic or inorganic bacterial carrier materials and/or bacterial aggregates present in the seed sludge. The operation of the system during the granulation process should be such that finely dispersed bacterial matter is rinsed out of the reactor, because only then can growth-in be delegated to a limited number of nuclei on mainly soluble types of wastewaters, and under very different environmental conditions, e.g., thermophilic (WIEGANT and DE MAN, 1986; WIEGANT et al., 1986; VAN LIER et al., 1994) and also psychrophilic conditions (VAN DER LAST and LETTINGA, 1991). Although very beneficial, the formation of granular sludge is not a prerequisite for a good performance of UASB systems; as a matter of fact thick well settling flocculent sludge of a high methanogenic activity will suffice in many cases.

Since large quantities of high quality excess anaerobic sludge (granular) are becoming available from existing full scale installations, the drawback of the "slow" (first) start-up is vanishing. This high quality excess sludge represents an ideal seed material for start-up of new installations, even in the case of wastewater significantly different in composition, strength, etc. The practical experiences so far with *secondary start-ups*, i.e., defined as the start-up with *non-adapted* granular sludge, generally are very satisfactory. However, in view of the fact that the sludge generally is not adapted to the new wastewater, in specific situations problems may occur, e.g.

- deterioration of the sludge granules, e.g., falling apart of the sludge aggregates;
- attachment of fast growing filamentous acidogenic bacteria; this may happen if too little time was invested for the sludge to adapt to the new substrate, and/or too little pre-acidification is applied;
- $CaCO_3$ scaling in case of a high Ca^{2+} concentration in the wastewater; problems can be prevented by recycling effluent to the influent in order to precipitate Ca as $CaCO_3$ (VAN LANGERAK et al., 1977).

Regarding its importance considerable emphasis is put on research in this field (e.g., ALPHENAAR, 1994).

4 Applicability of Anaerobic Sludge Bed Reactor Systems

Although initially investigated for the treatment of mainly soluble and medium strength wastewaters, soon after the full scale implementation of the UASB system in 1974, it was demonstrated that the system was also applicable to more complex, i.e., partially soluble wastewaters and later to low strength wastewaters like domestic sewage (LETTINGA et al., 1993; VAN HAANDEL and LETTINGA, 1994), possibly even at ambient temperatures as low as approximately 10 °C (DE MAN, 1990; VAN DER LAST and LETTINGA, 1991). On the other hand it also become clear that under such more severe conditions, the design and operation of the system needs specific improvements. In this respect *EGSB systems* and *staged* (compartmentalized) *reactor concepts* can be mentioned. Based on recent experimental evidence it now looks justified to state that EGSB systems in various cases permit AnWT at quite extreme conditions, such as treatment

- of very low strength wastewaters with COD_{biod} values <100 mg L^{-1} (KATO et al., 1994,1997; REBAC et al., 1995, 1996);
- at temperatures <10 °C, certainly for soluble but to some extent even for partially soluble low strength wastewaters, including cold and very low strength industrial wastewaters, like malting wastewater (LETTINGA et al., 1983a, b; VAN DER LAST and LETTINGA, 1991; REBAC et al., 1997; VAN LIER et al., 1997);
- of wastewaters containing toxic substrates like lauric and capric acid (RINZEMA et al., 1989, 1993a), but not compounds like oleic acid (HWU and LETTINGA, 1997; HWU et al., 1997), and (milk) lipids (RINZEMA et al., 1993b; PETRUY and LETTINGA, 1997; SANDERS, personal communication);
- for sulfate reduction under thermophilic and very likely mesophylic conditions. Very promising results have already been obtained using a compound like methanol as electron donor (WEIJMA et al., unpublished results).

Anaerobic treatment may also become a quite attractive option for thermophilic temperature conditions, although *staged sludge anaerobic reactor* (SSAR) *systems* should be applied in that case (VAN LIER et al., 1994, 1996; LETTINGA, 1995). The separate modules of such SSAR systems can consist of ASB modules. The main feature of the SSAR concept comprises the development of a balanced sludge in each separate compartment (module), which means a sludge that by definition is in balance with the specific conditions prevailing in that module, including the various intermediates dominating in the reactor medium in that specific module. Particularly for the degradation of various types of xenobiotics such a concept is of crucial importance because it will enable anaerobic degradation of "difficult" compounds, including partially soluble, inhibitory, poorly biodegradable compounds (FIELD et al., 1995). *Adaptation, maintenance of optimal environmental conditions* and *use of the proper reactor concept* are the key factors for the application of AnWT to "difficult" wastewaters. Consequently, sufficient time must be invested for developing properly balanced, adapted immobilized bacterial consortia; this would also be quite time consuming. Once a balanced microbial community has evolved, the anaerobic degradation of higher fatty acids and lower molecular weight aromatic compounds, like benzoic acid, phenols, cresols and TPA generally will proceed smoothly. Bacteria involved in degrading higher fatty acids and simple aromatics require methanogenic and sulfate reducing bacteria as a synergistic partner, favoring the thermodynamics of compound degradation. Various xenobiotic pollutants, with multiple chloro, nitro and azo groups are persistent to aerobic degradation, while they are readily reduced by anaerobic consortia to "end products" like lower chlorinated aromatics or aromatic amines. Highly toxic compounds such as chloroform, carbon tetrachloride, nitroaromatics and azo dyes are readily reduced to less toxic compounds by unadapted anaerobic sludge (VAN EEKERT et al., 1995; DONLON et al., 1996, 1997; RAZO-FLORES et al., 1997). Some toxic compounds like wood extractives present in pulping wastewater are not biodegradable under anaerobic conditions but are readily degraded in aerobic conditions. For these cases *upfront dilution* (diluting the incoming influent to subtoxic concentrations with the effluent of the aerobic post-treatment reactor) can be applied to make anaerobic treatment of concentrated pulping liquors feasible (KORTEKAAS et al., 1994).

5 Conclusions

Anaerobic wastewater treatment processes based on the upflow sludge bed principle like UASB and EGSB reactors belong to the category of "grown-up technologies". The understanding of the factors controlling sludge immobilization through granular sludge growth and/or deterioration can be considered as satisfactory for most practical applications, even although still more has to be understood of the formation of balanced bacterial ecocolonies. Such balanced bacterial communities are of crucial importance in the degradation of more difficult compounds. The EGSB system, presumably because virtually all retained sludge and immobilized bacterial consortia in this system are in contact with wastewater, offers great practical prospects, particularly for treating very low strength wastewaters (COD $\ll 1{,}000$ mg L^{-1}) and for application at lower temperatures than 10 °C. A distinct improvement in the anaerobic reactor technology can be obtained by staging the reactor, in order to prevent mixing up of the sludge to approach a plug flow pattern. A well-operated staged process will provide a higher treatment efficiency, e. g., for more difficult compounds, and also a higher process stability. In each separate module optimal conditions can be maintained. This is particularly true for thermophilic systems where substrate (intermediate) inhibition is a relatively important factor. Anaerobic sludge bed processes are presently successfully applied for a rapidly growing number of industrial effluents, and increasingly also for domestic sewage, particularly under tropical conditions. For low strength and cold wastewaters EGSB reactors equipped with sieve drum gas–solid separator devices look very promising.

6 References

ALPHENAAR, P. A. (1994), Anaerobic granular sludge: characterization and factors affecting its performance, *PhD-Thesis*, Wageningen: Agricultural University.

BINOT, B. A., BOT, T., NAVEAU, H. P., NYNS, E. J. (1983), Biomethanation by immobilized fluidized cells, *Water Sci. Technol.* **15**, 103–115.

DE MAN, A. W. A. (1990), Anaerobic treatment of raw sewage using a granular sludge UASB-reactor, *Final Report of an Investigation at Pilot Plant Scale Dealing with the Assessment of the Use of Anaerobic Treatment in the Netherlands*, Wageningen: Agricultural University, Department of Environmental Technology, March 1990.

DE ZEEUW, W. J. (1987), Granular sludge in UASB-reactors, in: Granular Anaerobic Sludge; Microbiology and Technology, *Proc. GASMAT-Workshop*, October 25–27, Lunteren, pp. 132–146.

DONLON, B. A., RAZO-FLORES, E., LETTINGA, G., FIELD, J. A. (1996), Continuous detoxification, transformation and degradation of nitrophenols in upflow anaerobic sludge blanket (UASB) reactors, *Biotechnol. Bioeng.* **51**, 439–449.

DONLON, B. A., RAZO-FLORES, E., LUIJTEN, M., SWARTS, H., LETTINGA, G., FIELD, J. A. (1997), Detoxification and partial mineralization of the azo dye Mordant Orange 1 in a continuous upflow anaerobic sludge blanket reactor, *Appl. Microbiol. Biotechnol.* **47**, 83–90.

EHLINGER, F. (1994), Anaerobic biological fluidized beds: experiences in France. Preprints, 7th *Int. Symp. Anaerobic Digestion, 80–99*, Cape Town, South Africa, January 23–27, pp. 315–323.

FIELD, J. A., STAMS, A. J. M., KATO, M., SCHRAA, G. (1995), Enhanced biodegradation of aromatic pollutants in cocultures of anaerobic and aerobic bacterial consortia, *Antonie Van Leeuwenhoek* **67**, 47–77.

FRANKIN, R. J., KOEVOETS, W. A. A., VAN GILS, W. M. A., VAN DER PAS, A. (1992), Application of the BIOBED® upflow fluidized bed process for anaerobic wastewater treatment, *Water Sci. Technol.* **25**, 373–382.

HEIJNEN, J. J., VAN LOOSDRECHT, M. C. M., MULDER, A., TIJHUIS, L. (1992), Formation of biofilms in a biofilm air-lift suspension reactor, *Water Sci. Technol.* **26**, 647–654.

HOLST, T. C., TRUC, A., PUJOL, R. (1997), Anaerobic fluidized beds: ten years of industrial experience. Proc. 8th Int. Conf. Anaerobic Digestion, Sendai, Japan, May 25–29, pp. 142–149.

HULSHOFF POL, L. W., LETTINGA, G. (1986), New technologies for anaerobic wastewater treatment, *Water Sci. Technol.* **18**, 41–53.

HULSHOFF POL, L. W., HEYNEKAMP, K., LETTINGA, G. (1987), The selection pressure as a driving force behind the granulation of granular sludge, in: Granular Anaerobic Sludge; Microbiology and Technology, *Proc. GASMAT-Workshop*, Lunteren, October 25–27, pp. 146–153.

HWU, C. S., LETTINGA, G. (1997), Acute toxicity of oleate to acetate-utilizing methanogens in mesophilic and thermophilic anaerobic sludges, *Enzyme Microbiol. Technol.* **21**, 297–301.

HWU, C. S., VAN BEEK, B., VAN LIER, J. B., LETTINGA, G. (1997), Thermophilic high rate treatment of wastewater containing long-chain fatty acids: effect of washed out biomass recirculation, *Biotechnol. Lett.* **19**, 453–456.

IZA, J., GARCIA, P. A., HERNANDO, S., FDZ-POLANCO, F., SANZ, I. (1988), Anaerobic fluidized bed reactors (AFBR): performance and hydraulic behaviour, in: *Advances in Water Pollution Control* (HALL, E. R., HOBSON, P. N., Eds.), pp. 155–165. Oxford, New York: Pergamon Press.

JEWELL W. J. (1983), Anaerobic attached film expanded bed fundamentals, in: *Proc. 1st Int. Conf. Fixed Film Biological Processes, 1982* (WU, Y. C., Ed.), pp. 17–42. University of Pittsburg, PA.

KATO, M. T., FIELD, J. A., VERSTEEG, P., LETTINGA, G. (1994), The feasibility of expanded bed granular sludge bed reactors for the anaerobic treatment of low strength soluble wastewaters, *Biotechnol. Bioeng.* **44**, 469–479.

KATO, M. T., FIELD, J. A., LETTINGA, G. (1997), The anaerobic treatment of low strength wastewaters, *Proc. 8th Int. Conf. Anaerobic Digestion*, Sendai, Japan, May 25–29, pp. 356–363.

KORTEKAAS, S., DOMA, H. S., POTAPENKO, S. A., FIELD, J. A., LETTINGA, G. (1994), Sequenced anaerobic-aerobic treatment of hemp black liquors, *Water Sci. Technol.* **29**, 409–419.

LETTINGA, G. (1995), Anaerobic digestion and wastewater treatment systems. *Antonie Van Leeuwenhoek* **67**, 3–28.

LETTINGA, G., HULSHOFF POL, L. W. (1986), Advanced reactor design, operation and economy, *Water Sci. Technol.* **18**, 12, 99–108.

LETTINGA, G., VAN VELSEN, A. F. M., HOBMA, S. W., DE ZEEUW, W. J., KLAPWIJK, A. (1980), Use of the upflow sludge blanket (USB) reactor concept for biological wastewater treatment, *Biotechnol. Bioeng.* **22**, 699–734.

LETTINGA, G., HOBMA, S. W., HULSHOFF POL, L. W., DE ZEEUW, W. J., DE JONG, P., GRIN, P. et al. (1983a), Design, operation and economy of anaerobic treatment, *Water Sci. Technol.* **15**, 177–195.

LETTINGA, G., HULSHOFF POL, L. W., WIEGANT, W., DE ZEEUW, W. J., HOBMA, S. W., et al. (1983b), Upflow sludge blanket processes, *Proc. 3rd Int. Symp. Anaerobic Digestion*, Boston, MA, August 14–19, pp. 139–158.

LETTINGA, G., DE MAN, A., VAN DER LAST, R. M., WIEGANT, W., VAN KNIPPENBURG, K. et al., (1993), Anaerobic treatment of domestic sewage and wastewater, *Water Sci. Technol.* **27**, 67–73.

LI, A., SUTTON, P. M. (1981), Dorr–Oliver Anitron system fluidized bed technology for methane production from dairy waste, in: *Proc. Whey Products Institute Annual Meeting*, Chicago, JL, April.

PETRUY, R., LETTINGA, G. (1997), Digestion of a milk fat emulsion, *Biores. Technol.* **61**, 141–150.

RAZO-FLORES, E., LUYTEN, M., DONLON, B., LETTINGA, G., FIELD, J. A. (1997), Biodegradation of selected azodyes under methanogenic conditions, *Proc. 8th Int. Conf. Anaerobic Digestion*, Sendai, Japan, May 25–29, pp. 429–436.

REBAC, S., RUSKOVA, J., GERBENS, S., VAN LIER, J. B., STAMS, A. J. M., LETTINGA, G. (1995), High-rate anaerobic treatment of wastewater under psychrophillic conditions, *J. Ferment. Bioeng.* **1**, 499–506.

REBAC, S., VISSER, A., GERBENS, S., VAN LIER, J. B., STAMS, A. J. M., LETTINGA, G. (1996), The effect of sulphate on propionate and butyrate degradation in a psychrophilic anaerobic expanded granular sludge bed (EGSB) reactor, *Environ. Technol.* **17**, 997–1005.

REBAC, S., VAN LIER, J. B., JANSSEN, M. G. J., DEKKERS, F., SWINKELS, K. T. M., LETTINGA, G. (1997), High-rate anaerobic treatment of malting wastewater in a pilot-scale EGSB system under psychrophilic conditions, *J. Chem. Technol. Biotechnol.* **68**, 135–146.

RINZEMA, A., ALPHENAAR, A., LETTINGA, G. (1989), The effect of lauric acid shock loads on the biological and physical performance of granular sludge bed UASB reactors digesting acetate, *J. Chem. Technol. Biotechnol.* **46**, 257–266.

RINZEMA, A., ALPHENAAR, A., LETTINGA, G. (1993a), Anaerobic digestion of long chain fatty acids in UASB-reactors and expanded granular sludge bed reactors, *Proc. Biochem.* **28**, 527--537.

RINZEMA, A., LETTINGA, G., VAN VEEN, H. (1993b), Anaerobic digestion of triglyceride emulsions in expanded granular sludge bed with modified separators, *Environ. Technol.* **14**, 423–432.

VAN DER LAST, A. R. M., LETTINGA, G. (1991), Anaerobic treatment of domestic sewage under moderate climatic conditions using upflow reactors at increased superficial velocities, *Paper, 6th Int. Symp. Anaerobic Digestion*, Sao Paulo, May 12–16.

VAN EEKERT, M. H. A., VEIGA, M. C., FIELD, J. A., STAMS, A. J. M., SCHRAA, G. (1995), Removal of tetrachloro-methane by granular methanogenic sludge, in: *Bioremedation of Chlorinated Solvents* (HINCHEE, R. E., LEESON, A., SEMPRINI, S., Eds.), pp. 139–145. Columbus, Batelle Press.

VAN HAANDEL, A. C., LETTINGA, G. (1994), *Anaerobic Sewage Treatment*, p. 226. Chichester: John Wiley & Sons.

VAN LANGERAK, A., HAMELERS, B., LETTINGA, G. (1997), Influent calcium removal by crystallization re-using anaerobic effluent alkalinity. *Proc. 8th Int. Conf. Anaerobic Digestion*, Sendai, Japan, May 25–29, pp. 126–133.

VAN LIER, J. B., BOERSMA, F., DEBETS, M. M. W. H., LETTINGA, G. (1994), High rate thermophilic anaerobic wastewater treatment in compartmentalized upflow reactors, *Water Sci. Technol.* **30**, 251–261.

VAN LIER, J. B., GROENEVELD, N., LETTINGA, G. (1996), Characteristics and development of thermophilic methanogenic sludge in compartmentalized upflow reactors, *Biotechnol. Bioeng.* **50**, 115–124.

VAN LIER, J. B., REBAC, S., LENS, P., VAN BIJNEN, F., OUDE ELFERINK, S. J. W. H. et al. (1997), Anaerobic treatment of partially acidified wastewater in a two-stage expanded granular sludge bed (EGSB) system at 8°C, *Water Sci. Technol.* **36**, 317–324.

VELLINGA, S. H. J., HACK, P. J. F. M., VAN DER VLUGT, A. J. (1986), New type high rate anaerobic reactor, *Proc. Water Treatment Conf. "Anaerobic Treatment a Grown-up Technology"*, Aquatech, Amsterdam, September 15–19, pp. 547–562.

VERSPRILLE, A. I., FRANKIN, R. J., ZOUTBERG, G. R. (1994), Biobed, a succesful cross-breed between UASB and fluidized bed, *Proc. 7th Int. Symp. Anaerobic Digestion*, Cape Town, South Africa, pp. 587–590.

WIEGANT, W. M., DE MAN, A. W. A. (1986), Granulation of biomass in thermophilic anaerobic sludge blanket reactors, *Biotechnol. Bioeng.* **28**, 718–727.

WIEGANT, W. M., HENNINK, M., LETTINGA, G. (1986), Separation of the propionate degradation to improve the efficiency of thermophilic anaerobic treatment of acidified wastewaters, *Water Res.* **20**, 517–524.

YSPEERT, P., VEREIJKEN, T. L. F. M., VELLENGA, S., DE VEGT, A. (1993), The IC-reactor for the anaerobic treatment of industrial wastewater, *Proc. 1993 Food Industry Environmental Conf.*, Atlanta, GA, pp. 14–16.

ZOUTBERG, G. R., HEYNEKAMP, K., VERSPRILLE, B. I. (1997), Anaerobic treatment of chemical wastewater in biobed EGSB-reactors, *Proc. 8th Int. Conf. Anaerobic Digestion*, Sendai, Japan, May 25–29, pp. 175–182.

26 Modeling of Biogas Reactors

HERBERT MÄRKL

Hamburg, Germany

List of Abbreviations

AC	acetic acid
BTR	biogas tower reactor
COD	chemical oxygen demand
DMG	dimethylglycine
DS water	desalinated water
HDPE	high density polyethylene
HPLC	high pressure liquid chromatography
TOC	total organic carbon
TSS	total suspended solids
UASB	upflow anaerobic sludge blanket

List of Symbols

A	area	m^2
a	interfacial area per volume	$m^2\ m^{-3}$
c_i	concentration of component i	$mol\ L^{-1}$
D	diffusion coefficient	$m^2\ s$
d_b	diameter, bubble	m
G	gas volume	m^3
H_i	Henry coefficient of component i	$mol\ L^{-1}\ Pa^{-1}$
K_i	kinetic constant of component i	$mol\ L^{-1}$
$K_{D,i}$	dissociation coefficient of component i	$mol\ L^{-1}$
k	proportional coefficient	h^{-1}
k_La	volume specific mass transport coefficient	h^{-1}
M_{prod}	produced biomass	$g\ h^{-1}$
$\dot{M}_{sedi,i}$	sedimenting mass flow from module i	$g\ h^{-1}$
p_i	partial pressure of component i	Pa
R	universal gas constant, $R = 8.314\ kJ\ kg^{-1}\ K^{-1}$	$kJ\ kg^{-1}\ K^{-1}$
SS_i	concentration of suspended solids in module i	$g\ L^{-1}$
T	absolute temperature	K
TOC	total organic carbon	$g\ L^{-1}$
u	velocity	$m\ s^{-1}$
w	velocity	$m\ s^{-1}$
V	volume	m^3
$\dot{V}_{exchange}$	exchange flow rate	$m^3\ h^{-1}$
$\dot{V}_{bubble}$	volumetric gas production rate	$m^3\ h^{-1}$
$\dot{V}_{feed}$	feed flow rate	$m^3\ h^{-1}$
$\dot{V}_{gas}$	gas flow rate	$m^3\ h^{-1}$
$\dot{V}_{recyc.}$	recycling gas flow rate	$m^3\ h^{-1}$
x	concentration of biomass	$g\ L^{-1}$
Y	yield coefficient	$g\ mol^{-1}$

Greek

γ_i	activity coefficient [–]
ε_G	gas hold up [–]
$\Delta\varepsilon$	difference of gas hold up [–]
τ	boundary layer renewal time s
μ	growth rate h^{-1}

Special Symbols of Chemical Components

Ac	total acetic acid
HAc	undissociated acetic acid
Ac^-	dissociated acetic acid
Pro	total propionic acid
HPro	undissociated propionic acid
Pro^-	dissociated propionic acid

1 Introduction

As early as 1973 GRAEF and ANDREWS (1973) published a mathematical model of the anaerobic digestion of organic substrates. The authors assumed that the conversion of volatile organic acids by the methanogenic bacteria to methane was the rate-limiting step in the sequence of biological reactions and that all volatile acids can be represented as acetic acid. Acetic acid is dissociated to a large extent at the relevant pH range between 6.6 and 7.4. The authors assume that only the undissociated form of acetic acid is the limiting substrate for the microbial production of methane.

The pH value in this mathematical model is calculated from the ion balance assuming electroneutrality in the fermentation broth. In this context the concentration of dissolved CO_2, which itself is partly dissociated to HCO_3^- and CO_3^{2-}, has also to be known. In the mathematical model of GRAEF and ANDREWS, the concentration of dissolved CO_2 is calculated on the basis of the mass transport from the liquid phase to the gas phase of the system. This transport phenomenon proved to be very important for the mathematical modeling of technical biogas reactors as will be shown later in this chapter.

The mathematical model of GRAEF and ANDREWS describes a homogenous completely mixed system as far as the liquid phase of the reactor is concerned.

1.1 Elements of the Mathematical Model

In this chapter it will be shown that the mentioned elements of the model of GRAEF and ANDREWS are an excellent basis for the quantitative analysis of modern biogas reactor systems. In the research work of different groups following this first publication it was proved that more knowledge is necessary, especially about the microbial kinetics of the biogas production (Sect. 3). Most of the real biogas reactors are far from being completely mixed systems as demonstrated by the example of a biogas tower reactor (BTR) (Fig. 1). This type of reactor is described in more detail by MÄRKL and REINHOLD (1994). The typical features of the BTR are its tower shape, the modular structure, and the internal installations. Gas collecting devices are used to withdraw the fermentation gas from different levels of the reactor. These gas collectors separate the reactor into modules along the height. By

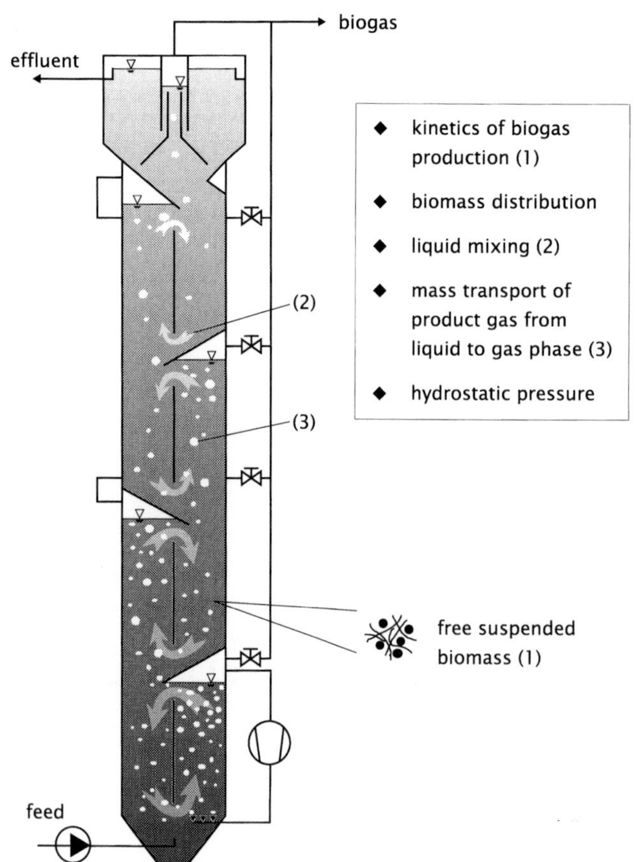

Fig. 1. Elements of a mathematical model for a biogas tower reactor (BTR).

means of these devices which are equipped with valves, the gas loading and the mixing intensity can be controlled separately in each module. To avoid flotation of active biomass due to excessive gas load, an effective biomass accumulation is generated within the reactor.

The biomass is represented in this reactor by free suspended microorganisms which are associated in the form of sedimentable more or less loose pellets (flocs). One very important point, when modeling such a reactor, will be to describe the local distribution of active biomass within the reactor (Sect. 4.2).

In a mathematical model the mixing behavior of the liquid phase is of similar importance (Sect. 4.1). The mixing of the BTR is caused by internal airlift loop units. Since the reactor is designed in the shape of a tower and is only fed at the bottom (Fig. 1), the question of mixing and supplying the microorganisms in each part of the reactor with nutrients is very important during the scale-up of such a system. Besides, hydraulic mixing is important to avoid toxic concentrations of substrate near the inlet of the reactor.

As already pointed out by GRAEF and ANDREWS (1973), the mass transport of product gas from the liquid phase to the gas phase is an essential element of the mathematical model. Local transport data as a function of the hydrodynamics for the example of the biogas tower reactor will be given for CO_2 and H_2S in Sect. 5 of this chapter. The last element of a mathematical model which is of high importance, especially when discussing tall reactors, is represented by the influence of the local hydrostatic pressure on the kinetics of biogas generation (Sect. 6).

1.2 Scale-Up Strategy

Parameters of the mathematical model elements must be identified by experiments, and the models have to be evaluated with respect to their suitability for the design of technical scale biogas reactors. The scale-up strategy is demonstrated in Fig. 2 with explanations in Tab. 1. Basic experiments are done in the laboratory. The hydrodynamic behavior and the

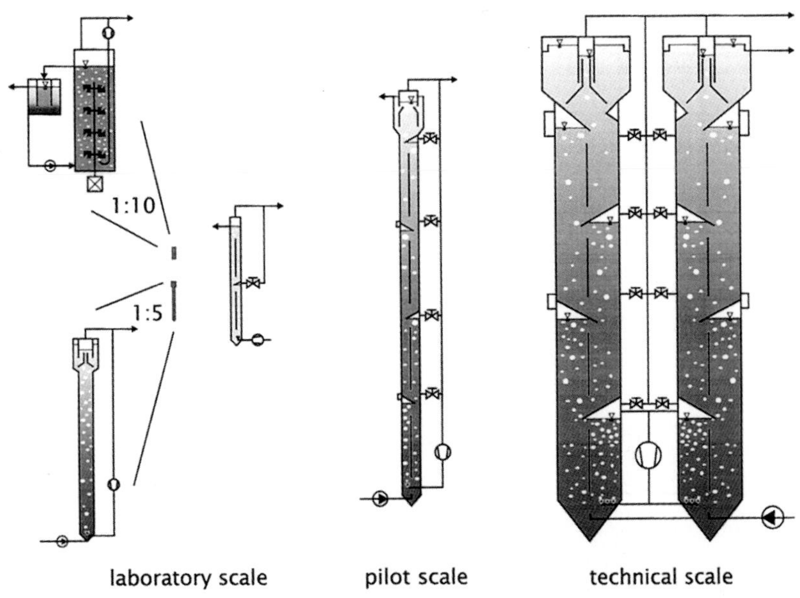

Fig. 2. Strategy for scale-up of the biogas tower reactor, specification of reactors according to Tab. 1.

Tab. 1. Specification of Biogas Reactors Used for the Experiments

Reactors	Experiments	Kinetics	Hydrodynamic and Liquid Mixing	Local Distribution of Biomass	Mass Transport of Product Gas	Influence of Hydrostatic Pressure
Laboratory	Stirred tank reactor, active volume 16 L	X			X	X
	Two UASB reactors with biogas recirculation, active volume 36.5 L and 39.5 L, height 1.7 m and 1.95 m respectively	X			X	
	Laboratory tower reactor, height 6.5 m, diameter 0.4 m, two modules		X	X		
Pilot	Pilot scale tower reactor, height 20 m, diameter 1 m, volume 15 m^3, four modules, photo of reactor shown in Fig. 3	X	X	X	X	X

sedimentation of non-active biomass were studied first in a small reactor unit with a height of 6.5 m. This laboratory reactor has a diameter of 0.4 m and consists of only 2 modules. To understand the influence of the height of a reactor, a pilot scale biogas reactor was built at the site of a company producing bakers' yeast (Fig. 3). The reactor was designed on the basis of the laboratory scale experiments and investigations on the kinetics of wastewater digestion from bakers' yeast production. The characteristic data of the wastewater are given in Tab. 2. The active volume of the reactor is 15 m^3. It is divided into 4 modules and at the top of the reactor a settling zone is integrated to maintain a high sludge concentration. The pilot reactor is well equipped with instruments for measuring concentrations and velocities along the height of the reactor. Because the pilot scale reactor reaches almost the height of a technical reactor (in Fig. 2 two reactor units with a height of 25 m and a diameter of 3.5 m are shown as examples) it is assumed that the mathematical model, which is evaluated on the pilot scale with respect to hydraulics and taking into account the influence of hydrostatic pressure, will permit prediction of the behavior of a technical scale reactor.

Fig. 3. Biogas tower reactor, pilot scale, at the site of Deutsche Hefewerke in Hamburg, built by Preussag Wassertechnik GmbH, Bremen.

2 Measuring Techniques

In order to understand the microbial degradation process and the system behavior of a biogas reactor and to establish reliable mathematical models, measuring techniques must be available – if possible – to on-line analyze special substances in the fermentation broth. Therefore, some new measuring techniques have been developed during the last years.

2.1 On-Line Measurement Using a Mass Spectrometer

The anaerobic treatment of wastewater loaded with high concentrations of sulfate is

Tab. 2. Characteristic Data of the Wastewater Used for the Experiments

	Average	Range	Unit
Total organic carbon (TOC)	13.8	11–19	$g\ L^{-1}$
Chemical oxygen demand (COD)	31	25–36	$g\ L^{-1}$
COD (incl. betaine)	44		$g\ L^{-1}$
Betaine	5.7	4–10	$g\ L^{-1}$
Sulfate	3.9	2.5–6.5	$g\ L^{-1}$
Acetic acid	58	0.6–120	$mmol\ L^{-1}$
Total suspended solids (TSS)	2.5	0.8–3	$g\ L^{-1}$
pH	4.5	4.3–5	–

often accompanied with problems due to the inhibition by hydrogen sulfide (H_2S). It is well established that the inhibition is caused by undissociated hydrogen sulfide present in the liquid phase, and not by the total sulfide content. This concentration of undissociated hydrogen sulfide depends on sulfate concentration in the wastewater, pH, and ionic strength. The concentration of undissociated hydrogen sulfide in the liquid phase undergoes great changes in the pH range used for anaerobic digestion, which is between 6.6 and 7.4. Since the actual concentration of dissolved H_2S cannot be calculated on the basis of the H_2S partial pressure in the biogas, because gas phase and liquid phase are not in equilibrium as will be demonstrated in Sect. 5, this concentration has to be measured.

Measurement on the basis of ion-sensitive electrodes proved to be very problematic especially with respect to their stability for a longer period of time. MEYER-JENS et al. (1995) developed a new type of probe connected to a mass spectrometer. The probe consists of a small tube (outer diameter 0.9 mm, inner diameter 0.7 mm) of polydimethylsilicon. The tube can directly be inserted into the fermentation liquid of the biogas reactor. The H_2S dissolved in the liquid phase penetrates through the membrane and will be transported by a carrier gas (technical nitrogen) to the mass spectrometer.

A very simple silicon membrane probe is shown in Fig. 4. The silicon membrane tube is arranged in a short bypass loop outside of the reactor. The fermentation liquid flows outside of the tube while the carrier gas passes inside. In Fig. 5 calibration of the measurement system is demonstrated. The calibration curve is perfectly linear within the relevant H_2S concentrations. The different gradient in the case of desalinated water and wastewater indicates the difference in the solubility (Henry coefficient) of H_2S in desalinated water (78.2 mmol L^{-1} bar^{-1}) and wastewater (68.5 mmol L^{-1} bar^{-1}) at 37 °C (data given by POLOMSKI, 1998).

The measurement system can also serve well for the detection of dissolved CO_2 in the biogas reactor. Fig. 6 shows the calibration curve for CO_2 in a sample of desalinated water. Also in the case of CO_2 the solubility (Henry coefficient) for wastewater ($2.21 \cdot 10^{-4}$ mmol L^{-1} Pa^{-1}) is 12% smaller than the literature data given for desalinated water ($2.52 \cdot 10^{-4}$ mmol L^{-1} Pa^{-1}). MATZ and LENNEMANN (1996) used a similar silicon membrane probe for the on-line monitoring of biotechnological processes by gas chromatographic mass spectrometric analysis. The sensitivity of the probe system could be improved considerably by an instationary heating of the membrane to a temperature of 200 °C. This quasi on-line measuring procedure has a sampling period of 15 min between two successive measuring cycles. Substances like dimethylsulfide, butanethiol, cresol, phenol, ethylphenol, and indole at concentrations between 10 and 200 ppm were detected.

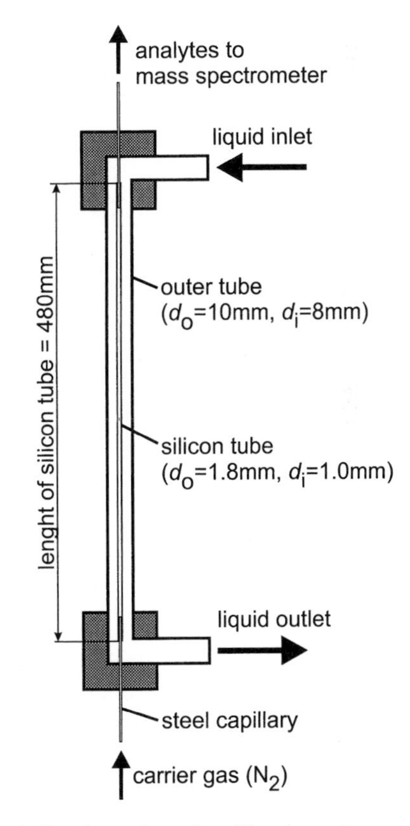

Fig. 4. Configuration of a silicon membrane probe operated in a short bypass at the reactor (after POLOMSKI, 1998).

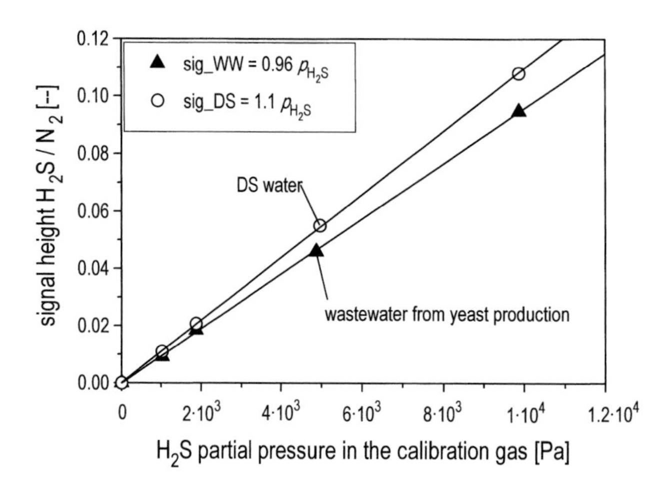

Fig. 5. Calibration of the silicon membrane probe coupled with a mass spectrometer measuring the concentration of dissolved H_2S in desalinated water (DS water) and in the wastewater of bakers' yeast production. The ordinate gives the value of the signal with reference to the N_2 signal of the carrier gas. At the abscissa the partial pressure of H_2S is denoted, with which the liquid is in equilibrium (after POLOMSKI 1998).

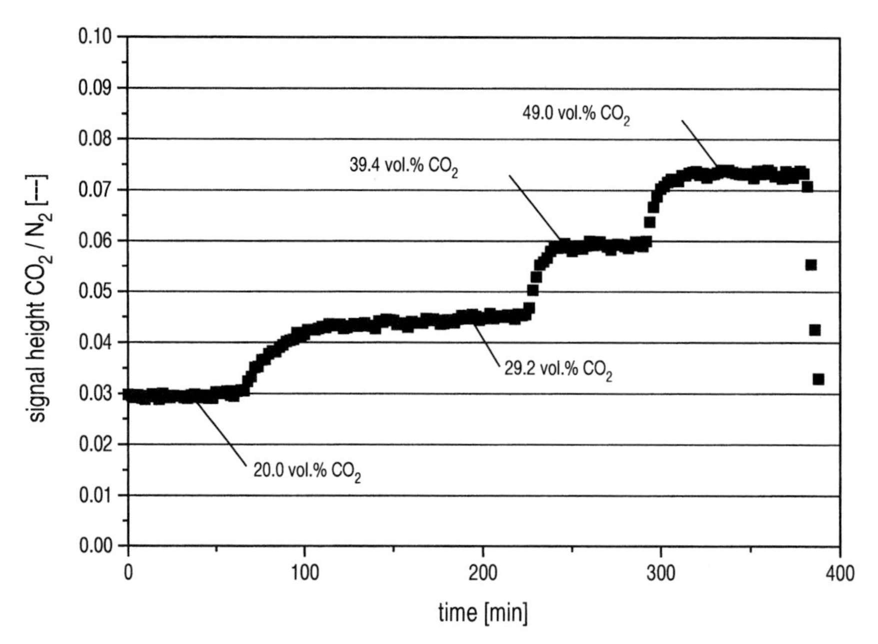

Fig. 6. Calibration of the silicon membrane probe coupled to mass spectrometer measuring the dissolved CO_2 in desalinated water (after POLOMSKI, 1998).

2.2 On-Line Monitoring of Organic Substances with High-Pressure Liquid Chromatography (HPLC)

ZUMBUSCH et al. (1994) reported the simultaneous on-line determination of betaine (trimethylglycine), N,N-dimethylglycine (N,N-DMG), acetic and propionic acid during the anaerobic fermentation of wastewater from bakers' yeast production. Ammonia and chloride could also be measured. The determinations were carried out by means of isocratic cation exchange chromatography. For the detection of betaine and N,N-DMG an ultraviolet detector was used. All other substances were detected by conductivity.

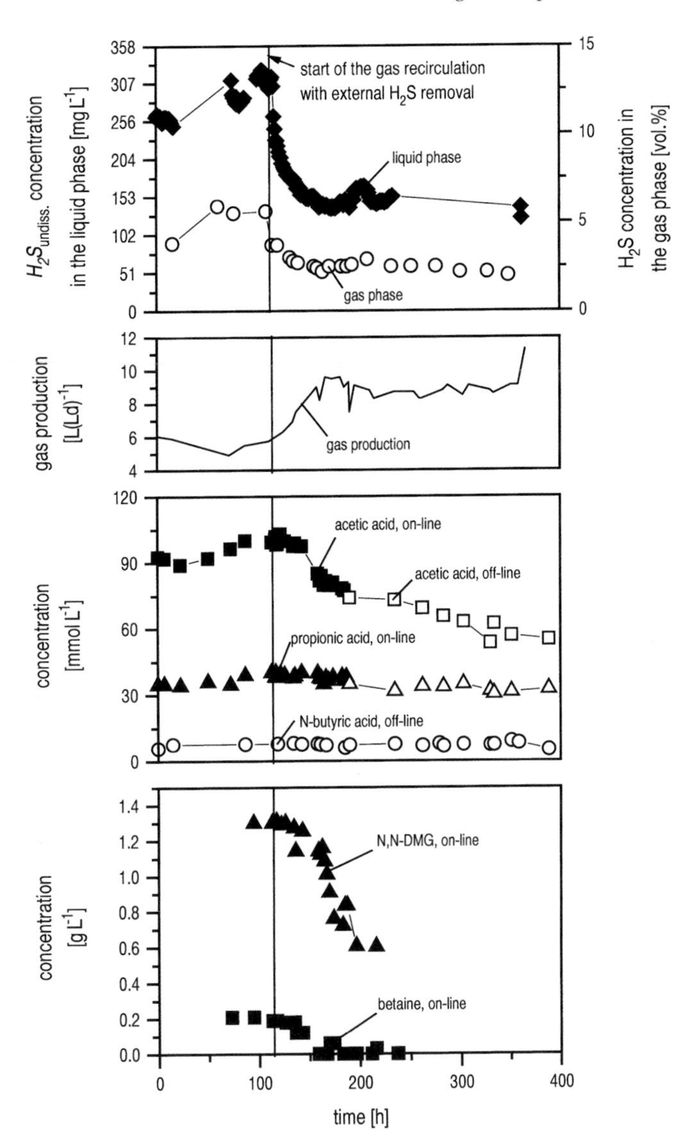

Fig. 7. Combined measurement with the silicon membrane probe mass spectrometer and on-line high-pressure liquid chromatography in a laboratory UASB reactor with and without elimination of H_2S in an external biogas recirculation loop over a period of some hundred hours (after POLOMSKI, 1998). On-line measurements were carried out in cooperation with ZUMBUSCH et al., 1994). H_2S concentrations were detected by the silicon membrane probe and mass spectrometer.

Besides other organic components such as acetic and propionic acid, betaine is the dominant organic substance in the wastewater. Up to 33% of the TOC is represented by betaine. The use of HPLC analysis combined with an automated ultrafiltration setup for on-line process monitoring gives interesting information about the dynamic behavior of the anaerobic process as demonstrated in Fig. 7. The graphs show the effect of hydrogen sulfide elimination out of the fermentation broth. Concentration of undissociated H_2S was decreased by a factor of 2 (from 300–150 mg L^{-1}). During the same time biogas production increased. The concentration of acetic acid decreased while the other acids measured remained more or less stable. Betaine was only completely degraded at lower H_2S concentrations. Interestingly, also the concentration of N,N-DMG, which is a metabolic product of the anaerobic degradation of betaine, is reduced to a lower level.

3 Kinetics

The anaerobic degradation of organic substances is performed in a sequence of biological reactions in a synthrophic cooperation of different microorganisms (Fig. 8). In a first step complex biopolymers (lipids, proteins, carbohydrates) are split into the monomers which can be assimilated by the microbial cell. This first breakdown is catalyzed by extracellular enzymes. In a second step fermentative bacteria form hydrogen, carbon dioxide, acetic acid, propionic acid and further intermediary products, predominantly volatile fatty acids.

Acetic acid, carbon dioxide, and hydrogen can directly be converted to methane and carbon dioxide (biogas) by methanogenic bacteria. Propionic acid as well as the other intermediates cannot directly be transformed to biogas. In an intermediate step the so-called acetogenic bacteria have to generate acetic acid, carbon dioxide, and hydrogen which then can be converted to biogas again. THAUER et al. (1977) showed that for thermodynamic reasons the degradation of propionic acid is only possible at very low concentrations of hydrogen (1 Pa). From this argument it is clear that the acetogenic reaction which generates hydrogen can only be performed by a direct and close synthrophic cooperation with methanogenic bacteria which consume hydrogen. Sulfate, a frequent component in almost all food industries' wastewater, is almost totally converted to hydrogen sulfide.

The dynamic behavior of the system is mainly controlled by the conversion of acetic acid to methane and carbon dioxide. This reaction is rather slow and, therefore, it is the bottleneck of the anaerobic digestion as demonstrated by experiments like those shown in Fig. 9. Whey which contains about 5% of lactose is continuously digested to biogas. At the beginning of the experiment the concentration of the different volatile acids denoted in Fig. 9 represent a steady state. In this situation the reactor was heavily overloaded during the next 2.5 h. As a result the concentration of acetic acid increased by a factor of 5. The concentrations of the other acids remained more or less constant. This finding supports the idea of GRAEF and ANDREWS (1973) that the conversion of acetic acid to methane and CO_2 is the rate-limiting step and that a mathematical model of the anaerobic degradation process can be reduced to this dominating reaction.

This simple idea is expressed in Fig. 10 and Eqs. (1 and 2) which describe the concentration of microorganisms x and the concentration of acetic acid Ac as a function of time.

$$\frac{\mathrm{dx}}{\mathrm{dt}} = \mu(c_{\mathrm{Ac}}, c_{\mathrm{i}})\,\mathrm{x} - \frac{\dot{V}_{\mathrm{feed}}}{V_{\mathrm{reactor}}}\,\mathrm{x} \tag{1}$$

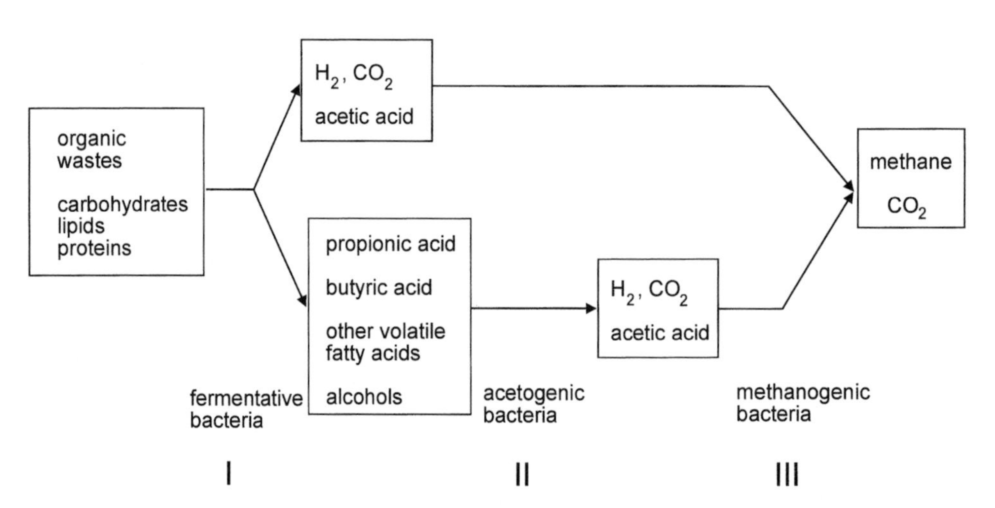

Fig. 8. Pathway of anaerobic biodegradation.

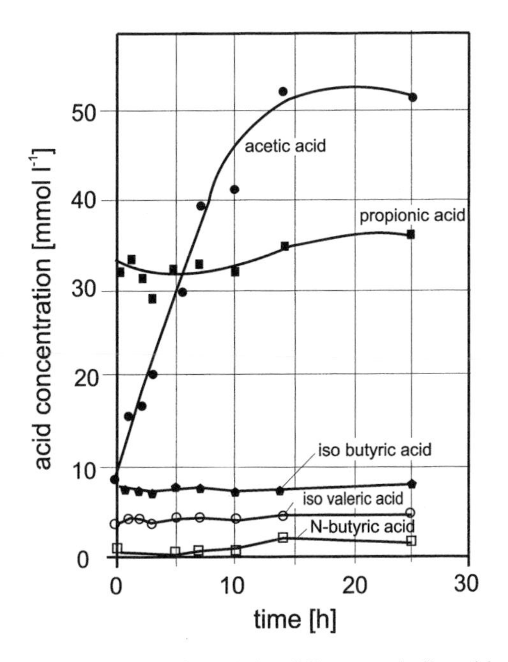

Fig. 9. Concentration of the different volatile acids as a function of the time during the continuous anaerobic degradation of whey, a by-product of cheese production (after MÄRKL et al., 1983). The experiments were performed in a 20 L stirred tank reactor. The graph shows the dynamic behavior of the system after being heavily overloaded with whey during the first 2.5 h of the experiment.

$$\frac{dc_{\mathrm{Ac}}}{\mathrm{dt}}=\frac{\dot{V}_{\mathrm{feed}}}{V_{\mathrm{reactor}}}(c_{\mathrm{Ac,e}}-c_{\mathrm{Ac}})-\frac{1}{\mathrm{Y}}\,\mu(c_{\mathrm{Ac}},c_{\mathrm{i}})\,\mathrm{x} \quad (2)$$

The reactor is directly fed by acetic acid $c_{\mathrm{Ac,e}}$. The growth rate of the microorganisms is denoted as $\mu(c_{\mathrm{Ac}}, c_{\mathrm{i}})$ and the biogas production can be assumed to be directly proportional to μ. At a first glance this model looks very simple. The real problem when solving the equation is that we need information on the kinetics of the reaction. The growth rate μ is a function of the acetic acid concentration c_{Ac} but may also depend on the concentration of other substances c_{i}. Before going into detail of formulating this kinetic equation one should understand more about the physico-chemical state of the fermentation broth.

Acetic acid is dissociated to a large extent in the relevant pH range between 6.6 and 7.4. The pH value is calculated on the basis of the activity of H^+, which equals $\gamma_1 \cdot c_{H^+}$.

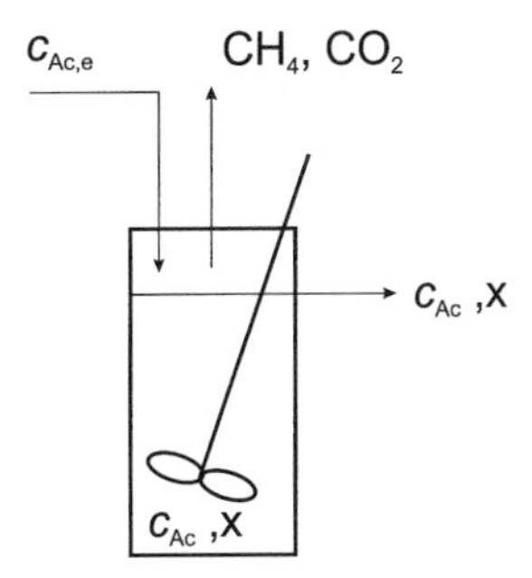

Fig. 10. Mathematical model of the anaerobic digestion in a stirred tank reactor. The reactor is directly fed by acetic acid (concentration in the inlet: $c_{\mathrm{Ac,e}}$). The concentration of acetic acid in the reactor is c_{Ac}, x denotes the concentration of biomass.

The concentration of H^+ ions c_{H^+} can be calculated according to Eq. (3) which represents the electroneutrality of the fermentation broth.

$$\begin{aligned} &c_{\mathrm{OH}^-}+c_{\mathrm{Ac}^-}+c_{\mathrm{Pro}^-}+c_{\mathrm{HCO}_3^-}\\ &\quad +c_{\mathrm{HS}^-}=c_{\mathrm{NH}_4^+}+c_{\mathrm{H}^+}+Z \end{aligned} \quad (3)$$

The amount of negative and positive ions must be balanced.

The most important ion concentrations in this calculation are the concentrations of dissociated acetic acid c_{Ac^-} and propionic acid c_{Pro^-}, OH^- ion c_{OH^-}, hydrocarbon ion $c_{\mathrm{HCO}_3^-}$, and dissociated hydrogen sulfide c_{HS^-}. Besides the H^+ ion, the ammonium ion with its concentration $c_{\mathrm{NH}_4^+}$ has a positive charge. Z represents the pool of not extra specified ions which are necessary to balance the equation and are assumed to be constant during the different stages of the fermentation.

The concentrations of these ions can be calculated according to Eqs. (4–14).

$$\mathrm{HAc}\leftrightarrows\mathrm{Ac}^-+\mathrm{H}^+ \qquad c_{\mathrm{Ac}}=c_{\mathrm{HAc}}+c_{\mathrm{Ac}^-} \quad (4)$$

$$\mathrm{HPro}\leftrightarrows\mathrm{Pro}^-+\mathrm{H}^+ \qquad c_{\mathrm{Pro}}=c_{\mathrm{HPro}}+c_{\mathrm{Pro}^-} \quad (5)$$

$$\begin{aligned} &(1)\quad \mathrm{H_2O}+\mathrm{CO_2}\leftrightarrows\mathrm{HCO_3^-}+\mathrm{H}^+\\ &(2)\quad \mathrm{HCO_3^-}\leftrightarrows\mathrm{CO_3^{2-}}+\mathrm{H}^+\\ &\qquad c_{\mathrm{C,tot}}=c_{\mathrm{CO_2}}+c_{\mathrm{HCO_3^-}}+c_{\mathrm{CO_3^{2-}}} \end{aligned} \quad (6)$$

$$\begin{aligned}&(1)\quad H_2S \leftrightharpoons HS^- + H^+\\&(2)\quad HS^- \leftrightharpoons S^{2-} + H^+\\&\qquad c_{S,tot} = c_{H_2S} + c_{HS^-} + c_{S^{2-}}\end{aligned} \tag{7}$$

$$\begin{aligned}&H_2O + NH_3 \leftrightharpoons NH_4^+ + OH^-\\&\qquad c_{N,tot} = c_{NH_3} + c_{NH_4^+}\end{aligned} \tag{8}$$

$$c_{OH^-} = \frac{1}{c_{H^+}} \cdot 10^{-14}\ \mathrm{mol^2\,L^{-2}} \tag{9}$$

$$c_{Ac^-} = \frac{K_{D,Ac}}{K_{D,Ac} + c_{H^+}\,\gamma_1^2}\,c_{Ac} \tag{10}$$

$$c_{Pro^-} = \frac{K_{D,Pro}}{K_{D,Pro} + c_{H^+}\,\gamma_1^2}\,c_{Pro} \tag{11}$$

$$c_{HCO_3^-} = \frac{K^I_{D,CO_2}}{K^I_{D,CO_2} + c_{H^+}\,\gamma_1^2}\,c_{C,tot} \tag{12}$$

$$c_{HS^-} = \frac{K^I_{D,H_2S}}{K^I_{D,H_2S} + c_{H^+}\,\gamma_1^2}\,c_{S,tot} \tag{13}$$

$$c_{NH_4^+} = \frac{K^I_{D,NH_3}}{K^I_{D,NH_3} + c_{OH^-}\,\gamma_1^2}\,c_{N,tot} \tag{14}$$

As an example the acetic acid HAc is dissociated to Ac^- and H^+ according to Eq. (4). The amount of total acetic acid (concentration of the total acetic acid: c_{Ac}) is calculated by the sum of dissociated (c_{Ac^-}) and undissociated species (c_{HAc}). From these two equations and the knowledge of the dissociation constant $K_{D,Ac}$ and the activity coefficient γ_1 the concentration of the dissociated species can be calculated according to Eq. (10). The activity coefficient is calculated by an approximation after DAVIS which can be found in LOEWENTHAL and MARAIS (1976). The activity coefficients depend on the ionic strength which was determined to be 0.235 mol L^{-1} in the case of wastewater from bakers' yeast production. The calculation of the dissociation constants was performed after BEUTIER and RENON (1978). Data for the discussed wastewater are given in Tab. 3.

The calculation in the case of the dissociation of CO_2 and H_2S is principally more complex because it is performed in two steps according to Eqs. (6 and 7) which are qualitatively shown in Fig. 11. But from these graphs it is clear that both, the concentrations of CO_3^{2-} and S^{2-} can be neglected in the relevant range of pH between 6.6 and 7.4. Therefore, calculation can be handled like a 1-step dissociation. Eqs. (3, 9–14) give the pH value in an implicit form. The pH can be evaluated by a stepwise iterative procedure.

3.1 Acetic Acid, Propionic Acid

Many experiments have been reported on the growth rate of acetic acid converting organisms with respect to the concentration of acetic acid in the fermentation broth. A reliable collection of those data is presented in Fig. 12. Because it is very difficult to measure the growth rate of methane producing organisms directly (in the case of a mixed culture it may be impossible) and because it is assumed that this growth rate is directly proportional to the methane production rate, usually the latter value is taken as a measure of bacterial activity. In Fig. 12 the methane production rate, related to the maximum methane production rate of a biogas reactor, is denoted at the ordinate of the graph.

At the abscissa the concentration of the total acetic acid measured in the fermentation broth is denoted. Data of different systems are collected. WITTY and MÄRKL (1986) reported the digestion of waste mycelium from antibiot-

Tab. 3. Equilibrium Constants for Wastewater of Bakers' Yeast Production, Temperature 37.4 °C

			Activity coefficients			
γ_1		0.73		monovalent ion		
γ_2		0.28		bivalent ion		
			Dissociation constant [mol L^{-1}]			
$K_{D,Ac}$	$K_{D,Pro}$	$K^I_{D,Ac}$	K^{II}_{D,CO_2}	K^I_{D,H_2S}	K^{II}_{D,H_2S}	K_{D,NH_3}
$1.17 \cdot 10^{-5}$	$1.3 \cdot 10^{-5}$	$4.94 \cdot 10^{-7}$	$5.79 \cdot 10^{-11}$	$1.44 \cdot 10^{-7}$	$2.7 \cdot 10^{-14}$	$6.83 \cdot 10^{-6}$

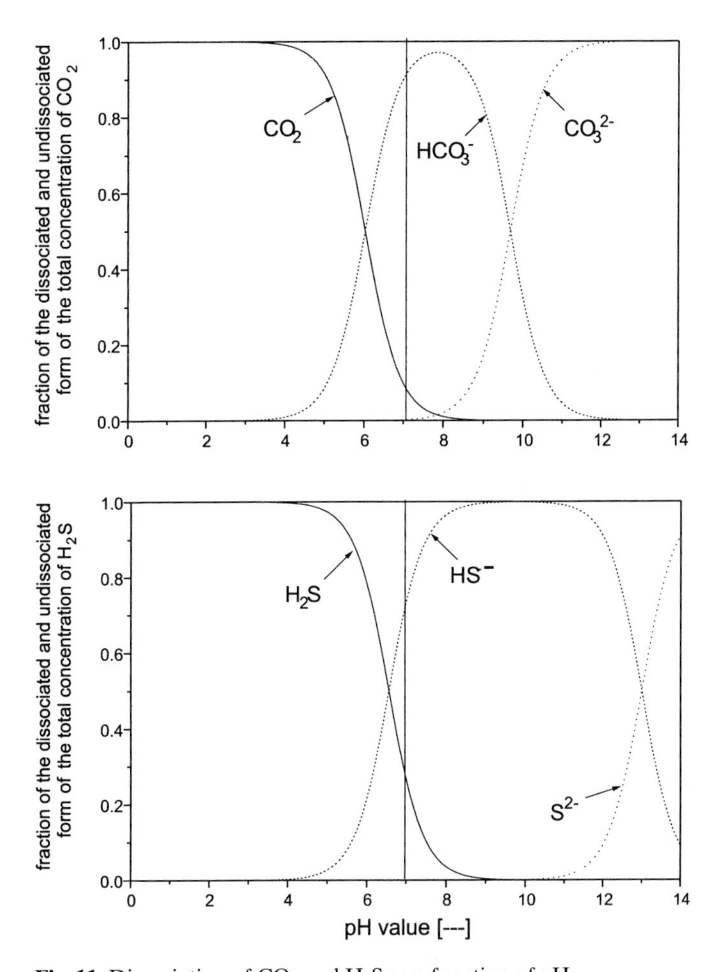

Fig. 11. Dissociation of CO_2 and H_2S as a function of pH.

ics production, THERKELSEN and CARLSON (1979) and THERKELSEN et al. (1981) digested a complex artificial substrate and AIVASIDIS et al. (1982) studied a pure culture of *Methanosarcina barkeri* which was directly fed with acetic acid. Due to the complexity of the synthrophic reaction system and also due to the instable nature of the biogas reaction the data points reflecting the experimental results of the different authors are somewhat scattered. But apart from this phenomenon, it can be stated, that the general behavior of the different systems differs to a large extent from one to the other.

In Fig. 13 exactly the same experimental data are presented. In this case at the abscissa only that part of the acetic acid is denoted which is not dissociated.

The undissociated acetic acid can be calculated with the help of Eq. (15) if the total acetic acid and the pH value are known.

$$c_{\mathrm{HAc}} = \frac{c_{\mathrm{H^+}}\,\gamma_1^2}{K_{\mathrm{D,Ac}} + c_{\mathrm{H^+}}\,\gamma_1^2}\,c_{\mathrm{Ac}} \tag{15}$$

It can easily be seen, that this parameter integrates all experimental data into a clear picture. From this finding it was derived that the microorganisms use and only "see" the undissociated form of acetic acid. The graph drawn in Fig. 13 has the form of a Michaelis–Menten kinetics, the K_s value is 0.07 mmol L^{-1}.

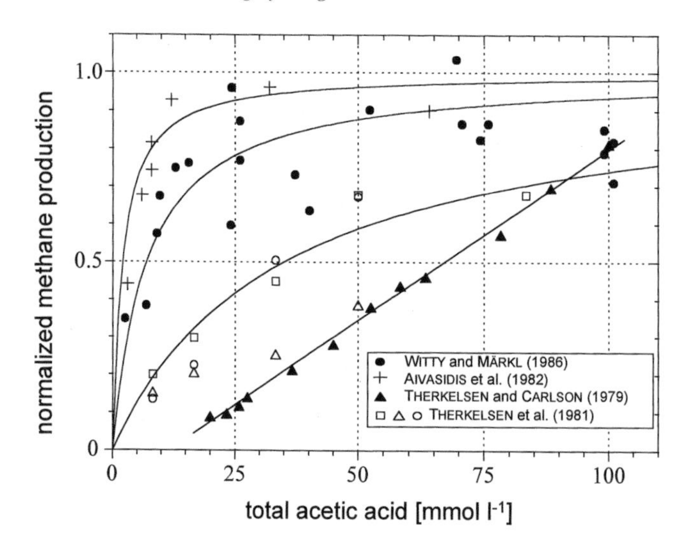

Fig. 12. Methane production as a function of the concentration of total acetic acid.

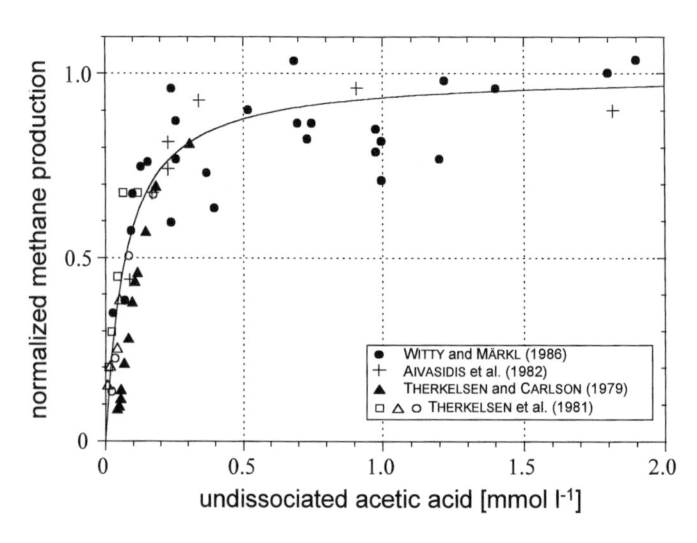

Fig. 13. Methane production as a function of the concentration of undissociated acetic acid.

The K_s value represents the concentration of substrate which corresponds to 50% of the maximal reaction rate. Because the undissociated acetic acid represents the relevant substrate, we may also write $K_s = K_{HAc}$.

To find out, if this K_{HAc} value is of general nature or if it depends on the type of microorganism which is active in the conversion of acetic acid to methane, some more kinetic data of authors who worked with more defined microbial systems are studied. These data are presented in Fig. 14. It can be seen that experiments performed with *Methanothrix* sp. result in a smaller K_{HAc} value ($K_{HAc} = 0.02$ mmol L^{-1}) than those performed with *Methanosarcina barkeri* which correspond to a somewhat higher value ($K_{HAc} = 0.12$ mmol L^{-1}). This result may be due to the fact that the relative surface of the filamentous *Methanothrix* sp. is larger compared to that of the rod-shaped *Methanosarcina* sp.

Altogether the K_{HAc} value, based on the undissociated acetic acid is roughly in the range between 0.02 mmol L^{-1} and 0.12 mmol L^{-1}. In the case of a mixed culture an average value of 0.07 mmol L^{-1} may be adequate. If this

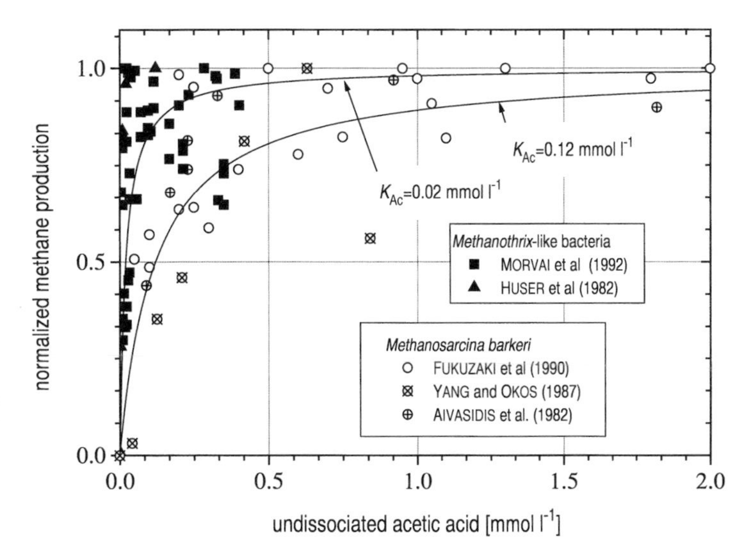

Fig. 14. Methane production as a function of the concentration of the undissociated acetic acid. Experimental data of cultures of *Methanothrix*-like sp. and *Methanosarcina barkeri*.

mixed culture contains mainly *Methanothrix* sp. to convert acetic acid, the value will be smaller. If *Methanosarcina* sp. dominate, the value of K_{HAc} should be higher. The author of this paper uses $K_{HAc}=0.07$ mmol L^{-1}.

WITTY and MÄRKL (1986) reported propionic acid inhibits the conversion of acetic acid to methane and CO_2. Experiments with different amounts of propionic acid resulted in an inhibition coefficient of $K_{HPro}=0.97$ mmol L^{-1}. Therefore, a concentration of 0.97 mmol L^{-1} of undissociated propionic acid causes an inhibition of 50% of the rate without propionic acid (Eq. 18).

$$c_{HPro}=\frac{c_{H^+}\,\gamma_1^2}{K_{D,Pro}+c_{H^+}\,\gamma_1^2}\,c_{Pro} \tag{16}$$

$$c_{H_2S}=\frac{c_{H^+}\,\gamma_1^2}{K_{D,H_2S}+c_{H^+}\,\gamma_1^2}\,c_{S,tot} \tag{17}$$

$$\mu=\mu_{max}\frac{c_{HAc}}{c_{HAc}+K_{HAc}}\cdot\frac{K_{HPro}}{K_{HPro}+c_{HPro}}\cdot\frac{K_{H_2S}}{K_{H_2S}+c_{H_2S}} \tag{18}$$

$K_{HAc}=0.07$ mmol L^{-1};
$K_{HPro}=0.97$ mmol L^{-1};
$K_{H_2S}=2.5$ mmol L^{-1}

3.2 Hydrogen Sulfide

The inhibition of methane generation from acetic acid by H_2S is well known. Because the medium sulfate concentration in the wastewater of a bakers' yeast production is as high as 3.9 g L^{-1} which is almost totally converted to hydrogen sulfide, the problem of H_2S inhibition is very substantial in this case. FRIEDMANN and MÄRKL (1994) reported an inhibition constant of 2.8 mmol L^{-1} (95.2 mg L^{-1}). Recent experiments of POLOMSKI (1998) on both the laboratory and pilot scale (for experimental apparatus see Fig. 2) resulted in an inhibition constant of 2.5 mmol L^{-1} (85 mg L^{-1}). The results of these experiments are shown in Fig. 15.

3.3 Conclusions

As far as the kinetics of the conversion of acetic acid to methane and CO_2 are concerned the results are summarized in Eq. (18). The data given are only valid near the steady state of the system; this means in a pH range between 6.6 and 7.4 and in the usual range of substrate concentrations. If the system was heavily overloaded for some time and was out of the described range, the microbial population might need some recovering time, which cannot be described by Eq. (18). Fig. 16 shows a sketch of different configurations in which

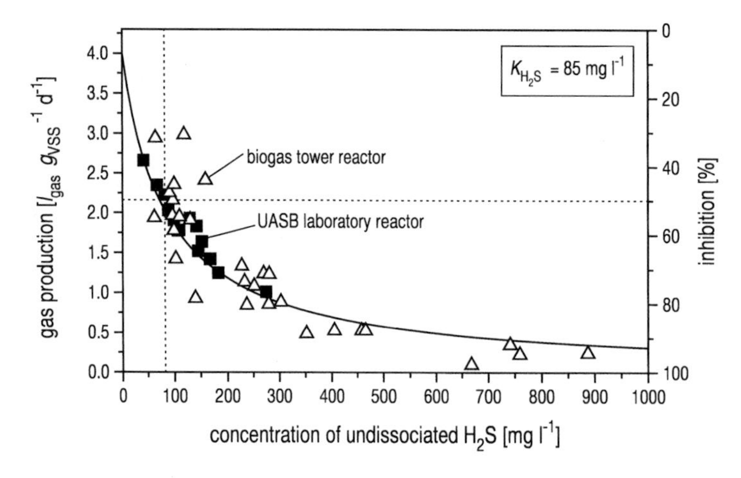

Fig. 15. Specific gas production of a UASB laboratory reactor and a biogas tower reactor on pilot scale (for both reactors, see Fig. 2 and Tab. 1) as a function of the concentration of undissociated H_2S. The data of gas production are corrected for the influence of acetic acid and propionic acid concentrations according to Eq. (18). VSS: volatile suspended solids (biomass).

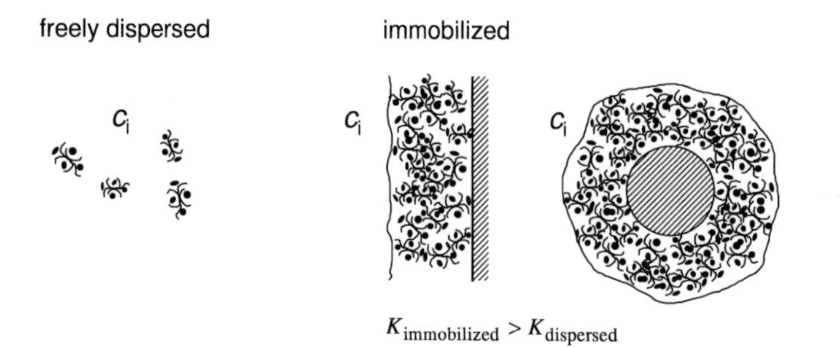

Fig. 16. Configuration of the active microbial system.

active microorganisms exist in biogas reactors. They can exist in a freely dispersed form, i.e., as single organisms, or as in the case of the biogas tower reactor in the form of loose pellets. The microbial system can also be immobilized in the form of a film at the wall or as a dense pellet. In the second case the conversion efficiency will also be influenced by transport phenomena. The data given in Eq. (18) hold only true for the case of freely dispersed organisms. When immobilized microorganisms dominate in a biogas reactor the constants (both saturation and inhibition constants) will be larger.

Eq. (18) can also be used to calculate the combined influence of two parameters as demonstrated in Fig. 17. This graph shows the gas production as a function of the concentration of total acetic acid. It can be seen that the behavior changes completely if it is compared to the graph of Fig. 12. Adding H_2S to the fermentation broth changes the kinetics from a Michaelis–Menten type to a substrate inhibition type. The effect can be explained by the continuous decrease of the pH value if the concentration of acetic acid is increased. At smaller values of pH more of the total hydrogen sulfide exists in the form of undissociated H_2S and the inhibitory effect increases along the abscissa.

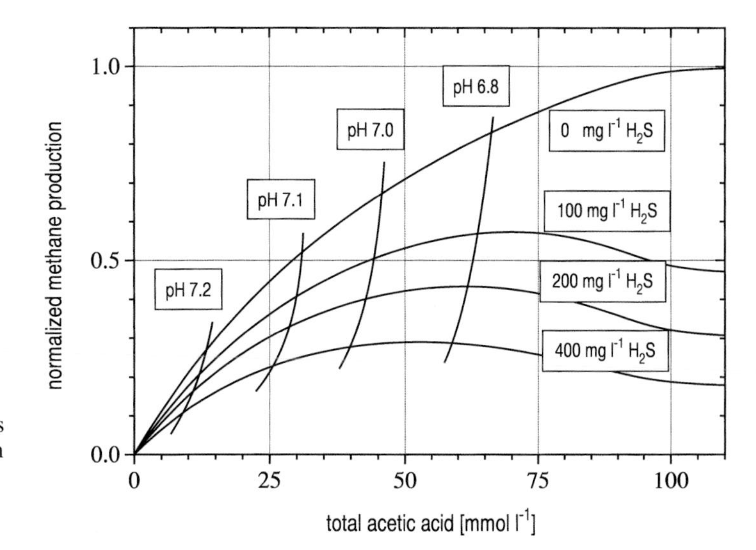

Fig. 17. Methane production as a function of the concentration of total acetic acid at different concentrations of total hydrogen sulfide ($c_{HPro}=0$).

4 Hydrodynamic and Liquid Mixing Behavior of the Biogas Tower Reactor

Mixing of the liquid phase with respect to the scale-up of a reactor has to be known to understand the nutrient supply of all of the active reactor regions and to avoid local overloading of the reactor. The biogas tower reactor of Fig. 1 consists of four modules. The active volume of one reactor unit, as shown in Fig. 2, is 280 m^3 (height 25 m, diameter 3.5 m). A settler at the top of the reactor helps to maintain high biomass concentrations within the reactor.

The principal function of one module is described in Fig. 18. Biogas produced below this module rises and is collected under the collecting device (1) from where it can be withdrawn. The quantity of gas withdrawn is controlled by a valve (2). If less gas is taken off than collected, the remaining gas rises through the open cross-sectional area (3) into the next module and into the next gas collecting device (5). The baffle (4) separates the space between two gas collecting devices into two channels joining at the top and the bottom of each module. Because of the rising gas on one side of the baffle, there is a difference in the gas holdup in two channels, which corresponds to a fluid circulation along the baffle (BLENKE, 1985). The upflow channel is called "riser" and the downflowing part is called "downcomer". The circulation velocity is strongly linked to the amount of gas passing through the open cross-section-

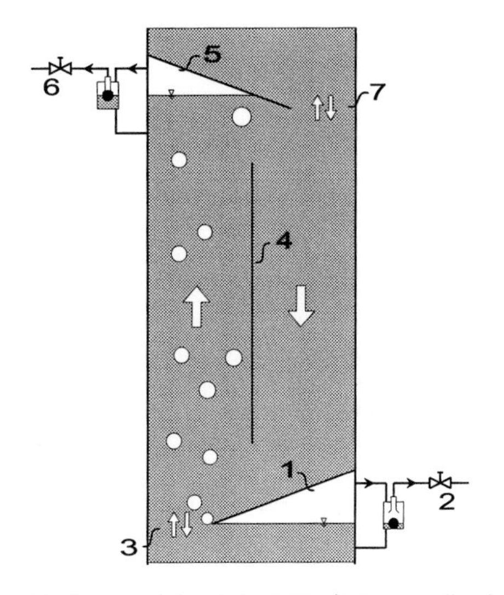

Fig. 18. One module of the BTR (1,5: gas collecting devices; 2,6: gas control units; 3,7: connecting cross-sectional areas; 4: baffle).

al area (3). Hence, by controlling the gas outlets (2) and (6) it is possible to control the liquid circulation velocity and the mixing time in a module. If no gas is withdrawn, both the circulation velocity and the mixing intensity are high. In contrast, if all collected gas is withdrawn, the circulation slows down and good settling conditions for biomass set in. In the lower zones of the reactor it might be advantageous to support the mixing characteristics, in the higher zones good settling conditions should be established. As in the case of mixing within one module, the mixing between two neighboring modules depends, as experiments showed, on the gas flow rate through the open cross-sectional area (3) between the two modules. If gas rises through this area, turbulence is generated, causing a convective transport between the two modules. If the gas flow rate through the connecting area (3) is low, the mass transport between the modules will decrease.

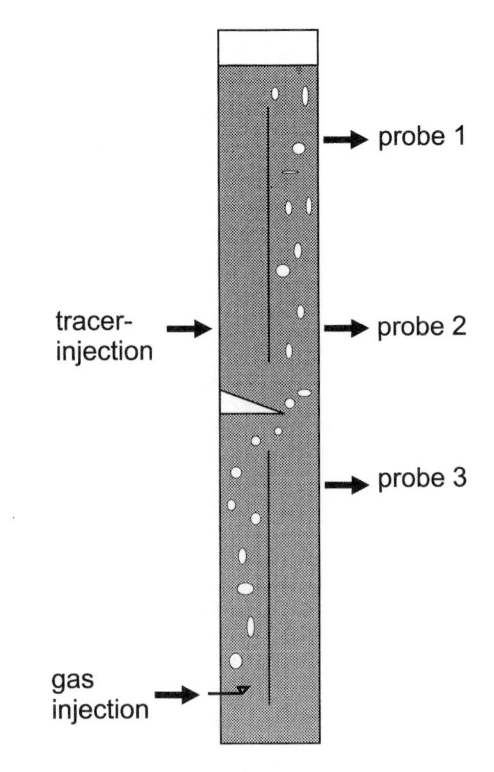

Fig. 19. Laboratory reactor representing a model of two modules of the BTR. It was filled with 0.1% ethanol in tap water, and air was injected through a sparger at the bottom to simulate biogas production (height: 6.5 m; diameter 0.4 m; volume 0.7 m^3; height of a module 3.5 m; length of the baffle 2.89 m).

4.1 Mixing of the Liquid Phase

REINHOLD et al. (1996) provide a theoretical and analytical analysis of the mixing behavior of a BTR. Experiments were performed on a laboratory (Fig. 19) and a pilot scale (Fig. 20). To study mixing, a saline solution was injected in the form of a Dirac impulse as a tracer at positions indicated in Figs. 19, 20. The course of the salt concentration was continuously monitored by conductivity probes at different positions. For the experiments the reactors were filled with tap water. To reduce the coalescence of bubbles and to obtain small bubbles as observed in biogas reactors, 0.1% of ethanol were added. Instead of biological gas production air was injected at the bottom of the reactors. It was assumed that the dominating effect on the circulating flow was caused by gas coming from the lower module. The results of the experiments can be summarized as follows: The mixing times for a 95% homogeneity after a tracer impulse are

(1) mixing within one module: $(1–2) \cdot 10^2$ s,
(2) mixing from one module to a neighboring module: $(0.5–2) \cdot 10^3$ s.

Because the heights of a module in the laboratory and in the pilot scale reactor do not differ much, the situation on the pilot scale is more or less the same as in the laboratory scale reactor.

Mixing within one module due to the axial dispersion in the circulating flow is shown in Fig. 21, presenting laboratory scale experiments. The concentration profiles over time for the four modules of the pilot scale reactor after applying a Dirac impulse in the third module are recorded in Fig. 22. The experimental behavior is simulated very well by a mathematical model.

The structure of the mathematical model developed is shown in Fig. 23. The structures of the two models A and B follow the modular structure of the reactor concept. Both models

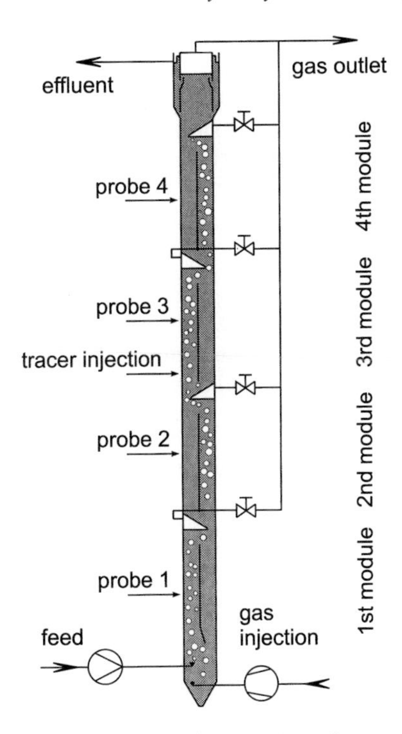

Fig. 20. Pilot scale BTR (height 20 m; diameter 1 m; volume 15 m^3; height of a module 4 m; length of the baffle 3 m).

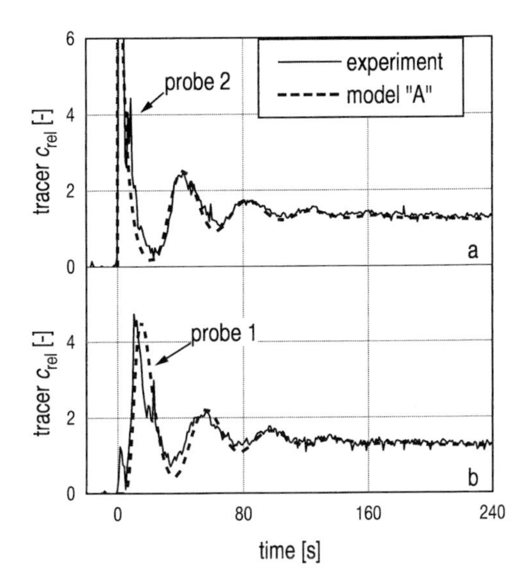

Fig. 21. Mixing of a Dirac impulse in the upper module of the laboratory scale reactor (Fig. 19).

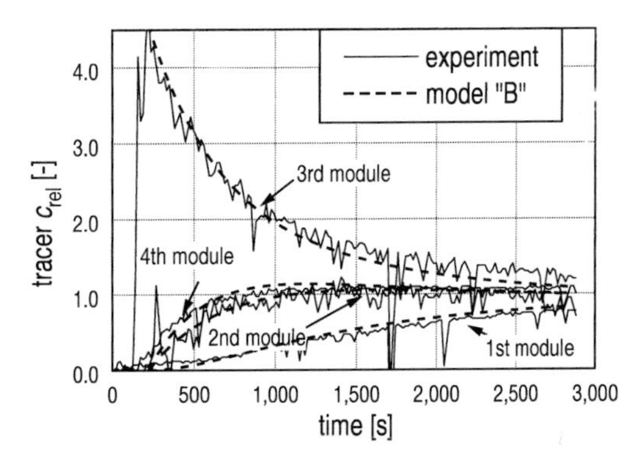

Fig. 22. Mixing of a Dirac impulse (3rd module) in the pilot scale BTR (Fig. 20).

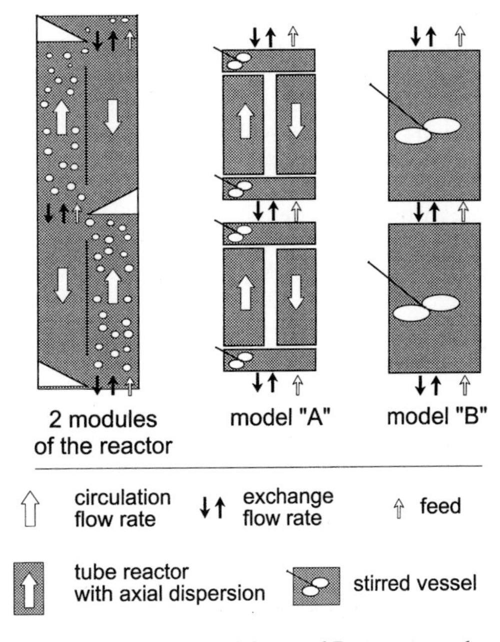

Fig. 23. Structure of models A and B as compared to the BTR for describing the mixing behavior of the liquid phase.

couple two neighboring modules by the flow of the feed ($\dot{V}_{\text{feed}}$) which is constant within the tower reactor and is directed from the bottom to the top of the reactor. The central idea of both models is to predict the coupling between two modules by the so called "exchange flow rate". This virtual flow rate ($\dot{V}_{\text{exchange}}$) accounts for the situation of a turbulent exchange of volume elements in the area between two modules ($A_{exchange}$). This turbulent flow is driven by the bubbles passing this area. $\dot{V}_{feed}$ is kept at zero during the experiment. Also, in reality the contribution of $\dot{V}_{feed}$ is usually small compared to the turbulent exchange flow $\dot{V}_{exchange}$.

4.1.1 Model A

Model A, which is a more detailed one, is able to describe mixing within a module as well as intermixing between modules. Each module is composed of four compartments. Riser and downcomer are mathematically replaced by tube reactors with axial dispersion. At the bottom and the top of each module, riser and downcomer join in mixing zones which are regarded as stirred vessels. The four zones are linked by the circulation flow rate $\dot{V}_{circ}$. Since the cross-sectional area of the riser A_{riser} and the downcomer $A_{downc.}$ as well as the flow rate through the riser $\dot{V}_{riser}$ and the downcomer $\dot{V}_{downc.}$ are equal, the mean circulation velocity w_m is given by Eq. (19).

$$w_m = \frac{\dot{V}_{riser}}{A_{riser}} = \frac{\dot{V}_{downc.}}{A_{downc.}} \tag{19}$$

Each real module of the reactor is thus replaced by a "ring structure" of four reactors. The transport and mixing mechanisms in this ring structure are convection and axial dispersion. The mathematical equations of model A can be derived from the mass balances of each of the four compartments. Model A needs three parameters for calculating the mixing behavior: the liquid circulation velocity w_m, the axial dispersion coefficient D_{ax}, and the exchange rate $\dot{V}_{exchange}$. All these parameters depend on the gas flow rate entering the module from the lower module. For more details of the model, the mathematical structure, and the procedure of solving the partial differential equations the publication of REINHOLD et al. (1996) should be consulted.

4.1.2 Model B

Model B, which is simpler, consists of only one stirred vessel per module (Fig. 23). This model is able to describe the mixing behavior of the whole reactor if the mixing intensity within one module is high compared to the tracer transport from one module to another. This assumption holds true in the real situation as shown earlier. Model B links the stirred vessels in the same way as model A. The intermixing between two neighboring modules is again modeled by an exchange flow rate $\dot{V}_{exchange}$ going up and down. Eq. (20) shows the material balance for one module i.

$$\begin{aligned} V_i \frac{dc_i}{dt} = &\dot{V}_{exchange,i-1}(c_{i-1} - c_i) \\ &+ \dot{V}_{exchange,i}(c_{i+1} - c_i) \\ &+ \dot{V}_{feed}(c_{i-1} - c_i) \end{aligned} \tag{20}$$

For describing the mixing of one compound in a BTR with model B, the number of the ordinary differential equations is equal to the number of modules. It is an initial-value problem which can be solved with a Runge–Kutta method. The only necessary parameter for calculating the mixing behavior is the exchange flow rate $\dot{V}_{exchange}$.

Experiments in both laboratory and pilot scale reactors are carried out at different gas flow rates $\dot{V}_{gas}$ to investigate the dependence of the hydrodynamic parameters on the gas loading. Although there are a lot of investigations on airlift loop reactors and bubble columns in general, the results cannot be applied to this type of reactor because the gas and liquid loading of the BTR are far below those of the airlift loop and bubble column reactors found in the literature. Since the present BTR has unique characteristics, it was necessary to determine the exchange flow rates $\dot{V}_{exchange}$. Consequently, studies on the circulation velocity w_m and the axial dispersion D_{ax} were carried out.

The characteristic parameters D_{ax}, w_m, $\dot{V}_{exchange}$, are a function of the gas loading of the system. The parameters can be obtained by using the least-square method to fit the simulation to the experimental data. The results are shown in Figs. 24–26.

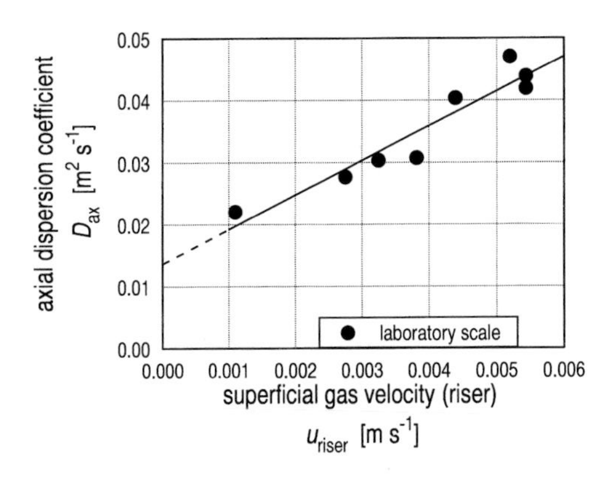

Fig. 24. Axial dispersion coefficient as a function of the superficial gas velocity in the riser.

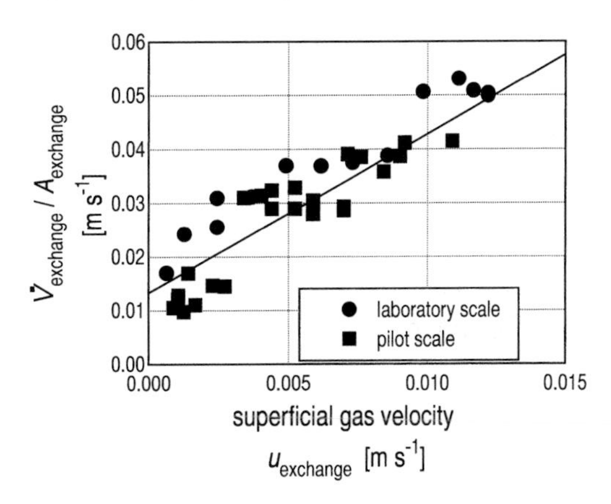

Fig. 26. Effect of the superficial gas velocity on the exchange flow rate between two modules.

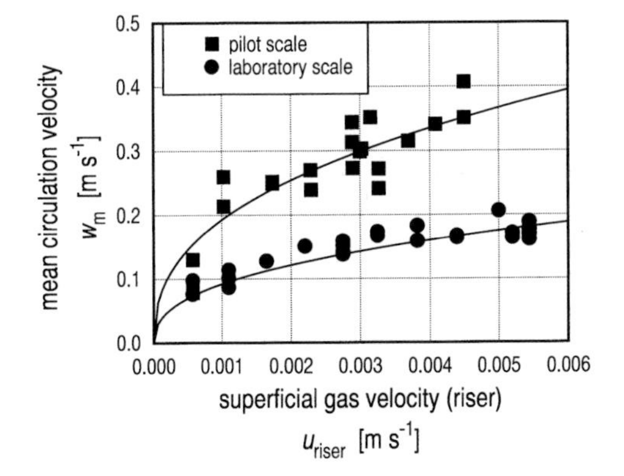

Fig. 25. Mean circulation velocity of the liquid as a function of the superficial gas velocity of the riser.

In Fig. 24 the axial dispersion coefficient is plotted against the superficial gas velocity of the riser u_{riser} which is given by Eq. (21).

$$u_{riser} = \frac{\dot{V}_{gas}}{A_{riser}} \tag{21}$$

This coefficient was only determined on the laboratory scale. Fig. 25 shows the mean circulation velocity w_m of the laboratory scale and the pilot scale reactors as a function of the superficial gas velocity of the riser u_{riser}, From a momentum balance the mean circulation velocity w_m can be calculated on the basis of the difference in gas holdup between riser and downcomer $\Delta\varepsilon$ and the knowledge of the total pressure loss (friction) coefficient ξ according to Eq. (22).

$$w_m = \sqrt{\frac{2\,g\,L}{\xi}\Delta\varepsilon} \tag{22}$$

L is the length of the baffle and g is the gravitational acceleration. The gas holdup can be correlated from the superficial gas velocity u_{riser} according to WEILAND (1978). Due to experiments of REINHOLD et al. (1996) it can be calculated according to Eqs. (23, 24).

$$\Delta\varepsilon = 0.27\,u_{riser}^{0.8} \quad \text{(laboratory scale)} \tag{23}$$

$$\Delta\varepsilon = 0.76\,u_{riser}^{0.8} \quad \text{(pilot scale)} \tag{24}$$

The pressure loss coefficients were determined to be $\xi = 6.9$ on the laboratory scale and to be $\xi = 4.8$ on the pilot scale. As expected, the total friction coefficient ξ slightly decreases during scale-up.

Fig. 26 shows the exchange flow rate $\dot{V}_{exchange}$ as a function of the superficial gas velocity $u_{exchange}$, which can be calculated according to Eq. (25).

$$u_{exchange} = \frac{\dot{V}_{gas}}{A_{exchange}} \tag{25}$$

It is quite remarkable that the findings of Fig. 26 are invariant with the scale of the reactor. Because $\dot{V}_{exchange}$ dominates the mixing behavior of the system (indeed, it is the only experimental parameter necessary for model B) this result is very important for the scale-up of biogas tower reactors.

4.2 Distribution of Biomass within the Reactor

The wastewater is passing the BTR from the bottom to the top of the reactor. With this feeding procedure the upflow velocity due to the feed increases – at a constant mean residence time of wastewater in the reactor – linearly with the height of the reactor. Therefore, in high reactors it is essential to have reliable mechanisms to retain the active biomass. These mechanisms and the resulting distribution of biomass within the reactor were studied by REINHOLD and MÄRKL (1997).

4.2.1 Experiments

Experiments were performed with a 2-modular laboratory tower reactor (height 6.5 m) and in a pilot scale tower reactor (height 20 m) as shown in Fig. 2 and described in Tab. 1. The laboratory tower reactor was filled with tap water and anaerobic, pelletized sludge. The sludge was biologically inactive as no substrate was present in the tap water, i.e., it did not produce any fermentation gas. Air was injected at the bottom of the column instead of the real biogas. As the reactor was built out of transparent material visual observations were possible in addition to measurements from samples. The following observations were made during these experiments:

(1) The suspended solid concentration in the lower module is always higher than in the upper module.
(2) The macroscopic liquid turbulence backmixing flow ($\dot{V}_{exchange}$) which causes the interaction between adjoining modules was visualized through the movement of the sludge pellets as indicated by the arrows (7) in Fig. 27.
(3) The most remarkable observation is also indicated in Fig. 27: At the lower end of the baffle (4) the suspension turns 180° (6) and from this flow a continuously downward directed solid mass flow is observed, falling at the downwards sloping flat blade (1) and from there sliding to the module below. This suggests to be a very effective mechanism of retaining active sludge in the reactor.

Measurements of sludge distribution were also performed in the pilot scale tower reactor digesting wastewater of the bakers' yeast company. The BTR was inoculated with pelletized sludge, and the space loading was increased in 75 d to 8.8 kg TOC m^{-3} d^{-1} with a TOC removal efficiency of 60%. The TOC conversion rate by the sludge was 1.2 kg TOC $(kgSS)^{-1}$ d^{-1} . The size of the pellets in the reactor

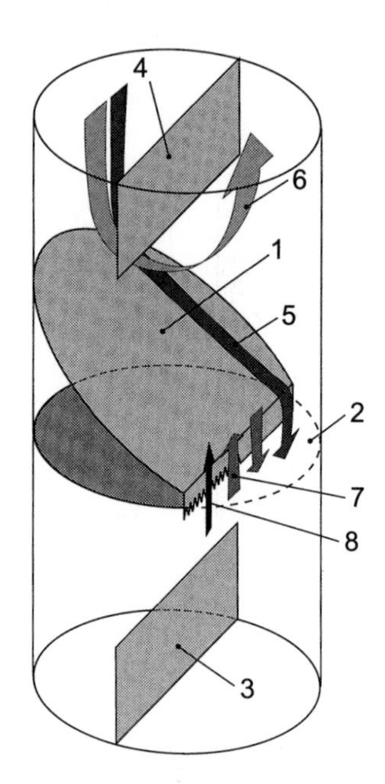

Fig. 27. Flows in the coupling zone of two adjacent modules: (1) gas collecting device; (2) connecting area between the modules; (3) upper end of the lower baffle; (4) lower end of the upper baffle; (5) sedimenting mass flow ($\dot{M}_{sedi}$); (6) circulating flow ($\dot{V}_{circ}$); (7) backmixing flow ($\dot{V}_{exchange}$); (8) feed flow ($\dot{V}_{feed}$).

ranged from 0.08–1 mm. Volatile suspended solids were about 50–60% of the suspended solids. Elementary analysis showed that 50% of the dry matter was carbon. Two main observations concerning the sludge distribution were made during these experiments on the pilot scale:

(1) There is no difference in the concentration of suspended solids between riser and downcomer. Each module can be regarded as a completely mixed reactor.
(2) The concentration of suspended solids in the BTR decreases gradually in upward direction from module to module.

4.2.2 Mathematical Modeling

The structure of the mathematical models is shown in Fig. 28. The mass balance of one module (i), including the interaction with an upper ($i+1$) and a lower module ($i-1$), leads to Eq. (26).

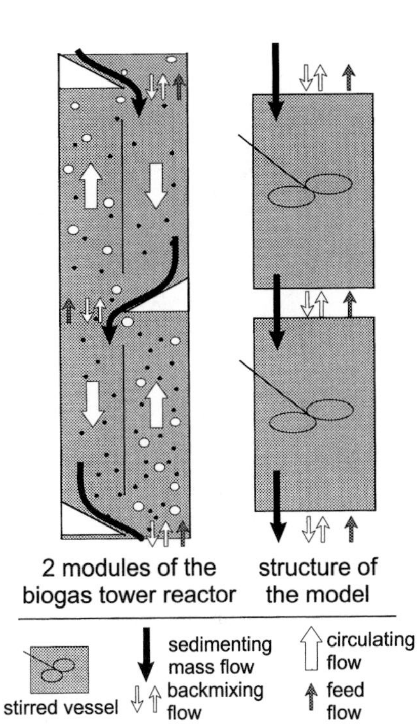

Fig. 28. Structure of the mathematical model compared to two modules of the BTR.

$$V_i \frac{dSS_i}{dt} = \dot{V}_{exchange,i-1}(SS_{i-1} - SS_i) + \dot{V}_{exchange,i}(SS_{i+1} - SS_i) + \dot{V}_{feed}(SS_{i-1} - SS_i) - \dot{M}_{sedi,i-1} + \dot{M}_{sedi,i} + \dot{M}_{prod} \frac{V_i\, SS_i}{\sum_{k=1}^{n} V_k \cdot SS_k} \quad (26)$$

In this equation V_i denotes the volume of a module, SS_i the concentration of suspended solids, $\dot{V}_{feed}$, the volumetric feed rate of the reactor and $\dot{M}_{sedi,i}$ the sedimenting mass flow between the modules. The produced biomass $\dot{M}_{prod}$ according to the growth of organisms could be calculated according to the TOC consumed (Eq. 27).

$$\dot{M}_{prod} = 0.1 \cdot \dot{V}_{feed}(TOC_{in} - TOC_{out}) \quad (27)$$

The volumetric exchange flow $\dot{V}_{exchange,i}$ is calculated with Eq. (28) on the basis of measurements of Fig. 26.

$$\frac{\dot{V}_{exchange}}{A_{exchange}} = 0.014 \text{ m s}^{-1} + 3.1 \cdot \frac{\dot{V}_{gas}}{A_{exchange}} \quad (28)$$

The only parameter which is not known, *a priori*, is the sedimenting mass flow rate $\dot{M}_{sedi}$. This value was determined by regression analysis from the experiments performed as shown in Fig. 29. It can clearly be seen, that the sedimentation capacity of active gassing sludge is much smaller compared to inactive (dead) sludge as used in the laboratory experiment. The sedimentation of active sludge follows Eqs. (29) and (30) where $A_{reactor}$ is the cross-sectional area of the reactor.

$$SS < 3 \text{ kg m}^{-3}: \frac{\dot{M}_{sedi}}{A_{reactor}} = 0.0025 \text{ m s}^{-1} \cdot SS \quad (29)$$

$$SS > 3 \text{ kg m}^{-3}: \frac{\dot{M}_{sedi}}{A_{reactor}} = 0.04 \text{ kg s}^{-1} \text{ m}^{-2} + 0.015 \text{ m s}^{-1} \cdot SS \quad (30)$$

In Fig. 30 results of experiments and the calculation on the basis of the mathematical model are compared. The agreement between simulated data and measured suspended solid concentration is quite good. The relative deviation

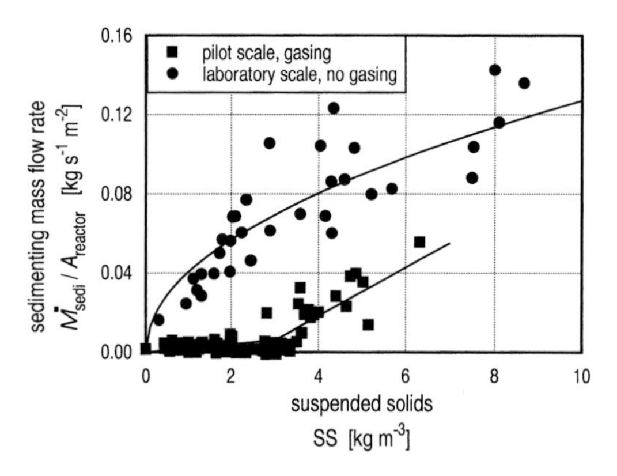

Fig. 29. Sedimenting mass flow rate between adjacent modules on the laboratory scale and pilot scale as a function of the suspended solid concentration in the upper module.

between measured and calculated values increases with decreasing concentration of suspended solids. The relative deviation ranges from 10–15% of the concentrations in the BTR. To investigate the scattering of the measured concentration of suspended solids, 10 samples were taken at one sampling point during a laboratory scale experiment. The detected suspended solid concentrations deviated by up to 8% from the average value. This shows that part of the deviation might be due to sampling errors.

REINHOLD and MÄRKL (1997) also demonstrated that the mathematical model allows for realistic predictions of the dynamic behavior of the system as was shown by simulating the start-up of a BTR.

5 Mass Transport from the Liquid Phase to the Gas Phase

As demonstrated in Sect. 3, it is necessary to have quantitative data for the concentration of the product gases, like CO_2, NH_3, and H_2S in the fermentation broth in order to calculate the pH value within the fermentation broth and get information on the kinetics of methane production. To calculate these concentra-

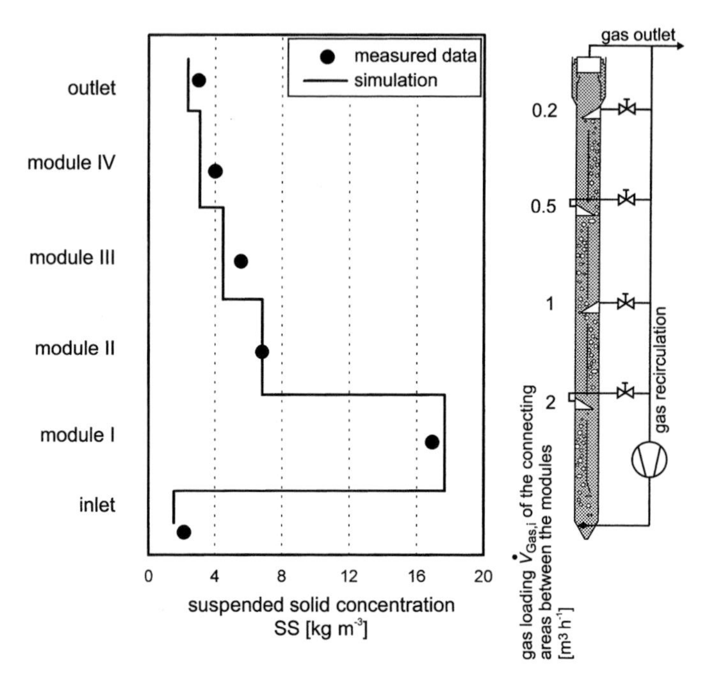

Fig. 30. Concentration of suspended solids as a function of reactor height in the 15 m^3 pilot scale BTR. $\dot{V}_{feed}$ = 315 L h^{-1}; total gas production of the reactor 2.3 m^3 h^{-1}; gas recirculated 1.8 m^3 h^{-1}; ΔTOC = 7.3 kg m^{-3}; gas loading $V_{gas,i}$ of the connecting areas between the modules as indicated.

tions it is necessary to have additional information about the transport of these gases from the liquid phase – where they are produced by the microorganisms – to the gas phase. Basic ideas of this calculation will be discussed with the example of H_2S because of its strong influence on biogas production. POLOMSKI (1998) performed experiments on the desorption of H_2S from the fermentation liquid broth in laboratory and pilot scale reactors.

The experimental setup of the laboratory experiment is shown in Fig. 31. The UASB reactor is supplied with wastewater at the bottom. A part of the generated biogas is withdrawn at the top of the reactor and recycled to the bottom where it is injected by means of a perforated tube. The recycled biogas is either recycled directly or over a H_2S adsorber. The H_2S concentration in the fermentation liquid and the partial presure of H_2S in the gas were measured by means of the silicon membrane probe and the mass spectrometer (see also Sect. 2.1).

The effect of gas recirculation can be explained by the example of one result of the experiment as shown in Fig. 32. At the ordinate on the left side the H_2S partial pressure is denoted, on the right side the H_2S concentration in the liquid phase is denoted which is in equilibrium with the partial pressure in the gas according to Eq. (31) and the Henry coefficient $H_{H_2S} = 2.330 \cdot 10^{-2}$ mg L^{-1} Pa^{-1} (equal to $6.85 \cdot 10^{-4}$ mmol L^{-1} Pa^{-1}).

$$c_i = p_i \cdot H_i \tag{31}$$

It can easily be seen that in this type of reactor without gas recirculation (first 23 h of the experiment) H_2S in the liquid and the gas phase is far from equilibrium. The concentration in the liquid phase is overconcentrated by 75% compared to the gas phase. After recirculating the gas and concurrently producing a larger interfacial area between liquid and gas phase the overconcentration is diminished to 18%. After starting recirculation of biogas 8 h are

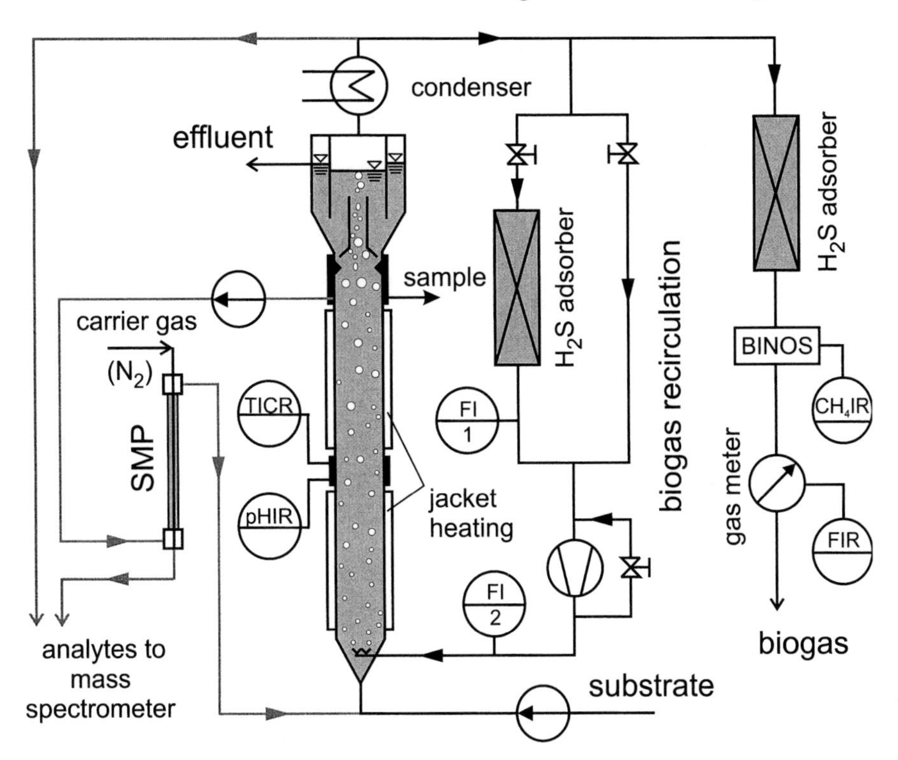

Fig. 31. Experimental set-up of a modified laboratory scale UASB reactor (active volume: 36.5 L) equipped with a bypass membrane probe and a mass spectrometer.

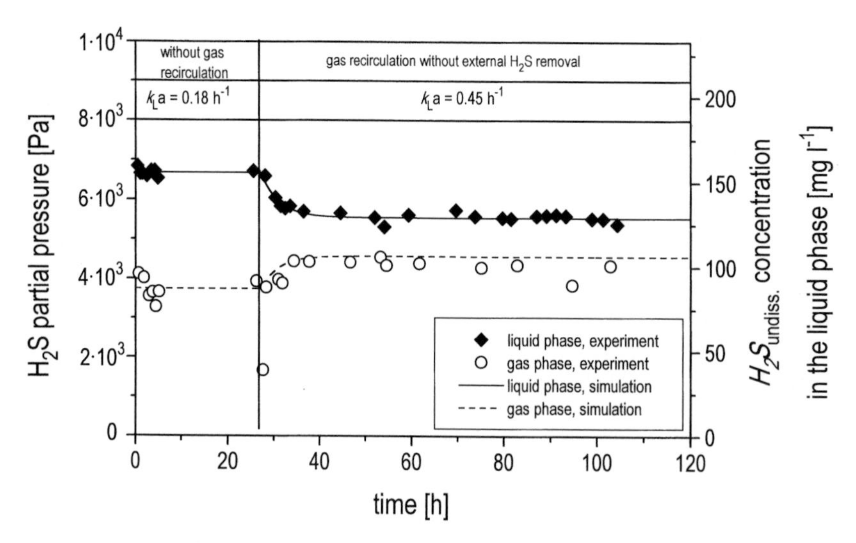

Fig. 32. Effect of recycling biogas on the H_2S concentration in the gas phase and the liquid phase in the laboratory UASB reactor. Experimental data are compared to results of the calculation with the mathematical model of Eqs. (32–34) (gas recirculation: 7.5 L h^{-1}; gas production: 7.45 L h^{-1}; pH value: 7.2 (constant); sulfate concentration in the wastewater: 3.85 g L^{-1}; residence time of wastewater: 2 d).

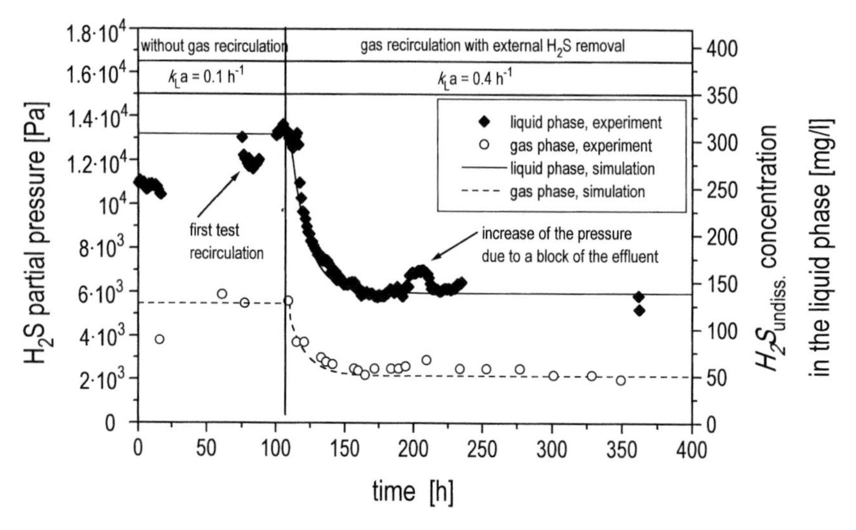

Fig. 33. Different from Fig. 32 an H_2S absorber was integrated in the recycled biogas (gas recirculation 27 L^{-1}; gas production: 7.6–13.7 L h^{-1}; pH value: 7.1–7.3; sulfate concentration: 5.13 g L^{-1}; mean residence time of wastewater in reactor: 1 d).

required to come from one steady state to the next. In Fig. 33 the result of a similar experiment is shown. Different from the first one, the recirculated biogas passes an H_2S absorber and is eliminated from the system. Although in this case the sulfate content in the wastewater is extremely high (5.13 g L^{-1}) the concentration of the H_2S in the liquid phase can be lowered to 130 mg L^{-1}.

In Figs. 32 and 33 it is also demonstrated that model calculations fit well with the experimental data. The H_2S content in the liquid and the gas phase can be calculated by using H_2S balances for the different compartments.

5.1 Liquid Phase

In Eq. (32) $V_{reactor}$ is the volume of the liquid in the reactor, $c_{S,tot}$ the concentration of the total sulfur in the liquid according to Eq. (7), $\dot{V}_{feed}$ is the volumetric feed rate of the reactor and $c_{S,tot,e}$ the concentration of the total sulfur in the reactor feed.

$$V_{reactor} \cdot \frac{dc_{S,tot}}{dt} = \dot{V}_{feed}(c_{S,tot,e} - c_{S,tot}) - V_{reactor} \cdot k_L a \, (c_{H_2S} - H_{H_2S} \cdot p_{H_2S,bubble}) \quad (32)$$

$k_L a$ denotes the transport capacity between the liquid phase and the gas phase (a: interfacial area per volume $m^2\ m^{-3}$), c_{H_2S} is the concentration of the undissociated portion of the H_2S in the liquid, and $p_{H_2S,bubble}$ the partial pressure of H_2S in the bubbles rising in the reactor liquid. H_{H_2S} is the Henry coefficient for H_2S.

5.2 Gas Bubbles

Eq. (33) gives the H_2S balance for the gas bubbles in the reactor suspension.

$$V_{bubble} \cdot \frac{dp_{H_2S,bubble}}{dt} = \dot{V}_{recyc.} \cdot p_{H_2S,recyc.} - \dot{V}_{bubble} \cdot p_{H_2S,bubble} + V_{reactor} \cdot k_L a \, (c_{H_2S} - H_{H_2S} \cdot p_{H_2S,bubble}) \cdot R \cdot T \quad (33)$$

V_{bubble} is the total volume of all the bubbles, $\dot{V}_{recyc.}$ denotes the volumetric flow rate of the biogas recycled from the top to the bottom of the reactor, $p_{H_2S,recyc.}$ the partial pressure of H_2S in this flow. $\dot{V}_{bubble}$ is the volumetric gas production rate of the reactor.

5.3 Head Space

Eq. (34) gives the H_2S (p_{H_2S}) partial pressure in the head space located over the reactor liquid, which is equal to the partial pressure of H_2S in the biogas outlet of the reactor.

$$\frac{dp_{H_2S}}{dt} = \frac{\dot{V}_{bubble}}{V_{gas}} (p_{H_2S,bubble} - p_{H_2S}) \quad (34)$$

V_{gas} is the volume of the head space. Assuming steady state, the partial pressure of H_2S in the gas bubbles rising in the biogas suspension $p_{H_2S,bubble}$ and the partial pressure of the produced biogas p_{H_2S} are equal.

The $k_L a$ value in Eqs. (32, 33) is not known *a priori*. It can be estimated by a regression analysis, fitting the model calculation to the experimental data. The resulting $k_L a$ values are denoted in the Figs. 32 and 33. It can be seen that recirculation of biogas in the experimental reactor of Fig. 31 increases the $k_L a$ value by a factor of 3–4 compared to the situation without gas recirculation.

Applying Eqs. (32–34) to each module of the biogas tower reactor and with additional regard to the liquid mixing which is described in Sect. 4.1, the H_2S content in the liquid phase and the gas phase in the different sections of the reactor can be simulated as shown in Fig. 34. It can be seen that the highest concentration of H_2S is found in the liquid phase at the bottom of the reactor (module I). The calculated $k_L a$ values in the pilot scale biogas tower reactor are larger by a factor of about 5–10 compared to those in the UASB reactor. Therefore, the differences between the liquid and the gas phase concentrations are much smaller. By recirculating biogas using an external H_2S adsorber values as small as 50 mg L^{-1} undissociated H_2S in the reactor liquid can be realized.

The $k_L a$ values identified are proportional to the gas holdup ε_G in the reactor suspension according to Eq. 35 as shown in Fig. 35.

$$k_L a = k \cdot \varepsilon_G \quad (35)$$

The proportional coefficient k depends largely on the size of the bubbles produced. The smallest interfacial area is found when bubbles are only generated by an overflow of the gas collection devices $k_{II} = k_{III} = k_{IV} = 100\ h^{-1}$. $k_L a$ is improved using a nozzle for the gas injection $k_{III,nozzle} = 200\ h^{-1}$. In the case of a gas injection over a frit made of HDPE $k_{I,frit} = 400\ h^{-1}$ is found.

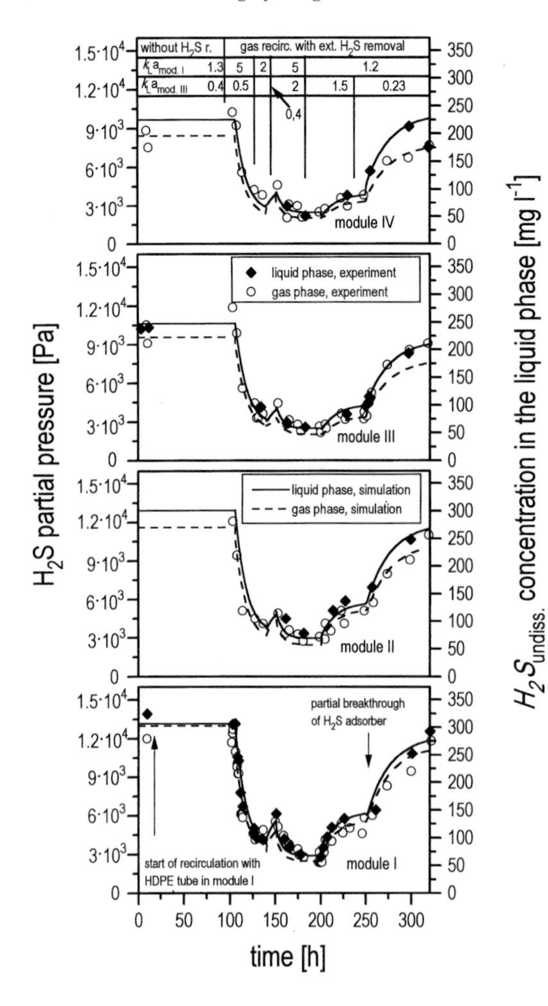

Fig. 34. Effect of recycling biogas on the H_2S concentration in the gas phase and the liquid phase in the pilot scale biogas tower reactor. Wastewater was simultaneously supplied to modules I, II, and III. Gas was recycled to module I: 2.3, 13, 5, 13, 1.8, 1.8 $m^3 h^{-1}$ and to module II: 0, 0, 2, 2, 3, 0 $m^3 h^{-1}$ in the sequence of the experiment according to the denoted time intervals; sulfate content of wastewater: 3.4–4.5 g L^{-1}; mean residence time of wastewater in the reactor: 2.1 d), r., recirc. recirculation, ext. external, mod. module.

6 Influence of the Hydrostatic Pressure on Biogas Production

Knowledge about the effect of higher hydrostatic pressure on the anaerobic digestion is important for the understanding of anaerobic microbiological processes in landfills, sediments of lakes, or deeper parts of high technical biogas reactors, especially of biogas tower reactors. Anaerobic reactors for the treatment of sewage sludge may also have a height up to 50 m in some cases.

In the literature very little has been published on this important effect. The reason may be found in the high complexity of performing experiments under steady state conditions and elevated pressure. A theoretical and experimental study was published by FRIEDMANN and MÄRKL (1993, 1994). It was shown that the most important effect of elevated pressure is due to the higher solubility of the product gases, like CO_2, H_2S, and NH_3. According to Eqs. (3–14) the pH may change because of a higher solubility of these gases and, therefore, the equilibrium of dissociation of all the relevant substances will shift with a significant effect on the kinetics of biogas production.

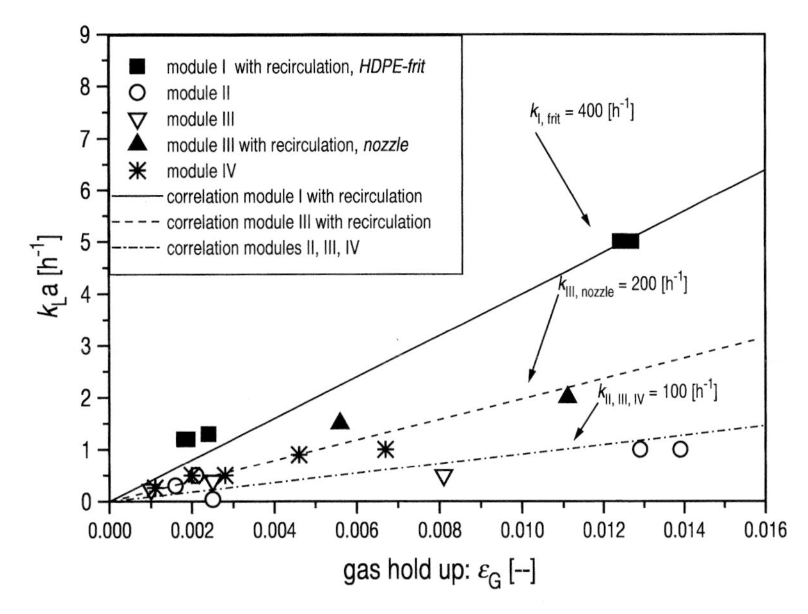

Fig. 35. k_La values as a function of gas holdup in the different reactor modules of the pilot scale biogas tower reactor. Gas was supplied to module I by means of a HDPE frit and to module III by means of a simple nozzle.

If a higher percentage of the produced gases is dissolved in the liquid phase of the reactor broth at higher pressures, a smaller amount of gas G will remain compared to the gas volume at a usual pressure of 10^5 Pa G_0. Because this effect is different for each gas component, the composition of the produced biogas will change when increasing hydrostatic pressure.

Assuming the number of gas bubbles will remain constant, increasing the pressure will lead to a smaller size of these bubbles (bubble diameter: d_B) according to Eq. (36).

$$d_b \sim \sqrt[3]{\frac{G}{G_0} \frac{1}{p_{hydrost}}} \tag{36}$$

The bubble diameter decreases because the relative content of dissolved gas is higher at higher pressures which results in a smaller G; furthermore, the pressure $p_{hydrost}$ itself reduces the bubble size. Highbie's penetration theory (Eq. 37) and Stoke's theory (Eq. 39) of bubble rising velocity (w_b) show the influence on the mass transfer coefficient k_L (Eq. 40).

$$k_L = 2\sqrt{\frac{D}{\pi\,\tau}} \tag{37}$$

$$\tau = \frac{d_b}{w_b} \tag{38}$$

$$w_b \sim d_b^2 \tag{39}$$

$$k_L \sim \sqrt{d_b} \tag{40}$$

It is assumed that the diffusion coefficient D in Eq. (37) is constant at the different pressures and the boundary layer renewal time τ can be calculated according to Eq. (38). Due to the fact that the interfacial contact area a between liquid and gas will also be smaller according to Eq. (41), the k_La value with respect to the hydrostatic pressure $p_{hydrost}$ can be calculated by Eq. (42).

$$a \sim d_b^2 \tag{41}$$

$$k_La \sim \left(\frac{G}{G_0} \frac{1}{p_{hydrost}}\right)^{0.83} \tag{42}$$

Altogether, at a higher hydrostatic pressure the transport capacity of the product gases from the liquid phase to the gas phase (Sect. 5, Eqs. 32 and 33) decreases and, therefore, the concentration of these gases in the liquid phase will increase additionally.

FRIEDMANN and MÄRKL (1993) give a solution of this complex physico-chemical system (39 nonlinear equations have to be solved simultaneously). Fig. 36 gives an example of the results showing the relative change of the composition of produced biogas as a function of pressure.

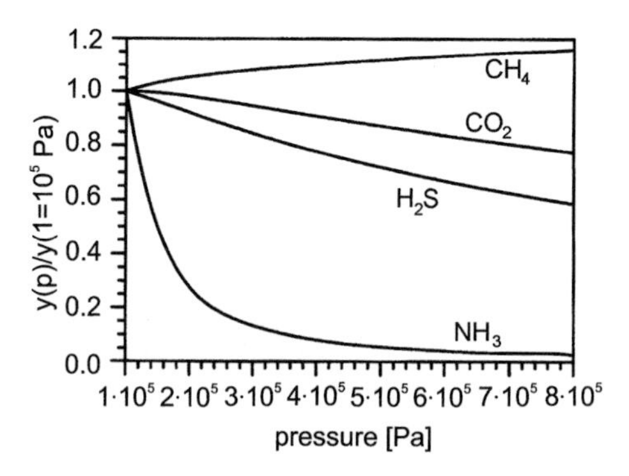

Fig. 36. Relative change of the composition of the produced biogas due to an alteration of the absolute pressure (model calculation for the digestion of wastewater of bakers' yeast production).

Experiments were performed in a 16 L stirred tank reactor. The reactor was equipped for continuous experiments under elevated pressures (up to $8 \cdot 10^5$ Pa). The pH value, the concentration of the undissociated H_2S in the liquid, and the H_2S content in the gas were measured *in situ* and continuously. Concentrations of the undissociated acetic acid and propionic acid were also detected (Fig. 37).

A comparison of the total experimental data investigated with model calculations is given in Fig. 38 showing a reasonable degree of agreement.

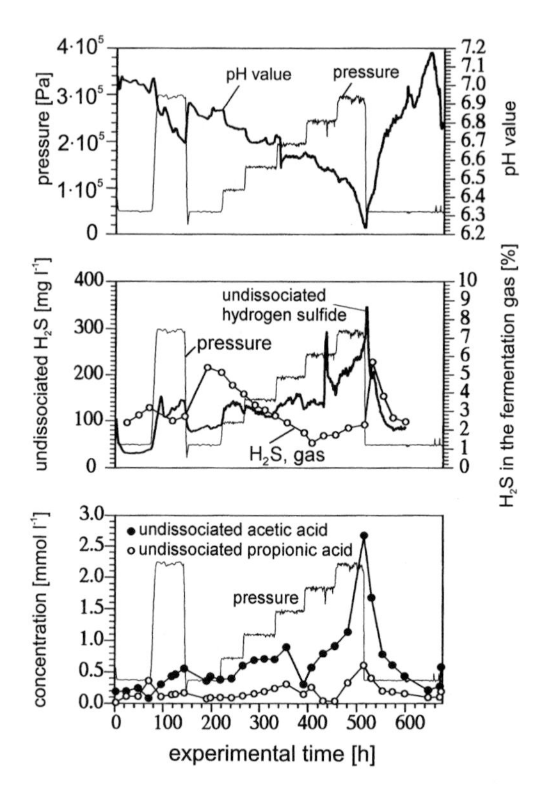

Fig. 37. Experiments digesting wastewater of bakers' yeast production at different pressures (pressure difference to normal pressure) in a stirred tank reactor (active volume 16 L). Continuous feeding of wastewater (TOC = 12 g L^{-1}, concentration of sulfate 2.8 g L^{-1}). Pressure was kept constant ±2,500 Pa within a certain time interval to achieve steady state conditions.

7 Reactor Control

The mathematical models described in Sects. 3–6 can be used to design controllers that stabilize the operation or automate the start-up of the reactor. Variables to be controlled include the pH value, the biomass concentration along the reactor height, and TOC. The controller influences the feed and the volume of the withdrawn biogas in all reactor modules.

The complexity of this control task results from the uncertainties of the reactor dynamics, which are reflected by the uncertainties of the models described so far. Further restrictions concern the availability of on-line measurable signals and the precision with which the control inputs can be operated. However, the structural properties of the reactor, which are reflected by the modular structure of the model, make it possible to use a modular controller structure, where each module is controlled by a decentralized control station, and a coordi-

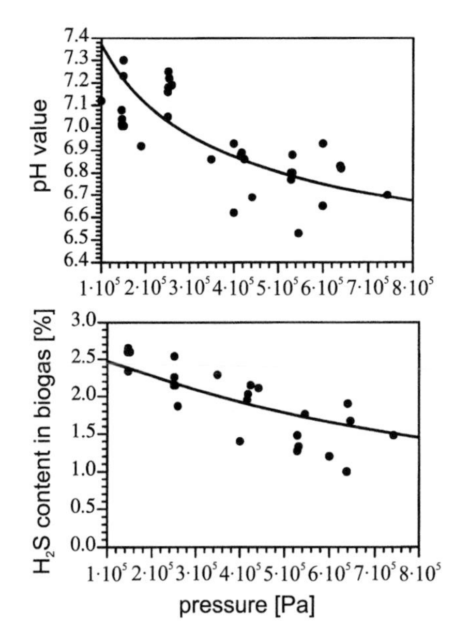

Fig. 38a, b. Comparison between model calculation and experiment. **a** pH as a function of the absolute pressure, **b** H_2S content in the biogas as a function of the absolute pressure.

nating layer takes account of the interconnection of the four independent control loops (Fig. 39).

The controller has been designed and tested by means of the dynamic model, which has been implemented in SIMULINK (simulation software, The Math Works Inc., USA) as shown in Fig. 40. The block diagram shows two kinds of blocks. The "modules" include the nonlinear differential equations that describe a reactor module, whereas the blocks called "links" are used to cope with the interconnections. Fig. 40 shows that all modules are strongly interconnected, which reflects the biochemical interdependence of all reactor elements. As described by PAHL et al. (1995), and PAHL and LUNZE (1998) different controllers have been selected by means of a linearized model which has been derived from the nonlinear model, but the properties of the closed loop systems have been studied by using the complete nonlinear model. An alternative way of controller design has been described by ILCHMAN et al. (1997), where an adaptive controller with high-gain feedback has been used. This control principle utilizes structural properties of the model, but it is robust in the face of parametric uncertainties.

Fig. 41 shows the implementation of the controller. The reactor is coupled by means of an Interbus-S to the control equipment in which the control laws are implemented. Separate personal computers were used to verify those

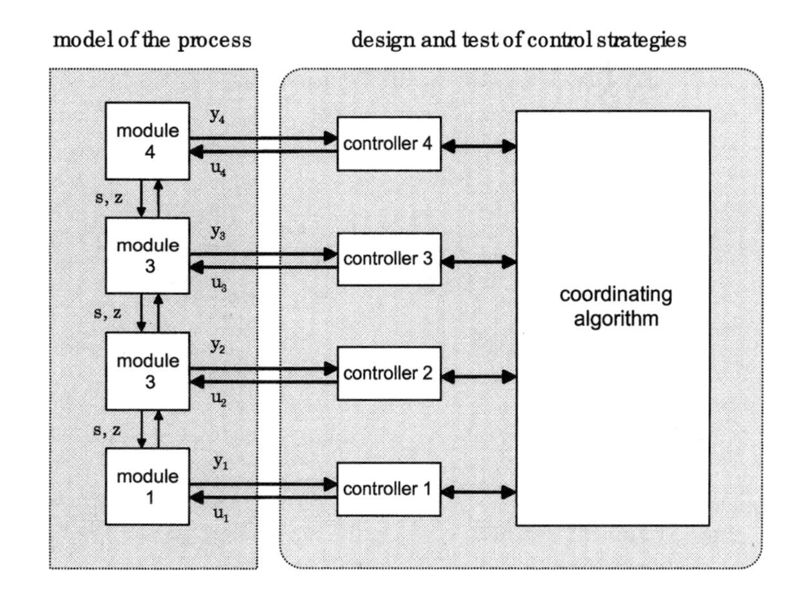

Fig. 39. Controller design. A model of the process reflecting the actual process structure is used to design and test control strategies off-line. (y_i ... measured variables, u_i ... control signals, s, z ... coupling signals).

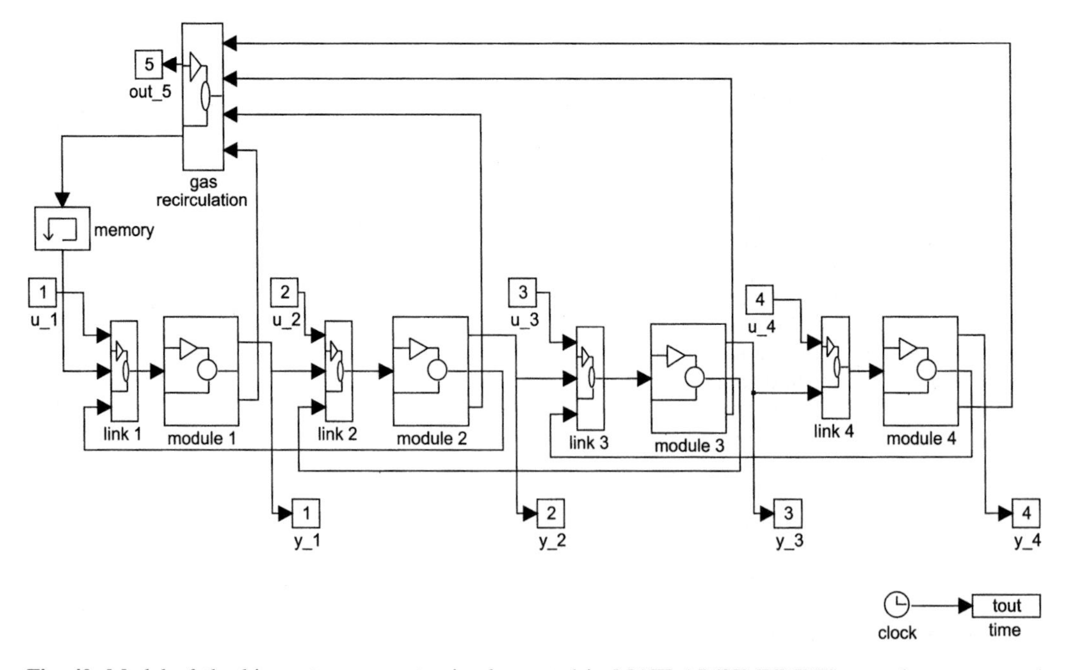

Fig. 40. Model of the biogas tower reactor implemented in MATLAB/SIMULINK, u and y are control signals and measured variables, respectively.

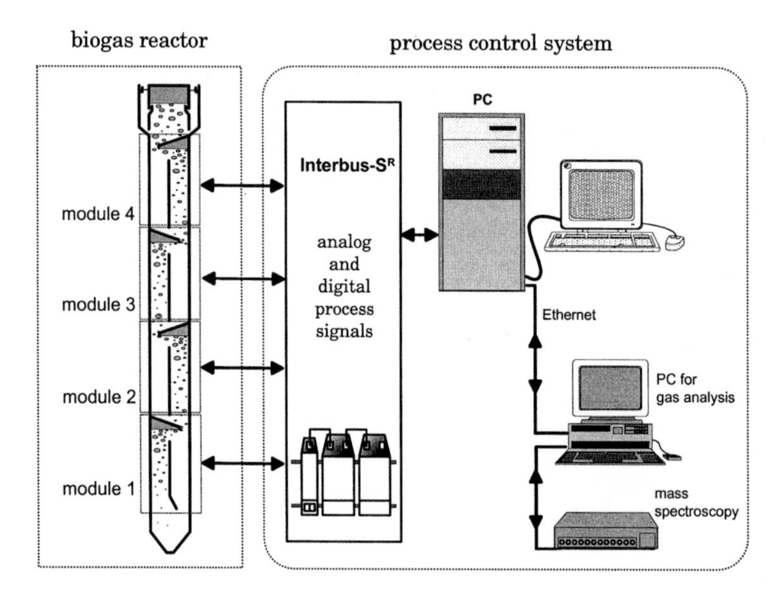

Fig. 41. Controller implementation.

controllers that use measurement of TOC by mass spectroscopy.

The experimental results verified the feasibility of the control principles used and showed that the dynamic models obtained have a sufficient accuracy for controller design.

8 Outlook

The mathematical model was developed and its use was demonstrated with the example of a biogas tower reactor. The hydrodynamic and liquid mixing behavior (Sect. 4) as well as the application of the mathematical model for reactor control (Sect. 7) are directly attributed to this reactor type.

But it should also be stated that the structure of the model and the basic understanding of the system behavior are of a more general nature and can be applied to all reactor types. This holds especially true for the measuring techniques (Sect. 2) and the microbial kinetics (Sect. 3). In each biogas reactor the transport of the gas phase (Sect. 5) will play an important role. The influence of gas oversaturation which is a function of that transport phenomenon is of particular importance in case of the production of gases which have an inhibitory effect on methane generation like H_2S and NH_3.

The influence of hydrostatic pressure on biogas production (Sect. 6) should be taken into consideration for most large technical scale bioreactors with a height of more than 5 m.

Acknowledgement
The research work reviewed in this chapter was mainly financed by the Deutsche Forschungsgemeinschaft within the Sonderforschungsbereich 238 for a period of 10 years. The intensive and fruitful cooperation with the colleagues Prof. G. MATZ, Prof. G. BRUNNER (both in the field of measuring techniques), Prof. J. LUNZE (control), and the support of the chairman of the Sonderforschungsbereich, Prof. H. TRINKS, and his successor Prof. J. WERTHER, are gratefully acknowledged. The author also wants to express his gratitude for the cooperation and financial support (pilot plant experiments) of the companies Deutsche Hefewerke, Hamburg (Dr. H.-G. KIRK, Dr. T. DELLWEG) and Preussag Wassertechnik GmbH, Bremen (Dipl.-Ing. H. BERBRICH).

9 References

AIVASIDIS, A., BASTIN, K., WANDREY, C. (1982), Anaerobic Digestion 1981, p. 361. Amsterdam: Elsevier.

BEUTIER, D., RENON, H. (1978), Representation of $NH_3-H_2S-H_2O$, $NH_3-CO_2-H_2O$ and $NH_3-SO_2-H_2O$ vapor-liquid equilibria, *Ind. Eng. Chem. Process Des.* **17**, 220–230.

BLENKE, H. (1985), Biochemical loop reactors, in: *Biotechnology* 1st Edn., Vol. 2, *Fundamentals of Biochemical Engineering*, pp. 465–517. Weinheim: VCH.

FRIEDMANN, H., MÄRKL, H. (1993), Der Einfluß von erhöhtem hydrostatischen Druck auf die Biogasproduktion, *Wasser-Abwasser gwf* **134**, 689–698.

FRIEDMANN, H., MÄRKL, H. (1994), Der Einfluß der Produktgase auf die mikrobiologische Methanbildung, *Wasser-Abwasser gwf* **6**, 302–311.

FUKUZAKI, S., NISHIO, N., NAGAI, S. (1990), Kinetics of the methanogenic fermentation of acetate, *Appl. Environ. Microbiol.* **56**, 3158–3163.

GRAEF, S. P., ANDREWS, J. F. (1973), Mathematical modeling and control of anaerobic digestion, *AIChE Symp. Series* **70**, 101–131.

HUSER, B. A., WUHRMANN, K., ZEHNDER, A. J. B. (1982), *Methanothrix soehngenii* gen. nov., a new acetotrophic non-hydrogen oxidizing methane bacterium, *Arch. Microbiol.* **132**, 1–9.

ILCHMANN, A., PAHL, M., LUNZE, J. (1997), Adaptive Regulation of a Biogas Tower Reactor, *Proc. Eur. Control Conf.*, Brussels 1997, paper 364 (CD by Belware Information Technology, Waterloo, Belgium).

LOEWENTHAL, R. E., MARAIS, G. V. R. (1976), *Carbonate Chemistry of Aquatic Systems: Theory and Application*, Vol. 1. Ann Arbor, MI: Science Publishers.

MÄRKL, H., REINHOLD, G. (1994), Biogas-Turmreaktor, ein neues Konzept in der anaeroben Abwasserreinigung, *Chem.-Ing. Tech.* **66**, 534–536.

MÄRKL, H., MATHER, M., WITTY, W. (1983), Meß- und Regeltechnik bei der anaeroben Abwasserreinigung sowie bei Biogasprozessen. *Münchener Beiträge zur Abwasser-, Fischerei- und Flußbiologie*, Vol. 36, pp. 49–64. München, Wien: R. Oldenbourg Verlag.

MATZ, G., LENNEMANN, F. (1996), On-line monitoring of biotechnological processes by gas chromatographic mass-spectrometric analysis of fermentation suspensions, *J. Chromatogr.* A **750**, 141–149.

MEYER-JENS, T., MATZ, G., MÄRKL, H. (1995), Online measurement of dissolved and gaseous hydrogen sulphide in anaerobic biogas reactors, *Appl. Microbiol. Biotechnol.* **43**, 341–345.

MORVAI, L., MIHÁLTZ, P., HOLLÓ, J. (1992), Comparison of the kinetics of acetate biomethanation by raw and granular sludges, *Appl. Microb. Biotechnol.* **36**, 561–567.

PAHL, M., LUNZE, J. (1998), Dynamic modeling of a biogas tower reactor, *Chem. Eng. Sci.* **53**, 995–1007.

PAHL, M., REINHOLD, G., LUNZE, J., MÄRKL, H. (1995), Modeling and simulation of complex biochemical processes, taking the Biogas Tower Reactor as an example, *Proc. Scientific Computing in der Chemischen Verfahrenstechnik*, Hamburg, 1–3 March. Heidelberg: Springer-Verlag.

POLOMSKI, A. (1998), Einfluß des Stofftransports auf die Methanbildung in Biogas-Reaktoren. *Thesis*, Technical University Hamburg-Harburg, Germany.

REINHOLD, G., MÄRKL, H. (1997), Model-based scale-up and performance of the Biogas Tower Reactor for anaerobic waste water treatment, *Water Res.* **31**, 2057–2065.

REINHOLD, G., MERRATH, S., LENNEMANN, F., MÄRKL, H. (1996), Modeling the hydrodynamics and the fluid mixing behavior of a Biogas Tower Reactor, *Chem. Eng. Sci.* **51**, 4065–4073.

THAUER, R. K., JUNGERMANN, K., DECKER, K. (1977), Energy conservation in chemotrophic anaerobic bacteria, *Bact. Rev.* **41**, 100–180.

THERKELSEN, H. H., CARLSON, D. A. (1979), Thermophilic anaerobic digestion of a strong complex substrate, *J. Water Pollut. Control. Fed. 51,* **7**, 1949–1964.

THERKELSEN, H. H., SÖRENSEN, J. E., NIELSEN, A. M. (1981), *Thermophilic Anaerobic Digestion of Wastewater Sludge and Farm Manure*, Project Brief of COWI-Consult in 45 Teknikerbyen, DK-2830 Virum.

WEILAND, P. (1978), Untersuchungen eines Airliftreaktors mit äußerem Umlauf im Hinblick auf seine Anwendung als Bioreaktor. *Thesis*, University of Dortmund, Germany.

WITTY, W., MÄRKL, H. (1986), Process engineering aspects of methanogenic fermentation on the example of fermentation of *Penicillium mycelium, Ger. Chem. Eng.* **9**, 238–245.

YANG, S. T., OKOS, M. R. (1987), Kinetic study and mathematical modeling of methanogenesis of acetate using pure cultures of methanogens, *Biotechnol. Bioeng.* **30**, 661–667.

ZUMBUSCH, P. VON, MEYER-JENS, T., BRUNNER, G., MÄRKL, H. (1994), On-line monitoring of organic substances with high-pressure liquid chromatography (HPLC) during the anaerobic fermentation of waste water, *Appl. Microbiol. Biotechnol.* **42**, 140–146.

27 Future Aspects – Cleaner Production

NORBERT RÄBIGER

Bremen, Germany

List of Abbreviations

DOC	dissolved organic carbon
ELU	environmental load units
EPS	environmental priority strategies
KEA	cumulative energy expenditure
MIPS	intensity of material consumption per service unit
VNCI	Vereniging van de Nederlandse Chemische Industrie (Association of the Dutch Chemical Industry)

1 Introduction

Since the mid-1970s a growing awareness of the emission of pollutants and their ecological impact has led to a comprehensive legislative initiative in the Federal Republic of Germany, aimed at stricter regulations for environmental protection. Legal reglementation oriented on the adherence to environmental quality standards for direct discharge into open waters has increased exponentially, and its growing influence on production-specific emissions in various industrial sectors can be observed. Nowadays the targets of environmental policies focus on the realization of prevention and recycling principles without relinquishing the above mentioned orientation on restrictive emission standards.

This orientation on standards is of two-fold significance for production-integrated environmental protection. On the one hand presently available technologies – in themselves characterized by the historical development of the legislation – can only partially be employed in production-integrated environmental protection; on the other hand it will be necessary to counteract the ensuing idea that ecological measures are imposed upon production from the outside and will in consequence be extremely expensive. Even though the guidelines of the Agenda 21 of Rio (Bundesministerium für Umwelt, 1993) imply opportunities for enterprises to improve their competitiveness on a global scale by introducing the principle of sustainability, this has not been recognized by all industrial sectors.

Today, according to Daly (1994) approximately 40% of all raw materials produced worldwide by photosynthesis are being exploited by man. In connection with the environmental impact resulting from the utilization of these raw materials this fact alone should make it obvious, that the introduction of production methods with a minimized demand on natural resources will be of great significance for the future. These methods of production will be characterized by measures such as:

(1) reducing the number and amount of raw materials needed and intermediate products, as well as reducing energy consumption,
(2) avoiding waste or producing ecologically harmless residues,
(3) recovery as well as recycling of useful residual materials, and
(4) establishing closed works internal water cycles.

This list is merely an example, but its links to production-integrated environmental protection are made evident by the similarity of targets. Measures of production-integrated environmental protection are thus an effective part of the chain of wealth creation of a production, and since the consumption of resources is minimized, they contribute to the overall financial cover. In contrast to end-of-pipe technologies cost calculations cannot solely be based on the fixed costs load, and consequently quite some interesting operational approaches to solutions for an ecologically sound production are conceivable.

From these goals a specific profile of demands can be derived for technologies that serve a production-integrated environmental protection, but at present only very few of the industrial processes developed and implemented so far meet these requirements, even though a suitable basic potential is given. In order to adapt them to the above described demands or to develop new processes, however, a basic conceptual change will be needed. Some first approaches have already been supplied by the development of novel processes in the chemical and metal processing industries.

2 Profile of Demands on Production-Integrated Measures

As can be seen in Fig. 1, any ecologically sound approaches to technical solutions that hold a promise for the future will be process and production-integrated measures and will, therefore, be aimed at minimizing the use of resources within production. One has to distinguish between two kinds of technical measures – those related to the product itself and those concerning the processes. Certainly, process-integrated environmental protection has to be paid highest priority and will contain all measures suited to minimize the consumption of resources at each step of the process, and to avoid emissions. The introduction of novel reaction systems as, e.g., described in (WIESNER, 1995) constitutes one of these measures.

Here, apart from the emissions linked to the energy demand, the product itself represents the decisive environmental burden. In this context, according to SCHNITZER (1998) a change in consumers' behavior will be necessary in order to achieve any further ecological benefits.

The mere meeting of consumers' needs rather than the amount of products sold would be at the center of interest. This, however, can only be accomplished if expenditures are not directed at the quantities produced, but at the usefulness of the products. Product leasing, a procedure already employed quite often today, is a first step into this direction. It has to be noted, however, that to reach process-integrated environmental protection will still need a long time.

Within production-integrated environmental protection the total production constitutes the balance limit for any conceivable zero emission. In order to minimize the use of resources all steps of the manufacturing process may be jointly considered, and this in turn will put the closing of works internal cycles into the center of interest. The resulting profile of demands on environmentally friendly approaches to process engineering will be demonstrated by means of Fig. 2.

The example given in Fig. 2 is based on the assumption that only an incomplete conver-

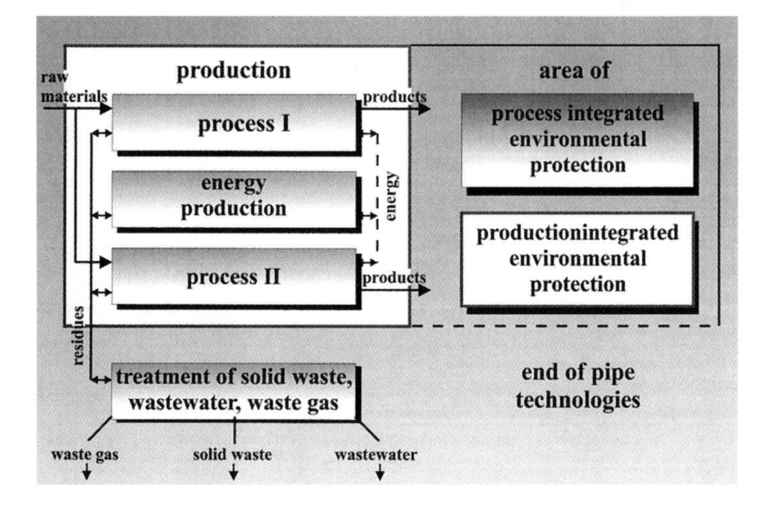

Fig. 1. Approaches to environmental protection in chemical production processes.

sion takes place within the reaction, whereby, apart from the educts A and B not only the product C, but also the by-product D will be present in the output. After product C has been separated from the effluent, there remains a multi-substance mixture, the treatment of which can be used to outline the demands on measures of production-integrated environmental protection.

In principle, the plant and its processes will have to be designed to operate at several stages, making it possible

(1) to recover useful materials in a first step and
(2) to treat wastewater in order to establish a closed cycle (E) in a second step.

It has to be pointed out for both steps of the process that quite often the concentrations of substances to be treated will be rather low, and accordingly, procedures with highly selective separating properties will be required. As far as the necessary water quality is concerned, this will be determined by the works internal user and not by the legislator. Thus, in these cases any technical orientation on legal standards – as in today's technologies – will loose in significance, so that currently there are only very few suitable techniques available that would serve these works internal purposes. Processes will have to be tailored to the specific demands of each production and will have to be employed at the ecologically weak points of production lines. Discovering these weak points will, therefore, be an indispensible part of production-integrated environmental protection.

3 Production-Integrated Water and Wastewater Treatment Technology

3.1 Procedures for the Recovery of Useful Materials

In principle, the recovery of useful materials from effluents is possible by combined appli-

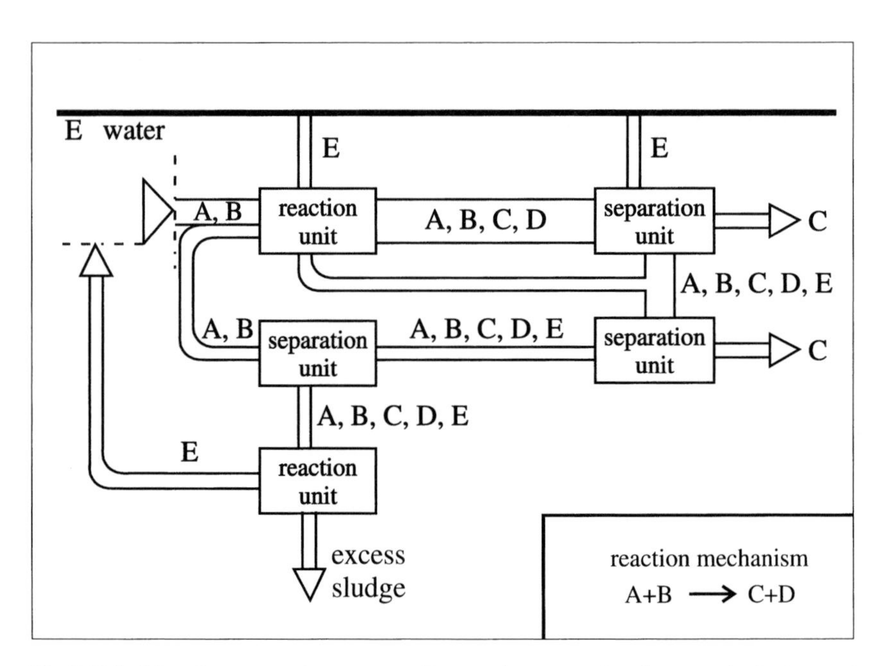

Fig. 2. Valuable substance and water recycling, environmental and economical advanced production.

cation of separation techniques, but today the necessary expenditure is often still too high. Therefore, the call for the development of better processes with a selective separating effect persists. In combination with wastewater treatment techniques and as part of the production integrated-environmental protection these processes will increase the yield of useful materials and thus, they hold much promise for the future. So far, however, only very few rare cases have come to grips with such tasks, still necessitating much further developmental work. In principle, a great number of techniques show sufficient potential for a selective separation of components in effluent mixtures. Prominent among these are:

(1) membrane technology,
(2) adsorption technology, and
(3) reactive extractions.

What makes these separating procedures so special is that they can be adapted to the specific selective separating task by adjusting the chemical potential at the phase–interface area which is responsible for the separation process.

In membrane technology various different membrane applications are known, ranging from microfiltration to high-pressure reverse osmosis (PETERS et al., 1997), and including dialysis or membrane technologies for solution diffusion (Gesellschaft für funktionelle Membranen- und Anlagetechnologien mbH., 1996). In this field the two latter processes are of special importance since their selective separating properties are already being exploited. First applications in the sectors of metal processing and food processing industries have already been described (BAUER and DANZ, 1997; KASCHEK et al., 1997); yet, their authors, also still stress the need for further developmental work.

One of the present developmental trends goes towards influencing the chemical potential at the phase–interface by measures that can externally be regulated and thus be adapted to the according requirements. Of special interest among these measures are externally imposed electromagnetic fields, with or without a previous separation of polar components.

In this method the chemical potential can either be influenced by the membrane material itself (e.g., dialysis) or by adding polar components. Such components might also be salt fractions, which are often present in wastewater.

In the example a nanofiltration membrane with 10 KDa was used for filtering a synthetic effluent with chlorobenzoic acid as organic component. A salinity-dependent rejection occurred which could reach values of up to 90%, whereas just the selection of a suitable mem-

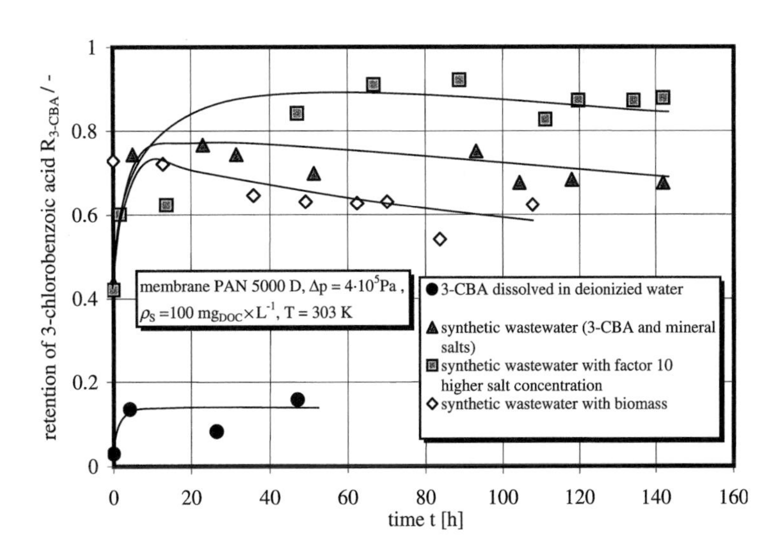

Fig. 3. Rejection of 3-chlorobenzoic acid in dependence on salinity in nanofiltration.

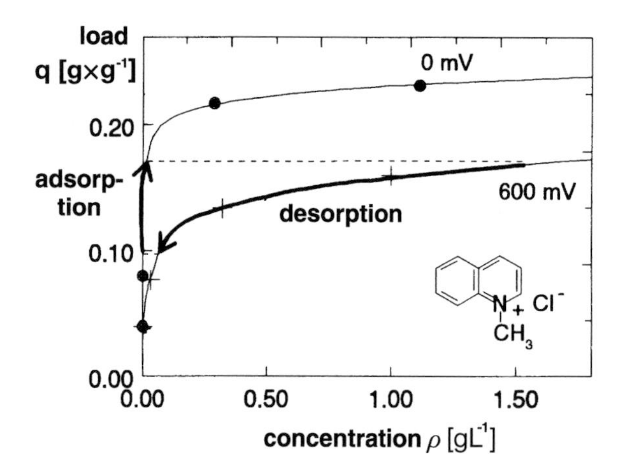

Fig. 4. Potential controlled equilibrium loading of methyl quinolinium chloride on activated carbon.

brane should only have achieved a 15% rejection. The result of such an investigation (BRAMBACH, 1997) – as indicated by Fig. 3 – clearly documents the above described possibility to influence the selective properties of a membrane depending on a given chemical potential.

In the present case a scaling layer enriched with salt components is formed at the inside of the membrane, causing an increased rejection of chlorobenzoic acid, a substance low in polarity. The degree of rejection depends on the salt concentrations and the composition of the salt fractions as well as on the inert fractions present in the phase boundary (e.g., the biofilm).

In adsorption technology superimposed electromagnetic fields also show an effect similar to that in membrane technology. In this case the chemical potential of the phase–interface area and the interaction of the molecules in the effluent mixture can be adjusted by alternating fields. Fig. 4 shows the results of an investigation into active carbon (BEN et al., 1996), which clearly illustrate the described effect. Not only hydrophobic components could be adsorbed, but in dependence on the strength of the field selected hydrophilic components could be adsorbed as well. The hysteresis observed is characteristic for such adsorption processes.

Of the three separation technologies favored in this context adsorption techniques certainly still require the greatest developmental efforts, since so far little has been found out about the effects of dislocation diffusion on the adsorption process. This disadvantage is, however, offset by the wide spread of adsorption technology, that could guarantee a fast implementation of advanced techniques with little investment into modifications.

In its application as a selective separation technique the reactive extraction method has experienced the greatest progress. The know-how in this field is mainly based on the selection of the solvent and its specific reactivity. This might consist of liquid ion exchangers, complex forming substances, ligands, etc., which selectively affect individual components. The extraction from the solvent can subsequently be carried out with known methods, whereby the thermal conditions dictated by the product must already be considered when choosing the appropriate system.

GERTH (1996) and SCHÄFER and SLUYTS (1996) cite reports on practice-oriented applications of this technology. It is remarkable that the high selectivity achieved seems to be decisive for an operational assessment of the various procedures. Even though the redemption times that have been achieved by the pharmaceutical industries (SCHÄFER and SLUYTS, 1996) when recovering active agents from their effluents by such a production-integrated method may certainly not be regarded as the average value, this technology seems to offer special advantages. Again, however, attention has to be drawn to the nesessary search for a suitable system of solvents with a specific reactivity.

3.2 Water Treatment Techniques

As has been illustrated so far, new approaches to environmentally friendly procedures have to be employed when in a next necessary step of the process the wastewater is treated after any useful materials have been recovered (Fig. 2). The present state of the technology is of no use or only of limited use to any production-integrated application. The demands on these new techniques are outlined in Fig. 5, where the physical separation procedures already described in Sect. 3.1 are neglected, and biochemical procedures are placed at the center of interest instead.

Fig. 5 uses the example of landfill seepage to demonstrate residence times necessary for biologically degrading 90% of different classes of substances by means of mixed bacterial populations. These residence times depend on the proportional mass distribution of the various classes of substances in the discharge. As the present technology is oriented on given standards, it mainly aims at eliminating a discharge characterized by a summation parameter. Thus, it is designed to achieve longer residence times either by being operated in several steps or by enlarging the reactive volume. Using this technology as an example the development of the legal restrictions can be illustrated, whereby one has to distinguish between three developmental stages, the first two of which comprise the elimination of

(1) BSB_5-(easily degradable DOC) and subsequently,
(2) the nutrient load (nitrogen and phosphate).

In addition it is required today to retain persistent compounds that are hard to degrade (harmful to water). Accordingly, the present approach is to run biological processes for establishing an adequate microbiocenosis in reactors with mostly rather large volumes and to eliminate the whole emission. The currently preferred biofilm technique (see Chapters 13, 17, 18, this volume) certainly offers one opportunity to adapt the population to specific requirements. For an application that would meet the new demands, however, it lacks the necessary reaction control.

This last mentioned task of retaining persistent compounds is already quite similar to production-integrated applications, since it already targets single substances. In contrast to the present state of the technology, however, new approaches in ecological process engineering should possess the ability to selectively eliminate or transform classes of persistent substances from effluent mixtures.

This is required because in principle even the smallest amount of pollutants in the effluent could impede the reuse of this water. Furthermore, if all contaminants were degraded within the first stage, this would lead to an accumulation of persistent compounds in the biosludge. As a consequence, the problem would only be shifted. Thus, if a selective elimination from effluent mixtures is to be accomplished the demands on new biological procedures in production-integrated applications have to insist on achieving a substrate-specific residence time behavior. Then, due to low concentrations of hard-to-degrade substances, one will succeed in achieving a selective transformation

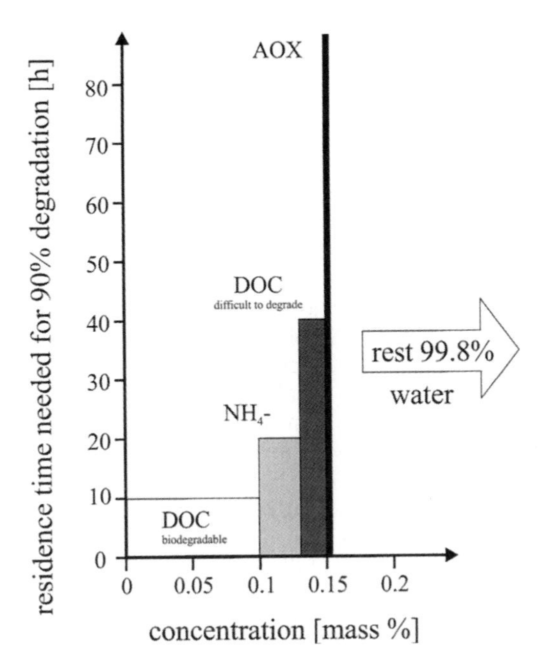

Fig. 5. Residence time needed for 90% degradation of substance classes in industrial wastewater.

process in reactors with small volumes as well. Furthermore, this process will lead to a biocenosis which is specifically aimed at the persistent component.

One of the possible solutions has been introduced by BRAMBACH (1997) and RÄBIGER and SCHIERENBECK (1997), and, as was to be expected, is based on combining a selective separation technique with a reaction controlling component. In this case, a membrane technique is used by which an extensive selective accumulation of the class of persistent substances is achieved in the reaction chamber. Classes of substances whose molecules fall short of the size at which the membrane cuts off the separation process will leave the reaction system as permeate, with only a short residence time. The permeate flow rate and the rejectivity of the membrane determine the selectivity and thus the throughput and performance of the reactor.

In the application illustrated the so-called suspension membrane reactor was designed as an airlift reactor (Fig. 6). In an aerated "upstream" assembly the membrane module forms the central unit. Three phases – water, air, and biomass, whether immobilized on particles or not – flow through this unit with sufficient velocity. After degassing at the top of the reactor the liquid, including the biomass, flows to the bottom of the reactor inside a "downstream" assembly, thus establishing a flow cycle. The driving force responsible for this is sustained and controlled by the gas supply.

The difference in gas content as well as in static pressure which arises between upstream and downstream assembly is proportional to the pressure drop across the circulating flow. Since the reactor for membrane filtration is under pressure ($3 \cdot 10^5 - 5 \cdot 10^5$ Pa), not the oxygen concentration, but the hydrodynamics constitute the factor that has to be regulated by aeration, decisively influencing the formation of the biofilm and consequently the permeation rate of the reactor.

The concentration profile shown in Fig. 6 corresponds to the solution diffusion model and depicts a rejection of specific classes of substances due to the chemical potential determined by the membrane. How this is influenced by the ingredients of the effluent from the previous step has already been described in Sect. 3.1. In this context it seems to be especially worth mentioning that the potential for an extended degradation of persistent and problematic compounds is created by an in-

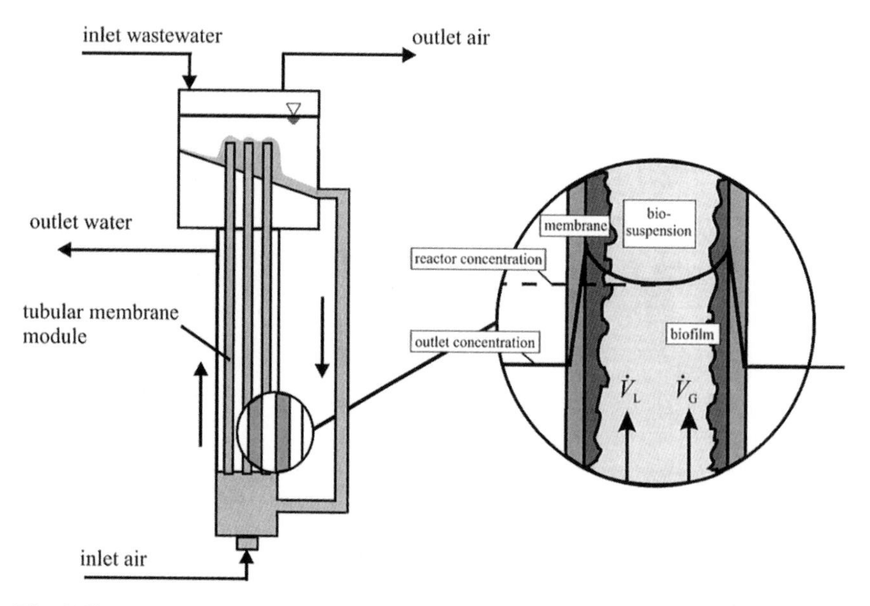

Fig. 6. The suspension membrane reactor – an example for a combined separation and reactor technology.

crease in their concentration. According to VAN UHDEN (1967) a number of substances exist whose removal will only become possible above a certain threshold of minimum concentrations (Fig. 7). Therefore, the rejection that can be attained by increasing the concentration not only effects an adjustment of the necessary substrate-specific residence time distribution, but also an increased conversion.

The results of the investigation shown in Fig. 8 – with 3-chlorobenzoic acid as organic component of a synthetic effluent – clearly demonstrate the great influence of a changed residence time behavior. In order to explain this a form of representation has been chosen that is known from reaction control, i.e., representing the residual portion of contaminants remaining in the effluent in dependence on the Damköhler number of first order. Results which have been obtained from an airlift reactor without a membrane module – corresponding to the systems currently employed – have to be regarded as describing the minimum performance, and thus the curve they form describes the limits of the process. Here the purification performance is determined by the hydrodynamically determined residence time distribution, which is that in an ideal stirred tank. Accordingly, big reactors with large volumes are required to achieve the residence time necessary for the degradation or removal of persistent components.

At large Damköhler numbers, however, the given conversion rate is restricted by mass transport, which would make no sense for persistent substances. In real applications, processes which are restricted in reactivity can only optimally be designed by cutting the link between the residence time determined by the substrate and that determined by the hydrodynamics. In the case that is regarded here, this has been achieved by retaining a certain class of substances by means of a membrane module. This prolongs the residence time relevant for the substrate, without prolonging the one determined hydrodynamically. Thus, at lower Damköhler numbers, i.e., in transformation processes with restricted reactivity and taking place in smaller volumes, a reactor performance is achieved which can be regulated by the rejectivity of the membrane.

The diagrams shown in Fig. 8 have been calculated by means of the dispersion model and describe the experimentally obtained results pretty well (RÄBIGER and SCHIERENBECK, 1997). It has to be pointed out, however, that

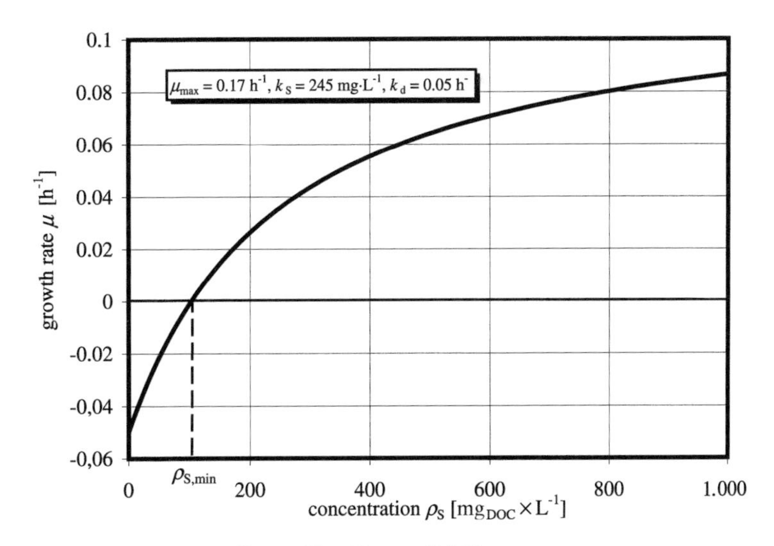

Fig. 7. Kinetics according to VAN UHDEN (1967).

$\mu = \mu_{max} \cdot p_S (p_S - k_S)^{-1} - k_D.$

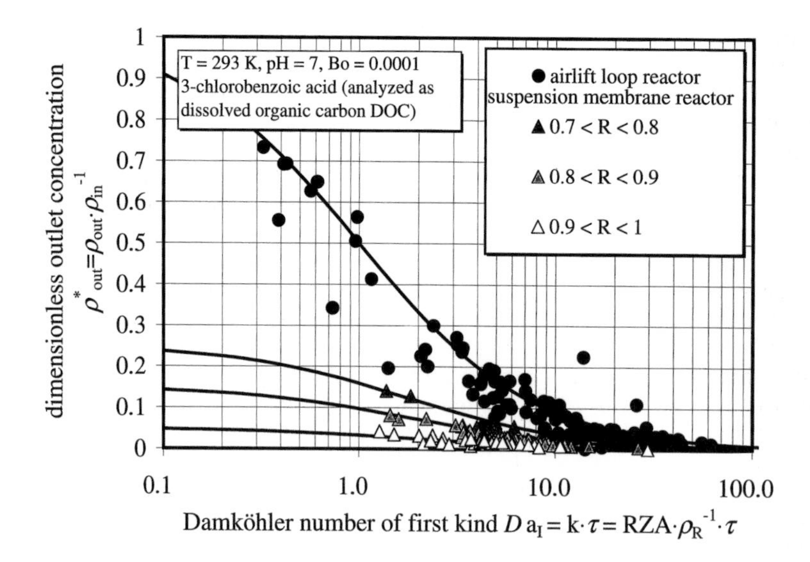

Fig. 8. Input and output analysis necessary for the factual inventory.

the active proportion of the biomass and its kinetics, both of which are important for the application of the Monod kinetics, have been experimentally determined. The high oxygen content caused by the operating pressure and the high age of the sludge obtained by the retention of the biomass lead to a very low production of excess sludge, about which very little is known up to now.

The reactor technology described above has also proven to work for multisubstance mixtures. Practical application, however, is limited by the availability of modern sensor technology and automatic controls (e.g., fuzzy logic), since these determine how close the process will be to the production process. The most significant technological developments of the future will be in the field of microsystems, as online measurements and preventive process control are decisive for a successful production integration. The production depth and range of production plants and machinery will expand considerably, so that a strong fluctuation in effluent loads has to be expected.

In principle, this kind of plants is already known from current end-of-pipe applications. Basically, any micro- (KRAFT and MENDE, 1997), ultra- (STAAB and GÜNDER, 1996) or nanofiltration (MELLIS, 1994) which follows the bioreactor process shows an identical potential for reaction control. But so far application of membrane technology has only been aimed at retaining the biomass, resulting in today's distinction between low pressure (KRAFT and MENDE, 1997) and high pressure (STAAB and GÜNDER, 1996; MELLIS, 1994) techniques. Apart from that current bioreactor technology is designed to eliminate the total emission load, thus being designed for large volumes. This in turn leads to a tranformation process with limited mass transport. As can be seen in Fig. 8, this permits no influence, or – due to the high concentration of biomass – only a rather limited influence of the membrance-specific rejection capacity on the conversion rate.

3.3 Materials Management and Weak Point Analysis of Production Lines

It has already been pointed out that it will be necessary to carry out a weak point analysis of production lines in order to be able to establish ecologically and economically optimized measures for a production based on resource minimization. The methodology suited to this approach has to live up to the demands on strategic management, according to which not only the ecological impact of the individual steps of the process will be made transparent, but also the creation of wealth within a production, and all potentials will become subject to evaluation. Such a combined analytical assessment seems to be imperative, as (SCHÄFER and SLUYTS, 1996) the economic behavior of manufacturers will have to undergo ever greater changes.

Within the framework of the extensive industrial growth of the past special attention was paid to regulating the demands in the market by means of a flexible supply. In future this will no longer be possible, since the demand will be of a clearly more critical nature. It will be subject to critical considerations which will not only apply to a simple selection from the products available, but will also extend to questions of their period of use, and thus will include an assessment of their overall functionality. The consumers in a modern service society are no longer merely passive customers, but pay their own contribution to the use of a product with all its ecological implications, e.g., by leasing. Demand will teach the manufacturer who invents and proposes new or novel products, what is to be done. In consequence, flexibly designed production plants will have to prove their optimal ecological relevance in order not to impair the success of the product (GIARINI and LIEDTKE, 1998).

In this context the assessment of any wealth creating potentials is of special significance. This holds true as well for products the service function of which is more important than their mere manufacturing function. The incremental value as a whole has to be measured and optimized with regard to the effectiveness of the system.

At present, there are many quantitative methods available to assess ecological impacts, such as

(1) the model of critical air and water volumes,
(2) the model of critical loads,
(3) the evaluation by ecopoints,
(4) the VNCI model (Netherlands),
(5) the EPS model (Sweden),
(6) the Tellus model (USA),
(7) the methodology of material intensity per product and service unit (MIPS),
(8) the methodology of accumulated energy expenditure,
(9) the WaterPinch analysis, and
(10) methodological approaches to the implementation of an environmental compatibility assessment.

Furthermore, there exist several qualitative evaluation methods (e.g., utility analysis, ABC analysis, verbal-argumentative evaluations).

Currently, the utility analysis represents one of the quite often employed assessment procedures; it offers a qualitative decision matrix for analyzing processes, products, and sites. The utility value of each criterion targeted and its alternatives are separately calculated by multiplying the estimated degree of meeting this target – determined by allocating certain scores – with some weighting factors (HOPFENBECK et al., 1996; KREIKEBAUM et al., 1994; GAHRMANN et al., 1993). Thus, environmental relevant factors within an enterprise can be assessed according to various different ecological criteria (e.g., energy consumption, current as well as expected environmental risks), and by simple grading methods they can be added up to an overall assessment. As a result of analyzing the various possible decisions the alternative with the highest total utility value will receive highest priority.

By their very nature all qualitative evaluation procedures give relatively large room to subjective assessments, since the ecological impact of various factors is graded in a "relative" way, ranging from factors that imply especially critical loads, to those that are regarded as less urgent and even minor or secondary. This relative evaluating procedure is supposed to make allowance for the fact that the evaluation of a company's ecological impacts can never be given with scientific precision, since environmental phenomena are always extremely complex and difficult to assess (HOPFENBECK et al., 1996; KREIKEBAUM et al., 1994; GAHRMANN et al., 1993).

Quantitative evaluation procedures on the other hand contain inflexible as well as rating assumptions which together constitute a precisely determined basis for calculating the future appraisal.

One example is *the model of critical air and water volumes*. Here the emission of each individual pollutant is put into relation to a so-called critical load (BECK, 1993; Umweltbundesamt, 1995; AHBE et al., 1990; HABERSATTER, 1991). This model is based on the assumption that for each pollutant emitted into the air or water a certain volume can be calculated which is loaded up to the legal limit (critical load) by the pollutant. For each medium the pollutant-specific characteristic values can be computed from these ratios and can be added up to give a single coefficient.

The *model of critical loads* constitutes an expansion of the model of critical air and water volumes. In addition to air and water this evaluation paradigm takes into account the overall energy consumption by means of energy equivalencies, and also the necessary dump site volume for solid wastes (BECK, 1993; Umweltbundesamt, 1995). The product itself, however, is not evaluated.

In continuation of the model of critical air and water volumes the *evaluation by ecopoints* takes up and modifies the orientation on "critical values" (BECK, 1993: Umweltbundesamt, 1995; AHBE et al., 1990; BRAUNSCHWEIG and MÜLLER-WENK, 1993). The calculation of ecopoints is based on the so-called ecological scarcity, which reflects the relation between the loading capacity of an environmental resource (the environmental medium) and the actual load (the loading value measured). Multiplying the environmental load with its according ecofactor results in ecopoints. The sum of all ecopoints for all environmental loads reveals the wheighted overall environmental burden.

The *VNCI model* from the Netherlands is based on 9 different environmental topics (global warming, ozone depletion, acidification, eutrophication, photochemical ozone formation, dispersion of toxic substances, disposal of waste, disruption depletion of natural resources), all of which can be found in the Dutch National Environmental Scheme (Umweltbundesamt, 1995). Each of these environmental effects is treated separately and characterized by indicators and equivalencies which in part are generally accepted, such as the CO_2 value corresponding to the greenhouse effect, but also by such that are contentious, e.g., the values corresponding to areas affected by noise, smell, and accidents. These coefficients indicating the environmental load are correlated to their respective impact across the Netherlands, then multiplied by the factor 1,000,000, and the result represents the "environmental impact index".

The result of the VNCI model is an "environmental result" for each of the 9 topics concerning environmental burdens. It is calculated by multiplying the environmental impact index by its respective weighting factor. The weighting factor is chosen from anything between 1 and 10, and it might arise accidentally, can be set by political decisions, or – depending on the circumstances – it may be assigned a new value time and again. These dimensionless "environmental results" can be added up; the sum will then determine the degree of ecological difference between two options.

In contrast to this the Swedish *EPS model* (environmental priority strategies) makes it possible to describe the environmental burden arising from a production system to quite some extent by using numerical values, the so-called environmental load units (ELU), which originate in the consumption of resources and energy as well as in the emission of pollutants (Umweltbundesamt, 1995). The Swedish parliament has agreed on areas worthy of special protection (biodiversity, production, human health, resources, aesthetic values) and the ELU which are derived from these protective targets are oriented on the degree of people's individual willingness to invest part of their capacity for work into influencing the quality of the environment ("willingness to pay").

The US *Tellus model* is used for assessing questions of waste management, whereby the alternatives for disposal are analyzed according to ecological aspects and are put into correlation with the framing economic conditions (Umweltbundesamt, 1995). The costs of the environmental load arising from the various options for disposal are added to the purely operational costs of disposal by internalizing external costs. For calculating the ecological costs the main focus is put on costs that arise from emissions into the air as well as into ground and surface water, whereby the monetary value put on these emissions is determined by the expenditure for retaining certain pollutants or preventing other environmental damage.

A new measure for the environmental impact of any kind of goods is defined by the methodology of investigating the materials expenditure per product and service (MIPS) (HOPFENBECK et al., 1996; Umweltbundesamt, 1995; SCHMIDT-BLECK, 1993). The measure for the degree to which the environment is burdended is derived from the intensity of material consumption per service unit or function throughout the whole life cycle of the product, i.e., MIPS. Thus, the total mass input forms the central parameter, whereby wastes and emissions are not separately considered, since they are already indirectly contained within the input values. The higher the number of services related to a certain consumption of resources, the smaller the MIPS value will be, and the better the environmental compatibility of the product.

The method of cumulative energy expenditure – KEA – registers all energy inputs into the system and all usable outputs, and relates them to the primary energy expenditure according to a certain set of rules (Umweltbundesamt, 1990). Thus, the KEA comprises all consumption of energy resources and all emissions resulting from generating and supplying this energy.

The WaterPinch analysis by LINNHOFF and DHOLE (1996) is a method applied in a great number of fields, if there are problems with wastewater or drinking water. In this method both the consumption of freshwater and of wastewater throughout a production are plotted in a diagram in which the horizontal axis represents the throughput of used water and the vertical axis the degree of water purity with respect to a single pollutant. The area in which the two curves of fresh water and of wastewater overlap tells something about the scope to which water can be reused, and the pinchpoint, the intersection between the two, represents the limits of the system.

The methodological approaches by GEBLER (1992) to assess environmental compatibility pursue the goal of a quantitative evaluation of ecological impacts of waste treatment systems. The ecological aspects of different system and planning versions are to be compared by determining certain numerical values, so-called toxicity equivalencies, which are directly connected to the environmental impact of a disposal system. This toxicity equivalency is determined by the mass of a pollutant emitted per ton of waste and a factor gained by weighting the pollutant according to its toxicological and ecotoxicological hazards. This weighting factor includes criteria of acute toxicological effects on aquatic and terrestrial organisms, carcinogenic and mutagenic criteria, persistence of substances in the air (abiotic decomposition) as well as in soil (biotic and abiotic decomposition) and the bioaccumulation potential. The *toxicity equivalency* devised by GEBLER (1992) tries to capture the potential threat to the environment by including basically all environmental impacts of pollutants.

If common features among the assessment procedures described above are to be found, it should be stated that the quantitative ones quite often avail themselves of effluent or emission standards, guidelines or similar regulations which are either ordered by law or issued by generally acknowledged institutions, obviously regarding them as generally acceptable criteria, or using them as basis for the calculation of their later evaluation (FLEISCHER, 1994). Most assessment methods belong into this category, including such as the model of critical air and water volumes, the model of critical loads, the assessment by ecopoints, the VNCI model, the EPS model, and the Tellus model. Hardly any or only a very limited ecological impact can be derived from them.

The lack of generally accepted methods for diagnosing environmental matters very often leads to the fact that environmental measures not always live up to the ecological targets called for and supported in the interest of the public. In consequence, it happens that the ecological effectiveness of measures is unintentionally misinterpreted. Comparing different measures aimed at an ecologically relevant improvement is impossible without a comprehensive kind of assessment. As long as evaluations are not standardized, different investigators will continue to arrive at different results (WIESNER, 1995). Apart from that insufficient insight into the ecological impacts of pollution has to be regarded as a further problem. For this reason the MIPS method and the methodological approach of GEBLER (1992) are a step in the right direction, as far as the environmental comparability of differing procedures is concerned.

Unfortunately not even a comparative analysis and representation of the applicability of these two methods will be possible since this would require special knowledge. For this reason only the method of GEBLER (1992) which has been further developed by the author will be discussed at this point.

First of all an environmental assessment of production processes requires detailed knowledge of the energy and material flows occurring throughout the whole production process. By means of an input–output analysis the input entities (raw materials, fuels, accessory materials, water, energy, possible intermediate products) as well as the output entities (products discharges, waste) of a production process are registered (Fig. 9).

Secondly, based on such factual inventory, each pollutant occurring during the production process and each product deriving from its degradation or conversion is assigned some functional value. This value directly corresponds to the toxicological effect and the environmental impact of the substance concerned, so that the material and energy flows present in the production process as well as the required services can be assessed by adding up the individual components. Each step of the process can thus be accorded either one functional value or a set of several values distinguishing between the various ingredients. An example of how this approach is applied to an actual case shall be given below. In this instance the described functional value for the environmental impact of the production was determined by a comprehensive recording of data. For the analysis of the weakest points the functional values found for each step of the process were plotted as a summation curve over the course of the whole production process. For each of the individual summation curves there holds: the steeper the curve rises, the greater is the potential threat of this process to the environment. Accordingly, the process steps for which the summation curve reaches maximum peaks are identified as ecolocically weakest points.

The product itself, the emission paths, and the energy contribute to the evaluation, and splitting the summation curve into their individual shares will permit a more precise pinpointing of the weakest points. This becomes very clear in Fig. 10. In this instance the results are represented by a related value (Te_0), which is the value derived from the previous production line and serves as starting point for the subsequent one, and they clearly highlight three weak points. Two of these (S2, S3) are related to the product, and one to emissions (S1). The product-related weak points are caused by coating processes for finishing the product. Weak point S1 on the other hand results from a degreasing process linked to surface treatment and the subsequent break-up of the emulsion in the discharge.

Of course the question immediately arises whether this impact assessment considers the significance of all the individual substances in

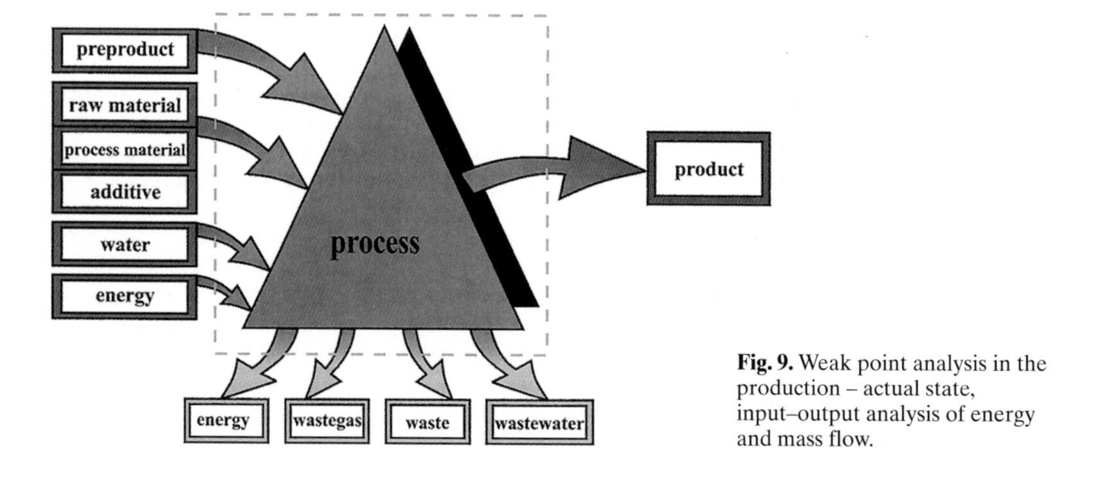

Fig. 9. Weak point analysis in the production – actual state, input–output analysis of energy and mass flow.

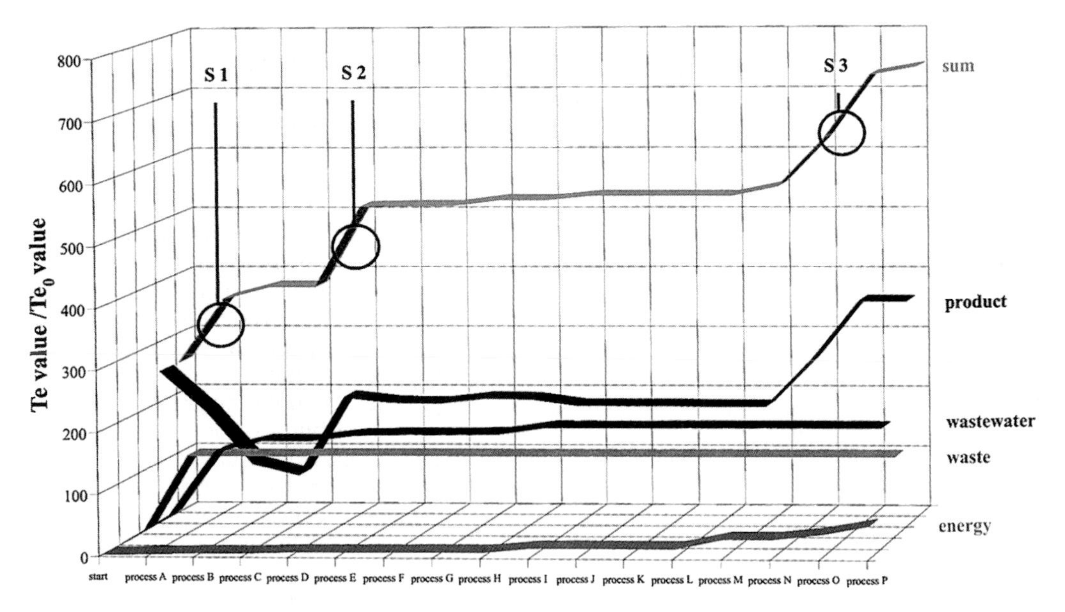

Fig. 10. Weak points (S) analysis in the production including production integrated measures.

a sufficiently comprehensive way. This cannot be completely confirmed, since the number of substances is too high and their environmental impact is not yet entirely known. On the other hand it should be pointed out that, even though a concentration on adding up just a few particularly hazardous ingredients might be a step in the right direction, it is by no means sufficient. The environmental impact of large amounts of very well-degradable pollutants can be identical to that of hazardous substances and should, therefore, not be neglected. In view of the methodology described this means that usually a 70% precision will suffice to identify the relevant weak points. Furthermore, it is possible to check the significance of the information obtained on particularly harmful ingredients by deriving the Te_0 value from the pollutant loads that would be caused by adherence to the legal emission standards.

Even though the environmental potentials of a production are subjectively weighted by this method, they still can clearly be demonstrated. The weakest points were found for the given example with the help of various functions, including the wealth creating potential of the production, and the effects of appropriate measures were tested. The result is shown in Fig. 11. It has to be noted, however, that in a first step only the weak point connected to the emission was considered. Finding solutions to the other two weak points (S2, S3) is closely linked to the development of a new product, something that can only be achieved in the long run. As a selective separating technique extraction has proved to be an especially suitable solution for weak point S1, because this procedure did not affect the properties of the recovered grease which were necessary for recycling and re-entering the substance into the production process. However, in this context it is of some significance that – according to this analysis – the ecological/economical optimum lay at a mere 85% separation of the grease from the effluent. Any more extensive recovery would have led to a considerable increase in energy consumption. Consequently, the individual functional summation value for the first weak point does not show all of the desired improvement, but the absolute value, and especially costs, have been improved decisively.

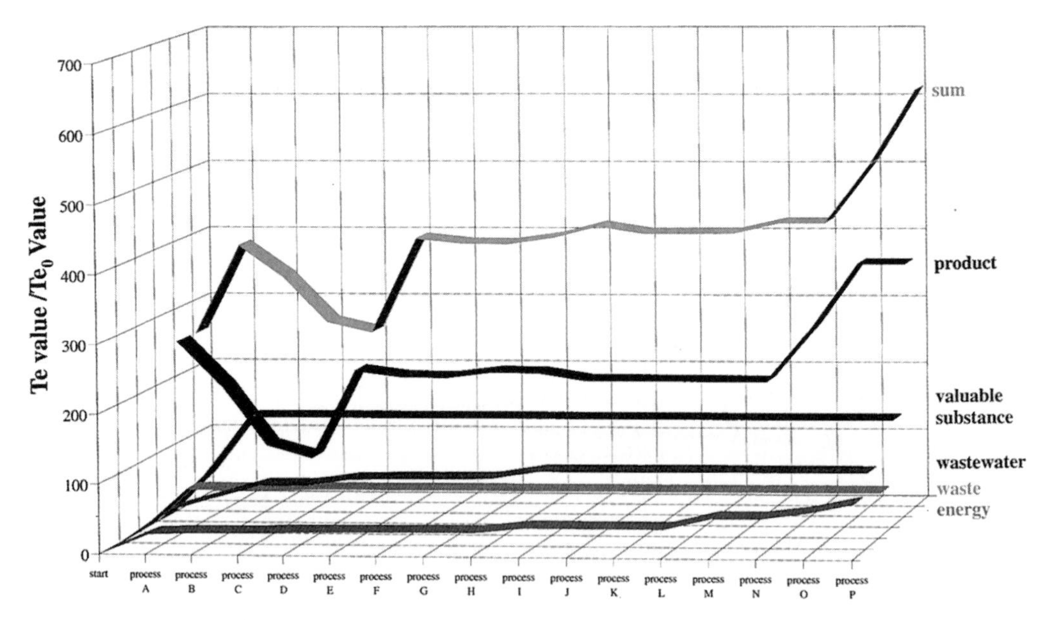

Fig. 11. Extraction module for resource recovery.

With the help of such a method environmental and wealth creating potentials of a production can be analyzed and the weak points as well as solutions for eliminating any deficits can be found. In all ecological solutions the closing of cycles will clearly remain at the center of interest, but this method offers further promising potential for the future development of production-integrated environmental protection.

4 References

AHBE, S., BRAUNSCHWEIG, A., MÜLLER-WENK, R. (1990), Methodik für Ökobilanzen auf der Basis ökologischer Optimierung, *Schriftenreihe Umwelt,* Nr. 133. Bern: Bundesamt für Umwelt, Wald und Landschaft (BUWAL).

BAUER, B., DANZ, K. (1997), Wasser- und Laugenmanagement in der Getränkeindustrie, in: *Preprints des Colloquium "Produktionsintegrierte Wasser-/Abwassertechnik"*, Bremen (GVC-VDI Düsseldorf, Ed.), C41–C58. Bremen: Institut für Umweltverfahrenstechnik.

BECK, M. (1993), *Ökobilanzierung im betrieblichen Management.* Würzburg: Vogel Buchverlag.

BEN, A., SCHÄFER, A., WENDT, H. (1996), Abwasserreinigung durch potentialgesteuerte Ad-/Desorption an Aktivkohle, in: *Preprints 3. GVC-Kongreß "Verfahrenstechnik der Abwasser- und Schlammbehandlung"*, 14–16 October, Würzburg (GVC-VDI Düsseldorf, Ed.), Vol. 3, pp. 285–289. Düsseldorf: GVC-VDI.

BRAMBACH, R. (1997), Reaktor zur biologischen Eliminierung schwer abbaubarer Kohlenwasserstoffe durch Einstellung substratspezifischer Verweilzeiten, *Fortschritt-Berichte VDI,* Reihe 15, Umwelttechnik, Nr. 183. Düsseldorf: VDI-Verlag.

BRAUNSCHWEIG, A., MÜLLER-WENK, R. (1993), *Ökobilanzen für Unternehmungen – Eine Wegleitung für die Praxis.* Stuttgart: Haupt Verlag.

Bundesministerium für Umwelt, Naturschutz und Reaktorsicherheit (1993), *Konferenz der Vereinten Nationen für Umwelt und Entwicklungen im Juni 1992 in Rio de Janeiro – Dokumente – Agenda 21, Umweltpolitik.* Bonn.

DALY, H. E. (1994), Die Gefahren des freien Handels, *Spektrum der Wissenschaft* **1**, 40–46.

FLEISCHER, G. (1994), *Produktionsintegrierter Umweltschutz*. Berlin: Verlag für Energie- und Umweltschutz GmbH.

GAHRMANN, A., HUMPFLING, R., SIETZ, M. (1993), *Bewertung betrieblicher Umweltschutzmaßnahmen, ökologische Wirksamkeit – ökonomische Effizienz*. Taunusstein: Eberhard Blottern Verlag.

GEBLER, W. (1992), Ökobilanzen in der Abfallwirtschaft – Methodische Ansätze zur Durchführung einer Programm-Umweltverträglichkeitsprüfung, *Stuttgarter Berichte zur Abfallwirtschaft*, Vol. 41. Berlin: Erich Schmidt Verlag.

GERTH, K. (1996), Aufbereitung von korrosiven Prozeßabwässern in einem pharmazeutischen Großunternehmen, in: *Preprints 3. GVC-Kongreß "Verfahrenstechnik der Abwasser- und Schlammbehandlung"*, 14–16 October, Würzburg (GVC-VDI Düsseldorf, Ed.), Vol. 2, pp. 847–864. Düsseldorf: GVC-VDI.

Gesellschaft für funktionelle Membranen- und Anlagentechnologien mbH (1996), *Brochure*. St. Ingbert, Germany.

GIARINI, O., LIEDTKE, P. M. (1998), *Wie wir arbeiten werden*. Hamburg: Hoffmann und Campe.

HABERSATTER, K. (1991), Ökobilanz von Packstoffen – Stand 1990, *Schriftenreihe Umwelt*, Nr. 133. Bern: Bundesamt für Umwelt, Wald und Landschaft (BUWAL).

HOPFENBECK, W., JASCH, C., JASCH, A. (1996), *Lexikon des Umweltmanagements*. Landsberg/Lech: Verlag Moderne Industrie.

KASCHEK, M., MAVROV, V., KLUTH, J. et al. (1997), Membranverfahren zur Laugenaufbereitung in der Getränkeindustrie mit dem Ziel der Kreislaufführung, in: *Preprints des Colloquium "Produktionsintegrierte Wasser-/Abwassertechnik"*, Bremen (GVC-VDI Düsseldorf, Ed.), C59–C78. Bremen: Institut für Umweltverfahrenstechnik.

KRAFT, A., MENDE, U. (1997), Das WABAG Submerged Membrane System für Prozeßwasserreinigung und -recycling am Beispiel von Mälzereiprozeßwasseraufbereitung, in: *Preprints des Colloquium "Produktionsintegrierte Wasser-/Abwassertechnik"*, Bremen (GVC-VDI Düsseldorf, Ed.), C79–C86. Bremen: Institut für Umweltverfahrenstechnik.

KREIKEBAUM, H., SEIDEL, E., ZABEL, H.-U. (1994), *Unternehmenserfolg durch Umweltschutz-Rahmenbedingungen, Instrumente, Praxisbeispiele*. Wiesbaden: Gabler Verlag.

LINHOFF, B., DHOLE, V. (1996), Freshwater and Wastewater Minimisation: From Concepts to Results. *Preprints 3. GVC-Kongreß "Verfahrenstechnik der Abwasser- und Schlammbehandlung"*, 14–16 October, Würzburg (GVC-VDI Düsseldorf, Ed.), Vol. 2, pp. 605–613. Düsseldorf: GVC-VDI.

MELLIS, R. (1994), Zur Optimierung einer biologischen Abwasserreinigungsanlage mittels Nanofiltration. *Thesis*, RWTH Aachen, Germany.

PETERS, T. A., FIEDLER, P., MARZINKOWSKI, J. M. et al. (1997), Realisierung eines geschlossenen Wasserkreislaufes am Beispiel der Textilveredelung durch Mikrofiltration des biologisch behandelten Abwassers und dessen Wiederverwendung als Prozeßwasser, in: *Preprints des Colloquium "Produktionsintegrierte Wasser-/Abwassertechnik"*, Bremen (GVC-VDI Düsseldorf, Ed.), A35–A46. Bremen: Institut für Umweltverfahrenstechnik.

RÄBIGER, N., SCHIERENBECK, A. (1997), Selektive Eliminierung von schwer abbaubaren Stoffen mit Membran-Bioreaktoren, *Membran Hit 97* **103**, 1–23 (Institut für Siedlungswasserwirtschaft und Abfalltechnik, Hannover).

SCHÄFER, J.-P., SLUYTS, D. (1996), Aufbereitung eines Prozeßabwassers mittels Reaktivextraktion, *GVC-VDI Jahrbuch*, pp. 285–295. Düsseldorf: VDI-Verlag.

SCHMIDT-BLECK, F. (1993), *Mips – Das Maß für ökologisches Wirtschaften*. Berlin: Birkhäuser Verlag.

SCHNITZER, H. (1998), Die auf einer Stoffanalyse basierende Implementierung von vorsorgendem integriertem Umweltschutz, *Chem. Ing. Techn.* **1**, 64–73.

STAAB, K. F., GÜNDER, B. (1996), Aerobe Abwasserreinigung ohne biologische Überschußschlammproduktion, *Preprints 3. GVC-Kongreß "Verfahrenstechnik der Abwasser- und Schlammbehandlung"*, 14–16 October, Würzburg (GVC-VDI Düsseldorf, Ed.), Vol. 3, pp. 61–65. Düsseldorf: GVC-VDI.

Umweltbundesamt (1995), *Methodik der produktbezogenen Ökobilanzen – Wirkungsbilanz und Bewertung*, Texte 23. Berlin.

VAN UHDEN, U. (1967), Transport-limited growth in the demostat and its competitive inhibition, a theroretical treatment, *Arch. Mikrobiol.* **58**, 145–154.

WIESNER, J. (Coordinator) (1995), Production-integrated environmental protection, in: *Ullmann's Encyclopedia of Industrial Chemistry*, Vol. B8 (ARPE, H. et al., Eds.). Weinheim: VCH.

Index

N

O

X

Y

Z